1 MONTH OF
FREE
READING

at

www.ForgottenBooks.com

By purchasing this book you are eligible for one month membership to ForgottenBooks.com, giving you unlimited access to our entire collection of over 1,000,000 titles via our web site and mobile apps.

To claim your free month visit:

www.forgottenbooks.com/free918382

ISBN 978-0-266-97710-0
PIBN 10918382

U. S. DEPARTMENT OF COMMERCE
R. P. LAMONT, Secretary

BUREAU OF STANDARDS
GEORGE K. BURGESS, Director

BUREAU OF STANDARDS
JOURNAL OF RESEARCH

July, 1932

Vol. 9, No. 1

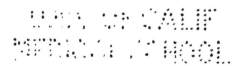

UNITED STATES
GOVERNMENT PRINTING OFFICE
WASHINGTON : 1932

THE USE OF α-BENZOINOXIME IN THE DETERMINATION OF MOLYBDENUM

By H. B. Knowles

ABSTRACT

Present methods for the determination of molybdenum in any considerable amount require a number of tedious and time-consuming operations prior to the actual determination of that element. α-benzoinoxime, advocated as being specific for copper, has been found to precipitate molybdenum quantitatively and to isolate it from most of the more commonly encountered elements. As a result of the present study a procedure has been developed by which molybdenum can be determined in ores, steels, and other products in much less time than by present methods and with all the accuracy of the best methods now in use.

CONTENTS

I. INTRODUCTION

In an investigation of some reagents proposed as being specifics for copper, consideration was given α-benzoinoxime, which has been recommended by Feigl,[1] and for which he has proposed the name "Cupron." Briefly, it is stated by Feigl that complete precipitation of copper results when an alcoholic solution of the reagent is added to a hot, "clear blue" ammoniacal solution of copper. In the presence of iron, aluminum, lead, etc., precipitation is made in an ammoniacal tartrate solution, while in the presence of nickel, a tartaric acid solution buffered with tartrate is recommended. The resulting precipitate is said to be insoluble in water, alcohol, dilute ammonium hydroxide, acetic, and tartaric acids; easily soluble in mineral acids and slightly soluble in concentrated ammonium hydroxide. After appropriate washing it is dried and weighed as

$$\text{Cu } (C_6H_5\text{—CH—C—}C_6H_5)$$
$$\overset{|}{O} \quad \overset{\|}{N}O$$

containing 22.02 per cent of copper.

---

[1] F. Feigl, Ber., vol. 56, II, p. 2083, 1923.

30200

The desirability of using α-benzoinoxime for the precipitation of copper in the presence of other elements, notably molybdenum, led to a study of the behavior of the reagent toward such elements. Preliminary tests disclosed the fact that while the reagent was not as specific for copper as could be wished, it did present interesting possibilities in its reactions with molybdenum. This element was not only quantitatively precipitated in an acetic acid solution buffered with acetate, but also in cold mineral acid solutions containing as much as 20 per cent by volume of sulphuric acid.

II. EXPERIMENTAL

1. PRELIMINARY CONSIDERATIONS

(a) PERMISSIBLE ACIDITY

Experiments showed that to prevent the interference of certain elements the solution must be distinctly acid with either sulphuric, hydrochloric, or nitric acid. A solution containing 5 per cent by volume of sulphuric acid is preferable, although good precipitations were also obtained from solutions containing as much as 20 per cent by volume of sulphuric acid. Solutions containing 5 per cent of hydrochloric or nitric acid gave excellent results in instances, in which sulphuric acid was objectionable, such as in the presence of tin or lead. Successful precipitations were also obtained in solutions containing 5 per cent by volume of phosphoric acid. Precipitations made in the presence of tartaric acid were not quite complete, while those made in solutions containing hydrofluoric acid indicated that this acid must be absent.

(b) TEMPERATURE OF THE SOLUTION

Precipitations at 80° to 90° C., as well as those performed at room temperature showed that there was danger of reducing sexivalent molybedenum before it was precipitated. Experiments showed that this condition was avoided best by working with a cold solution and adding sufficient bromine water to faintly tinge the solution after the addition of the reagent.

(c) AMOUNT OF REAGENT REQUIRED

When slightly more than the theoretical quantity of the reagent was used, as determined by the relation 1 Mo to 3 $C_6H_6 \cdot CH(OH) \cdot C:(NOH) \cdot C_6H_5$, incomplete precipitation resulted. Similar unsatisfactory results prevailed when ten times the theoretical quantity of reagent was employed. This effect may have been caused, in part, by the large quantity of alcohol added with the reagent. It was found that from two to five times the theoretical amount of reagent provided an adequate excess.

(d) TIME NECESSARY FOR COMPLETE PRECIPITATION

Experiments to determine the time necessary for complete precipitation showed that although filtration could be begun almost immediately following the final addition of the reagent, equally satisfactory results were obtained when filtrations were made at the end of 10 minutes, during which time ample opportunity was afforded to permit intermittent thorough stirring. Low results were obtained when the precipitates were allowed to remain in contact with the solutions for 30 minutes.

(e) WASHING THE PRECIPITATE

It was found that water, absolute ethyl alcohol, and various mixtures of diluted alcohol containing a little sulphuric acid were not suitable for washing the precipitate. The most promising wash solution was found to be a cold, diluted sulphuric acid (1 per cent by volume) containing a small amount of α-benzoinoxime.

(f) FINAL TREATMENT OF THE PRECIPITATE

Experiments in which it was sought to weigh or titrate the molybdenum precipitate, after appropriate washing and drying, gave no immediate indication of success, and further attempts were abandoned. Difficulties, such as are experienced in the direct ignition of some organic compounds, were not encountered, for it was found that the precipitate could be ignited directly to molybdic oxide after a short drying.

2. PROCEDURE RECOMMENDED FOR GENERAL USE

Prepare a solution containing 10 ml of sulphuric acid (specific gravity 1.84) in a volume of 200 ml and not more than 0.15 g of sexivalent molybdenum. If vanadates or chromates are present add sufficient freshly prepared sulphurous acid to reduce them and heat to boiling. Continue the boiling until the odor of sulphur dioxide can no longer be detected. Chill the solution to a temperature of 5° to 10° C. Stir and slowly add 10 ml of a solution of 2 g of α-benzoinoxime in 100 ml of alcohol and 5 ml extra for each 0.01 g of molybdenum present. Continue to stir the solution, add just sufficient bromine water to tint the solution a pale yellow and then add a few milliliters of the reagent. Allow the beaker and contents to remain in the cooling mixture 10 to 15 minutes with occasional stirring, stir in a little macerated filter pulp and filter through a paper of close texture, such as S. & S. No. 589 Blue Band. Filtration can be greatly facilitated by using a coarser filter, such as S. & S. No. 589 Black Band, but it is then requisite that the filtrate be very carefully examined and the first portions refiltered if they are not absolutely clear. Wash the precipitate with 200 ml of a cold, freshly pr are solution containing 25 to 50 ml of the prepared reagent and 10 ml of sulphuric acid in 1,000 ml. On standing, the filtrate will deposit needlelike crystals if sufficient reagent has been employed.

Transfer the washed precipitate to a weighed platinum crucible, cautiously dry; char, without flaming, over a very low gas flame and then ignite to constant weight in an electric muffle at 500° to 525° C. In umpire analyses of materials containing silica it is best to remove that constituent before proceeding with the precipitation of molybdenum rather than to treat the final precipitate with sulphuric and hydrofluoric acids, because of the uncertainty of completely decomposing molybdenum sulphate at the temperature of ignition. If the oxide contains no impurities, except tungsten, it should dissolve completely in warm dilute ammonium hydroxide. If an insoluble residue remains, it must be separated by filtration, ignited, weighed, and the weight subtracted. If tungsten may be present, the clear ammoniacal solution should be acidified with hydrochloric acid and treated with cinchonine as in II, 4(b), "analysis of ores and commercial products."

Results obtained by use of the recommended procedure with pure solutions of molybdenum are given in Table 1.

TABLE 1.—*Determination of molybdenum in pure solution*

Experiment No.	Mo added	Mo found	Difference
1	0.0001	0.0001	0.0000
2	.0010	.0009	−.0001
3	.0051	.0051	.0000
4	.0103	.0101	−.0002
5	.0205	.0205	.0000
6	.0513	.0513	.0000
7	.1026	.1027	+.0001

3. BEHAVIOR OF OTHER ELEMENTS IN RECOMMENDED PROCEDURE

In a survey of the behavior of other elements in the procedure recommended in II, 2, the following elements were studied: Silver, lead, mercury, bismuth, copper, cadmium, arsenic, antimony, tin, selenium, tellurium, aluminum, iron, titanium, zirconium, chromium, vanadium, silicon, tungsten, tantalum, columbium, cerium, uranium, rhenium, nickel, cobalt, manganese, zinc, and the members of the platinum group—ruthenium, rhodium, palladium, osmium, iridium, and platinum.

The only elements that give precipitates in mineral acid solutions with α-benzoinoxime are tungsten, palladium, sexivalent chromium, quinquevalent vanadium, and tantalum. The precipitation of tungsten and palladium is seemingly quantitative and the use of α-henzoinoxime in quantitative determinations of these elements is being studied. For example, in a determination of molybenum in the Bureau of Standards standard sample of chrome-vanadium steel No. 72 the result obtained was too high. The error was afterwards found to be caused by tungsten, the presence of which had not been previously noted. The precipitation of sexivalent chromium and quinquevalent vanadium was not studied because satisfactory methods for their determination are available, and it was found that they cause no interference when reduced to lower valences (Cr^{III} and V^{IV}). The precipitate with tantalum appeared more like the hydrated acid than a compound with the reagent, and the reaction was not studied further because tantalum occurs but seldom, and most of it would be removed before precipitation of molybdenum would be attempted.

Silver, lead, mercury, bismuth, copper, cadmium, arsenic, antimony, tin, aluminum, iron, titanium, zirconium, trivalent chromium, quadrivalent vanadium, cerium, uranium, nickel, cobalt, manganese, and zinc are not precipitated either when alone or when associated with molybdenum. Experiments dealing with mixtures of these elements are shown in Table 2. The separation of molybdenum from antimony is of particular interest because good methods for the separation of these elements are lacking.

Selenium, tellurium, rhenium, ruthenium, rhodium, osmium, iridium, and platinum are not precipitated when they occur alone. Their behavior when associated with molybdenum was not studied.

Columbium and silicon, in addition to the already mentioned sexivalent chromium, quinquevalent vanadium, palladium, tungsten, and tantalum contaminate the precipitate and must be removed before precipitation of molybdenum is attempted, or else determined in the weighed precipitate and deducted.

The results given in Table 2 were obtained when a single precipitation of molybdenum was made in the presence of other elements. With the exception of experiment No. 3, tests of the final precipitates showed no evidence of the presence of contaminating elements. The filtrate and washings from each experiment, after evaporation with nitric and sulphuric acids to destroy organic matter and testing colorimetrically by treating with potassium thiocyanate and stannous chloride [2] showed less than 0.1 mg of molybdenum.

TABLE 2.—*Determination of molybdenum in the presence of other elements*

Experiment No.	Mo added	Mo found	Difference	Remarks
	g	*g*	*g*	
1...............	0.0103	0.0101	−0.0002	Precipitated in 5 per cent HCl+1 per cent HNO₃.
2...............	.0103	.0102	−.0001	Precipitated in 5 per cent HCl.
3...............	.0513	.0632	+.0119	Precipitated in presence of 0.05 g V^v.
4...............	.0513	.0514	+.0001	Precipitated in presence of 0.05 g V^{iv}.
5...............	.0513	.0514	+.0001	Precipitated in presence of 0.05 g Sn^{iv} in 5 per cent HCl.
6...............	.0513	.0514	+.0001	Precipitated in presence of 0.05 g each of Ni, Co, Mn, Cr^{iii}, Fe^{iii}, Zn, Cu.
7...............	.0513	.0514	+.0001	Precipitated in presence of 0.05 g each of Tl, Zr, Ce^{iii}, Al, U.
8...............	.0513	.0510	−.0003	Precipitated in presence of 0.05 g each of Pb, Sb, As, after removal of Pb as sulphate. PbSO₄ not examined for Mo.
9...............	.0513	.0516	+.0003	Precipitated in presence of 0.05 g each of Ag, Bi, Cd, Hg^{ii}.
10...............	.0513	.0516	+.0003	Precipitated in presence of 0.05 g Sb^v in 5 per cent acid (50 per cent H₂SO₄: 50 per cent HCl).

It was found that if correct colorimetric determinations of molybdenum are to be obtained the complete absence of nitric acid and platinum must be assured. The removal of nitric acid is readily accomplished by heating to fuming with an excess of sulphuric acid, while the absence of platinum is best assured by conducting all necessary fusions and evaporations in either silica or porcelain laboratory ware.

4. APPLICATION OF THE PROCEDURE

(a) ANALYSIS OF STEEL

To study the applicability of the procedure to the determination of small amounts of molybdenum in steel, samples of the Bureau of Standards standard sample No. 72 and of No. 11d with small additions of molybdenum were used. One-gram samples of the steel were dissolved in 50 ml of diluted sulphuric acid (1+6) and the solution treated with a minimum amount of nitric acid (specific gravity 1.42) to decompose carbides and oxidize the molybdenum. The solutions were filtered, if not perfectly clear, diluted to 100 ml with water, cooled, treated with sufficient ferrous ammonium sulphate to reduce vanadic and chromic acids, and then cooled to 5° to 10° C. Five to ten milliliters of the α-benzoinoxime reagent was added, followed by the addition of bromine water and a few more milliliters of the reagent. After standing 10 to 15 minutes, the precipitates

---

[2] Lundell, Hoffman, and Bright, Chemical Analysis of Iron and Steel, p. 323, John Wiley & Sons, New York, 1931.

were filtered, washed, and ignited as in the procedure already described. The ignited oxides were examined for any insoluble residue and tungsten they might have contained.

The results obtained by the above procedure are shown in Table 3.

TABLE 3.—*Determination of molybdenum in steel*

Experiment No.	Material	Percentage of molybdenum		Difference	Remarks
		Added or present	Found		
		Per cent	*Per cent*	*Per cent*	
1	Chrome molybdenum steel No. 72.	0.149	0.166	+0.017	Gravimetric determination.
2	do	.149	.160	+.011	Do.
3	B. O. H. steel No. lld..	.0051	.0054	+.0003	Colorimetric determination after solution of the precipitate.
4	do	.0103	.0096	−.0007	Do.
5	do	.51	.52	+.01	Gravimetric determination.
6	do	5.13	5.15	+.02	Do.

(b) ANALYSIS OF ORES AND COMMERCIAL PRODUCTS

To study the applicability of the method to large amounts of molybdenum, determinations were made on the Bureau of Standards standard sample of calcium molybdate No. 71 and samples of wulfenite and molybdenite ore. The procedure adopted in the analysis of these materials consisted of an initial attack of 0.2 to 0.5 g of the material with either hydrochloric acid, nitric acid, or a mixture of both. Molybdenite is perhaps more readily attacked by treating with a mixture of fuming nitric acid (specific gravity 1.49) and bromine. Following the preliminary decomposition the solution was diluted with water, treated with 25 ml of diluted sulphuric acid (1 + 1) and evaporated until fumes of sulphuric acid appeared. After the addition of 100 ml of water, the solution was heated to dissolve soluble sulphates, filtered, and washed with diluted sulphuric acid (2 + 100). The insoluble residue obtained at this stage contained a small amount of molybdenum which was determined by fusing the residue with sodium carbonate, extracting with water, and making a colorimetric test of the water extract. The filtrate, at room temperature, was then diluted to 200 ml and treated with a few drops of tenth normal potassium permanganate, enough to produce a permanent pink tinge, to insure complete oxidation of the molybdenum. Freshly prepared sulphurous acid was then added to reduce vanadates and chromates, and the solution boiled until no odor of sulphur dioxide could be detected. After thorough cooling, the solution was treated with an excess of α-benzoinoxime, and the resulting precipitate filtered, washed, dried, ignited, and weighed. The ignited oxide was dissolved in the least possible amount of warm dilute ammonium hydroxide, filtered, washed with warm water, and the filter with its contents ignited and weighed. The ammoniacal extract containing all the molybdenum was acidified with hydrochloric acid, treated with cinchonine, digested overnight, and any precipitate of tungsten was filtered, washed, ignited at 525° C. and weighed. The weight of this residue, together with that of the residue insoluble in ammonium hydroxide, was deducted from the weight of the molybdenum oxide.

Results of the experiments are given in Table 4.

TABLE 4.—*Determinations of molybdenum in ores and commercial products*

Experiment No.	Material	Mo found	Remarks
1	Calcium molybdate	35.22 per cent Mo	Certificate value 35.3 per cent.
2	do	35.26 per cent Mo	
3	Wulfenite ore	18.77 per cent MoO_3	11 analysts reported values between 18.29 and 21.86 per cent by various procedures.
4	do	18.84 percent MoO_3	Do.
5	Molybdenite ore	70.4 per cent MoO_3	13 analysts reported values between 68.62 and 74.79 per cent by various procedures.
6	do	70.1 per cent MoO_3	Do.

In some further tests on the wulfenite ore, a 5 g sample was treated as described, and the solution diluted to exactly 500 ml after decomposition and the removal of the lead as sulphate. Direct precipitation with α-benzoinoxime in 50 ml aliquots indicated 18.73, 18.67, and 18.73 per cent of molybdenum trioxide. Tests on similar aliquots indicated 18.77 and 18.75 per cent after the long and tedious process of eliminating arsenic and vanadium by triple precipitation with ammonium hydroxide in the presence of ferric iron, separation from tungsten by precipitation of the molybdenum as sulphide from an acid solution containing tartaric acid, recovery of the molybdenum which escaped the sulphide precipitation and other time-consuming attendant operations. The removal of copper, which does not interfere in the α-benzoinoxime procedure, is a requisite of the longer procedure before attempting either a gravimetric or volumetric determination of molybdenum.

In connection with the process of removing arsenic and vanadium by precipitating with ammonium hydroxide in the presence of ferric iron, it is to be noted that if lead is present in the original material, and escapes separation as sulphate, it will combine with and retain molybdenum in the precipitate of ferric hydroxide. The amount of molybdenum thus held may be conveniently determined by dissolving the ferric hydroxide in sulphuric acid, treating with potassium thiocyanate and stannous chloride, and comparing the resulting color with that of a standard molybdenum solution by means of a colorimeter.

III. ACKNOWLEDGMENTS

The author desires to express his gratitude to G. E. F. Lundell, under whose direction this investigation has been conducted, and also to R. Gilchrist for helpful assistance in the investigation dealing with the platinum metals.

WASHINGTON, April 29, 1932.

EFFECT OF ZINC COATINGS ON THE ENDURANCE PROPERTIES OF STEEL

By W. H. Swanger and R. D. France

ABSTRACT

The effect of the surface alterations, resulting from the application and presence

It is generally agreed that the character of the surface of a metal is an important factor in determining its resistance to repeated stresses. If an endurance limit is accepted as an intrinsic property of a metal, this limit is correctly determined only when smoothly polished specimens with generous fillets are used. The damaging effects of surface corrosion and of mechanically produced notches have formed the subject of numerous investigations on "fatigue of metals." The information gained from these investigations has shown the necessity for avoiding "notch effects" in highly stressed members subjected to repeated stresses. Careful removal of tool marks, protection from corrosion and use of adequate fillets at abrupt changes of section, aid materially in realizing in practice the normal endurance strength of metals.

Metallic coatings are frequently used on iron and steel to protect against corrosion. It is a matter of considerable interest to know what effect such metallic coatings may have upon the fatigue limit of metals when damage by corrosion is not involved. From a mechanical standpoint the presence of a metallic coating on a specimen of iron or steel introduces factors which complicate this problem. There are two surfaces, the free surface of the coating and that of the underlying steel, the characteristics of which may influence the fatigue limit of the composite specimen. Another factor is the endurance strength of the coating itself. Very little is known about this property of the various protective metallic coatings in general use, but it is probably low in comparison with the endurance strength of steels. The nature of the bond or interface between coating and steel is believed to have a very important influence on the endurance properties of the composite specimen. The nature of the surface of the steel, the kind of coating and the manner in which it is applied largely determine the character of the bond between coating and steel.

This investigation was restricted to a study of the effect of hot-dipped galvanized and electroplated zinc coatings on the endurance properties of low carbon open hearth iron and two carbon steels.

II. MATERIALS

Zinc coatings were chosen because they are the most commonly used protective metallic coatings on ordinary structural grades of iron and steel. Both hot-dipped galvanized and electroplated coatings were used because of the known difference in the nature of the bond between steel and zinc coating of these two types. Sherardized, "galvannealed," and sprayed zinc coatings were not studied. It is believed that the difference in the nature of the bond of hot-dipped galvanized coatings and sherardized or "galvannealed" coatings, and of electrodeposited coatings and sprayed zinc coatings, is one of degree rather than of kind. It is, of course, possible that each of the above-mentioned types of zinc coatings might affect the endurance properties of a given steel to a different degree.

The open-hearth iron and the two carbon steels were purchased from jobbers and were not specially made for this investigation. The chemical compositions of the three materials (ladle analyses) are given in Table 1.

TABLE 1.—*Chemical composition of steels*

	Carbon	Manganese	Phosphorus	Sulphur	Silicon
	Per cent	*Per cent*	*Per cent*	*Per cent*	*Per cent*
Open-hearth iron	0.02	0.03	0.042	0.005	
0.45 per cent C steel	.45	.60	.015	.040	0.18
0.72 per cent C steel	.72	.31	.017	.019	.24

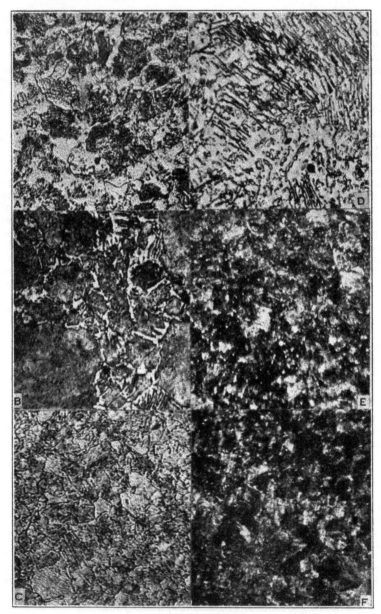

FIGURE 1.—*Structure of the carbon steels. Specimens were etched with nital (alcohol containing 2 per cent nitric acid).* × 450

a, 0.45 per cent carbon steel, normalized at 875° C., annealed at 800° C.; *b*, 0.45 per cent carbon steel, normalized at 875° C., annealed at 800° C., quenched in oil from 830° C.; *c*, 0.45 per cent carbon steel, normalized at 875° C., annealed at 800° C., quenched in oil from 830° C., tempered at 595° C.; *d*, 0.72 per cent carbon steel, normalized at 795° C., annealed at 765° C.; *e*, 0.72 per cent carbon steel, normalized at 795° C., annealed at 765° C., quenched in oil from 775° C.; *f*, 0.72 per cent carbon steel, normalized at 795° C., annealed at 765° C., quenched in oil from 775° C., tempered at 450° C.

The open-hearth iron test specimens were machined from the center of 1-inch diameter hot-rolled bars and the specimens of the two carbon steels from the center of corresponding bars of three-quarter inch diameter.

All of the specimens were carefully machined on a lathe and finish ground to size. They were then polished longitudinally until all traces of circumferential tool marks were eliminated with emery papers of successively finer grit, ending with 0000 paper.

Endurance limits of the two carbon steels, coated and uncoated, were determined with the steels in the normalized and annealed condition, in the oil-quenched condition, and in the tempered condition. The details of the heat treatments are given in Table 2.

To minimize any decarburization effect, an atmosphere of illuminating gas was maintained in the furnace during the heat treatments. As a further precaution the hardened specimens were machined oversize and a layer 0.005 inch thick was ground off the test length after the heat treatments.

The open-hearth iron was used in the "as rolled" condition. The microstructures of the two carbon steels in the three conditions of heat treatment are shown in Figure 1.

TABLE 2.—*Heat treatment of carbon steels*

	Temperature for—			
	Normalizing [1]	Annealing [2]	Quenching [3]	Tempering [4]
	°F. °C.	°F. °C.	°F. °C.	°F. °C.
0.45 per cent carbon steel:				
Annealed	1,607 875	1,472 800	......	......
Quenched	1,607 875	1,472 800	1,526 830	......
Tempered	1,607 875	1,472 800	1,526 830	1,103 595
0.72 per cent carbon steel:				
Annealed	1,463 795	1,409 765	......	......
Quenched	1,463 795	1,409 765	1,427 775	......
Tempered	1,463 795	1,409 765	1,427 775	842 450

[1] ¾-inch rods, heated with furnace, held 20 minutes, air cooled.
[2] ¾-inch rods, heated with furnace, held 40 minutes, cooled with furnace.
[3] Machined test bars, heated with furnace, held 20 minutes, quenched in oil.
[4] Machined test bars, heated with furnace, held 60 minutes, cooled with furnace.

The galvanized coatings were applied by the research division of the New Jersey Zinc Co. (of Pa.) by a method approximating commercial practice for hot-dip galvanizing. The specimens to be galvanized were first polished to the same degree as the specimens tested in the uncoated condition. They were then dipped in a hydrochloric acid solution (2 parts water to 1 part hydrochloric acid, specific gravity 1.19) for two minutes and immediately into the zinc bath held at 440° C. (824° F.). A high-grade zinc (containing 99.94+ per cent Zn) was used. The weight of coating obtained varied from 1.6 to 2.0 oz./ft.^{2} The galvanized coatings varied from 0.0017 to 0.0035 inch in thickness. This variation in thickness was probably caused by the fact that some of the specimens had to be dipped more than once to obtain a complete coating. The length of time in the zinc bath, accordingly, varied from 45 to 100 seconds.

In order to distinguish between the effect of the acid pickling and the combined effect of pickling and galvanizing, fatigue tests were made on specimens of each material which, after final polishing, had been dipped for two minutes into hydrochloric acid of the same strength used for the galvanized specimens.

The specimens of the quenched 0.45 and 0.72 per cent carbon steels which were not galvanized were dipped into a lead bath for 45 seconds at 440° C. (824° F.) so that they had the same heat treatment as the corresponding hot-dipped galvanized specimens. A final polish with 0000 emery paper was given to the lead-dipped specimens before they were dipped into the acid or tested.

The electroplated zinc coatings were applied at the Bureau of Standards.[1] The procedure was as follows: (a) cathode-electrolytic cleaning, two minutes, at 90° C.; (b) hot-water rinse; (c) cold-water rinse; (d) pickled in sulphuric acid (2 N), two minutes at 50° C.; (e) hot-water rinse; (f) hot alkali dip without current, two minutes; (g) scrubbed with cleaning solution, bristle brush; and (h) plated in acid zinc bath, 24 minutes 1.5 amperes, 35° C.

The electrolytic cleaner was made up as follows: Sodium carbonate, 30 g per liter; trisodium phosphate, 30 g per liter; and sodium hydroxide, 7.5 g per liter.

The zinc anodes for the electroplating process were of the same order of purity as the zinc used for the hot-dipped coatings. The thickness of the electrodeposited coatings varied from 0.0021 to 0.0031 inch, which is roughly equivalent to a 2-ounce coating.

III. TESTING PROCEDURE

The endurance limit determinations were made by both the rotating beam and the axial loading methods of stressing. The rotating beam tests were made on R. R. Moore machines and the axial loading tests were made on Haigh alternating stress testing machines, in which the specimens were subjected to alternating equal tensile and compressive stresses. The form of the specimens, methods of calibration of the testing machines, and testing procedure followed have been described previously by one of the authors.[2]

Axial loading tests were not made on zinc-plated specimens as it was believed that the effect of the electrodeposited coating on the endurance limit determined by this method, would be of the same order as was found for the rotating beam tests. As a further economy in number of specimens, axial loading tests were not made on the 0.45 and 0.72 per cent carbon steels in the quenched condition because these steels are seldom used in this condition. Rotating beam tests of electroplated specimens of the 0.72 per cent carbon steels in the quenched condition and of the 0.45 per cent carbon steel in the annealed and in the quenched conditions were also omitted.

Usually nine specimens were used in the determination of each endurance limit. One specimen of each series, except the pickled specimens, tested on the Moore machines was subjected to 25,000,000 cycles of reversed stress at the endurance limit and then restressed at a value 5,000 lbs./in.² above the endurance limit. An annealed, a quenched, and a tempered specimen of the 0.45 per cent carbon steel and an annealed and a tempered specimen of the 0.72 per cent carbon

[1] The plating was done by the electrochemistry laboratory under the supervision of Dr. W. Blum.
[2] R. D. France, Endurance Testing of Steel: Comparison of Results Obtained with Rotating-Beam versus Axially-Loaded Specimens, Proc., Am. Soc. Testing Materials, vol. 31, pt. 2, p. 176, 1931.

steel were stripped of their galvanized coatings, after which they were tested at a stress just under the fatigue limit determined on the acid-pickled specimens of the corresponding materials.

The stresses applied to the acid pickled and the coated specimens were calculated on the diameter of the polished specimen before it was pickled or coated. The diameters were measured with a special micrometer capable of a precision of plus or minus 0.0001 inch. The change in diameter caused by either the acid treatment alone or the acid treatment and the application of the zinc coating was in all cases less than 0.0002 inch.

The tensile strengths of the three materials were determined on standard 0.505 inch diameter test bars, heat treated in the same way as the endurance specimens. Hardness determinations were made on the ends of the tensile and endurance test bars.

IV. RESULTS

The results of the fatigue limit determinations are given in Table 3, together with the tensile strength and hardness of the steels and the per cent change in fatigue limits caused by the pickling, by the pickling and galvanizing, and by the electroplating. The fatigue limits are also shown graphically in Figure 2.

Conventional S-N diagrams for all of the fatigue limit determinations are given in Figures 3 to 9.

TABLE 3.—*Physical properties and results of endurance tests*

Material (per cent carbon)	Heat treatment	Tensile strength	Hardness		Rotating beam								Axial loading			
			Brinell hardness number	Rockwell number	Uncoated		Pickled		Galvanized		Electroplated		Uncoated		Galvanized	
					Endurance ratio [1]	Endurance limit	Fatigue limit	Change in fatigue limit	Fatigue limit	Change in fatigue limit	Fatigue limit	Change in fatigue limit	Endurance ratio [1]	Endurance limit	Fatigue limit	Change in fatigue limit
		Lbs./in.²				*Lbs./in.²*	*Lbs./in.²*	*Per cent*	*Lbs./in.²*	*Per cent*	*Lbs./in.²*	*Per cent*		*Lbs./in.²*	*Lbs./in.²*	*Per cent*
0.02	Hot rolled	44,000	90	51 B	0.61	27,000	22,000	−18.5	26,000	−4.0	27,500	+1.8	0.59	26,000	22,900	−13.5
.45	Annealed	81,000	153	84 B	.44	36,000	33,000	−8.0	27,000	−25.0			.31	28,500	24,500	−4.0
.45	Quenched	121,500	248	28 C	.65	80,000	60,000	−25.0	44,000	−42.5						
.45	Tempered	102,000	207	98 B	.45	46,500	46,500	0	40,000	−14.0	49,000	+4.8	.45	46,000	20,000	−34.0
.72	Annealed	92,000	192	93 B	.38	35,000	31,500	−10.0	20,500	−13.0	35,000	0	.31	29,000	28,500	−2.0
.72	Quenched	176,000	340	39 C	.70	124,500		−40.0	75,000	−40.0						
.72	Tempered	168,500	332	37 C	.55	94,000	57,000	−7.5	84,500	−12.0	105,000	+11.0	.37	62,500	47,000	−24.0

[1] Endurance ratio—$\dfrac{\text{Endurance limit}}{\text{Tensile strength}}$.

V. DISCUSSION

The fatigue limits of the specimens that had been dipped in acid were lower than the fatigue limits of the polished uncoated specimens of the corresponding materials. The decrease was not uniform for the different materials but ranged from zero for the tempered 0.45

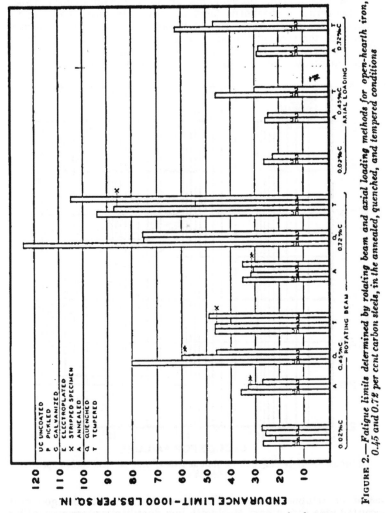

FIGURE 2.—*Fatigue limits determined by rotating beam and axial loading methods for open-hearth iron, 0.45 and 0.72 per cent carbon steels, in the annealed, quenched, and tempered conditions*

per cent carbon steel to 40 per cent for the quenched 0.72 per cent carbon steel. This result was clearly a manifestation of the "notch effect" caused by the acid treatment and was of the same nature as the corrosion effect which has been shown by McAdam (1, 2),[3] to have a pronounced influence on the endurance properties of metals.

---

[3] The numbers in parentheses here and throughout the text refer to the papers listed in the selected bibliography appended to this paper.

The variations in the effect of the acid treatment on the fatigue limits are undoubtedly associated with different solubility rates of the materials in the three conditions of heat treatment. The difference in the surface contours of the steel in different specimens of any one series was of about the same magnitude as the difference between the three series of specimens (acid pickled, galvanized, and electroplated). Figure 10 shows, in longitudinal section, typical surface contours of the steel of the specimens after the various treatments.

The decrease in fatigue limit resulting from the acid treatment was much greater for the quenched carbon steels than for the an-

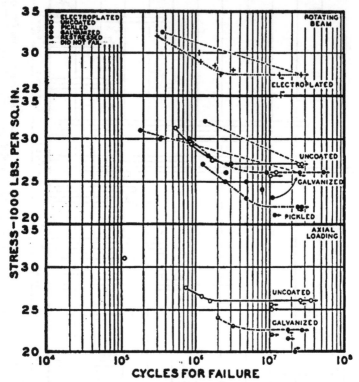

FIGURE 3.—*S–N diagrams for fatigue tests of 0.02 per cent carbon open-hearth iron*

nealed or tempered steels. This is in accord with the generally accepted idea that a hard steel with low ductility is more susceptible to notch effects than a softer and more ductile steel.

There was a marked decrease in the fatigue limit of the galvanized materials as determined by the rotating-beam method except for the open-hearth iron, for which there was little if any difference (4.0 per cent). For the carbon steels the decrease ranged from 13 to 42.5 per cent, and was greater for the quenched and the tempered steels than for the annealed steels. By the axial loading method of test the decrease for the open-hearth iron was 13.5 per cent and for the carbon steels the decrease was much greater in the tempered steels than in the annealed steels.

Except for the open-hearth iron and the quenched 0.72 per cent carbon steel, the decrease in fatigue limit was greater for the galvanized than for the acid-pickled material. It might be considered that this further decrease was caused by an increased pitting or notch effect on the surface of the steel by the action of the zinc in the galvanizing process. That this was not the only cause is indicated by the fact that galvanized specimens of the annealed and the

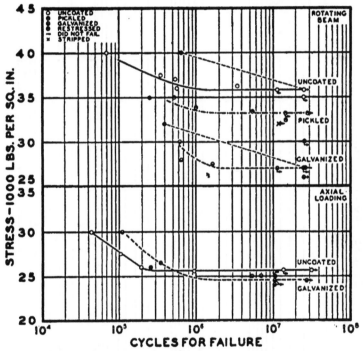

FIGURE 4.—*S–N diagrams for fatigue tests of 0.45 per cent carbon steel, annealed*

quenched 0.45 per cent carbon steels, and of the annealed 0.72 per cent carbon steels, when stripped of the zinc coating[4] did not fail in 10,000,000 cycles in the rotating beam machines at stresses just under the fatigue limits of the acid-pickled materials. The stripped specimen of the tempered 0.45 per cent carbon steel failed after 3,500,000 cycles which indicated that its fatigue limit was not much lower than the stress at which it failed. The stripped specimen of the tempered 0.72 per cent carbon steel failed after 300,000 cycles, but at a stress 32,000 lbs./in.² higher than the fatigue limit of the galvanized material.

Hence it is believed that the conclusion is warranted that the presence of a hot-dip galvanized coating causes a serious lowering

[4] The zinc coatings were dissolved in hydrochloric acid (specific gravity 1.19) containing 1 ml of antimony chloride solution [32 g of SbCl₃ in 1,000 ml HCl (specific gravity 1.19)] to 100 ml of acid. A.S.T.M. specification A 90-30.

of the fatigue limit of carbon steels below that which would be obtained in the same steels in the polished but uncoated condition.

A similar conclusion can be drawn from the results of investigations by Harvey (3, 4), Haigh (5), and Fuller (6) of the protection against corrosion fatigue afforded to steels by galvanized or other types of metallic coatings. Although their results showed that the endur-

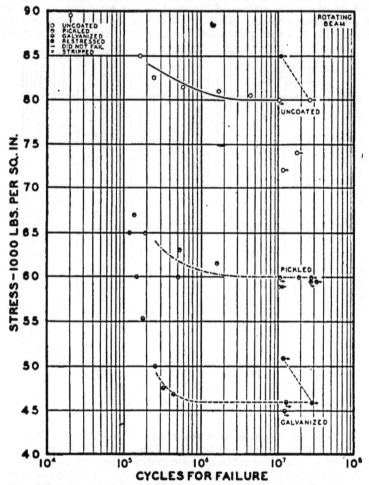

FIGURE 5.—*S–N diagrams for fatigue tests on 0.45 per cent carbon steel, quenched*

ance properties of the coated metals were definitely better than for the uncoated metals, when subjected to simultaneous stress and corrosion, the corrosion-fatigue limits of the galvanized materials were at the same time lower than the endurance limits of the uncoated materials not subjected to corrosion.

In marked contrast to the lower fatigue limit of the galvanized specimens, the fatigue limits of the zinc-plated specimens of the softer steels were equal to those of the corresponding uncoated specimens. The fatigue limits of the zinc-plated specimens of the tempered steels were higher than those of the corresponding uncoated specimens by 5.5 per cent for the medium carbon and 11 per cent for the higher carbon steel.

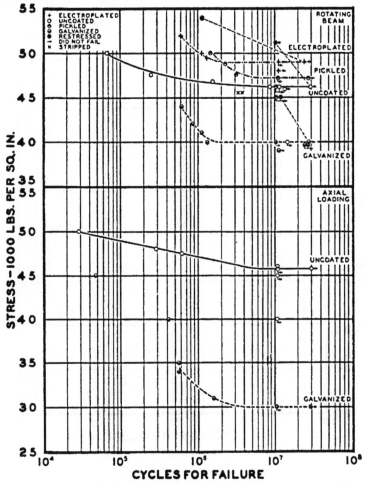

FIGURE 6.—*S–N diagrams for fatigue tests on 0.45 per cent carbon steel, tempered*

Both the hot-dip galvanized and the electroplated specimens had been subjected to comparable acid treatments before the coatings were applied. Furthermore, the steels of each series were identical in composition and had received the same heat treatments. Hence any differences in fatigue properties of the galvanized and electroplated specimens of the same series can not be ascribed to differences in the steels themselves, particularly to any suspected differences in

ra₊e of notch propagation. It is believed that the increased fatigue
limits of the electroplated coatings were not the result of a strengthening effect of the zinc coating. Numerous cracks were found in
both types of coatings in the more highly stressed portions of the
specimens after they were removed from the testing machines.
This indicates that the maximum fiber stresses in the coatings were
above their endurance limits. It is reasonable to expect that the
endurance strength of the galvanized coating was higher than that
of the electrodeposited zinc. Consequently any strengthening effect

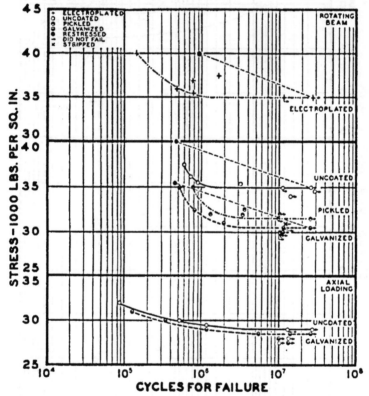

FIGURE 7.—*S–N diagrams for fatigue tests on 0.72 per cent carbon steel, annealed*

from the coating would be expected to be derived from the galvanized rather than from the electrodeposited coating.

The two types of coating are different in many respects. The
electrodeposited coatings, as would be expected, were homogeneous
throughout. The bond between zinc and steel may have been a
"molecular bond," but at the interface there was little, if any, diffusion of iron into the zinc. Figure 11 (a) shows a typical cross
section of the electroplated steel specimen.

The structure of hot-dipped galvanized coatings on steel has been
studied by a number of investigators (7, 8, 9). Without going into

a detailed discussion of the composition of the layers it suffices at
this time to state that the galvanized coatings on the specimens used
in this investigation consisted of at least three layers. The outer
layer was predominantly pure zinc, the innermost layer consisted
largely of the more or less well-known iron-zinc intermetallic compounds. The intermediate layer consisted largely of intermetallic
compound or compounds interspersed in a zinc matrix. On the
annealed steel and the open-hearth iron specimens the intermediate
layer was relatively much thicker than the innermost layer, whereas,
on the quenched and on the tempered steels the thickness of the
innermost layer approached that of the intermediate layer.

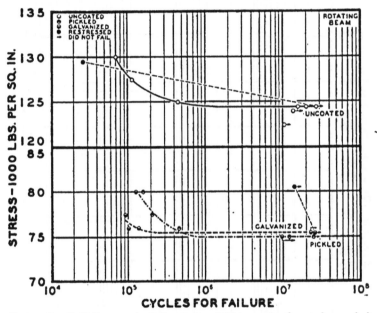

FIGURE 8.—*S–N diagrams for fatigue tests on 0.72 per cent carbon steel, quenched*

Micrographs (c) and (d) Figure 11 show in cross-section the galvanized coatings on an annealed and on a quenched 0.45 per cent
carbon steel. The difference in thickness of the inner, iron rich,
layers is marked. Possibly a difference in solubility in zinc of the
heat-treated steel, and the annealed steel or open-hearth iron is
responsible for the difference in thickness of the alloy layers.

The scratch hardness of the two types of coating indicated that the
outer layer of the galvanized coatings was of the same order of hardness as the electrodeposited coatings. The intermediate and innermost layers of the galvanized coatings were increasingly harder as
the steel surface was approached. This is illustrated in the micrographs (e) and (f) of Figure 11. The alloy layer adjacent to the steel
(e) appears to be even harder than the steel itself.

As stated before, there were numerous cracks in both types of
coating of the tested specimens. In the electrodeposited coatings

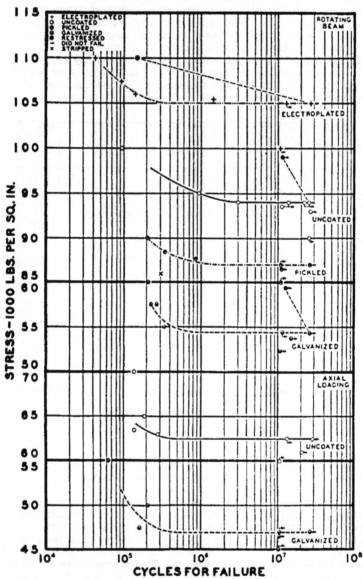

FIGURE 9.—*S–N diagrams for fatigue tests on 0.72 per cent carbon steel, tempered*

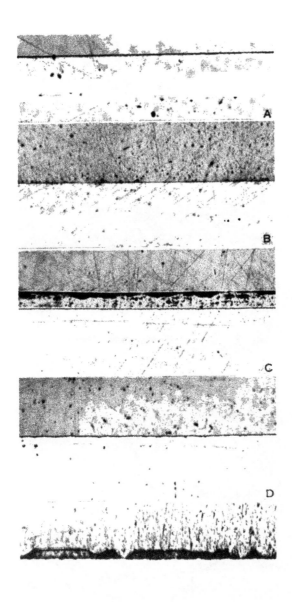

FIGURE 10.—*Typical contours of the steel surface of specimens unetched.* × 50

a, Polished; b, polished and then acid pickled; c, polished, acid pickled, and then galvanized; d, polished, acid pickled, galvanized, and then stripped; and e, polished and then electroplated.

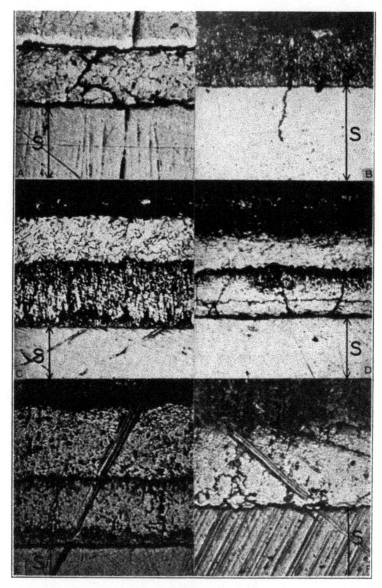

FIGURE 11

a, Structure of electrodeposited zinc coating on 0.72 per cent carbon steel, annealed; b, crack in galvanized coating extending into steel, 0.45 per cent carbon, annealed; c, structure of galvanized coating on 0.45 per cent carbon steel, annealed. Note the relatively thick intermediate layer, the thin innermost layer, and the light colored outer layer of relatively pure zinc; d, galvanized coating on 0.45 per cent carbon steel tempered. Note the thickness of the inner, hard, brittle layer adjacent to the hardened steel as compared with the thin layer of similar composition next to the softer steel of photograph c; e, scratch made with Bierbaum microcharacter across galvanized coating on tempered 0.45 per cent carbon steel. Note differences in width of scratch in the outer zinc layer and the intermediate and innermost iron-zinc alloy layers. Note that the innermost alloy layer appears to be harder than the steel; f, scratch made with Bierbaum microcharacter across electrodeposited zinc coating on tempered 0.72 per cent carbon steel. Discontinuity of scratch between zinc and steel was caused by difference in elevation between zinc and steel. Etched with chromic acid solution containing sodium sulphate (20 g CrO$_3$, 1.5 g Na$_2$SO$_4$ in 100 ml H$_2$O). $\times$ 275. Arrows indicate steel base.

the cracks were irregular (a) Figure 11, and appeared to be intergranular. In the galvanized coatings there were many more cracks in the intermediate and innermost layers than in the outer layer. Many of the cracks appeared to have started in the layer adjacent to the steel and to have progressed toward the outer surface, and in some instances the surface was not quite reached. In many instances the cracks undoubtedly originated at the surface and progressed inwardly toward the steel.

Of the broken specimens which were sectioned for examination under the microscope, two were found in which there was a crack in the steel which was a continuation of a crack in the innermost layer of the galvanized coatings. One of these is shown in (b) Figure 11. Although there were many cracks in the galvanized coatings, the evidence indicated that only a few had advanced into the steel. The probability is that none of the cracks in the electrodeposited coatings had extended into the steel, and that the cracks which caused failure of the electroplated specimens in the fatigue test originated in the steel itself.

The explanation of the lower fatigue limits of the galvanized specimens as compared with the uncoated or electroplated specimens is believed to lie: (a) In the difference in the stress conditions at the bottom of a crack in the inner, relatively hard, layers of the galvanized coating and those in a similar crack in the softer electrodeposited zinc; and (b) in the difference in the nature of the bond between zinc and steel in the two types of coatings.

It has been shown (10) that in relatively soft and ductile metals subjected to repeated stresses, slip lines form either previously to, or subsequently to, the appearance of cracks and that the cracks advance in the direction of the slip lines. The slip lines are an indication of plastic deformation under an applied stress which decreases when the deformation occurs. It is believed that the zinc of the electrodeposited coatings had sufficient ductility to deform around the bottom of an advancing crack, and that the resulting decrease in stress concentration when the crack had advanced to the steel was sufficient to stop the crack at that point. The discontinuity between zinc and steel was an additional aid in halting further advance of the crack. Consequently, the normal endurance limit of the steel was attained.

In the case of the galvanized specimens, a crack advancing into the relatively hard and very brittle inner layers did not meet with any conditions conducive to a decrease of stress concentration. The crack progressed to the outer steel fibers with undiminished stress. The intimate bond between coating and steel offered no obstacle to the advance of the crack into the steel. Naturally not every crack produced in the coating in the course of the fatigue test penetrated into the steel. A fortuitous combination of maximum stress concentration and conditions at the surface of the steel most favorable to the propagation of the stress, determined the location of the crack which led to failure of the specimen. Consequently since the presence of a hot-dipped galvanized coating promotes stress concentrations, the fatigue limit of such a coated specimen was appreciably lower than the normal endurance limit of the steels.

The data obtained on the specimens which were restressed at 5,000 lbs./in.^{2} above the fatigue limit, after they had been subjected to 25,000,000 cycles of stress at the fatigue limit, indicated that only the quenched and the tempered steels were appreciably strengthened by the previous understressing. All of the annealed carbon steel and open-hearth iron specimens tested failed to "run" at the higher stress. The 0.72 per cent carbon steel specimen, uncoated and the two electroplated specimens of the 0.72 and the 0.45 per cent carbon steel, in the tempered condition, also failed to run.

The interesting observation was made that distinct spangles were developed on the surface of the zinc of the galvanized specimens shortly after they were placed in operation in the rotating beam machines.

VI. ACKNOWLEDGMENT

Grateful acknowledgment is made by the authors to P. R. Kosting for his aid in the preparation of the micrographs of the zinc coatings and of the scratch hardness tests.

VII. SELECTED BIBLIOGRAPHY

1. D. J. McAdam, jr., Influence of Water Composition on Stress Corrosion, Proc., Am. Soc. Testing Materials, vol. 31, pt. 2, p. 259, 1931; and additional papers by McAdam listed at the end of this paper.
2. D. J. McAdam, jr., Stress Corrosion of Metals, Proc., Zurich Congress of the New International Association for the Testing of Materials, 1931.
3. W. E. Harvey, Zinc as a Protective Coating Against Corrosion-Fatigue of Steel, Metals, and Alloys, vol. 1, p. 458, April, 1930.
4. W. E. Harvey, Cadmium Plating v. Corrosion-Fatigue. Pickling v. Corrosion-Fatigue, Metals and Alloys, vol. 3, p. 69, March, 1932.
5. B. P. Haigh, Chemical Action in Relation to Fatigue of Metals, Trans. Inst. Chem. Engrs., vol. 7, p. 29, 1929.
6. T. S. Fuller, Some Aspects of Corrosion Fatigue, Trans. Inst. of Metals Division, A. I. M. E., p. 47, 1929.
7. H. S. Rawdon, Structure of Commercial Zinc Coatings, Proc., Am. Soc. Testing Materials, vol. 18, pt. 1, p. 216, 1921.
8. W. H. Finkeldey, The Microstructure of Zinc Coatings, Proc., Am. Soc. Testing Materials, vol. 26, pt. 2, p. 304, 1926.
9. W. Guertler, Structure of Galvanized Iron, Zeitschrift für Metallographie, vol. 1, p. 353, 1911.
10. H. F. Moore and Tibor Ver, A Study of Slip Lines, Strain Lines, and Cracks in Metals under Repeated Stress, Bull. No. 208, Eng. Expt. Sta., Univ. Illinois, June 3, 1930.
11. H. W. Gillett, The Need for Information on Notch Propagation, Correlated Abstracts, Metals and Alloys, vol. 1, p. 114, September, 1929.
12. H. W. Gillett, What is This Thing Called Fatigue? Metals and Alloys, vol. 2, p. 71, February, 1931.

WASHINGTON, April 30, 1932.

POWER INPUT AND DISSIPATION IN THE POSITIVE COLUMN OF A CÆSIUM DISCHARGE

By F. L. Mohler

ABSTRACT

The power input per centimeter is equal to the voltage gradient times the current. Two recognized sources of power dissipation are the recombination of ions on the tube walls and the atomic radiation.

The voltage gradient is measured between two probes at the axis of the dis-

tions were determined by a photo-electric cell. The first doublet of the principal series at 8,521 and 8,944 A contributes most of the radiation.

Results show that with low currents and pressures the radiation and recombination losses account for nearly the entire power input. With increasing currents and pressures there is an increasing balance of power which is unaccounted for. All the power terms change very slowly with pressure. The recombination loss is negligible at low currents and the most important factor at high currents.

CONTENTS

Conditions in the positive column are determined by the nature

the voltage gradient times the current, and it follows from the nature of the column that the power dissipated in each element of length is equal to the input. The dissipation of power appears as heating of the tube walls, either directly or by conduction and radiation from

---

[1] Tonks and Langmuir, Phys. Rev., vol. 34, p. 876, 1929.

the gas, and as selective radiation of the atoms which is transmitted by the tube walls. The general statement is made in the Handbuch der Physik that the total radiation is a small fraction of the power input, but the experiments cited deal with gases which have their strongest radiation in the far ultra-violet and this is completely absorbed by the tube walls. Thus Crew and Hulburt [2] find that the total radiation emitted by a hydrogen discharge is about 1 per cent of the input, but they point out that about nine times the observed radiation is in the first Lyman line alone.

The alkali vapor spectra are the only spectra in which all the strong lines are transmitted by glass. The possibility of obtaining a high radiant efficiency is utilized in the Osram sodium lamp.[3] Under normal operating conditions (current density 1 amp. per cm^2) the lamp radiates about 9 per cent of the power input. This radiation is predominantly D lines and the luminous efficiency is probably higher than any other light source, though the figure for radiation efficiency is certainly not impressive. The fact is that the lamp has to be overloaded to a point of low efficiency to maintain a high tube temperature and vaporize the sodium. With one-tenth of the normal current and external heating a radiation efficiency of 70 per cent is observed.

There is no published work in which radiation measurements have been combined with detailed electrical measurements in the study of a discharge. Other phases of power dissipation have been treated by Langmuir and his associates in their comprehensive theoretical [4] and experimental [5] studies of the mercury positive column. An important factor is the flow of ions to the walls and their recombination on the walls. The fraction of the power dissipated in this way depends on the current and is small for low currents. Elastic collisions between electrons and atoms will give some energy to the atoms, but Killian has evaluated this loss and finds it negligibly small for the mercury arc. He concludes that power dissipated in inelastic collisions between electrons and atoms and radiated by the atoms must account for the balance of the power loss not accounted for by recombination on the walls. The conclusion seems entirely reasonable, though there is a possibility that the excited atoms dissipate some energy by collisions with normal atoms. The probability of dissipation of energy by collision (quenching) during the life of an excited state is undoubtedly small under most discharge conditions; but since the resonance radiation is very strongly absorbed, the radiation diffuses slowly through the gas by repeated absorption and emission. For this reason the small probability of quenching during the radiation life must be multiplied by a large and unknown factor to give the probability of quenching during the diffusion time.

This paper reports measurements of power input, radiation loss, and wall recombination loss in the positive column of a cæsium discharge. Tubes of one diameter 1.8 cm have been used, and the effect of the two remaining independent variables, vapor density, and current, have been studied. This work has been done in conjunction with a study of collision processes in the positive column which will be reported in another paper.

[2] Crew and Hulburt, Phys. Rev., vol. 29, p. 843, 1927.
[3] Kreft, Pirani, and Rompe, Tech. Wissenschaft Abhandl aus dem Osram Konzern, vol 2, p. 24, 1931.
[4] Tonks and Langmuir, General Theory of the Plasma of an Arc, Phys. Rev., vol. 34, p. 876, 1929.
[5] Killian, Uniform Positive Column of an Electric Discharge in Mercury Vapor, Phys. Rev., vol. 35, p. 1238, 1930.

II. ELECTRICAL MEASUREMENTS

Figure 1 illustrates the type of discharge tube used. A thermionic discharge from an oxide-coated platinum strip cathode was employed. Vapor pressure was controlled by a separate heater around the side tube containing the cæsium. Tubes were outgassed by baking and by running a discharge for four or five hours before sealing off from the pumps.

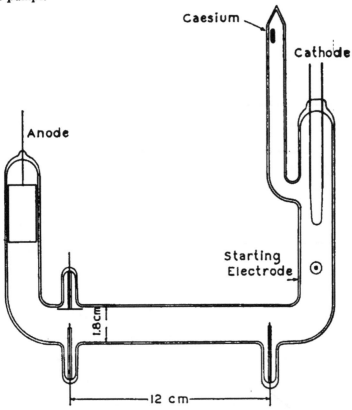

FIGURE 1.—*Discharge tube used in making power input and dissipation measurements*

The voltage gradient was measured between two small probes near the axis of the tube and 12 cm apart. For the lower pressure and current conditions the probes were platinum wires 0.4 mm in diameter and 2 mm long. For the higher currents the platinum wires were cut off flush with the insulating tube. In cæsium vapor difficulty is encountered from leakage over the insulating surfaces, and this was greatly reduced by avoiding a close fit between the insulating tube and the wire. Melting off of the small probes by the striking of an arc was a source of trouble. If a probe is too large the current may rise to a destructive value even at potentials negative to the space potential. The random space current becomes so large that an electrode less than 0.01 cm² in area may rob the discharge and suddenly become the anode. The ion current to the walls was measured by a disk of platinum 1 cm in diameter which was bent to fit against the wall.

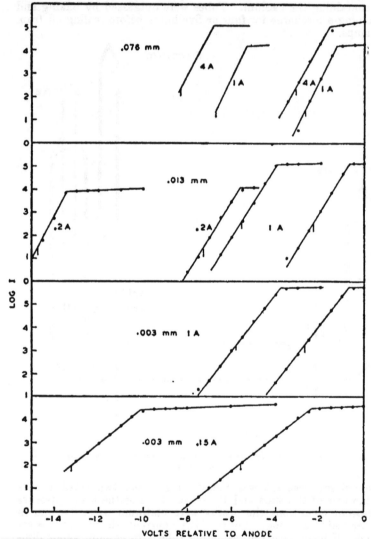

FIGURE 2.—*Log current voltage curves for the two small probes at the axis of the tube and 12 cm apart*

Vertical dashes mark the potentials of zero current.

Figure 2 gives plots of the log of the electron current to the small probes versus potential relative to the anode for several typical conditions. The plots are accurately linear for all conditions. The slope of the left branch measures the electron energy; the intersection of the two branches measures the space potential. The slopes as measured by the two probes are equal within experimental error. The electron current to the disk also rises at the same rate, showing that the electron energy is the same at the center and at the wall. This has been shown before by Killian.[6]

The voltage difference between the probes can be obtained from the difference between the space potentials, or since the curves are parallel, it can be measured between the potentials of zero current. The two intervals agree within about 0.1 volt, while the difference is several volts. This difference, divided by 12 and multiplied by the current, gives the input in watts per centimeter of the column.

The power dissipated by flow of ions to the tube walls is equal to the ion current in amperes times the sum of the ionization potential (3.9 volts) plus the potential drop through which the ions fall, plus the energy in electron volts of the electrons. The leakage to the disk at negative potentials was always measurable, but in general small enough so that an extrapolation of the current voltage curve to the wall potential should involve no serious error. This current multiplied by the geometrical factor, 7.2, gives the ion current to 1 cm of the tube. The assumption that the current, corrected for leakage, measures pure ion current requires some justification. In many discharges the electron emission from negative electrodes is an important part of the current. This emission comes from photo-electric effect and from collision of excited atoms with the metal surface and in either case depends on whether the energy of the excited states is greater than the work function of the metal. The energy of the cæsium resonance state is 1.45 electron volts. Ives and Olpin[7] find that the minimum work function of a pure metal surface in cæsium is about equal to this, while the results of Boeckner and the author[8] on probe radiation indicate that platinum probes in a cæsium discharge have a high work function compared to other metals (well above 2 volts, but not measured). The consistency of the present measurements of power input and loss gives further evidence that no serious error has been made in interpreting the current at negative potentials as an ion current.

In the discharge space the random electron current is about a thousand times the ion current, and the insulating wall must take up a potential which draws the full ion current to it and limits the electron current to an equal value. The potential difference is limited to a thin sheath a fraction of a millimeter thick over the surface of the tube. There is also a small potential difference in the same direction between the space potential near the wall and at the center of the tube. Since the electrons have a Maxwell distribution of velocity, the small fraction of fast electrons which pass through the retarding field will reach the wall with the same distribution of velocity as the electrons in the field free space. The contribution of the electorn kinetic energy in electron volts (0.2 to 0.3 volt) is obtained from the semilog. plot.

$$V_0 = (V_2 - V_1)/(ln I_1 - ln I_2)$$

A rigorous measurement of the potential through which the ions fall

[6] Killian, Phys. Rev., vol. 35, p. 1238, 1930.
[7] Ives and Olpin, Phys. Rev., vol. 34, p. 117, 1929.
[8] Mohler and Boeckner, B. S. Jour. Research, vol. 7, p. 751, 1931.

would require a small probe at the wall which could be brought to the space potential. (The disk is much too big for this.) Then the difference between the potential of zero current and the space potential would give the potential drop in the sheath (about 2 volts). The average ion also falls through a potential drop somewhat less than the difference in space potentials near the walls and at the center of the tube, a matter of a few tenths of a volt. From a consideration of Killian's measurements it was evident that the difference between the space potential and potential of zero current at the center of the the tube was slightly larger than the potential drop across the wall sheath and less than the drop between the center of the tube and the wall. The error in using this potential drop instead of the true potential difference can only be 0.1 or 0.2 volt; and as the recombination energy is 6 or 7 electron volts, the use of another probe at the wall seemed unnecessary.

III. RADIATION MEASUREMENTS

The predominant lines in an alkali metal spectrum are the first doublet in the principal series, and in cæsium these lines fall in the near infra-red at 8,521 and 8,944 A. Nearly all other lines fall in a range transmitted by glass and water except the first doublet of the diffuse series near 36,000 and 35,000 A. As the tube is operated in a furnace at 250° to 300° C., the use of a water cell in making thermopile measurements was unavoidable and this far infra-red doublet was completely filtered out. The G. E. cæsium-oxygen-silver photoelectric cell is very sensitive to the resonance lines, and it was useful in evaluating the transmission of the water cell, tube walls, etc. A disk thermopile 0.5 cm in diameter was used.

The procedure was as follows: The thermopile was calibrated without the water cell or any window against a radiation standard. It was then exposed through a water cell and furnace window to the cæsium radiation from the full width and a 10 cm length of the positive column. The galvanometer deflection of 3 or 4 cm could be measured with sufficient precision by averaging many observations. It was inconveniently small for routine measurements, so the deflection of the direct radiation was compared with the deflection produced when the radiation from a 1 cm length of the tube was focused by a short focus lens so that the entire image fell on the thermopile. The transmission of the tube wall, furnace window, and water cell for the cæsium radiation were measured by the photo-electric cell. It was assumed that the radiation from the discharge was equal in all directions. Two independent calibrations before and after the series of experiments gave values of 2.13×10^{-3} and 2.18×10^{-3} watts per centimeter galvanometer deflection per centimeter of the column for the total radiation.

An experiment was made on the absorption of the cæsium radiation by a 4 cm cell of cæsium vapor. With the discharge vapor pressure 0.004 mm, the cell containing vapor at about ten times that pressure reduced the photo-electric effect to less than one-tenth. As the vapor is perfectly transparent to all lines except the principal series lines, this shows that most of the energy is in the resonance lines. Photographs of the spectrum indicate that these lines remain predominant at all pressures. Radiation measurements were made before and after a series of electrical measurements with no evidence of any error appreciable in comparison with the electrical measurements or with the error in optical adjustments when tubes were changed.

IV. RESULTS

Discharges operated with a voltage drop ranging from 5 to 15 volts. It was found that a large change in the battery voltage with a compensating change in the series resistance to give equal current did not change the radiation by 0.1 per cent. Also a change in the voltage drop in the tube produced by changing the cathode temperature had no effect. Currents ranged from 0.15 to 4 amperes; pressures from 0.001 to 0.3 mm. The low pressure limit was set by the dark space approaching the probe nearest the cathode.

Figures 3 and 4 show the power input; P, the power dissipated by recombination; W, the radiation; R, and P-W as functions of the pressure and the current. With currents of 0.2 amp. practically all the power input is radiated. With currents of 1 amp. or less and pressures less than 0.03 mm, P-W is nearly equal to R; that is, the power input is nearly all accounted for. The mean value for all measurements in this range gives

$$R/(P\text{-}W) = 0.95 \pm 0.04$$

where 0.04 is the mean error and does not include the calibration error. With increasing current and pressure there is a rapidly increasing difference between P-W and R. A plot of this unexplained residual is given as a function of the current for various pressures in Figure

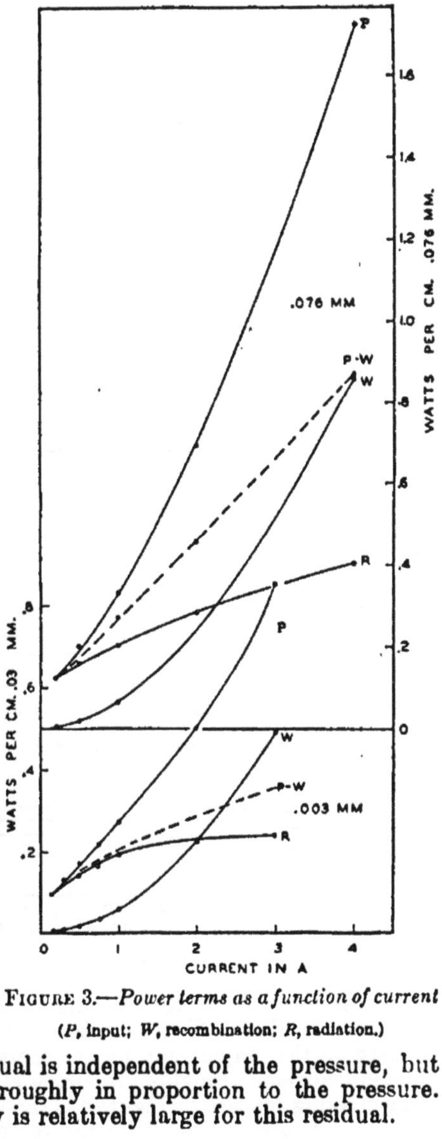

FIGURE 3.—*Power terms as a function of current*

(*P*, input; *W*, recombination; *R*, radiation.)

5. Up to 0.013 mm the residual is independent of the pressure, but above this point increases roughly in proportion to the pressure. The experimental uncertainty is relatively large for this residual.

V. CONCLUSIONS

Little can be said as yet about the factors determining the form of the curves of Figures 3, 4, and 5. With increasing current the gradient at first decreases and then increases and correspondingly the input curve of Figure 3 changes. The rapid increase at higher currents

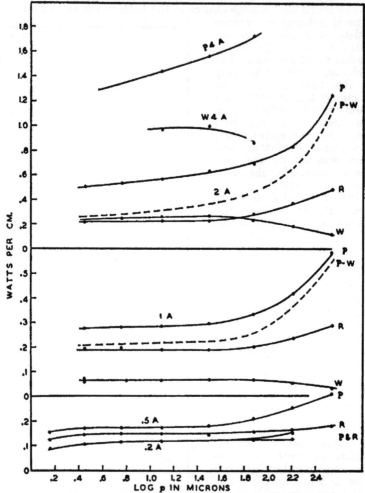

FIGURE 4.—*Power terms as a function of the log of the pressure in microns*

comes in part from the recombination loss and in part from the unknown residual. The variation in recombination loss, W, depends largely on the ion current, for the energy per ion changes very slowly. The ion current is roughly proportional to the electron concentration. The striking feature of Figure 4 is the very slow variation of all power terms with pressure. The measurements cover a 300-fold range of pressure, and over half of this range the variation is scarcely more than the experimental error. An appreciable variation is evident with high pressures and high currents, and the unknown residual effect is an important factor in this. The fact that there is an unexplained balance in the power equation is not surprising. We have mentioned in the introduction several other sources of power loss; the unmeasured far infra-red radiation, the dissipation of energy by elastic collisions, and the quenching of resonance radiation by interatomic collisions, and there are undoubtedly other possibilities. The evidence is as yet insufficient to choose between them. It is reassuring to find that over a considerable range the power loss is predominantly radiation of the resonance lines and recombination on

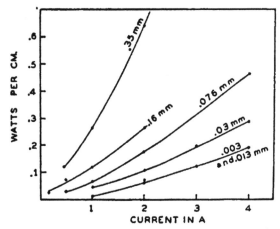

FIGURE 5.—*Balance of the power loss not accounted for by recombination and radiation (P-W-R) as a function of the current*

the walls. It suggests that an approximate theory can be developed which considers production of ions and of the first excited state exclusively and that other phenomena can be treated independently. In this connection it is of interest to note that the visible spectrum of cæsium involving higher excited states shows large variations in intensity and intensity distribution in contrast to the very gradual change of the total radiation which is largely resonance radiation.

The published figures on the radiation efficiency of the Osram sodium lamp referred to in the introduction are consistent with our results, though the data refer to the entire discharge rather than the positive column, so that the values are lower. The cæsium positive column, with a current density of 1 amp. per cm^2 at low pressure, radiates 28 per cent of the power and at 0.1 amp. per cm^2 the radiation is 96 per cent as compared with the overall efficiencies of 9 and 70 per cent for the sodium discharge. Evidently the alkali discharges are qualitatively similar in their characteristics.

Alkali vapor discharges of the type used in this work may have useful applications as sources of intense monochromatic radiation. The intensity of this radiation is insensitive to pressure and current variations, so that a very accurate control of intensity is possible. The fact that the efficiency of production of radiation can be made almost unity is of little practical value, since the high efficiency is attained only by keeping the power input and brightness low.

WASHINGTON, April 11, 1932.

RP456

THERMAL EXPANSION OF SOME SILICATES OF ELEMENTS IN GROUP II OF THE PERIODIC SYSTEM

By R. F. Geller and Herbert Insley

ABSTRACT

A ceramic body having the unusually low average coefficient of linear thermal expansion of 0.53×10^{-6} in the range 0° to 200° C., reported by W. M. Cohn, was investigated, and it was found that this material is composed essentially of the mineral cordierite. The investigation was amplified to include thermal expansion determinations for some compounds of other elements in Group II of the periodic system. The data obtained indicate that zinc orthosilicate, celsian and beryl appear worthy of consideration by those interested in the development of bodies having low coefficients of thermal expansion.

A ceramic body having the unusually low coefficient of linear therma.

Cohn's findings are duplicated. The study was amplified to include linear thermal expansion determinations for some silicates of other elements in Group II of the Periodic Table.

The specimens of series A (Table 1) were made of a body composed of 43 per cent talc (approximately 95 per cent $3MgO \cdot 4SiO_2 \cdot H_2O$), 35 per cent Florida kaolin (93 to 95 per cent kaolinite), and 22 per cent electric furnace corundum (99 per cent Al_2O_3) dry sieved through a No. 200 sieve. This body was intended as a duplicate of

---

[1] Temperatures are given in °C., throughout this report.
[2] Ber. der Deut. Ker. Gesell., vol. 10, p. 271, June, 1929, and vol. 11, p. 62, February, 1930, and Ann. Physik,

the body described by Singer [3] as typical of those investigated by Cohn.

The specimens of series B (Table 1) were made of a body composed of 39.5 per cent talc, 47.2 per cent Georgia kaolin (96 to 98 per cent kaolinite), and 13.3 per cent anhydrous Al_2O_3 obtained by heating reagent quality aluminum nitrate.

The specimens of series C (Table 1) were made of a wet ground mixture of 24 per cent of reagent quality basic magnesium carbonate $(Mg(OH)_2 \cdot (MgCO_3)_4)$, 45 per cent of potters' flint less than 10μ in size, and 31 per cent of Al_2O_3 obtained as described for series B.

No attempt was made to produce specimens containing only one crystal phase. Specimens containing a minimum of 90 per cent of the crystalline compound investigated, as determined by petrographic examination, were considered of sufficient purity to establish the approximate thermal expansion of that compound.

Unless otherwise stated, the various heat treatments described in Tables 1 and 2 and in the text were made in electrically heated furnaces, the resistors being 80 per cent platinum-20 per cent rhodium wire.

Expansion determinations were made by the interferometer method.[4] The specimens of series A and B (Table 1) were molded and heated in the form of disks about 2 cm in diameter, 0.3 cm in thickness, and with a hole through the center about 0.7 cm in diameter. Three small cones or "feet" protruded from opposite points on each side. The over-all height varied between 0.50 and 0.75 cm. Irregularly shaped pieces, approximately pyramidal in form and from 0.50 to 0.75 cm in height, were used for expansion determinations of the other materials investigated; three such pieces were used for each determination. The rate of temperature change during both heating and cooling tests did not as a rule exceed 3° per minute, and for most of the tests was maintained at from 1° to 2° per minute.

[3] Ber. der Deut. Ker. Gesell., vol. 10, p. 209, June, 1929.
[4] Fizeau, Ann. d. Phys., vol. 128, p. 564, 1866. C. G. Peters and C. H. Cragoe, B. S. Sci. Paper No. 393, 1920. G. E. Merritt, B. S. Sci. Paper No. 485, 1924.

TABLE 1.—*Thermal expansion determinations of bodies composed essentially of "cordierite," $2MgO·2Al_2O_3·5SiO_2$*

[All temperatures in °C.]

Specimen No.	Composition, determined petrographically	Preparation	Test No.	Linear thermal expansion, expressed as average coefficient [1]			
				R.T.[2] to 100° (×10⁻⁴)	R.T.[2] to 200° (×10⁻⁴)	Miscellaneous R.T. to— (°C.)	Coefficient (×10⁻⁴)
A-1[3]	A, corundum, inclusions in phase C. B, crystalline grains with tendency to elongated forms and corroded surfaces. Index between 1.60 and 1.65; probably forsterite. C cordierite, $2MgO·2Al_2O_3·5SiO_2$. Constitutes the ground mass of the body; low double refraction, twinning common; mean index about 1.53.	Body A. Held at 1,350° for 1 hour	1	0.7	1.1	500	1.9
A-15	Measurably less corundum than in A-1; it occurs as inclusions in the cordierite which constitutes the balance of the body.	Body A. Held at 1,350° for 16 hours	1	.5	.8	660	1.6
A-15a	do.	Specimen A-15. Reheated at 950° to 960° for 13 days. Maximum temperatures reached in preceding tests: 310°, 203°, 313°, 505°.	5	.4	.8	400	1.3
A-17	do.	Body A. Held at 1,300 to 1,350° for 67 hours. Maximum temperatures reached in preceding tests: 323°, 535°.	3	.2	.6	365 75 400	.9 −.2 +.9
A-3	Approximately 90 per cent cordierite and 10 per cent inclusions, largely corundum.	Body A. Heated in a gas-fired kiln at approximately 1,350° for 3 days and permitted to cool in the kiln.	1			75–400	1.0
A-3a[4]	do.	Rods approximately 1¼ cm in diameter and 30 cm long. Preparation same as A-3.	1			75 400	−.3 +.8 +1.0
B-2	90 per cent or more cordierite, balance corundun and a small amount of what appears to be enstatite.	Body B. Heated at 1,345° for 45 hours	1	.5	.9	308	1.1
B-2a	A, cordierite 85 to 90 per cent of total; index 1.53+. Present as well developed grains, larger than in preceding sample. B, spinel occurring as inclusions in phase A. C, very small amount of mullite as groups of fine needles in phase A.	Specimen B-2. Reheated at 1,410° for 17 hours. Maximum temperature reached in test No. 1, 211°.	2		.5	55 55–100 604	−.1 +.2 1.3

[1] Values derived from observations on heating.

[2] R. T.=Room temperature; varied from 20° to 28°.

[3] Chemical analysis by J. F. Klekotka showed SiO_2, 40.7 per cent; R_2O_3, 43.1 per cent; MgO, 15.2 per cent; CaO, 0.06 per cent; ignition loss, 0.15 per cent.

[4] Expansion determined by P. Hidnert using the apparatus and method described in Bureau of Standards Scientific Paper No 524 (1926) by W. Souder and P. Hidnert. R. T.=25°.

TABLE 1.—*Thermal expansion determinations of bodies composed essentially of "cordierite," $2MgO \cdot 2Al_2O_3 \cdot 5SiO_2$—Continued*

[All temperatures in °C.]

Specimen No.	Composition, determined petrographically	Preparation	Test No.	Linear thermal expansion, expressed as average coefficient [1]		Miscellaneous	
				R.T. to 100°	R.T. to 200° [2]	R.T. to—	Coefficient
C-1	Composed of at least 90 per cent cordierite; index 1.53. The balance is made up of forsterite in irregular, partly dissolved crystals and of corundum in minute rounded grains.	Reagent quality materials. Heated at 1,425° for 20 minutes, ground to pass 80 mesh and reheated at 1,425° for 2 hours.	1	X10⁻⁴ 0.5	X10⁻⁴ 0.9	°C. 400	X10⁻⁴ 1.3
C-1a	do	Portion of C-1. Reheated at 750° to 800° for 2 weeks.	1		.9		
C-1b	do	Portion of C-1. Reheated at 900° for 4 days. Maximum temperature of test No. 1, 215°.	2		2.4		
C-2a	do	Portion of C-1. Reheated at 900° to 920° for 24 hours.	1		1.6		
C-2b	do	Specimen C-2a. Reheated at 1,025° to 1,050° for 3 days.	1		1.9		
C-2c	do	Specimen C-2b. Reheated at 1,120° to 1,150° for 2 days.	1		2.0		
C-2d	do	Specimen C-2c. Reheated at 1,300° to 1,380° for 2 days.	1		1.0		
C-3	Composed of cordierite of index 1.52+ and approximately 20 per cent of an undetermined constituent of index above 1.56 present as fibrous bundles (mullite?) growing from nuclei (corundum?) and always occurring as inclusions in the cordierite.	C-1. Melted at, and quenched from, 1,550° C. Devitrified at 1,100°, time 64 hours. Maximum temperature of test No. 1, 225°.	2	.6	1.0	285	1.1
C-4	Similar to C-3 except that the fibrous crystals appear to be somewhat better developed.	Portion of glass from C-3 devitrified at 1,355°, time 70 hours. Maximum temperatures of preceding test, 362°, 231°, and 446°.	4	-.5	{	100-200, 100-330	.3, .6
C-5	Practically complete devitrification to "unstable" $2MgO \cdot 2Al_2O_3 \cdot 5SiO_2$ of index 1.54+. The ternary compound constitutes 90 to 95 per cent of the total; balance fibrous growths similar to those in C-3.	Portion of glass from C-3 devitrified at 900°; time 4½ days.	1	3.4	4.0	317	4.0
D	Some mineral matter other than cordierite present. Occurs as extremely minute prismatic crystals of relatively high index, frequently concentrated along runlets.	Mineral from Ortherfol, Finland. Supplied by U. S. National Museum, specimen No. 78326. Material is not uniform. Values reported were selected from three duplicate tests.	1	.5	.9	300	1.2

TABLE 2.—*Thermal expansion determinations of some compounds containing elements in Group II (periodic table), including compounds of magnesium other than cordierite*

[All temperatures in °C.]

Specimen No.	Composition, determined petrographically	Preparation	Test No.	Linear thermal expansion, expressed as average coefficient [1]			
				R. T.[2] to 100° ($\times 10^{-4}$)	R. T.[2] to 200° ($\times 10^{-4}$)	Miscellaneous R. T. 10— (°C.)	Miscellaneous Coefficient ($\times 10^{-4}$)
F	Forsterite, 80 to 83 per cent $2MgO \cdot SiO_2$	Fusion in electric arc of talc and magnesium carbonate. Maximum temperature of test No. 1, 138°.	2	7.2	8.0	304	8.3
F	Spinel, at least 90 per cent $MgO \cdot Al_2O_3$, balance glass and birefracting material, probably clinostatite or forsterite.	Specimen cut from commercial brick and twice reheated to 1,800° C. Expansion determined by R. A. Heindl (B. S. Technical News Bulletin No. 152, December, 1929).				1,400	8.6
G	Beryl, indices, ω, 1.580; ϵ, 1.573 ($3BeO \cdot Al_2O_3 \cdot 6SiO_2$).	Mineral from San Jose de Bujauba, Minas Geraes, Brazil. U. S. National Museum specimen No. 94218. Maximum temperature of test No. 1, 460°.	2			48 / 225	−.2 / +.5
G-a	do	Mineral from San Jose de Bujauba, Minas Geraes, Brazil. U. S. National Museum specimen No. 94218. Maximum temperature of preceding tests, 460° 225°.	3	1.0		235 / 300	2.1 / 2.3
H	do	Mineral from Royalston, Mass. U. S. National Museum specimen No. 1662x. Maximum temperature of test No. 1, 146°.	2	.3	.7	290	.9
I	do	Mineral from Brazil. U. S. National Museum specimen No. 44386. Maximum temperature of test No. 1, 183°.	2	.3	.7	265	1.0
J	Beryl, indices, ω, 1.588; ϵ, 1.580.	Mineral from San Diego County, Calif. U. S. National Museum specimen No. 93926.	1	.9	1.3		
J-a	do	Mineral from San Diego County, Calif. U. S. National Museum specimen No. 93926. Maximum temperature of preceding test, 200°.	2	1.6		225 / 300	2.2 / 2.2
K	Zinc orthosilicate, approximately 95 per cent $2ZnO \cdot SiO_2$, balance glass.	Prepared by E. N. Bunting (B. S. Jour. Research, vol. 8 (2) (R. P. 413), p. 279, 1932), and obtained by devitrification. Melting point, 1,510° C.	1	.8	1.2	350	1.6
L	Zinc aluminate, 38 per cent ZnO, 53 per cent Al_2O_3, balance mostly oxides of iron.	Mineral supplied by U. S. National Museum. Maximum temperature of test No. 1, 210°.	2	6.4	7.3	525	7.7
M	Celsian, approximately 90 per cent $BaO \cdot Al_2O_3 \cdot 2SiO_2$.	$BaCO_3$ and kaolin heated for 2½ hours at from 1,550° to 1,560° C.	1	2.7	3.1	400	3.4
N	Anorthite, approximately 95 to 97 per cent $CaO \cdot Al_2O_3 \cdot 2SiO_2$.	Repeated heating of reagent quality materials (at temperatures about 50° below the melting point of anorthite) and regrinding. Prepared in the cement section of the Bureau of Standards.	1	3.5	4.0	360	4.3
O	Dicalcium aluminum silicate, approximately 65 per cent $2CaO \cdot Al_2O_3 \cdot SiO_2$ in well-formed crystals.	Reagent quality materials wet ground. Reheated twice, maximum temperatures 1,300° and 1,470°, respectively. Prepared by H. Insley.	1	2.3	8.0	400	8.1
P	Calcium metasilicate, >95 per cent = $CaO \cdot SiO_2$.	Same as Specimen N.	1	10.9	11.7	600	7.9
Q	Corundum, over 99 per cent Al_2O_3.	Commercial electric furnace product.	1	5.9	7.1		5.3
R	Mullite, almost entirely $3Al_2O_3 \cdot 2SiO_2$; a small percentage of glass and corundum present.	Fusion of alumina and silica. Expansion determined by R. A. Heindl (B. S. Tech. News Bull. No. 139, November, 1928).				1,100	

[1] Derived from observations on heating. [2] R. T. = Room temperature. [3] R. T. = Room temperature; varied from 20° to 25°. [3] Parallel to c axis. [4] Perpendicular to c axis.

TABLE 3.—*Values indicating reproducibility of results*

Values given are the total expansion in μ per meter from room temperature to t.

Specimen No.	t, °C.	First test	Second test	Third test	Fourth test
		μ	μ	μ	μ
A-17 {	100	5	15	15	15
	155	60	60	50	45
	235	170	150	140	

t, °C.	A-1[1]		A-3[1]		C-1[1]	
	(a)	(b)	(a)	(b)	(a)	(b)
100	50	45	10	5	40	40
200	200	195	80	90	150	160
300	365	380	180	190	295	305
400	590	610			480	500
560	1,000	1,020				

[1] Duplicate specimens.

The reproducibility of results is indicated by the data in Table 3.

III. RESULTS

1. CORDIERITE

The first expansion determinations were made on specimens of body A. Results are summarized in Table 1 and shown graphically in Figure 1. Since cordierite $(2MgO \cdot 2Al_2O_3 \cdot 5SiO_2)$[5] was the predominant phase in all of the specimens of the A series, the next series (B and C, Table 1) were prepared of mixtures in the proper proportions to produce this compound. However, in no case did the product contain more than approximately 90 to 95 per cent of cordierite. When heated at about 1,350° the principal impurities in the talc bodies (A-3, Table 1) were corundum (Al_2O_3) and forsterite $(2MgO \cdot SiO_2)$, while heating at higher temperatures for long periods of time (B-2a) evidently produced some decomposition of the ternary compound with the resultant formation of small amounts of spinel $(MgO \cdot Al_2O_3)$ and mullite $(3Al_2O_3 \cdot 2SiO_2)$. Prolonged heating of the specimens of series C also promoted the growth of bundles of fibrous crystals, apparently mullite, which originated from minute rounded grains of corundum.

Thermal expansions of the minor constitutents (forsterite, spinel, corundum, and mullite) were determined by test or taken from values given in the literature (Table 2). Their expansions are all considerably higher than the expansions of the specimens which are composed predominantly of "cordierite" (as distinguished from the "unstable" forms of $2MgO \cdot 2Al_2O_3 \cdot 5SiO_2$ by refractive indices and other optical properties)[6] and there was no indication of contraction during heating below 100° for these materials. It is believed that the low expansions observed are characteristic of bodies composed essentially of cordierite.

[5] G. A. Rankin and H. E. Merwin, Am. J. of Sci., vol. 45, p. 301, 1918. E. A. Wulfing and L. Oppenheimer, Ber. der Heidl. Akad. der Wissen, No. 10, 1914.
[6] See footnote 5.

The difference in expansion of specimens A–3 and C–1 (Table 1), both of which contain about 90 per cent cordierite, can not be explained satisfactorily as due to the effect of the impurities alone. That there may be undetermined differences in the crystal structure

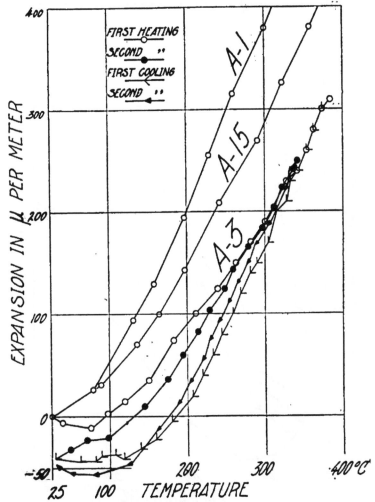

FIGURE 1.—*Thermal expansion curves for cordierite specimens of series A heated at 1,350° C. for one hour (A–1), 16 hours (A–15), and for 3 days (A–3)*

or composition of the cordierite is indicated by Rankin and Merwin,[7] who have observed considerable difference in indices of refraction of the cordierite from samples which appear optically homogeneous, but which vary somewhat in composition. This they attribute to solid solution of SiO_2 in the cordierite. Slight difference in the indices

[7] See footnote 5, p. 40.

of refraction were observed in some of the samples of cordierite used in this investigation, as is evidenced by B-2a and C-3.

The "unstable" form (C-5) with the composition $2MgO \cdot 2Al_2O_3 \cdot 5SiO_2$ was obtained by the method used by Rankin and Merwin;[5] that is,

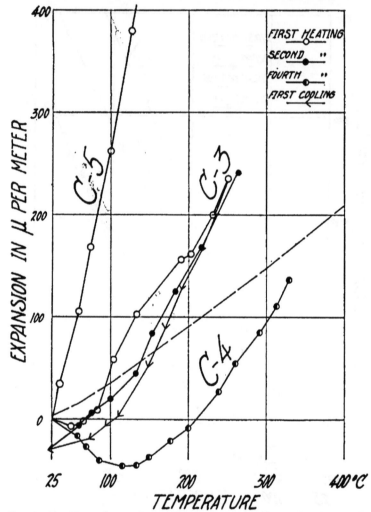

FIGURE 2.—*Thermal expansion curves for cordierite specimens of series C prepared from the glass by devitrification at 900° C. (C-5), 1,100° C. (C-3, and 1,355° C. (C-4)*

The dashed line is for fused quartz and is drawn from data in B. S. Sci. Paper No. 524.

by heating a glass of the same composition at about 900° for a considerable time. The product obtained corresponds in indices of refraction and other optical properties to that described by them. Its

---

[5] See footnote 5, p. 40.

B. S. Journal of Research, RP456

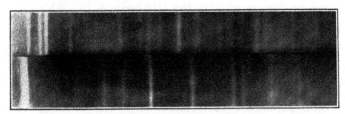

FIGURE 3.—*X-ray patterns of the stable (top) and unstable (bottom) forms of $2MgO \cdot 2Al_2O_3 \cdot 5SiO_2$ (cordierite)*

thermal expansion (C-5, Table 1 and fig. 2) is unquestionably considerably higher than that of cordierite. The increased expansion of specimens C-1b and C-2c after reheating at from 900° to 1,150° could not be duplicated with the talc body A-15. This increased expansion can not be attributed to the unstable form of the 2:2:5 compound, because neither petrographic examinations nor X-ray patterns of the reheated specimens C-1b and C-2c showed its presence. That the X-ray photograph would have shown the pattern of the unstable compound, if present in sufficient quantity to affect the expansion, seems probable from the marked difference in the two patterns. (Fig. 3.)

The small permanent contraction of about 40μ per meter evidently caused by the first heating (specimen C-3 in fig. 1) is characteristic of this form and in one case was noted to a lesser degree (5μ to 10μ per meter) after a specimen (A-17) had been heated and cooled four times. It was also noted that the cordierite specimens may have a slightly higher rate of contraction during cooling than of expansion during heating, in the temperature range above 100°, even though there is no measurable permanent length change. This is typified by the second heating and second cooling curves for A-3. (Fig. 1.)

The results do not permit the presentation of thermal expansion curves, or of coefficients, as correct for a compound of the definitely determined composition $2MgO \cdot 2Al_2O_3 \cdot 5SiO_2$. It is fairly evident, however, that (*a*) the crystalline phase observed as constituting the bulk of the specimens of series A, B, and C is the same as the stable form of the ternary compound identified by Rankin and Merwin [9] having the approximate composition 2:2:5; (*b*) the thermal dilatation of this stable form of the ternary compound is remarkably low; it may contract slightly during heating from about 20° to 80°; and the average linear expansion, when heated from 20° to 400°, is about 1μ per meter per °C. (equivalent to a coefficient of 0.000001); (*c*) the average linear change for the temperature range of 20° to 100° is much lower than for any other equivalent range; and (*d*) the expansion of the specimen of the unstable form tested is unmistakably higher than that of the stable.

(a) GENERAL COMMENTS

The specimens of series A and B were uniformly light buff to gray in color and of a firm but open texture. The gain in weight (water absorption) of an A-3a specimen (Table 1) resulting from autoclaving it in water for one and one-half hours at a steam pressure of 150 lbs./in.[2] was 25.8 per cent; there was no measurable length change or "moisture expansion" over a 10 cm length. The specimens of series C prepared either by devitrification, or by heating at 1,425°, resembled opaque white glass which was very difficult to pulverize with mortar and pestle, and pieces could be heated to a red heat and plunged into cold water without being visibly affected. Specimens of the glass (series C) heated at 900° for five hours or less showed no devitrification. Heating at 900° for four and one-half days produced the unstable, or high expanding, form (C-5, fig. 2) and heating at 950° for two days (or at higher temperatures, C-3 and C-4, fig. 2) produced the stable form. At 1,380° the glass devitrified completely to the stable form in 15 minutes.

[9] See footnote 5, p. 40.

2. COMPOUNDS OTHER THAN CORDIERITE

Of the other compounds investigated (Table 2 and figs. 4 and 5) the three having the lowest average expansion are beryl (*G*, *H*, *I*, and *J*), zinc orthosilicate (*K*), and celsian (*M*). The data for the four specimens of beryl are presented in some detail, partly because of their very low expansions and partly because beryl has received some attention, as a material of low thermal expansion, from previous investigators.[10] Both cordierite and beryl may contract slightly when heated from 20° to 100° and their average coefficients are not greatly different. However, on first heating and cooling beryl it is likely to show a permanent expansion of 30 to 40 μ per meter. The expansion

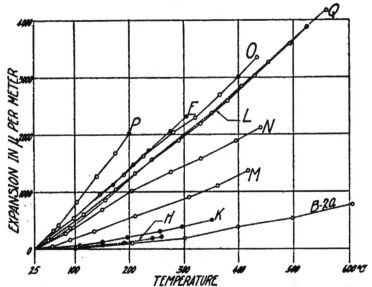

FIGURE 4.—*Thermal expansion curves of some of the materials described in Table 2, E, forsterite; K, zinc orthosilicate; L, zinc aluminate; M, celsian; N, anorthite; O, dicalcium aluminum silicate; P, calcium metasilicate*

Curves for a cordierite specimen (B-2a, Table 1) and for a beryl specimen (*H*) are included for comparative purposes.

curves for the "stabilized" specimens (that is, curves obtained after two or more tests) are given in Figure 5. Worthy of note are (*a*) the similarity in behavior of specimens *G*, *H*, and *I*, which have the same indices (Table 2); (*b*) the somewhat higher rate of expansion of *J*, which has higher indices; and (*c*) the fact that the expansion perpendicular to the *C* axis is relatively much higher than the expansion parallel to the same axis and is practically identical for both *G* and *J* specimens.

Several attempts to produce a specimen composed chiefly of some one crystalline silicate of cadmium (which has the same crystal habits as zinc, magnesium, and beryllium) were unsuccessful. The proper

[10] H. Fizeau, Compt. rend., T. LXVI, p. 1005, 1868. A. V. Bleininger and F. H. Riddle, J. Am. Ceram. Soc., vol. 2, p. 564, 1919. Robert Twells, jr., J. Am. Ceram. Soc., vol. 5, p. 226, 1922.

mixture of CdO and SiO₂ to form $CdO \cdot SiO_2$, when heated at 1,425° showed no evidence of reaction; after heating at 1,550° there was evidence of reaction, but the material rapidly fell to a powder when removed from the furnace. A mix to form $2CdO \cdot SiO_2$ was heated to a

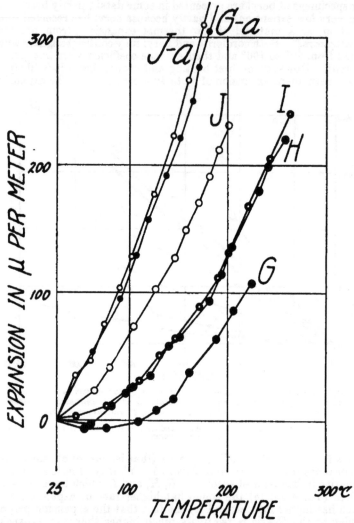

FIGURE 5.—*Thermal expansion curves for specimens of beryl described in Table 2*

J-a and G-a were determined perpendicular, the others were determined parallel, to the C axis.

maximum temperature of 1,600°. The product is described in Table 4. This material, even when quenched in water from 1,600°, also fell to a powder and no thermal expansion observations were attempted. Glasses having the calculated compositions $CdO \cdot Al_2O_3 \cdot 2SiO_2$ and

$2CdO \cdot 2Al_2O_3 \cdot 5SiO_2$ were obtained when the respective mixtures were heated at 1,450°. They were devitrified at approximately 1,250° and the products are described in Table 4. The average coefficient of expansion for the temperature range 25° to 300° was found to be 7.0×10^{-6} for the 1:1:2 glass and 6.5×10^{-6} for the 2:2:5 glass. Substituting these values in the formula given by Hall [11] one may calculate the "expansion factor" for CdO. The calculated values obtained in this way were 97 and 104×10^{-9}, the average being practically 100×10^{-9}. This average value for CdO would not apply to temperature ranges other than 25° to 300°.

TABLE 4.—*Properties of some cadmium silicates*

	Composition, determined petrographically
Calculated composition, $2CdO \cdot SiO_2$ .... *Preparation*—Wet ground mixture of CdO and SiO_2 heated at 1,550° to 1,600° C. for 45 minutes and cooled rapidly.	Two phases, neither of which predominates: Phase A. Index just above 1.7 (about 1.71) with very low or no birefringence. Phase B. Isotropic and occurs as inclusions in phase A; the index is much lower than 1.7. (NOTE.—In Sprechsaal, vol. 52, p. 256, 1919, F. M. Jaeger and H. S. van Klooster report the following values: $2CdO \cdot SiO_2$, melting point 1,243° to 1,252° C.; indices N_1 and $N_2 > 1.739$. $CdO \cdot SiO_2$, melting point $1,242 \pm 0.5°$ C.; indices N_1 and $N_2 > 1.739$)
Calculated composition, $CdO \cdot Al_2O_3 \cdot 2SiO_2$ .. *Preparation*—Wet ground mixture of CdO and Georgia kaolin melted to a clear glass at 1,450° C. and devitrified at 1,250° C.	Three phases present: Phase A. Probably the most abundant, occurs as irregular crystals of very low double refraction (<0.005) with a mean index of about 1.65. Phase B. Either glass or isotropic crystals occurring usually as interstitial material between grains of phase A. Index lies between 1.61 and 1.62. Phase C. Minute rounded grains of high index and high double refraction occurring as inclusions in phase A.
Calculated composition, $2CdO \cdot 2Al_2O_3 \cdot 5SiO_2$. *Preparation*—Same as for $CdO \cdot Al_2O_3 \cdot 2 SiO_2$.	Three phases present: Phase A. Most abundant and occurs with some faces (probably prismatic) well developed. Crystals are biaxial, very large optic axial angle (approximately 90°) and indices: $\alpha = 1.575 \pm 0.003$, $\gamma = 1.590 \pm 0.003$. Phase B. Isotropic, probably glass. Somewhat less abundant than phase A and index variable but between 1.615 and 1.620. Phase C. Least abundant; occurs as irregular crystalline growths showing a tendency to dendritic forms. Double refraction very low (probably 0.005 or less), mean index 1.65 ± 0.005.

IV. SUMMARY

The low expansion of a magnesia body reported by W. M. Cohn [12] is a characteristic of cordierite (the "stable" form of a compound having the approximate formula $2MgO \cdot 2Al_2O_3 \cdot 5SiO_2$). This form may be prepared by sintering a mixture of talc, clay, and corundum at 1,350°, although the time required is considerably shortened by heating at from 1,400° to 1,425°. At temperatures above 1,425° to 1,450° decomposition of the cordierite will take place. It may be formed also by devitrification of a glass of the proper composition at temperatures above 950° and below about 1,425°.

Zinc orthosilicate, celsian or barium feldspar, and beryl also appear worthy of further consideration by those interested in the development of bodies having low coefficients of thermal expansion.

WASHINGTON, February 19, 1932.

[11] J. Am. Ceram. Soc., vol. 13 (3), p. 182, March, 1930.   [12] See footnote 2, p. 35.

AN AUTOMATIC REVERBERATION METER FOR THE MEASUREMENT OF SOUND ABSORPTION

By W. F. Snyder

ABSTRACT

The usual electrical recording method for measuring sound absorption in a reverberation room requires the determination of the rate of decay of sound. The automatic reverberation meter described in this paper works through a range of about 70 decibels, the rate of decay being measured over intervals as short as desired through the whole range.

An automatic control of the apparatus has been developed which allows measurements to be made with but little attention on the part of the observer.

The working ranges of the four methods [1] that have been used at the Bureau of Standards for measuring the rate of decay of sound in a room are shown diagrammatically in Figure 1. The approximate

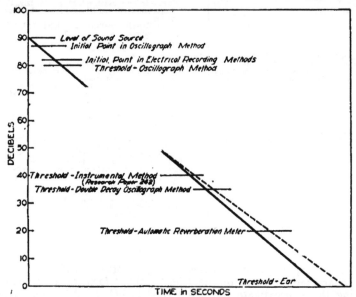

FIGURE 1.—*Working ranges of various methods used at the Bureau of Standards to measure reverberation time*

range through which the rate of decay is measured by each method is indicated by the difference in decibels between the initial point and the final threshold. The ordinary oscillograph method is very limited in range, being about 7 decibels. The other methods described were

---

[1] See reference 6 in the bibliography.

used through ranges of 40 to 50 decibels. The present automatic reverberation meter works through a range of about 70 decibels. The ear, of course, covers even a greater range since in reverberation measurements the ear must start at the initial sound level in the room and must follow the decay to the threshold of hearing.

More important, however, than a measurement over a wide range of intensity, is the accurate determination of the rate of decay of the sound energy throughout the measured range. Such a procedure is essential when measuring highly absorbent materials in a reverberation room or whenever "two-room" conditions exist.[3] Under these conditions a double rate of decay may exist as shown by the dotted

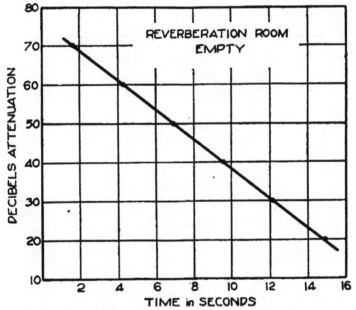

FIGURE 2.—*Curve showing the rate of decay at 512 cycles made over a range of 50 decibels*

It is the result of 60 observations.

line in Figure 1. The ear and instrumental methods previously described give but two points on the decay curve, the initial starting point and the threshold point, and we may be unaware of any change in the rate of decay between the points measured. The same is true in the double-decay oscillograph method. There is, however, in this case a short range at both the initial and threshold points where the rate is accurately known. The two short decay curves photographed on the oscillograms sometimes showed different rates of decay, indicating that the rate was not always uniform through the whole range.

Measurements with a reverberation meter as shown in Figure 2 yield a number of points along the decay curve at definite intervals, and the rate of decay may be determined throughout the working range.

[3] See reference 9 in the bibliography.

B. S. Journal of Research, RP457

FIGURE 3.—*Equipment used in making absorption measurements with automatic reverberation meter*

Since the decay of sound in a room is of a nonuniform character we must resort to a statistical method of measurement, each measurement being repeated at least ten times. The procedure in making these measurements is as follows: Sound radiated from a source such as a loud speaker is suddenly cut off and a timing mechanism started at the same time. The sound as it dies away is picked up by a microphone and when the sound reaches a certain level in the room an electrical arrangement stops the timer. The timer has then recorded the length of time the sound takes to decay from the initial intensity to the threshold at which the instrument is set. Shifting the threshold of the recording instrument by a definite number of decibels, a different time is measured. Knowing the time it takes the sound to decay through definite ranges of sound levels, a curve can be plotted similar to Figure 2 in which the slope gives the rate of decay. From the rate of decay the absorption of the room can be calculated by the reverberation equation.[*]

Making statistical measurements in such fashion becomes quite laborious. The present design of the reverberation meter used at the Bureau of Standards has made much of the operation automatic.

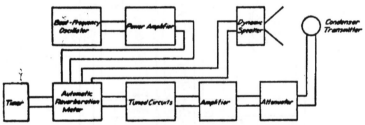

FIGURE 4.—*Schematic diagram of entire apparatus*

Figure 3 shows the recording equipment and Figure 4 is a schematic diagram of the entire apparatus. Alternating current in the audio-frequency range is generated by a beat-frequency oscillator, strengthened by a power amplifier and finally transformed into sound energy by a dynamic speaker. A condenser transmitter transforms the sound back into electrical energy. The weak electrical current is then amplified, the threshold of the instrument being controlled by an attenuator. Immediately following the amplifier is a group of tuned circuits which aid in reducing the level of extraneous noises. These tuned circuits attenuate approximately 15 decibels at the first octave above and below the frequency to which they are tuned. The automatic reverberation meter with its associated timer terminates the recording apparatus.

Figure 5 shows in detail the electrical circuit contained within the reverberation meter which controls the timer. There is one stage of amplification transformer coupled to a screen-grid tube. The screen-grid tube which acts as a rectifier is directly coupled through a 0.5 megohm resistor to a pentode tube. The pentode tube is used as an oscillator and was so chosen because of its high plate current when connected in series with a high resistance relay. Some difficulty has been found in getting a pentode tube to oscillate under the particular

---

[*] See reference 6 in the bibliography.

circuit requirements which are necessary for the desired operation of the equipment. The frequency of oscillation is about 700 kc per second.

The grid bias of the pentode tube depends upon the potential across the plate resistor of the rectifier tube which in turn is a function of the sound energy picked up by the microphone. When the sound level is high, as at the beginning of the decay, the grid bias on the oscillator is sufficiently negative to prevent it from oscillating and the plate current is nearly zero. As the sound dies away the grid bias becomes more positive. This continues until a sufficiently positive bias sets the tube into oscillation with a sudden increase in plate current. In consequence, a relay in series with the tube opens a set of contacts, stopping the timer. The sudden increase in current

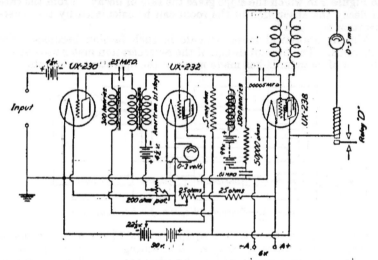

FIGURE 5.—*Electrical circuit diagram of apparatus which stops the electric timer*

through the relay winding gives a positive control to the operation. The point at which this goes into operation, which has been called the threshold of the instrument, depends upon the amount of amplification used and the setting of the attenuator.

The 1,500-henry choke in series with the grid circuit of the pentode tube and the 0.01 μf by-pass condenser serve as a filter for the audio frequencies. This is a necessary precaution to prevent the relay from humming when low frequencies are measured. The circuit is so designed that once the tube goes into oscillation a negative shift in the grid bias caused by an increase in the sound energy amounting to about 4 decibels is necessary to stop oscillations, thereby restarting the timer. This more or less helps to integrate out a variable sound decay and at the same time prevents the timer from following variations less than 4 decibels which would cause excessive starting and stopping of the timer.

Figure 6 shows the arrangement of relays which makes the operation of the reverberation meter almost automatic. The apparatus consists of three relays, a small alternating current motor with a 1,100 to 1 reduction gear coupled by a shaft to a cam operating a 4-pole switch and a self-stopping device. Large condensers across the contacts of the relays prevent sparking and setting up electrical disturbances in the near-by amplifying equipment.

The operation of this device is as follows: A rod connected to the armature of relay A is lifted, closing the motor circuit. When no sound is on, the lower contact at relay D (fig. 6) is closed, allowing current to flow through the windings of relay A, thus holding the armature of relay A after it is lifted.

Connected to the same armature is a small pin fitting into cam B, which when lifted allows relay A to operate properly, since the armature is now close to the pole piece, the pin riding around on top of cam B. A 4-pole switch operated by the segments along cam A is closed after the cam begins to rotate. This switch closes the loud-speaker circuit and closes a circuit through the winding of relay A which

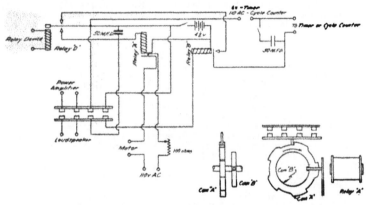

FIGURE 6.—*Relays and cams which provide automatic control for the reverberation meter*

normally opens at relay D as the sound comes on. It also closes the circuit of the winding of relay B which prevents the timer from operating. After a further rotation of the cam shaft the 4-pole switch suddenly springs open breaking the loud-speaker circuit, opening the motor circuit and allowing the contact of relay B to close, which starts the timer at the beginning of the decay. While the sound is on or as it is decaying the timer contact of relay D is closed. As the sound decays to the threshold of the instrument and the plate current suddenly increases in the pentode tube, relay D operates stopping the timer. At the same time the other contact on relay D closes and the motor is started. The whole process is repeated. This continues for five measurements. At the fifth measurement the pin on the armature of relay A drops back into the slot of cam B and stops the entire mechanism.

An average of 10 observations is usually taken at six different sound levels spaced at 10 decibel intervals making a range of 50 decibels. This is sufficient for most absorption measurements. The apparatus has a range, however, of about 70 decibels.

The apparatus described above offers a statistical method of dealing with a type of measurement that presents certain difficulties, since we are determining a rate of decay that is seldom uniform. The method averages a series of observations in a mechanical fashion and is almost entirely automatic. This being true, the human equation in reverberation and absorption measurements has been practically eliminated.

BIBLIOGRAPHY OF PUBLISHED ARTICLES REFERRING TO MECHANICAL AND ELECTRICAL METHODS OF MEASURING REVERBERATION TIME

1. E. Meyer and Paul Just, On the Measurement of Reverberation Time and Sound Absorption, Elek. Nach. Tech., vol. 5, pp. 293–300, 1928.
2. E. Meyer, Automatic Reverberation Measurement, Zeit. f. Tech. Physik., vol. 11, pp. 253–259, 1930.
3. M. J. O. Strutt, Automatic Reverberation Measuring Instrument, Elek. Nach. Tech., vol. 7, pp. 280–282, July, 1930.
4. V. L. Chrisler, The Measurement of Sound Absorption by Oscillograph Records, J. Acoustical Soc. of Am., vol. 1, No. 3, pp. 418–421, April, 1930.
5. V. L. Chrisler and W. F. Snyder, Recent Advances in Sound Absorption Measurements, J. Acoustical Soc. Am., vol. 2, No. 1, pp. 123–128, July, 1930.
6. V. L. Chrisler and W. F. Snyder, The Measurement of Sound Absorption, B. S. Jour. Research, vol. 5, pp. 957–972, October, 1930.
7. V. L. Chrisler and W. F. Snyder, Measurements with a Reverberation Meter, J. Motion Picture Engrs., vol. 18, No. 4, pp. 479–487. April, 1932.
8. E. C. Wente and E. H. Bedell, A Chronographic Method of Measuring Reverberation Time, J. Acoustical Soc. Am., vol. 1, No. 3, pp. 422–427, April, 1930. Also Bell Telephone Laboratory Reprint B-500.
9. Carl F. Eyring, Reverberation Time Measurements in Coupled Rooms, J. Acoustical Soc. Am., vol. 3, No. 2, pp. 181–206, October, 1930.
10. F. L. Hopper, The Measurement of Reverberation Time and Its Application to Acoustic Problems in Sound Pictures, J. Acoustical Soc. Am., vol. 2, No. 4, pp. 499–505, April, 1931.
11. F. L. Hopper, The Determination of Absorption Coefficients for Frequencies up to 8,000 cycles, J. Acoustical Soc. Am., vol. 3, No. 3, pp. 415–427, January, 1932.
12. Harry F. Olson and Barton Kreuzer, The Reverberation Time Bridge, J. Acoustical Soc. Am., vol. 2, No. 1, pp. 78–82, July, 1930.
13. R. F. Norris and C. A. Andree, An Instrumental Method of Reverberation Measurement, J. Acoustical Soc. Am., vol. 1, No. 3, pp. 366–372, April, 1930.
14. R. F. Norris, Application of Norris-Andree Method of Reverberation Measurement to Measurements of Sound Absorption, J. Acoustical Soc. Am., vol. 3, No. 3, pp. 361–370, January, 1932.

WASHINGTON, April 26, 1932.

NOTE ON THE PROBABLE PRESENCE OF 2, 2-DIMETHYL-PENTANE IN A MIDCONTINENT PETROLEUM [1]

By Johannes H. Bruun [2] and Mildred M. Hicks-Bruun

ABSTRACT

This paper presents evidence which indicates that 2, 2-dimethylpentane is present in small amount (not more than a few hundredths of 1 per cent) in a midcontinent petroleum. The evidence consists in the approximate agreement of the measured properties of a petroleum fraction containing 46 mole per cent of cyclohexane with those computed on the assumption that the remaining 54 per cent is 2, 2-dimethylpentane.

The properties compared were initial freezing point, boiling point, refractive index, molecular weight and eutectic temperature. The evidence thus obtained was confirmed by comparing the infra-red absorption spectrum of the petroleum fraction with that of a synthetic mixture having the assumed composition.

CONTENT

I. INTRODUCTION

From a Caucasian petroleum, Markownikoff (1) [3] obtained a fraction boiling between 78.5° and 79° C. From the boiling range and the specific gravity of this fraction, as well as from its unusual stability toward strong nitric acid, he concluded that 2,2-dimethylpentane was probably present in the Caucasian petroleum. With the exception of Markownikoff, it seems that no one has reported any indication of the presence of 2,2-dimethylpentane in any petroleum.

II. EXPERIMENTAL PROCEDURE AND RESULTS

The present investigation deals with the same fractions of an Oklahoma petroleum which were used for the isolation and determination of cyclohexane (2).

It was found that a large cut (5.5 kg) boiling between 80° and 80.5° C. consisted of about 94 mole per cent of cyclohexane and had a refractive index of 1.422 which is 0.003 unit lower than that of pure cyclohexane. It was therefore apparent that some compound

---

[1] Financial assistance has been received from the research fund of the American Petroleum Institute. This work is part of project No. 6, The Separation, Identification, and Determination of the Constituents of Petroleum.

[2] Research Associate, representing the American Petroleum Institute.

[3] Figures in parenthesis here and elsewhere in the text indicate references given in the bibliography at the end of this paper.

or compounds with lower refractive index than that of cyclohexane must also be present in this fraction. The object of the present investigation was to concentrate this unknown constituent for possible identification.

While working on the isolation of cyclohexane, it was noted that if the cut boiling between 80° and 80.5° C. was subjected to fractionation by equilibrium melting, fractions of rapidly decreasing refractive indices were obtained. Thus by a single fractionation the following changes in the physical constants were obtained:

	Stage I Stage II
$n \frac{20}{D}$ changed from......................	1.422 to 1.417
Freezing point changed from...............	−12° to −28° C.
Boiling point changed from...............	80.4° to 80.0° C.
Molecular weight changed from..........	85.1 to 86.1

In view of this apparently efficient separation it was decided to subject the material to systematic equilibrium melting. The progress of the fractionation is shown by the chart in Figure 1. The large circle at the upper left corner (R. I. 1.416) represents the original fraction (obtained by distillation) which had the boiling range 79° to 80° C. The other nine large circles at the top represent fractions which were obtained by subjecting the 80.0° to 80.5° cut to equilibrium melting. (This fractionation is described in the cyclohexane paper (2).) The numbers inside the large circles indicate the refractive indices and the initial freezing points of the fractions. As indicated by the solid lines, the large fractions were separated by equilibrium melting into a number of smaller fractions, indicated by the smaller circles. The refractive indices of these fractions are given by the numbers inside the circles. The fractions having similar refractive indices were then mixed for further fractionation. This is indicated by the broken lines in Figure 1. As shown in the chart, the molecular weight was determined for some of the fractions.

As a result of the fractionation shown in Figure 1, the material was separated into comparatively large fractions consisting mainly of cyclohexane (fractions with high refractive indices) and into small quantities of material with low refractive index and low freezing point. Fractionation of this material, which contained the unknown constituent, was continued until its refractive index had been brought down to 1.392 and its freezing point had reached −124.6° C. (Stage III.) At this point, however, the available quantities of this material were too small to attempt further purification.

The time-temperature cooling curve of the final fraction is shown in Figure 2. Table 1 shows some of the physical constants of synthetic 2, 2-dimethylpentane, of cyclohexane, and of the petroleum fractions during three stages of the fractionation. Stage I represents the large cut boiling between 80° and 80.5° C. The material in Stage II was obtained by a single fractionation of the mixture in Stage I by equilibrium melting. (See p. 54 and upper left of fig. 1.) Stage III represents the fraction which was obtained as a final result of the systematic equilibrium melting. (See bottom of fig. 1.)

TABLE 1.—*Some physical constants of 2, 2-dimethylpentane, of cyclohexane, and of the petroleum fractions during three stages of the fractionation*

| | Cyclo-hexane (C$_6$H$_{12}$) synthetic [a] | The petroleum fractions | | | 2, 2-di-methyl-pentane (C$_7$H$_{16}$) synthetic [b] |
		Stage I	Stage II	Stage III	
n $\frac{20}{D}$	1.426	1.422	1.417	1.392	1.382
Molecular weight	84.09	[c] 85.1	[c] 86.1	[c] 93.7	100.12
Normal boiling point in ° C.	80.8	80.4	80.0	79.5	78.9
Freezing point in ° C.	6.4	−12	−28	−124.6	−125.6

[a] J. Timmermans, J. chim. phys., vol. 23, p. 760, 1926.
[b] G. Edgar and G. Calingaert, J. Am. Chem. Soc., vol. 51, p. 1544, 1929.
[c] For method, see M. M. Hicks-Bruun, B. S. Jour. Research, vol. 5, pp. 575-583, 1930.

As the fractionation proceeded a definite increase was noted in the molecular weight. Thus, while the molecular weight for Stage I was 85.1, the material in Stages II and III was found to have molecular weights of 86.1 and 93.7, respectively. This is unmistakable evidence of the presence of a constituent having a higher molecular weight than that (84.09) of cyclohexane.

The material in Stage III had a distillation range of about 0.5° C. It did not react with bromine or iodine and it was insoluble in water. Furthermore this fraction exhibited an unusual stability when heated with fuming nitric or with chlorosulphonic acid.

By correlating the values of the molecular weight (93.7), the refractive index and the boiling point of this fraction with the corresponding constants of known hydrocarbons, it was concluded that a heptane was probably present in the cyclohexane fraction.

It should be further noted that the boiling point of the material in Stage I (which contained 94 mole per cent of cyclohexane) was 80.4° C. as compared with 80.8° C. for pure cyclohexane. As a result of further fractionation by equilibrium melting fractions of still lower boiling points (80.0° C. in Stage II and 79.5° C. in Stage III) were obtained. This fact indicates clearly that the presumed heptane is one with a boiling point below that of cyclohexane or that the mixture is azeotropic with a boiling point minimum. The only heptane with a boiling point below 80.8° is 2,2-dimethylpentane.

The above-mentioned great stability of the material in Stage III against chemical reagents is in accordance with the structure of 2,2-dimethylpentane, because of all the isomeric heptanes, only the following three do not possess a reactive tertiary carbon atom: 2,2-dimethylpentane (boiling point 78.9° C.); 3,3-dimethylpentane (boiling point 86° C.); and *n*-heptane (boiling point 98.4° C.).

III. ANALYSIS OF THE FINAL FRACTION

For the purpose of determining the approximate composition of the material (Stage III, Table 1) having the lowest refractive index (1.392), the following procedure was used. The freezing point of cyclohexane was determined with a Beckmann thermometer. A solution of known concentration of the material (Stage III) in the cyclohexane was then made up, after which the lowering of the freezing point was determined. It is evident that the cyclohexane present in the material (Stage III) will have no effect upon the freezing point, and that the total lowering will be caused by the other con-

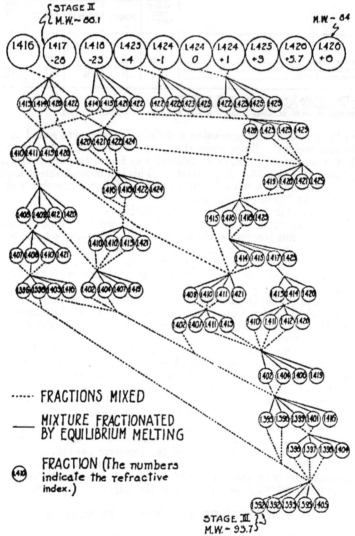

Figure 1.—*Chart showing the fractionation by means of equilibrium melting*

stituents in this material. From the lowering thus obtained the mole percentage of cyclohexane for the material in Stage III could be calculated. However, the two most recent values reported in the literature for the heat of fusion (L_{f_0}) of cyclohexane differ by more than 20 per cent. Hence calculations were made using each value, with these results:

L_{f_0} = 5.87 cal./g (3).
45.9 mole per cent of cyclohexane and
54.1 mole per cent of other constituents.
L_{f_0} = 7.4 cal./g (4).
30.8 mole per cent of cyclohexane and
69.2 mole per cent of other constituents.

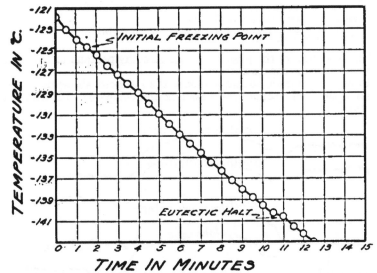

FIGURE 2.—*Time-temperature cooling curve of the final fraction.* (*Stage III*)

IV. CORRELATION OF THE PROPERTIES OF THE FINAL FRACTION WITH ITS COMPOSITION

The amount of the final fraction (Stage III) was so small that further fractionation for the purpose of isolating a pure sample of the presumed heptane was impossible. Under these circumstances the best that can be done is to compare the measured properties of the final fraction with those of a mixture consisting of 45.9 moles of cyclohexane and 54.1 moles of 2,2-dimethylpentane. This comparison is shown in Table 2. The calculated values are based on the assumption that the mixture is an ideal solution.

It will be observed that the agreement is good except in the case of the refractive index, where the observed value is low by 0.008 unit, which might be caused by the presence of a small amount of a third constituent, such as normal hexane, for example.

TABLE 2.—*Observed and calculated properties of the final fraction*

Property =	F. P.[a]	B. P.[a]	Molecular weight	n_D^{20}	Eutectic halt
	°C.	*°C.*			*°C.*
Observed values	−124. 6	79. 5	93. 7	1. 392	[b]−140. 2
Values calculated from L_{fe}=5.87 cal./g[e]	−124. 6	79. 8	92. 8	[d]1. 399	[e]−138. 2
L_{fe}=7.4 cal./g[f]	−136. 8		95. 1	1. 394	[e]−136. 0

[a] Initial.
[b] See Figure 2.
[c] See reference (3) in bibliography.
[d] A synthetic mixture was found to have n_D^{20}=1.400.
[e] For method see Washburn and Read, Proc. Nat. Acad., vol. 1, p. 191, 1915.
[f] See reference (4) in bibliography.

The evidence for the presence of 2,2-dimethylpentane is therefore strong but by no means decisive.

In order to obtain further evidence the infra-red absorption spectrum of the final fraction was measured and compared with that of a synthetic mixture.

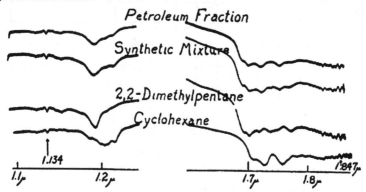

FIGURE 3.—*The infra-red absorption spectra of the final petroleum fraction (Stage III), the synthetic mixture, the 2,2-dimethylpentane, and the cyclohexane*

Measured by U. Liddel, of the Fixed Nitrogen Research Laboratory of the Department of Agriculture.

Through the courtesy of G. Edgar and G. Calingaert, of the Ethyl Gasoline Corporation, a small sample of synthetic 2,2-dimethylpentane was obtained. This sample was mixed with pure cyclohexane (and a small amount of *n*-hexane) so as to yield a mixture of approximately the same composition as that calculated for the material in Stage III. The spectra of the two mixtures as well as those of the pure samples of cyclohexane and of 2,2-dimethylpentane were then measured by U. Liddel, of the Fixed Nitrogen Research Laboratory of the United States Department of Agriculture, and are shown in Figure 3. The two upper curves reveal a striking resemblance between the synthetic mixture and the final fraction (Stage III), which was presumably of the same composition.

The characteristic minimum at about 1.19μ is clearly evident in the spectrum of the petroleum fraction. This evidence when taken together with that shown in Table 2 renders it highly probable that 2,2-dimethylpentane is a major constituent of the petroleum fraction

(Stage III). Based upon the crude oil the amount present can not, however, exceed a few hundredths of 1 per cent. With larger amounts of material to work with, there should be no difficulty in isolating the pure hydrocarbon.

V. ACKNOWLEDGMENT

The authors acknowledge the technical advice and suggestions of E. W. Washburn, director of American Petroleum Institute project No. 6.

VI. BIBLIOGRAPHY

1. W. Markownikoff, Ber., vol. 33, p. 1908, 1900.
2. J. H. Bruun and M. M. Hicks-Bruun, B. S. Jour. Research, vol. 7, p. 607, 1931.
3. M. Padoa, Atti Accad. Lincei, vol. 28, p. 240, 1919.
4. G. S. Parks and H. M. Huffman, Ind. Eng. Chem., vol. 23, p. 1139, 1931.

WASHINGTON, May 3, 1932.

NOTES ON THE ORIFICE METER; THE EXPANSION FACTOR FOR GASES

By Edgar Buckingham

ABSTRACT

The discharge coefficient of an orifice meter, determined with water, is applicable when the meter is used for measuring the flow of a gas, provided that the differential pressure is so small that the accompanying change of density is insignificant. But if the differential is a considerable fraction of the absolute static pressure, the water coefficient must be multiplied by an "expansion factor" which allows for the effects of change of density.

The paper contains a discussion of recent experimental data which show how the expansion factor depends on the form of the meter, the ratio of downstream to upstream pressure, and the specific heat ratio of the gas. The conclusions are summarized in an empirical equation which may be used for computing the value of the expansion factor in certain practically important cases.

A theoretical method of computing the expansion factor is developed and is shown to agree reasonably well with the facts observed under conditions that are approximately in accordance with those postulated by the theory.

CONTENTS

I. INTRODUCTION

The type of meter to which the following notes refer is illustrated diagrammatically by Figure 1, which shows some of the notation to be used, as well as certain limitations on the relative dimensions of the parts. It may be assumed that the readers to whom these notes are addressed would find a detailed description of the orifice meter superfluous.

Since the pressures p_1 and p_2, observed at the upstream and downstream side holes or pressure taps, depend on the locations of the holes, it is necessary to specify the distances, l_1 and l_2, from the

61

center of each hole to the nearer face of the orifice plate, and some one of the following four schemes is usually adopted:

(a) Pipe taps, $l_1 = 2.5D$, $l_2 = 8D$.
(b) Throat taps, $l_1 = D$, $l_2 = 0.5D$.
(c) Flange taps, $l_1 = l_2 = 1$ inch for all sizes of pipe.
(d) Corner taps, the side holes are at the faces of the plate, or the pressures are taken off through narrow circumferential slits between the plate and the flanges, as illustrated in the lower half of Figure 1.

Combinations (a), (b), and (c) are in common use in the United States, and (d) has been adopted as standard by the Society of German Engineers (1).[1]

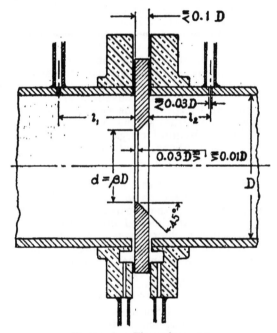

FIGURE 1.—*The orifice meter*

Side holes are shown above and ring slits below

There are advantages in adopting an arrangment such that meters of different sizes shall be geometrically similar as regards location of the side holes, a condition satisfied by (a), (b), and (d), but not by (c).

When the pipe diameter, D, is 8 inches or more and the orifice diameter, d, is not more than about $0.75D$, arrangements (c) and (d) give nearly the same pressure readings and may be regarded as equivalent, except in work of high precision. But with pipes as small as 4 inches in diameter, this assumption is no longer safe; and even with larger pipes, the two arrangements may give appreciably different results if the diameter ratio is as large as 0.8.

---

[1] Figures in parenthesis here and throughout the text indicate references given at the end of this paper.

II. THE ORIFICE METER EQUATION

The indications of an orifice meter are usually, and most conveniently, interpreted by means of some equation which is substantially equivalent to

$$M = N\,C\,A\sqrt{2\rho(p_1 - p_2)} \tag{1}$$

in which

M = the required rate of mass flow, or the mass discharged per unit time;

A = the area of the orifice;

ρ = the density of the fluid being metered;

p_1, p_2 = the pressures at the upstream and downstream taps;

N = a numerical constant dependent on the units; and

C = the discharge coefficient of the orifice, a number which does not depend on the units.

If the fluid is a gas, its density must be referred to some specified pressure and temperature, and these are most commonly taken to be the upstream pressure p_1 and the temperature t_1 of the gas approaching the orifice. We shall adopt this convention and denote the density under these conditions by ρ_1. For the sake of simplicity, it will also be supposed that all quantities are measured by a system of normal units, such as "British absolute" or cgs, because we then have $N = 1$; and with these two conventions, equation (1) takes the form

$$M = CA\sqrt{2\rho_1(p_1 - p_2)} \tag{2}$$

Before an orifice can be used as a flow meter, the value of C must be known, and this value depends on the rate of flow, the properties of the gas, the dimensions of the apparatus, the location of the pressure taps, etc. In the experimental investigations needed for the elucidation of this subject, observations of pressure and temperature at the orifice are combined with measurements of the rate of discharge by some independent method, and the equation is used in the form

$$C = \frac{M}{A\sqrt{2\rho_1(p_1 - p_2)}} \tag{3}$$

which may be regarded as a definition of C.

It may be remarked that ρ_1 denotes the true density at p_1, t_1; and if the gas in question is one that shows large departures from Boyle's law under the anticipated working conditions, the use of the familiar equation $pv = RT$, in computing the value of ρ_1 from the results of a density determination under laboratory conditions that are very different from the working conditions, may lead to large errors (2).

III. RESTRICTION TO HIGH VALUES OF THE REYNOLDS NUMBER

Let R_d be the Reynolds number defined by the equation

$$R_d = \frac{4M}{\pi d\eta} \tag{4}$$

in which η is the viscosity of the fluid.

If R_d is large, say $R_d > 200,000$, the value of C found by testing an orifice with water, or other liquid, is sensibly independent of the rate of flow (3), and this shows that the effects of viscosity have become negligible. But if the same orifice is tested with a gas, such as air, the value obtained for C varies with the rate of flow, even though R_d be high enough to make the effects of viscous forces insignificant; for the decrease of density as the pressure falls from p_1 to p_2, in contra-distinction to the constancy of density of a liquid, introduces a new element into the phenomena of flow (5, 6, 7).

The condition that R_d shall have the required high value is nearly always satisfied in the commercial metering of gases, and since the object of this paper is to discuss the changes of C which are due to compressibility alone, it will be assumed, from this point onward, that the requirement is fulfilled.

IV. THE EXPANSION FACTOR

In discussing the effects of compressibility, it will be convenient to employ the following notation:

$\beta = d/D =$ the diameter ratio of the orifice, or

$m = \beta^2 =$ the area ratio;

$K =$ the value found for C when the orifice is tested with a liquid at high values of R_d: so long as the installation remains unchanged, K is a constant of the orifice;

$y = p_2/p_1 =$ the pressure ratio at which the discharge coefficient determined by experiments on the gas has the value C;

$\gamma = C_p/C_v =$ the specific heat ratio of the gas; and

$Y = C/K =$ the expansion factor.

If the fall of pressure at the orifice is made so small that the accompanying decrease of density is insignificant, the gas must behave very nearly like a liquid; and experiment confirms the conclusion that $C \doteq K$ when $y \doteq 1$.

We therefore write

$$C = KY \qquad (5)$$

in which the expansion factor, Y, describes, or allows for, the varying effect of compressibility on the discharge coefficient; and $Y \doteq 1$ when $y \doteq 1$.

In many important cases, the relation $Y = f(y)$ is very nearly linear, as is illustrated by the simultaneous values of y and C given in Table 1 and plotted in Figure 2.

TABLE 1.—*Relation of C to y*

Diameter ratio 0.6209. Specific heat ratio 1.283. Throat taps

$\frac{p_2}{p_1} = y$	C observed	$\frac{0.6669 - 0.23}{(1-y)}$	Difference	$\frac{p_2}{p_1} = y$	C observed	$\frac{0.6669 - 0.23}{(1-y)}$	Difference
0.549	0.5618	0.5632	−0.0014	0.797	0.6211	0.6202	+0.0009
.553	.5660	.5652	+.0008	.904	.6444	.6448	−.0004
.560	.5654	.5657	−.0003	.911	.6469	.6464	+.0005
.647	.5866	.5857	+.0009	.913	.6486	.6469	+.0017
.651	.5872	.5866	+.0006	.947	.6543	.6547	−.0004
.650	.5879	.5864	+.0015	.948	.6527	.6549	−.0022
.700	.5973	.5979	−.0006	.953	.6557	.6561	−.0004
.703	.5979	.5986	−.0007	.975	.6589	.6612	−.0023
.711	.5999	.6004	−.0005	.977	.6620	.6616	+.0004
.808	.6250	.6227	+.0023	.977	.6605	.6616	−.0011
.799	.6213	.6207	+.0006				

This is one of a large number of results of experiments carried out under the direction of H. S. Bean, of the Bureau of Standards, for the Committee on Gas Measurement of the Natural Gas Department of the American Gas Association. In this instance, the experimental work was done at Los Angeles in 1929, with a natural gas of specific heat ratio $\gamma = 1.283$, and the pressures were taken at throat taps.

The Los Angeles experiments included tests of 23 orifices in pipes of 4, 8, and 16 inches nominal diameter, and Table 2 gives a list of the orifices, together with the number of tests on each and the lowest value of y, for throat taps, to which the tests extended.

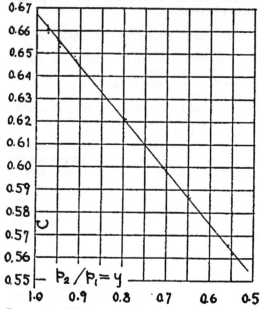

Figure 2.—*Relation of discharge coefficient to pressure ratio*

From Table 1

The orifice plates were one-eighth inch thick, with the edges of the orifice square and sharp at both faces. They were installed in commercial steel pipes which had been selected for smoothness, and a short nest of smaller pipes was placed in each of the three pipes at a distance of $10D$ to $15D$ ahead of the orifice, to insure straightness of flow. Pipe, throat, and flange taps were provided, and pressures were read at all three pairs. The absolute pressure was never more than 2.6 atmospheres so that departures from Boyle's law could be ignored. Since it was not practicable to measure the rates of flow by means of a gasometer, they were determined by passing the discharge from the orifice under test through standard reference orifices, of which any number up to 6 could be used in parallel. Details of the experiments and their results will be described in a later publication.

The linear relationship shown by Table 1 and Figure 2 is characteristic of the orifices for which $0.2 \lessgtr \beta \lessgtr 0.75$. In general, the departures of the plotted points from the best straight line that could be drawn among them by simple inspection were somewhat greater than in this series, but they were not systematic.

For diameter ratios of 0.8 or more, when the range of y was long enough to give a well defined band of points, the axis of the band was concave upward, the curvature increasing with β; and the tests of the orifice for which $\beta = 0.1241$ indicated that a curve, slightly convex upward, would be a little better than a straight line, a result in accordance with other observations on orifices of diameter ratios below about 0.2. But the general conclusion may be drawn from the Los Angeles observations with throat taps, that, over the ranges $0.2 \lesssim \beta \lesssim 0.75$ and $1.0 > y > 0.5$, the expansion factor can be represented, within the experimental errors, by the linear equation

$$Y = 1 - \epsilon(1 - y) \qquad (6)$$

in which the slope coefficient, ϵ, is constant for any one orifice but increases with β.

TABLE 2.—*List of orifices tested at Los Angeles, 1929*

Number of orifice	Pipe diameter D	$\dfrac{d}{D} = \beta$	$\beta^4 = m^2$	Number of tests	Lowest value of y
	Inches				
1		0.3724	0.0192	24	0.533
2		.4967	.0609	21	.541
3		.6207	.1484	20	.516
4	4.03	.7449	.3079	31	.544
5		.8069	.4239	25	.503
6		.8691	.5705	24	.607
7		.1241	.0002	21	.543
8		.3105	.0093	21	.529
9		.4967	.0609	21	.515
10	8.05	.6209	.1486	21	.549
11		.6829	.2175	28	.495
12		.7450	.3080	31	.479
13		.8070	.4241	31	.486
14		.8693	.5711	29	.550
15		.1951	.0014	26	.472
16		.3901	.0231	30	.523
17		.5527	.0933	25	.597
18		.6501	.1786	22	.741
19	15.38	.6989	.2386	21	.798
20		.7477	.3125	14	.875
21		.7963	.4021	12	.887
22		.8289	.4721	18	.933
23		.8615	.5508	12	.952

When the measurements of pressure were made with flange taps ($l_1 = l_2 = 1$ inch instead of $l_1 = D$, $l_2 = 0.5D$), the results were similar to those for throat taps, the only difference being that the values of ϵ were a trifle larger and that the linear relationship persisted up to higher values of β.

Experiments by R. Witte (3, 4), with corner taps, have also given a linear relation of Y to y for air and nitrogen; and the same was true of his more extensive experiments on superheated steam, except for an as yet unexplained anomaly at the highest values of y, which may be connected with the phenomena of delayed condensation. The remaining points lay along straight lines. (See reference 4, figs. 14 and 15, p. 295.)

The experimental data cited above seem to be the most extensive and trustworthy available, though not the only ones. For example,

J. L. Hodgson (5) has published curves representing the results of experiments with corner taps on air for diameter ratios $\beta = 0.421$, 0.632, and 0.843, and they are all convex upward; but for the two lower diameter ratios one can not be certain, from the small figure published, that straight lines would not do as well as the curves. For $\beta = 0.843$ there is no doubt about the curvature, but, unfortunately, there are no other published data with which this curve can be compared, so that it stands alone.

In the Los Angeles experiments with flange taps, the orifices numbered 5, 13, and 14, in Table 2 ($\beta = 0.8069$, 0.8070, and 0.8693) gave well-defined straight lines, while the observed points for number 6 ($\beta = 0.8691$) lay along a curve that was strongly concave upward. There is, however, no necessary inconsistency between these observations and Hodgson's. For when the edge of the orifice is as near the wall of the pi e as it is with these large values of β, the pressure at the wall in the vicinity of the plate varies rapidly with distance from the plate; and flange taps, which are merely near the plate, may well give quite different results from taps right in the corners. The fact that plate 14 gave a straight line, while plate 6, with the same value of β, gave a curve, was doubtless due partly, if not wholly, to the fact that the tap distances were $D/8$ in the one case and $D/4$ in the other.

The results of the early work of the Bureau of Standards on compressed air (6) were also presented as curves that were convex upward; but the observed points were so much scattered that straight lines might equally well have been used, for all but the lowest values of β. The experimental conditions at Los Angeles were more favorable, both in the steadiness of the gas supply and in the longer range of values of y that could be covered, and they permitted of so much higher precision that the Los Angeles results may be regarded as superseding the earlier ones.

Many experiments with natural gas from other fields have been carried out under the direction of H. S. Bean for the committee named above, and have given results like those already described; but the experimental conditions were generally less satisfactory and a detailed discussion of the results would not change or invalidate the conclusions drawn from the more precise data gathered at Los Angeles. These additional experiments will not be further considered here, nor will an exhaustive review of the literature be attempted; but one excellent set of observations remains to be mentioned, although the arrangement of the apparatus did not correspond exactly to any of the four usual pressure tap combinations.

H. Bachmann (7), working with air, determined the discharge coefficient of an orifice on the end of a pipe, with the jet issuing into the atmosphere. The dimensions were D 82.5 mm (3.25 inches), $d = 20.032$ mm, $\beta = 0.2428$, and $l_1 = 1.82D$; and p_2 was taken to be the barometric pressure.

In view of the low value of β, this arrangement must have gi n very nearly the same results as if the pipe had been continued downstream and the pressures had been measured at throat taps.

The values of C from 17 experiments covering the range $0.996 \gtreqless y \gtreqless 0.535$ are reproduced without systematic error by the equation

$$C = 0.600\,[1 - 0.302\,(1 - y)] \tag{7}$$

the greatest departure being -0.6 per cent, and the mean ± 0.2 per cent.

One experiment, at $y=0.997$, gave $C=0.5907$ as compared with 0.5995 from equation (7); but having regard to the admirable consistency of the other values, it seems fair to assume that this one experiment was affected by an unusually large error of some sort.

From the foregoing brief review of experimental data, it appears that, over the range $1.0 > y > 0.5$, the expansion factor may be represented, within the present accuracy of orifice meter testing, by linear equations of the form of equation (6), provided that: (a) The tap distances are not greater than $l_1 = D$ and $l_2 = 0.5D$; and (b) the diameter ratio is within the limits $0.2 \gtrless \beta \gtrless 0.75$.

V. COORDINATION OF THE EXPERIMENTAL VALUES OF ϵ

Admitting the substantial correctness of the general form of equation (6), we have next to intercompare the values of the slope coefficient ϵ that fit the various sets of observations.

1. THROAT TAPS

For each of the orifices tested at Los Angeles the values of $C\sqrt{1-m^2}$ were plotted as ordinates[2] against y as abscissa; and for each orifice for which $\beta < 0.75$, the result was a more or less definite and straight band of points. By stretching a fine thread along the band and making readings at $y=0.5$ and $y=1.0$, two straight lines, of greatest and least slope, were determined, between which any line that could reasonably be drawn to represent the band must lie. The values of ϵ for these lines were computed, and their mean is recorded as ϵ_{obs} in column 5 of Table 3.

TABLE 3.—*Slope coefficient of Y for throat taps or free discharge*

Observer	D	$\frac{d}{D}=\beta$	$\beta^4=m^2$	ϵ_{obs}	$\delta\epsilon_{obs}$	$\frac{0.41+0.33m^2}{\gamma}$ $=\epsilon_{calc}$	ϵ_{obs} $-\epsilon_{calc}$	Col. umn 8 —col umn 6	K_{obs}
1	2	3	4	5	6	7	8	9	10
	Inches								
		0.3724	0.0192	0.336	±0.006	0.334	+0.012	+0.006	0.614
	4.03	.4967	.0609	.342	.002	.335	+.007	+.005	.630
		.6207	.1464	.363	.003	.358	+.005	+.002	.668
		.7449	.3079	.386	.006	.399	−.013	−.007	.748
		.1241	.0002	.313	.003	.320	−.007	−.004	.595
		.3105	.0093	.329	.003	.322	+.007	+.004	.606
H. S. Bean, throat taps, $\gamma=1.283$	8.05	.4967	.0609	.331	.004	.335	−.004	....	.634
		.6209	.1486	.344	.003	.358	−.014	−.011	.667
		.6829	.2175	.383	.006	.376	+.007	+.001	.696
		.7450	.3080	.403	.005	.399	+.004	....	.737
		.1951	.0014	.304	.007	.320	−.016	−.009	.599
		.3901	.0231	.300	.022	.326	−.026	−.004	.613
	15.28	.5537	.0933	.307	.032	.344	−.037	−.005	.643
		.6801	.1786	.337	.017	.365	−.028	−.011	.679
		.6989	.2286	.379	.023	.381	−.002	....	.705
		.7477	.3125	.386	.022	.400	−.014	....	.733
H. Bachmann, free discharge, $\gamma=1.40$	3.25	.2428	.0035	.302	.010	.294	+.008	....	.600

[2] Values of C might equally well have been used.

The number in column 6 is, in each case, one-half the difference between the two extreme values and gives a rough estimate of the uncertainty of the value of ϵ_{obs}.

The foregoing procedure evidently involves a considerable exercise of personal judgment which might have been avoided by utilizing the method of least squares. But there was no satisfactory method for weighting the separate points—which were certainly not all of equal weight—and the result of any arbitrary assignment of weights would have been no more authoritative or probable than that obtained, as described, by simple inspection.

If tests were carried out on a series of orifices which differed only in diameter ratio, the values found for ϵ should evidently lie along some smooth curve, $\epsilon = f(\beta)$, within the errors of experiment; and it appears that the relation would be approximately linear in β^4 or m^2. Column 7 of Table 3 contains values computed from the empirical equation

$$\epsilon_{calc.} = \frac{0.41 + 0.33m^2}{\gamma} \tag{8}$$

and column 8 contains the values of $(\epsilon_{obs.} - \epsilon_{calc.})$. Column 9 shows the positive or negative excess of $(\epsilon_{obs.} - \epsilon_{calc.})$ over the estimated uncertainty of $\epsilon_{obs.}$, given in column 6. For 5 of the 17 orifices, including Bachmann's, $(\epsilon_{obs.} - \epsilon_{calc.})$ is within the estimated uncertainty, while for the other 12 it is outside, by amounts up to 0.011.

It is quite possible that the errors in determining ϵ were larger than the admittedly rough estimates shown in column 6; and there may also have been differences of finish between the different plates, so that even if there had been no experimental errors, the points would not have lain on a smooth curve or a straight line. But in any event, the departures are not so important as might appear at first sight. An error of 0.010 in ϵ changes C by 0.4 per cent at $y = 0.6$, or by 0.2 per cent at $y = 0.8$, which is a lower value of y than is often encountered in practice; and it seems probable that when an orifice meter for gas is used with throat taps, the equation

$$Y_{calc.} = 1 - \frac{0.41 + 0.33m^2}{\gamma}(1 - y) \tag{9}$$

will always give values of Y that are accurate enough for ordinary commercial purposes.

2. FLANGE AND CORNER TAPS

In Table 4, with the same notation as Table 3, the data in the upper part refer to Bean's observations with flange taps on the Los Angeles natural gas, the values of $\epsilon_{obs.}$ and $\delta\epsilon_{obs.}$ having been found from the observations in the manner described above for throat taps. The lower part of the table refers to Witte's (4) observations with corner taps on superheated steam, air, and nitrogen: and the values of $\epsilon_{obs.}$ were obtained by readings from the published plots of Y against y. (See reference 4, figs. 14 and 15, p. 295.)

TABLE 4.—*Slope coefficient of Y for flange and corner taps*

Observer	D	$\frac{d}{D}=\beta$	$\beta^4=m^2$	$\epsilon_{obs.}$	$\delta\epsilon_{obs.}$	$\frac{0.41+0.37m^2}{\gamma}$ $=\epsilon_{calc.}$	$\epsilon_{obs.}-$ $\epsilon_{calc.}$	Column 8−column 6
1	2	3	4	5	6	7	8	9
	Inches							
		0.3724	0.0192	0.333	±0.009	0.325	+0.008	------
		.4967	.0609	.341	.003	.337	+.004	+0.001
	4.03	.6207	.1484	.367	.003	.362	+.005	+.002
		.7449	.3070	.387	.015	.408	−.021	−.006
		.8069	.4239	.411	.005	.442	−.001	------
		.1241	.0002	.310	.006	.320	−.010	−.004
		.3105	.0093	.335	.004	.322	+.013	+.009
		.4967	.0609	.329	.005	.337	−.008	−.003
Bean, flange taps, γ=1.283	8.05	.6209	.1486	.350	.004	.392	−.012	−.008
		.6829	.2175	.393	.003	.382	+.011	+.008
		.7450	.3080	.412	.005	.408	+.004	------
		.8070	.4241	.443	.002	.442	+.006	+.004
		.8693	.5711	.491	.003	.483	+.008	+.005
		.1951	.0014	.309	.006	.320	−.011	−.005
		.3901	.0231	.304	.012	.326	−.022	−.010
	15.38	.6501	.1786	.349	.015	.371	−.023	−.007
		.6989	.2386	.409	.018	.398	+.021	+.003
		.7477	.3125	.433	.030	.410	+.023	------
		.20	.0016	.328	-------	.313	+.015	------
		.4935	.0595	.336	-------	.330	+.026	------
Witte, corner taps, γ=1.31	3.94	.58	.1136	.352	-------	.345	+.007	------
		.70	.2401	.382	-------	.381	+.001	------
		.76	.3329	.414	-------	.407	+.007	------
		.152	.0005	.275	-------	.293	−.018	------
	.79	.326	.0112	.332	-------	.296	+.036	------
Witte, corner taps, γ=1.40	.197	.606	.1347	.312	-------	.328	−.016	------

In Witte's experiments on steam, the rate of flow was determined by condensation and weighing, and the experimental accuracy was probably higher than could be attained with natural gas. On the other hand, in his experiments on air and nitrogen, the flow was measured by a small wet-drum meter; and while these measurements may have been accurate, the orifices were too small for exact reproduction, and comparison with larger orifices of ostensibly the same geometrical shape is of little significance. The most important result of these small-scale experiments is their satisfactory confirmation of the linear relationship between Y and y.

Since Witte's values of Y are published in the form of small plots, from which it is difficult to make accurate readings, the values of $\epsilon_{obs.}$ given in Table 4 may not do justice to the accuracy of the original data. No attempt has been made to estimate the uncertainty denoted by $\delta\epsilon_{obs.}$ It is impossible to assign definite weights to the 26 values of $\epsilon_{obs.}$; but the 5 values for the largest pipe at Los Angeles and Witte's 3 values for air and nitrogen seem to be considerably less trustworthy than the others.

The values in column 7 of Table 4 were computed from the equation

$$\epsilon_{calc.}=\frac{0.41+0.37m^2}{\gamma} \tag{10}$$

and columns 8 and 9 have the same meanings as in Table 3.

As with equation (9) for throat taps, so here it appears that the slightly modified equation

$$Y_{calc.} = 1 - \frac{0.41 + 0.37m^2}{\gamma}(1-y) \tag{11}$$

represents the facts to an approximation sufficient for ordinary commercial metering, under the following conditions: (a) for flange taps, up to $\beta = 0.8$ when $l_1 = l_2 \leqq D/4$, or up to $\beta = 0.87$ when $l_1 = l_2 \leqq D/8$; and (b) for corner taps up to $\beta = 0.76$, the highest value for which Witte gives data.

When an orifice for which $\beta = 0.869$ was tested at Los Angeles in the 4-inch pipe, with flange taps, the resulting band of points was strongly concave upward, whereas an orifice of the same diameter ratio in the 8-inch pipe gave a well-defined straight line. The simple linear relation persisted to a higher value of β when the pressure taps were relatively closer to the orifice plate; and in Witte's measurements with corner taps, values of β above 0.76 would probably still have given the linear relation described by the general equation (6) or, in particular cases, by equations (7), (9), and (11).

VI. THEORETICAL COMPUTATION OF Y

Some orifice-metering devices work with more than the critical pressure drop, but in the normal meter the range of pressure is less than 2 to 1, and usually very much less. Even with gases that show considerable departures from Boyle's law over the range from 1 atmosphere up to the high pressures at which they may be metered, the departures are nearly always neglible over the range of pressure in an orifice meter; and although the use of the ideal gas equation, $pv = RT$, for computing the density at p_1 from the density at atmospheric pressure, might lead to serious errors, it is permissible to treat the expansion through the orifice as subject to this equation. It may also be stated, without discussing the details of the experimental evidence, that when R_d is large, the flow is very nearly isentropic, at least as far as the vena contracta. The changes of density in the jet may therefore, without appreciable error, be treated as conforming to the thermodynamic equations for isentropic expansion of an ideal gas.

Let us now suppose that the pressure taps are so situated that p_1 is the static pressure in the approaching stream of gas just before it has begun to converge toward the orifice, and p_2 is the static pressure in the jet at the vena contracta, where the flow has become straight and the pressure in the jet sensibly uniform and equal to the static pressure of the gas in the surrounding space.

The area of the orifice being A, let μ_a be the contraction coefficient, so that the cross section of the jet at the vena contracta is $A\mu_a$. Then by the usual, familiar train of reasoning we arrive at the equation

$$M = A\mu_a \sqrt{\frac{2\gamma}{\gamma - 1}p_1\rho_1 \frac{y^{\frac{2}{\gamma}} - y^{\frac{\gamma+1}{\gamma}}}{1 - \mu_a^2 m^2 y^{\frac{2}{\gamma}}}} \tag{12}$$

And upon comparing this with equation (2) in the form

$$M = KYA\sqrt{2\rho_1(p_1 - p_2)} \tag{13}$$

and introducing the abbreviation

$$\frac{\gamma}{\gamma - 1} \frac{y^{\frac{2}{\gamma}} - y^{\frac{\gamma+1}{\gamma}}}{1 - y} \equiv Z \tag{14}$$

we get the equation

$$Y = \frac{1}{K}\sqrt{\frac{Z}{\frac{1}{\mu^2_a} - m^2 y^{\frac{2}{\gamma}}}} \tag{15}$$

from which Y may be computed, for any given values of m, K, γ, and y, if the value of μ_a can be determined.

In default of a solution of the equations of motion, μ_a can be found only by recourse to some plausible, simplifying assumption. In an earlier paper (8) it was assumed that, at any given mass flow, the force exerted by the gas on the upstream face of the orifice plate was the same, whether the subsequent flow through the orifice was isentropic or went on without change of density, as for liquids. If the jet issues into a space in which the static pressure is uniform, and is therefore the same all over the boundary of the jet and the downstream face of the plate as in the vena contracta, the assumption makes it possible to apply the momentum principle and obtain a relation between the contraction coefficient μ_a, and the contraction coefficient, μ, for a jet of liquid from the same orifice. The latter may readily be shown to satisfy the equation

$$\mu = \frac{K}{\sqrt{1 + m^2 K^2}} \tag{16}$$

so that μ may be computed from the diameter ratio of the orifice and the value of K, which is accessible to measurement, either by experiments with a liquid or as the limiting value of C in experiments with a gas.

The relation in question (equation (20) of reference 8) may be put into the form

$$\mu_a = \frac{Z}{y^{\frac{1}{\gamma}}B}\left(1 - \sqrt{1 - \frac{y^{\frac{2}{\gamma}}B}{Z^2}}\right) \tag{17}$$

where

$$B = \left(m^2 + \frac{2}{\mu} - \frac{1}{\mu^2}\right)Z - m^2 y^{\frac{2}{\gamma}} \tag{18}$$

and Z is defined by equation (14).

The value, or lack of value, of the assumption on which equation (17) is based is to be determined by comparing the resulting "theoretical" values of Y with values found by experiments with a gas on an orifice which is so installed that the conditions regarding p_1 and p_2 are satisfied.

The value of K, found either by testing with a liquid or by extrapolation to $y=1$ from the experiments with the gas, is substituted in equation (16), together with the measured value of $\beta^2 = m$, to give the value of μ; a value of y is selected, and with the given value of γ the value of Z is computed from equation (14); and after these preliminaries, B, μ_a, and Y are computed, successively, from equations (18), (17), and (15). The value of Y may then be compared with the value found by experiment at the selected value of y.

VII. COMPARISON OF THEORETICAL AND EXPERIMENTAL VALUES OF Y

In the deduction of equation (12), p_1 and p_2 represent the static pressure in the stream just before it begins to converge toward the orifice, and the static pressure in the jet at the vena contracta; and if the theory is to be tested by comparing values of Y from equation (15) with values obtained by experiment, the pressure taps in the experimental apparatus must be so placed as to conform to the requirements of the theory. For diameter ratios up to 0.75, the former condition may be satisfied, within the precision of all but the most refined measurements, by placing the upstream side hole anywhere within the limits $0.5D \lesssim l_1 \lesssim 2D$; but the location of the downstream side hole requires more care.

Visual observations with orifices installed in glass pipes have shown that the vena contracta occurs at about the same cross section of the pipe as the minimum static pressure at the wall, and it is commonly assumed that this minimum pressure is identical with the pressure in the jet at the vena contracta. No direct experimental proof of this is known to the present writer, but there is no obvious reason for doubting that the assumption is substantially correct, and it will be accepted here.

For low values of β, the downstream minimum of pressure is about one pipe diameter from the orifice, but is too flat to be located accurately. As β is increased, the minimum becomes more pronounced and moves closer to the orifice, but its position also depends to some extent on the rate of flow, being blown farther downstream if the speed of the jet is raised (9). Nevertheless, the pressure in a fixed side hole at the distance $l_2 = 0.5D$ from the orifice plate is only very slightly higher than the minimum pressure, unless β is large; and up to $\beta = 0.75$ the difference is not more than 0.005 (p_1-p_2), which corresponds to a change in C of only 0.25 per cent. It is therefore evident that measurements of p_1 and p_2 with throat taps will come very close to satisfying the conditions presupposed in the deduction of equation (15).

The deduction of equation (17) is subject to the further condition that the static pressure on the downstream face of the orifice plate, and over the bounding surface of the jet as far as the vena contracta, shall be uniform and equal to the pressure inside the jet at the vena contracta. This requirement is satisfied when the jet discharges into the open air, as in Bachmann's experiments (7), or in the ordinary installation when β is small. As β is increased, the static pressure in the region about the jet becomes less uniform, if we may judge by observations at a series of small side holes distributed along the wall of the pipe, and the conditions for the validity of equation (17) are less nearly satisfied.

1. THROAT TAPS

It is impossible to form a quantitative estimate of the effects of the departures from the theoretical conditions just considered, and the comparison of theory with experiment will therefore be carried up to $\beta = 0.75$, which is as far as the Los Angeles experiments with throat taps gave well-determined values of Y.

To cover this range of diameter ratios six of the Los Angeles orifices were selected as having particularly well determined values of ϵ_{obs}; they are marked with asterisks (*) in column 5 of Table 3. Values of Y were computed from equation (15) at $y = 0.5---0.9$, with $\gamma = 1.283$ and the values of K shown in column 10 of Table 3, which were found graphically at the same time as the values of ϵ_{obs}.

The first result to be noted is that the theoretical curves, $Y = f(y)$, are slightly convex upward, the curvature being greatest for low values of β. For $y \gtrless 0.6$, the computed points are not far from the straight line drawn from the point $(y = 1, Y = 1)$ through the point computed for $y = 0.7$. The slope of this line will be denoted by $\epsilon_{0.7}$, and the ordinates by

$$Y_{0.7} = 1 - \epsilon_{0.7} (1 - y) \qquad (19)$$

The value of $(Y - Y_{0.7})$ at any value of y, is the amount by which the computed theoretical curve is above the straight line at that value of y, and these amounts are shown in Table 5.

TABLE 5.—*Curvature of the computed curve $Y = f(y)$*

γ	D	β	K_{obs}	Values of $(Y - Y_{0.7})$.				
				$y = 0.5$	0.6	0.7	0.8	0.9
	Inches							
	8.05	0.1241	0.595	−0.0093	−0.0080	±0.0000	+0.0012	+0.0011
	8.05	.3105	.606	−.0090	−.0028	±.0000	+.0011	+.0011
1.283	4.03	.4967	.630	−.0071	−.0026	±.0000	+.0010	+.0008
	4.03	.6207	.668	−.0059	−.0020	±.0000	+.0008	+.0006
	8.05	.6829	.696	−.0049	−.0016	±.0000	+.0006	+.0004
	8.05	.7450	.737	−.0036	−.0011	±.0000	+.0003	+.0001
1.40	8.25	.2428	.600	−.0076	−.0028	±.0000	+.0012	+.0011

The fourth figure in Y is not certain, but the table suffices to give an idea of the degree of curvature and its regular increase as β decreases. As already noted, experiments indicate that when β is small the true curve is slightly convex upward, although for larger values of β it is sensibly a straight line.

The computations were also carried out for Bachmann's (7) orifice with $\beta = 0.2428$, $K = 0.600$, and $\gamma = 1.40$; and Table 6 contains values of the following quantities for each of the seven orifices:

Y computed from equation (15);

$Y_{0.7}$ computed from equation (19); and

Y_{obs} computed from equation (6), with ϵ_{obs} taken from column 5 of Table 3.

TABLE 6.—*Computed and observed values of Y for throat taps (Bean, $\gamma = 1.285$) and free discharge (Bachmann, $\gamma = 1.40$)*

$\epsilon_{0.7}$ ϵ_{obs} ϵ_{calc}	β	K		$y=0.5$	0.6	0.7	0.8	0.9
0.303 .313 .320	0.1241	0.595	Y...... $Y_{0.7}$... Y_{obs}...	0.841 .849 .844	0.876 .879 .875	0.9094 .909 .906	0.941 .940 .937	0.971 .970 .969
.313 .329 .322	.3105	.606	Y...... $Y_{0.7}$... Y_{obs}...	.836 .844 .836	.872 .875 .868	.9062 .906 .901	.939 .937 .934	.970 .969 .967
.233 .342 .335	.4967	.630	Y...... $Y_{0.7}$... Y_{obs}...	.826 .834 .829	.864 .867 .863	.9000 .900 .897	.934 .933 .932	.968 .967 .966
.365 .363 .358	.6207	.668	Y...... $Y_{0.7}$... Y_{obs}...	.812 .818 .819	.852 .854 .855	.8904 .890 .891	.928 .927 .927	.964 .963 .964
.387 .383 .376	.6829	.696	Y...... $Y_{0.7}$... Y_{obs}...	.801 .807 .809	.843 .845 .847	.8838 .884 .885	.923 .923 .923	.962 .961 .962
.419 .408 .399	.7450	.737	Y...... $Y_{0.7}$... Y_{obs}...	.797 .791 .799	.831 .832 .839	.8742 .874 .879	.916 .916 .919	.958 .958 .960
.283 .302 .294	.2428	.600	Y...... $Y_{0.7}$... Y_{obs}...	.851 .859 .849	.884 .887 .879	.9151 .915 .909	.945 .943 .940	.973 .972 .970

In the first column, values of $\epsilon_{0.7}$ are given for comparison with those of ϵ_{obs} and ϵ_{calc}, repeated from columns 5 and 7 of Table 3.

Small discrepancies between Tables 6 and 5 are due to the dropping of subsequent figures.

2. CORNER TAPS

Observations of pressure at corner taps do not quite satisfy the conditions for which equation (15) was deduced, and if the results of such observations are to be used for testing the value of that equation, the experimental values of K must first be reduced to what they would have been if the pressures had been observed at throat taps, which conform more nearly to the requirements of the theory. This can not, at present, be done with any great accuracy, but in order not to neglect the opportunity offered by the publication of Witte's (4) observations on steam, the reduction will be attempted. It might be effected by means of Witte's observations on the longitudinal distribution of pressure at the wall of the pipe near the orifice plate; but the uncertainty of readings from the rather small-scale curves by which the results are represented (see reference 4, Pt. II) has led me to prefer using the somewhat similar data obtained at Chicago in 1924 (9).

Letting K_c denote the value of K for corner taps and writing

$$K = b \, K_c \tag{20}$$

values of the reduction factor b were found, by interpolation in Table 19 of reference 9, from the equation

$$b = \frac{C(24,12)}{C(1,1)} \tag{21}$$

in which $C(24,12)$ and $C(1,1)$ represent the values of C for an orifice installed in a smooth pipe of 23.3 inches inside diameter, when the distances from the orifice plate to the side holes were, respectively, 24 and 12 inches, and 1 and 1 inch, the ratio being deduced from the observed longitudinal distribution of pressure.

The identification of $C(1,1)$ with the value $C(0,0)$ that would be obtained with the side holes right at the faces of the plate instead of $D/23.3$ away, is of course not exact; the difference is small but not yet accurately known, and it varies with the diameter of the side holes (4). Furthermore, the pressure distribution is not entirely independent of the pressure ratio. The values of b found as described above are therefore slightly uncertain, but in default of a detailed tabulation of Witte's measurements it appears that we can do no better at present.

Table 7 refers to the five orifices for which Witte gives values of $Y = f(y)$ determined with superheated steam (see reference 4, fig. 15): the notation and arrangement are the same as in Table 6, with the addition of two columns containing K_c, as given by Witte, and b, obtained as already described. The values of Y were computed from equation (15) with $\gamma = 1.31$ and $K = bK_c$; and those of ϵ_{obs} and ϵ_{calc} in the first column are repeated from Table 4.

TABLE 7.—*Computed and observed values of Y for corner taps (Witte, $\gamma = 1.31$)*

$\epsilon_{0.7}$ ϵ_{obs} ϵ_{calc}	β	K_c	b	$bK_c = K$		$y=0.5$	0.6	0.7	0.8	0.9
1	2	3	4	5	6	7	8	9	10	11
0.304 .326 .313	0.20	0.604	0.9984	0.6030	Y_____ $Y_{0.7}$_____ Y_{obs}_____	0.840 .848 .836	0.875 .878 .869	0.9087 .909 .902	0.940 .939 .934	0.971 .970 .967
.318 .356 .330	.4935	.622	.9991	.6215	Y_____ $Y_{0.7}$_____ Y_{obs}_____	.833 .841 .822	.870 .873 .858	.9046 .905 .893	.937 .936 .929	.969 .968 .964
.336 .352 .345	.58	.644	.9998	.6439	Y_____ $Y_{0.7}$_____ Y_{obs}_____	.825 .832 .824	.863 .866 .859	.8992 .899 .894	.934 .933 .929	.967 .966 .965
.378 .382 .381	.70	.692	1.0081	.6976	Y_____ $Y_{0.7}$_____ Y_{obs}_____	.805 .811 .809	.847 .849 .847	.8866 .887 .885	.925 .924 .924	.963 .962 .962
.416 .414 .407	.76	.730	1.0202	.7447	Y_____ $Y_{0.7}$_____ Y_{obs}_____	.798 .792 .793	.832 .834 .834	.8753 .875 .876	.917 .917 .917	.969 .968 .969

A question might arise here concerning the values of ϵ_{obs}. Let p_1' and p_2' be the pressures measured at corner taps, while p_1 and p_2 are the upstream and downstream minimum pressures dealt with by the theory. To be comparable with values of Y computed from the reduced values of K, the observed values of Y should be plotted against the simultaneous values of p_2/p_1; and if, by an oversight, the abscissa in the plot were p_2'/p_1', the values of ϵ_{obs} read from the plot would need slight corrections, which would, however, be negligible except for the two largest orifices. In reality, the abscissa in the figure is stated to be p_2/p_1, and since it must be assumed that the statement is correct, no further reduction has been undertaken.

In view of the uncertainties involved in the foregoing reduction of Witte's experimental data, to say nothing of the difficulty of making accurate readings from his published figure, the surprisingly close agreement of Y and Y_{obs} for the two largest orifices is not to be taken too seriously. Nevertheless, it appears that, over the range $0.6 < y < 1.0$ and $0.2 \eqslantgtr \beta \eqslantgtr 0.76$, equation (15), developed by theoretical reasoning from an initial approximating assumption, does give a fairly good representation of the best established experimental facts for steam ($\gamma = 1.31$), as well as for natural gas ($\gamma = 1.283$) and for air ($\gamma = 1.40$).

VIII. SUMMARY

1. NOTATION

With all quantities expressed in terms of normal units, such as "British absolute" or cgs, let C be the discharge coefficient of an orifice meter of the type illustrated by Figure 1, as defined by the equation

$$M = CA\sqrt{2\rho_1(p_1 - p_2)} \tag{A}$$

in which:

$M =$ the mass discharged per unit time;
$A =$ the area of the orifice;
$p_1, p_2 =$ the static pressures observed at the upstream and down-stream side holes or pressure taps; and
$\rho_1 =$ the density of the gas at p_1 and the upstream temperature.

In addition to the foregoing notation, let

$d =$ the diameter of the orifice;
$D =$ the diameter of the pipe in which it is installed;
$\beta = d/D =$ the diameter ratio, or
$m = \beta^2 =$ the area ratio;
$l_1, l_2 =$ the distances from the orifice plate to the centers of the upstream and downstream side holes;
$y = p_2/p_1 =$ the pressure ratio;
$\gamma = C_p/C_v =$ the specific heat ratio of the gas;
$\eta =$ the viscosity of the gas;
$R_d = 4M/\pi d\eta =$ the Reynolds number;
$K =$ the value obtained for C when the meter is tested with water or other liquid under conditions that make $R_d > 200{,}000$—M, ρ_1, and η now referring to the liquid;
$Y = C/K =$ the expansion factor for the gas; and
$\epsilon = (1-Y)/(1-y)$, so that

$$C = KY \tag{B}$$

and

$$Y = 1 - \epsilon(1-y) \tag{C}$$

2. CONCLUSIONS

The following statements and conclusions are subject to the restriction that $R_d > 200{,}000$, a condition which is nearly always satisfied in the commercial measurement of gas by orifice meters except when the orifices are very small.

1. The water coefficient, K, is sensibly constant for any one orifice when installed and operated in a prescribed manner.
2. When the meter is used for gas, $C \doteq K$ or $Y \doteq 1$, when $y \doteq 1$.

These two facts are already familiar from the published work of Witte and others. The condition $R_d > 200,000$ results from Witte's experiments (3). The following statements are conclusions from the discussion in the present paper of experimental data obtained with pressure taps located within the limits $l_1 \lessgtr 2D$ and $l_2 \lessgtr 0.5D$: they may or may not be true outside those limits.

3. When $\beta < 0.2$, the curve $Y = f(y)$ is slightly concave toward the y axis, but the data for low values of β are scanty and no more specific statement is possible.

4. When $0.2 \lessgtr \beta \lessgtr 0.75$, $Y = f(y)$ is linear within the present accuracy of orifice meter measurements, at least as far down as the critical value of y; in other words, ϵ is a constant for any one orifice meter.

5. When $\beta > 0.75$, the linearity of $Y = f(y)$ may or may not persist, according to the location of the pressure taps.

(a) For $l_1 = D$ and $l_2 = 0.5D$ (throat taps), the curve is convex toward the y axis at $\beta = 0.8$ and still more so when $\beta = 0.87$. Presumably, the linear relationship ceases to hold soon after β exceeds 0.75.

(b) For $l_1 = l_2 = D/4$ (flange taps in a 4-inch pipe), $Y = f(y)$ is still represented by a straight line at $\beta = 0.8$; but at $\beta = 0.87$ the curve is strongly convex toward the y axis.

(c) For $l_1 = l_2 = D/8$ (flange taps in an 8-inch pipe) the linear relationship still persists at $\beta = 0.87$.

(d) For corner taps there are no satisfactory data above $\beta = 0.76$; but (b) and (c), above, indicate that $Y = f(y)$ would still be linear at considerably higher values of β.

6. Within the limits $0.2 \lessgtr \beta \lessgtr 0.75$, where each of the foregoing arrangements of the pressure taps gives a linear change of Y with y, or a constant ϵ for each orifice, the values of ϵ vary systematically with β and γ, and the values of Y are given approximately by the equation

$$Y = 1 - \frac{0.41 + 0.35m^2}{\gamma}(1-y) \tag{D}$$

The available observations may be slightly better represented by using separate equations for throat, and for corner and flange taps; but the difference is little, if at all, greater than the experimental uncertainties. And it seems probable that when the pressure ratio is greater than 0.8, as it is in the vast majority of practical metering operations, the mean equation (D) will always give the value of Y correctly within 0.5 per cent, and usually much closer than that, provided that the pressure taps are located within the limits $l_1 \lessgtr 2D$ and $l_2 \lessgtr 0.5D$.

Further accumulation of experimental data may require some modification of the numerical coefficients of equation (D), but it seems improbable that the changes will be of serious importance to gas engineers.

Although the variations of the limiting or liquid coefficient, K, have not been discussed in this paper, it may be stated here that the values of Y, or of the slope coefficient ϵ, are much less sensitive to changes of tap location or roughness of the pipe than the values of K.

7. In continuation of an earlier paper (8), a theoretical method for computing Y has been developed, and has been shown to be in fair agreement with the experimental facts in a number of typical cases.

IX. REFERENCES

1. Regeln für die Durchflussmessung mit genormten Düsen und Blenden, V. D. I. Verlag, Berlin, 1930.
2. H. S. Bean, An Apparatus and Method for Determining the Compressibility of a Gas and the Correction for "Supercompressibility," B. S. Jour. Research, vol. 4 (RP170), p. 645, May, 1930.
3. R. Witte, Durchflusszahlen von Düsen und Stauründern; Techn. Mech. u. Thermodynamik, vol. 1, Nos. 1, 2, 3, 1930. (Forschung, vol. 1.)
4. R. Witte, Die Strömung durch Düsen und Blenden; Forschung, vol. 2, pp. 245 and 291, July and August, 1931.
5. John L. Hodgson, The Orifice as a Basis of Flow Measurement; Inst. Civ. Eng., 1925, Selected Engineering Papers, No. 31.
6. B. S. Jour. Research, vol. 2 (RP49), p. 561, March, 1929.
7. H. Bachmann, Beitrag zur Messung von Luftmengen; Dissertation, Darmstadt, 1911.
8. Edgar Buckingham, Note on Contraction Coefficients of Jets of Gas, B. S. Jour. Research, vol. 6 (RP303), p. 765, May, 1931; or Beitrag zur Berechnung der Kontraktionszahl, Forschung, vol. 2, p. 185, May, 1931.
9. Howard S. Bean, M. E. Benesh, and Edgar Buckingham, Experiments on the Metering of Large Volumes of Air; B. S. Jour. Research, vol. 7 (RP335), p. 93, July, 1931.

WASHINGTON, May 23, 1932.

RP460

THEORY OF VOLTAGE DIVIDERS AND THEIR USE WITH CATHODE RAY OSCILLOGRAPHS

By Melville F. Peters, George F. Blackburn, and Paul T. Hannen

ABSTRACT

Four requirements are given for voltage dividers, and equations have been developed for the most general type of capacitance voltage divider. Special cases of the most general type are discussed. These equations show that if the capacitance of the voltage divider is small, the capacitance to ground can not be neglected if the deflection of the cathode beam is to be proportional to the applied voltage. A number of oscillograms of spark discharge using five types of circuits confirm the equations. It was found that resistors had to be placed in the voltage divider to damp out oscillations set up in the divider. Equations are given for resistance voltage dividers expressing the relation between the divided voltage and the applied voltage in terms of the resistance and self-capacitance of the resistors. The equations show that only under special conditions will the deflection of the cathode beam be proportional to the applied voltage. Oscillograms taken at frequencies of 20 to 1,000 kc confirm the equations as closely as can be expected with the simple assumptions used in deriving them. Methods are given for the determination of the sensitivity of the cathode beam at the photographic plate, as well as the determination of the reduction ratio of the voltage divider.

CONTENTS

I. INTRODUCTION

A fundamental requirement for an instrument designed to measure potential difference in an electric circuit is that the measuring instrument shall not introduce more than a negligible change in the character of the circuit under consideration. This sometimes imposes limiting conditions upon the instrument, especially in the case of measuring the voltage across a spark gap with a cathode ray oscillograph under conditions found in practice.

81

In automotive ignition systems the capacitance and inductance are low. Measurements made on aircraft engines showed the capacitance of the leads to vary from 40 to 180 $\mu\mu f$. On radio-shielded engines the capacitance is usually greater than this and has been found to be as high as 400 $\mu\mu f$. The proximity of the leads to the engine causes the inductance to change with frequency. At high frequencies the iron has very little importance and computations made for the shorter leads, disregarding the effect of iron, showed the inductance to be approximately 1 μh.

If the voltages are sufficiently low so that they may be applied directly to the deflection plates of the oscillograph, the latter increases the capacitance of the circuit by approximately 20 $\mu\mu f$. If a short lead is used from the magneto to the spark gap the combined capacitance of the circuit need not exceed 40 $\mu\mu f$, which enables the experimenter to set up a test circuit which is a sufficiently close approximation to the actual circuit as regards capacitance.

If the voltage is too high to be applied directly to the deflection plates a voltage divider must be used. In a study of magneto discharges it was found that a voltage divider was required, and one was designed which introduced a capacitance of 20 $\mu\mu f$. This voltage divider proved unsatisfactory and the present investigation was undertaken to study various possible arrangements and to determine which would be the most suitable for measuring the voltage characteristics of magneto spark discharges.

II. GENERAL REQUIREMENTS FOR A VOLTAGE DIVIDER

The oscillograph records the difference of potential between the two deflection plates at any instant. In order that the voltage recorded by the oscillograph shall be a true representation of the voltage to be measured it is necessary that the following conditions be satisfied.

1. Introduction of the voltage divider must not affect materially the character of the circuit.

2. The potential difference across the deflection plates must at every instant be proportional to that across the divider, and the ratio must remain constant at all frequencies. This insures that the two potential differences will be in phase.

3. The voltage between the deflection plates and ground must not be sufficient to cause a breakdown in the deflection tube or introduce spurious effects on the cathode beam.

4. There must be no natural oscillation in any part of the voltage divider which will produce spurious effects.

In the following discussion these requirements will be referred to by number.

III. ANALYSIS OF VOLTAGE DIVIDER CHARACTERISTICS

1. CAPACITANCE VOLTAGE DIVIDER

The most general type of capacitance voltage divider is shown diagrammatically in Figure 1. Referring to the figure let A and A' be two points, which may be the electrodes of a sphere gap or spark plug, having potentials V_A and V_A' with respect to ground. Let V_s and V_s' be the potentials with respect to ground of the two deflection plates as indicated. Let C_1 and C_1' be the capacitances of the two series capacitors, and C_3' the capacitance of the capacitor shunted across the deflection plates of the oscillograph. The capacitance of these plates is C_3. Let C_2 and C_2' be capacitances to ground, which are the effective capacitances to ground of the portion of the circuit metallically connected to C_2 and C_2', respectively. By an inspection of Figure 1 equations (1) and (2) may be written

$$(V_A - V_a)C_1 = V_a C_2 + (V_a - V_a')(C_3 + C_3') \tag{1}$$

$$(V_A' - V_a')C_1 = V_a'C_2' + (V_a' - V_a)(C_3 + C_3') \tag{2}$$

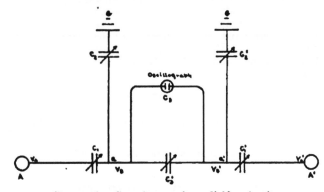

FIGURE 1.—*Capacitance voltage divider circuit*

Subtracting (2) from (1) an expression is obtained for

$$V_A - V_A' = V_a\left[1 + (C_3 + C_3')\left(\frac{1}{C_1} + \frac{1}{C_1'}\right) + \frac{C_2}{C_1}\right]$$
$$- V_a'\left[1 + (C_3 + C_3)\left(\frac{1}{C_1'} + \frac{1}{C_1'}\right) + \frac{C_2'}{C_1'}\right] \tag{3}$$

In a similar manner an expression may be obtained for $V_a - V_a'$ which reduces to

$$V_a - V_a' = \frac{(V_A - V_A')\left[C_1 C_1' + \frac{C_1 C_2' + C_1' C_2}{2}\right] + \frac{(V_A + V_A')}{2}\left[C_1 C_2' - C_1' C_2\right]}{(C_1 + C_2)(C_1' + C_2' + C_3 + C_3') + (C_1' + C_2')(C_3 + C_3')} \tag{4}$$

Equation (4) shows that the deflection which corresponds to $V_a - V_a'$ will, in general, depend not only on the potential difference $V_A - V_A'$ which is to be measured, but also on the average potential $\frac{V_A + V_A'}{2}$ of the points A and A' above ground. This latter effect can be eliminated by adjusting the circuit to make

$$\frac{C_2}{C_1} = \frac{C_2'}{C_1'} \tag{5}$$

When this adjustment is made equations (3) and (4) may be written

$$\frac{V_s - V_s'}{V_A - V_A'} = \frac{1}{K} \qquad (6)$$

where

$$K = \left[1 + (C_3 + C_3') \left(\frac{1}{C_1} + \frac{1}{C_1'} \right) + \frac{C_2}{C_1} \right]$$

and will be referred to as the reduction factor. Equation (6) shows that if the capacitances of the capacitors are independent of frequency the voltage divider will satisfy the second general requirement.

The method of adjustment of the capacitors to satisfy equation (5) is suggested by equations (3) and (4), for if the potential V_A is made equal to V_A', there must be no deflection of the cathode beam. This is true for all values of $V_A = V_A'$ only when equation (5) is satisfied.

To obtain an expression for V_s solve equation (1) for $V_s - V_s'$ and substitute in equation (5). This gives

$$V_s = \frac{V_A\, C_1}{C_1 + C_2} - \frac{(V_A - V_A')(C_3 + C_3')}{(C_1 + C_2)\left[1 + (C_3 + C_3')\left(\frac{1}{C_1} + \frac{1}{C_1'} \right) + \frac{C_2}{C_1} \right]} \qquad (7)$$

In a similar manner an expression is obtained for V_s' from equations (2) and (5)

$$V_s' = \frac{V_A'\, C_1'}{C_1' + C_2'} + \frac{(V_A - V_A')\,(C_3 + C_3')}{(C_1' + C_2')\left[1 + (C_3 + C_3')\left(\frac{1}{C_1} + \frac{1}{C_1'} \right) + \frac{C_2'}{C_1'} \right]} \qquad (8)$$

Equations (7) and (8) give the potentials of the two deflection plates to ground in terms of the applied potential to ground and the electrical constants of the voltage divider. The maximum value which V_s and V_s' may have to satisfy the third requirement depends upon the limitations imposed by the oscillograph. For this reason they must be estimated or determined experimentally, in which case, it may be more desirable to solve equations (7) and (8) for $\frac{C_2}{C_1}$ and $\frac{C_2'}{C_1'}$ respectively, in terms of $\frac{(C_3 + C_3')}{C_1}$, $\frac{(C_3 + C_3')}{C_1'}$. This solution leads to the following relations:

$$\frac{C_2}{C_1} = \frac{1}{2}\left[\frac{V_A}{V_s} - (C_3 + C_3')\left(\frac{1}{C_1} + \frac{1}{C_1'} \right) - 2 \right]$$

$$\pm \frac{1}{2} \sqrt{ (C_3 + C_3')^2 \left(\frac{1}{C_1} + \frac{1}{C_1'} \right)^2 + 2\,(C_3 + C_3')\left(\frac{1}{C_1} + \frac{1}{C_1'} \right)\frac{V_A}{V_s} - 4\frac{(C_3 + C_3')}{C_1}\frac{(V_A - V_A')}{V_s} + \frac{V_A^2}{V_s^2} } \qquad (9)$$

$$\frac{C_3'}{C_1'} = \frac{1}{2}\left[\frac{V_A'}{V_B'} - (C_3 + C_3')\left(\frac{1}{C_1} + \frac{1}{C_1'}\right) - 2\right]$$

$$\pm \frac{1}{2}\sqrt{(C_3 + C_3')^2\left(\frac{1}{C_1} + \frac{1}{C_1'}\right)^2 + 2(C_3 + C_3')\left(\frac{1}{C_1} + \frac{1}{C_1'}\right)\frac{V_A'}{V_B'} + 4\frac{(C_3 + C_3')}{C_1'}\frac{(V_A - V_A')}{V_B} + \frac{V_A'^2}{V_B'^2}} \tag{10}$$

If the maximum potentials V_A, V_A', V_B and V_B' are known approximately, the problem is to determine the five variable capacitors so as to obtain the desired deflection of the cathode beam and at the same time satisfy the four requirements. Methods of making calculations will now be given, and to simplify the presentation four cases will be considered governed by the maximum potentials of V_A and V_A', and a fifth case pertaining to the elimination of oscillations in the voltage divider. The five corresponding circuits will be designated hereafter as circuits 1, 2, 3, 4, and 5. They are shown diagrammatically in Figure 2.

An experimental study of these five circuits has been made at frequencies from 20 to 1,000 kc per second. It was found that at frequencies above 1,000 kc care must be taken in the design and selection of the capacitors and that the drop in the leads from the source to the voltage divider, as well as in the leads in the voltage divider itself become important. It was also found that resonance occurred in the voltage divider when voltage at 10^4 kc frequency was applied across A and A'. This was first observed when sparking occurred across the plates of the capacitors C_1 and C_1', and later verified by connecting a milliammeter between A and C_1. At resonance the milliammeter indicated a large current.

To eliminate repetition, certain characteristics of the circuits in the five cases are given.

1. To satisfy requirement 1 the maximum capacitance of C_1 was taken as 10 $\mu\mu$f.

2. The cathode beam was deflected 1 cm at the photographic plate for each 250 volts applied to the horizontal deflection plates. The maximum deflection which can be measured is approximately 5.6 cm, so that $V_B - V_B'$ must not be greater than 1,400 volts.

3. With a potential difference of 60,000 volts applied to the cathode tube the average potential gradient between cathode and anode is $\frac{60,000}{17.5} = 3,500$ volts per centimeter. The distance between the two pairs of deflection plates is 11 mm, and to insure against appreciable leakage from one pair to the other the average potential gradient between them was arbitrarily limited to half of that between cathode and anode. The potential difference between the pairs of deflection plates was accordingly limited to $\frac{3,500}{2} \times 1.1 = 1,925$ volts. Since the maximum potential of the oscillator plate to ground may be ± 700 volts but was usually 500 volts, the greatest potential V_B or V_B' should have is $1,925 - 500 = 1,425$ volts. This is also the potential of V_B when $V_B' = 0$ and the deflection of the cathode beam at the photographic plate is 5.6 cm.

4. The capacitance C_3 in this treatment signifies the capacitance of the circuit metallically connected to one plate of the capacitor C_3 with respect to the circuit metallically connected to the other plate of C_3 with C_3' omitted. The capacitance C_3 was estimated to be 25 $\mu\mu$f.

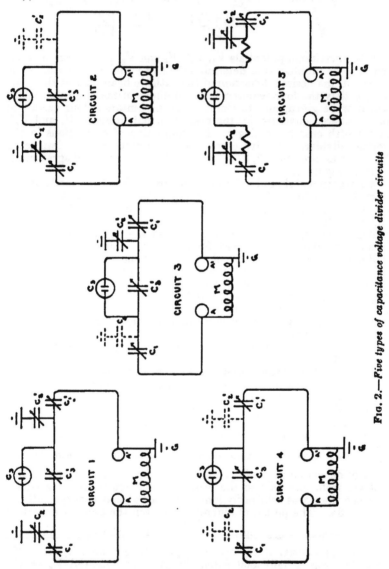

FIG. 2.—*Five types of capacitance voltage divider circuits*

5. The capacitance C_2 signifies the capacitance to ground of the circuit consisting of the ungrounded plate of C_2 and the parts metallically connected to it, and C_2' has the corresponding meaning. It was estimated that when C_1 and C_1' had their maximum values of 10

$\mu\mu$f each, the capacitances designated as C_2 and C_2' would not be less than 25 $\mu\mu$f each, even though the capacitors C_2 and C_2' were removed.

Circuit 1. V_A and V_A' large.—Let the maximum potential of A be 10,000 volts and suppose that at the same instant A' has a maximum potential of 6,000 volts. To simplify the problem make the circuit symmetrical; that is, $C_1 = C_1'$, $C_2 = C_2'$.

If the value $V_A - V_A' = 4,000$ volts be substituted in equation (6) it is found that in order to obtain a deflection of the cathode beam corresponding to 1,400 volts, K must be equal to 2.86.

It does not follow that it is permissible to adjust the circuit so that K shall have this value, and it is found that with this value of K, it is not possible to satisfy the third requirement, concerning the maximum potentials of V_s and V_s'. It is therefore necessary to use a larger value of K; that is, to work with less than the full deflection of the cathode beam.

To find the smallest value of K which makes it possible to satisfy the first three requirements it is necessary to use equation (9). Substituting in equation (9) the values

$$\frac{V_A}{V_s} = \frac{10,000}{1,400} = 7.14, \quad \frac{V_A - V_A'}{V_s} = \frac{10,000 - 6,000}{1,400} = 2.86, \quad C_1 = C_1'$$

and plotting $\frac{C_2}{C_1}$ versus $\frac{C_3 + C_3'}{C_1}$ the curve shown at (a) in Figure 3 was obtained. From this curve may be obtained the value which $\frac{C_2}{C_1}$ must have when $\frac{C_3 + C_3'}{C_1}$ is selected, or vice versa, in order that V_s shall be equal to 1,400 volts when V_A and V_A' have their maximum values. If K is computed from curve (a) and the value plotted versus $\frac{C_3 + C_3'}{C_1}$ the curve shown at (b) is obtained. This curve shows that the smallest value of K is obtained when $\frac{C_3 + C_3'}{C_1} = 0$, and the value of K increases as $\frac{C_3 + C_2'}{C_1}$ increases. To make $\frac{C_3 + C_3'}{C_1}$ as small as possible C_3' is made zero, and C_1 is given its largest value of 10 $\mu\mu$f. C_3 is constant and equal to 25 $\mu\mu$f so that the smallest value of $\frac{C_3 + C_3'}{C_1}$ is $\frac{0 + 25}{10} = 2.5$. The corresponding value of $\frac{C_2}{C_1}$, curve (a), is 5.52, hence $C_2 = 55.2$ $\mu\mu$f and $K = 11.52$. Substituting this value of K in equation (6), $V_s - V_s' = 347$ volts, which is the maximum possible potential difference across the deflection plates when all the requirements are complied with.

Since this corresponds to a deflection of less than 2 cm at the photographic plate it is worth while to consider whether this deflection could be appreciably increased by removing the restrictions on the values of V_s and V_s'. Plotting V_s and V_s' as given by equations (7) and (8) with C_2 as the independent variable the two curves shown at (a) and (b) in Figure 4 were obtained. Determining $V_s - V_s'$ from curves (a) and (b) for a number of values of C_2, the curve shown at (c) was po . This curve shows that with the value $C_1 = C_1' = 10$ $\mu\mu$f, $C_3 + C_3' = 25$ $\mu\mu$f, and the smallest value of $C_2 = C_2' = 25$ $\mu\mu$f, the greatest deflection which can be obtained at the photographic plate corresponds to 450 volts at the deflection plates. Curve (a) shows however that V_a would be 2,500 volts, and since, as has been mentioned previously, the other pair of deflection plates may have a potential to ground of opposite sign and amounting to 500 volts, the

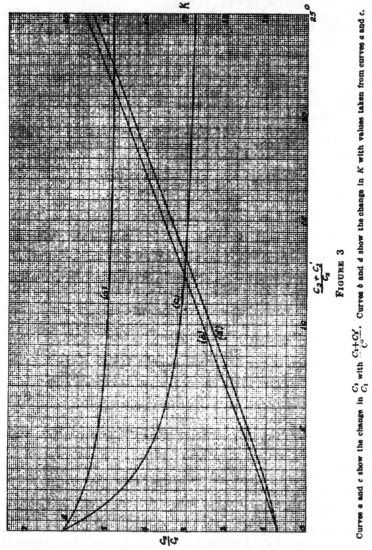

FIGURE 3

Curves a and c show the change in $\frac{C_1}{C_1}$ with $\frac{C_3+C_3'}{C_3}$. Curves b and d show the change in K with values taken from curves a and c.

potential difference between the pairs of plates might be sufficient to cause a discharge to take place between them.

The manner in which the capacitors C_1, C_1', and C_3', may be varied to increase the deflection are obvious and need not be further considered.

If V_A' becomes negative to V_A the greatest voltage which may be measured is $K(V_s - V_s') = 11.5 \times 1,400 = 16,100$. Thus, if the absolute value of V_A is not greater than 10,000 volts, V_A' may have any absolute value not exceeding 6,100 volts. Since the arrangement is symmetrical the absolute potential V_A' may be 10,000 volts if the absolute potential of V_A does not exceed 6,100 volts.

Circuit 2. V_A large, $V_A' = 0$.—If computations for circuit 1 are made for $V_A = 10,000$ volts and $V_A' = 0$, instead of 6,000 volts, the maximum

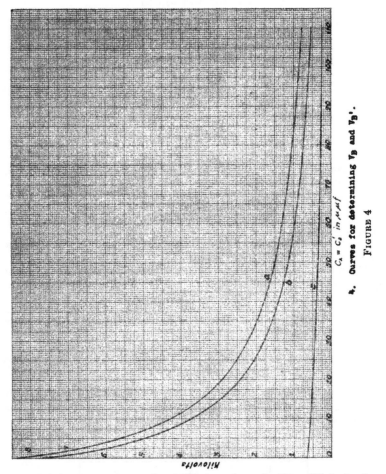

FIGURE 4

potential difference across the deflection plates is 870. With the data given C_3 may be determined from equation (9). Curves (c) and (d) are shown in Figure 3 with $V_A' = 0$, so that a comparison may be made with curves (a) and (b) which were calculated with $V_A = 6,000$ volts. If C_2, C_2' and $C_3 + C_3'$ are given their smallest capacitance, 25 μμf each, and C_1, C_1', their largest capacitance, 10 μμf each, the greatest potential difference $V_s - V_s'$ is $\frac{10,000}{10.2_5} = 975$ volts. In order to get a greater deflection it is customary to make $C_1' = \infty$. Equation (3) becomes

$$V_A - V_A' = V_s\left[1 + (C_3 + C_3')\left(\frac{1}{C_1}\right) + \frac{C_2}{C_1}\right] - V_s'\left[1 + (C_3 + C_3')\left(\frac{1}{C_1}\right)\right] \quad (11)$$

and since $V_s' = V_A' = 0$

$$\frac{V_s}{V_A} = \frac{C_1}{C_1 + C_2 + C_3 + C_3'} \quad (12)$$

For all values of V_s' other than zero the relation expressed by (12) does not hold, and requirement 2 is not satisfied. The capacitance $C_2 + C_3 + C_3'$ is now the capacitance of V_s to ground and may be denoted by C_G. Equation (12) may be written

$$\frac{V_s}{V_A} = \frac{C_1}{C_1 + C_G} \quad (13)$$

and the capacitance

$$C_G = C_1\left(\frac{V_A}{V_s} - 1\right) \quad (14)$$

For example, if C_1 is limited to 10 $\mu\mu$f and $V_s = 1{,}400$ volts for the maximum deflection, then $C_G = 61.4$ $\mu\mu$f.

Circuit 3. V_A large, V_A' small.—Let the maximum potential difference of $V_A - V_A' = 10{,}000$ volts, where V_A' may take any value from 0 to 1,000 volts. Consider two possible arrangements of circuit 2 designated (a) and (b):

(a) In this arrangement $C_1 = 10$ $\mu\mu$f, $C_3 = 25$ $\mu\mu$f, $C_3' = 0$ and $C_2 = C_G - C_3 = 61.4 - 25 = 36.4$ $\mu\mu$f.

(b) In this arrangement $C_2 = 25$ $\mu\mu$f, which is the capacitance of the leads and capacitor plates to ground. $C_1 = 10$ $\mu\mu$f. Since $C_G = 61.4$, $C_3 + C_3' = 61.4 - 25 = 36.4$ $\mu\mu$f.

If $V_A' \pm 0$, equation (12) can not be used. Putting $C_1' = \infty$ in equation (4) the following relation is obtained for the voltage across the deflection plates.

$$V_s - V_s' = \frac{V_A C_1 - V_A'(C_1 + C_2)}{C_1 + C_2 + C_3 + C_3'} \quad (15)$$

If the potentials to ground shown in columns 1 and 2 of Table 1 for V_A and V_A' are substituted in equation (15), values for V_s and $V_s - V_s'$ are obtained for the two arrangements.

TABLE 1

V_A	$V_A' = V_s'$	(a)		(b)	
		V_s	$V_s - V_s'$	V_s	$V_s - V_s'$
10, 000	0	1, 400	1, 400	1, 400	1, 400
10, 200	200	1, 490	1, 290	1, 530	1, 330
10, 400	400	1, 590	1, 190	1, 660	1, 260
10, 600	600	1, 690	1, 090	1, 790	1, 190
10, 800	800	1, 790	990	1, 920	1, 120
11, 000	1, 000	1, 890	890	2, 050	1, 050

These computations show that unless V_A' remains at ground potential the deflection is not proportional to the applied potential difference.

When the magneto is used as a source of voltage V_A' is usually small compared to V_A until the spark gap breaks down. If peak-voltage measurements are to be made the capacitance C_1' may be made large in which case C_2' is also increased so that $\frac{C_2}{C_1} = \frac{C_2'}{C_1'}$. Table 2 gives the values for V_s and $V_s - V_s'$ using circuit 1 with the following capacitances.

(c) $C_1 = C_1' = 10$ μμf, $C_2 = C_2' = 44.3$ μμf, $C_3 = 25$ μμf and $C_3' = 0$, which are the same values given in circuit 2.

(d) $C_1 = 10$ μμf, $C_1' = 60$ μμf, $C_2 = 38.5$ μμf, $C_2' = 231$ μμf, $C_3 = 25$ μμf, and $C_3' = 0$.

TABLE 2

V_A	V_A'	(c)		(d)	
		V_s	$V_s - V_s'$	V_s	$V_s - V_s'$
10,000	0	1,400	960	1,400	1,290
11,000	1,000	1,580	960	1,600	1,290

In both (c) and (d) the deflection is proportional to the applied potential difference. At low frequencies, arrangement (d) is superior to arrangement (c). At high frequencies the large capacitance of C_1', C_2' to ground which is $\frac{C_1' \times C_2'}{C_1 + C_2}$ will allow large currents to flow and the drop in the leads from A' to C_1' and C_1' to C_2' will affect the voltage across the deflection plates. At frequencies not greater than 10^3 kc this should not be serious when V_A' is small. This circuit can be used to advantage when it is uncertain as to whether or not V_A' remains at zero potential.

Circuit 4. V_A *and* V_A' *both small.*—When V_A and V_A' are both small, but $V_A - V_A'$ exceeds 1,400 volts, the capacitances C_2 and C_2' are made as small as practicable by omitting the two capacitors C_3 and C_2'.

Substituting in equation (8) the limiting values previously mentioned $C_1 = C_1' = 10$ μμf, $C_2 = C_2' = 25$ μμf, and $C_3 + C_3' = 25$ μμf, $\frac{V_s - V_s'}{V_A - V_A'} = 1/8$, which is the largest value $1/K$ may have with this arrangement under the conditions specified. If C_1 and C_1' are increased, $\frac{C_2}{C_1}$, $\frac{C_2'}{C_1'}$ and $(C_3 + C_3')\left(\frac{1}{C_1} + \frac{1}{C_1'}\right)$ become less, and $1/K$ increases. Rogowski, Wolff, and Klemperer [1] have described this circuit for which the capacitances of C_1, $(C_3 + C_3')$, and C_1' are much greater.

Circuit 5. Aperiodic voltage divider circuit.—If the voltage divider is used to measure spark discharges, oscillations may be expected in any circuit where $\frac{r}{4L} < \frac{1}{LC}$. Such oscillations were present and the cir-

[1] Rogowski, Wolff, and Klemperer: "Die Spannungsteilen Kathodenoszillographen." Archiv für Elektrotechnik, vol. 22, p. 579, 1929–30.

cuit was made aperiodic by inserting resistors at (a) and (a'), (fig. 1 or 2) so as to make $\dfrac{R^2}{4L^2} \!>\! \dfrac{1}{LC}$, where L is the inductance of the lead C_1C_3, and C the smaller of the two capacitances C_1 and C_3. In this manner requirement 4 is satisfied.

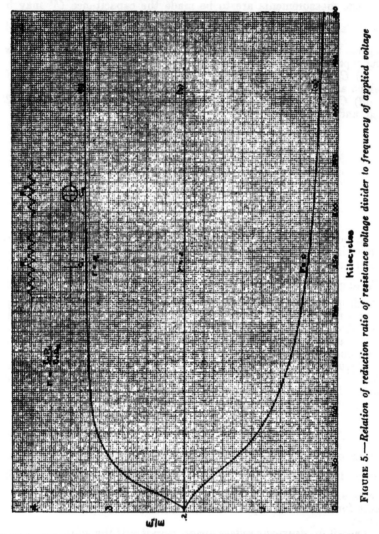

FIGURE 5.—*Relation of reduction ratio of resistance voltage divider to frequency of applied voltage*

2. RESISTANCE VOLTAGE DIVIDERS

The resistance voltage divider arrangement is shown in Figure 5. At low frequencies the impedance of the deflection plates is so high compared with that of the portion R_2 of the voltage divider that only a small part of the current is shunted to the plates, so that the ratio of the voltage to be measured to that across the deflection plates is independent of frequency. The difference in phase between the two is also small. Computations which follow show that in order to keep the reduction ratio constant as the frequency is increased, resistance of the divider must be decreased, and at frequencies found in spark discharges a suitable resistance would be so low as to prevent a spark passing across the gap. If the resistance has a small capacitance and inductance the relation becomes more complicated. When grid leaks are used for resistance the inductance is small, whereas the capacitance across the grid leaks when supported in holders becomes important at high frequencies. Neglecting the inductance of the resistors and their capacitance to ground, an equation showing the relation between the voltage applied to the divider and the voltage across the deflection plates may be determined in terms of the resistance and capacitance of each portion of the voltage divider.

Let R_1 and R_2 designate the resistances of the resistors shown in the diagram of Figure 5, and let C_1 represent the capacitance of the resistor R_1, and C_2 the combined capacitance of the resistor R_2 and the oscillograph deflection plates. The impedance of each branch may be written

$$Z_1 = \frac{R_1}{1+j\omega C_1 R_1} \text{ and } Z_2 = \frac{R_2}{1+j\omega C_2 R_2}$$

where ω is 2π times the frequency η and $j = \sqrt{-1}$. The total impedance for the circuit is

$$Z = \frac{R_1}{1+j\omega C_1 R_1} + \frac{R_2}{1+j\omega C_2 R_2}$$

and since $E = IZ$, where E is the voltage across the divider and I the current, the relation becomes

$$\frac{E}{E_2} = \frac{\dfrac{R_1}{1+j\omega C_1 R_1} + \dfrac{R_2}{1+j\omega C_2 R_2}}{\dfrac{R_2}{1+j\omega C_2 R_2}}$$

which reduces to

$$\frac{E_2}{E} = \frac{1}{\sqrt{\left[\dfrac{R_1(1+\omega^2 C_1 C_2 R_1 R_2)}{R_2(1+\omega C_1 R_1^2)}+1\right]^2 + \left[\dfrac{\omega R_1(C_2 R_2 - C_1 R_1)}{R_2(1+\omega C_1 R_1^2)}\right]^2}} \tag{16}$$

which expresses the relation between the voltage across the deflection plates and that applied to the voltage divider.

If the resistance has no capacitance $C_1 = 0$ and equation (16) reduces to

$$\frac{E_2}{E} = \frac{1}{\sqrt{\left[\dfrac{R_1}{R_2}+1\right]^2 + [\omega C_2 R_1]^2}} \tag{17}$$

If $C_1 R_1 = C_2 R_2$, equation (16) becomes independent of frequency since it reduces to

$$\frac{E_2}{E} = \frac{1}{\dfrac{R_1}{R_2}+1} = \frac{R_2}{R_1 + R_2} \tag{18}$$

These equations show that whether $\frac{E_2}{E}$ will increase, decrease, or not change with frequency depends upon the values of R_1, R_2, C_1 and C_2. Putting $r = \frac{C_1 R_1}{C_2 R_2}$, $R_1 = 8 \times 10^5$ ohms, $R_2 = 2 \times 10^5$ ohms, $C_2 = 20 \ \mu\mu f$, and assigning to C_1 the values of 0, 5, and 10 $\mu\mu f$ so that r takes the corresponding values of 0, 1, and 2 the curves shown in Figure 5 were obtained. In these curves (*a*) was plotted from equation (17), showing that the ratio decreases very rapidly with increase in frequency. As ω increases without limit $\frac{E_2}{E}$ approaches 0. Curve (*b*) is a plot of equation (18) which shows that the voltage divider is independent of frequency when $r = 1$. Curve (*c*) is a plot of equation (16) and shows that the ratio $\frac{E_2}{E}$ increases with an increase in ω. As ω approaches ∞, $\frac{E_2}{E}$ approaches the limiting value $\frac{C_1}{C_1 + C_2}$.

If the capacitance C_1 is assumed to be so small it may be neglected, computations from equation (17) show that in order that $\frac{E_2}{E} = 0.2$ shall not diminish by more than 5 per cent (or not be less than 0.19), $\omega^2 C^2 R_1^2$ must not be greater than 2.56. Since $C_2 = 20 \times 10^{-12} f$ (which includes the leads as well as the deflection plates), $\omega^2 R_1^2 = 2.56 \times \frac{10^{22}}{4}$ $= 6.4 \times 10^{21}$. If $\omega = 10^7$ or $\eta = \frac{10^7}{2\pi}$, $R_1^2 = \frac{6.4 \times 10^{21}}{10^{14}} = 6.4 \times 10^7$. Therefore, $R_1 = 8,000$ ohms, $R_2 = 2,000$ ohms and $R_1 + R_2 = 10,000$ ohms, which is the maximum resistance which may be used at a frequency of $\frac{10^7}{2\pi}$ or approximately 1,600 kc. Such a resistance across the gap would prevent sparking. Silsbee[2] has shown that a resistance of 5×10^4 ohms is so low as to seriously affect the character of the discharge and might actually prevent the occurrence of a spark. For high-compression engines such as are in use to-day the resistance must be much greater than this. Such low resistances are objectionable in this work because they affect the character of the spark. At times it may happen that fouling of plugs will reduce the resistance across a spark plug to 5×10^4 ohms, but this is a special condition, and while an interesting one, should not be a limiting condition of the problem.

The remainder of this report is devoted to an experimental study of the theory developed in this and the preceding sections, together with a discussion of oscillograph technique and the calibration of the voltage divider.

IV. DESCRIPTION OF APPARATUS

The cathode ray oscillograph was of the high-voltage cold-cathode type. The cathode was operated at approximately 60,000 volts, the supply being the secondary of a high tension transformer. The primary was connected through rotary switches to a 60-cycle 110-volt

[2] Silsbee, N. A. C. A. Report No. 241, Electrical Characteristics of Spark Generators for Automotive Ignition, p. 27.

supply, the switches being operated by a synchronous motor driven from the same source. The switches were constructed so that the 60,000 volts was applied to the cathode on the negative half cycle, either once in 20 cycles or once in 100 cycles. For the present investigation the slower rate of excitation was used.

The cathode tube (*ED*, fig. 6) was modified so that the cathode could be removed and polished. This was made possible by fusing the tungsten rod supporting the cathode to the movable portion of the ground joint as shown in the figure. The diaphragm *D* was mounted in the glass tube so that both the cathode and diaphragm moved together.

The deflection tube has two pairs of deflection plates, one pair horizontal and the other pair vertical. They are shown at *A* and *B* in Figure 6. *C* is a coil which produces a magnetic field deflecting the cathode beam in a horizontal direction. If observations are to be made at low frequencies the voltage is applied to plates *A*, producing a vertical deflection. If, at the same time, the cathode beam is swept across the field by the magnetic sweep *C*, a record is obtained which

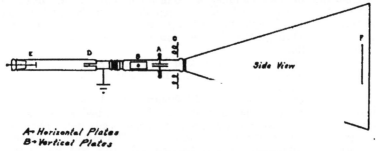

A— *Horizontal Plates*
B— *Vertical Plates*

FIGURE 6.—*Diagram of cathode ray oscillograph*

shows the variation with time of the voltage across the deflection plates. An oscillator with a range of 20 to 1,000 kc was supplied with the oscillograph. If this oscillator is connected to plates *A* and the sweep actuated, a sine wave is obtained. For high frequencies the voltage to be measured is connected to plates *B*, and superimposes upon the sine wave a horizontal deflection proportional to the voltage across the plates. The sine wave then gives the time coordinate.

With 60,000 volts applied to the cathode the deflection of the beam at the photographic plate is approximately 1 cm for each 250 volts applied to the horizontal deflection plates. The maximum cathode beam deflection which can be photographed is 5.6 cm, thus limiting the useful voltage across the deflection plates to about 1,400.

The longest time required to sweep the beam across the photographic plate is 1,200 microseconds, and when necessary a faster sweep is used. If the spark discharge is to be photographed in the center of the film the magneto has to be timed accurately. This is accomplished by driving the magneto with the synchronous motor through a set of gears. Adjustment of the gears allows for approximate timing, and the final adjustment is made with the timing device on the magneto. The magneto is radio shielded and was designed for use with 12-cylinder aircraft engines.

122486—32——7

Figure 7 is a photograph showing the arrangement of the five capacitors C_1, C_1', C_2, C_2', and C_2'. C_1 and C_1' are two high-voltage radio transmitting capacitors. The range of capacitance is from 10 to 80 $\mu\mu f$. C_2 and C_2' are two high-voltage radio transmitting capacitors having a range of capacitance from 40 to 300 $\mu\mu f$. Since the maximum voltage across C_2 and C_2' is 1,400, it is not necessary to use high-voltage capacitors. It is necessary to keep the leakage of the portion of the circuit called C_2 and C_2' in Section III small, and this requires that the capacitors in the circuit have a high insulation resistance. Since the high-voltage capacitors had a much higher insulation resistance than the low-voltage capacitors which were available, the former were used. C_2' was made by removing alternate plates from a low-voltage capacitor and supporting as shown in the photograph. The range of capacitance was 10 to 80 $\mu\mu f$.

To further reduce leakage the bakelite binding posts were removed and the lead wires brought to the oscillograph through 1-inch holes in the case. This at first seemed an unnecessary precaution, but it was found that the resistance of the binding posts to ground was much less than that of the remainder of the insulated circuit. In magneto discharges a small potential difference remains across the gap between discharges, and although the discharges alternate in sign, this additional leakage through the binding posts was sufficient to noticeably affect the results.

The five circuits may be obtained by simply removing or short circuiting the necessary capacitors. In circuit 5 the resistors are inserted at (a) and (a').

The leads shown at AD and $A'D'$ should be fixed when the voltage divider is adjusted so that they will not be disturbed while making measurements. Changing the position of the leads will unbalance the circuit and the results will be misleading. The method of balancing has been mentioned under Section III, and will be further discussed in Section V.

Figure 8 is a sketch of another arrangement used for circuit 3. The brass cylinder B was threaded and screwed to the binding post protruding through the bakelite bushing BH. This bushing passed through the case of the oscillograph. A is another brass cylinder larger than B, and the capacitance between A and B was changed by sliding the brass rod C through the bakelite bushing E. F was a grounded brass cylinder surrounding the capacitor. Leakage to ground through BH is objectionable for reasons previously given.

A better arrangement is to place the brass cylinder B, together with the other parts of the capacitor inside of the oscillograph case. The small leakage through the bakelite bushing will cause very little change in the spark discharge.

The resistance type of voltage divider was made of a number of resistances connected in series. The advantage of using a number of low resistances in series rather than a single high resistance is that a breakdown is less likely to occur and the capacitance of the unit is less, provided the same type is used for both the high and low resistance. However, this increases the capacitance to ground. This capacitance to ground may be reduced by suspending the resistors away from all metal by means of silk threads.

The spark gap used in most of the preliminary work consisted of two spheres in air. The spheres were 2 cm in diameter with a screw

B. S. Journal of Research, RP460

FIGURE 7.—*Arrangement of capacitors for voltage divider*

FIGURE 9   FIGURE 10   FIGURE 11

Oscillograms showing effect of unbalanced voltage divider.

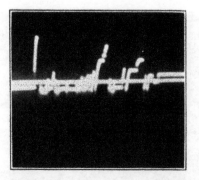

FIGURE 13.—*Slope of traces shows the reduction ratio of circuit 4 is the same at 20 and 1,000 kc*

FIGURE 14.—*Magneto spark discharge in CO_2 at 75 pounds pressure (circuit 1)*

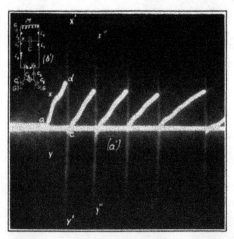

FIGURE 15.—*a', Magneto spark discharges in CO_2 at 15 pounds pressure; b', circuit arrangement used in obtaining oscillogram*

adjustment to change the gap opening. Later, measurements were made using an aviation spark plug in an atmosphere of CO_2 at pressures of 15, 75, and 135 lbs./in.² (absolute). The spark plug was screwed into the end of a brass cylinder about 4 inches long and 2 inches in diameter which was provided with a quartz window for the purpose of observing the spark.

Measurements of the high voltages used in calibrating the oscillograph were made with an electrostatic voltmeter having a range of 0 to 10 kv. A kenotron with an upper range of 30,000 volts was placed in series with the spark gap and voltmeter to ground.

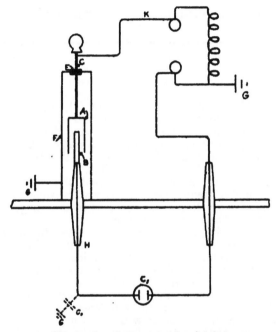

FIGURE 8.—*Arrangement for circuit 2*

V. EXPERIMENTAL RESULTS WITH CIRCUITS 1 TO 5

1. METHOD OF BALANCING VOLTAGE DIVIDER TO SATISFY EQUATION (5)

Before oscillograms showing the variation in voltage are taken it is necessary to balance the capacitance voltage divider as previously explained, so that the condition $\frac{C_2}{C_1} = \frac{C_2'}{C_1'}$ is satisfied. Since the method of balancing is the same for circuits 1, 3, 4, and 5 only one set of oscillograms showing the circuit unbalanced and balanced will be given.

With the two lead wires from the divider (fig. 1) connected to the insulated side of the sphere gap, and the horizontal deflection plates in use, the oscillogram shown in Figure 9 was obtained for circuit 4.

For correct adjustment of the position of the tube there would be no overlap of the beam as shown at (*a, b*). This adjustment will be discussed later.

Figure 10 is an oscillogram showing the apparent difference in potential between the two lead wires when C_1 is considerably greater than C_1'. Figure 11 shows the result when the condition $\frac{C_2}{C_1} = \frac{C_2'}{C_1'}$ is satisfied.

It will be understood that subsequent oscillograms pertaining to these four circuits were taken after the circuits had been balanced. It is not necessary to balance circuit 2.

2. EXPERIMENTAL VERIFICATION OF REQUIREMENT 2 AND EQUATION (6)

Requirement 2, which is expressed by equation (6), states that if the voltage divider is balanced and a potential difference is applied to the leads A and A' (fig. 1) a plot of peak voltage versus cathode beam deflection should give a straight line passing through the origin. This is shown to be the case for circuits 1, 3, 4, and 5 by the straight lines shown in Figure 12, at (*a*), (*b*), (*c*), and (*d*), respectively. These lines were obtained by applying a known potential difference to the terminals of the voltage divider and measuring the deflection. The voltage was supplied by the magneto over a range of 2,000 to 10,000 at a frequency of about 20 kc. The details as to the manner of making the calibration will be given in Section VIII.

Equipment was not available to extend the measurements of Figure 12 to higher frequencies, but it was possible to determine that the ratio of the voltages was constant in this range by the use of the oscillator. The procedure follows:

The leads from the oscillator were connected to the horizontal plates, to give a vertical deflection. Another pair of leads was connected from the oscillator to the voltage divider, the divided voltage being applied to the vertical deflection plates so as to produce a horizontal deflection. If the ratio of the applied voltage to the reduced voltage is constant, the trace of the cathode beam must be a straight line. If the ratio is to be independent of frequency the slope of the line must not change with frequency. In Figure 13 are shown two traces of the cathode beam. The straight line (*a*) was photographed at 20 kc. and (*b*) was photographed at 1,000 kc. As closely as measurements can be made these lines are parallel. These traces were photographed with circuit 4, but similar results were obtained for circuits 1, 3, and 5. Circuit 2 was not tested.

In each of these four circuits, C_1 and C_1' were selected so that the voltage divider did not materially change the characteristics of the circuit (requirement 1). Figures 12 and 13 show that the deflections of the cathode beam are proportional to the applied voltage and that this proportionally does not change with frequency (requirement 2). By proper selection of C_2 and C_2' the potentials of the deflection plates to ground were kept within the required limits (requirement 3). The absence of oscillations in the circuit (requirement 4) may best be determined by examining oscillograms of the spark discharge.

3. OSCILLOGRAMS SHOWING THE VARIATION OF VOLTAGE IN MAG-
NETO SPARK DISCHARGES

In Figure 12 straight lines were drawn showing the relation between $V_A - V_A'$ and the corresponding deflection of the cathode beam. These calibrations were made before the oscillograms shown in the

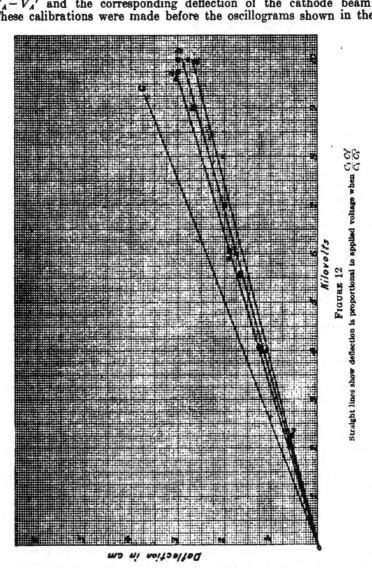

FIGURE 12

remainder of this section were taken. For this reason values of K computed from Figure 12 may not agree with those which follow, since in the latter they were selected so as to give a large deflection when the spark discharge took place at a pressure of 135 pounds. With this setting of the capacitors, oscillograms were taken with circuits 1, 3, 4, and 5, to show the variation of voltage across a spark plug in CO_2 gas during discharge. Pressures of 15, 75, and 135 pounds were used. All pressures are expressed in pounds per square inch absolute. The oscillograms for circuit 2 were taken with the spark plug in CO_2 gas at 15 pounds pressure as well as with a sphere gap in air at atmospheric pressure.

In the case of circuit 4 the capacitors were purposely selected so that V_B was greater than 1,400 volts at the higher pressures. This was done to show the necessity of keeping V_B small.

Throughout this section only representative oscillograms are given for each circuit. It will be understood that for oscillograms similar to that shown in Figure 14 the magnetic sweep provides the time coordinate. For oscillograms similar to that shown in Figure 17 the oscillator provides the time coordinate.

Circuit 1.—The approximate capacitances of the six capacitors follow:·

$C_1 = C_1' = 10$ $\mu\mu f$, $C_2 = C_2' = 65$ $\mu\mu f$, $C_3 = 15$ $\mu\mu f$ (this was found by measurement to be nearer the true capacitance than 25 $\mu\mu f$, which was assumed in the computations), $C_3' = 10$ $\mu\mu f$.

Figure 14 is an oscillogram showing a spark discharge in CO_2 at a pressur of 75 pounds. This figure shows traces of the cathode beam below the sweep line which correspond to a potential of more than twice the breakdown voltage of the gap. These traces are due to oscillations in the potential divider. From this it follows that circuit 1 does not satisfy requirement 4.

Circuit 2.—(*a*) With the arrangement shown in Figure 2, circuit 2, the capacitances of the six capacitors were approximately $C_1 = 10$ $\mu\mu f$, $C_1 = \infty$, $C_2 = 25$ $\mu\mu f$, $C_2' = 15$ $\mu\mu f$, $C_3 = 15$ $\mu\mu f$, $C_3' = 0$. Since C_2' is small the voltage drop between V_4' and the deflection plate may be neglected.

In Figure 15, (*a'*) is an oscillogram taken with circuit 2 to show the variation in voltage across a spark plug in CO_2 gas at 15 pounds pressure using the arrangement shown at (*b'*). The capacitor C shunted across the spark plug and the inductances L_1, L_3, and L_1' had a capacitance of 2,800 $\mu\mu f$. The inductances designated by L_1 and L_1' represent the inductance of the leads in the parts of the circuit indicated. The inductance of each L_1 and L_1' was approximately 1 μh. L_3 was a coil having an inductance of 3 μh.

This oscillogram was taken with the horizontal deflection plates connected to the voltage divider. At (*a*) the voltage began to increase and continued to increase until it reached (*d*), where a discharge took place across the spark plug. During the discharge the voltage drop across the spark plug should be small because the resistance of the gap is small. The fainter traces at (*x'*, *y'*) etc., indicate a voltage drop across the plug much greater than the breakdown voltage of the gap. An explanation of these large deflections follows.

If during the discharge $V_4 - V_4'$ is small the traces shown by (*x'*, *y'*) etc., can not come from the voltage drop across the spark gap, and should, therefore, appear if both leads of the voltage divider are connected to the same side of the spark gap. Figure 16 (*b'*) shows diagrammatically the circuit arrangement used in obtaining the oscillogram shown in Figure 16 (*a'*). The traces (*x'*, *y'*) etc., are present, and while measurements show that they are slightly larger in Figure 15 (*a'*) than in Figure 16 (*a'*) this is to be expected because in the first case V_A differs by a small amount from V_A' and in the second $V_A = V_A'$. That this is so becomes apparent from equation (15),

$$V_s - V_s' = \frac{V_A C_1 - V_A'(C_1 + C_3)}{C_1 + C_2 + C_3 + C_3'}$$

for if $V_A = V_A'$ the voltage $V_s - V_s'$ across the deflection plates will be greater or less than

$$\frac{V_A' C_3}{C_1 + C_2 + C_3 + C_3'}$$

which is the value resulting from equation (15) when $V_A = V_A'$.

An inspection of the circuit diagram shows that during the discharge of the capacitor C, the drop in $(L_1' + L_2)$ will make V_A' different from ground potential. V_A' has been measured and found to be 2,300 volts. Substituting this value of V_A' in equation (15) for the circuit arrangement shown in Figure 16 (b')

$$V_s - V_s' = \frac{-V_A' C_3}{C_1 + C_2 + C_3 + C_3'} = \frac{-2,300 \times 25}{50} = -1,150 \text{ volts.}$$

This corresponds to a deflection of $\frac{1,150}{250} = 4.6$ cm, which agrees with the measured value of 4.7 cm for (y'). The measured length of (x') is 4.5 cm.

The faint trace (x, y) is attributed to conditions arising from the breakdown of the gap in the distributor. Figure 15 (a') shows that when the gap breaks down the potential of the circuit increases so rapidly that the cathode beam leaves only a faint trace. This is shown by the separation at (a). This sudden application of voltage causes an appreciable difference of potential between ground and V_A'. The measured deflection of the cathode beam is 2.0 cm (fig. 16 (a')), which corresponds to a voltage across the deflection plates of $2.0 \times 250 = 500$. Substituting in equation (15), $500 = \frac{-V_A' \times 25}{500}$ or $V_A' = -1,000$ volts at the instant the distributor gap breaks down.

Figure 16 (a') shows that (a, d) is horizontal, and indicates that V_A' is zero before the spark gap breaks down. Equation (12) may then be used to compute $V_A - V_A'$ from Figure 15 (b').

Figure 17 is an oscillogram showing the traces (x', y') etc., spread out to show the nature of the oscillations. The oscillator operating at 20 kc was used to provide a time coordinate. The oscillogram was taken by connecting the deflection plates directly across a small portion of the lead L_1'. The average frequency obtained from measurements made on this and similar oscillograms was 1.1×10^3 kc. Neglecting the effect of the magneto the computed frequency of the circuit

$$(C, V_A, V_A', L_2, L_1) \text{ is } \eta = \frac{1}{2\pi\sqrt{C(L_1 + L_1' + L_2)}} = \frac{10^3}{6.28\sqrt{28 \times 5}}$$
$$= 1.25 \times 10^3 \text{ kc}$$

which agrees with the measured value since the inductors L_1, L_1', and L_3 were only approximately determined. The maximum current in the circuit is

$$I = E \sqrt{\frac{C}{(L_1 + L_1' + L_3)}} = 3,000 \sqrt{\frac{2,800 \times 10^{-12}}{5 \times 10^{-6}}} = 71 \text{ amperes}$$

where E is taken as 3,000 volts, since it is the peak potential difference across the spark plug.

The drop in voltage between (g) and A_4' using the computed frequency (since the computed current was also used) is

$$V_A' - V_g = \omega L I = 2\pi \times 1.25 \times 10^6 \times 4 \times 10^{-6} \times 71 = 2,230 \text{ volts}$$

which agrees with the measured value of 2,300 volts if (g) is assumed to be at ground potential.

It may be well to mention at this time that the portion of the circuit shown grounded, is the magneto. During the discharge of the capacitor the magneto base may not remain at ground potential although grounded to a water pipe with a copper strip. This departure from ground potential becomes more pronounced as the capacitance and inductance of the circuit are decreased. With circuit constants sufficiently small, so that $\eta = 10^4$ kc, the maximum potential of the magneto to ground was found to be 750 volts. No measurements were made of the potential difference between the magneto base and ground, or between (g) and ground, with the circuit shown in Figures 15 (b') and 16 (b'). At this relatively low frequency, 1.25×10^3 kc, the drop in voltage between the magneto base and ground, or (g) and ground would be much less.

If better agreement were desired between the computed and measured values of V_A', the potential of (g) to ground would have to be determined and the result added to 2,230 volts.

(b) The capacitance of the arrangement shown in Figure 8 measured from A to B was approximately 20 $\mu\mu$f. The measurement included C_1 as well as the capacitance to ground which in this type of capacitor is relatively large as compared to C_1. Figure 18 is an oscillogram taken with the same circuit arrangement as shown in Figure 15 (b'), except a sphere gap was used instead of a spark plug. It will be noted that the deflections (x', y') are not vertical. This will be discussed later.

Circuit 3.—The approximate capacitances of the six capacitors used in circuit 3 are: $C_1 = 10$ $\mu\mu$f, $C_1' = 70$ $\mu\mu$f, $C_2 = 25$ $\mu\mu$f, $C_2' = 175$ $\mu\mu$f, $C_3 = 15$ $\mu\mu$f, and $C_3' = 25$ $\mu\mu$f.

The oscillogram shown in Figure 19 represents a spark discharge in CO_2 at a pressure of 135 pounds. It shows very clearly just what is taking place between the secondary of the magneto, the gap in the distributor, the lead from the magneto distributor, and the spark plug.

The sweep line KL was taken with the four deflection plates grounded. The horizontal deflection plates were then connected to the voltage divider and the cathode beam swept across the photographic plate until it reached (a) where the contact in the primary circuit was broken.

After the contact was broken the series gap in the distributor of the magneto broke down, and this raised the potential difference across the gap to the voltage indicated at (b). This increase was so

B. S. Journal of Research, RP460

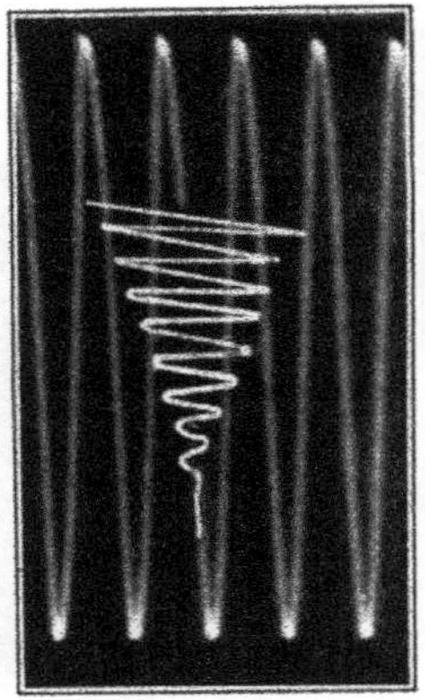

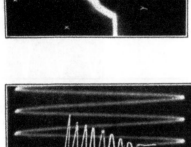

Figure 17.—Waveform of the superimposed oscillations shown at xy, x'y', etc. (Figs. 15 and 16)

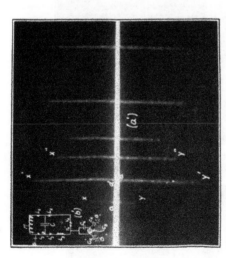

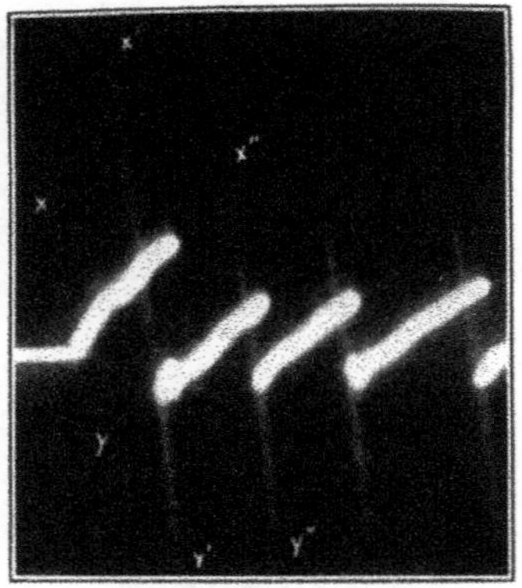

Figure 16.—a', Magneto spark discharge in CO_2 at 15 pounds pressure; b', circuit arrangement used in obtaining oscillogram

Figure 18.—Magneto circuit arrangement similar to that used in Figure 15 with sphere gap in air replacing spark plug in CO_2

The sloping lines indicate improper adjustment of the cathode tube.

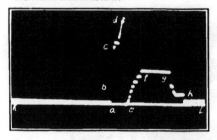

FIGURE 19.—*Magneto spark discharge in CO₂ at 135 pounds pressure (circuit 3)*

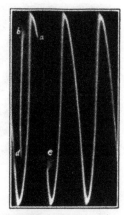

FIGURE 20.—*Magneto spark discharge in CO₂ at 15 pounds pressure (circuit 4)*

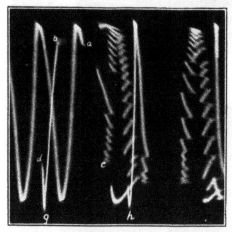

FIGURE 21.—*Magneto spark discharge in CO₂ at 75 pounds pressure (circuit 4)*

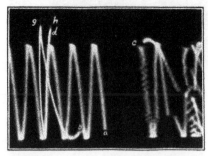

FIGURE 22.—*Magneto spark discharge in CO₂ at 135 pounds pressure (circuit 4)*

FIGURE 23.—*Magneto spark discharge in CO_2 at 75 pounds pressure (circuit 5)*

FIGURE 24.—*Magneto spark discharge in CO_2 at 135 pounds pressure (circuit 5)*

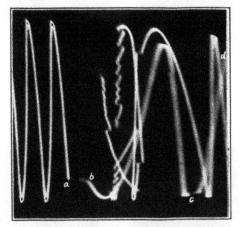

FIGURE 25.—*Magneto spark discharge in CO_2 at 135 pounds pressure (circuit 5)*

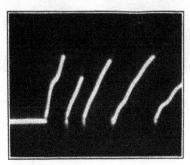

FIGURE 26.—*Magneto spark discharge in CO_2 at 15 pounds pressure (circuit 5)*

B. S. Journal of Research, RP460

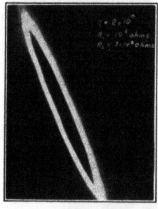 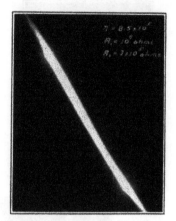

FIGURE 27   FIGURE 28

Change in reduction ratio and phase of applied voltage to the divided voltage using resistance voltage divider.

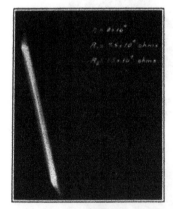 

FIGURE 29   FIGURE 30

Change in reduction ratio and phase of applied voltage to the divided voltage using resistance voltage divider.

B. S. Journal of Research, RP460

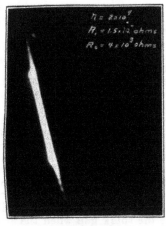

FIGURE 31   FIGURE 32

Change in reduction ratio and phase of applied voltage to the divided voltage using resistance voltage divider.

FIGURE 33

Deflections (a) and (b) show the effect of improper cathode tube alignment. The vertical deflection at (c) indicates a correct alignment.

FIGURE 34.—*Symmetrical wave form resulting from proper cathode tube alignment*

rapid that the cathode beam left no trace. The gap remained broken and the voltage increased to (c) when the series gap broke down, healed, and broke down again until the voltage across the gap was indicated by (d), where it healed again. The series gap again broke down and with this small increase in voltage the spark plug gap broke down. The voltage then dropped so rapidly that again the cathode beam did not leave a trace. At (e) there is an indication below the sweep line of a small voltage. The spark gap in the magneto broke, and a series of discharges took place across the gap until the voltage indicated at (f) was reached. The voltage in the magneto probably increased slightly, but not enough to break down the gap in the magneto. The voltage across the spark plug remained constant until finally the magneto voltage began to fall, and a discharge took place from the spark plug circuit to the magneto. This continued until (h) was reached where the potential difference between the magneto and the spark plug circuit was too small to further break down the gap and the potential at (h) remained on the spark plug. This voltage is also shown to exist across the spark gap when the sweep first appears on the left side of the oscillogram.

Circuit 3 shows no oscillation, so it satisfies requirement 4. It may be used where the large capacitances C_1' and C_2' are not objectionable.

Circuit 4.—The approximate capacitance of the six capacitors are $C_1 = C_1' = 20$ μμf, $C_2 = C_2' = 25$ μμf, $C_3 = 15$ μμf, $C_3' = 75$ μμf.

The capacitances of C_1 and C_1' were increased to 20 μμf to bring out the effect of increasing V_s and V_s'.

Figures 20, 21, and 22 are oscillograms representing spark discharges at pressures of 15, 75, and 135 pounds, respectively, in CO_2 gas. For these photographs the oscillator provided a time axis. The voltage to be measured was applied to the vertical plates. The photographs are lettered in the same manner as in Figure 19. These photographs show that oscillations are set up in the voltage divider and requirement 4 is not satisfied.

In Figure 21 the peak potential of V_s is approximtely 1,770 volts and in Figure 22 it is approximately 2,750 volts. The effect of this increased voltage is shown at (g) and (h). These traces indicate voltages on the deflection plates connected to the oscillator greater than the voltage of the oscillator. They are similar to those which appeared in Figures 9, 10, and 11 on the horizontal plates. These deflections are spurious, and are probably due to a potential field set up between the deflection plates B and their surroundings.

Circuit 5.—The approximate capacitances of the six capacitors used in circuit 5 were: $C_1 = C_1' = 10$ μμf, $C_2 = C_2' = 65$ μμf, $C_3 = 15$ μμf, $C_3' = 0$. Five hundred ohm resistors were inserted at (a) and (a').

Figure 23 is an oscillogram showing the variation in voltage, the discharge taking place in CO_2 gas at 75 pounds pressure. The long faint traces which appear in Figure 14 are not present in this photograph. The faint traces appearing below the heavy traces are probably due to slow electrons, since similar faint traces may be found at maximum voltage of the oscillator. (See fig. 25.)

Figures 24 and 25 are oscillograms representing spark discharges at 135 pounds pressure. Faint traces due to oscillations in the voltage divider are absent in these photographs. Figure 26 is an oscillogram taken with circuit 5 under the same sparking conditions as the oscillogram shown in Figure 15 with circuit 2. The oscillations in the voltage divider which are pronounced with circuit 2 are not present with circuit 5. The faint traces appearing at peak voltage are due to the slow electrons in the cathode beam.

Figure 25 shows a progressive shortening of the oscillator amplitude as the voltage across the vertical deflection plates increases. This will be discussed in Section VII.

VI. EXPERIMENTAL RESULTS WITH RESISTANCE VOLTAGE DIVIDERS

It was shown theoretically in Section III that on account of the effective capacitance of resistors and their supports at high frequencies, the reduction ratio of a resistance divider, in general, changes with the frequency. Oscillograms to illustrate the change in reduction ratio as the frequency is varied are shown in Figures 27 to 32. The oscillator voltage was applied directly to the horizontal plates and the reduced voltage to the vertical plates. In each case the total resistance was $R_1 + R_2$, and the vertical deflection plates were connected across R_2.

With R_1 consisting of two 500,000-ohm grid leaks in series and R_2 of two 150,000-ohm grid leaks in series, the oscillogram shown in Figure 27 was taken at 20 kc and that in Figure 28 at 850 kc. The reduction ratio measured from Figure 27 is 0.35 while from Figure 28 it is 0.47. The low frequency reduction ratio, $\frac{R_2}{R_1 + R_2}$, is 0.23.

The oscillograms shown in Figures 29 and 30 were taken at 20 and 1,000 kc, respectively, with $R_1 = 15,000$ ohms (a 10,000-ohm and a 5,000-ohm grid leak in series) and $R_2 = 4,000$ ohms (two 2,000-ohm grid leaks in series). $\frac{R_2}{R_1 + R_2} = 0.21$. The reduction ratios measured from the oscillograms are 0.20 and 0.30, respectively.

The oscillograms shown in Figures 31 and 32 were taken with wire-wound resistors, with $R_1 = 75,000$ ohms (a single resistor) and $R_2 = 15,000$ ohms (a 10,000-ohm resistor in series with one of 5,000 ohms). $\frac{R_2}{R_1 + R_2} = 0.17$. The measured reduction ratios are 0.17 at 20 kc and 0.29 at 1,000 kc.

It is generally impossible to arrive at these reduction ratios by using equation (16), since neither C_1 nor C_2 are accurately known. Furthermore, equation (16) does not take into account capacitances to ground. It is sufficient for the purposes of this discussion to show on the theoretical side that the ratio $\frac{E_2}{E}$ may either increase or decrease with increase in frequency of the applied voltage and to show on the experimental side that this ratio does vary with the frequency.

VII. OSCILLOGRAPH TECHNIQUE

In connection with the study of voltage dividers, certain details in the technique of cathode-ray oscillograph operation were found to be important. Some of these are mentioned in the literature of the cathode-ray oscillograph, but because of their importance it seems desirable to consider them more in detail at this time.

1. ALIGNMENT OF THE CATHODE TUBE

The usual method of aligning the cathode tube is to place the bright spot in the center of the photographic plate. This means that the active spot on the electrode, the small opening or diaphragm through which the electrons enter the deflection tube and the center of the photographic plate are in a straight line. In the deflection tube the cathode beam must travel between two sets of deflection plates. When a voltage is applied to the plates an electrostatic field exists with respect to the surroundings. If the deflection plates and surroundings are symmetrical with respect to the cathode beam, the electrostatic field will have a kind of symmetry that will produce no deflection of the ray except that which is normal to, and caused by, the difference of potential between each pair of plates. If the field is not symmetrical, the voltage applied to each pair of deflection plates will produce a deflection made up of two components, one normal and one parallel to the plates.

If the two vertical deflection plates B and one horizontal plate A shown in Figure 6 are connected to ground and a voltage applied to the second horizontal plate shown at A, the deflection with the usual adjustment of the cathode tube will generally show a negligible horizontal component. If plates A are used with a voltage divider, such as circuit 4, the difference of potential between the plates may be small in comparison with the difference of the potential between the plates and ground. In this case, the horizontal component may be as great as, or greater than, the vertical component. This effect is shown in Figures 9 and 10, where the tip of the deflection is not vertically above or below the break in the sweep. Figure 11 shows a horizontal deflection although the indicated potential difference between the plates is zero. By swinging the cathode tube horizontally a position was found where the tip of the deflection did not extend more than 1 mm beyond the break in the sweep line. With circuit 4 it was impossible to eliminate this effect entirely when V_B and V_B' were large.

Another example of incorrect alignment of the cathode tube is shown in Figure 18. The lines (x', y') etc., may be made normal to the horizontal deflection plates by swinging the cathode tube horizontally. The traces (x', y') etc., in Figure 15 (a') shows the tube alignment greatly improved. To determine whether or not the cathode tube is correctly aligned for a particular set of conditions the following procedure may be used.

With the four deflection plates at ground potential and the sweep demagnetized the cathode tube is adjusted until the spot appears in the center of the fluorescent screen. The horizontal plates are connected to the voltage divider and the divider balanced as previously described, after which the leads to the divider are connected across the spark gap. If now a spark discharge takes place across the gap the electrostatic field between the deflection plates and the surroundings is that which will exist during the photographing of the discharge. By swinging the cathode tube horizontally, a position will be found where the deflection is vertical. In Figure 33, (a) and (b) show the tube out of alignment while (c) shows the tube aligned horizontally. The oscillograms were taken with circuit 4. A corresponding procedure is followed for the vertical alignment. It is well to check both alignments after the adjustments are completed.

Figure 34 was taken after alignment of the tube by this method. It will be noted that point (*O*) falls midway between (*K*) and (*L*). This method of alignment may not always be the most desirable, because if the horizontal plates are not parallel to the trace produced by the magnetic sweep, or the vertical plates not perpendicular to this trace, the adjustment will displace the cathode tube so that the bright spot is considerably displaced from the center of the photographic plate. For measuring potential differences a small slope is not objectionable unless it is desired to determine a relation between the rate of increase and decrease of the voltage. In most cases it is possible to correct for the small slope. When it is desired to determine the instantaneous relation between current and voltage, using one pair of plates to measure the voltage and the other pair to measure the potential difference across some reactance, the pattern photographed is usually so complicated that it is difficult to make corrections for both pairs of plates. It is this latter problem that led to the method given above for the adjustment of the cathode tube. In Figure 34 the bright spot is above and to the left of the center of the photographic plate.

2. SENSITIVITY OF THE CATHODE BEAM

It is well known that the sensitivity of the cathode beam to either an electrostatic or an electromagnetic field changes with the cathode voltage. The maximum voltage which may be applied to the cathode before a discharge takes place is limited by the pressure and composition of the gases in the cathode tube. Indirectly, then, the sensitivity of the cathode beam depends upon the pressure and composition of gases. In measuring voltages it is therefore necessary to know that the electrostatic sensitivity at the time of photographing the phenomena is the same as at the time of calibration, or else to have available a ready means of determining the sensitivity.

Measurements made with a hot-wire pressure gauge can not be relied upon to determine sensitivity if the photographic film has not been thoroughly out gassed. Unless this out gassing is carried out in a separate container before the film is placed in the oscillograph a great deal of time is lost. The procedure generally followed until this portion of the work was undertaken was for the observer to reduce the pressure until the sweep line appeared dotted and then allow the pressure to increase until the line appeared solid. At this time the photographs were taken. Working in this manner with a film which had not been out gassed the sweep first appeared dotted with a gauge reading of 54 on film 1 and a few minutes later the sweep first appeared dotted with a gauge reading of 42 on film 6. In common with other investigators the practice was adopted of photographing the sweep with a known voltage across the horizontal plates immediately after photographing the phenomena, and using this line as a measure of sensitivity.

This same idea was used to indicate visually the sensitivity before photographing. The deflection plates, oscillator, and a source of known voltage were connected to a double-pole double-throw switch so that either the oscillator or known voltage could be applied to the deflection plates. About 1 centimeter below the top of the fluorescent screen a narrow line was drawn with willemite. When a vacuum was obtained which gave a good solid sweep the known voltage was adjusted until the sweep passed through the willemite line.   This line appeared green against a bluish background.

To control the sensitivity it is only necessary to apply this voltage to the deflection plates and adjust the vacuum until the sweep passes

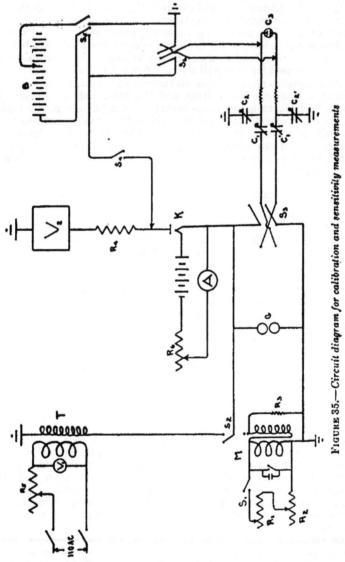

FIGURE 35.—*Circuit diagram for calibration and sensitivity measurements*

through the willemite line.   The voltage was supplied by a small transformer T and rectified by the kenotron K as shown in Figure 35. The double-throw switch S_2 was arranged so that the transformer or magneto could be connected to the kenotron. V_1 is an alternating current voltmeter used to indicate the voltage drop across the primary of the transformer. By changing the resistance R_5 to compensate for changes in the 110-volt supply a constant voltage was supplied to the primary. It was assumed this maintained a constant secondary voltage. The voltmeter V_2 was not sensitive enough for this work, since it was graduated in kilovolts. With S_5 open and S_4 closed, the voltage was applied to the deflection plates, and its polarity reversed, through S_6. Unless the line voltage is constant radio B batteries will give better results.

The need for such control is shown by Figures 36 and 37. These figures show only the left and right halves of oscillograms taken with potential differences of 0, ± 218, ± 437, and ± 692 volts applied to the horizontal deflection plates from the B battery through S_5 and S_6. (Fig. 35.) The hot-wire gauge readings were the same, but on account of changes in the composition of the gases the gauge readings did not indicate the same pressures in the oscillograph. This change in gas composition resulted from different out-gassing of the films. The average deflection computed from Figure 36 is 275 volts per centimeter and from Figure 37 is 215 volts per centimeter. The measurements were made at the center of the film.

Figures 38 and 39 show the left and right halves of two oscillograms taken after the sensitivity indicated by the cathode beam appeared the same. The average deflection computed from the center of Figure 38 is 250 volts per centimeter and from the center of Figure 39 is 255 volts per centimeter. The average rounded value has been taken as 250 volts per centimeter when the deflection is measured at the center of the film using the slowest electromagnetic sweep. The two extreme traces on the oscillogram are photographs of the constant voltage supplied by the transformer and kenotron to the deflection plates. None of the constant potential lines are symmetrical with respect to the center of the photograph because the cathode tube had been adjusted as given under VII, 1.

These lines appear curved since the voltage applied to the cathode tube is taken from the peak of a 60-cycle wave and therefore varies with the time of application. The voltage reaches a maximum at the center of the film which is the position of least sensitivity. The curvature of the lines becomes less as the speed of the cathode beam across the film is increased, because in this case there is a smaller change in the cathode voltage in the interval. Observations show that as the speed at which the cathode beam is swept across the film is increased the same potential difference on the deflection plates produces a slightly greater deflection at the center of the film. This is to be expected if the cathode beam is swept across the film at such times as to include different segments of the calibration lines shown in Figures 38 and 39.

By referring to Figure 6 it will be seen that the deflection of the cathode beam at the photographic film is greater for a given potential difference across plates B than for the same potential difference across plates A. Figure 40 is an oscillogram taken with voltages of 0, ± 218, ± 437, and ± 692 applied across the vertical deflection plates. The vertical deflection was obtained by applying the oscillator voltage to the horizontal deflection plates. The alignment of the cathode tube was the same as for Figures 38 and 39. Measurements made between the sharp edges starting from the far central ruled lines gave an average deflection of 195 volts per centimeter.

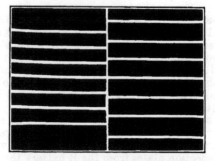

FIGURE 36   FIGURE 37

Two sets of calibration lines with sensitivity matched
with thermocouple pressure gauge.

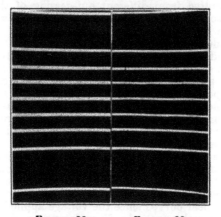

FIGURE 38   FIGURE 39

Two sets of calibration lines with sensitivity matched
with known voltage and willemite mark.

FIGURE 40.—*Calibration lines using
vertical plates*

B. S. Journal of Research, RP460

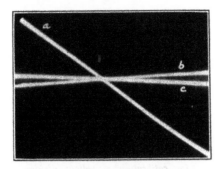

FIGURE 41

Line *a* indicates the ratio of the sensitivity between the
horizontal and vertical deflecting plates.

FIGURE 42.—*Effect of nonuniform
electrostatic fields*

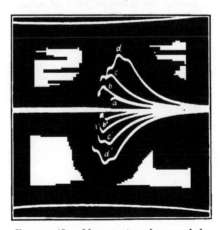

FIGURE 43.—*Magneto impulses used for
the calibration of voltage dividers*

A 1-megohm resistor was shunted across the magneto
terminals.

B. S. Journal of Research, RP460

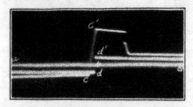

FIGURE 44.—*Magneto impulses used for calibration of voltage divider without resistor shunted across magneto terminals*

FIGURE 46.—*Oscillogram used for determining the reduction ratio of voltage divider using oscillator*

At the time this calibration was made precautions had not been taken to control the sensitivity of the cathode beam.

A second method of determining the sensitivity of the vertical deflection plates is to determine the ratio of the sensitivity of the two pairs of deflection plates and multiply the sensitivity of the horizontal deflection plates by this ratio. If s_1 be the sensitivity of the cathode beam to plates A (volts/cm) and s_2 that to plates B, then the ratio $v = \frac{s_1}{s_2}$ must be constant. This ratio is independent of the voltage applied to the cathode tube as well as the composition and pressure of the gases in the oscillograph. This ratio is conveniently found by applying the oscillator voltage directly to both pairs of plates. The slope of the resulting line, shown at (a) in Figure 41, is the ratio v, which is thus found to have the constant value 1.39 for this deflection tube. Only the straight portion of this line near the center was used because as will be shown presently a small distortion may take place with large deflections of the cathode beam. Thus if s_1 is known, s_2 is found from $s_2 = \frac{s_1}{v}$. For $s_1 = 250$ volts per centimeter, $s_2 = 180$ volts per centimeter.

To explain the slight curvature at the extremities of line (a) in Figure 41 as well as the progressive shortening of the oscillator amplitude in Figure 25 it is necessary to consider the space variation of electrostatic field intensity transversely across the field between the deflection plates. This variation is indicated in Figure 42. The traces (g) and (h) were obtained in the same manner as the two extreme calibration traces shown in Figures 38 and 39. The trace (a) was obtained by applying a constant voltage across the horizontal plates (fig. 6) and the oscillator voltage across the vertical plates. Between the vertical plates the cathode beam oscillated in a horizontal central plane at a frequency of 20 kc. The constant voltage deflected the oscillating cathode beam toward one of the horizontal deflection plates, depending upon the polarity. If the field traversed by the cathode beam is uniform the trace on the photographic film will be parallel to the plates, and if not, it will travel along a path showing the variation in field intensity. The inner edge of (a) was recorded at maximum cathode voltage, which may be considered constant in this very short interval of 2.5×10^{-5} seconds. The lighter portion above this dark line is the record of many oscillations at lower cathode voltage. The trace (c) was obtained by reversing the polarity of the constant voltage. Trace (f) was obtained in a similar manner after grounding the horizontal deflection plates.

The traces (b) and (d) were obtained by applying the constant voltage across the vertical deflection plates and the oscillating voltage across the horizontal plates. The cathode beam is first deflected to the right or left by the vertical plates, depending on the polarity of the applied voltage. The cathode beam, deflected away from the central axis horizontally, then enters the oscillating field of the horizontal plates, and oscillates at a frequency of 20 kc. With the constant voltage deflecting the beam to the right the trace (b) was photographed, and with voltage reversed, (d) was photographed. Trace (e) was taken in a similar manner with the vertical plates grounded. The most intense portion, of the traces, which are the inner edges of (b) and (d), were recorded at maximum cathode voltage.

If it is assumed that the maximum oscillator voltage remains constant during the photographing of (b), (e), and (d), as well as the sensitivity as determined by the pressure and gas composition, lines joining the ends should show the change in sensitivity with transverse distance from the center of the horizontal plates. These lines would be parallel to (a) and (c), respectively. They would be difficult to draw because the slow electrons striking the middle of the photographic film fall in the same vertical plane as the fast electrons, whereas at (b) and (d) they are deflected to the side.

The curvature of lines (a), (b), (c), and (d) can be attributed to the existence of potential fields whose intensities are not uniform throughout.

It has been shown that an oscillogram taken with the oscillator voltage deflecting the cathode beam vertically and the slow magnetic sweep deflecting it horizontally, with the vertical plates grounded, would show a variation of oscillator amplitude depicted by the lines (g) and (h) of Figure 42. The amount of this variation diminishes as the speed of the sweep is increased. If at the same time a voltage is applied to the vertical deflection plates, rapidly increasing to an amount sufficient to cause the horizontal deflection (e) to (b), the path of the beam between the horizontal plates is suddenly transferred from a field of maximum intensity (a') to a field of less than maximum intensity (a). If now the combined action of the electrostatic field of the vertical plates and the electromagnetic field of the sweep has not caused the beam to be deflected entirely off the photographic film, the resulting oscillogram will show a change in the amplitude of the oscillation. This change in amplitude will depend upon the horizontal displacement of the cathode beam by the voltage across the vertical deflection plates as well as the instantaneous cathode voltage. These two effects may either counteract or reinforce each other. Effects of this kind are believed to be the cause of the progressive shortening of the amplitude of the oscillation at (a), (b), (c), and (d) shown in Figure 25. In order that the shortening in Figure 25 be explained quantitatively by Figure 42 it is necessary to assume that the cathode beam in the former is displaced more to the right than the displacement shown in the latter figure.

While it is possible to explain qualitatively, at least, the lengthening of (g) and (h) in Figures 21 and 22 in the same manner, it is believed that these spurious effects were caused by V_B and V_B' becoming large. This seems all the more reasonable since a lengthening appeared in Figure 11 with no potential difference between the deflection plates. It is also now apparent why the extremities of line (a) in Figure 41 are slightly curved.

It is not intended to discuss these effects in detail or make a study of the variation of voltage throughout the whole field, but peculiar effects were obtained and their interpretation was difficult until the oscillogram shown in Figure 42 was taken.

VIII. CALIBRATION OF THE CAPACITANCE VOLTAGE DIVIDER

If the capacitance of each capacitor and lead wire were accurately known, the quantity represented by K, which has been called the reduction factor of the voltage divider, could be calculated. If this method is followed, care must be taken in measuring the capacitances of the six capacitors as well as the leads between the capacitors. The disadvantage of this method is that it requires a series of careful measurements with each change in the settings of the capacitors.

It is more feasible, therefore, to calibrate experimentally, and this may be done in two ways: (1) By applying a known alternating voltage across the divider and photographing the resulting cathode beam deflection, and (2) by determining the ratio between the cathode beam deflection when an alternating voltage is applied directly to one pair of plates and the reduced voltage to the other pair.

1. CALIBRATION BY APPLICATION OF KNOWN VOLTAGES

The leakage which takes place through the insulation of the capacitors prevents the use of a constant voltage for calibrating a capacitance voltage divider. A convenient source of alternating voltage is found in the magneto, since the polarities of consecutive impulses are of opposite sign. The circuit connections for calibrating are shown in Figure 35. The electrodes G were separated to prevent a spark discharge across the gap. The voltage induced in the secondary of the magneto M was controlled by shunting variable resistances R_1 and R_2 across the primary breaker contacts. This voltage was applied to the divider through switches S_2 and S_3, the latter switch controlling the polarity applied to the deflection plates. The kenotron K acted as a check valve and allowed the voltmeter V_2 to indicate the peak voltage. The magneto, which was synchronized with the oscillograph, generated 36 surges per second, 18 positive and 18 negative with respect to ground.

Figures 43 and 44 are oscillograms showing the application of this method with and without the resistor R_3 shunted across the secondary of the magneto. Voltage divider circuit 5 was used. In Figure 44 the peak voltages indicated by the voltmeter were 2,000 and 6,000, and are, respectively, represented by (bd') and (bc'), where (b) lies on the line of zero potential difference. The corresponding voltages of opposite polarity remaining from the previous surge are (db) and (cb).

The advantage of using the 1-megohm grid leak R_3 is shown by Figure 43. In this case the potential difference across the gap is quickly brought to zero. The surges were all of the same polarity with respect to the magneto, but their polarity with respect to the deflection plates was reversed on each voltage by switch S_3. (Fig. 35.) The peak voltages represented by (a) and (a'), (b) and (b'), (c) and (c'), and (d) and (d') are, respectively, $\pm 2,350$, $\pm 4,075$, $+5,900$ and $-6,075$, and $+9,600$ and $-9,750$. These deflections determine the lines (a) and (b) shown in Figure 45. Since no two surges are exactly alike and the voltmeter indicates the average of 18 surges per second, a variation between observations is to be expected. Before photographing the surges shown in Figure 43, the sensitivity was checked in the manner described in Section VII.

With a smaller reduction ratio the straight line (d) in Figure 45 was determined. The plotted points represent the average of six independent observations. They were taken to show that the reduction factor of circuit 5 is constant, and that the deviations of the plotted points in Figure 12 become less as a greater number of observations are averaged. These deviations are caused by the magneto since the peak voltages of two sparks are not necessarily the same.

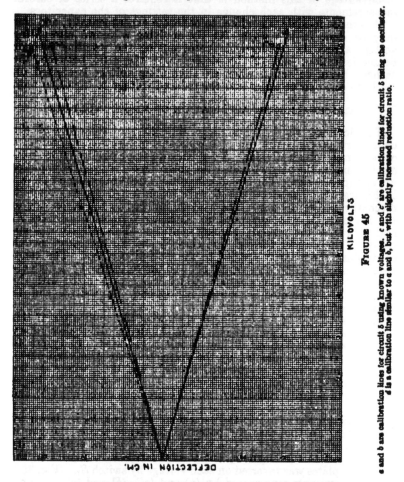

FIGURE 45

a and *b* are calibration lines for circuit 5 using known voltages. *c* and *c'* are calibration lines for circuit 5 using the oscillator. *d* is a calibration line similar to *c* and *b*, but with slightly increased reduction ratio.

2. CALIBRATION BY USE OF THE OSCILLATOR

A method for using the oscillator supplied with the oscillograph to determine the relative sensitivity of the two pairs of deflection plates has already been discussed. This method, which consists simply in applying the undivided oscillator voltage to one pair of plates and the divided voltage to the other pair, is also useful for calibration. The slope of the resulting line, of which Figures 13 and 46 are examples, determines the ratio

$$\frac{V_B - V_B'}{V_A - V_A'}$$

after the ratio of sensitivity between the two pairs of plates has been determined.

In Figure 46 the straight line (*a*) was obtained by applying the oscillator voltage directly to the vertical deflection plates and the divided voltage (using circuit 5) to the horizontal plates. The same capacitor settings were used as for Figure 43, the reversal of slope being obtained by reversing the leads to the voltage divider. The exposures shown at (*b*) and (*c*) in Figure 41 were made after making a small change in the capacitor settings. With the divided voltage applied to the horizontal plates, as in this instance, the reduction factor K is equal to $\frac{v\delta_2}{\delta_1}$, where δ_1 is the mean of two deflections measured along two vertical ruled lines, δ_2 is the total horizontal deflection measured between these lines, and $v = \frac{s_1}{s_2} =$ the ratio of sensitivity for plates A (horizontal) in volts per centimeter to that for plates B (vertical). The constant v is 1.39. Measurements made on Figure 46 gives $\delta_1 = 0.55$ cm and $\delta_2 = 10.75$ cm. Hence, $\frac{1}{K} = 0.071$.

On account of the small slope produced, when K becomes greater than 15 or 20 the method just described may not be found practicable. In such cases it will be necessary to calibrate by the method of applying known voltages. But the convenience of using the auxiliary equipment supplied with the oscillograph makes the second method preferable when the reduction ratio is of such magnitude as to allow its use. It is then unnecessary to obtain a source of high alternating voltage or high-voltage measuring instruments. (The term "high voltage" is relative. The maximum voltage used in this investigation was 10,300.)

It is interesting to note that method (2) gives

$$\frac{1}{K} = \frac{V_B - V_B'}{V_A - V_A'}$$

absolutely, without regard to the sensitivity of the cathode beam.

To compare the results obtained by the two methods of calibration it is only necessary to substitute this value of K in the expression,

$$D = \frac{V}{Ks}$$

where D is the deflection in centimeters, V is an arbitrary applied voltage and s is sensitivity in volts per centimeter. Let $V = 10,000$. From the previous sensitivity measurements (Sec. VII) it is known that

$$s_1 = 250 \text{ volts per centimeter}$$

and

$$s_2 = 180 \text{ volts per centimeter}$$

and for the same capacitor settings $\frac{1}{K}$ was found above to be 0.0712. Therefore

$$D_1 = 0.071 \times \frac{10,000}{250} = 2.85 \text{ cm for plates A}$$

and

$$D_2 = 0.071 \times \frac{10,000}{180} = 3.95 \text{ cm for plates B}$$

The deflection D_1 is indicated on Figure 45 by the lines (c) and (c'), and agrees closely with the calibration by the method of known potentials. If the leads had been reversed and another straight line obtained the average of the two would give a better value for v than was obtained from the single line (a) in Figure 41. This should lead to a better agreement between the two methods.

IX. CONCLUSION

The results of this investigation show that if it is desired to measure voltages in spark discharges whose peak voltages lie within the range found in the average automotive engine ignition system, voltage divider circuit 5 will probably be found satisfactory for use with a cathode ray oscillograph of the type described. However, it is well to call attention to the fact that some of the other circuits may be equally satisfactory for other types of measurements. The faint traces which were observed in oscillograms due to oscillations in the voltage divider circuit can probably be eliminated by inserting resistances as was done in circuit 5. One of the chief merits of the capacitance voltage divider is its adaptability to a wide range of conditions of voltage measurement.

Resistance voltage dividers are unsatisfactory for frequencies commonly found in spark discharges unless special care is taken to balance the capacitance and inductance of the resistance. In the development of the equation for resistance voltage dividers only the simplest assumptions were made as to capacitance, and the inductance of the resistors was completely neglected. This equation, therefore, represents only in a very rough way what actually occurs. The oscillograms obtained in testing the resistance divider are evidence in themselves of the unsuitability of this type of divider for high-frequency work.

In the literature of cathode ray oscillography there has been a disappointing lack of detail regarding the theory and technique of measurements. It is hoped that the details set forth in this paper will be of value to those who are using the instrument in work of a similar nature.

Another point that can hardly be overemphasized is the fact that in this work no phenomena have been recorded by the oscillograph that have not been susceptible of rational explanation. It has been the repeated experience of the authors that recorded phenomena which are most baffling when first observed on an oscillogram can usually be interpreted after careful study and experimental investigation.

X. ACKNOWLEDGMENTS

This work was supported by the Bureau of Aeronautics of the Navy Department and the National Advisory Committee for Aeronautics.

The authors wish to thank Doctors Mueller, Silsbee, and Brickwedde for their many helpful suggestions.

WASHINGTON, April 8, 1932.

O

U. S. DEPARTMENT OF COMMERCE
R. P. LAMONT, Secretary

BUREAU OF STANDARDS
LYMAN J. BRIGGS, Acting Director

BUREAU OF STANDARDS JOURNAL OF RESEARCH

August, 1932

Vol. 9, No. 2

UNITED STATES
GOVERNMENT PRINTING OFFICE
WASHINGTON : 1932

For sale by the Superintendent of Documents, Washington, D. C. · · · · · · · · $3 per year on subscription

A VACUUM TUBE AMPLIFIER FOR FEEBLE PULSES

By L. F. Curtiss

ABSTRACT

A resistance-capacity coupled 5-stage amplifier is described which is suitable for automatic registration of the primary ionization pulses produced when corpuscular rays pass through a shallow ionization chamber. A discussion of an improved type of ionization chamber is given.

CONTENTS

I. INTRODUCTION

It has long been recognized that a direct method of recording current pulses produced by ionizing particles, such as α or H rays, would be of great value in the study of this type of radiation. By such a procedure, in which the current pulse occurring in an ionization chamber is amplified linearly and recorded, a decided advance over the now familiar Geiger counter is achieved in that the particle is not only counted, but also the ionization which it produces as it passes through the ionization chamber is instantaneously measured and recorded. In view of the fact that an α ray passing through a chamber 5 mm in depth would produce on the average only about 10^4 ions, the rise in potential of the insulated collector of the ionization chamber can not be much greater than 10^{-4} volt for any practicable value of the capacity of the collector. The problem is therefore to amplify this potential pulse up to the order of 20 volts before applying it to the grid of the output tube of the amplifier. This is very difficult or impossible to do with ordinary radio tubes, since poor grid insulation, large grid currents, large grid capacity, and other undesirable characteristics serve to obliterate entirely the effect of the arrival of such a small charge to the grid. It is not surprising, therefore, that this form of registration has not rapidly replaced the Geiger counter in which the primary ionization is amplified by impact ionization 10^7 times in the counter itself. This augmented ionization current is then easily handled by a very simple amplifier. Thus, once given a satisfactory counter, it is a comparatively simple matter to record the entrance of particles into the counter. However, in the counter the primary ionization serves only as a trigger to set off a much larger impact ionization which can never be more than very roughly proportional to the primary ionization.

Greinacher [1] first demonstrated that it was possible to amplify directly the primary ionization of a single particle. For this purpose he used an ionization chamber with electric field only sufficiently intense to insure rapid and complete collection of the primary ions. Later developments of this arrangement led to vacuum-tube circuits of many stages of amplification which were not only difficult to operate but risked a serious sacrifice of the linear characteristics required to make the amplifier of value.

More recently the development of amplifiers which meet the requirements of linearity and reasonable convenience of operation have been undertaken by Wynn-Williams and Ward [2] and by Leprince-Ringuet.[3] Although accomplishing the same result the two circuits differ considerably in details. Unfortunately neither circuit is readily available to investigators working in this country, since the tubes used are of foreign manufacture, therefore it is difficult to procure or to replace them, or to secure information to enable one to select from tubes available here those which would have similar characteristics. For these reasons the writer, in undertaking to construct an amplifier, was faced with the necessity of working out an arrangement based on the principles outlined by Wynn-Williams and Ward, but making use of entirely different types of tubes with the exception of the output tube. In doing this it was found possible to use commercial radio tubes in all stages except the first. The tube used here is so simple in construction that it can be made in almost any laboratory, and full details of its construction are given.

Since a complete discussion of the principles involved in the design of such circuits is given by Wynn-Williams and Ward and the theory of the operation is very adequately discussed by Ortner and Stetter,[4] these matters are not discussed here.

II. DESCRIPTION OF THE AMPLIFIER

As pointed out by Leprince-Ringuet, the essential difficulty met with in amplifying these primary ion pulses is that of conserving the small quantity of charge and converting the sudden rise in potential produced when this charge is driven on to the collecting system of the ionization chamber into a current pulse which can be readily amplified. No ordinary radio tube is suitable for this purpose, since the grid in such tube is not insulated sufficiently to deal successfully with such small quantities of charge. Since neither the tube used by Wynn-Williams and Ward nor by Leprince-Ringuet was available, the writer constructed a simple tube from the description given by Leprince-Ringuet. Figure 1 shows a sketch of this tube. It is a 3-electrode tube which departs from the usual arrangement in that the filament is located between the grid G and the plate P, the grid in this case being a plane electrode like the plate. The filament (an oxide-coated nickel ribbon operating on 0.4 ampere and 0.6 volt) is stretched parallel to the plane of these electrodes. The filament and plate leads are taken out

[1] H. Greinacher, ZS. f. Phys., vol. 36, p. 364, 1926.
[2] C. E. Wynn-Williams and F. A. B. Ward, Proc. Roy. Soc., vol. 131, p. 391, 1931.
[3] L. Leprince-Ringuet, Annales des P. T. T., vol. 20, p. 480, 1931.
[4] G. Ortner and G. Stetter, ZS. f. Phys., vol. 54, p. 449, 1929.

through a press seal at the lower end of the tube, the grid lead being brought out separately at the top. This arrangement provides for better insulation of the grid which is further increased by making the tube of pyrex glass. Other features of this tube which make it particularly suited to this work are a a low grid-ground capacity and a filament emission limited by space charge. The high insulation and low capacity of the grid, a plate about 1 cm square, make possible the rapid and sufficient rise of potential of the grid required to produce a current pulse in the plate circuit. The space-charge characteristics of this tube are of great value in cutting down the shot effect. With several succeeding stages of amplification, the shot effect becomes one of the limiting factors in the operation of the amplifier. This effect is so large in ordinary tubes that they can not be used in the first stage since the current pulses resulting from it mask entirely the pulses which come from the ionization chamber.

It is quite possible that other types of electrometer tubes, for example the FP54, would serve here, but the tube described is much simpler in construction and operation and likewise less expensive.

Having determined the type of tube for the input stage, the remainder of the amplifier becomes more or less routine radio practice as related to audio-frequency resistance-capacity coupled amplifiers. The complete circuit is shown in the wiring diagram reproduced in Figure 2. It has five stages, four of which use commercial radio tubes. In order to obtain as much amplification per stage as possible, thus reducing the required number of stages, screen grid alternating current amplifier tubes (type 224) were selected for the three amplifying

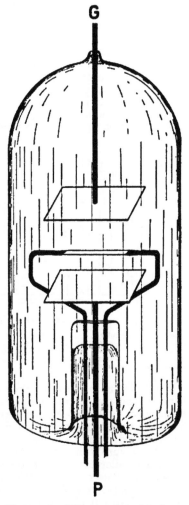

FIGURE 1.—*Diagram of special electro meter tube*

G=grid lead; *P*=plate lead

stages. Following Wynn-Williams and Ward, a pentode was used in the output stage and the time constants of all these stages were made large. The amplifying constant of the 224 tubes connected as shown is 1,000. Naturally the amplification per stage does not attain such a high value, but with the resistances here used it is reasonable to expect an over-all amplification of about 100 per stage. This high degree of amplification is necessary since, as Leprince-Ringuet points out, the input tube does not amplify the potential applied to it. In fact, the potential surge produced at the plate end of the plate circuit resistor by the plate current of this tube is probably of the order of one one-bundredth of the rise of potential of its grid. However, this is a potential surge in a battery-fed current which can be amplified readily without unusual precautions regarding insulation.

The constants for all parts of the circuit are given below the diagram in Figure 2. Those condensers not designated by a letter are 2 microfarad by-pass condensers to shield the amplifier from extraneous pulses picked up by battery leads. Thorough shielding of the whole amplifier by a divided metal box is of course essential. The amplifier is very sensitive to high-frequency oscillations which ruin its effectiveness by setting up relaxation oscillations in the coupling condensers.

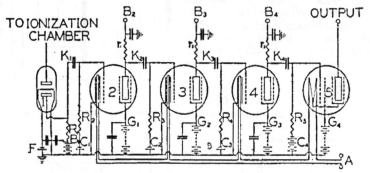

FIGURE 2.—*Wiring diagram of amplifier*

1 = Electrometer tube; 2, 3, 4 = 224 tubes; 5 = 247 tube; r_1, r_2, r_3, = 250,000 ohms; R_2, R_4, R_5 = 1 megohm. R_3 = 2 megohms; R_1 = 250,000 ohms; B_1 = 4.5 volts; B_2, B_3, B_4 = 250 volts; G_1, G_3, G_3 = 25 volts; G_4 = 250 volts C_1, C_2, C_3 = 1.5 volts; C_4 = 25 volts; K_1 = 0.0005 microfarad; K_2, K_3 = 0.1 microfarad; K_4 = 2 microfarad. F (fila ment battery for electrometer tube) = 0.6 volt. A = 10 volts (1.75 amperes). Condensers not designated are 2 microfarad by-pass condensers.

Particular note is to be taken of the coupling of the electrometer tube to the rest of the amplifier. This is the "distorting" stage mentioned by Wynn-Williams and Ward. It is very important to have the time constant of this circuit small. The value used here is 0.001 second. Since a higher value of grid leak was required than that used by Wynn-Williams and Ward, this small constant is secured by using a smaller coupling condenser (0.0005 microfarad). For some reason that does not seem self-evident, Leprince Ringuet was able to use a comparatively slow circuit at this point (0.04 second). However, he does not give a diagram of his complete circuit so that he may have introduced the "disotrtion" at a later stage. It may be well to emphasize that it is not safe to depend merely on the values of the resistances and capacities given and to construct an amplifier without testing its performance thoroughly by means of an oscillograph. This is particularly important, of course, when other forms of recording are to be used. The writer's experience indicates that only by taking careful oscillograph records can it be made certain that the amplifier is functioning properly.

FIGURE 4.—*Oscillograph records*

a, α particles in presence of β and H particles; *b*, zero line in absence of radiation.

III. THE IONIZATION CHAMBER

Wynn-Williams and Ward describe the precautions to be taken in designing ionization chambers for use with an amplifier of this kind. The guard-ring type of chamber is particularly suited for this use, but in the form which they describe the "parasitic" capacity has not been reduced to a minimum. Their design is shown in Figure 3(a). The capacity of this system consists of (1) that between the window, W, and the electrode, E; (2) that of the cylindrical condenser formed between the guard ring, G, and the electrode, E; and (3) the capacity of the lead, L, mainly located at the point where it passes through the insulator, S. The capacity (1), is that of the ionization chamber itself and can not be reduced without reducing the effective volume of the chamber. The other two, however, are not associated with the volume of the chamber and any means of reducing them will increase the sensitivity of the arrangement. Of these, (2), that represented by the cylindrical condenser between the electrode, E,

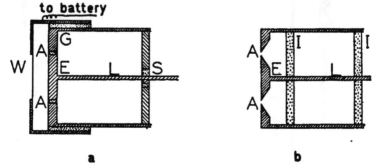

FIGURE 3.—*Section of ionization chambers*

and the guard ring, G, is by far the largest. In fact, it may be several times larger than the rest of the capacity of the system, particularly when, as in the chamber described by Wynn-Williams and Ward, insulating material is inserted in the annular gap to support the electrode, E. This difficulty may be greatly reduced by modifying the internal parts of the ionization chamber as shown in Figure 3 (b). The electrode, E, and the guard ring here have tapered edges which reduce the capacity at this point to a minimum. The insulator is removed entirely. The lead, L, is supported by two large insulators, II, at points where the capacity is a minimum. This requires the lead, L, to be stiffer and, therefore, of larger diameter, but the actual increase in capacity between, say, 0.5 millimeters and a 1.5-millimeter rod at this point is small compared with the reduction of capacity effected by the modification.

As examples of the kind of records that may be obtained with this amplifier and ionization chamber, reproductions of oscillograph records are shown in Figure 4 (a) and (b). Figure 4 (a) is a record of α particles obtained in the presence of β and H radiation and indicates the ease with which different types of radiation may be identified. Figure 4 (b) a blank record on the same time scale, indicates the nature of the zero line in the absence of radiation. The unsteadiness in this line is made up of a combination of effects, such as shot effect in the first tube, sound vibrations in the ionization chamber, and mechanical shock, all of which may set up small pulses in the amplifier. The ionization chamber with a thin metal window is a sensitive microphone which must be shielded as far as possible from all sound. The zero line is much less steady than that shown if reasonable precautions are not taken to shield the complete amplifier from noise and mechanical shock.

The tallest peaks in Figure 4 (a) are due to α particles and vary slightly in height as a result of differences in speeds and in lengths of paths of the particles as they traverse the ionization chamber in different directions. The shortest sharp peaks are produced by H particles and are roughly of one-fourth the height of the peaks due to α particles. Since a high-speed α particle produces about the same ionization as a low-speed H particle, there are some peaks of intermediate height which could not be identified merely by a study of the record. Since a β particle has only about one two-hundredth of the ionizing power of an α particle, the ionization of the β particles merge with the extraneous disturbances and can not be identified with certainty from the record. This makes it clear why α and H particles can be recorded by this method in presence of γ radiation. This is not possible with the Geiger counter since the secondary β rays produce pulses which can not readily be differentiated from those due to α particles.

IV. ACKNOWLEDGMENTS

The writer wishes to acknowledge aid from consultations with Dr. A. V. Astin and Dr. G. W. Gardiner regarding problems which arose in the course of constructing the amplifier, and from B. W. Brown for help in building and testing it. Mr. R. G. Kenedy, jr., prepared the oxide-coated filament for the electrometer tube. The writer is particularly grateful to Dr. F. West and to Dr. C. F. Burnam, of the Howard A. Kelly Hospital of Baltimore, for the material from which the polonium source used in testing the amplifier was extracted.

WASHINGTON, May 16, 1932.

RP462

INFRA-RED SPECTRA OF HELIUM

By William F. Meggers and G. H. Dieke [1]

ABSTRACT

Employing Xenocyanine photographic plates to explore the infra-red spectra of helium emitted by a Geissler tube used "end-on" about 120 new lines were recorded in the spectral interval 8,361 to 11,045 A. Analysis of these data shows that 32 of the lines belong to the arc spectrum of neutral He atoms (He I) and the remainder are ascribed to the band spectrum of the He_2 molecule. The He I lines are accounted for as combinations of previously known terms and as extensions to the fundamental series, while the He_2 lines for the most part represent hitherto lacking intercombinations between groups of terms which account for the visible bands of helium.

CONTENTS

I. INTRODUCTION

The spectra of helium have been investigated [2] so extensively that they would appear to have been exhausted. However, the recent discovery of a remarkable new infra-red sensitizer, Xenocyanine, by the Eastman Kodak Research Laboratories [3] has greatly extended the photographic range of spectra, and use of these plates at the Bureau of Standards has revealed many new lines in the emission spectra of chemical elements. Up to the present, the longest published wave length observed photographically from helium was 7,281 A, but with Xenocyanine plates 120 lines have been recorded in the infra-red between 8,361 and 11,045 A. In this range, only the reasonance line at 10,830 A was known, since its detection radiometrically by Paschen [4] in 1908. Analysis shows that 32 of the above lines belong to the He I spectrum, while the remainder are ascribed to the band spectrum of the He_2 molecule.

II. EXPERIMENTAL

The source of the infra-red He spectra described in this paper was a Geissler tube made by Robert Goetze in Leipzig. It had a narrow capillary and was designed for end-on projection of the light into the

[1] Associate Professor of Physics, The Johns Hopkins University.
[2] H. Kayser, Handbuch der Spectroscopie, vol. 5, p. 508, 1912. H. Kayser, and H. Konen, Handbuch der Spectroscopie, vol. 7, p. 606, 1930.
[3] C. E. K. Mees, J. Opt. Soc. Am., vol. 22, p. 204, 1932.
[4] F. Paschen, Ann. der Phys. (4), vol. 27, p. 537, 1908.

spectrograph. The tube was operated with uncondensed discharges from a 40,000-volt transformer, the primary of which was fed by 5 or 6 amperes, 60 cycle, alternating current, 110 volts.

A concave grating spectrograph in Wadsworth mounting[5] was employed in recording the spectra on Xenocyanine plates, the He exposure being made in the first order spectrum in which the dispersion is 10.26 Å per millimeter. Comparison spectra of the iron arc were photographed in the second order of the grating, above and below the He exposure, and wave-length measurements were based on the internationally adopted secondary standards.[6] The duration of exposure averaged 16 hours for He and 2 minutes for Fe.

The appearance of such a large number of new lines in the He exposures first suggested the presence of impurities or spurious lines, but careful examination showed that the O I triplet at 9,260.9, 9,262.8, and 9,266.1 Å was the only impurity, and no evidence of Lyman ghosts was found. It was immediately suspected that most of the lines in this region must be due to the He_2 molecule, but the production of this band spectrum under the conditions of excitation employed was somewhat surprising. It is well known that the ordinary Geissler tube of He excited by uncondensed discharges shows only the so-called arc spectrum, He I. The band spectrum of the He_2 molecule is excited with great intensity by using a higher pressure of He in a discharge tube with a large bore in place of the usual capillary, and passing through it discharges from an induction coil or transformer with a condenser in parallel and a small spark gap in series with the tube,[7] but it appears also with uncondensed discharges in tubes with a very fine constriction in the bore.[8] The latter conditions are approximated by the tube described above, and probably account for the production of the observed mixture of He I and He_2 spectra.

III. RESULTS

1. THE He I SPECTRUM

The Bohr model of a neutral He atom consists of a nucleus and two electrons so that its characteristic spectrum must resemble that of alkaline earths with two valence electrons. Its energy states must comprise a system of singlet terms and a system of triplet terms, among which singlet S (1^1S_0) represents the normal state of the atom.

Analysis of the structure of the He I spectrum has been summarized in treatises on spectral series[9] in which singlet and doublet systems are described. In the last treatise, it is shown that the doublet system is in reality a triplet system in which the level separations are so small that they are difficult to detect. Partial resolution of the ^3P terms gave the impression of doublets in the spectrum. Since the new infra-red (He I) lines are all accounted for as combinations of known terms or extensions of known series we are presenting our results in Table 1 in Grotrian's modern notation. Table 1 may,

[1] W. F. Meggers and K. Burns, B. S. Sci. Paper (No. 441), vol. 18, p. 191, 1922.
[2] Trans. Internat. Astron. Union, vol. 3, p. 86, 1928.
[3] W. E. Curtis, Proc. Roy. Soc. A 89, p. 146, 1913.
[4] T. R. Merton and J. G. Pilley, Proc. Roy. Soc. A 109, p. 267, 1925.
[5] A. Fowler, Series in Line Spectra, p. 91, Fleetway Press, London, 1922. F. Paschen and R. Goetze, Seriengesetze der Linienspektren, p. 26, Julius Springer, Berlin, 1922. W. Grotrian, Handbuch der Astrophysik, vol. 3, p. 555, Julius Springer, Berlin, 1930.

however, be regarded as an extension to the He tables in any of the above-mentioned reports.

TABLE 1.—*He I*

Fundamental series $3^1D = 12,305.09$. $3^1D - m^1F$

λ_{air}	Intensity	ν_{vac}	m	m^1F
18,693.4	2	5,348.0	4	6,857.1
12,792.2	1	7,815.1	5	4,390.0
10,917.0	3	9,157.5	6	3,047.6
10,031.16	15	9,966.20	7	2,038.9
9,529.27	4	10,491.10	8	1,714.0
9,213.1	1	10,851.1	9	1,354.0

Fundamental series $3^1D = 12,209.10$. $3^1D - m^1F$

λ_{air}	Intensity	ν_{vac}	m	m^1F
18,684.2	3	5,350.71	4	6,858.39
12,784.6	1	7,819.89	5	4,389.21
10,912.95	6	9,160.92	6	3,049.82
10,027.73	40	9,969.61	7	2,239.49
9,526.17	10	10,494.52	8	1,714.58
9,210.28	6	10,854.46	9	1,354.64
8,996.7	2	11,112.1	10	1,097.0
8,845.0	1	11,302.7	11	906.4

Combinations

λ_{air}	Intensity	ν_{vac}	Combination
10,233.0	2	9,769.6	$3^1P - 7^1S$
9,682.4	1	10,325.2	$3^1P - 8^1S$
11,045.1	1	9,051.3	$3^1P - 6^1D$
10,138.50	10	9,860.69	$3^1P - 7^1D$
9,625.80	3	10,385.90	$3^1P - 8^1D$
9,303.2	2	10,746.0	$3^1P - 9^1D$
9,085.6	1	11,003.4	$3^1P - 10^1D$
11,013.1	1	9,077.6	$3^1S - 5^1P$
9,603.50	6	10,410.01	$3^1S - 6^1P$
8,914.7	2	11,214.4	$3^1S - 7^1P$
10,667.62	4	9,371.60	$3^3P - 6^3S$
9,702.66	10	10,303.63	$3^3P - 7^3S$
9,346.2	2	10,896.6	$3^3P - 8^3S$
10,311.32	40	9,695.43	$3^3P - 6^3D$
9,516.70	30	10,504.97	$3^3P - 7^3D$
9,063.40	6	11,030.36	$3^3P - 8^3D$
10,072.10	2	9,925.70	$3^3D - 7^3P$
9,552.8	1	10,465.3	$3^3D - 8^3P$
9,463.66	60	10,563.84	$3^3S - 5^3P$
8,361.7	3	11,956.0	$3^3S - 6^3P$

The so-called fundamental (hydrogenlike) series of He have been extended from 2 to 6 members in the singlet system and from 2 to 8 in the triplet system, the first two lines of each series representing the radiometric observations of Paschen, so that the intensities of these are not to be compared with the photographic estimates of the remainder. The third line of each series lies close to 11,000 A, and to compensate for declining photographic sensitivity their estimated intensities should be multiplied by 10 or more to make them comparable with the succeeding lines.

2. THE He₂ BANDS

The known energy levels of the helium molecule[10] may be regarded as originating from a single valence electron and may therefore, be classified as $s\Sigma$, $p\Sigma$, $p\Pi$, $d\Delta$, etc., levels in which the s, p, d mean, as in atomic spectra, that the orbital moment of the valence electron is 0, 1, 2, whereas the symbols Σ, Π, Δ designate the component Λ of this orbital momentum along the internuclear axis. The low levels which are the final states in the transitions which give rise to the visible helium bands are $2s\Sigma$, $2p\Pi$, and $2p\Sigma$. The negative s and d levels combine with the $2p\Sigma$ and $2p\Pi$ levels, whereas the positive p levels combine with the $2s\Sigma$ level. This fact divides the terms of the helium molecule into two groups, the $2s\Sigma$ level and those terms which combine with it on the one hand and the $2p\Sigma$ and $2p\Pi$ levels and those terms which combine with them on the other hand. The relative term values in each group can be determined with the accuracy of the wave-length measurements, but the relative position of the two groups can only be determined by calculating the limits of two Rydberg series. This is due to the fact that all the intercombinations between the two groups which are allowed by the selection rules must fall in the infra-red and were not observed so far.

The new lines in the infra-red given in the present paper are partly due to just such transitions.[11] The most prominent band in this region has its head at 9,123 A and is due to a transition $2p\Sigma \rightarrow 2s\Sigma$. (See Table 2.)

TABLE 2.—*He₂*

Band at 9,123 A　　$2p\Sigma \rightarrow 2s\Sigma$

K	P branch			R branch		
	λ	I	ν	λ	I	ν
1	9,193.85	4	10,874.21	9,158.92	7	10,915.32
3	9,222.96	10	839.53	42.23	12	935.25
5	57.50	20	799.09	30.66	10	949.11
7	97.14	20	753.05	24.29	8	956.75
9	9,341.91	15	701.51	23.17	6	958.10
11	91.90	10	644.55	27.34	5	953.09
13	9,447.12	8	582.34	37.00	5	941.51
15	9,507.74	8	514.86	52.23	5	923.30
17	73.90	7	442.20	73.18	5	898.36
19	9,645.68	7	364.49	9,200.00	5	866.59
21	9,723.33	6	281.72	33.03	3	827.71
23	9,807.13	5	193.87	72.60	2	781.50
25	97.47	3	100.82			
27	9,994.6	1	002.7			

Band at 9,346 A　　$2p\Sigma^1 \rightarrow 2s\Sigma^1$

K	P branch			R branch		
	λ	I	ν	λ	I	ν
1				9,376.6	1	10,661.9
3	9,441.6	2	10,588.5	60.67	2	680.07
5	77.72	3	548.17	50.50	2	691.69
7	9,519.39	3	501.99	46.2	2	696.6
9	67.08	3	449.64	48.0	2	694.5
11	9,620.88	2	391.21	55.70	2	685.74
13	80.54	2	327.17	69.8	1	669.7
15	9,746.64	2	257.13	90.3	1	646.4
17	9,819.18	2	181.36			
19	98.67	2	099.60			
21	9,985.43	1	011.85			

[10] All the bands discussed in this paper belong to the triplet system. The triplet separation, however, is so small that the lines appear single, except in some cases under very high dispersion. The singlet system is weaker and much less completely known. No singlet bands were found in the region under investigation.

[11] On photographs sent to one of us two years ago by Prof. T. Takamine which were taken on Neocyanine plates under low dispersion, the presence of bands in the infra-red could be seen. The dispersion was not large enough, however, to permit an analysis.

Both the initial and the final state of this band are already known.
The final state is that of the so-called main series [12] ($2s-np$ in the
old notation) which contains the most prominent bands of the
spectrum. The initial state $2p\Sigma$ is the final state of a group of
bands, the first of which was discovered by Merton and Pilley [13]
and which are described in full elsewhere.[14]

The combination relations given in Table 3 prove the identity of
the band. The second column (marked new) gives the rotational
differences of the final state

$$R(K-1)-P(K+1)=2s\Sigma(K+1)-2s\Sigma(K-1)$$

The next column gives the average of the same differences obtained
from the bands of the main series.[15] The agreement is very good
which proves that $2s\Sigma$ must be the final state of the band.

In the same way we prove from the agreement of the differences

$$R(K)-P(K)=2p\Sigma(K+1)-2p\Sigma(K-1)$$

given in column 7 for the new band and in column 8 for the previously
known bands which have $2p\Sigma$ as final state that $2p\Sigma$ is the initial
state of the new band. The band has only a P and R branch as
must be expected for a $\Sigma \rightarrow \Sigma$ transition.

There is a fainter band of exactly the same structure with its head
at 9,346 A. (See Table 2.) It is the $2p\Sigma^1 \rightarrow 2s\Sigma^1$ band which has
the same electronic transition as the band at 9,123 A mentioned
above, but a $1 \rightarrow 1$ vibrational transition (instead of $0 \rightarrow 0$ for λ 9,123).
The columns 4 and 5 of Table 3 prove in the same way as explained
above for the main band that $2s\Sigma^1$ is actually the final state. The
initial state $2p\Sigma^1$ was not known before.

TABLE 3.—*Combination relations*

K	$2s\Sigma(K+1)-2s\Sigma(K-1)$				K	$2p\Sigma(K+1)-2p\Sigma(K-1)$		
	$v=0$		$v=1$			$v=0$		$v=1$ (new)
	New		New			New		
2	75.79	75.80	73.4	73.39	1	41.11	41.07	--------
4	136.16	136.20	131.90	131.87	3	95.72	95.69	91.57
6	196.06	196.06	189.70	189.84	5	150.02	150.00	143.52
8	255.24	255.22	247.0	247.08	7	203.70	203.69	194.61
10	313.55	313.49	303.3	303.47	9	256.59	256.61	248.86
12	370.75	370.76	358.57	358.63	11	308.54	308.56	294.53
14	426.65	426.63	412.6	412.60	13	359.17	359.14	342.53
16	481.10	481.07	465.0	465.03	15	408.44	408.22	389.27
18	533.87	533.85	--------	--------	17	456.16	456.45	--------
20	584.87	584.91	--------	--------	19	502.10		--------
22	633.84	633	--------	--------	21	545.99		--------
24	680.68	680	--------	--------	23	587.63		--------

Besides the two bands given above, we must expect in this region
the bands $3s\Sigma \rightarrow 2p\Sigma$ and the whole $3d \rightarrow 2p\Sigma$ complex. Only a few
lines of the $3s\Sigma \rightarrow 2p\Sigma$ band could be identified near 10,400 A. The
sensitiveness of the photographic plate in this region is already too

[12] W. E. Curtis and R. G. Long, Proc. Roy. Soc., A 108, p. 513, 1925.
[13] T. R. Merton and F. G. Pilley, Proc. Roy. Soc., A 109, p. 267, 1925.
[14] G. H. Dieke, S. Imanishi, and T. Takamine, ZS. f. Phys., vol. 57, p. 305 1929
[15] Mainly from unpublished data in the possession of one of the authors

weak to allow also the weaker lines to be recorded. The $3d \rightarrow 2p\Sigma$ complex is completely present, it is given in Table 4. The calculated frequencies are obtained with the help of the known differences between the $2p\Sigma$ and the $2p\Pi$ states. This band complex shows the same characteristics as the $4d \rightarrow 2p\Sigma$ complex.[16] The P branch of $3d\Sigma \rightarrow 2p\Sigma$ is practically absent, and the R branch of $3d\Pi \rightarrow 2p\Sigma$ can not be found at all. This is due to the decoupling of the orbital momentum. The band $3d\Delta \rightarrow 2p\Sigma$ is entirely absent. It is forbidden if the orbital momentum is coupled to the internuclear axis, as Λ would change by two units. For the $3d\Delta$ states the decoupling is still very small. For the $4d\Delta$ and $5d\Delta$ states it is much bigger, and, therefore, the transitions from these states to the $2p\Sigma$ state are present, though with faint intensity.

Also the $5p\Pi$, $6p\Pi$, etc., $\rightarrow 3s\Sigma$ transitions have to be expected in this region. But they are apparently much weaker than the bands mentioned above and were not found.

TABLE 4.

$3d\ \Sigma \rightarrow 2p\Sigma$

K	P branch			R branch		
	ν calc.	I	ν obs.	ν calc.	I	ν obs.
0				10, 667.76	1	10, 667.6
2	10, 626.67	....	....	71.75	3	71.58
4	576.05		10, 576.3	65.26	4	65.17
6	515.25	¹ 2d	514.86	53.90	5	53.72
8	450.20	¹ 8	449.64	40.11	4	40.09
		¹ 3				
10	383.51	1	383.38	26.02	4	26.03
12	317.47	1-	317.5	12.13	3	12.05
14	252.97	1-	252.8	599.29	3	599.03
16	190.86		....	87.44	2	87.6
18	....		....		2	76.3

¹ Confused with R_{16}.
² Masked by P_{18} of 9,123.   ³ From P_2 of λ 9,346.

$3d\ \Pi \rightarrow 2p\Sigma$

K	P branch			Q branch			R branch		
	ν calc.	I	ν obs.	ν calc.	I	ν obs.	ν calc.	I	ν obs.
0				10, 863.77	2	10, 863.8	10, 897.18	......	....
2	10, 853.09	....	....	60.16	2	60.02	932.61	......	....
4	37.24	1	10, 837.14	54.84	¹ 6	54.46	990.40	......	....
6	40.42	....	(¹)	48.36	2	48.32	11, 050.84	......	....
8	46.96	2	46.8				110.86	......	....
10	54.42	¹ 6	54.46	41.57	1	41.2	160.68	......	....
12	60.98	1	60.9	34.69	1	34.44	226.19	......	....
14	67.05	....	(¹)	28.43	1	28.4	280.75	......	....
16	72.10	1	71.9	23.17	1	23.2	333.25	......	....
18	....		....		1	19.2			

¹ Too close to P_2 of λ 9,123.   ² P_{16}, Q_6, and 3^2D-9^1F.   ³ Too close to R_{16}.

3. CONSTANTS

The constants which can be calculated from the bands at 9,123 and 9,346 Å are

$$\lambda\ 9{,}123 : 2s\Sigma \rightarrow 2p\Sigma \qquad \nu_0 = 10{,}889.59$$
$$\lambda\ 9{,}346 : 2s\Sigma^1 \rightarrow 2p\Sigma \qquad \nu_0 = 10{,}637.8$$

16 G. H. Dieke, Zs. f. Phys., vol. 57, p. 71, 1929.

From these values for the origins of the bands it follows that the difference between the first and second vibrational state of the initial electronic level is

$$\text{for } 2p\Sigma \qquad \omega = 1{,}481.0$$

if the value of ω for $2s\Sigma$ is 1,732.8.

The rotational constants are

$$2p\Sigma \quad \nu = 0 \qquad B = 6.851 \qquad \beta = 5.42 \cdot 10^{-4}$$
$$\nu = 1 \qquad B = 6.55$$

in which B and β are the constants in the formula

$$B(K + \tfrac{1}{2})^2 - \beta(K + \tfrac{1}{2})^4$$

for the rotational energy.

It is interesting to note that the values B and ω for the $2p\Sigma$ state are the smallest of all levels observed so far. This means that the bond which holds the two atoms together is weakest for the $2p\Sigma$ state. It is weaker than for the He_2^+ ion for which the constants are

$$B = 7.10 \,[17]$$
$$\omega = 1{,}627.2 \,[18]\,[19]$$

As stated above it is possible now with the help of the $2p\Sigma \rightarrow 2s\Sigma$ band to calculate all the relative term values of the triplet system with the accuracy of the wave-length measurements. In order to obtain the absolute energy values it is necessary to calculate the limit of one Rydberg series. We take

$$2s = 34{,}301.8$$

which is the value which Curtis and Long [20] calculated from the limit of the series $np\Pi \rightarrow 2s\Sigma$.[21] Then we get for the other low terms

$$2p\Pi = 29{,}533.7$$
$$2p\Sigma = 23{,}412.2$$

All other terms can then be easily calculated from the zero lines of the bands which have the level in question as initial level.

The $2p\Pi$ term can also be directly calculated as head of the $ns\Sigma \rightarrow 2p\Pi$ series. The value obtained in this way is $2p\Pi = 29{,}530.7$.[22] The difference of only 3 cm^{-1} between this value and the value given above shows how well the terms can be represented by a Rydberg series.

Finally, all of the infra-red He lines observed photographically are presented in order of increasing wave length in Table 5, in which the column headings are either self-explanatory or given by the key to the classifications.

[17] See footnote 16, p. 126.
[18] W. Weizel and E. Pestel, ZS. f. Phys., vol. 56, p. 197, 1929.
[19] S. Imanishi, Sc. Pa. Inst. Phys. Chem. Res., vol. 11, p. 139, 1929.
[20] See footnote 12, p. 125.
[21] If we assume the value calculated in the usual way from the head of a Rydberg series we assume that the actual levels which have half a quantum of vibrational energy follow the Rydberg law. It looks more plausible to assume that the actual electronic levels; that is, the levels extrapolated to no vibrational energy are the ones which obey the Rydberg formula more closely. Imanishi (see footnote 19) has calculated the limit of the $np\Pi \rightarrow 2s\Sigma$ series under this assumption and found a somewhat different value for $2s\Sigma$. But also this procedure is open to objections as the different orientation for the orbital momentum cause a shift and splitting up of the levels. (The $s\Sigma$ terms would be free from this objection.) On account of this complication, and because the absolute values of the energies are of very little importance for most purposes anyway, we have chosen the old value of Curtis and Long rather than that of Imanishi.
[22] G. H. Dieke, Phys. Rev., vol. 38, p. 646, 1931.

TABLE 5.—*Infra-red He spectra*

Key to the classification of the He₂ lines:

A: 2pΣ→2sΣ.
B: 2pΣ¹→2sΣ¹.
C: 3dΠ→2pΣ.
D: 3dΣ→2pΣ.
E: 3sΣ→2pΣ.

Intensity	λ_{air} I. A.	ν_{vac} cm⁻¹	He I	He₂ A	B	C	D	E
3	8,361.7	11,956.0	2¹S–6¹P					
3	8,776.09	403.41						
1	9,845.0	302.7	3¹D–11¹F					
2	8,914.7	214.4	3¹S–7¹P					
2	8,996.7	112.1	3¹D–10¹F					
1–	8,997.6	111.0						
6	9,063.40	030.36	3¹P–9¹D					
1	82.9	006.7						
1	85.6	11,003.4	3¹P–10¹D					
1	9,094.90	10,992.16						
6	23.17	958.10		R₄				
8	24.29	956.75		R₇				
5	27.34	953.09		R₁₁				
10	30.66	949.11		R₅				
5	37.00	941.51		R₁₃				
12	42.23	935.25		R₃				
5	52.23	923.30		R₁₅				
7	58.92	915.32		R₁				
5	73.18	898.36		R₁₇				
2	74.6	896.7	3¹P–8¹S					
4	93.55	874.21		P₁				
1	9,195.5	871.9				P₁₆		
5	9,200.00	866.59		R₁₉				
2	02.4	863.8				Q₂		
1	04.8	860.9				P₁₄		
2	05.56	860.02				Q₄		
6	10.28	854.46	3¹D–9¹F			Q₆; P₁₀		
1	213.1	851.1	3¹D–9¹F					
2	15.49	848.32				Q₆		
2	16.8	846.8				P₉		
1	21.5	841.2				Q₁₀		
10	22.96	839.53		P₃				
1	25.00	837.14				P₄		
1	27.29	834.44				Q₁₃		
1	32.5	828.4				Q₁₄		
3	33.03	827.71		R₂₁				
1	36.9	823.2				Q₁₅		
1	40.3	819.2				Q₁₈		
1	49.6	809.3						
20	57.50	799.09		P₅				
2	60.91	795.11						
3	62.78	792.94						
4	66.10	789.07						
2	72.60	781.50		R₂₂				
20	9,297.14	753.05		P₇				
1	9,303.2	746.0	3¹P–9¹D					
15	41.91	701.51		P₉				
2	46.2	696.6			R₇			
2	48.0	694.5			R₄			
2	50.50	691.69			R₆			
2	55.70	685.74			R₁₁			
2	60.67	680.07			R₃			
3	68.12	671.58					R₁	
1	69.8	669.7			R₁₃			
1	71.6	667.6					R₆	
4	73.75	665.17					R₄	
1	76.6	661.9			R₁			
5	83.82	653.72					R₆	
1	9,390.3	10,646.4			R₁₅			

TABLE 5.—*Infra-red He spectra*—Continued

Intensity	λ_{air} l. A.	ν_{vac} cm⁻¹	He I	He II				
				A	B	C	D	E
10	9,391.90	10,644.55		P_{11}				
4	9,395.84	640.09					R_8	
4	9,408.27	626.03					R_{10}	
1	17.73	615.36						
3	20.67	612.05					R_{11}	
2	32.24	599.03					R_{14}	
2	41.6	588.5			P_3			
2	42.4	587.6					R_{15}	
8	47.12	582.34		P_{15}				
2	52.5d	576.3					P_6; R_{13}	
60	63.66	563.84	3^2S-5^2P					
3	9,477.72	548.17			P_5			
8	9,507.74	514.86		P_{16}	P_5		P_6	
30	16.70	504.97	3^1P-7^1D					
3	19.39	501.99			P_7			
10	26.17	494.52	3^2D-6^2F					
4	29.27	491.10	2^1D-8^1F					
1	52.8	466.3	3^2D-8^2P					
3	57.08	449.64			P_9		P_8	
7	9,573.90	442.20		P_{17}				
6	9,603.50	410.01	3^1S-6^1P					
2	20.88	391.21			P_{11}			
3	25.80	385.90	3^1P-8^1D					
1	28.14	383.38					P_{10}	
7	45.68	364.49		P_{12}				
2	80.54	327.17			P_{13}			
1	82.4	325.2	3^1P-8^1S					
1	9,667.6	317.5					P_{12}	
10	9,702.66	303.63	3^2P-7^2S					
6	22.33	281.72		P_{31}				
2	46.64	257.13			P_{15}			
1	50.8	252.5					P_{14}	
1	9,782.7	219.3						
5	9,807.13	193.87		P_{23}				
2	19.18	181.36			P_{17}			
2	97.47	100.82		P_{24}				
2	9,898.67	099.60			P_{19}			
1	9,985.43	011.85			P_{21}			
1	9,994.6	10,002.7		P_{27}				
40	10,027.73	9,969.61	3^2D-7^2F					
15	31.16	966.20	3^1D-7^1F					
2	72.10	925.70	3^1D-7^1P					
10	10,138.50	860.69	3^1P-7^1D					
2	10,233.0	769.6	3^1P-7^1S					
1	258.8	745.1						R_{13}
40	10,311.32	695.43	3^3P-6^3D					
1	50.9	658.4						R_8
1	393.3	618.9						R_6
1	10,433.1	582.2						R_4
1	10,599.2	432.1						P_6
4	10,667.62	371.60	3^1P-6^1S					
2	10,714.4	330.7						
100	10,829.09	231.86	$2^3S-2^3P_0$					
500	10,830.30	230.83	$2^3S-2^3P_{1,\,2}$					
6	10,912.95	160.92	3^1D-6^1F					
3	917.0	157.5	3^1D-6^1F					
1	11,012.1	077.6	3^1S-6^1P					
1	11,045.1	9,051.3	3^1P-6^1D					

WASHINGTON, May 28, 1932.

127984—32——2

TESTS OF CELLULAR SHEET-STEEL FLOORING

By J. M. Frankland[1] and H. L. Whittemore

ABSTRACT

The strength and elastic properties of a new type of cellular steel structural floor were determined. The floor panels were made from 12, 14, 16, and 18 United States standard gage special corrugated sheets which were spot-welded together into panels 2 feet wide and about 11 feet long. The weights ranged from 8 to 16 lbs./ft.[2] The specimens were tested under transverse loading at the quarter points, measurements of strain and deflection being taken at mid-span. It was found that the Euler-Bernoulli theory commonly used in designing beams could be applied satisfactorily to predict the elastic behavior. On 10-foot spans, the floor panels behaved elastically under loads equal to or greater than those which would produce a deflection of one-three hundred and sixtieth of the span. The maximum load was in all cases considerably higher than the limit of elasticity of the panels and assured a considerable margin of safety against overloading. The sections were very stable under the concentrated loads and reactions.

The tensile properties and Young's modulus of elasticity were determined for the sheet steel used in the floor specimens, as was also the shearing strength of the spot welds.

CONTENTS

I. INTRODUCTION

1. GENERAL

Modern practice in building construction often provides a structural steel frame on which is carried the walls and floors. The walls provide protection against weather, the floors carry the actual weight of the occupancy. The floor beams inclose areas usually 15 to 25 feet square

---

[1] Research Associate representing the American Institute of Steel Construction, New York, N. Y.

which are filled in with a variety of constructions. A concrete slab is often used, with or without the use of steel or concrete secondary beams. Tile arches with reinforcement are sometimes used, or else a combination of concrete and tile construction. For all these constructions the weight of the floor is usually greater than the live load it is designed to carry, often considerably so. The cost of the steel frame and the foundations is therefore greatly influenced by the weight of the floor construction. Consequently many engineers and manufacturers are endeavoring to design and produce a load-carrying member for a floor which will be lighter and more convenient than these already in use. It would be desirable to have a lightweight unit which would lend itself to the possibilities of mass production and at the same time could be economically shipped to the site and laid rapidly in place, affording workmen at once a working floor and a place for storing materials. At present the greatest possibilities seem to lie in metal construction.

The advantages to be gained by the use of a lightweight floor in buildings are apparent. For bridges also such a floor would be desirable. As W. H. Thorpe has shown [2] for a bridge of given general type, the load-carrying capacity per pound of structure depends primarily on the span and the ratio of live load to weight of the floor system. For long-span bridges and tall buildings the weight of floors has a cumulative effect. If the dead loads due to the floor system can be reduced, it will result in a much greater reduction in the total weight of the structure, making possible very appreciable economies in material in these two fields.

2. THE CELLULAR FLOOR

Sheet steel, which is now one of the leading products of the steel industry, can be readily formed into corrugated sheets in thicknesses up to 7 gage.[3] The load-carrying properties of these sheets may be materially increased by adding another sheet, flat or corrugated, to form a cellular construction. The two sheets may be either riveted or welded together, the latter process being particularly adapted to large-scale production. A panel made in such a way would be characterized by longitudinal stiffening cells. Such a construction lends itself to an economic use of the material, as much of it as possible being stressed to the allowable working stress, resulting in a beam of high strength in proportion to its weight.

Panels of this sort seem well suited for the construction of floors. They could be shop fabricated and shipped to the construction site where they would be available for rapid placing to form a continuous working floor as soon as the beams were erected to carry them. The cellular structure provides also a system of ducts which appears to offer possibilities of considerable economy in running plumbing, heating and ventilating, electrical, and compressed-air lines.

Experiments by a manufacturer resulted in the development of two types of sheet steel cellular floor panels which appeared to have sufficient strength. These are shown in Figures 1 and 2. The panels are 2 feet wide and any desired length up to 12 feet. The sheets are joined by spot welds.

---

[2] Steel Bridge Weights, Engineering, vol. 120, p. 534, 1925.
[3] 0.188 inch. United States standard gage for sheet, plate, iron, and steel.

B. S. Journal of Research. RP463

FIGURE 1.—*End view of floor types A, B, and C*

The manufacturer's experiments indicated that panels weighing in the neighborhood of 10 lbs./ft.^{2} would be adequate for most floor loads. The dead load of such a floor would consequently be due for the major part to the fireproofing and floor finish. It has been estimated that with this construction there can be saved about one-half of the dead load due to the floor compared to constructions now in common use, this saving varying somewhat with the live load and being subject to modification according to local requirements for fireproofing.

Before putting such a new construction into service, it was considered advisable to make a thorough study of the panels under load

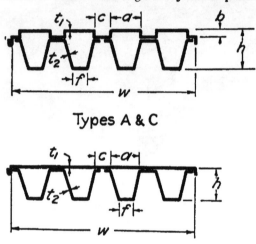

Types A & C

Type B

FIGURE 2.—*Sections of floor panels*
The average numerical values of the dimensions indicated by letters are given in Table 2 for each specimen.

to see how closely their performance could be predicted by conventional engineering theory.

II. ACKNOWLEDGMENTS

The American Institute of Steel Construction sponsored the investigation, which was made under the research associate plan at the Bureau of Standards. The specimens were designed and manufactured by the H. H. Robertson Co., Pittsburgh, Pa. Dr. A. H. Stang and L. R. Sweetman, of the bureau staff, assisted in the analysis of the problem and in making the tests.

III. DESCRIPTION OF THE SPECIMENS

The specimens were of three types, sections of which are shown in Figure 2. Types A and C consisted of two corrugated sheets and type B of one corrugated sheet and a flat sheet, the bottom sheets being of the same dimensions for each type. The corrugated upper surface was designed to bind with a surface slab of concrete or similar material.   The troughs in the top sheet also offer certain advantages in laying pipes and ducts.   If a concrete slab is not used, the troughs can be covered with sheet-metal channels to give a smooth upper panel surface.   Cork, rubber tile, or linoleum may be laid directly on the flat top sheet of the panels of type B.   Different strengths were obtained by varying the thicknesses of top and bottom sheets. Type C differed from type A in having the lower corrugations cut away for a distance of 6 inches from each end, the projecting portion of the top sheeting being reinforced in the manner shown in Figure 1. There was no reinforcement of the welds at the ends of the bottom corrugations.   The object of this design was to enable this type of panel to be framed into the floor beams with the top sheet lying directly on the upper flange, thus reducing the distance from the finished floor surface to the ceiling immediately below.

Where the top and bottom sheets of a panel were in contact between the cells, they were joined by two rows of three-eighths-inch spot welds, the welds being about 1½ inches apart in each row.   The horizontal portion of the assembled section where these welds occur will be spoken of subsequently as the web.   In all specimens the bottom sheet was made in two parts separated down the center web. In all the specimens numbered X through XXIV the top sheet was in one piece, this representing the latest manufacturing method. The top sheets of specimens I through IX were made in two pieces with the separation occurring down the center web, this type of construction being clearly shown in the type A panel of Figure 1. In all type B specimens the top sheets were in one piece.

The top sheet of the panels was bent down on either side for a lateral connection to adjacent panels when erected.

The specimens were uncoated showing the mill scale on the sheets except for numbers I, VI, X, XI, and XII, which had been given one coat of paint, inside the cells as well as on the outside surface.

All specimens were 24 inches wide and either 10 or 11 feet long. The cells were spaced 6 inches on centers and in types A and C were 5⅜ inches deep, in type B, 4⅜ inches deep.

A specimen will be described hereafter by its type letter, followed by the gage numbers of the top and bottom sheets, respectively. The gage used is the United States standard gage for sheet, plate, iron, and steel.   The nominal thicknesses in inches, together with the maximum and minimum values observed may be found in Table 1. The relative variation of thickness in any one sheet was much less. When the bottom corrugations were made from material lighter than 16 gage, preliminary experiments by the manufacturer had shown that the cell bottoms are liable to crumple over the support.   To avoid this the cell bottoms of the A16–18 specimens tested were reinforced at the ends by pieces of 18 gage sheet of the same shape and 6 inches long which were welded to the webs, giving in effect a double cell bottom over the supports.

The first group of specimens, numbered I to XII, were manufactured on a power cornice brake, the dimensions being obtained by hand setting of the sheet in the machine.   These will be referred to later as "hand-set specimens."   They were characterized by a certain irregularity in the dimensions and a lack of evenness in the

load-bearing surfaces. Specimens XII to XXIV were formed in the
same machine over dies which gave much more uniform dimensions.

TABLE 1.—*Thickness of sheets*

Gage No.	Nominal thickness	Actual Thickness	
		(Minimum)	(Maximum)
	Inch	*Inch*	*Inch*
12	0.107	0.097	0.114
14	.077	.069	.083
16	.061	.056	.066
18	.049	.044	.051

The sheets were join in all cases by resistance spot welding with
three-eighths inch diameter welds. In the form of spot welding used,
projection welding, small projections are formed on one of the sheets
at the places where the welds are to be made. A hand-operated
machine making one weld at a time and with the current controlled
by a time switch was used on the first 12 specimens. The remaining
specimens were welded on a multiple machine with full automatic
control. Specimens XV, XXII, XXIII, and XXIV were welded on
the automatic machine after its operation had been improved.

The dimensions of each cell and web were taken at both ends of a
specimen and the results averaged to obtain data from which the
moment of inertia and neutral axis location could be calculated. The
measurements were taken to the nearest sixteenth of an inch, and the
averaged results expressed to the nearest hundredth. Average dimensions and descriptions of the specimens are given in Table 2.
Since the deviations in the dimensions of the most irregular panels
did not produce differences of more than 2 per cent in the moments
of inertia calculated cell by cell and from the averaged dimensions,
it was not felt worth while to list the dimensions of the particular parts.

TABLE 2.—*Description and dimensions of floor specimens*

[The dimensions designated by letters are shown in fig. 2]

Panel	Type	a	b	c	f	h	w	t_1	t_2	Length	Nominal weight
		Inches	*Inches*	*Inches*	*Inches*	*Inches*	*Inches*	*Inch*	*Inch*	*Feet*	*Lbs./ft.*[1]
I	A14-16	3.86	1.29	2.04	2.00	5.64	24.78	0.074	0.060	11	10.50
II	A14-16	3.91	1.29	2.16	2.16	5.65	24.44	.075	.062	11	10.50
III	A14-16	3.90	1.32	2.07	2.13	5.69	24.31	.074	.061	11	10.50
IV	A16-18	3.89	1.29	2.12	2.13	5.64		.060	.049	11	8.50
V	A16-18	3.88	1.31	2.15	2.12	5.60	24.50	.060	.047	11	8.50
VI	A16-18	3.90	1.28	2.03	2.11	5.66	24.44	.061	.049	11	8.50
VII	C14-16	3.88	1.29	2.14	2.18	5.66	24.47	.075	.062	11	10.50
VIII	C14-16	3.84	1.29	2.09	2.12	5.66	24.50	.075	.062	11	10.50
IX	C14-16	3.88	1.31	2.10	2.16	5.66	24.03	.070	.062	11	10.50
X	B16-16	3.76	0	2.10	2.16	4.35	24.22	.057	.060	11	8.25
XI	B16-16	3.80	0	2.07	2.12	4.35	24.34	.060	.039	11	8.25
XII	B16-16	3.72	0	2.11	2.18	4.37	24.13	.060	.060	11	8.25
XIII	A14-16	3.94	1.26	2.05	2.06	5.70	24.00	.076	.065	11	10.50
XIV	A16-18	3.81	1.23	2.22	2.12	5.64	24.58	.063	.049	11	8.50
XV	B16-16	3.82	0	2.12	2.20	4.34	24.00	.061	.060	11	8.25
XVI	A14-14	3.88	1.24	2.10	2.17	5.71	24.50	.077	.079	11	11.50
XVII	A14-14	3.91	1.27	2.10	2.15	5.71	24.12	.077	.079	11	11.50
XVIII	A14-14	3.91	1.25	2.11	2.14	5.70	24.12	.077	.079	11	11.50
XIX	A12-12	3.91	1.24	2.15	2.32	5.53	24.25	.108	.105	10	16.10
XX	A12-12	3.92	1.25	2.11	2.28	5.57	24.00	.113	.100	10	16.10
XXI	A12-12	3.88	1.25	2.11	2.27	5.61	24.06	.108	.101	10	16.10
XXII	A14-16	3.92	1.25	1.98	2.19	5.64	23.88	.080	.064	11	10.50
XXIII	A14-16	3.88	1.24	2.00	2.12	5.66	23.75	.080	.061	11	10.50
XXIV	A14-16	3.87	1.24	2.02	2.14	5.66	23.81	.079	.064	11	10.50

[1] Calculated from average height of hand-set A14-16 specimens.

IV. TENSILE TESTS OF THE MATERIAL

1. METHODS OF TEST

The principal requirements of the usual commercial grades of black sheet steel are that the sheets should have a good surface and a high ductility to undergo various forming operations. They are not primarily intended for structural purposes. It was, therefore, desirable to investigate the structural properties, tensile strength, yield point and Young's modulus of elasticity of the material used in these floor panels.

A screw-power beam testing machine having a capacity of 20,000 pounds was used, the machine having two load ranges, one up to 2,000 pounds and the other up to 20,000 pounds. The former range was used for all specimens except the thickest (12 gage).

The specimen used is illustrated in Figure 3 and conforms to the requirements of the sheet tensile specimen recommended by the American Society for Testing Materials in their Tentative Methods

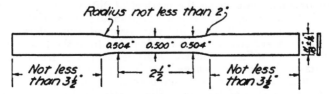

FIGURE 3.—*Sheet tensile specimen*

of Testing of Metallic Materials (Specification E8–27T, Proc., vol. 27, I, p. 1067). The gage length was 2 inches.

One hundred and twelve specimens were tested, 9 of 12 gage, 26 of 14 gage, 65 of 16 gage, and 12 of 18 gage. Stress-strain measurements were taken for 7 specimens of 14 gage, 6 of 16 gage, and 6 of 18 gage.

The tensile specimens were taken from coupons cut from the end of each sheet prior to forming. Tests were made both in the direction of rolling and transverse to this direction, these specimens being marked respectively *L* and *T*. Ninety-three specimens were taken from panels I through XII and furnished information as to the properties of practically all the sheets entering into those panels. Nineteen specimens were taken from the remaining panels.

The stress-strain curves were obtained with a Ewing extensometer of 2-inch gage length mounted on the edges of the specimen. The load was maintained until any drift in extensometer readings appeared to cease and then the reading recorded.

For the remainder of the specimens only yield point, tensile strength, and elongation were determined. A slow machine speed was used as it was found that the values obtained were to a considerable degree dependent on the rate of extension. In some cases a preliminary load of roughly one-half the yield point was applied more rapidly and the rest of the test continued at the slower speed. The speeds adopted were 0.01 inch per minute until the load began to pick up after the yield point, and 0.09 inch per minute for the remainder of the test. These speeds were measured on the movable crosshead of the machine under no load. The yield points were determined by "drop of beam."

The cross section of the specimens was measured at the ends and at the center of the gage length. As will be seen from Figure 3, the width of the specimen is less at the center to assure fracture near the middle of the gage length. The minimum of the three areas measured was used to compute yield point and tensile strength. The effective area for the determinations of Young's modulus was obtained by assuming that the area could be represented by a quadratic function of the position in   gage length. The effective area will then be given very closely by

$$A = \tfrac{1}{6} (A_1 + 4A_0 + A_2)$$

where A_0 is the area at the center and A_1 and A_2 the areas at the ends of the gage length. The formula will be recognized as Simpson's rule. The area so obtained is a closer approximation than would be obtained by the use of the average area.

From the stress-strain data were determined Young's modulus and the proportional limit, using the method of differences proposed by Dr. L. B. Tuckerman, of the Bureau of Standards.[4] A trial modulus of elasticity of 30,000,000 lbs./in.[2] was assumed and differences between the observed strain and that calculated from the trial modulus were plotted in a difference curve. A straight line was drawn through the points on the difference curve in such a way as to include the most points within a range of one hundred-thousandth inch per inch strain on either side of the line. Slight irregularities at the lowest loads were disregarded. From the slopes of these lines were derived the actual moduli of the specimens. The proportional limit was taken as the stress at which the difference began to depart from the straight line by more than one-hundred-thousandth inch per inch strain. In a few cases, the permanent set was measured after each load.

2. RESULTS

Figures 4 and 5 are typical stress-strain curves up to the yield point. Though these figures show results for two transverse specimens, the curves are representative of the results found for all specimens. Table 3 gives the results derived from the stress-strain measurements. The values of the proportional limit there assigned are to be considered as dependent on the width of the error band that was assumed. The measurements indicated that the material was not perfectly elastic at low loads, but these departures from elasticity were of the same order as the random error in the readings, so that it remained uncertain to what extent this inelasticity was real. In any case, inelastic behavior before the yield point is of little significance in estimating the suitability of a material for structural purposes. The yield point may therefore be taken as a measure of the safe working stress of the material.[5]

---

[4] See Determination and Significance of the Proportional Limit in the Testing of Metals, R. L. Templin, and discussion, Proc., Am. Soc. Testing Materials, vol. 29, II, p. 523, 1929. Discussion, p. 538.
[5] See note 4.

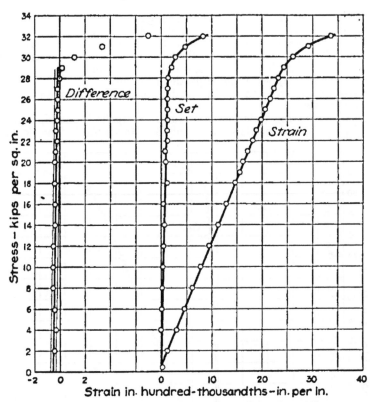

FIGURE 4.—*Stress-strain curves for tensile specimen 9T*

TABLE 3.—*Results of stress-strain measurements on tensile specimens*

Specimen	Thickness	Proportional limit	Yield point	Tensile strength	Young's modulus	Elongation in 2 inches
	Inch	*Kips/in.²*	*Kips/in.²*	*Kips/in.²*	*Kips/in.²*	*Per cent*
1 L [1]	0.078	14.7	21.6	39.7	27,600	38
2 T	.080	17.0	22.7	44.7	29,600	33
3 L [1],[2]	.069	8.6	19.0	39.9	27,500	
4 T	.073	10.0	21.2	43.3	29,100	33
5 L	.063	20.1	21.0	40.2	28,700	31
6 L	.061	12.4	21.1	39.4	28,700	32
7 T	.064	17.1	29.0	43.7	28,900	27
8 L	.048	27.4	35.1	51.4	28,400	25
9 T	.050	28.0	33.9	56.0	29,800	25
10 T	.073	20.2	24.6	45.8	29,400	38
11 L	.075	16.0	21.0	40.7	28,000	36
12 T	.076	20.6	27.3	45.0	29,900	35
13 L	.059	20.0	30.4	43.7	28,800	31
14 T	.061	29.7	35.2	45.1	29,300	40
15 L	.062	18.0	32.9	46.2	29,900	36
16 L [1]	.045	20.1	25.1	46.0	28,400	
17 T	.047	15.4	31.3	55.8	31,400	25
18 L	.049	25.0	25.0	47.6	28,800	33
19 T	.050	21.6	28.5	51.4	30,600	30

[1] Considerable set at low loads.   [2] Broke near gage mark.

The average for 10 specimens of Young's modulus in the direction of rolling was 28,500,000 lbs./in.^{2}, and transverse to the direction of rolling the average of nine tests was 29,700,000 lbs./in.^{2} A value of 28,700,000 lbs./in.^{2} was taken for the longitudinal modulus, this being arrived at by neglecting two specimens (1L and 3L), for which the elastic limit appeared to be particularly low.

Based on strength and yield-point determinations, the material was of three types:

1. Sheets having yield points of 20 to 28 kips/in.^{2} (one kip equals 1,000 pounds) and tensile strengths of 38 to 48 kips/in.^{2} The elonga-

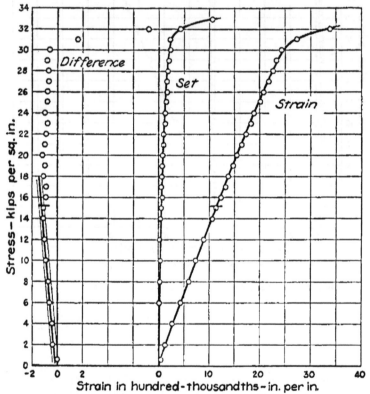

FIGURE 5.—*Stress-strain curves for tensile specimen 17T*

tion in 2 inches ranged from 30 to 40 per cent. At the yield point there was a sharp drop of beam, the load falling off considerably;

2. Sheets having yield points of 35 to 47 kips/in.^{2} and tensile strengths of 46 to 50 kips/in.^{2} There was a marked drop in load at the yield point. The elongation was as great as for the first type;

3. Sheets having tensile strengths of 50 up to 70 kips/in.^{2} and a lower elongation than the preceding types. The yield points ranged from 29 to 52 kips/in.^{2} The elongations were in the neighborhood of 25 per cent. At the yield point the drop of beam was not so pronounced and occasionally the load remained nearly constant.

Only a small proportion of the specimens were of the third type and these were all from 16 and 18 gage material. The higher strength appears to be due to cold working at the mill. The second type of sheet was made from open-hearth steel rolled on a continuous strip mill.

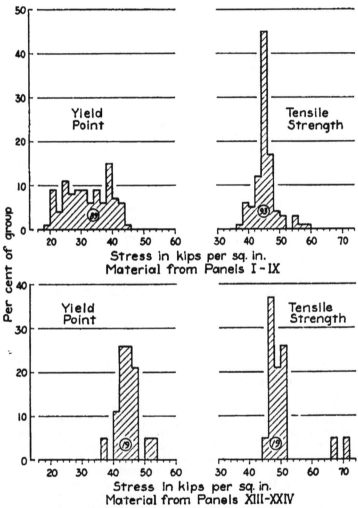

FIGURE 6.—*Distribution of yield point and tensile strength among the specimens*

The first type of material was found only in the first 12 panels, which also included some of the second type. All but one of the tensile coupons taken from panels XII through XXIV were of the second type, the exception being a 16-gage sheet of the third type.

Figure 6 shows graphically how the yield point and tensile strength were distributed among the specimens.

The ordinate represents the per cent of the group which falls within the range of 2 kips/in.^{2} indicated on the horizontal scale. The ringed number is the total of specimens in each group.

V. SHEARING STRENGTH OF THE WELDS

1. GENERAL

The loads applied to the top surface of the floor panels caused shearing forces on the welds joining the top and bottom elements. It is obviously of importance to know what shears may be safely applied to spot welds such as those used in the cellular floor. A series of tests were conducted to obtain data on this point.

2. METHODS OF TESTING

The test of a spot-welded joint in shear alone presents certain difficulties. If a joint with a single lap is pulled in tension, the eccentric loading produces a couple tending to rotate the join of the two parts. In sheet metal this tends to split the joint apart and to

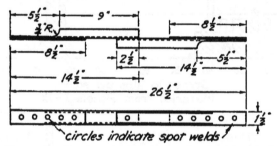

FIGURE 7.—*Specimen for determining shearing strength of welds*

produce buckles about the weld. If a double lap is used, the strength of the joint is considerably increased and fracture may occur outside of the welds; in fact, double lap joints in 14, 16, and 18 gage sheet containing two ⅜-inch diameter welds were tested to destruction and all broke by tearing of the strip inside the cover plates.

A variety of joints containing 1, 2, and 3 welds were obtained and tested in tension. From these results a satisfactory specimen was designed as shown in Figure 7. It consists essentially of two channel-shaped sections from the steel sheet lapped back to back and welded with a single spot. In order to grip the specimen in the jaws of the testing machine, the channel flanges were cut away at the ends. It was found necessary to reinforce the gripped portion of the specimen by welding on a strip along the back of each channel. The specimens were projection welded using a technique as close as possible to the welding methods used in fabricating floor. Such specimens in 14, 16, and 18 gage sheet containing one ⅜-inch diameter spot weld fractured in the weld without appreciable bending of the channels or buckling of the sheet about the weld.

The specimens were tested in tension in a 100,000-pound Amsler machine using the 10,000-pound range. The autographic recording apparatus supplied with the machine was used on an 8-inch gage length, so that a load-extension curve was obtained in each case.

3. RESULTS

The specimens described in the preceding section are noted below by the type letter G, followed by the gage number of the sheet, and an individual specimen number to distinguish those of the same gage. Six specimens each of 14, 16, and 18 gage sheet were tested. The results are summarized in Table 4.

The gross area of the fractured surface of welds was estimated by assuming an elliptical shape and measuring two perpendicular diameters to the nearest 0.01 inch. The net area was obtained from the gross area by deducting for the larger blowholes. Both figures are, of course, only approximate, as are also in corresponding degree the stresses derived from them. The maximum load is definitely the more significant figure.

The autographic records showed that failures were preceded by little or no plastic deformation. The single exception to this was G18-3, which failed by the weld tearing out of the sheet. The ultimate strength of the weld is therefore a measure of the permissible design loads.

TABLE 4.—*Strength of welds*

Specimen	Type of failure	Maximum load	Gross area of weld	Net area of weld	Shearing strength on gross area	Shearing strength on net area
		Kips	*Square inch*	*Square inch*	*Kips/in.²*	*Kips/in.²*
G14-1	Negative bearing	4.59				
G14-2	Shear	3.39	0.105	0.102	32.3	33.2
G14-3	do	4.02	.119	.117	33.8	34.4
G14-4	do	3.86	.101	.099	38.2	39.0
G14-5	do	3.64	.095	.087	38.3	41.8
G14-6	do	3.75	.108	.104	34.7	36.1
G16-1	do	2.94	.075	.074	39.2	39.7
G16-2	Negative bearing	2.61				
G16-3	do	2.57				
G16-4	do	2.24				
G16-5	do	2.43				
G16-6	Shear	2.49	.058	.055	42.9	45.3
G18-1	Negative bearing	2.91				
G18-2	Shear	2.83	.104	0.057-0.090	27.2	31.4-49.6
G18-3	Negative bearing	2.81				
G18-4	do	3.09				
G18-5	do	2.87				
G18-6	do	3.14				

Those specimens in which the weld metal itself did not shear failed by the weld spot tearing out of the sheet. On the side of a weld toward the gripped end of that portion of the specimen there exists a concentrated tensile "negative bearing stress," analogous to the compressive bearing stress in the plate of a riveted joint. Where compressive failure occurs in the plate in the riveted joint, the material in each sheet parts in tension in the welded joint. Failure progresses by the welded regions twisting and tearing out of the sheets. Figure 8 shows late stages in the failure in "negative bearing" of two 16-gage specimens, one single-lap and one channel type. Some of the failures of this kind appeared to start in defective material near the edge of the weld and then spread to the edge of the unfused material.

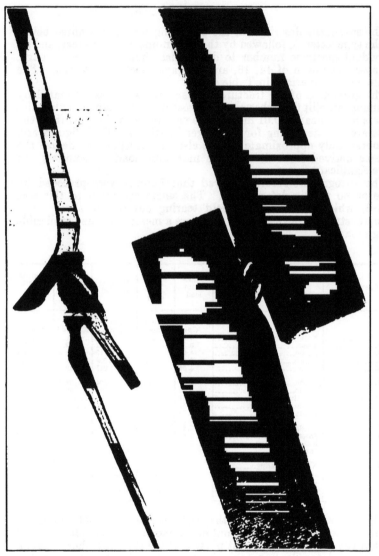

FIGURE 8.—*Typical failures of spot welds in negative bearing for channel-type (see fig. 7) and single-lap specimens*

FIGURE 9.—*View of a floor panel ready for test*

The fractures of welds that broke in shear presented a variety of appearances. The welding of specimens G16-1 and G16-6 was defective, the former containing three weld spots of gross areas 0.019, 0.008, and 0.048 square inch, and the latter containing two spots of gross areas 0.013 and 0.045 square inch. Apparently the sheets were in good electric contact at more than one place due to a poorly made projection or to uneven surfaces. The welds in both specimens contained oxide inclusions. The weld in G18-2 contained coarse radial blowholes about the periphery, there being also a central region with large voids. Tarnish in the center suggests that there was poor adhesion in that portion, but the tarnish may possibly have occurred following fracture. In the 14-gage specimens all the sheared welds contained blowholes, in some cases only one at the center, while in others there were many fine radial blowholes. In the latter cases the surface of fracture showed considerable curvature through points of weakness. Where the weld contained only a few circular blowholes, the fracture presented the appearance of a broken piece of coarse-grained ductile material.

Application of conclusions drawn from these tests must necessarily be restricted to material of the same grade of sheet steel and to welds of the same size made under similar conditions. Higher-strength material would be expected to show higher values, at least in negative bearing.

It is evident that the gage and the tensile strength of the sheet determine whether shearing or negative bearing failure will occur. Below a certain thickness the concentration of tensile stress at the edge of the weld will produce negative bearing failure before the shearing strength of the weld is reached. Above this thickness the load required to shear the weld will not produce the requisite tensile stress to produce a failure in negative bearing. The specimens may be distinguished on this basis. The 16 and 18 gage specimens all broke in negative bearing with the exception of G16-1, G16-6, and G18-2, in which the welds were defective. The 14-gage specimens broke by shearing of the weld, the sole exception to this being G14-1, in which the weld was evidently of much greater shearing strength. In general, then, material of 14 gage or heavier may be expected to develop the full shearing strength of these welds while lighter gages will fail at lower loads in negative hearing.

The shearing strength developed in a welded joint thus depends upon the gage of the sheet when the gage is below a certain critical thickness. Above that thickness the strength should be independent of the gage. Though the tensile properties of the sheet used in the weld specimens were not determined, there was general evidence through the tests that the 16-gage material was of rather low strength, so that the low figures for the 16-gage specimens compared to the 18-gage specimens is probably to be explained on these grounds. The specimen of lowest strength was G16-4 which broke at 2.24 kips in negative bearing, this being a lower breaking load than observed from the specimens with defective welds previously described. Two kips would thus be a conservative figure for the strength of a single one of these spot welds in shear.

VI. METHOD OF TESTING FLOOR PANELS

1. INFORMATION DESIRED

The most important points to be ascertained in the tests of the floor panels themselves were: First, how closely do the deflections and stresses agree with those calculated by the Euler-Bernoulli theory, which is usually used in designing engineering structures; and second, under what range of loads does the floor behave elastically? It is further of importance to know how failure occurs and what are the points of structural weakness with their effects on the response of the panel to load.

It was decided therefore to make transverse tests to destruction of the individual floor panels, measuring deflections and strains at suitable locations in order to correlate these with the loads.

2. TEST PROCEDURE

The test panels were mounted at the ends on roller supports. The A12–12 specimens were 10 feet in length and were tested on a span of 9 feet 8 inches; the remaining panels were 11 feet long and were tested when possible on a span of 10 feet, it being necessary to use a span of 10 feet 6 inches for the type C specimens and the type A specimens with reinforcing at the ends of the cell bottoms (A16–18). In each case the specimen had a length of bearing of 4 inches. Equal loads were applied at the quarter points of the span in a screw power-testing machine of 600,000-pound capacity, using a poise giving a 300,000-pound range. By loading at the quarter points the bending moment between the loads was uniform and equal to the maximum moment which would be produced by the load uniformly distributed. The maximum shear is likewise equal to the maximum shear produced by uniform loading. The computed deflection at mid-span, however, is 10 per cent greater than would have been produced by uniform loading. Since the maximum bending moment is developed over the middle half of the span and the maximum shear is developed over the portions of the specimen between the loads and the supports, inhomogeneities in the structure will be more liable to discovery than if the panel were uniformly loaded.

Steel bearing plates with pads of ¼-inch canvas belting were used to distribute the loads and reactions to the specimen. The plates were 4 inches wide and one-half inch thick and extended across the specimen, the pads being cut to the same size and placed between the specimen and the bearing plate. The load was transmitted to the specimen from the movable cross head of the machine by an I beam carried on a spherical bearing, rollers being placed between the loading beam and the bearing plates. This arrangement can be seen in the typical set-up shown in Figure 9. In the figure it will be noticed that the end bearings are free to accommodate themselves to small twists in the specimen by means of a cylindrical seating.

The deflection at mid-span was measured at each load by means of dial micrometers accurate to 0.002 inch. These dials were mounted on a stiff frame supported at the horizontal webs of the panel immediately over the supports. The frame was carried on 3 steel balls in such a way that 1 foot was free to rotate about a point, 1 to move along a line, and 1 to move in a plane, the support being therefore

B. S. Journal of Research. RP463

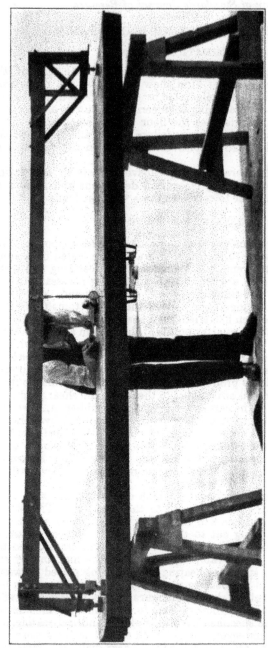

FIGURE 10.—*A floor panel showing apparatus for taking strain and deflection measurements*

B. S. Journal of Research, RP463

FIGURE 11.—*Bottom of a floor panel showing arrangement of strain gages*

kinematically nonredundant. This frame and its mounting is shown
in Figures 9 and 10. Such a mounting assures the frame remaining
unstrained during all portions of the test. The deflection was therefore measured unambiguously relative to a plane close to the neutral
surface of the panel, this neutral surface being computed to lie seven-
eighths to 1 inch below the web for specimens of types A and C, and
1½ inches below the web for type B specimens. The deflection was
taken on the webs at mid-span at two points 6 inches on either side
of the center line.

The strains on a 10-inch gage length in the top and bottom of each
cell were measured, also at mid-span, by the use of Whittemore hand
strain gages.[6] These gages read to a ten-thousandth of an inch and
the readings can be estimated to a hundred-thousandth of an inch,
which corresponds to a stress of about 30 lbs./in.[2] The error in the
strain gages is believed not to exceed two divisions (600 lbs./in.[2] equivalent stress), this error being due for the most part to irregularities in
the dial mechanism. The readings on the top of the panel, the compression side, were taken by the customary hand application of the
gage as illustrated in Figure 10, while the readings on the bottom
were taken with instruments attached to the specimen with rubber
bands. Details of the method of attaching the bottom gages can be
seen in Figure 11. The gage lengths on the bottom were staggered
two inches alternately on either side of the mid-span for convenience
in attaching the instruments.

According to the conventional engineering theory, the stresses in
the extreme fibers should be constant between the loads, this being due
to the constancy of bending moment and section modulus. The
stresses were computed by multiplying the strain by the average value
of Young's modulus (28,700,000 lbs./in.[2]) found in the tensile tests of
the material.

The height of each cell at the two ends of the specimen was measured
to the nearest hundredth of an inch by means of internal calipers at
each load for which the top strain gages were read.

The usual procedure in a test consisted of taking cell heights, top
and bottom strain-gage readings, and deflection measurements with
no load on the panel. These readings were repeated at 2,000-pound
intervals. The deflection dials and bottom strain gages were read at
each 1,000-pound interval. In the case of the heaviest panels, XIX,
XX, and XXI, of the A12–12 type, readings were made every 1,500
pounds; for all other specimens readings were taken every 1,000
pounds. The first group of 12 panels contained three specimens of
each type. One of these was loaded in 1,000-pound increments
straight up to the point at which it would support no added load.
Strain-gage readings were taken until the change in gage length
indicated that the material was well beyond its elastic range. The
deflection measurements were taken over the full range of the apparatus (about 1¾ inches), which sufficed to carry the readings nearly
to the maximum load. A second specimen of the same type was loaded
until a "limiting deflection" was reached approximately equal to one
three hundred and sixtieth of the span plus 10 per cent. As previously
mentioned, the loading in the tests should produce a deflection 10

---

[6] See Arch Dam Investigation, vol. I, Am. Soc. Civil Engrs., p. 64, November, 1927.

per cent greater than the same load uniformly distributed. In order
to avoid cracking plaster ceilings, floors are generally restricted to a
maximum deflection when uniformly loaded of one three hundred and
sixtieth of the span. The limiting deflection of the tests consequently
corresponds to the deflection limits set on floors in practice. At this
load the specimen was held for one hour to observe any creep that
might occur. Following this halt, the test was continued to failure as
before. The third panel of the group was loaded up to the limiting
deflection with the same load increments as before, then unloaded in
steps to zero, readings being taken at the same loads as in the first
portion of the test. The rest of the test then continued as for the

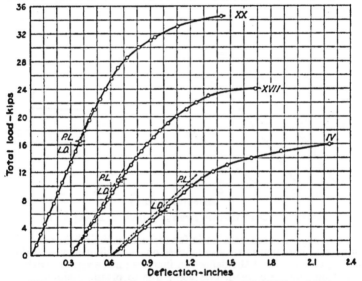

FIGURE 12.—*Load-deflection curves for panels IV, XVII, and XX*

first specimen. The object of this repeated loading was to observe
hysteresis effects as well as permanent set on removing the load. Any
effects of prestressing should show up on the second loading. Since
secondary effects were found to obscure any real hysteresis that may
have been pr s n , the procedure followed for the last 12 specimens
was to load to the limiting deflection, unload to zero to observe permanent set, no intermediate readings being taken during unloading,
then reload to the limiting deflection and continue the test as before.
To observe creep, one specimen of each group was held for an hour
at the limiting deflection load before unloading.
    Careful watch was maintained to observe the development of
buckling failure in the various parts of the specimen, the principal
points of interest being the cell tops at and between the loads and the
side walls of the cells at the supports.

VII. RESULTS OF THE PANEL TESTS

The curves of Figures 12 and 13 show typical relations of load to average deflection at mid-span. It will be observed that up to a certain point on the curves, the deflection is proportional to the load within the experimental error. This was found to be true for all the specimens. The load at this point will be spoken of subsequently as the proportional limit of the floor panel. The proportional limit for the panel is indicated on the curves by P. L. and the limiting deflection load by L. D. A deviation of 0.002 inch from proportionality was taken as criterion for determining this limit. The proportional limit of the panel so defined and the load at the limiting deflection

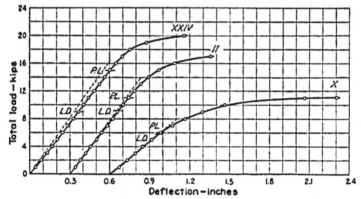

FIGURE 13.—*Load-deflection curves for panels II, X, and XXIV*

of the panels are used as a basis for discussing their behavior. For this reason these points are also marked on succeeding curves. Above the proportional limit the deflection increases more rapidly with successive increments of load, this being due to plastic yielding and to buckling. This yielding progresses until a point is reached where the panel will support no added load. Specimens remained intact, though deformed, after the test. Figures 14 to 17 show the appearance of some typical panels after testing.

Table 5 summarizes the strength properties of the specimens and gives also the spans on which they were tested and the permanent set indicated by the deflection dials when unloaded completely from the limiting deflection. The columns headed equivalent uniform loads per square foot give the distributed loads which were computed to give the same maximum bending moment on a 10-foot span as the observed loads.

TABLE 5.—*Strength properties of panels*

Panel No.	Description	Span	Limiting deflection load	Proportional limit	Maximum load	Equivalent uniform load on 10-foot span			Set after loading to limiting deflection
						Limiting deflection load	Proportional limit	Maximum load	
		Ft. in.	*Kips.*	*Kips.*	*Kips.*	*Lbs./ft.²*	*Lbs./ft.²*	*Lbs./ft.²*	*Inch*
I	A14-16	10 0	8.5	9.0	18.61	425	450	930	---
II	A14-16	10 0	9.2	11.0	17.65	460	550	882	---
III	A14-16	10 0	9.9	9.0	16.99	495	450	850	0.008
IV	A16-18	10 6	6.2	10.0	17.25	326	525	906	---
V	A16-18	10 6	5.7	11.0	17.00	299	578	892	.010
VI	A16-18	10 6	6.5	6.5	15.77	341	341	828	---
VII	C14-16	10 6	7.5	6.0	13.67	394	315	718	---
VIII	C14-16	10 6	8.0	8.2	14.35	420	430	753	.016
IX	C14-16	10 6	8.4	3.8	15.30	441	200	803	---
X	B16-16	10 0	5.7	6.0	11.54	285	300	577	---
XI	B16-16	10 0	5.0	5.0	11.06	250	250	552	---
XII	B16-16	10 0	5.2	7.4	13.22	260	370	666	.026
XIII	A14-16	10 0	9.7	8.0	20.45	485	400	1,022	.014
XIV	A16-18	10 6	6.5	10.0	18.06	341	525	948	.004
XV	B16-18	10 0	5.6	8.0	14.00	280	400	700	.002
XVI	A14-14	10 0	11.0	11.0	25.65	550	550	1,282	.016
XVII	A14-14	10 0	10.6	11.0	26.15	530	550	1,308	.011
XVIII	A14-14	10 0	11.0	11.0	25.65	550	550	1,282	.016
XIX	A12-12	9 8	15.3	19.5	38.15	739	942	1,844	.008
XX	A12-12	9 8	15.8	16.5	38.20	764	797	1,846	.005
XXI	A12-12	9 8	16.4	16.5	34.14	793	797	1,747	.008
XXII	A14-16	10 0	9.3	14.6	20.10	465	.730	1,005	.003
XXIII	A14-16	10 0	9.9	14.8	21.50	495	740	1,090	.001
XXIV	A14-16	10 0	9.0	14.6	21.00	450	730	1,050	.005

For use in interpreting these results, section properties of the panels were calculated from the dimensions in Table 2. The moment of inertia, position of the neutral axis and section moduli were computed for each panel. In order to determine whether the use of average dimensions was justified in calculating these properties, the moment of inertia and position of the neutral axis was calculated cell by cell for some of the most irregular specimens. No appreciable differences, however, were found between the two methods of calculation. The dotted lines shown in the load-deflection and load-strain curves of the panels indicate the deflection calculated from these section properties.

The moment of inertia was also derived from the slope of the load-deflection curve below the proportional limit. The slope was determined by the method of least squares which takes here a particularly simple form on account of the equal increments of load.

Table 6 gives the computed values of the section properties and also the observed values of the moments of inertia taken from the load-deflection curves.

B. S. Journal of Research, RP463

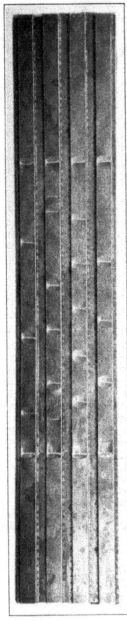

FIGURE 14.—*Top view of an Al₄ 16 panel after testing*

FIGURE 15.—*Side view of an A16 18 panel after testing*

B. S. Journal of Research, RP463

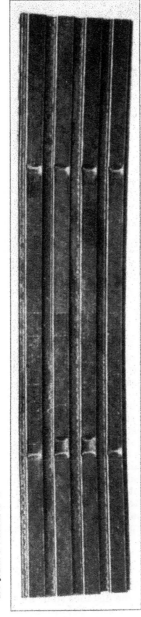

FIGURE 16.—*Top view of an A12-12 panel after testing*

FIGURE 17.—*Side view of a B16-16 panel after testing*

TABLE 6.—*Section properties of panels*

S_c = compressive section modulus for panel.
S_t = tensile section modulus for panel.
v_c = distance from neutral axis to extreme compression fiber.
v_t = distance from neutral axis to extreme tension fiber.

$$r_{calc.} = \frac{v_c}{v_c + v_t}.$$

I = moment of inertia of panel.

Panel	Type	S_c	S_t	v_c	v_t	r (calc.)	I (calc.)	I (obs.)
		*In.*³	*In.*³	*Inches*	*Inches*		*In.*⁴	*In.*⁴
I	A14-16	10. 1	6. 1	2. 13	3. 51	0. 385	21. 5	20. 5
II	A14-16	10. 4	6. 4	2. 15	3. 50	. 381	22. 4	21. 5
III	A14-16	10. 3	6. 3	2. 17	3. 52	. 381	22. 3	24. 5
IV	A16-18	8. 4	5. 1	2. 11	3. 52	. 374	17. 8	16. 9
V	A16-18	8. 2	4. 8	2. 08	3. 52	. 371	17. 0	16. 2
VI	A16-18	8. 7	5. 1	2. 09	3. 57	. 369	18. 1	17. 4
VII	C14-16	10. 5	6. 4	2. 15	3. 51	. 380	22. 5	20. 4
VIII	C14-16	10. 4	6. 4	2. 20	3. 54	. 383	22. 8	20. 8
IX	C14-16	10. 6	6. 5	2. 16	3. 52	. 380	23. 0	26. 3
X	B16-16	8. 7	4. 6	1. 51	2. 84	. 347	13. 2	13. 9
XI	B16-16	8. 5	4. 5	1. 50	2. 85	. 345	12. 7	13. 4
XII	B16-16	9. 2	4. 8	1. 49	2. 88	. 341	13. 7	13. 8
XIII	A14-16	10. 8	6. 6	2. 17	3. 53	. 381	23. 4	23. 4
XIV	A16-18	8. 5	5. 0	2. 10	3. 54	. 372	17. 8	17. 1
XV	B16-16	9. 0	4. 7	1. 48	2. 86	. 341	13. 4	13. 4
XVI	A14-14	12. 0	8. 1	2. 30	3. 41	. 403	27. 6	27. 0
XVII	A14-14	11. 8	8. 1	2. 32	3. 39	. 406	27. 3	25. 4
XVIII	A14-14	11. 6	8. 0	2. 33	3. 38	. 405	27. 1	26. 2
XIX	A12-12	15. 9	10. 6	2. 22	3. 31	. 402	35. 2	33. 7
XX	A12-12	16. 5	10. 4	2. 15	3. 42	. 386	35. 5	35. 9
XXI	A12-12	15. 9	10. 4	2. 22	3. 39	. 396	35. 4	36. 1
XXII	A14-16	11. 1	6. 6	2. 11	3. 53	. 374	23. 4	22. 4
XXIII	A14-16	11. 1	6. 6	2. 11	3. 55	. 373	23. 5	22. 9
XXIV	A14-16	11. 0	6. 6	2. 12	3. 54	. 375	23. 4	21. 8

Figures 18 to 21 show some typical load-strain curves for the four cells of various panels. The curves for Panel III, Figure 18, show the strains in the pan on the second loading, the dashed lines showing the calculated values of the strains are drawn through the set obtained at zero load after loading to the limiting deflection.

From the compressive and tensile strain data the neutral axis could be located at different loads on the assumption that plane sections remain plane. It was found desirable to express the location of the neutral axis in terms of a proportion of the total height of the cell, thus minimizing the effect of irregularities in the heights of the cells. As will appear later, the relation to load of the apparent position of the neutral axis given by the strain data is of value in determining the way in which the specimen begins to fail at mid-span and also in judging whether the various parts of the section are acting integrally. The location of the neutral axis was expressed in terms of a variable defined as follows:

$$r = \frac{e_c}{e_c + e_t}$$

where e_c and e_t are the compressive and tensile strains respectively. This r is the ratio to the total height of the distance from the neutral axis to the top of the cell. Even when plastic deformation has occurred, r continues to represent quite closely the position of the neutral axis [7] provided that the upper and lower elements composing the

---

[7] Bach and Baumann, Elastizität und Festigkeit, 9th ed., p. 266. Eugen Meyer, Berechnung der Durchbiegung von Stäben, deren Material den Hookeschen Gesetz nicht folgt. Zeit. des Verein Deutscher Ingenieure, p. 167, 1908.

cell act integrally. Values of r are determined very simply by graphical methods and from these values plots of r against load were made

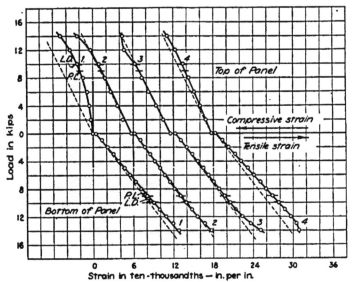

FIGURE 18.—*Load-strain curves for Panel III*

for each cell. For a sturdy beam acting according to the usual simple theory, these values of r should be identical with the values of r_{calc}. given in Table 6.

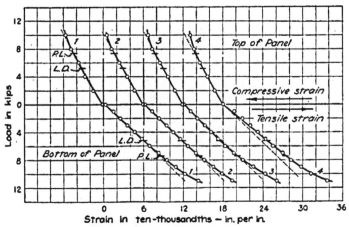

FIGURE 19.—*Load-strain curves for Panel XII*

Figures 22 to 24 give characteristic curves showing the values of r at the various loads. The values of r_{calc}. are indicated by the dashed lines.

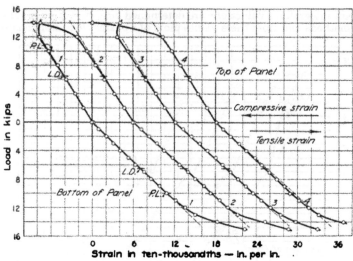

FIGURE 20.—*Load-strain curves for Panel XIV*

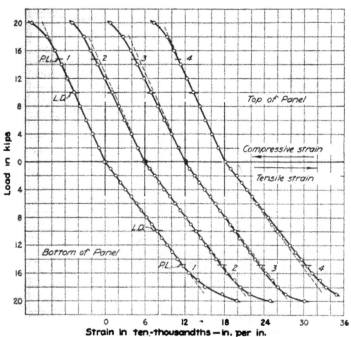

FIGURE 21.—*Load-strain curves for Panel XXIII*

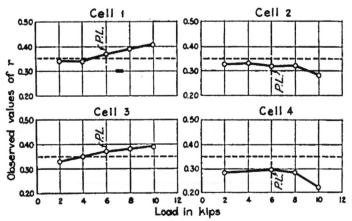

FIGURE 22.—*Values of r for Panel X*

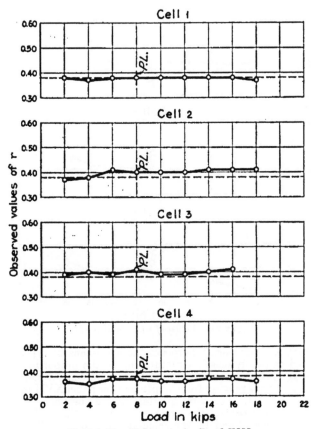

FIGURE 23.—*Values of r for Panel XIII*

Some typical curves showing the reduction in height of the cells over the supports as the load increases are given in Figures 25 to 27.

VIII. DISCUSSION OF THE TEST RESULTS OF THE PANELS

1. SPECIMENS OF TYPES A AND B

(a) ELASTIC BEHAVIOR

The close agreement shown in Table 6 between the calculated moment of inertia and that derived from the load-deflection data

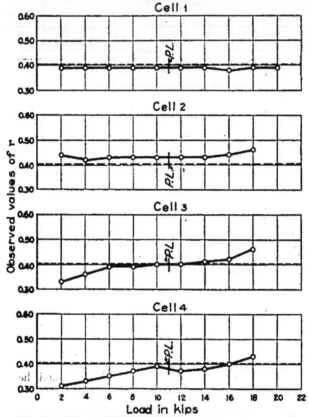

FIGURE 24.—*Values of r for Panel XVIII*

shows that for engineering purposes the specimens of types A and B behaved as purely elastic structures up to the proportional limit. The average ratio of observed to calculated moment of inertia is 98 per cent, the ratios ranging from 93 per cent for panel XVII to 110 per cent for Panel III. For three-quarters of the specimens the ratios were within 4 per cent of the average value. In the load-deflection curves of Figures 12 and 13, the calculated deflection will be seen to be close to the observed value, the observed value for all the panels being in general a little greater. The sets in deflection after unloading from the limiting deflection, given in Table 5, were negligible. They were due to readjustment of the specimen near the loads and supports and to local yielding. It is believed that additional applications of load would produce only slight changes in the values of the set. The creep in deflection observed after holding at constant load for one hour at the limiting deflection was negligible, the maximum value being 0.002 inch for Panel XI.

When some of the specimens formed by hand setting of the brake (panels I to XII) were placed on the supports, there were gaps of

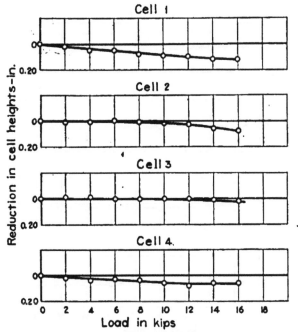

FIGURE 25.—*Reduction in cell heights, Panel I*

as much as a quarter of an inch between cell bottoms and the bearing pads. Furthermore, the loads were not distributed evenly over the cells, since the cell tops were not all quite in the same horizontal plane, though these inequalities were smaller than those observed for the cell bottoms at the supports. When the load was applied the horizontal webs bent, forcing the cell tops into the same plane under the load and tending to close up the gaps over the supports, but even at the limiting deflection the loads and reactions were not evenly distributed over the cells because of the differences in the heights of the cells. Because of this condition, the strain and deflection often increase more rapidly for the first increments of load than for succeeding increments. Examples of this may be seen in the load-deflection curves for Panels IV and X (figs. 12 and 13) and in the load-strain curves for Panel III (fig. 18).

On the other hand, the die-braked specimens (XII to XXIV) were much more uniform in dimensions and were flatter. For these panels the inequalities of the kind mentioned produced no appreciable effects on the deflections or strains.

The load-strain curves of typical panels given in Figures 18 to 21 show that the observed strain is in no case much greater than that computed by the conventional theory. The sets in the strain readings after loading to the limiting deflection were small, Panel III as a whole showing the largest values of these sets.

For all specimens of types A and B, except Panel III, Table 5 shows that the proportional limit was equal to or higher than the load

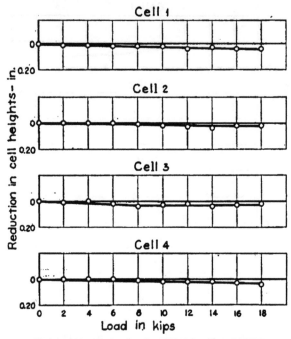

FIGURE 26.—*Reduction in cell heights, Panel XIII*

at the limiting deflection. Panel III was a product of the earlier manufacturing technique.

Buckling under the loading plates was present in all panels before the limiting deflection was reached and could be observed distinctly looking along the interior of the cell illuminated with a flash light. After removing the limiting deflection load, slight permanent buckles remained in about two-thirds of the panels which were examined for permanent buckles. In one case three of the eight areas of contact of the loading plates with the specimen were buckled, in another case two, and in the remaining five cases only one. It appeared that the uniformity of the die-braked panels resulted in less severe indentation under load and in fewer permanent buckles. For loads below the limiting deflection, the buckling under the loading plates

is almost entirely elastic. Under repeated loading the presence of
slight permanent buckles did not produce any measurable change in
the elastic properties.

(b) NEUTRAL AXIS

The position of the neutral axis as determined by the strain measurements at the middle of the specimens was usually in good agreement
with the values calculated from the dimensions. The close agreement of the observed and calculated moments of inertia indicates
also that the actual nuetral axis was close to its calculated position.
For some cells, however, the observed values of r do not check well

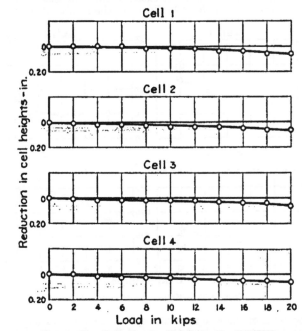

FIGURE 27.—*Reduction in cell heights, Panel XXIV*

with the calculated values. An extreme example of this may be
seen in Figure 28, showing the values of r for Panel IV. This was
due to the top and bottom of the cells not acting integrally in portions
of the specimen near the gage lines.

The integral action of the upper and lower elements depends
upon shearing forces being transmitted by the welds. When welds
fail under load, or if the welding should fail to join portions of the
web, there is a lack of integral action between the top and bottom
of the panel. What happens then may best be visualized by considering the action of the top and bottom elements when placed together
as in the panel, but unconnected by any welding. The bending
moment then is distributed between the two elements in proportion
to their respective moments of inertia. Since the moment of inertia
of the bottom element is much greater than that of the top, most

of the bending moment will be carried by the lower element. It follows that for a given bending moment on a panel the compressive stress at the top of an unwelded panel is less, while the tensile stress at the bottom is greater than if the panel had been welded. Strain-gage readings on the upper and lower surfaces of such a panel would indicate the position of the neutral axis to be above that for a welded panel. The values of r observed in such a case would be less than those calculated on the assumption of integral action of the upper and lower elements of the panel.

Actual cases of incomplete integration in the webs are intermediate between the fully welded and the unwelded case described. The values of r in Figure 28 show an extreme effect of this kind. Cells 3 and 4 of Panel XVIII (fig. 24) illustrate the results of poorly inte-

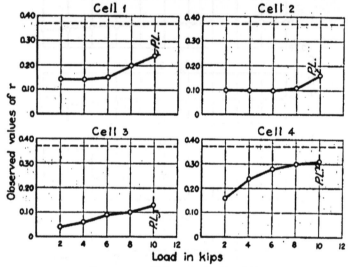

FIGURE 28.—*Values of* r *for Panel IV*

grated portions of the web near mid-span. As the load increased on the panel, the intact welds at the end of the defective portion began to take up the shearing forces that otherwise would have been trans-mitted by that portion. This tightening-up process brings the value of r nearer the calculated value as the load increases, as can be seen in the curves. The compressive-strain curve for the first cell of Panel III (fig. 18) illustrates the effect of defective integration on the strains. The tensile-strain curve for this cell shows an increase over the calculated value, but the percentage change is not as great as for the compressive strain. In general, it may be said that defective integration in the web reduced the compressive stresses in the upper cells to a greater degree than it increased the tensile stresses in the lower cells. For Panel IV the compressive stresses observed at the proportional limit were markedly lower than the computed values, while the tensile stresses observed were not much greater than the computed values, the maximum value of the observed stress being

33.6 kips/in.² compared with a computed stress of 31.2 kips/in.². At the limiting deflection the maximum excess of observed over computed tensile stress amounted to 4.3 kips/in.², or 18 per cent, for the hand-set Panel III, as can be seen in Figure 18 (cell 4). For the more accurately dimensioned panels formed over dies and welded automatically, the maximum excess was 1.9 kips/in.², or 9 per cent for Panel XIX.

(c) WELDS

The shearing forces on the welds may be computed from the formula

$$s = \frac{3}{16}\frac{SQ}{I}$$

where

$s =$ the shearing force in kips on a single weld,
$S =$ the total vertical shear in kips per panel,
$Q =$ the static moment in in.³ about the neutral axis of that portion of the section lying above the welds.
$I =$ the moment of inertia in in.⁴ of a single panel.

The coefficient three-sixteenths inch takes account of the spacing of the welds. For the panels tested, s ranged from $0.032S$ to $0.036S$. At the proportional limit, the panels with the exception of those of the A12–12 type gave a minimum factor of safety of the welds in shear of 10, based on the conservative strength of two kips per weld that was recommended in Section V. The A12–12 specimens had a minimum factor of safety of 5.5. The failure of welds below the proportional limit, which is indicated in some cases by the decrease of r with load, can not, therefore, be ascribed to shear alone. As the neutral axis was below the horizontal web for all the panels, it is probable that under the compressive stresses the sheets composing the web separated by buckling, causing tension in the welds in addition to shearing forces.

Certain specimens showed poor joining of the webs as indicated by the change of r with load, this being true of many of the first group of panels received (I to XII) and also of the A12–12 panels, XIX to XXI. No failure could be ascribed to defective welds alone, although undoubtedly the increased tensile stress produced yielding in the panels at earlier loads. Any significant effects of poor or inadequate welding would be shown by a lowering of the proportional limit.

(d) CELL HEIGHTS AT THE SUPPORTS

The pressure of the reactions reduced the height of the cells over the supports as shown by Figures 25, 26, and 27. This reduction of the height was shown to be almost entirely elastic up to the limiting deflection load by the absence of appreciable permanent changes in the cell heights. Except for Panel III, the maximum observed permanent reduction, after applying the limiting deflection load, was 0.03 inch for one cell of Panel XXII. Two cells in Panel III decreased 0.05 inch in height, one cell 0.04 inch, and three cells 0.03 inch. This anomalous behavior of Panel III is due apparently to irregularity in sizes of the cells, this panel having been formed on a hand-set brake. The average permanent reduction in cell heights after unloading from the limiting deflection was 0.007 inch.

The maximum reductions in cell height observed at the limiting deflection were 0.08 inch for one cell of Panel I, 0.07 inch for one cell of Panel III, and 0.05 inch for one cell of Panel XVI. The high values for Panel I and III are due to the irregularity of dimensions characteristic of the panels formed on the hand-set brake. Values for the other panels were all less than 0.05 inch and averaged 0.02 inch. The reduction in cell height for die-braked panels is there-' fore negligible at the limiting deflection load.

(c) FAILURE OF THE PANELS

As the load was increased above the proportional limit, buckles appeared in the cell tops between the loads and increased in size and number up to the maximum load. At the same time the indentations in the cell tops under the loading plates increased slowly in depth and in a few panels buckles appeared in the upper portion of the side walls of the cells. Failure in the upper portion of the specimen appeared to take place entirely by buckling, though the A12–12 specimens (see fig. 16) showed strain lines on the cell tops running about 45° to the axis of the panel, which indicated that the yield point in compression had been reached.

In no cases did any crumpling or buckling start from irregularities of shape or from dents which had occurred in manufacture or shipment. Even in the panels having the thinnest bottom sheets (18 gage), the reinforcing at the ends of these panels inhibited completely any crumpling of the cell bottoms over the supports.

For most of the specimens sharp cracks were heard as the maximum load was approached and occasionally at loads below the proportional limit. It is believed that these sounds were due to the fracturing of welds. Many weld fractures, however, were undoubtedly inaudible.

Departure from elastic behavior in the specimens may be due to the following causes: (1) Yielding of the cell bottoms in tension, (2) buckling of the cell tops in compression, (3) failure of welds, and (4) effect of the concentrated loads.

The first three points will be discussed in the light of the observations at mid-span, 30 inches from the loads.

The stress at which compression failure by buckling began to occur is dependent on the yield point of the material, since the elastic buckling is negligible with respect to the inelastic. For 14 and 16 gage sheet having a yield point of 20 to 25 kips/in.^{2}, inelastic buckling seemed to set in at a stress of around 12 kips/in.^{2} for specimens of type A. The 14-gage material with a yield point of 41 to 46 kips/in.^{2} in type A buckled at a stress of about 22 kips/in.^{2} For type B, the buckling stress appeared to be a little lower with respect to the yield point, 16-gage material with a yield point of 21 kips/in.^{2} beginning to show marked buckling at stresses of 10 to 12 kips/in.^{2}

The dimensions of panels of the A14–16 and A16–18 types were such that simultaneous compression and tension failures occurred at mid-span provided that the observed values of r were in fair agreement with the computed values. The compression failures in the A16–18 panels, once they started, progressed more rapidly, however, than in the heavier top sheets of the A14–16 type. When the observed values of r agree with the computed values, the ratio of compressive to tensile stress in these two types is about 0.60. When this ratio becomes a little greater, as the A14–14 and A12–12 types, for which it is about 0.67, failure tends to begin by buckling and then develops by combined buckling and yielding in tension.

In the type B panels the restraint offered to the buckling by the webs caused tensions in the welds which, in conjunction with the shearing forces, were sometimes enough to cause failure of the welds when buckling was well advanced. In Figure 29 is shown a B16–16 specimen in which the top sheet separated from the bottom sheet at a load near the maximum, thus permitting the buckles to spread from one cell top to another across the full width of the panel.

The effects of the welding on the failures have already been discussed.

The effect of the concentration of the loads on the proportional limits of the panels may best be judged by comparing the stresses at mid-span at the proportional limit with the yield point of the bottom sheets given either by the tensile tests or by the load-strain curves. On the average, the proportional limit occurs at a mid-span stress of around 0.7 of the yield poin , though some specimens had proportional limits which develop the yield point stresses at the center (V, XXII, and XXIII). A closer examination of the results on this point is of little use, since in many cases materials of widely different yield points were used for the bottom sheets.

The local effects produced by the concentrated loads may be three, as follows: (1) Indentation and buckling of the cell tops under the loading plates; (2) buckling of the web producing weld failures near the loads and consequently increased tensile stress in the cell bottom; and (3) possibly an increase in stress under the concentrated loads considerably above that calculated from the conventional theory. Such stresses have been predicted by Schnadel.[8] For box girders with a length of 6.3 times the width and loaded at the quarter points, he has calculated the maximum longitudinal stress in the top and bottom plates at the quarter points to be 70 per cent in excess of that at mid-span.

The disappearance of buckles under the loading plates on unloading from the limiting deflection and the fact that there was no evidence of reduction in the proportional limit on repeated loading justifies the conclusion that indentation of the cell tops by the concentrated loads is of small importance up to the proportional limit. It would be reasonable to expect, however, that the indentations would lower the compressive resistance of the panel at these points. The effect predicted by Schnadel should be verified by some independent means, but must be accepted as a possibility. In any case, it may be concluded that under uniform loading and on similar spans the panels will behave satisfactorily over a range at least equal to that given by the proportional limit of the present tests.

Had the dimensions of the first 12 specimens been uniform and had the material and welding been likewise more uniform, the results would, undoubtedly, have not shown so much scatter. The tests have shown that the three stages of manufacturing technique have resulted in panels progressively better in strength properties and in uniformity of dimensions.

---

[8] Georg Schnadel, Die mittragende Breite in Kastenträgern und in Doppelboden, Werft-Reederei-Hafen, Mar. 7, 1928, Heft 5, p. 92.

B. S. Journal of Research, RP463

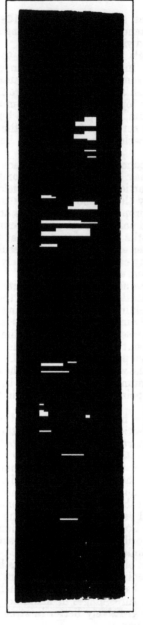

FIGURE 29.—*Top view of a B18-16 panel after testing*

2. SPECIMENS OF TYPE C

The agreement between the observed and computed moments of inertia in Table 6 is poor for the specimens of type C. No reason was found for the high moment of inertia observed for Panel IX. The proportional limits (Table 5) were low with respect to the limiting deflection loads and the maximum loads were lower than for any specimens of the A14–16 type. The panels failed from an inherent defect in design, the sheets composing the webs splitting apart at the ends due to combined tension and shear on the welds. The tension on the welds was produced by the panel being supported at the ends only by the projecting top element.

IX. CONCLUSIONS

The loading tests on sheet steel floor panels having longitudinal stiffening cells and either flat or corrugated top surfaces justify the following conclusions for types A and B (fig. 1).

1. The elastic range of the panels was equal to or in some cases in excess of the maximum working range set by the usual deflection requirements (deflection less than one three hundred and sixtieth of the span) in 10-foot floor panels. This elastic range is the fundamental criterion of the usefulness of the panels.

2. The maximum load carried by these panels showed a considerable and satisfactory margin of safety against overloading.

3. The method in common use for designing beams affords a satisfactory basis for predicting the elastic behavior of these floor sections. The stiffness of the panels calculated from average dimensions was in excellent agreement with that calculated from the deflection. The distribution of longitudinal stress in the die-braked panels can be satisfactorily predicted by the usual methods of design if an allowance of 10 per cent is made to cover the possible irregularities in the distribution of tensile stress.

4. The spot welds used in joining the sheets were amply strong in shear, but some may have failed by buckling apart of the two sheets in the web between the welds. These failures, however, had no appreciable effect on the behavior of the panels within the elastic range.

5. The location of the neutral axis is in accord with the calculated location except where imperfect integration by the welds of the top and bottom elements of the section may lead to a displacement. It is therefore desirable that controlled automatic welding be used in the manufacture of such panels.

6. The thin-walled sections showed a considerable stability against secondary failure, even under concentrated loads.

Panels of type C (fig. 1) showed an inherent defect in design, failure being due to tension on the end welds of the web, there being no provision to strengthen the ends of the webs to withstand these forces.

WASHINGTON, June 4, 1932.

127984—32——4

THE INFLUENCE OF TEMPERATURE ON THE EVOLUTION OF HYDROGEN SULPHIDE FROM VULCANIZED RUBBER

By A. D. Cummings

ABSTRACT

Evolution of hydrogen sulphide from vulcanized rubber containing 8 to 32 per cent sulphur has been measured when the temperature of the specimens was raised step by step from 105° to 265° C., and the time intervals kept equal. The same sample of each compound was employed throughout the whole temperature range. The rate of decomposition increases as the temperature is raised, except in the case of the compounds containing the higher percentages of sulphur, when it passes through a maximum.

The loss of hydrogen sulphide has also been determined for four different compounds containing 4, 10, 18, and 32 per cent sulphur when samples were heated for 200 hours at 136° C., and also when other samples of these compounds were maintained for the same length of time at 220° C. In each case, the rate of decomposition decreases rapidly at first, but after several days' heating, the decrease becomes relatively slow. In general, the rate increases with temperature and with increasing sulphur content. Other products evolved are moisture and organic compounds.

The data presented in this paper supplement and confirm previous results on the evolution of hydrogen sulphide from vulcanized rubber and indicate at what temperatures direct thermal decomposition of different rubber-sulphur compounds will become significant. No mechanism to explain the chemical changes involved is suggested. This question is worthy of further investigation.

CONTENTS

I. INTRODUCTION

This paper presents the results of measurements of the evolution of hydrogen sulphide from rubber-sulphur compounds heated at various temperatures. Determination of the amount of hydrogen sulphide produced under different conditions was used to measure the degree of decomposition of the rubber. Two types of experiments were carried out: (1) Samples of vulcanized rubber containing 8 to 32 per cent sulphur were heated for 8-hour intervals at 13 temperatures between 105° and 265° C., and (2) samples having sulphur contents of 4, 10, 18, and 32 per cent were maintained for about 200 hours at constant temperature, one set of specimens at 136° and another group at 220° C.

These measurements of deterioration were undertaken in connection with an investigation on the electrical properties of vulcanized rubber at relatively high temperatures. During these experiments, samples of rubber-sulphur compounds had been subjected to a wide range of temperatures. Time of exposure to each temperature had been about eight hours. The purpose of the present work was to determine when the sulphur content of a specimen had changed sufficiently to affect its dielectric constant and power factor by a measurable amount. In order to approximate the conditions under which the electrical tests were made, it was necessary to determine the amount of decomposition when rubber vulcanized with 8 to 32 per cent sulphur was heated for successive intervals of eight hours each at temperatures changed in unequal steps from 105° to 265° C. To make this information more complete and to obtain additional data which could be compared with previous investigations, the work was extended to include determinations of the loss of hydrogen sulphide from vulcanized rubber heated for a long time at constant temperature. The electrical properties of the whole series of rubber-sulphur compounds is the subject of a separate investigation at this bureau, and will be reported in another paper.

Many investigators have shown that hydrogen sulphide is associated with the vulcanization of rubber and with its subsequent deterioration during aging. Stevens and Stevens have noted the evolution of hydrogen sulphide during the vulcanization of ebonite at temperatures above 70° C. Webster, Fry, and Porritt have shown that ebonite evolves hydrogen sulphide at ordinary temperatures as well as when heated and have measured the rate of decomposition at several temperatures. Wolesensky has found that both soft and hard rubber lose hydrogen sulphide at all temperatures above 25° C. For more details of previous investigations on this problem during the last few years, the reader is referred to the papers by the authors mentioned.[1] A review of the literature up to 1929 is given in the paper by Wolesensky.

The results described in this paper represent a single set of measurements of the evolution of hydrogen sulphide when compounds of purified rubber and sulphur were heated under different conditions of time and temperature. Possible effects or variations which might be encountered with rubber samples made at different times or prepared in different forms for exposure to the heat, or with other changes in experimental conditions were not investigated. The results show the loss of hydrogen sulphide when rubber-sulphur compounds representative of the whole series from soft to hard rubber were heated step by step from 105° to 265° C., and also when heated for 200 hours at 136° and at 220° C.

[1] Edward Wolesensky, B. S. Jour. Research, vol. 4, p. 501, 1930; Rubber Chem. Tech., vol. 3, p. 386, 1930. J. D. Fry and B. D. Porritt, India Rubber J., vol. 78, p. 307, 1929. D. M. Webster and B. D. Porritt, India Rubber J., vol. 79, p. 239, 1930; Rubber Chem. Tech., vol. 3, p. 618, 1930. H. P. Stevens and W. H. Stevens, J. Soc. Chem. Ind., vol. 50, p. 397T, 1931.

II. PREPARATION OF SPECIMENS

The rubber specimens were prepared from protein-free rubber hydrocarbon and sulphur. The rubber hydrocarbon was obtained from latex by digestion with hot water and extraction with water and alcohol, in accordance with the method described by McPherson,[2] which gives a purified material containing about 99.5 per cent hydrocarbon. A master batch of rubber and sulphur was made and portions of this were blended with fresh rubber hydrocarbon to give any sulphur content desired. The specimens were prepared by pressing samples of these rubber-sulphur mixtures between thin aluminum plates separated by a spacer 1.25 mm in thickness. The specimens were then placed in an autoclave and vulcanized for 40 hours at 141° C. under pressure of carbon dioxide. The long period of vulcanization

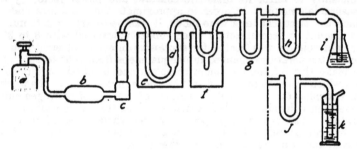

FIGURE 1.—*Apparatus used for measurement of decomposition of vulcanized rubber*

a, cylinder of commercial hydrogen to provide an inert atmosphere and sweep out products evolved while the rubber is heated.

b, concentrated alkaline pyrogallol solution on glass wool to absorb traces of oxygen.

c, tower containing calcium chloride and magnesium perchlorate to dry the gas stream.

d, glass tube containing the rubber sample cut into strips.

e, oil bath, electrically heated and thermostatically controlled.

f, trap in ice-salt freezing mixture to collect any volatile liquids.

g, U tube containing magnesium perchlorate to absorb moisture given off from the rubber.

h, U tube containing saturated potassium hydroxide solution on glass wool to absorb hydrogen sulphide, protected at each end by drying agent to prevent loss of moisture from the tube.

i, protective drying tube and bubble counter.

NOTE.—In some cases lead acetate solution was used as an absorber for hydrogen sulphide. This is indicated by the gas-washing bottle, *k,* preceded by the drying tube, *j,* to prevent back diffusion of moisture into the weighed drying tube, *g.*

resulte'd in a product which contained practically no free sulphur in the soft and medium hard rubber ranges. In the case of pure hard rubber, analyses on different samples of stock containing 32 per cent sulphur showed from 0.5 to 0.8 per cent free sulphur. In the present experiments, the 32 per cent sample was the only one to show a deposit of sublimed sulphur.

III. APPARATUS AND PROCEDURE FOR DETERMINING THE HYDROGEN SULPHIDE DISSOLVED

The essential features of the apparatus were a supply of inert gas, a tube to hold the rubber sample, a constant temperature bath, and an absorber for the hydrogen sulphide. The individual parts of the apparatus are indicated and described in Figure 1.

[2] A. T. McPherson, A Method for the Purification of Rubber and Properties of the Purified Rubber B. S. Jour. Research, vol. 8 (RP449), p. 751, 1932.

The general method chosen for measuring the deterioration of vulcanized rubber when heated was determination of the amount of hydrogen sulphide evolved. This determination was carried out in two ways: (1) the hydrogen sulphide was absorbed in lead acetate solution and the lead sulphide collected and weighed, and (2) the hydrogen sulphide was absorbed by saturated potassium hydroxide solution contained in a tube which could be weighed directly. Both methods were equally satisfactory under appropriate conditions. Absorption in lead acetate is preferable when a very small amount of hydrogen sulphide is to be measured, but when much lead sulphide is formed this method becomes exceedingly time-consuming on account of slow filtration.

1. MEASUREMENT OF THE EVOLUTION OF HYDROGEN SULPHIDE DURING STEP-BY-STEP HEATING

The decomposition in this part of the work was carried out with a simplified apparatus consisting of gas supply, sample tube and heater, trap, and hydrogen sulphide absorber. In order to duplicate more nearly the conditions during the electrical test when the inert gas was not purified, the purifying train was omitted. The drying tube, *g*, was not used because it was not intended to account for all products formed. Only sections *a*, *d*, *e*, *f*, and *k* of the apparatus were employed.

The tubes containing 5-gram samples of rubber in strips cut from the specimens previously described were placed in the oil bath and the stream of hydrogen [3] started. The bath was heated from room temperature up to 105° C. where it was held for eight hours. The absorber containing lead acetate solution was then changed, provided any precipitated lead sulphide was visible, and the temperature was raised to 115°. The lead sulphide formed was collected by filtration through a Gooch crucible, washed, dried, and weighed. This procedure was repeated step by step every eight hours at 13 temperatures between 105° and 265°, with samples containing from 8 to 32 per cent sulphur. All samples were heated for eight hours at each temperature before being raised to the next higher temperature. It was already known that the electrical properties of rubber containing less than 8 per cent sulphur were not changed a significant amount by loss of sulphur when heated under these conditions, consequently the experiments were begun with the 8 per cent compound. Temperatures below 100° did not cause a change in sulphur content sufficient to be detected in the electrical measurements on a specimen containing the maximum amount, 32 per cent, of sulphur, therefore 105° was selected as the initial temperature.

2. MEASUREMENT OF THE EVOLUTION OF HYDROGEN SULPHIDE AT CONSTANT TEMPERATURE

The apparatus was used complete as described for this part of the investigation, and both methods for the absorption of the hydrogen sulphide were employed. The specimen tubes containing 20 to 50 g

---

[3] No significant error was introduced by reaction of the hydrogen with any small amount of free sulphur in the rubber. The rate of reaction of hydrogen with sulphur is relatively slow at the highest temperature used in the present experiments. (See J. W. Mellor, A Comprehensive Treatise on Inorganic and Theoretical Chemistry, vol 10, pp. 117 ff., 1930.) Furthermore, an experiment in which steam was used as the inert gas gave substantially the same rate of evolution of hydrogen sulphide as was obtained with hydrogen (unpublished work of A. T. McPherson).

of rubber were set up at room temperature outside the oil bath. While the bath was being heated to the temperature desired, the apparatus was swept with hydrogen to displace air. The tubes for absorption of moisture and of hydrogen sulphide, when potassium hydroxide was used, were removed, weighed, and then replaced. The tubes containing the rubber samples were next immersed in the hot oil, and the hydrogen stream was adjusted to about one bubble per second. The hydrogen sulphide was measured at intervals by weighing the U-tube absorbers directly or by determining the amount of lead sulphide formed. The amount of moisture and volatile oils was determined only at the end of the experiment. This served to account for the entire loss in weight of the samples. The loss of hydrogen sulphide was determined for four rubber compounds containing 4, 10, 18, and 32 per cent sulphur when samples were heated for about 200 hours at 136° C., and also when other samples were heated at 220° C. for the same length of time. The temperature of the oil bath was held within ± 1° at the former temperature and ± 3° at the latter.

IV. RESULTS OF DECOMPOSITION DURING STEP-BY-STEP HEATING

The evolution of hydrogen sulphide from vulcanized rubber containing 8 to 32 per cent sulphur when heated in the 8-hour stepwise manner increases with rising temperature, with the exception that the loss from the compounds containing the higher percentages of sulphur passes through a maximum. In view of the large change in sulphur content during heating, this is not surprising. The amount of hydrogen sulphide evolved in an 8-hour interval at each temperature is shown in Figure 2.

The change in sulphur content calculated from the loss of sulphur as hydrogen sulphide between 105° and each higher temperature is given in Figure 3. These values for the sulphur content are not identical with those which would be determined by analysis of a sample of rubber taken at each temperature, because there was some loss in weight of the sample due to distillation of volatile liquids. However, the error is not great. This is indicated by the agreement between the final sulphur content calculated from the data of the present experiments and that obtained by analysis of samples used in the electrical measurements mentioned previously. This comparison, printed in Table 1, shows that the heating in the present investigation had satisfactorily paralleled the conditions during the electrical test and that the procedure and calculations were adequate to furnish the information desired.

V. RESULTS OF DECOMPOSITION AT CONSTANT TEMPERATURE

In addition to determination of hydrogen sulphide, measurements were made of the amounts of other decomposition products and moisture which resulted from heating vulcanized rubber containing 4, 10, 18, and 32 per cent sulphur at 136° and 220° C. for 200 hours.

1. RATE OF LOSS OF HYDROGEN SULPHIDE

The results show the effects of time, temperature, and composition of samples on the rate of evolution of hydrogen sulphide.

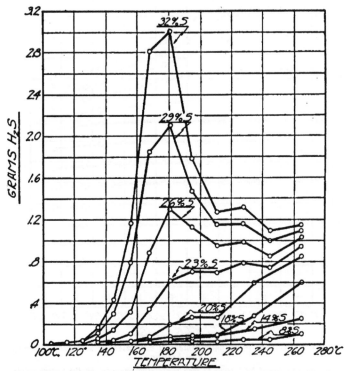

FIGURE 2.—*Evolution of hydrogen sulphide from vulcanized rubber during step-by-step heating*

Grams of hydrogen sulphide liberated from 100 g vulcanized rubber heated for eight hours at each temperature indicated. Every sample was started at 105° and the same sample was carried through to 265°.

(a) EFFECT OF TIME

For a given composition at each temperature, the rate of loss of hydrogen sulphide, high at first, decreases rapidly during the first few days' heating, but subsequently the decrease becomes slow.

(b) EFFECT OF TEMPERATURE

For a given composition and period of time, the rate increases with rising temperature.

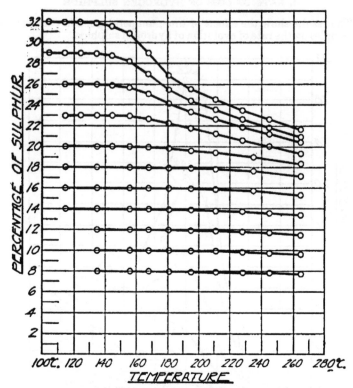

FIGURE 3.—*Change in sulphur content of vulcanized rubber during step-by-step heating*

Samples were heated for 8 hours at each temperature indicated by the various points. Every sample was started at 105° and the same sample was carried through to 265°.

TABLE 1.—*Analysis of electrical test specimens for sulphur compared with sulphur content calculated from the present experiments*

Original sulphur	Sulphur after heating calculated from present measurements	Total sulphur in specimens after high temperature electrical test
Per cent	Per cent	Per cent
32	21.6	22.0
29	20.9	21.5
26	20.4	20.5
23	19.3	19.6
20	18.3	18.1
18	17.0	16.8
16	15.4	15.2
14	13.4	13.4
12	11.4	11.1
10	9.6	9.5
8	7.7	7.8

(c) EFFECT OF COMPOSITION OF SAMPLES

For a given temperature at equal times, the rate increases as the sulphur content of the rubber becomes greater, with one exception, namely, that there is a noticeable similarity between the 10 and 18 per cent compounds in loss of hydrogen sulphide at 136°. This observation was satisfactorily repeated.

The rate of evolution of hydrogen sulphide from the samples investigated is shown in Figure 4. These curves are started at the point of

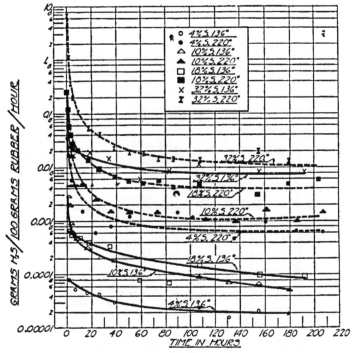

FIGURE 4.—*Rate of evolution of hydrogen sulphide from vulcanized rubber heated at 136° and 220° C., plotted on a semilogarithmic scale*

maximum evolution, an initial increase in rate being disregarded because it must be due only to the heating up of the sample and the approach to a quasi-steady state throughout the apparatus. The percentage of the original weight lost as hydrogen sulphide plotted against the time is given in Figure 5.

2. OTHER DECOMPOSITION PRODUCTS AND MOISTURE

Along with the hydrogen sulphide, volatile oils having slight terpenelike odors were formed. At 220°, these oils made up 10 to 40 per cent of the total loss in weight of the samples during 200 hours' heating. There was also a trace of some compound having a faint onionlike odor which was never condensed or absorbed by potassium hydroxide or lead acetate. No attempt was made to identify this or any of the other organic products. Some moisture[4] was evolved from the rubber and was measured at the end of each group of experiments in order to be able to account for the entire loss in weight of the samples. The percentage of the original weight obtained in the products from a typical experiment with a 32 per cent compound at 136° is given in Table 2, which shows that the total loss was satisfactorily accounted for.

VI. CHARACTERISTICS OF THE RUBBER AFTER HEATING

The specimens all changed materially in physical properties during the exposure to heat. Compounds which had initially contained up

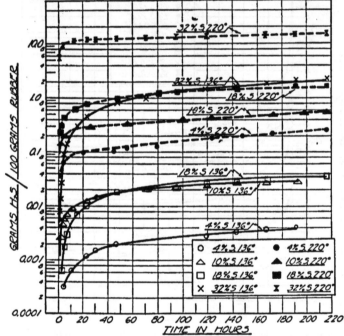

FIGURE 5.—*Percentage loss of hydrogen sulphide at 136° and 220° C., plotted on a semilogarithmic scale*

to 26 per cent sulphur became translucent and showed a reddish color when samples 1 mm in thickness were examined in sunlight, whereas all above 12 per cent of sulphur were originally opaque. At the end of the heating, the intermediate and hard rubbers were brittle at room temperature, but became flexible and tacky when warmed. A similar change in properties, less in degree, was evident in the samples of soft rubber. The conversion of vulcanized rubber into thermoplastic materials has been noted by Kemp,[5] who heated rubber mixed with

[4] The moisture may have been an impurity or may have come from the thermal decomposition of a small amount of oxidized material in the rubber. For a discussion of the products formed by oxidation of rubber, see Oxidation Studies of Rubber, Gutta-Percha, and Balata Hydrocarbons by A. R. Kemp, W. S. Bishop, and P. A. Lassalle, Ind. Eng. Chem., vol. 23, p. 1444, 1931.

[5] A. R. Kemp, U. S. Patent 1638335, Aug. 9, 1927. For a general review of the field of thermoplastic products made from rubber, including patent references, the reader is directed to the Chemistry of Rubber, by Harry L. Fisher, Chem. Rev., vol. 7, No. 1, pp. 94, 112–123, March, 1930.

8 to 16 per cent sulphur to 200° to 280° C., and the manufacture of such products from raw rubber is now a commercial process.

TABLE 2.—*Percentage of original sample accounted for after a 32 per cent compound was heated 200 hours at 136° C.*

	Per cent
Residual rubber	95. 55
Oils	. 01
Sulphur	. 02
Moisture	. 18
Hydrogen sulphide	2. 17
Total	99. 93

VII. DISCUSSION

The course followed by the loss of hydrogen sulphide when vulcanized rubber is heated is well established and agrees with the results

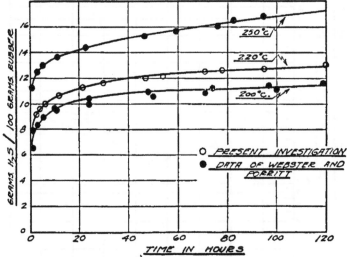

FIGURE 6.—*Comparison between present data and those of Webster and Porritt.* (See footnote 1, p. 164.)

Loss of hydrogen sulphide from hard rubber at 200°, 220°, and 250° C.

of previous investigations so far as can be told from consideration of the data taken under different conditions. The curve for evolution of hydrogen sulphide from the 32 per cent compound at 220° lies between similar curves published by Webster and Porritt [6] for ebonite heated at 200° and 250° C. This comparison is shown in Figure 6.

Calculation of the ratio of sulphur lost at the end of 190 hours to the initial sulphur content shows that the percentage of the original sulphur lost at 136° is greater for the compound containing 10 per cent sulphur than for that containing 18 per cent, and at 220° is greater for the compound containing 4 per cent sulphur than for that having 10 per cent. On the other hand, the weight of sulphur actually lost per gram of rubber compound was nearly equal for the

---

[6] See footnote 1, p. 164.

10 and 18 per cent compounds at 136°, and was less for the 4 per cent sample than for that containing 10 per cent sulphur at 220°. These calculations are given in Table 3 and the proportionate loss of sulphur is plotted in Figure 7. The similarity in rate and in total percentage of loss at 136° for the samples containing 10 and 18 per cent sulphur led to a repetition of this experiment over the first 80 hours using

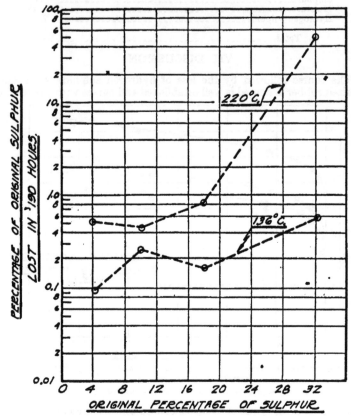

FIGURE 7.—*Percentage of original sulphur lost in 190 hours at 136° and 220° C., plotted on a semilogarithmic scale as a function of initial sulphur content*

other specimens containing the same percentages of sulphur. This determination checked the original observations satisfactorily. Also, an inspection of the data from the earlier experiments showed a distinct similarity in rate of evolution from compounds containing 10 to 18 per cent sulphur at the lower end of the temperature range. These observations can be interpreted as indicating that there may be a range of compositions, varying with the temperature, where rates of decomposition are similar, and where the proportionate loss of sulphur may grow less while the original sulphur content of the specimens increases. There may be a fairly sharp break at the end of this range, above which the loss of hydrogen sulphide may become much more rapid and follow the course expected for the order of ascending sulphur content. The data of the present experiments do not show whether the cases at hand are part of a general phenomenon or only isolated instances.

TABLE 3.—*Comparison between loss of hydrogen sulphide and proportionate loss of sulphur*

Original sulphur	H_2S lost at end of 190 hours as per cent of original compound		Sulphur lost in 190 hours as per cent of original sulphur	
	136°	220°	136°	220°
Per cent				
4	0.004	0.23	0.094	5.29
10	.029	.51	.26	4.88
18	.033	1.56	.17	8.15
32	1.96	13.92	5.77	40.93

The chemical nature of the changes in the rubber molecule during thermal decomposition has not been thoroughly explained, although change in unsaturation as a result of loss of hydrogen sulphide and the effect of heat on vulcanized rubber has been commented upon by Winkelmann,[7] and by the other investigators previously mentioned.[8] A few determinations of unsaturation made on the rubber after heating in the present experiments have not given any additional insight into the mechanism of the chemical changes involved. This phase of the problem may be investigated further at some future time.

Since this manuscript was prepared, a paper on the pyrolysis products of ebonite by Midgley, Henne, and Shepard has been published.[9]

The author wishes to express his thanks to A. H. Scott for assistance with some of the measurements reported in this paper.

WASHINGTON, June 1, 1932.

[7] H. A. Winkelmann, Ind. Eng. Chem., vol. 18, p. 1163, 1926.
[8] See footnote 1, p. 164, and footnote 5, p. 171.
[9] T. Midgley, Jr., A. L. Henne, and A. F. Shepard, J. Am. Chem. Soc., vol. 54, p. 2953, 1932.

RP465

SOME OF THE FACTORS WHICH AFFECT THE MEASURE-MENT OF SOUND ABSORPTION

By V. L. Chrisler and Catherine E. Miller

ABSTRACT

It has been found that air has an appreciable absorption for sound at frequencies as low as 512 cycles per second. This absorption varies with the temperature, the moisture content, and the barometric pressure. Curves are given showing such changes in absorption in the reverberation room at the Bureau of Standards.

Attention is called to the fact that when a highly absorbent sample is placed in a very reverberant room the decay curve may not be logarithmic.

Measurements of sound absorption on the same samples made by different observers in different reverberation rooms have in the past shown variations difficult to explain. Some of the factors responsible for these variations are now beginning to be recognized and understood. Humidity, temperature, and even barometric pressure have been found to cause quite perceptible changes in sound absorption measurements. P. E. Sabine [1] first called attention to the fact that decreased humidity materially increased the absorption of an empty reverberation room for frequencies above 2,000 cycles. Later Knudsen [2] described some pioneer work on this effect and determined coefficients for the sound absorption of air containing varying amounts of water vapor. Additional experimental work shows that not only humidity, but also temperature and pressure cause changes in the sound absorption of air.

For the most part the effect of these factors (in rooms of 10,000 to 20,000 cubic feet) is perceptible only at the higher frequencies of 1,000 cycles or more, but under favorable conditions this effect can be measured at 512 cycles. In large rooms of 1,000,000 cubic feet or more this effect should be noticeable at all frequencies.

These hitherto unexplained changes were noticed by us over two years ago and records of temperature, humidity, and pressure have been kept for the last year in the hope that a correlation would become evident. Knudsen's paper gave the first suggestion as to how the data might be interpreted. He determined the reverberation time of two rooms which had the same lining material but different volumes. The temperature in these two rooms was maintained approximately constant, but the relative humidity varied.

[1] P. E. Sabine, The Measurement of Sound Absorption Coefficients, J. Franklin Inst., vol. 207, p. 347.
[2] V. O. Knudsen, The Effect of Humidity upon the Absorption of Sound in a Room, and a Determination of the Coefficients of Absorption of Sound in Air, J. Acoustical Soc. Am., July, 1931.

As a result of his investigation, Knudsen proposed that the customary formula for the reverberation time be written with a corrective term depending upon humidity, as follows:

$$T = \frac{0.05V}{-S \log_e (1-\alpha) + 4mV} \tag{1}$$

where
  T = reverberation time,
  V = volume of room in cubic feet,
  S = surface in square feet,
  α = average coefficient of sound absorption, and
  m is an attenuation constant, measuring the decay of the sound
    intensity with the distance traveled. The observed values of
    T, S, and V were substituted in equation (1) thus obtaining
    two equations from which a and m were calculated.

In this way Knudsen determined m for different relative humidities and at four different frequencies for temperatures between 21° and 22° C. It was found that a was a constant within the limits of experimental error, but that m varied with the humidity.

To fit Knudsen's results to the present work, the assumption was made that the effect of water vapor on sound absorption of air depended solely upon vapor pressure and was independent of temperature and barometric pressure. On this basis Knudsen's curves for m, given in terms of relative humidity, were replotted in terms of vapor pressure and the corresponding values of $4mV$ subtracted from the measured total absorption of our reverberation room. The residual absorption was then plotted against temperature. The scatter of these points suggested that Knudsen's values of m did not adequately represent the effect of water vapor in our experiments. In addition, Knudsen's figures did not cover the full range of our vapor pressures. His curves were, therefore, modified by successive trial and error until the curves for residual absorption showed a minimum scatter of points from a smooth curve, the average variation being less than 1 per cent of the initial measured absorption and the maximum variation being only a little over 2 per cent.

The resulting empirical curves for m at 2,048 and 4,096 cycles are shown in Figure 1 and the residual absorption-temperature curves in Figures 2 and 3.

The question of sound absorption of air has been considered theoretically by Rayleigh[3] and others, and the conclusion was reached that the sound absorption should be proportional to the square of the frequency. This relation appears to hold good in the empirical curves for m shown in Figure 1, as the value for m at 4,096 cycles is four times that at 2,048 cycles. An exception to this is found when the vapor pressure is about 0.15 inch of mercury or less. Here the ratio seems to be about 3.9.

In drawing the curves in Figures 2 and 3 the term $4mV$ was calculated from the curves in Figure 1 and subtracted from the measured total absorption. The remainder has hitherto been assumed to depend only on the absorption of the surface of the room, as indicated by its form $-S \log (1-a)$. But after correction has thus been made

[3] Rayleigh, Theory of Sound, vol. 2, pp. 315–316.

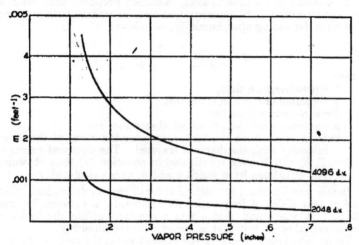

FIGURE 1.—*Values of m for air for different vapor pressures, at frequencies of 2,048 and 4,096 cycles*

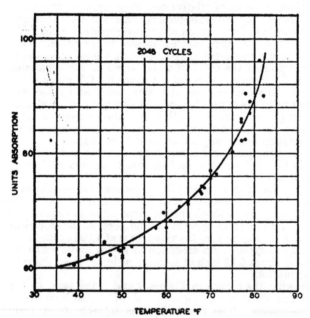

FIGURE 2.—*Change of absorption of air with change of temperature at 2,048 cycles*

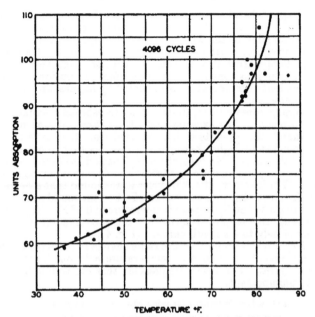

FIGURE 3.—*Change of absorption of air with change of temperature at 4,096 cycles*

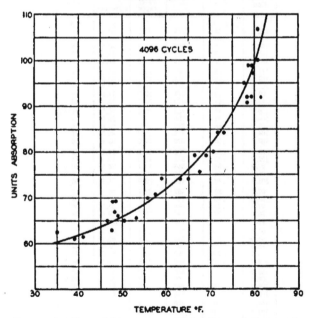

FIGURE 4.—*Curve of Figure 3 corrected for changes in barometric pressure*

for water vapor, the absorption shows a very large variation with temperature, much more than can reasonably be ascribed to any change in absorption of the wall surfaces. It appears, then, that an additional correction must be made for the absorption of air as a function of temperature if a constant absorption coefficient for the walls of the room is assumed. That this surface absorption is practically independent of temperature is indicated by all measurements on absorbing materials.

On examining these curves, drawn so as to represent as fairly as possible the mean of the experimental observations, it was noticed that, in general, those points which departed from the curve on one side corresponded to higher barometric pressures than those on the other side of the curve. No theoretical explanation has been found for this fact, but it is possible to deduce from these deviations an empirical correction. For instance, at 4,096 cycles it was found that a decrease of 0.1 inch in barometric pressure has an effect on the sound absorption equivalent to that of an increase in temperature of 0.8° F. Replotting the curve with this empirical correction the experimental points fit much more closely to a smooth curve. (Fig. 4.)

At 2,048 cycles it is found that the same change in barometric pressure is equivalent to a change in temperature only one-fourth that with 4,096 cycles. For lower frequencies this correction becomes practically unimportant.

TABLE 1.—*Corrections for water vapor at 2,048 cycles*

Total absorption	Vapor pressure	$m \times 10^6$	Correction for water vapor	Total absorption minus correction	Temperature
	Inches Hg				° F
102.1	0.54	365	21.7	80.4	75
101.7	.69	303	18.0	83.7	77
108.6	.68	307	18.2	90.4	78
105.2	.62	331	19.6	85.6	77
107.6	.67	312	18.5	89.1	79
113.4	.68	307	18.2	95.2	81
107.5	.72	295	17.5	90.0	82
101.8	.62	331	19.6	82.2	77
95.0	.52	375	22.2	72.8	68
99.4	.47	400	23.7	75.7	70
100.2	.46	403	23.9	76.3	71
104.4	.71	295	17.5	86.9	79
105.0	.65	320	19.0	86.0	78
104.7	.66	315	18.7	86.0	77
102.4	.65	320	19.0	83.4	78
96.3	.52	375	22.2	74.1	68
102.0	.62	331	19.6	82.4	74
97.9	.56	356	21.1	76.8	70
98.7	.44	415	24.6	74.1	68
101.3	.46	403	23.9	77.4	65
97.1	.36	467	27.7	69.4	59
96.6	.40	440	26.1	70.5	63
97.5	.32	504	29.9	68.6	56
96.2	.39	447	26.6	69.6	59
94.4	.314	510	30.6	63.8	52
94.0	.298	540	32.1	61.9	50
94.9	.278	552	32.8	62.1	50
98.3	.251	593	35.3	63.0	49
102.6	.200	690	41.0	61.6	43
102.9	.228	630	37.4	65.5	44.5
103.2	.213	660	39.2	64.0	45
96.6	.273	560	33.3	63.3	50.5
102.2	.223	640	38.0	64.2	46
106.3	.173	705	41.9	62.2	44
102.5	.204	680	40.4	62.1	42
129.5	.134	1,160	68.9	60.6	39
113.6	.153	905	53.8	59.8	36.3
114.6	.159	880	52.3	62.3	37.5

TABLE 2.—*Corrections for water vapor at 4,096 cycles*

Total absorption	Vapor pressure	$m \times 10^6$	Correction for water vapor	Total absorption minus correction	Observed temperature	Corrected temperature
	Inches Hg				° *F.*	° *F.*
163	0.69	121	72	91	77	78.3
173	.68	122	73	100	78	81
173	.62	132	78	95	77	77.5
173	.67	124	74	99	79	79.6
179	.69	121	72	107	81	80.7
167	.71	118	70	97	82	79.8
170	.62	132	78	92	77	78.4
168	.52	150	89	79	68	68.7
176	.47	161	96	80	70	70.8
180	.46	162	96	84	71	71.8
168	.71	118	70	98	79	79
167	.65	127	75	92	78	78.9
165	.66	125	74	91	77	----------
168	.65	127	75	93	78	----------
165	.52	150	89	76	68	67.
162	.62	132	78	84	74	73.1
168	.56	143	85	83	70	68.4
172	.45	165	98	74	68	65.4
174	.46	162	96	78	65	66.4
182	.36	186	111	71	59	57.7
179	.40	176	105	74	63	63
189	.32	200	119	70	56	56.2
180	.39	178	106	74	59	58.7
185	.314	202	120	65	52	50.4
172	.387	179	106	66	56.8	53.3
195	.289	216	128	67	50	49.1
201	.278	222	132	69	50	47.9
205	.251	239	142	63	49	48.1
228	.200	281	167	61	43	38.7
221	.228	256	152	69	44.5	48.1
225	.213	269	160	65	45	46.6
200	.273	226	134	66	50.5	53
222	.223	260	155	67	46	48.4
227	.204	273	165	62	42	40.9
313	.134	424	250	63	39	----------
269	.159	346	206	63	37.5	35.1

The original measurements at 2,048 and 4,096 cycles, and the corrections made for the moisture content of the air are shown in Tables 1 and 2. In Table 2 the corrected temperature in the last column is obtained from the observed temperature by applying the above-mentioned change of temperature empirically equivalent to the variation in barometric pressure. This correction at 2,048 cycles is quite small and has been neglected in Table 1.

Figure 5 shows the relation between m and vapor pressure for 1,024 and 512 cycles and Figures 6 and 7 show the corresponding absorption-temperature curves. These curves were calculated from the curves for 2,048 and 4,096 cycles by Rayleigh's law and fitted the data so well that no further empirical adjustment was made.

One of the most striking illustrations of the use of these curves occurred on a day of cold, windy weather in Washington in March, 1932. It has been our custom to measure the absorption of the empty room either immediately before or immediately after the measurements on the sample of the material. On this particular day the measurements were made on the sample first. As soon as the sample was removed from the room, measurements were made of the absorption of the empty room at 2,048 and 4,096 cycles. Much to our surprise it was found that at 4,096 cycles the empty room had more absorption than when it contained the sample.

By referring to the wet and dry bulb thermometer measurements made before and after the sample was removed it was found that there had been a considerable change in the humidity and a very small change in temperature, caused by the high wind changing the air in the room while the door was opened for removing the sample. When corrections taken from the curves of Figures 1, 2, and 3 were made for the changes in humidity and temperature, the sample was found to have a reasonable degree of absorption. On a later day when conditions (in respect to humidity and temperature) were approximately the same inside and outside the reverberation room, measurements were repeated on this same sample, and the coefficients

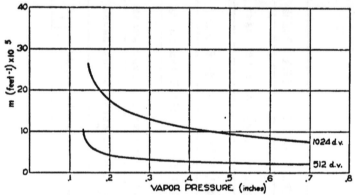

FIGURE 5.—*Values of m for different conditions of vapor pressure, for frequencies of 512 and 1,024 cycles*

of absorption which were obtained were found to agree almost exactly with those previously found.

The fact that air may have considerable absorption for sound is of interest in other fields. For instance, the distance that sound signals can be heard is of vital importance to shipping. Until quite recently it was thought that wind direction and velocity, layers of air of different densities which might cause reflection and refraction, and noise due to a storm were the principal factors which affected the distance at which a sound signal could be heard. More recent work shows that both temperature and humidity are important factors. Horner[4] gives some of the results of a study in which it was found that the distance at which a sound signal might be heard depends upon the humidity and the temperature of the air. He states that when the humidity is high, distant sounds can be heard with abnormal loudness, while under very low humidity these same sounds may become completely inaudible. He also stated "the worst acoustical conditions were almost invariably found in the type of weather commonly known as oppressive." Here the temperatures were high and evidently the increased absorption due to the high temperature was the predominating factor.

---

[4] Horner, Effect of Meteorological Condition on Sound Transmission at Sea, Nautical J., 1927.

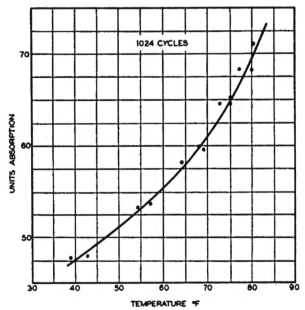

FIGURE 6.—*Change of absorption of air with change of temperature at 1,024 cycles*

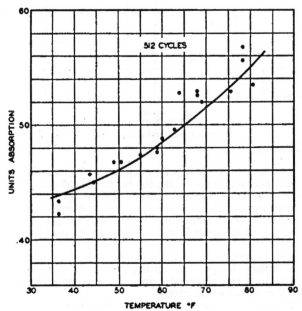

FIGURE 7.—*Change of absorption of air with change of temperature at 512 cycles*

Recent measurements made near Boston by the Lighthouse Service also show some interesting facts. A comparison was being made between a siren and an electric oscillator. In both cases the fundamental note was about 180 cycles per second. A considerable percentage of the energy of the siren was in overtones while the electric oscillator gave practically a pure note. When the observer was close the siren sounded the louder, but at a distance of 2 or 3 miles the oscillator was the louder, showing that the air has considerably less absorption for low-pitched notes.

In all sound-absorption measurements the assumption has heretofore been made that the decay of sound energy in a reverberation

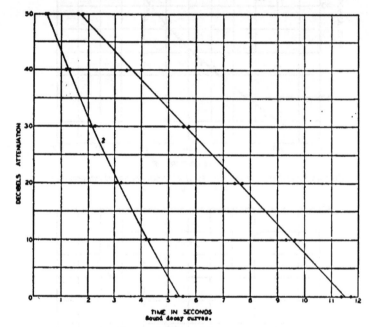

TIME IN SECONDS
Sound decay curves.

FIGURE 8.—*Sound-decay curves*

1, empty room.
2, highly absorbent sample in room.

room is logarithmic. With improved methods of measurement now available it is possible to determine the form of the decay curve with considerable accuracy. Figure 8 shows the decay curve at 1,024 cycles plotted logarithmically, for an empty reverberation room and for the same room containing a highly absorbent sample of material. With the sample in the room it will be noticed that the rate of decay is not uniform.

In the case of curve 2 of Figure 8, while the rate of decay is variable it shows no sudden change. In another case, illustrated in Figure 9, the distribution of the observed points is fitted more closely by a broken line than by a smooth curve.

Since the slope of the decay curve is dependent in part upon the absorption of the sample, it might be supposed from the form of curve 2 (fig. 8) and of the curve in Figure 9 that the coefficient of absorption of the sample varied with the intensity of the sound; but by repeating the experiment of Figure 9, starting from an initial level of sound intensity some 20 or 30 db lower, it is found that the knee is not fixed in position, but suffers a corresponding shift downward. We must, therefore, conclude that the change in slope is not due to change in coefficient with the intensity of the sound, but is rather to be ascribed to nonuniform distribution of the sound energy. Considerations of conditions in the reverberation room indicate that

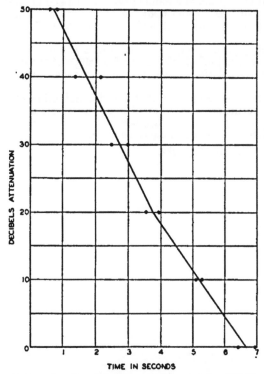

TIME IN SECONDS

FIGURE 9.—*Sound-decay curve with a sample of highly absorbent material in the reverberation room*

this is the probable explanation. In the empty room the absorption coefficient of the walls, etc., is only about 1 per cent, and is approximately uniform over the whole interior, but when a sample of highly absorbent material is placed in the room the rate of absorption over the surface of this sample may be 60 or more times as great as over the wall surfaces. Moreover, this sample is usually of an area which is not negligible as compared to the boundaries of the room.

The source of sound is designed so as to give as nearly as possible a a uniform initial distribution of sound energy in the room, but after the source has been stopped the extreme heterogeneity of absorption present makes it probable that the distribution of sound energy does not remain uniform as it decays.

The question then arises as to how the absorption should be calculated, as in all reverberation methods a straight line decay has been assumed. If the reason advanced above for the change of slope is correct, the slope of the initial portion of the curve would give the better value for the absorption. With a curve of varying slope the ear method, necessarily involving the whole curve, is incapable of giving correct results, as it measures only the average slope of the whole curve.

SUMMARY

The total absorption of a room appears to depend upon the amount of water vapor present and upon the temperature. The calibration of the room is therefore not definite unless these factors are kept constant.

The coefficient of absorption of a sample of material will depend upon whether the initial, average, or final slope of the decay curve is used in the calculation. The ear method necessarily employs an average.

It may now be recognized that the determination of the sound-absorption coefficient of a material is not as simple a matter as has been hitherto supposed, but appears to depend upon a number of factors which are now beginning to be understood.

WASHINGTON, May 20, 1932.

RP466

THE RECIPROCAL SPHERICAL ABERRATION OF AN OPTICAL SYSTEM INCLUDING HIGHER ORDERS

By Harold F. Bennett

ABSTRACT

The aberration present when an axial point is imaged by a centered system of spherical optical surfaces may be expressed by any one of a number of power series. In this paper the reciprocal of the distance between the intersection of a ray with the axis and a fixed point on the axis is expressed as a series in h, where h is the perpendicular distance from the ray to the fixed point.

Formulas are derived by which the constant coefficients of the series expressing the aberration in the image space of a single spherical surface may be computed if the corresponding coefficients for the object space are known, the fixed point of reference being the center of curvature of the surface. These formulas can not be used in the case of a plane surface since there is no center of curvature. Accordingly, after developing transfer formulas by which the aberration may be referred to a new point of reference, a second set of formulas is derived in which the point of reference is taken as the vertex of the surface. A final set of formulas expresses the longitudinal in terms of the reciprocal aberration.

As a numerical example, the computation of the aberration of the third, fifth, and seventh orders of an ordinary photographic "landscape" lens is given in full. Its results compare favorably with those of a trigonometric tracing of rays in which seven place tables were used.

In conclusion the convergence of these series is briefly discussed and some relations to diffraction theory are pointed out.

CONTENTS

I. INTRODUCTION

1. SPHERICAL ABERRATION

Rays of monochromatic light proceeding from a luminous point and passing through a lens are generally focussed not in a single image

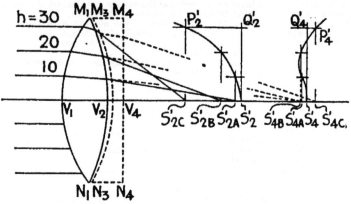

FIGURE 1.—*Spherical aberration of a single lens and of a doublet*

The paths of three rays when the concave lens is not present are shown in full lines. The concave lens and the paths of the rays if both lenses are present are shown in dotted lines. All parts of this figure are drawn in proportion to values obtained by an actual calculation. ($n_{1,1}=1.54$; $n_{1,4}=1.63$.)

point, but at varying distances from the lens according to the zone of the lens traversed. In the case of an object point on the axis of a centered system of spherical surfaces (the plane being considered a sphere of infinite radius) this defect in imagery is called "spherical aberration." Points off the axis are affected by the same aberration, but they are also affected by other aberrations, the presence of which makes analysis more difficult. The term "spherical aberration" is sometimes used more broadly to include these extra-axial aberrations and sometimes rather loosely in referring to the axial aberration where nonspherical surfaces are involved. In the present paper "spherical aberration" in the stricter sense of the word is discussed and a method is given for computing and expressing it quantitatively.

As an example of spherical aberration consider the portion of Figure 1 drawn in full lines. Light from an infinitely distant object point is incident upon the lens M_1N_1, and the three rays shown are refracted at the two surfaces so as to intersect the axis at S'_{2A}, S'_{2B}, and S'_{2C}, which are the foci for these respective zones. If perpendiculars to the axis are drawn at these points, their intersections with lines parallel to the axis through the points of incidence on the first surface form the locus of the curve $S'_2P'_2$ which may be considered as representing both in direction and magnitude the spherical aberration in the image. For example, the distance $Q'_2P'_2$ is the measure of the longitudinal spherical aberration of the ray at $h_1=30$. The dotted lines illustrate the method of correcting the spherical aberration by the insertion of a negative component, M_3. $.N_4$, of suitable design.

2. SERIES EXPRESSIONS OF LONGITUDINAL SPHERICAL ABERRATION

The aberration of a lens may be expressed quantitatively by giving the location of one or more points on the curve or by an equation for the curve. Such an equation is customarily written as an infinite series in which the terms beyond a certain order are neglected; thus the aberration represented by the curve $S'_4P'_4$ (fig. 1) may be expressed as follows:

$$\text{Long. spher.} = S' - s' = A'h^2 + B'h^4 + C'h^6 + \ldots \qquad (1)$$

where $S' - s'$ represents the abscissa and h the ordinate, and where s', A', B', C', $\ldots$, are constants. The odd powers of h are absent because of axial symmetry.

The number of terms of the series which it is necessary to retain depends upon the conditions of the problem in hand. The straight line, $S'_4Q'_4$, given by the equation $S' = s'$ is a first approximation which is exact only for infinitesimal apertures. A value of the constant A' may then be found such that the parabola $S' = s' + A'h^2$ gives a sufficiently close approximation to the curve up to some aperture depending upon the accuracy required. Each of the terms of higher order has the property of remaining quite small for small apertures and then beginning to increase rapidly at some finite value of h. Accordingly, if the accuracy is not to diminish as larger and larger apertures are considered, the fifth order [1] term and then the seventh order [1] term must be taken into account, and so on.

In the present example A' is negative while B' and C' are positive, causing the upper portion of the curve to turn in the positive direction.

The same aberration, it is evident, could quite as legitimately be represented by one of a number of other curves and its corresponding

---

[1] The longitudinal spherical aberration which varies as the ith power of the aperture ($i=2, 4, 6 \ldots$) corresponds to an angular aberration (see equation (3), p. 194) which varies as the $(i+1)$th power of the aperture and also to a path difference which varies as the $(i+2)$th power of the aperture. The corresponding reciprocal spherical aberration (see Sec. I, 4) varies also as the ith power of the aperture, but the lateral aberration varies as the $(i+1)$th power. These differences of the expressions which refer in different manners to the same aberration give rise to a difficulty in nomenclature. To avoid this the use of the terms primary, secondary, tertiary, etc., has been proposed. These designations not only become somewhat awkward for the higher-order aberrations treated in this paper, but some confusion might occur because primary and secondary already have special meanings in connection with astigmatism and curvature of field.

In the original papers of L. Seidel the aberrations are measured by their lateral value so that the terms for spherical aberration assume the orders 3, 5, 7, etc. In view of this and the general acceptance of Seidel's work it has been considered as desirable to designate a particular term of spherical aberration by the order which the corresponding lateral geometric aberration assumes. The expression "the third order longitudinal (or reciprocal) aberration" may then be interpreted as referring to "the longitudinal (or reciprocal) aberration which corresponds to the third order lateral aberration," and similarly for the terms of higher order.

It is to be noted that the constant coefficient $B^{(i+1)}$ of reciprocal aberration has the dimensions $L^{-(i+1)}$ and thus corresponds exactly to the above designation of order $(i+1)$.

equation, for instance by a curve the ordinates of which are equal to the heights of incidence on the last surface instead of the first. For any particular ray the total aberration would necessarily be the same according to either equation but, since the two values of h are not directly proportional, different portions of the total aberration would be attributed to the different orders. It can be shown that the various orders differ, even if only a single surface is considered, according as the ordinate is defined as the height of incidence upon the surface, as the height of intersection with a plane tangent at the vertex, as the distance from the vertex or from the center of curvature of the surface measured along a perpendicular to the ray, as the arc, tangent, or sine of the angle of inclination of the ray with the axis, or as one of a number of other functions of the aperture. Upon investigation it is found that the third order coefficient, A', is identical in each group in which (as for the first three cases mentioned) the ordinate reduces to the same infinitesimal for very small apertures. For large apertures, however, the third order for a particular ray differs in the several equations insofar as h^2 is slightly different. Thus it may be seen that, if the expressions "third order," "fifth order," etc., are to have exact significance, the exact definition of the aperture must be indicated. Moreover, direct comparisons between different lenses can be made only when the aberrations of both are expressed in exactly the same manner.

3. METHODS OF EVALUATING THE CONSTANTS

It has been pointed out that successively closer approximations to the aberration curve can be obtained by evaluating additional terms of the series (equation (1)). The usual methods of evaluating these coefficients are of three general types. First, if the lens has actually been constructed, a number of narrow beams of light may be singled out, as in the Hartmann test, by a diaphragm with several small holes, and their positions after traversing the lens may be determined photographically or otherwise. Second, from the specifications of the lens system the theoretical position of several rays may be found by trigonometric ray tracing. In either case an empirical curve or equation is fitted to the discrete points thus obtained. Third, trigonometric relations, such as those used in ray tracing, are expanded as Taylor's or similar series in ascending powers of whatever function of the aperture is chosen as the parameter. The aberration coefficients are thereby expressed as functions of the object distance and of the constants of the lens system. The present investigation is of this third type, which is usually called the algebraic or analytical method.

Ray tracing gives directly the total longitudinal aberration of one or more individual rays for any given lens. The algebraic method, however, is often more useful to the lens designer in that it reveals more fully the portion of the final aberration which is contributed by each surface of the lens and suggests in what manner the design should be altered. Moreover, very often when the design is altered only a small part of the computation need be repeated. The algebraic method has been considered objectionable because the third order equations are not sufficiently exact while if higher order terms are included the computation becomes too laborious. It is believed, however, that frequently the additional information is worth the labor involved, and, accordingly, formulas are given for enough orders so that sufficient accuracy may be attained in practically all cases.

4. RECIPROCAL SPHERICAL ABERRATION

The frequent occurrence of reciprocal distances in optical equations has suggested that the spherical aberration formulas might be simplified in form if the aberration instead of being expressed as a difference between the focal lengths of different zones were expressed as a difference between the powers of the reciprocals of the focal lengths.

An idea of the relation of the reciprocal aberration to the longitudinal may be gained as follows: Take for a moment the back image distance $V_4S'_4 = s'_4$ (fig. 1) as the unit of length. Plot a curve with the same ordinates as $S'_4P'_4$, but with abscissas equal to the reciprocals of those of this curve when measured from the vertex, V_4, as the origin. Where the abscissa of the old curve is slightly less than one that of the new is slightly more than one. Thus it is easily seen that the new curve will be very similar to $S'_4P'_4$ but reversed with with respect to $S'_4Q'_4$. Since the aberration is measured by the departure from the vertical straight line, the reciprocal aberration in any case is opposite in sign to the longitudinal aberration. The statements regarding equations expressing aberration, which were made in the two preceding sections (2 and 3), apply directly to the new curve except that other symbols, to be introduced later (see equations (2) and (2')) should replace the ones in equation (1). If a different unit of length is used the reciprocal aberration curve will merely be changed in scale in the horizontal direction. If, however, a different origin be selected, the curve will also be changed in shape.

5. HISTORICAL SKETCH AND BIBLIOGRAPHY

The theory of third-order aberrations is treated in the majority of books on geometrical optics. Among these Conrady[2] is unexcelled from a practical standpoint, while Von Rohr[3] gives a more elegant mathematical development and also includes an extensive bibliography.

Fifth-order terms for rays in the axial plane were published by Keller[4] and by Bauer.[5] Their expressions are not strictly accurate, however, in that h is not exactly defined. Kerber[6] derived the fifth order reciprocal aberration of a single surface defining h as the height of incidence. Schupmann[7] used a fifth order term which he credited to Kerber. Von Rohr[8] and König derived the fifth order term by Abbe's method of invariants with the angle of inclination of the ray as the variable. Risco[9] extended these equations to apply to aspherical surfaces. Smart[10] published terms of the fifth order, but unfortunately a number of his equations appear to be in error.

[2] A. E. Conrady, Applied Optics and Optical Design, pt. 1, 518 pp., Oxford Univ. Press, 1929.
[3] M. Von Rohr, editor, Die Bilderzeugung in optischen Instrumenten. J. Springer, Berlin, 1904. Also translation by R. Kanthack, His Majesty's Stationery Office, London, 1920.
[4] G. A. Keller, Zur Dioptrik, Entwicklung der Glieder fünfter Ordnung. 24 pp., C. R. Schurich, Munich, 1865.
[5] K. L. Bauer, Carl's Reportorium f. Phys. Tech., vol. 1, pp. 219–241, 1886.
[6] A. Kerber, Centr. Z. f. Opt. u. Mech., vol. 7, pp. 217–218, 1886.
[7] L. Schupmann, Die Medial-Fernrohre. B. G. Teubner, Leipzig, 1899.
[8] M. Von Rohr, pp. 238–244. See footnote 3.
[9] M. Martinez Risco, Estudios Generales sobre Aberración Esférica de Orden Superior, Anales Soc. Espanola Fis. y Quim vol. 25 pt. 1, pp. 100–136, 1927.
[10] E. H. Smart, Phil. Mag., vol. 20, pp. 82–91, 1910.

Early investigators, notably Petzval, may have been in possession of formulas of higher orders than the fifth,[11] but it is not known that any of these have been published.   It is true, however, that investigations of aberrations from somewhat different standpoints have been extended to fifth and higher orders.   Most of these are based upon the eikonal function, which depends upon the variation of path length through the lens.

6. NOTATION AND SIGN CONVENTION

A thoroughly consistent and complete system of notation for optical equations is, unfortunately, not available.   In the present paper the aims have been to follow, in general, one of the systems already in use and to use as small a number of symbols as is compatible with clarity and brevity in presentation.   In particular may be mentioned the use of the Greek characters β and ζ with superscripts ($\beta^{(3)}$, $\beta^{(5)}$, etc.) as constant coefficients of the various orders of reciprocal spherical aberration.   (See equation (2) and footnote 1, p. 189.)

The original direction of the light is assumed to be from left to right.   The refracting and reflecting surfaces are numbered in the order in which they are encountered by the light and are indicated by numerical subscripts.   The well-known idea of regarding reflection as a special case of refraction in which the index ratio is -1 has been extended to include absolute indices (and indices relative to air), the index of any medium being considered as negative in sign if the light travels through it in a negative direction.[12]   Nevertheless, any other self-consistent sign convention may be used if preferred.

A list of the characters with their meanings is given herewith. Unprimed letters denote magnitudes in the object space, primed letters the corresponding magnitude in the image space.   If the two are identical, as in the case of r or ϕ, the character is always written unprimed.   In the following summary the primed character is omitted in most cases, it being assumed that its meaning is easily deduced from the definition of the corresponding unprimed character.

ENGLISH LETTERS

A, B, C, ... =the third order, fifth order, seventh order, ..., coefficients of longitudinal spherical aberration, positive if the aberration tends to make the rim ray cross the axis to the right of the paraxial focus.

c=a subscript meaning "referred to the center of curvature" of the appropriate surface.

d=the distance from the old point of reference to the new one to which the transfer is made, positive if to the right.

h=(in Introduction) linear aperture, subject to a variety of definitions.

---

[11] See, for instance, Trans. Opt. Soc. London, vol. 22, p. 214; 1920–21.   Also Berek, Grundlagen der praktischen Optik, p. 43.   Walter de Gruyter Co., Berlin, 1930.

[12] This convention is also applicable in path length equations as may be seen from the following simple example.   Suppose a ray from some point A is reflected back to A from a mirror at B.   The first distance, A to B or AB, being positive, the distance B to A or BA, after reflection must be negative.   Adding the distances together would give zero for the path length.   However, if the index be considered negative after an odd number of reflections and, as is universally done, the distance in each medium be multiplied by the index of refraction to get the equivalent path length in air the result is:

$$\text{Equiv. path length} = nAB + n'BA$$
$$= nAB + (-n)(-AB)$$
$$= 2nAB$$

which is obviously correct.

An alternative sign convention in which some of the angles are considered as obtuse after reflection has been proposed by T. Smith in Trans. Opt. Soc. London, vol. 27, pp. 312–318, 1925–26.

h=the distance from the point of reference or origin of coordinates to a ray, measured along a perpendicular to the ray
and positive if upward.

i, j=generalized integers.

n=the index of refraction of a medium, usually taken relative
to air, to be given a negative sign if the light traverses it
from right to left.

r=the radius of curvature of a spherical surface, positive if the
surface is concave to the right.

s=the distance from the vertex of a surface to the paraxial object
point for that surface, positive if to the right.

$t=s-r$=the distance from the center of curvature of a surface to
the paraxial object point for that surface, positive if to the
right.

S=the distance from the vertex of a surface to the intersection
of a ray in the object pencil with the axis; reduces to s as
the aperture approaches zero.

$T=S-r$=the distance from the center of curvature of a surface
to the intersection of a ray in the object pencil with the
axis; reduces to t as the aperture approaches zero.

v=a subscript meaning "referred to the vertex" of the appropriate surface.

GREEK LETTERS

α=the acute angle between a ray and the axis, positive if the ray lies above
the axis to the left of the point of intersection.

$\beta^{(i)}$=the coefficient of reciprocal spherical aberration of the ith order, in the
series referred to the center of curvature, usually of the sign opposite
that of the corresponding longitudinal aberration coefficient.

$\zeta^{(i)}$=the coefficient of reciprocal spherical aberration of the ith order in the
series referred to the vertex of the lens surface, usually of the sign
opposite that of the corresponding longitudinal aberration coefficient.

Θ=the reciprocal of T ($q. v.$).

ϑ, ϑ'=the angle of incidence, the angle of emergence; that is, the acute angle
between a ray and the normal to the surface at the point of incidence,
positive if the ray lies above the normal to the right of the surface.

κ=the constant of transfer to a new point of reference.

  =the ratio of the new paraxial reciprocal object (or image) distance to the
old paraxial reciprocal image distance.

κ, κ' in Part IV take on the special meaning that the transfer is made from the
vertex to the center of curvature of a single surface, κ applying to the
object pencil and κ' to the image pencil.

μ=relative index=n/n'.

μ'=relative index=n'/n.

ρ, σ, τ=the reciprocals of r, s, and t.

Σ=the reciprocal of S.

ϕ=the central angle or the angle between the radius drawn to the point of
incidence and the axis, positive if the radius lies above the axis to the
left of the center of curvature.

Ψ=reciprocal object distance from any point of reference on the axis.

II. THE RECIPROCAL ABERRATION OF A SINGLE SPHER-ICAL SURFACE REFERRED TO THE CENTER OF CURVA-TURE

1. PRELIMINARY STATEMENT OF PROBLEM

A set of formulas will first be derived which express the reciprocal
aberration coefficients of the refracted or image pencil of any spherical
surface when the corresponding constants of the incident or object
pencil for that surface are known, the point of reference being chosen
at the center of curvature of the surface.

In Figure 2, VS represents the axis of symmetry of a lens system of which one surface, with vertex at V and center of curvature at C, is shown in cross section. In a symmetrical pencil of rays incident upon this surface, consider the ray QP, incident at the point P. Let this ray (extended if necessary) intersect the axis at the point S, removed from the center of curvature by a distance T, positive if to the right. Let h_e denote the perpendicular distance from the center of curvature to the ray, positive if upward, and r the radius of curvature of the surface, positive if the surface is concave toward the right.

Let $\beta^{(3)}$, $\beta^{(5)}$, $\beta^{(7)}$, etc., be constants, such that if h_e is given, T is determined to as close an approximation as may be desired by retaining enough terms of the series:

$$\frac{1}{T}=\frac{1}{t}+\beta^{(3)}h_e{}^2+\beta^{(5)}h_e{}^4+\beta^{(7)}h_e{}^6+ \ \ldots \qquad (1)$$

Here t is the limiting value of T as h_e approaches zero; that is, the paraxial value of T, $\beta^{(3)}h_e{}^2$ is the third order term of reciprocal aberra-

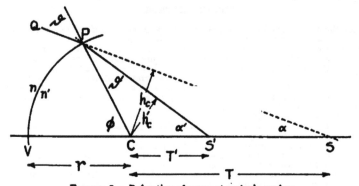

FIGURE 2.—*Refraction of a ray at a single surface*

tion, $\beta^{(5)}h_e{}^4$ is the fifth order, etc. For convenience in writing the equation, let $\frac{1}{T}=\theta$ and $\frac{1}{t}=\tau$, then

$$\theta=\tau+\beta^{(3)}h_e{}^2+\beta^{(5)}h_e{}^4+\beta^{(7)}h_e{}^6+\beta^{(9)}h_e{}^8+ \ \ldots \qquad (2)$$

The angles α, ϑ, and ϕ are shown in Figure 2. They are defined by the following equations, apparent from the geometry of the figure

$$\sin \alpha=\frac{h_e}{T}=\theta h_e$$

$$=\tau h_e+\beta^{(3)}h_e{}^3+\beta^{(5)}h_e{}^5+ \ \ldots \qquad (3)$$

$$\sin \vartheta=\frac{h_e}{r}=\rho h_e \qquad (4)$$

$$\phi=\alpha+\vartheta$$

where ρ is written for $\frac{1}{r}$.

Let the ray pass from a medium of index n into one of index n' and write for convenience $\dfrac{n'}{n} = \mu'$.[13] Denote the various functions of the refracted ray, PS', by priming those symbols which when unprimed represent the corresponding functions of the incident ray. Then:

$$\theta' = \tau' + \beta'^{(3)}h'_e{}^3 + \beta'^{(5)}h'_e{}^5 + \ldots \tag{2'}$$

$$\sin \alpha' = \tau'h'_e + \beta'^{(3)}h'_e{}^3 + \beta'^{(5)}h'_e{}^5 + \ldots \tag{3'}$$

$$\sin \vartheta' = \rho h'_e \tag{4'}$$

and

$$\alpha' + \vartheta' = \phi = \alpha + \vartheta \tag{5}$$

2. DERIVATION OF THE FORMULAS

By Snell's law and equations (4) and (4')

$$\sin \vartheta = \rho h_e = \mu' \rho h'_e \tag{6}$$

From this $h_e = \mu' h'_e$; then substituting in equation (3):

$$\sin \alpha = \mu' \tau h'_e + \mu'^3 \beta^{(3)}h'_e{}^3 + \mu'^5 \beta^{(5)}h'_e{}^5 + \mu'^7 \beta^{(7)}h'_e{}^7 + \mu'^9 \beta^{(9)}h'_e{}^9 + \\ \mu'^{11} \beta^{(11)}h'_e{}^{11} \tag{7}$$

The sines of α, ϑ, and ϑ' have now been expressed as functions of the same variable, h'_e. Of the possible methods of combining these to give $\sin \alpha'$, the unknown angle of equation (5), the following one using trigonometrical addition formulas and series expansions has been thought the most convenient:[14]

$$\cos \vartheta = 1 - \frac{1}{2}\sin^2 \vartheta - \frac{1}{8}\sin^4 \vartheta - \ldots$$

by equation (6)

$$= 1 - \frac{1}{2}\mu'^2 \rho^2 h'_e{}^2 - \frac{1}{8}\mu'^4 \rho^4 h'_e{}^4 - \frac{1}{16}\mu'^6 \rho^6 h'_e{}^6$$

$$- \frac{5}{128}\mu'^8 \rho^8 h'_e{}^8 - \frac{7}{256}\mu'^{10}\rho^{10}h'_e{}^{10}$$

$$\cos \vartheta' = 1 - \frac{1}{2}\rho^2 h'_e{}^2 - \frac{1}{8}\rho^4 h'_e{}^4 - \frac{1}{16}\rho^6 h'_e{}^6 - \ldots$$

$$\sin(\vartheta - \vartheta') = \sin \vartheta \cos \vartheta' - \cos \vartheta \sin \vartheta'$$

$$= (\mu' - 1)\rho h'_e - \frac{1}{2}(\mu' - \mu'^2)\rho^3 h'_e{}^3$$

$$- \frac{1}{8}(\mu' - \mu'^4)\rho^5 h'_e{}^5 - \frac{1}{16}(\mu' - \mu'^6)\rho^7 h'_e{}^7$$

$$- \frac{5}{128}(\mu' - \mu'^8)\rho^9 h'_e{}^9 - \frac{7}{256}(\mu' - \mu'^{10})\rho^{11}h'_e{}^{11}$$

---

[13] The sense of the index ratio and the direction of the ray are thus mutually defined. If the ray passes through the refracting surface from left to right, then the index n pertains to the medium on the left and vice versa. The sign convention for distances and angles is independent of the direction of propagation of light along the ray. The index is considered as negative if the light travels from right to left. (See footnote 12, p. 192.)

[14] It is to be remembered, of course, if one is interested only in the third order or the third and fifth orders, that the higher orders may be ignored entirely. Moreover, it will doubtless be found that the method of derivation of the formulas is more easily followed if only the first one or two terms of each equation are taken into consideration.

TABLE 1.—$Cos\ \vartheta\ cos\ \vartheta'$

1	$h'_c{}^2$	$h'_c{}^4$	$h'_c{}^6$	$h'_c{}^8$	$h'_c{}^{10}$
1	$-\frac{1}{2}\mu'^2\rho^2$	$-\frac{1}{8}\mu'^2\rho^4$	$-\frac{1}{16}\mu'^2\rho^6$	$-\frac{5}{128}\mu'^2\rho^8$	$-\frac{7}{256}\mu'^2\rho^{10}$
	$-\frac{1}{2}\rho^2$	$+\frac{1}{4}\mu'^2\rho^4$	$+\frac{1}{16}\mu'^4\rho^6$	$+\frac{1}{32}\mu'^4\rho^8$	$+\frac{5}{256}\mu'^2\rho^{10}$
		$-\frac{1}{8}\rho^4$	$+\frac{1}{16}\mu'^2\rho^6$	$+\frac{1}{64}\mu'^2\rho^8$	$+\frac{1}{128}\mu'^6\rho^{10}$
			$-\frac{1}{16}\rho^6$	$+\frac{1}{32}\mu'^2\rho^8$	$+\frac{1}{128}\mu'^4\rho^{10}$
				$-\frac{5}{128}\rho^8$	$+\frac{5}{256}\mu'^2\rho^{10}$
					$-\frac{7}{256}\rho^{10}$

$$\cos(\vartheta-\vartheta') = \cos\vartheta\cos\vartheta' + \sin\vartheta\sin\vartheta'$$
$$= \text{the expansion [15] in Table } 1 + \mu'\rho^2 h'_c{}^2$$
$$= 1 - \frac{1}{2}(\mu'-1)^2\rho^2 h'_c{}^2 - \frac{1}{8}(\mu'^2-1)^2\rho^4 h'_c{}^4$$
$$- \frac{1}{16}(\mu'^2-1)^2(\mu'^2+1)\rho^6 h'_c{}^6$$
$$- \frac{1}{128}(\mu'^2-1)^2(5\mu'^4+6\mu'^2+5)\rho^8 h'_c{}^8$$
$$- \frac{1}{256}(\mu'^2-1)^2(7\mu'^6+9\mu'^4+9\mu'^2+7)\rho^{10} h'_c{}^{10}$$

$$\cos\alpha = 1 - \frac{1}{2}\sin^2\alpha - \frac{1}{8}\sin^4\alpha - \ \ldots$$
$$= \text{the expansion [15] in Table } 2$$

TABLE 2.—$Expansion\ of\ cos\ \alpha$

[The coefficient of each power of h'_c is the sum of all the quantities below it in the same column]

1	$h'_c{}^2$	$h'_c{}^4$	$h'_c{}^6$	$h'_c{}^8$	$h'_c{}^{10}$
1	$-\frac{1}{2}\mu'^2\gamma^2$	$-\mu'^4\gamma\beta(3)$	$-\mu'^6\gamma\beta(5)-\frac{1}{2}\mu'^6\beta(3)^2$	$-\mu'^8\gamma\beta(7)-\mu'^8\beta(3)\beta(5)$	$-\mu'^{10}\gamma\beta(9)-\mu'^{10}\beta(3)\beta(7)$
					$-\frac{1}{2}\mu'^{10}\beta(5)^2$
	$-\frac{1}{8}\mu'^4\gamma^4$	$-\frac{1}{2}\mu'^4\gamma^3\beta(3)$	$-\frac{1}{2}\mu'^6\gamma^3\beta(5)-\frac{3}{4}\mu'^6\gamma^2\beta(3)^2\cdot2$	$\frac{1}{2}\mu'^{10}\gamma^3\beta(7)-\frac{3}{2}\mu'^{10}\gamma^3\beta(3)\beta(5)-$
					$\frac{1}{2}\mu'^{10}\gamma^3\beta(5)^2$
		$-\frac{1}{16}\mu'^4\gamma^4$	$-\frac{3}{8}\mu'^6\gamma^2\beta(3)$	$-\frac{3}{8}\mu'^{10}\gamma^2\beta(5)-\frac{15}{16}\mu'^{10}\gamma^2\beta(3)^2\cdot3$
			$-\frac{5}{128}\mu'^6\gamma^3$	$-\frac{5}{16}\mu'^{10}\gamma^3\beta(3)$
				$-\frac{7}{256}\mu'^{10}\gamma^{10}$

$$\sin\alpha' = \sin[\alpha+(\vartheta-\vartheta')]$$
$$= \sin\alpha\cos(\vartheta-\vartheta') + \cos\alpha\sin(\vartheta-\vartheta')$$
$$= \text{the expansion [15] in Table } 3$$
$$= \tau'h'_c + \beta'^{(3)}h'_c{}^3 + \beta'^{(5)}h'_c{}^5 + \beta'^{(7)}h'_c{}^7 + \ldots \tag{8}$$

the last term being obtained from equation (3').

---

[15] This member of the equation has been written in tabular form for clearness and convenience. In Tables 1 and 2 the quantities in a single column are to be considered as inclosed in brackets and multiplied by the power of h at the top. In Tables 3 and 4 the quantity in each small rectangle in the body of the table is to be multiplied by the power of h at the top of the column and also by the β—coefficient at the left end of the row. The quantity represented by the whole table is the sum of all these products. The products may be grouped either by row or by column (in this case the latter is more convenient) and the common factors taken out.

TABLE 3.—*Expansion of* $\sin [\alpha + (\vartheta - \vartheta')]$

[Each quantity in the body of the table is to be multiplied both by the power of h' at the top and by the β-coefficient at the left]

	h'_0	h'_1	h'_3	h'_5	h'_7	h'_9	h'_{11}
1	$(\mu'-1)\rho$	$-\tfrac{1}{2}\mu'(\mu'-1)\rho\tau$	$-\tfrac{1}{8}\mu'(\mu'^2-1)\rho\tau^2$	$-\tfrac{1}{16}\mu'^2(\mu'^2-1)\tau\cdot(\mu'^2+1)\rho\tau$	$-\tfrac{1}{128}\mu'^2(\mu'^2-1)\tau\cdot(5\mu'^4+6\mu'^2+5)\rho\tau\tau$	$-\tfrac{1}{256}\mu'^2(\mu'^2-1)\tau\cdot(7\mu'^4+0\mu'^4+0\mu'^2+7)\rho\tau\tau$
	$+(\mu'-1)\rho$	$-\tfrac{1}{2}(\mu'-\mu'^3)\rho\tau$	$-\tfrac{1}{8}(\mu'-\mu'^5)\rho\tau^3$	$-\tfrac{1}{16}(\mu'-\mu'^7)\rho\tau^4$	$-\tfrac{5}{128}(\mu'-\mu'^9)\rho\tau\tau\tau$	$-\tfrac{7}{256}(\mu'-\mu'^{11})\rho\tau^{13}$
		$-\tfrac{1}{2}(\mu'^3-\mu'^5)\rho\tau^3$	$+\tfrac{1}{8}(\mu'^3-\mu'^5)\rho\tau^4$	$+\tfrac{1}{16}(\mu'^3-\mu'^7)\rho\tau^4$	$+\tfrac{1}{64}(\mu'^3-\mu'^9)\rho\tau^6$	$+\tfrac{5}{256}(\mu'^3-\mu'^{11})\rho\tau^8$
			$-\tfrac{1}{8}(\mu'^5-\mu'^7)\rho\tau^4$	$+\tfrac{1}{16}(\mu'^5-\mu'^7)\rho\tau\tau\tau$	$+\tfrac{1}{32}(\mu'^5-\mu'^9)\rho\tau^6$	$+\tfrac{1}{128}(\mu'^5-\mu'^{11})\rho\tau^8$
				$-\tfrac{1}{16}(\mu'^7-\mu'^9)\rho\tau\tau\tau$	$+\tfrac{5}{128}(\mu'^7-\mu'^{11})\rho\tau^8$
β'	$+\mu'^3$	$-\tfrac{1}{2}\mu'^2(\mu'^2-1)\rho\tau$	$-\tfrac{1}{8}\mu'^2(\mu'^2-1)\rho\tau^4$	$-\tfrac{1}{128}\mu'^2(\mu'^2-1)\tau\cdot(\mu'^2+1)\rho\tau^4$	$-\tfrac{1}{128}\mu'^2(\mu'^2-1)\tau\cdot(5\mu'^4+6\mu'^2+5)\rho\tau^3$
		$-(\mu'^3-\mu'^5)\rho\tau$	$+\tfrac{1}{2}(\mu'^3-\mu'^5)\rho\tau^4$	$+\tfrac{1}{8}(\mu'^3-\mu'^5)\rho\tau^5$	$+\tfrac{1}{16}(\mu'^3-\mu'^7)\rho\tau^4$
			$-\tfrac{1}{2}(\mu'^5-\mu'^7)\rho\tau^4$	$+\tfrac{3}{8}(\mu'^5-\mu'^9)\rho\tau^7$
				$-\tfrac{5}{16}(\mu'^7-\mu'^{11})\rho\tau^7$

TABLE 3.—*Expansion of* $\sin[\alpha + (\delta - \delta')]$—Continued

Since, to the order of approximation attained, this is true for any value of h'_e, the coefficients of like powers of h'_e may be equated and are readily reduced to the following form:

$$\tau' = \mu'\tau + (\mu'-1)\rho \tag{9a}$$

$$\beta'^{(3)} = \mu'^3\beta^{(3)} - \tfrac{1}{2}\mu'\rho[(\mu'-1)^2\rho\tau + (\rho^2+\mu'^2\tau^2) - \mu'(\rho^2+\tau^2)] \tag{9b}$$

$$\begin{aligned}
\beta'^{(5)} = \mu'^5\beta^{(5)} &- \tfrac{1}{2}\mu'^3\beta^{(3)}(\mu'-1)\rho[(\mu'-1)\rho+2\mu'\tau] \\
&- \tfrac{1}{8}\mu'\rho[(\mu'^2-1)^2\rho^3\tau + (\rho^2-\mu'^2\tau^2)^3 - \mu'^3(\rho^2-\tau^2)^2]
\end{aligned} \tag{9c}$$

$$\begin{aligned}
\beta'^{(7)} = \mu'^7\beta^{(7)} &- \tfrac{1}{2}\mu'^6\beta^{(3)2}(\mu'-1)\rho - \tfrac{1}{2}\mu'^5\beta^{(5)}(\mu'-1)\rho[(\mu'-1)\rho+2\mu'\tau] \\
&- \tfrac{1}{2}\mu'^3\beta^{(3)}\rho[\tfrac{1}{4}(\mu'^2-1)^2\rho^3 - \mu'^2\tau(\rho^2-\mu'^2\tau^2) + \mu'^3\tau(\rho^2-\tau^2)] \\
&- \tfrac{1}{16}\mu'\rho[(\mu'^2-1)^2(\mu'^2+1)\rho^5\tau + (\rho^2-\mu'^2\tau^2)^3(\rho^2+\mu'^2\tau^2) \\
&\quad - \mu'^5(\rho^2-\tau^2)^3(\rho^2+\tau^2)]
\end{aligned} \tag{9d}$$

$$\begin{aligned}
\beta'^{(9)} = \mu'^9\beta^{(9)} &- \mu'^8\beta^{(3)}\beta^{(5)}(\mu'-1)\rho - \tfrac{1}{2}\mu'^7\beta^{(7)}(\mu'-1)\rho[(\mu'-1)\rho+2\mu'\tau] \\
&+ \tfrac{1}{4}\mu'^7\beta^{(3)3}\rho[(\rho^2-3\mu'^2\tau^2) - \mu'(\rho^2-3\tau^2)] \\
&- \tfrac{1}{2}\mu'^6\beta^{(5)}\rho[\tfrac{1}{4}(\mu'^2-1)^2\rho^3 - \mu'^2\tau(\rho^2-\mu'^2\tau^2) + \mu'^3\tau(\rho^2-\tau^2)] \\
&- \tfrac{1}{2}\mu'^3\beta^{(3)}\rho[\tfrac{1}{8}(\mu'^2-1)^2(\mu'^2+1)\rho^5 - \mu'^2\tau(\rho^2-\mu'^2\tau^2)(\rho^2+3\mu'^2\tau^2) \\
&\quad + \mu'^5\tau(\rho^2-\tau^2)(\rho^2+3\tau^2)] \\
&- \tfrac{1}{128}\mu'\rho[(\mu'^2-1)^2(5\mu'^4+6\mu'^2+5)\rho^7\tau + (\rho^2-\mu'^2\tau^2)^2\cdot \\
&\quad \cdot(5\rho^4+6\mu'^2\rho^2\tau^2+5\mu'^4\tau^4) - \mu'^7(\rho^2-\tau^2)(5\rho^4+6\rho^2\tau^2+5\tau^4)]
\end{aligned} \tag{9e}$$

$$\begin{aligned}
\beta'^{(11)} = \mu'^{11}\beta^{(11)} &- \tfrac{1}{2}\mu'^{10}[\beta^{(5)2}+2\beta^{(3)}\beta^{(7)}+\tau\beta^{(3)3}](\mu'-1)\rho \\
&- \tfrac{1}{2}\mu'^9\beta^{(9)}(\mu'-1)\rho[(\mu'-1)\rho+2\mu'\tau] \\
&+ \tfrac{1}{4}\mu'^9\beta^{(3)}\beta^{(5)}\rho[(\rho^3-3\mu'^2\tau^2) - \mu'(\rho^2-3\tau^2)] \\
&- \tfrac{1}{2}\mu'^7\beta^{(7)}\rho[\tfrac{1}{4}(\mu'^2-1)^2\rho^3 - \mu'^2\tau(\rho^2-\mu'^2\tau^2) + \mu'^3\tau(\rho^2-\tau^2)] \\
&- \tfrac{1}{16}\mu'^7\beta^{(3)3}\rho[(\rho^4+6\mu'^2\rho^2\tau^2-15\mu'^4\tau^4) - \mu'^3(\rho^4+6\rho^2\tau^2-15\tau^4)] \\
&- \tfrac{1}{2}\mu'^6\beta^{(5)}\rho[\tfrac{1}{8}(\mu'^2-1)^2(\mu'^2+1)\rho^5 - \mu'^2\tau(\rho^2-\mu'^2\tau^2)(\rho^2+3\mu'^2\tau^2) \\
&\quad + \mu'^5\tau(\rho^2-\tau^2)(\rho^2+3\tau^2)] \\
&- \tfrac{1}{16}\mu'^3\beta^{(3)}\rho[\tfrac{1}{8}(\mu'^2-1)(5\mu'^4+6\mu'^2+5)\rho^7 - \mu'^2\tau(\rho^2-\mu'^2\tau^2)\cdot \\
&\quad \cdot(\rho^4+2\mu'^2\rho^2\tau^2+5\mu'^4\tau^4) + \mu'^7\tau(\rho^2-\tau^2)(\rho^4+2\rho^2\tau^2+5\tau^4)] \\
&- \tfrac{1}{256}\mu'\rho[(\mu'^2-1)^2(7\mu'^6+9\mu'^4+9\mu'^2+7)\rho^9\tau \\
&\quad + (\rho^2-\mu'^2\tau^2)^2(7\rho^6+9\mu'^2\rho^4\tau^2+9\mu'^4\rho^2\tau^4+7\mu'^6\tau^6) \\
&\quad - \mu'^9(\rho^2-\tau^2)^2(7\rho^6+9\rho^4\tau^2+9\rho^2\tau^4+7\tau^6)]
\end{aligned} \tag{9f}$$

If, instead, the quantities in Table 3 be factored the following equations (written only to the ninth order) result:

$$\tau' = \mu'(\tau+\rho) - \rho \text{ or } (\tau'+\rho) = \mu'(\tau+\rho) \tag{10a}$$

$$\beta'^{(3)} = \mu'^3\beta^{(3)} + \frac{1}{2}\mu'(\mu'-1)\,\rho\,(\rho+\tau)\,(\rho-\mu'\tau) \tag{10b}$$

$$\begin{aligned}
\beta'^{(5)} = \mu'^5\beta^{(5)} &- \frac{1}{2}\mu'^3\beta^{(3)}(\mu'-1)\rho[(\mu'-1)\rho+2\mu'\tau] + \frac{1}{8}\mu'(\mu'-1)\rho(\rho+\tau) \\
&(\rho-\mu'\tau)\,[(\mu'^2+\mu'+1)\,\rho^2 - \mu'(\mu-1)\,\rho\tau + \mu'^2\tau^2]
\end{aligned} \tag{10c}$$

$$\begin{aligned}
\beta'^{(7)} = \mu'^7\beta^{(7)} &- \frac{1}{2}\mu'^6\beta^{(3)2}(\mu'-1)\rho - \frac{1}{2}\mu'^5\beta^{(5)}(\mu'-1)\rho[(\mu'-1)\rho+2\mu'\tau] \\
&- \frac{1}{8}\mu'^3\beta^{(3)}(\mu'-1)\rho[(\mu'+1)(\mu'^2-1)\rho^3 + 4\mu'^2\tau(\rho^3+\mu'^2\tau^2)] \\
&+ \frac{1}{16}\mu'(\mu'-1)\,\rho(\rho+\tau)(\rho-\mu'\tau)[(\mu'^4+\mu'^3+\mu'^2+\mu'+1)\rho^4 \\
&- \mu'(\mu'^2-1)\rho^3\tau + \mu'^2(\mu'^2+1)\rho^2\tau^2 - \mu'^2(\mu'-1)\rho\tau^3 + \mu'^4\tau^4]
\end{aligned} \tag{10d}$$

$$\beta'^{(9)} = \mu'^{9}\beta^{(9)} - \mu'^{8}\beta^{(3)}\beta^{(5)}(\mu'-1)\rho - \frac{1}{2}\mu'^{7}\beta^{(7)}(\mu'-1)\rho[(\mu'-1)\rho + 2\mu'\tau]$$

$$- \frac{1}{4}\mu'^{7}\beta^{(3)2}(\mu'-1)\rho[\rho^{2}+3\mu'\tau^{2}] - \frac{1}{8}\mu'^{5}\beta^{(5)}(\mu'-1)\rho\cdot$$

$$\cdot[(\mu'+1)(\mu'^{2}-1)\rho^{3}+4\mu'^{2}\tau(\rho^{2}+\mu'\tau^{2})] - \frac{1}{16}\mu'^{3}\beta^{(3)}(\mu'-1)\rho\cdot$$

$$\cdot[(\mu'+1)(\mu'^{4}-1)\rho^{5}+2\mu'^{3}(\mu'^{2}+\mu'+1)\rho^{4}\tau+4\mu'^{4}\rho^{2}\tau^{3}+6\mu'^{6}\tau^{5}]$$

$$+ \frac{1}{128}\mu'(\mu'-1)\rho(\rho+\tau)(\rho-\mu'\tau)[5(\mu'^{6}+\mu'^{5}+\mu'^{4}+\mu'^{3}+\mu'^{2}+\mu'+1)\rho^{6}$$

$$- \mu'(\mu'-1)(5\mu'^{4}+6\mu'^{3}+7\mu'^{2}+6\mu'+5)\rho^{5}\tau$$

$$+ \mu'^{2}(5\mu'^{4}+\mu'^{3}+3\mu'^{2}+\mu'+5)\rho^{4}\tau^{3}-\mu'^{3}(\mu'-1)(5\mu'^{2}+4\mu'+5)\rho^{3}\tau^{3}$$

$$+ \mu'^{4}(5\mu'^{2}-\mu'+5)\rho^{2}\tau^{4}-5\mu'^{5}(\mu'-1)\rho\tau^{5}+5\mu'^{6}\tau^{6}] \tag{10e}$$

If preferred, $(\tau'+\mu'\tau)$ may be substitued for $[(\mu'-1)\rho+2\mu'\tau]$ throughout these equations. In most numerical applications the computing of the higher powers of μ' may be avoided by dividing each equation by the appropriate power of n', remembering that $\mu'=\frac{n'}{n}$. Equations (9) then take the following form:

$$\frac{\tau'}{n'}=\frac{\tau}{n}+\left(\frac{1}{n}-\frac{1}{n'}\right)\rho \tag{11a}$$

$$\frac{\beta'^{(3)}}{n'^{3}}=\frac{\beta^{(3)}}{n^{3}}-\frac{1}{2}\rho\left[\left(\frac{1}{n}-\frac{1}{n'}\right)^{2}\frac{\tau}{n}\rho+\frac{1}{n}\left(\frac{\rho^{2}}{n'^{2}}+\frac{\tau^{2}}{n^{2}}\right)-\frac{1}{n'}\left(\frac{\rho^{2}}{n^{2}}+\frac{\tau^{2}}{n^{2}}\right)\right] \tag{11b}$$

$$\frac{\beta'^{(5)}}{n'^{5}}=\frac{\beta^{(5)}}{n^{5}}-\frac{1}{2}\frac{\beta^{(3)}}{n^{3}}\left(\frac{1}{n}-\frac{1}{n'}\right)\rho\left[\left(\frac{1}{n}-\frac{1}{n'}\right)\rho+2\frac{\tau}{n}\right]-\frac{1}{8}\rho\left[\left(\frac{1}{n^{2}}-\frac{1}{n'^{2}}\right)^{2}\frac{\tau}{n}\rho^{3}\right.$$

$$\left.+\frac{1}{n}\left(\frac{\rho^{2}}{n'^{2}}-\frac{\tau^{2}}{n^{2}}\right)^{3}-\frac{1}{n'}\left(\frac{\rho^{2}}{n^{2}}-\frac{\tau^{2}}{n^{2}}\right)^{3}\right] \tag{11c}$$

$$\frac{\beta'^{(7)}}{n'^{7}}=\frac{\beta^{(7)}}{n^{7}}-\frac{1}{2}\frac{\beta^{(3)2}}{n^{6}}\left(\frac{1}{n}-\frac{1}{n'}\right)\rho-\frac{1}{2}\frac{\beta^{(5)}}{n^{5}}\left(\frac{1}{n}-\frac{1}{n'}\right)\rho\left[\left(\frac{1}{n}-\frac{1}{n'}\right)\rho+2\frac{\tau}{n}\right]$$

$$-\frac{1}{2}\frac{\beta^{(3)}}{n^{3}}\rho\left[\frac{1}{4}\left(\frac{1}{n^{2}}-\frac{1}{n'^{2}}\right)^{2}\rho^{3}-\frac{1}{n}\frac{\tau}{n}\left(\frac{\rho^{2}}{n'^{2}}-\frac{\tau^{2}}{n^{2}}\right)+\frac{1}{n'}\frac{\tau}{n}\left(\frac{\rho^{2}}{n^{2}}-\frac{\tau^{2}}{n^{2}}\right)\right]$$

$$-\frac{1}{16}\rho\left[\left(\frac{1}{n^{3}}-\frac{1}{n'^{3}}\right)^{2}\left(\frac{1}{n^{2}}+\frac{1}{n'^{2}}\right)\frac{\tau}{n}\rho^{5}+\frac{1}{n}\left(\frac{\rho^{2}}{n'^{2}}-\frac{\tau^{2}}{n^{2}}\right)^{3}\left(\frac{\rho^{2}}{n'^{2}}+\frac{\tau^{2}}{n^{2}}\right)\right.$$

$$\left.-\frac{1}{n'}\left(\frac{\rho^{2}}{n^{2}}-\frac{\tau^{2}}{n^{2}}\right)^{3}\left(\frac{\rho^{2}}{n^{2}}+\frac{\tau^{2}}{n^{2}}\right)\right] \tag{11d}$$

If parallel light is incident upon a lens surface, $\tau=\beta^{(3)}=\beta^{(5)}=\ldots=0$. If also $n=1$ then equations (11) are greatly simplified, as follows:

$$\frac{\tau'}{n'}=\left(1-\frac{1}{n'}\right)\rho \tag{12a}$$

$$\frac{\beta'^{(3)}}{n'^{3}}=\frac{1}{2}\left(1-\frac{1}{n'}\right)\rho\frac{\rho^{2}}{n'} \tag{12b}$$

$$\frac{\beta'^{(5)}}{n'^{5}}=\frac{1}{8}\left(1-\frac{1}{n'^{3}}\right)\rho^{3}\frac{\rho^{2}}{n'} \tag{12c}$$

$$\frac{\beta'^{(7)}}{n'^{7}}=\frac{1}{16}\left(1-\frac{1}{n'^{5}}\right)\rho^{5}\frac{\rho^{2}}{n'} \tag{12d}$$

The higher-order equations in these last two sets may be written in a similar manner.

3. DISCUSSION

(a) INHERENT AND PROPAGATED ABERRATION

If the incident pencil is free from aberration, as is the case when it arises from a point on a material object, then $\beta^{(3)} = \beta^{(5)} = \ldots = 0$ and all terms in the aberration of the refracted pencil vanish except those in the upper row of Table 3; that is, all except the last term in each of the equations of (9), (10), or (11). This remaining term in each order will be referred to as the inherent aberration of that order, since it is independent of the aberration due to the other surfaces of the system and is entirely due to the surface in question. It is, of course, affected by a change of object distance.

If aberration is present in the incident pencil, then the terms of the image aberration which have a β coefficient do not, in general, vanish. These terms are to be considered as representing the aberration due to the preceding surfaces of the system after its propagation through the surface under consideration, and will be designated in the following manner:

The term in the fifth order equation ((9c), (10c), or (11c)) which contains the factor $\beta^{(3)}$ will be referred to as the third order term of the fifth order or the $\beta'^{(5)}$ $(\beta^{(3)})$ term. Similarly, the third order squared term of the seventh order denotes the $\beta'^{(7)}$ $(\beta^{(3)2})$ term, and the third-fifth order term of the ninth order denotes the $\beta'^{(9)}$ $(\beta^{(3)} \beta^{(5)})$ term, etc. These distinctions will apply directly to the system of equations to be developed later (Pt. III) except that the symbol $\digamma$ will replace β.

(b) NOTES ON NUMERICAL APPLICATIONS

There are a number of similarities among different terms of these equations which make the computation easier than would appear at first sight. The $\beta'^{(i)}(\beta^{(j)})$ term is identical with the $\beta'^{(i+2)}(\beta^{(j+2)})$ term, the $\beta'^{(i+4)}(\beta^{(j+4)})$ term, etc., except for factors in μ'^2. Moreover, the $\beta'^{(i)}(\beta^{(3)2})$, $\beta'^{(i+2)}(\beta^{(3)}\beta^{(5)})$, $\beta'^{(i+4)}(\beta^{(3)}\beta^{(7)})$, $\beta'^{(i+4)}(\beta^{(5)2})$, and the $\beta'^{(i+4)}(\beta^{(3)3})$ terms differ only by factors in μ'^2, $\frac{1}{2}$, and in one case τ. (i and j may be 3, 5, 7, or 9; $i \gtrless j$.)

It is, of course, optional which form of the equations is to be used in computing. The third order of (10b) is doubtless simpler than that of (9b). However, if orders higher than the third are also to be computed, the equations (9) have their advantages, notably in the marked similarities among the following bracketed quantities:

$$[(\mu'-1)^2 \rho \tau + (\rho^2 + \mu'^2 \tau^2) - \mu'(\rho^2 + \tau^2)]$$

$$[(\mu'^2-1)^2 \rho^3 \tau + (\rho^2 - \mu'^2 \tau^2)^2 - \mu'^3(\rho^2 - \tau^2)^2]$$

$$\left[\frac{1}{4}(\mu'^2-1)^2 \rho^3 - \mu'^2 \tau(\rho^2 - \mu'^2 \tau^2) + \mu'^3 \tau(\rho^2 - \tau^2)\right]$$

$$[(\mu'^2-1)^2(\mu'^2+1)\rho^5 \tau + (\rho^2 - \mu'^2 \tau^2)^2(\rho^2 + \mu'^2 \tau^2) - \mu'^5(\rho^2 - \tau^2)^2(\rho^2 + \tau^2)]$$

Each of these quantities is the sum of three terms. Each term in the last bracket (seventh order) contains as factors the corresponding terms in the first two brackets, except for (μ'^2+1) instead of $(\mu'-1)^2$. Also the terms in the third bracket contain factors of the terms in the second bracket. All this reduces considerably the labor involved in

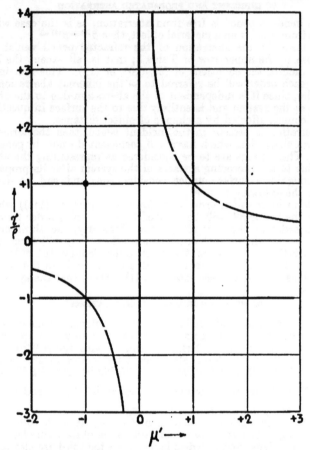

FIGURE 3.—*Zero reciprocal spherical aberration contour*

The point at $\mu'=-1$, $\frac{r}{\rho}=+1$ pertains only to orders higher than the third.

computing. Either of these sets may be written in the reduced form, as equations (11).

(c) ROOTS, ANOMALOUS POINTS, AND SPECIFIC EXAMPLES

It is interesting to note the conditions under which the aberration of a surface vanishes; that is, to find the root of the inherent terms of the equations (9) or (10) when set equal to zero. In equations (10) the factors common to all orders may be written thus:

$$\rho^3\left[\mu'(\mu'-1)\left(1+\frac{\tau}{\rho}\right)\left(1-\mu'\frac{\tau}{\rho}\right)\right]=0$$

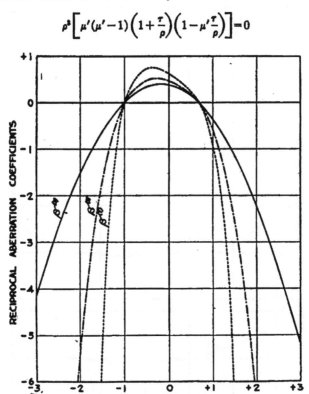

FIGURE 4.—*Reciprocal aberration coefficients for surface of unit radius, $\mu'=1.5$*

When ρ has a finite value not zero the roots of this are as follows:

$\mu'=0$ Physically impossible.
$\mu'=1$ Same index on both sides; no refraction.

or
$\left.\begin{array}{l}\tau/\rho=-1\\[4pt]\tau=-\rho\end{array}\right\}$ Object and image coincide at the vertex.

or
$\left.\begin{array}{l}\mu'\dfrac{\tau}{\rho}=\ 1\\[4pt]\tau=\rho/\mu'\end{array}\right\}$ The well-known aplanatic point.

These roots are shown in Figure 3. If $\rho=0$, the surface is a plane and there is no center of curvature from which to measure the distances. If the object coincides with the center of curvature then

$\tau = \infty$ and the series is nonconvergent. (See Part VII, sec. 1.) Another set of formulas will be derived which may be used in these cases.

The orders higher than the third have the additional roots given by equating to zero the quantities in brackets in the inherent terms of equations (10). The only root yet found for these is the isolated point

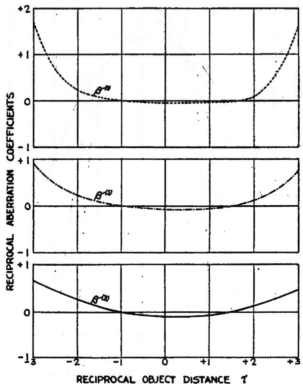

FIGURE 5.—*Reciprocal aberration coefficients for surface of unit radius, $\mu' = 1/1.5$*

$\left(\dfrac{\tau}{\rho} = +1, \ \mu' = -1 \right)$ indicated in Figure 3. Since the value $\mu' = -1$ indicates a reflection, it may be noted that in the image formed by a spherical mirror the reciprocal aberration of all orders higher than the third vanishes when the object is at a distance $2r$ from the vertex.

The advisability of computing and tabulating the values of these coefficients for different values of μ' and τ is being considered. Figures 4 and 5 show, for the values $\mu' = 1.5$ and $\mu' = 1/1.5$, the form of the functions which would result. These values would have to be multiplied by the power of ρ corresponding to the order of the aberration to obtain the actual value of any particular coefficient.

III: TRANSFER FORMULAS

1. DERIVATION

The refracted or image pencil from one surface is, of course, identical with the incident or object pencil of the succeeding surface. However, when the point of reference is changed, the aberration coefficients are also altered. Formulas will now be developed for computing the new coefficients.

Let the subscripts i and j refer, respectively, to a surface whose aberration is known and to the succeeding surface, whose aberration is to be computed. Then, as before, see (2') and (3') for the known surface,

$$\Theta'_i = \tau'_i + \beta'_i{}^{(3)}h'{}_i^2 + \beta'_i{}^{(5)}h'{}_i^4 + \cdots$$
$$\sin \alpha'_i = \Theta'_i h'_i = \tau'_i h'_i + \beta'_i{}^{(3)}h'{}_i^3 + \cdots \tag{13}$$

and for the succeeding surface,

$$\Theta_j = \tau_j + \beta_j{}^{(3)}h_j^2 + \beta_j{}^{(5)}h_j^4 + \cdots \tag{14}$$
$$\sin \alpha_j = \Theta_j h_j = \tau_j h_j + \beta^{(3)}{}_j h^3_j + \cdots \tag{15}$$

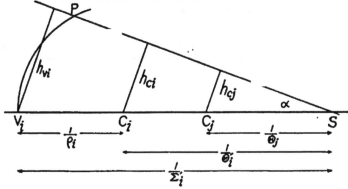

FIGURE 6.—*Diagram showing aperture, h, as measured from different points on the axis*

Let C_i and C_j (fig. 6) be the respective centers of curvature, and let the distance from C_i to C_j be d, positive if to the right. Then, for any ray, as PS, the adding of axial distances gives:

$$\frac{1}{\Theta'_i} = \frac{1}{\Theta_j} + d \tag{16}$$

Similarly, for a paraxial ray

$$\frac{1}{\tau'_i} = \frac{1}{\tau_j} + d, \text{ or } \frac{1}{\tau_j} = \frac{1}{\tau'_i} - d \tag{17}$$

then:

$$\frac{\tau_j}{\tau'_i} = 1 + d\tau_j = \frac{1}{1 - d\tau'_i} = \kappa \tag{18}$$

where κ is a new symbol introduced for convenience.

Manifestly
$$\sin \alpha'_i = \sin \alpha_j$$

or, by equations (13) and (15):

$$h'_t = \frac{1}{\Theta'_t}\Theta_f h_f$$

Then, substituting in turn from equations (16), (14), and (18):

$$h'_t = (1+d\Theta_f)h_f$$
$$= (1+d\tau_f)h_f + d\beta_f^{(3)}h^3_f + \ldots$$
$$= \kappa h_f + d\beta_f^{(3)}h^3_f + d\beta_f^{(5)}h^5_f + \ldots$$

This last expression for h'_t may be substituted in the right-hand member of equation (13), and then since the latter is equal to equation (15)

$$\sin \alpha_f = \tau_f h_f + \beta_f^{(3)}h^3_f + \beta_f^{(5)}h^5_f + \ldots$$
$$= \text{the series expanded [16] in Table 4.}$$

Since this equation is true for any value of h_f the coefficients of like powers of h_f may then be equated.

The first order equation is

$$\tau_f = \kappa\tau'_t \tag{18a}$$

[16] See footnote 15, p. 198.

TABLE 4.—sin α_j

[Each term in the body of the table is to be multiplied both by the power of h_j at the top of the column and by the β'_j—coefficient at the left]

	h_j	h_j^3	h_j^5	h_j^7	h_j^9	h_j^{11}
γ'_j	z	$+4\beta'_j{}^{(1)}$	$+4\beta'_j{}^{(3)}$	$+4\beta'_j{}^{(5)}$	$+4\beta'_j{}^{(7)}$	$+4\beta'_j{}^{(9)}$
$\beta'_j{}^{(1)}$		$+z^3$	$+3z^2 4\beta'_j{}^{(1)}$	$+3z^2 4\beta'_j{}^{(3)}+3z^2 4\beta'_j{}^{(5)\,2}$	$+3z^2 4\beta'_j{}^{(5)}+6z^2 4\beta'_j{}^{(1)}\beta'_j{}^{(3)}+4^2\beta'_j{}^{(1)\,3}$	$+3z^2 4\beta'_j{}^{(7)}+6z^2 4\beta'_j{}^{(1)}\beta'_j{}^{(5)}+3z^2 4\beta'_j{}^{(3)\,2}$
$\beta'_j{}^{(3)}$			$+z^5$	$+5z^4 4\beta'_j{}^{(1)}$	$+5z^4 4\beta'_j{}^{(3)}+10z^3 4\beta'_j{}^{(1)\,2}$	$+5z^4 4\beta'_j{}^{(5)}+20z^3 4\beta'_j{}^{(1)}\beta'_j{}^{(3)}+10z^2 4\beta'_j{}^{(1)\,3}$
$\beta'_j{}^{(5)}$				$+z^7$	$+7z^6 4\beta'_j{}^{(1)}$	$+7z^6 4\beta'_j{}^{(3)}+21z^5 4\beta'_j{}^{(1)\,2}$
$\beta'_j{}^{(7)}$					$+z^9$	$+9z^8 4\beta'_j{}^{(1)}$
$\beta'_j{}^{(9)}$						$+z^{11}$

The third order, after combining terms, is

$$(1-d\tau'_t)\beta_J{}^{(3)} = \kappa^2\beta'_t{}^{(3)}$$

or, substituting from equation (18)

$$\beta_J{}^{(3)} = \kappa^4\beta'_t{}^{(3)} \tag{18b}$$

The fifth order, likewise, is

$$(1-d\tau_t')\beta_J{}^{(5)} = 3\kappa^2 d\beta_J{}^{(3)}\beta_t{}'{}^{(3)} + \kappa^3\beta'_t{}^{(5)}$$

Substituting for $(1-d\tau_t')$ and $\beta_J{}^{(3)}$ from equations (18) and (18b):

$$\beta_J{}^{(5)} = \kappa^6(\beta'_t{}^{(5)} + 3\kappa d\beta'_t{}^{(3)2}) \tag{18c}$$

The higher orders may be reduced in a similar manner and the transfer formulas summarized as follows:

$$\kappa = \frac{1}{1-d\tau'_t} \tag{18}$$

$$\tau_J = \kappa\tau'_t \tag{18a}$$

$$\beta_J{}^{(3)} = \kappa^4\beta'_t{}^{(3)} \tag{18b}$$

$$\beta_J{}^{(5)} = \kappa^6(\beta'_t{}^{(5)} + 3\kappa d\beta'_t{}^{(3)2}) \tag{18c}$$

$$\beta_J{}^{(7)} = \kappa^8(\beta'_t{}^{(7)} + 8\kappa d\beta'_t{}^{(3)}\beta'_t{}^{(5)} + 12\kappa^2 d^2\beta'_t{}^{(3)3}) \tag{18d}$$

$$\beta_J{}^{(9)} = \kappa^{10}(\beta'_t{}^{(9)} + 10\kappa d\beta'_t{}^{(3)}\beta'_t{}^{(7)} + 5\kappa d\beta'_t{}^{(5)2} + 55\kappa^2 d^2\beta'_t{}^{(3)2}\beta'_t{}^{(5)}$$
$$+ 55\kappa^3 d^3\beta'_t{}^{(3).4}) \tag{18e}$$

$$\beta_J{}^{(11)} = \kappa^{12}[\beta'_t{}^{(11)} + 12\kappa d(\beta'_t{}^{(3)}\beta'_t{}^{(9)} + \beta'_t{}^{(5)}\beta'_t{}^{(7)}) + 78\kappa^2 d^2(\beta'_t{}^{(3)2}\beta'_t{}^{(7)}$$
$$+ \beta'_t{}^{(3)}\beta'_t{}^{(5)2}) + 364\kappa^3 d^3\beta'_t{}^{(3)3}\beta'_t{}^{(5)} + 273\kappa^4 d^4\beta'_t{}^{(3)6}] \tag{18f}$$

2. NOTES ON GENERALITY

These transfer formulas are more general than has been indicated By definition, the series

$$\Theta = \tau + \beta^{(3)}h_c^2 + \beta^{(5)}h_c^4 + \ldots$$

refers exclusively to the center of curvature of a lens surface. It is evident, however, that a similar series may be written referring to any point on the axis and that the transfer formulas will apply in transferring to any other point on the axis if Θ, τ, β, and h_c are replaced by the corresponding symbols from the new series. Moreover, these formulas apply as well to some of the series with different definitions of h mentioned in the introduction, the condition for their validity being

$$\Psi h = f(\alpha)$$

where Ψ corresponds to Θ above, but may refer to any point on the axis and $f(\alpha)$ is any function of α the use of which results in a convergent series throughout the required range of h for each case in question.

3. TRANSFER FORMULAS INVOLVING POWERS OF n

If the reduced form of the aberration equations (equations (11) or (12)) is used, then the transfer formulas must also be divided by the corresponding powers of n. They are then as follows:

$$\kappa = \frac{1}{1 - n_i' d \frac{\tau_i'}{n_i'}} \tag{19}$$

$$\frac{\tau_j}{n_j} = \kappa \frac{\tau_i'}{n_i'} \tag{19a}$$

$$\frac{\beta_j^{(3)}}{n_j^3} = \kappa^4 \frac{\beta_i'^{(3)}}{n_i'^3} \tag{19b}$$

$$\frac{\beta_j^{(5)}}{n_j^5} = \kappa^6 \left(\frac{\beta_i'^{(5)}}{n_i'^5} + 3\kappa n_i' d \frac{\beta_i'^{(3)2}}{n_i'^6} \right) \tag{19c}$$

$$\frac{\beta_j^{(7)}}{n_j^7} = \kappa^8 \left(\frac{\beta_i'^{(7)}}{n_i'^7} + 8\kappa n_i' d \frac{\beta_i'^{(3)} \beta_i'^{(5)}}{n_i'^8} + 12\kappa^2 n_i'^2 d^2 \frac{\beta_i'^{(3)3}}{n_i'^9} \right) \tag{19d}$$

Other reduced forms of aberration equations are to be given later. The various forms of the transfer formulas are collected here for convenience in reference.

Center of curvature to vertex:

$$n\kappa = \frac{1}{-\frac{1}{n} - d \frac{\tau_i'}{n_i'}} \tag{20}$$

$$n_j \sigma_j = n (n\kappa) \frac{\tau_i'}{n_i'} \tag{20a}$$

$$n_j \varsigma_j^{(3)} = (n\kappa)^4 \frac{\beta_i'^{(3)}}{n_i'^3} \tag{20b}$$

$$n_j \varsigma_j^{(5)} = (n\kappa)^6 \left[\frac{\beta_i'^{(5)}}{n_i'^5} + 3 (n\kappa) d \frac{\beta_i'^{(3)2}}{n_i'^6} \right] \tag{20c}$$

$$n_j \varsigma_j^{(7)} = (n\kappa)^8 \left[\frac{\beta_i'^{(7)}}{n_i'^7} + 8 (n\kappa) d \frac{\beta_i'^{(3)} \beta_i'^{(5)}}{n_i'^8} + 12 (n\kappa d)^2 \frac{\beta_i'^{(3)3}}{n_i'^9} \right] \tag{20d}$$

Vertex to center of curvature:

$$\frac{\kappa}{n} = \frac{1}{n - dn_i' \sigma_i'} \tag{21}$$

$$\frac{\tau_j}{n_j} = \frac{1}{n} \frac{\kappa}{n} n_i' \sigma_i' \tag{21a}$$

$$\frac{\beta_j^{(3)}}{n_j^3} = \left(\frac{\kappa}{n} \right)^4 n_i' \varsigma_i'^{(3)} \tag{21b}$$

$$\frac{\beta_j^{(5)}}{n_j^5} = \left(\frac{\kappa}{n} \right)^6 \left[n_i' \varsigma_i'^{(5)} + 3 \frac{\kappa}{n} d (n_i' \varsigma_i'^{(3)})^2 \right] \tag{21c}$$

$$\frac{\beta_j^{(7)}}{n_j^7} = \left(\frac{\kappa}{n} \right)^8 \left[n_i' \varsigma_i'^{(7)} + 8 \frac{\kappa}{n} d (n_i' \varsigma_i'^{(3)}) (n_i' \varsigma_i'^{(5)}) \right.$$
$$\left. + 12 \left(\frac{\kappa}{n} d \right)^2 (n_i' \varsigma_i'^{(3)})^3 \right] \tag{21d}$$

Vertex to vertex:

$$\kappa = \frac{1}{1 - \dfrac{d}{n} n_i' \sigma_i'} \tag{22}$$

$$n_j \sigma_j = \kappa n_i' \sigma_i' \tag{22a}$$

$$n_j \zeta_j'^{(3)} = \kappa^4 n_i' \zeta_i'^{(3)} \tag{22b}$$

$$n_j \zeta_j'^{(5)} = \kappa^6 \left[n_i' \zeta_i'^{(5)} + 3\kappa \frac{d}{n} (n_i' \zeta_i'^{(3)})^2 \right] \tag{22c}$$

$$n_j \zeta_j'^{(7)} = \kappa^8 \left[n_i' \zeta_i'^{(7)} + 8\kappa \frac{d}{n} (n_i' \zeta_i'^{(3)})(n_i' \zeta_i'^{(5)}) \right.$$

$$\left. + 12 \left(\kappa \frac{d}{n} \right)^2 (n_i' \zeta_i'^{(3)})^3 \right] \tag{22d}$$

Higher orders are derived from equations (18e) and (18f) in a similar manner.

IV. THE RECIPROCAL ABERRATION OF A SINGLE SURFACE REFERRED TO THE VERTEX

A second set of formulas will now be investigated in which the point of reference is chosen at the vertex of the surface. These are to be used in the cases where the formulas referred to the center of curvature are not applicable (see p. 203), although in many cases either set may be used.

1. DERIVATION OF THE FORMULAS

As has been stated, aberration series of the form (2) or (2') may be written referring to any point on the axis. Let the two which refer to the vertex of a surface (as V_i, fig. 6) be

$$\Sigma = \sigma + \zeta^{(3)} h_s^2 + \zeta^{(5)} h_s^4 + \zeta^{(7)} h_s^6 + \ldots \tag{23}$$

for the incident pencil and

$$\Sigma' = \sigma' + \zeta'^{(3)} h_s'^2 + \zeta'^{(5)} h_s'^4 + \zeta'^{(7)} h_s'^6 + \ldots \tag{23'}$$

for the refracted pencil. Probably the easiest way to establish a relation between these two series is to transfer the aberration to the center of curvature by means of equations (18) so that the resulting coefficients are identical with those of series (2) and (2') and then to substitute these coefficients in equations (10).

Let κ and κ' be the constants of transfer in the object and image spaces, respectively. Then substituting the appropriate values for the object space in the formulas (18), the following relations are derived:

$$\kappa = \frac{1}{1-r\sigma} = 1 + r\tau = \frac{\rho}{\rho-\sigma} = \frac{\rho+\tau}{\rho} \tag{24}$$

$$\tau = \kappa\sigma \tag{24a}$$

$$\beta^{(3)} = \kappa^4 \zeta^{(3)} \tag{24b}$$

$$\beta^{(5)} = \kappa^6 (\zeta^{(5)} + 3\kappa\tau\zeta^{(3)2}) \tag{24c}$$

$$\beta^{(7)} = \kappa^8 (\zeta^{(7)} + 8\kappa\tau\zeta^{(3)}\zeta^{(5)} + 12\kappa^2\tau^2\zeta^{(3)3}) \tag{24d}$$

$$\beta^{(9)} = \kappa^{10} (\zeta^{(9)} + 10\kappa\tau\zeta^{(3)}\zeta^{(7)} + 5\kappa\tau\zeta^{(5)2} + 55\kappa^2\tau^2\zeta^{(3)2}\zeta^{(5)} + 55\kappa^3\tau^3\zeta^{(3)4}) \tag{24e}$$

By a similar substitution and also by reason of equations (10a) and (24)

$$\kappa' = \frac{\rho}{\rho-\sigma} = \frac{\rho+\tau}{\rho} = \frac{\mu'(\rho+\tau)}{\rho} = \mu'\kappa \tag{25}$$

in the case of the refracted pencil. Continuing the substitution

$$\tau' = \kappa'\sigma' = \mu'\kappa\sigma' \tag{25a}$$

$$\beta'^{(3)} = \mu'^4\kappa^4\zeta'^{(3)} \tag{25b}$$

and similarly the higher orders differ in form from (24c to 24e) only by the use of $\mu'\kappa$ instead of κ.

Substituting these values throughout equations (10), dividing each equation by the power of $\mu'\kappa$ appearing as a factor of the left member, and writing $\frac{1}{\mu'} = \frac{n}{n'} = \mu$

$$\sigma' = \left(\sigma + \frac{\rho}{\kappa}\right) - \mu\frac{\rho}{\kappa} \tag{26a}$$

$$\zeta'^{(3)} = \mu\zeta^{(3)} + \frac{1}{2}\mu(1-\mu)\frac{\rho}{\kappa^2}\left(\frac{\rho}{\kappa} + \sigma\right)\left(\mu\frac{\rho}{\kappa} - \sigma\right) \tag{26b}$$

$$(\zeta'^{(5)} + 3\mu'\kappa\tau\zeta'^{(3)2})$$
$$= \mu(\zeta^{(5)} + 3\kappa\tau\zeta^{(3)2}) - \frac{1}{2}\mu\zeta^{(3)}(1-\mu)\frac{\rho}{\kappa} \cdot$$
$$\cdot[(1-\mu)\frac{\rho}{\kappa} + 2\sigma] + \frac{1}{8}\mu(1-\mu)\frac{\rho}{\kappa^2}\left(\frac{\rho}{\kappa} + \sigma\right)\left(\mu\frac{\rho}{\kappa} - \sigma\right) \cdot$$
$$\cdot[(1 + \mu + \mu^2)\frac{\rho^2}{\kappa^2} - (1-\mu)\frac{\rho}{\kappa}\sigma + \sigma^2] \tag{26c}$$

$$(\zeta'^{(7)} + 8\mu'\kappa\tau\zeta'^{(3)}\zeta'^{(5)} + 12\mu'^2\kappa^2\tau^2\zeta'^{(3)3})$$
$$= \mu(\zeta^{(7)} + 8\kappa\tau\zeta^{(3)}\zeta^{(5)} + 12\kappa^2\tau^2\zeta^{(3)3}) - \frac{1}{2}\mu\zeta^{(3)2}(1-\mu)\rho$$
$$- \frac{1}{2}\mu(\zeta^{(5)} + 3\kappa\tau\zeta^{(3)2})(1-\mu)\frac{\rho}{\kappa}[(1-\mu)\frac{\rho}{\kappa} + 2\sigma]$$
$$- \frac{1}{8}\mu\zeta^{(3)}(1-\mu)\frac{\rho}{\kappa}[(1 + \mu + \mu^2 + \mu^3)\frac{\rho^3}{\kappa^3} + 4\mu\frac{\rho^2}{\kappa^2}\sigma + 4\sigma^3]$$
$$+ \frac{1}{16}\mu(1-\mu)\frac{\rho}{\kappa^2}\left(\frac{\rho}{\kappa} + \sigma\right)\left(\mu\frac{\rho}{\kappa} - \sigma\right)[(1 + \mu + \mu^2 + \mu^3 + \mu^4)\frac{\rho^4}{\kappa^4}$$
$$- (1-\mu^3)\frac{\rho^3}{\kappa^3}\sigma + (1+\mu^2)\frac{\rho^2}{\kappa^2}\sigma^2 - (1-\mu)\frac{\rho}{\kappa}\sigma^3 + \sigma^4] \tag{26d}$$

$$(\zeta'^{(9)} + 10\mu'\kappa r \zeta'^{(3)}\zeta'^{(7)} + 5\mu'\kappa r \zeta'^{(5)2}$$
$$+ 55\mu''\kappa^2 r^2 \zeta^{(3)2}\zeta'^{(5)} + 55\mu'^3\kappa^3 r^3 \zeta'^{(3)4})$$
$$= \mu(\zeta'^{(9)} + 10\kappa r \zeta^{(3)}\zeta^{(7)} + 5\kappa r \zeta^{(5)2} + 55\kappa^2 r^2 \zeta^{(3)2}\zeta^{(5)}$$
$$+ 55\kappa^2 r^3 \zeta^{(3)4}) - \mu(\zeta^{(3)}\zeta^{(5)} + 3\kappa r \zeta^{(3)3})(1-\mu) - \frac{1}{2}\mu\cdot$$
$$\cdot(\zeta^{(7)} + 8\kappa r \zeta^{(3)}\zeta^{(5)} + 12\kappa^2 r^2 \zeta^{(3)3})(1-\mu)\frac{\rho}{\kappa}\left[(1-\mu)\frac{\rho}{\kappa} + 2\sigma\right]$$
$$-\frac{1}{4}\mu\zeta^{(3)2}(1-\mu)\rho\left[\mu\frac{\rho^2}{\kappa^2} + 3\sigma^2\right] - \frac{1}{8}\mu(\zeta^{(5)} + 3\kappa r \zeta^{(3)2})(1-\mu)\frac{\rho}{\kappa}\cdot$$
$$\cdot\left[(1 + \mu - \mu^2 - \mu^3)\frac{\rho^3}{\kappa^3} + 4\mu\frac{\rho^2}{\kappa^2}\sigma + 4\sigma^3\right] - \frac{1}{16}\mu\zeta^{(3)}(1-\mu)\frac{\rho}{\kappa}\cdot$$
$$\cdot\left[(1 + \mu - \mu^4 - \mu^5)\frac{\rho^5}{\kappa^5} + 2(\mu + \mu^3 + \mu^3)\frac{\rho^4}{\kappa^4}\sigma + 4\mu\frac{\rho^2}{\kappa^2}\sigma^3 + 6\sigma^5\right]$$
$$+\frac{1}{128}\mu(1-\mu)\frac{\rho}{\kappa^2}\left(\frac{\rho}{\kappa} + \sigma\right)\left(\mu\frac{\rho}{\kappa} - \sigma\right)\left[5(1 + \mu + \mu^2 + \mu^3 + \mu^4 + \mu^5 + \mu^6)\frac{\rho^6}{\kappa^6}\right.$$
$$- (5 + \mu + \mu^2 - \mu^3 - \mu^4 - \mu^5)\frac{\rho^5}{\kappa^5}\sigma + (5 + \mu + 3\mu^2 + \mu^3 + 5\mu^4)\frac{\rho^4}{\kappa^4}\sigma^2$$
$$\left. - (5 - \mu + \mu^2 - 5\mu^3)\frac{\rho^3}{\kappa^3}\sigma^3 + (5 - \mu + 5\mu^2)\frac{\rho^2}{\kappa^2}\sigma^4 - (5 - 5\mu)\frac{\rho}{\kappa}\sigma^5 + 5\sigma^6\right] \tag{26e}$$

Equation (26b) is used in eliminating $\zeta'^{(3)}$ from equations (26c) to (26e). Following this step, the resulting form of equation (26c) is used in eliminating $\zeta'^{(5)}$ from the higher order equations, and so on. Remembering at the same time that, from equation (24), $\frac{\rho}{\kappa} = (\rho - \sigma)$ and $\left(\frac{\rho}{\kappa} + \sigma\right) = \rho$. The equations may be brought into the following form:

$$\sigma' = \rho - \mu(\rho - \sigma) = \mu\sigma - (\mu - 1)\rho \tag{27a}$$

$$\zeta'^{(3)} = \mu\zeta^{(3)} - \frac{1}{2}\mu(\mu - 1)(\rho - \sigma)^2[\mu(\rho - \sigma) - \sigma] \tag{27b}$$

$$\zeta'^{(5)} = \mu\zeta^{(5)} + \frac{1}{2}\mu\zeta^{(3)}(\mu - 1)(\rho - \sigma)[(5\mu + 1)(\rho - \sigma) - 4\sigma] - \frac{1}{8}\mu(\mu - 1)\cdot$$
$$\cdot(\rho - \sigma)^2[\mu(\rho - \sigma) - \sigma]\,[(7\mu^2 - 5\mu + 1)(\rho - \sigma)^2 - (5\mu - 5)(\rho - \sigma)\sigma + \sigma^3] \tag{27c}$$

$$\zeta'^{(7)} = \mu\zeta^{(7)} - \frac{1}{2}\mu\zeta^{(3)2}(\mu - 1)[(7\mu + 4)(\rho - \sigma) - 3\sigma]$$
$$+\frac{1}{2}\mu\zeta^{(5)}(\mu - 1)(\rho - \sigma)[(7\mu + 1)(\rho - \sigma) - 6\sigma]$$
$$+\frac{1}{8}\mu\zeta^{(3)}(\mu - 1)(\rho - \sigma)[(63\mu^3 - 33\mu^2 - 7\mu + 1)(\rho - \sigma)^3$$
$$- (96\mu^2 - 68\mu - 8)(\rho - \sigma)^2\sigma + (40\mu - 32)(\rho - \sigma)\sigma^2 - 4\sigma^3]$$
$$-\frac{1}{16}\mu(\mu - 1)(\rho - \sigma)^2[\mu(\rho - \sigma) - \sigma(33)[\mu^4 - 47\mu^3 + 25\mu^2 - 7\mu + 1)(\rho + - \sigma)^4$$
$$- (47\mu^3 - 80\mu^2 + 40\mu - 7)(\rho - \sigma)^3\sigma + (25\mu^2 - 40\mu + 17)(\rho - \sigma)^2\sigma^2$$
$$- (7\mu - 7)(\rho - \sigma)\sigma^3 + \sigma^4] \tag{27d}$$

$$\zeta'^{(9)} = \mu\zeta^{(9)} - \mu\frac{1}{\rho-\sigma}\zeta^{(3)3}(\mu-1)[(411\mu+81)(\rho-\sigma)-330\sigma] - \mu\zeta^{(3)}\zeta^{(5)}.$$

$$\cdot(\mu-1)[(9\mu+5)(\rho-\sigma)-4\sigma] + \frac{1}{2}\mu\zeta^{(7)}(\mu-1)(\rho-\sigma)[(9\mu+1)(\rho-\sigma)$$

$$-8\sigma] - \frac{1}{8}\mu\zeta^{(3)2}(\mu-1)[(198\mu^3-42\mu^2-80\mu-3)(\rho-\sigma)^3$$

$$-(240\mu^2-116\mu-75)(\rho-\sigma)^2\sigma+(70\mu-51)(\rho-\sigma)\sigma^2+41\sigma^3]$$

$$+\frac{1}{8}\mu\zeta^{(5)}(\mu-1)(\rho-\sigma)[(99\mu^3-61\mu^2-9\mu+1)(\rho-\sigma)^3-(160\mu^2-124\mu$$

$$-10)\cdot$$

$$\cdot(\rho-\sigma)^2\sigma+(70\mu-60)(\rho-\sigma)\sigma^2-6\sigma^3] + \frac{1}{16}\mu\zeta^{(3)}(\mu-1)(\rho-\sigma)\cdot$$

$$\cdot[(429\mu^5-531\mu^4+140\mu^3+20\mu^2-9\mu+1)(\rho-\sigma)^5$$
$$-(960\mu^4-1,372\mu^3+428\mu^2+38\mu-10)(\rho-\sigma)^4\sigma$$
$$+(770\mu^3-1,190\mu^2+1,320\mu+20)(\rho-\sigma)^3\sigma^2-(280\mu^2-394\mu+130)\cdot$$

$$\cdot(\rho-\sigma)^2\sigma^3+(50\mu-40)(\rho-\sigma)\sigma^4-4\sigma^5] - \frac{1}{128}\mu(\mu-1)(\rho-\sigma)^2\cdot$$

$$\cdot[\mu(\rho-\sigma)-\sigma][(715\mu^6-1,525\mu^5+1,335\mu^4-665\mu^3+215\mu^2-35\mu+5).$$
$$\cdot(\rho-\sigma)^6-(1,525\mu^5-3,671\mu^4+2,709\mu^3-1,489\mu^2+371\mu-35)(\rho-\sigma)^5\sigma$$
$$+(1,265\mu^4-3,189\mu^3+2,873\mu^2-1,099\mu+165)(\rho-\sigma)^4\sigma^2$$
$$-575\mu^3-1,369\mu^2+1,049\mu-275)(\rho-\sigma)^3\sigma^3+(215\mu^2-371\mu+165)\cdot$$
$$\cdot(\rho-\sigma)^2\sigma^4-(35\mu-35)(\rho-\sigma)\sigma^5+5\sigma^6] \qquad (27e)$$

It may, at times, be preferable to multiply these equations by n' and write them in the following reduced form

$$n'\sigma' = n\sigma - (n-n')\rho \qquad (28a)$$

$$n'\zeta'^{(3)} = n\zeta^{(3)} - \frac{1}{2}n(\mu-1)(\rho-\sigma)^2[\mu(\rho-\sigma)-\sigma] \qquad (28b)$$

$$n'\zeta'^{(5)} = n\zeta^{(5)} + \frac{1}{2}n\zeta^{(3)}(\mu-1)(\rho-\sigma)[(5\mu+1)(\rho-\sigma)-4\sigma]$$
$$-\frac{1}{8}n(\mu-1)(\rho-\sigma)^2[\mu(\rho-\sigma)-\sigma][(7\mu^2-5\mu \qquad (28c)$$
$$+1)(\rho-\sigma)^2-(5\mu-5)(\rho-\sigma)\sigma+\sigma^2]$$

and similarly for higher orders. In computations by means of these equations it is not necessary to find the ζ coefficients explicitly, but only the "reduced coefficients," $n\zeta^{(3)}$, $n\zeta^{(5)}$, etc.

2. DISCUSSION

It is evident that these formulas are considerably more complex in form than those which refer to the center of curvature. They are included here as supplementary formulas to be applied where the other system is unsuitable. The eleventh order has not been included partly because of its unwieldiness and partly because in the cases where these equations will be most frequently applied, namely, surfaces of small or zero curvature or where the object is near the center of curvature, the inherent aberration will be small and fewer orders will, in general, be required than for the other surfaces of the system.

The zero points common to all orders of inherent aberration are given by

$$\mu(\mu-1)\left(\frac{\sigma}{\rho}-1\right)^2\left[\mu\left(\frac{\sigma}{\rho}-1\right)+\frac{\sigma}{\rho}\right]\rho^2=0$$

$\mu=0$, physically impossible.

$(\mu-1)=0$, the same root as in the other system.

$\left(\frac{\sigma}{\rho}-1\right)^2=0$, a double root at the center of curvature.

$(\mu+1)\left(\frac{\sigma}{\rho}-1\right)=-1$, the aplanatic point again.

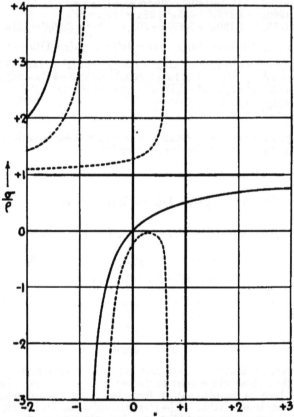

FIGURE 7.—*Zero reciprocal spherical aberration contour*

The solid lines indicate zero values of all orders. The dotted curve indicates an additional root of the fifth order. Additional roots of higher orders are not shown.

These roots are shown for $\rho=1$ in Figure 7, as is also an additional root of the fifth order. The additional roots of the higher orders have not been investigated.

Figures 8 and 9, corresponding to Figures 4 and 5, show the values of the inherent aberration coefficients for $\mu = 1/1.5$ and $\mu = 1.5$ when $\rho = 1$. It is to be remembered that $\mu = 1/1.5$ corresponds to $\mu' = 1.5$ in Figure 4.

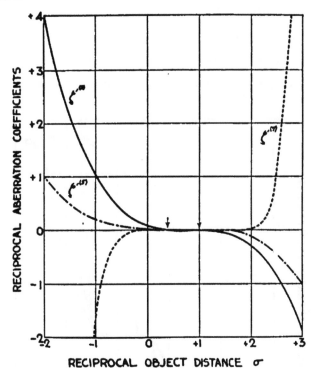

FIGURE 8.—*Reciprocal aberration coefficients for single surface of unit radius, $\mu = 1/1.5$*

V. INVERSION TO LONGITUDINAL SPHERICAL ABERRATION

Given the reciprocal aberration of a pencil of rays, the longitudinal aberration may be derived by expanding the brackets in the following equation

$$S = \frac{1}{\Sigma} = \frac{1}{\sigma + \zeta^{(3)}h^2 v + \zeta^{(5)}h^4 v + \ldots}$$

$$= \frac{1}{\sigma\left(1 + \frac{1}{\sigma}\zeta^{(3)}h^2{}_\bullet + \frac{1}{\sigma}\zeta^{(5)}h^4 v + \ldots\right)}$$

$$= s[1 + (s\zeta^{(3)}h^2{}_\bullet + s\zeta^{(5)}h^4{}_\bullet + s\zeta^{(7)}h^6{}_\bullet + s\zeta^{(9)}h^8{}_\bullet + s\zeta^{(11)}h^{10}{}_\bullet)]^{-1}$$

according to the formula $(1+x)^{-1} = 1 - x + x^2 - x^3 + \ldots$ If the longitudinal aberration be expressed as before. (See Introduction.)

$$S = s + Ah^2 + Bh^4 + Ch^6 + Dh^8 + Eh^{10} \tag{1}$$

the following formulas are found for converting reciprocal into longitudinal aberration:

$$A = -s^3 \zeta^{(3)} \tag{29a}$$

$$B = -s^2(\zeta^{(5)} - s\zeta^{(3)2}) \tag{29b}$$

$$C = -s^2(\zeta^{(7)} - 2s\zeta^{(3)}\zeta^{(5)} + s^2\zeta^{(3)3}) \tag{29c}$$

$$D = -s^2[\zeta^{(9)} - s(2\zeta^{(3)}\zeta^{(7)} + \zeta^{(5)2}) + 3s^2\zeta^{(3)2}\zeta^{(5)} - s^3\zeta^{(3)4}] \tag{29d}$$

$$E = -s^2[\zeta^{(11)} - 2s(\zeta^{(3)}\zeta^{(9)} + \zeta^{(5)}\zeta^{(7)}) + 3s^2(\zeta^{(3)2}\zeta^{(7)} + \zeta^{(3)}\zeta^{(5)2}) \\ -4s^3\zeta^{(3)3}\zeta^{(5)} + s^4\zeta^{(3)4}] \tag{29e}$$

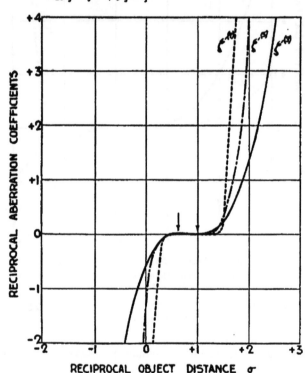

FIGURE 9.—*Reciprocal aberration coefficients for single surface of unit radius, $\mu=1.5$*

These formulas are quite general. If the aberration refers to some point other than the vertex of a surface, it is necessary only to replace these symbols by the corresponding ones in the new series. The parameter may be taken as any function of the aperture and of any constants (as, for instance, the constants of a surface of the lens) so long as a convergent series result both for the reciprocal and for the longitudinal aberrations.

VI. NUMERICAL EXAMPLE

1. DETAILS OF CONSTRUCTION OF THE LENS

As an illustration of the application of the formulas derived in the preceding parts of the paper, the reciprocal spherical aberration of the third, fifth, and seventh orders has been computed for an ordinary landscape lens such as is found in most low-priced cameras. The details of construction of the lens as given by Conrady [17] and the necessary reciprocals are as follows, except that the decimal point has been changed in order to bring it near the first significant figure both in the lengths and in their reciprocals:

$$r_1 = -0.5556 \qquad\qquad\qquad\qquad\qquad \rho_1 = -1.7999$$
$$n_{1,\,2} = 1.63487 \qquad d_{1,\,2} = 0.01 \qquad n^{-1}{}_{1,\,2} = 0.61167 \qquad \rho_2 = +0.67024$$
$$r_2 = +1.492$$
$$n_{2,\,3} = 1.54712 \qquad d_{2,\,3} = 0.03 \qquad n^{-1}{}_{2,\,3} = 0.64636$$
$$r_3 = -0.2506 \qquad\qquad\qquad\qquad\qquad \rho_3 = -3.9904$$

The formulas (12), (11), and (28) will be used at the respective surfaces. The final longitudinal aberration will then be derived and compared with the results of trigonometric ray tracing.

2. FIRST SURFACE; INFINITE OBJECT DISTANCE

Since $\tau_1 = 0$ and $n_1 = 1$, the simplified formulas (12) may be used at the first surface. The numerical work is conveniently arranged in four columns, one for the paraxial values and one each for the three orders of aberration computed, as follows:

	Paraxial	Third order	Fifth order	Seventh order
$\frac{1}{n'j}$	0.61167	0.61167	0.22885	0.06562
$\left(1 - \frac{1}{n'j}\right)$	.38833	.38833	.77115	.91438
ρ'_1	−1.7999	−1.7999	−5.8310	−18.890
$\frac{\rho}{n'}$	··········	1.9816	1.9816	1.9816
		½	⅛	¹⁄₁₆
$\frac{\tau}{n'};\ \frac{\beta'^{(t)}}{n'^t}$	−0.69896	−0.69253	−1.11381	−2.1392

where $j = 1, 1, 3$, and 5 in the respective columns.

The first line in this computation need not be written down if the computer so prefers. The next four lines contain the factors of the coefficients of the respective orders. The reciprocal of the intersection distance measured from the center of curvature could now be found for a ray at any height h'_{c1} from the center of curvature after refraction at this surface, this quantity being given by the series

$$\theta'_1 = \left(\frac{\tau_1}{n'_1}\right)n'_1 + \left(\frac{\beta'_1{}^{(3)}}{n'_1{}^3}\right)n'_1{}^3 h'_{c1}{}^2 + \left(\frac{\beta'_1{}^{(5)}}{n'_1{}^5}\right)n'_1{}^5 h'_{c1}{}^4 + \left(\frac{\beta'_1{}^{(7)}}{n'_1{}^7}\right)n'_1{}^7 h'_{c1}{}^6 - \cdots$$
$$= -0.69896\,n'_1 - 0.69253\,n'_1{}^3 h'_{c1}{}^2 - 1.11381\,n'_1{}^5 h'_{c1}{}^4$$
$$- 2.1392\,n'_1{}^7 h'_{c1}{}^6 - \cdots$$

---

[17] Conrady, Applied Optics, p. 316. See footnote 2, p. 191.

3. TRANSFER FROM CENTER OF CURVATURE TO CENTER OF CURVATURE; FIRST SURFACE TO SECOND

Formulas (19), Section III, 3, are used in transferring the point of reference to the center of curvature of the second surface. The numerical work is most conveniently carried out in two parts, first computing the various terms and factors in the formulas, and second combining them to find the new coefficients, as follows:

1	d	+2.0576	κ	+0.29640	$\dfrac{\beta'^{(1)} 1}{\pi'^{1}}$	+0.47980
2	$n'd$	+3.3639	κ^2	.089043		
3	$\kappa n'd$	+1.00379	κ^4	.79287·10⁻¹	$\dfrac{\beta'^{(1)}\beta'^{(1)}}{\pi'^{1}}$	+.77135
4			κ^5	.70800·10⁻¹		
5	$(\kappa n'd)^2$	+1.00759	κ^6	.62864·10⁻¹	$\dfrac{\beta'^{(1)} 2}{\pi'^{1}}$	−.33214

		Paraxial	Third order	Fifth order	Seventh order
6	$\dfrac{r'_1}{\pi'_1}; \dfrac{\beta'_1{}^{(i)}}{\pi'_1{}^i}$.	−0.69896	−0.69253	−1.11381	−2.1392
7	κ−term...			+1.44425	+6.1942
8	κ^3−term...				−4.0159
9	Sum......			+.33044	+.0391
10	$\dfrac{r_2}{n_2}; \dfrac{\beta_2{}^{(i)}}{n_2{}^i}$..	−.20857	−.54909·10⁻¹	+.23329·10⁻¹	+.0246·10⁻¹

In this computation the value of d is found from the original data $(d_{c1\,c2} = d_{v1v2} - r_1 + r_2)$, then κ and its powers and multiples are computed, $n'd$ being found incidentally to the computation of κ. Line 6 is copied from the last line of the computation in the preceding section. The "sums" in line 9 are the numerical values of the quantities in brackets in the formulas, which, when multiplied by the proper power of κ give the coefficients for the object pencil at the second surface.

4. SECOND SURFACE AND TRANSFER TO THIRD; CENTER OF CURVATURE TO VERTEX

The aberration at the second surface is computed by means of formulas (11). As may be seen below, the computation falls readily into three parts. In lines 1 to 10 the various monomial and binomial quantities and their powers which appear in the formulas are computed. In lines 11 to 14 are computed the four bracketed quantities which appear in the $\dfrac{\beta'^{(3)}}{n'^3}$−inherent, the $\dfrac{\beta'^{(5)}}{n'^5}$−inherent, the $\dfrac{\beta'^{(7)}}{n'^7}\left(\dfrac{\beta^{(3)}}{n^3}\right)$, and the $\dfrac{\beta'^{(7)}}{n'^7}$−inherent terms respectively.[18] In each case the three products composing the quantity are written in the first three columns and their sum is written in the last column. In lines 15 to 20 the various terms in the formulas are evaluated and added together to find the constants of the refracted pencil. The inherent aberration appears in line 16, the terms containing the factor $\dfrac{\beta^{(3)}}{n^3}$ are in line 17, and so on.

---

[18] These designations are explained in Sec. II, 3; p. 201.

In the transfer to the vertex of the third surface, the first part of the computation, lines 25 to 29, has been written below the second part, lines 20 to 24 in order to eliminate the recopying of line 20. Moreover, the computation differs slightly from that in the preceding section in the use of formulas (20) instead of (19).

The numerical work, then, is as follows:

	i	ρ^i	$\frac{1}{n^i}$	$\frac{1}{n'^i}$	$\left(\frac{1}{n^i}-\frac{1}{n'^i}\right)$
1	1	+0.67024	0.61167	0.64636	−0.03469
2	2	+.44922	.37414	.41778	−.04364
3	3	+.30109			
4	5	+.13825			

		$\left(\frac{r}{n}\right)^i$	$\left(\frac{\rho}{n}\right)^i$	$\left(\frac{\rho}{n}\right)^i$	$\left(\frac{\rho}{n}-\frac{\rho}{n'}\right)+(i-1)\frac{r}{n}$
5	1	−0.20857	+0.43322	+0.40997	−0.02325
6	2	+.043501	+.18768	+.16808	−.23182
7	3				−.44039

		$\left(\frac{1}{n^i}-\frac{1}{n'^i}\right)^2$	$\left(\frac{\rho^2}{n'^i}-\frac{r^2}{n^i}\right)$	$\left(\frac{\rho^2}{n^i}-\frac{r^2}{n^i}\right)$	
8	1	+0.001203	+0.14418	+0.12456	
9	2	+.001904	+.020788	+.015520	

		$\left(\frac{1}{n^i}+\frac{1}{n'^i}\right)$	$\left(\frac{\rho^2}{n'^i}+\frac{r^2}{n^i}\right)$	$\left(\frac{\rho^2}{n^i}-\frac{r^2}{n^i}\right)$	
10	1	+0.79192	+0.23118	+0.21158	

		Bracketed quantities			Sum
11	$\beta'(1)$	−0.00017	+0.14141	−0.13676	+0.00448
12	$\beta'(i)$	−.000120	+.012715	−.010031	+.002564
13	$\beta'(1)(\beta(1))$	+.000143	+.018394	−.016795	+.001742
14	$\beta'(i)$	−.0000425	+.0029395	−.0021225	+.0007745

		Paraxial	Third order	Fifth order	Seventh order
15	$\frac{r_2}{n_2}$; $\frac{\beta_2(i)}{n_2'}$	−0.20857	−0.54909·10⁻²	+0.23329·10⁻³	+0.0246·10⁻⁴
16	Inher.	−.02325	−.1501 ·10⁻²	−.2148 ·10⁻³	−.3244·10⁻⁴
17	$(\beta(2))$			+.0281 ·10⁻³	+.0321·10⁻⁴
18	$(\beta(1))$				−.0119·10⁻⁴
19	$(\beta(1))$				+.0035·10⁻⁴
20	$\frac{r_3'}{n'_3}$; $\frac{\beta_3(i)}{n'_3 r^i}$	−.23182	−.6992 ·10⁻²	+.0466 ·10⁻³	−.2761·10⁻⁴
21	(z)			−.6975 ·10⁻⁴	+.1239·10⁻⁴
22	(z^2)				−.9275·10⁻⁴
23	Sum			−.6509 ·10⁻⁴	−1.0797·10⁻⁴
24	$n \alpha z$; n_3^c	−1.16660	−.7827	−.7709	−1.3529

		Paraxial	Third order	Fifth order	Seventh order
25	d	−1.462	$n\alpha$	+3.2527	$\frac{\beta(1)^2}{n^6}$	+0.48688·10⁻⁴
26	$1/n'$	.64636	$(n\alpha)^2$	10.5801		
27	$n'd$	−4.7554	$(n\alpha)^4$	1.11939·10⁵	$\frac{\beta(1)\beta(2)}{n^6}$	−.03258·10⁻⁴
28	$(n'd)^2$	+22.614	$(n\alpha)^6$	1.18433·10⁵		
29	σ	−.75404	$(n\alpha)^6$	1.25303·10⁶	$\frac{\beta(1)^2}{n^6}$	−.3418·10⁻⁴

5. THIRD SURFACE AND INVERSION TO LONGITUDINAL ABERRATION

The aberration due to the third surface of the lens might also be computed by the same method as was used at the second surface. However, it is desired to illustrate the use of formulas (28), and moreover it is preferable to have the final aberration referred to the back vertex without the additional transfer from the center of curvature to the vertex. Since the image is in air, then n' is unity, $\mu = n$, equations (28) are identical with equations (27), and the "reduced" coefficients, $n'\sigma'$, $n'\zeta'^{(3)}$, . . ., are equal to the actual coefficients, σ', $\zeta'^{(3)}$, . . .

In lines 1 to 4 of the computation the preliminary data are prepared. The bracketed quantities computed in lines 5 to 22 are given the designation of the term in which they appear. The various sums of the powers of n appear in the first column of numbers. These are multiplied by the appropriate powers of $(\rho - \sigma)$ and σ, and the products are written in the next column. Finally, these products are added together to give the total value appearing in the last column.

The coefficients of longitudinal spherical aberration are computed in lines 28 to 34 by formulas (29), Part V, in much the same manner as a transfer is made to a new point of reference.

These computations are as follows:

	i	n^i	$(\rho - \sigma)^i$	σ^i		
(1)	1	1.54712	−3.2364	−0.75404		
(2)	2	2.39358	+10.4743	+.56858		
(3)	3	3.70316	−33.899	−.42873		
(4)	4	5.7292	+109.711	+.32328		

			Bracketed quantities			Sum
(5) (6)	$\zeta'^{(4)}(\zeta^{(0)})$	$(5n+1)$	8.7356	$\cdot(\rho-\sigma)$ -4σ	−28.272 +3.016	−25.256
(7) (8) (9)	$\zeta'^{(4)}(Ink.)$	$(7n^2 -. .)$ $-(5n-5)$	10.0195 −2.7356	$\cdot(\rho-\sigma)^2$ $\cdot(\rho)\sigma$ $+\sigma^2$	+104.947 −6.6759 +.5686	+98.840
(10) (11)	$\zeta'^{(7)}(\zeta^{(1)})$	$(7n+4)$	14.8298	$\cdot(\rho-\sigma)$ -3σ	−47.995 +2.2621	−45.733
(12) (13)	$\zeta'^{(7)}(\zeta^{(0)})$	$(7n+1)$	11.8298	$\cdot(\rho-\sigma)$ -6σ	−38.286 +4.524	−33.762
(14) (15) (16) (17)	$\zeta'^{(7)}(\zeta^{(5)})$	$(63n^2 -. .)$ $-(96n^2 -.)$ $(40n-32)$	144.481 −116.579 29.8848	$\cdot(\rho-\sigma)^3$ $\cdot(\rho)^2\sigma$ $\cdot(\rho)\sigma^2$ $-4\sigma^3$	−4,897.8 +920.75 −54.99 +1.71	−4,030.3

(18)(19)(20)(21)(22)	ʃ'⁽¹⁾ (Inh.)	$(33n^4-.\ .)$ $-(47n^3-.\ .)$ $(25n^3-.\ .)$ $-(7n-7)$	65.0247 −37.446 14.9547 −3.8298	$\cdot(\rho-\sigma)^4$ $\cdot(\rho)^4\sigma$ $\cdot(\rho)^4\sigma^2$ $\cdot(\rho)\sigma^3$ $+\rho^4$	+7,133.92 −957.16 +89.06 −5.31 +.32	+6,260.8
			Paraxial	Third order	Fifth order	Seventh order
(23)(24)(25)(26)(27)	$n\sigma_3$; $n\Gamma'_3$⁽ⁱ⁾ Inher. (ʃ⁽³⁾) (ʃ⁽⁵⁾) (ʃ⁽³⁾¹)		−1.16660 +2.1832	−0.7827 +18.854	−0.7709 +465.88 −27.08	−1.3529 +14,755. −1,080.2 −35.7 +11.9
(28)(29)(30)	σ'₃;Γ'₃⁽ⁱ⁾ s-term s²-term		+1.0166	+18.071	+438.03 −321.23	+13,650 −15,573 +5,710
(31)	Sum		....	....	+116.80	+3,787
(32)	σ', A', B', C'		+.98367	−17.486	−113.02	−3,664
(33)(34)	s^2 $\Gamma'^{(1)^2}$		0.96761 +326.56	ʃ'⁽¹⁾ ʃ'⁽¹⁾ ʃ'⁽¹⁾³	+7,915.6 +5,901.3

It is more or less customary to change the dimensions of a lens proportionately so that the focal length is 1,000 mm when plotting the aberrations. This is done to a satisfactory approximation in the present case by merely changing the decimal point. The intersection distance measured from the back vertex for an image ray at height h'_3 is then given by the series

$$S'_3 = s'_3 + A'h'^2_3 + B'h'^4_3 + C'h'^6_3$$

$$= 983.67 - 0.017486\ h'^2_3 - 0.00011302\ h'^4_3 - 0.000003664\ h'^6_3$$

6. COMPARISON WITH RAY TRACING

In comparing the results of algebraic and trigonometric computations it is well to keep in mind the following essential difference between the two: In the trigonometric work the longitudinal spherical aberration is found as the difference between the intersection lengths of a paraxial and another ray. If either length is determined with an uncertainty of 0.01 mm, the uncertainty in the aberration thus determined is at least as large as that. In the algebraic series, on the other hand, the spherical aberration is expressed as one or more correction terms to be applied to the paraxial focal length, and accordingly the aberration may sometimes be found with an uncertainty of 0.001 mm or less when the intersection lengths are not known to within 0.1 mm.

For the lens in our example Conrady gives the following values obtained by tracing a ray with the use of five place tables:

$$h_1 = 30.0 \qquad \alpha'_3 = 1.8253°$$

$$s'_3 = +983.67 \qquad S'_3 = 967.13$$

$$\text{Long. Spher.} = -16.54$$

For purposes of comparison, the algebraic aberration must be determined for the same ray, that is

$$h'_3 = S'_3 \sin \alpha'_3 = 967.13 \sin 1.8253° = 30.805$$

This value substituted into the final series in section 5, above, gives:

$A'h'^3_3 = -16.593$	$s'_3 = +983.67$
$B'h'^4_3 = -.1018$	Long. Spher. $= -16.698$
$C'h'^6_3 = -.0031$	$S'_3 = +966.972$

Total $= -16.698$

These results are shown in Figure 10.

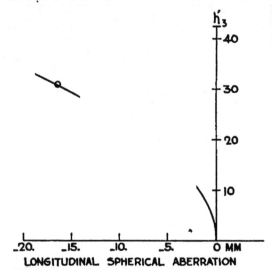

LONGITUDINAL SPHERICAL ABERRATION

FIGURE 10.—*Spherical aberration of "landscape" lens*

The curve shows the aberration of this lens as given by the accompanying calculation. The circle indicates a value obtained by trigonometric ray tracing using 5 place tables. The curve as shown corresponds to a back focal length of 983.6 mm. The horizontal scale, however, is twice the vertical.

In order to determine whether the discrepancy of 0.158 mm between the two values of Long. Spher. is due more largely to the trigonometric or to the algebraic computation, a seven-figure trigonometric computation has been carried out. Its results are as follows:

$s'_3 = 983.591$	$a'_3 = 1.82576°$
$S'_3 = 966.888$	Long. Spher. $= -16.703$

When this more precise value is taken as a standard of comparison, it may be seen that the algebraic computation, even if only the third spherical order term be considered, gives a better value for the spherical aberration than does the five-figure trigonometric computation. Moreover, if the fifth order term is included the resulting aberration differs but little from that given by the seven-place work. The seventh order term is negligible in the present case, but with larger relative apertures it would become more prominent.

It is interesting to note that Conrady computes the third order spherical aberration to be −16.702, which is virtually identical with the total aberration given by the 7-place computation, and then states that this value "agrees very well with the more exact trigonometric amount, −16.54, showing that the higher spherical aberration is small." The difference between the two values of the third order aberration may be considered as due to the causes mentioned in Part I, section 2. On the other hand, the close agreement between the first two paraxial image distances mentioned above appears to be largely fortuitous.

Although a final appraisal of the values of algebraic *v.* trigonometric computations should not be based upon one numerical example alone, still the following estimate, based upon this computation along with others which the author has made, may well be given here. The algebraic method is believed to be capable of giving as dependable results as the trigonometric in all practical cases. As a general rule, an exactness comparable to that given by 5-figure ray tracing may be obtained for apertures up to approximately $f/10$ by computing the third order only, and for apertures up to approximately $f/6$ by computing the third and fifth orders. These stated apertures will vary widely, tending to be smaller for meniscus than for symmetrical lenses. The labor involved in computing the paraxial and third order quantities by the method described in this paper is about the same as that of tracing one paraxial and one rim ray, that involved in computing also the fifth order is about the same as that of tracing one paraxial and two other rays. However, the possibility exists of compiling tables of the series aberrations, for instance as functions of the index of refraction and the reciprocal object distance, and thus of reducing the work to a minimum.

VII. DISCUSSION

1. CONVERGENCE OF ABERRATION SERIES

The validity of series expressions of aberration has been questioned on the ground that the magnitude of the terms which are neglected is not known. In fact Baker and Filon[19] have pointed out two cases in which these series actually become nonconvergent. There are analogous cases in the system of equations developed in the foregoing paragraphs, and these will be briefly discussed here.

In one case of nonconvergence the image distance for rays at some finite distance from the axis is infinite and, accordingly, the longitudinal aberration is infinite and can be expressed by no convergent series. In reciprocal aberration this case can be seen merely to be one of a zero value of the series. Nevertheless, analagous difficulty arises in the reciprocal series when a ray passes through the point of reference so that the reciprocal distance, and hence the reciprocal aberration, becomes infinite. This difficulty, however, can easily be avoided by a proper choice of the point of reference.

The second case mentioned by Baker and Filon concerns aberration series which are written in terms of the inclination of the ray in the

[19] Baker and Filon, On an Empirical Formula for the Longitudinal Spherical Aberration of a Thick Lens; Trans. Opt. Soc. London, vol. 20, pp. 67–92, 1918–19. See particularly pp. 74–76.

image space and involves longitudinal and reciprocal series alike. It was shown that two rays in the image pencil may have equal inclinations but unequal aberrations so that the aberration is a double-valued function of the independent variable. Manifestly no series can be convergent for both values.

In the present system of equations there is the following analagous case: Let ST (fig. 11) be the caustic curve of a pencil of rays. If the origin be chosen as at O so that a normal, N, to the caustic curve passes through it, then for every ray, R_BS_B, tangent to the curve beyond N, there is another ray, R_AS_A, such that the normals, OQ and OP, of the two rays are equal. There are, then two values of the aberration for a single value of the variable, h, and consequently no series can be convergent throughout this interval.

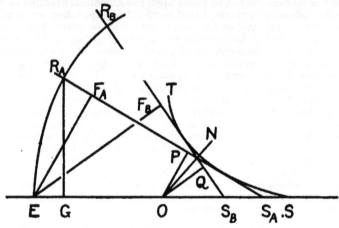

FIGURE 11.—*Diagram illustrating nonconvergence of aberration series*

However, by choosing a suitable point of reference, this difficulty is avoided, or at least the interval of convergence may be sufficiently extended for practical purposes.

Without stopping for the proof for other types of series, it may be stated as a more or less general rule that, if any aberration series be written using as a variable a length measured from a point on the axis to a point on the ray in some specified manner, than an absolute limit to the interval of convergence of the series is reached when the specified point on the ray coincides with the point of tangency of the ray with the caustic curve. In practice, however, long before this limit is reached the series will become so slowly convergent as to be of no practical value. It is to be observed, however, that nonconvergence can be avoided in the proposed system by a proper choice of the point of reference, while Baker and Filon found it necessary to use a function of the incident ray rather than of the refracted ray as the variable. This is objectionable on account of the indefiniteness inherent in using a function of the incident ray. For example, if h is given in S_4P_4 (fig. 1), S'_4 may be found from equation (1). However, this determines only one point on the ray since the ray might traverse the lens by a different path and still intersect the axis at the same point. Even if the height of incidence on the last surface is used as the ordinate and the axial point determined as before, the exact direction of the ray is not known unless the radius of the surface is given also. On the other hand, in the proposed system when the axial point is found by equation (1), the position of the ray is competely determined by a simple geometrical construction. Moreover, the ratio (or in the case of reciprocal aberration, the product) of these two quantities equals the sine of the angle of inclination of the ray while in the two systems just mentioned there is no direct relation.

2. RELATION TO PATH DIFFERENCE EQUATIONS

If it is desired to find the path difference between the two rays at some point on the axis, in order to work out the diffraction effects according to formulas such as those published by L. C. Martin,[30] a further advantage is gained in measuring the aperture along a line perpendicular to the ray rather than to the axis. In Martin's formula the independent variable is the height of intersection of the ray with a wave front through a fixed axial point. If $ER_A R_B$ (fig. 11) is a wave front, then $R_A G$ is such a height. Plainly an axial aberration series could not easily be written using this variable since the shape of the wave front is not usually known. If the wave front were exactly spherical then it is evident that $R_A G$ would be exactly equal to EF which is the variable used in the present paper. Now, since wave fronts do not ordinarily depart from sphericity by more than one or two wave lengths, or 0.001 mm, when h equals, say, 20 mm, the two lengths could be considered equal in most cases without appreciable error, so that the coefficients of the series derived in this paper could be substituted directly into Martin's equations.

A further investigation of the relation between reciprocal aberration and path length differences has been begun, and exact equations have been derived. It is expected that these will be published soon.

3. CONCLUSION

Besides the theoretical interest which attaches to the equations derived in this paper, there also is believed to be a considerable possibility of their practical application. No optical computer to-day questions the value of the third order aberrations developed by Seidel, and yet Berek [21] states that "L. Seidel himself did not consider the aberration formulas of the third order which he published between 1853–1856 suitable for practical purposes and, therefore, furnished trigonometrical tracing formulas to the C. A. Steinheil Optical Co. (1866)." Further, it may be stated that the mathematical tools furnished by Seidel lay idle for about 60 years, and only recently their usefulness, in spite of their restrictions, has been recognized. Although the fifth order, up to the present time, has been considered too unwieldy to be useful, still it is not impossible that it, like the third order, will in time be brought into a simpler form adaptable to numerical work. In addition to this there is the possibility of the compilation of tables of these aberrations which would shorten considerably the work of computing.

WASHINGTON, June 16, 1932.

---

[30] L. C. Martin, A Physical Study of Spherical Aberration, Trans. Opt. Soc., London, vol. 23, pp. 63–92 1921–22.
[21] Berek, p. 41, see footnote 11, p. 192.

AN ATTACHMENT FOR TURNING APPROXIMATELY SPHERICAL SURFACES OF SMALL CURVATURE ON A LATHE

By I. C. Gardner

ABSTRACT

This device replaces the compound rest on a lathe and was particularly designed for the production of the convex or concave surfaces of lens-grinding tools although it is well adapted for the production of such surfaces for any purpose. It is intended to be used only for the production of surfaces of which the radius of curvature is too great to permit the use of a tool mounted on the end of a radius rod. The mathematical theory underlying the design and a detailed description of the instrument as built are given in detail. The attachment, as constructed, permits disks under 300 mm in diameter to be faced and any radius of curvature, greater than 500 mm may be obtained.

The surfaces produced are not accurately spherical, but for many purposes the approximation is entirely satisfactory. The approximation becomes poorer as either the curvature or diameter of the surface is increased. For a surface 200 mm in diameter with a radius of curvature of 1,000 mm the maximum departure from sphericity is 0.02 mm. If the diameter of the tool is 300 mm and the radius of curvature is 500 mm, the departure from sphericity is approximately 0.3 mm. This last value is the maximum departure for any surface lying within the working range of the instrument.

CONTENTS

I. INTRODUCTION

In the construction of machines or tools it is often necessary to form a small portion of a spherical surface of which the radius of curvature is very long. This is particularly the case when making lens-grinding tools. The tool employed for grinding or polishing the surface of a lens may be described as a circular disk, usually of brass or cast iron, with a hub at the center of one face by which it can be mounted on a spindle and with the other face, in special cases plane, but generally convex or concave, the radius of curvature being the same in magnitude but opposite in sense to that of the surface desired on the finished lens.

When the radius of curvature is not too great such surfaces can be conveniently generated by mounting the cutting tool of a lathe on the end of a radius rod of the desired length. However, if the radius of curvature is much greater than 2,000 mm such a method is not

227

suitable, and convenient methods for producing such surfaces are not commonly available in the machine shop. There is a machine tool by which spherical surfaces of small curvature can be generated, but its method of operation requires the use of a milling cutter with its diameter at least as great as that of the lens tool to be made. It follows that a machine of this type must be large and strongly built. It is necessarily expensive and can not be used economically unless the demand for spherical surfaces is such that it can be kept in operation for long periods of time.

There are other devices on the market for producing tools for the manufacture of ophthalmic lenses. Some of these are not fitted for making a surface with the radius of curvature greater than 2 or 3 m. Others, which permit surfaces of less curvature to be made, employ templates, and hence do not permit any desired curvature to be obtained except by the construction of a special template.

Accordingly, when the radius of curvature is large, the difficulty of readily producing such a surface, by methods hitherto available, is so great that, in some instances, the lens designer actually avoids the use of surfaces of small curvature although this course is not satisfactory except for certain types of optical systems. In other cases, arcs are described on sheet metal by a marker on the end of the required length of piano wire. Two templates are filed from the metal—the one convex, the other concave—and these are ground together to eliminate the smaller irregularities. These templates are used to control the form of the tool which is turned by hand on a lathe. Hand turning would not be sufficiently precise except that lens tools are usually required in pairs, convex and concave, and the two tools can be ground together before use. The turning of tools by such a method is not only costly and tedious, but the cost is often further increased by the excessive amount of grinding which is required to prepare hand-turned tools for use.

Because of these difficulties a new method of producing approximately spherical surfaces of small curvature has been developed. By this method the surfaces may be produced on a lathe by substituting a relatively simple attachment for the compound rest. An adjustment can be readily made to give any radius of curvature within the range of the attachment. The sharper curvatures can not be generated by this device, but this offers no serious disadvantages as simple and satisfactory methods are already available for such curves. The surfaces generated are not truly spherical but in many cases, particularly for the larger radii of curvature, or for tools of small diameter, the departure from sphericity is so small that the tools may be used as turned. For the sharper curvatures falling within the range of the instrument it may be necessary to grind the members of a pair of tools together, but the time of grinding will be small as compared with that required for tools turned by hand.

II. GENERAL DESCRIPTION OF THE ATTACHMENT

According to this method the spherical surface to be produced is turned on a lathe by a single-tooth tool mounted in a special tool holder which is guided to generate the desired surface by a simple linkage system. A plan view of the fundamental parts is shown diagrammatically in Figure 1. The line KN is the axis of the lathe spindle and the arc KL is a meridian section of the curved surface which is turned by the cutting tool indicated at K. At A and B are two pins which are fixed to the lathe carriage. The triangle CDE, on which the cutting tool is rigidly mounted, is pushed across the lathe by a cross-feed and, as it advances, the two sides remain in contact with the pins A and B, thereby causing the triangle to rotate

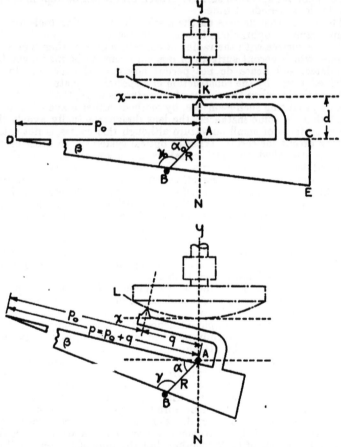

FIGURE 1.—*Diagrammatic sketch of linkage*

in such a manner that the cutting tool generates the arc KL. When the lathe spindle is revolving, therefore, a surface of revolution with the arc KL as a meridian section is produced.

Under the guidance of this linkage, the tool generates an arc which is not truly circular, but the approximation is satisfactory over the entire range of curvatures for which the use of a radius bar is unsatisfactory. Therefore the two methods supplement each other and, together, provide for the production of any curvature. The amount of curvature is determined by the magnitude of the angle between lines DE and DC and by the length AB. It is convenient to hold AB constant and provide means for varying the angle. The instrument may be calibrated to permit a setting to be readily made for any desired curvature. It is an advantage of this linkage that the constraints at A and B do not change their directions of action during the production of a given surface and accordingly inaccuracies, which may arise from lost motion, are substantially eliminated.

III. DERIVATION OF THE EQUATION OF THE CURVE GENERATED BY THE LINKAGE

In order to determine the exact form of the surface generated by this attachment it is desirable to derive the equation of the meridian section, which is the plane curve generated by the linkage system. In Figure 1 the essential portions of the linkage are illustrated in two positions. In the upper drawing the triangle CDE is shown in its initial position with DC at right angles to the bed of the lathe and with K, the generating point, in the prolongation of the axis of the spindle of the lathe. As this triangle is advanced across the bed of the lathe the sides DC and DE remain in contact with the fixed constraints A and B, thus causing the triangle in the course of its motion to assume the position shown in the lower drawing of Figure 1. Two parameters, p, the distance from D to A, and a, the angle between the lines AB and CD, serve to define the position of the triangle. The angle DBA will be denoted by γ. The initial values of these are p_0, a_0, and γ_0 as indicated in the upper drawing of Figure 1. The lengths AB and AK, designated, respectively, as R and d are constants of the instrument. The angle CDE, denoted by β, remains constant during the generation of any desired surface, but is varied in order to produce different curvatures. It is convenient to write

$$p - p_0 = q \tag{1}$$

where q is the extent of travel of DC past A.

The $x\ y$ system of coordinates indicated on Figure 1 is assumed fixed with respect to the bed of the lathe and the parts A and B. Then, for any point on the generated curve,

$$\begin{aligned} x &= q \cos (a - a_0) - d \sin (a - a_0) \\ y &= q \sin (a - a_0) + d \left[\cos (a - a_0) - 1\right] \end{aligned} \tag{2}$$

but for the more convenient application of these equations it is desirable that the trigonometric functions of $(a - a_0)$ be replaced by equivalent expressions in terms of β. If, by definition

$$\begin{aligned} P &= \frac{p}{R} \sin \beta \\ Q &= (1 - P^2)^{1/2} \end{aligned} \tag{3}$$

then

$$\begin{aligned} \sin \gamma &= P \\ \cos \gamma &= -Q \end{aligned} \tag{4}$$

Since

$$a = 180° - (\beta + \gamma)$$

FIGURE 3.—*Front view of attachment mounted on a lathe*

FIGURE 4.—*Back view of attachment mounted on a lathe*

$$\sin \alpha = \sin (\beta + \gamma) = P \cos \beta - Q \sin \beta \quad \text{and} \qquad (5)$$
$$\cos \alpha = -\cos (\beta + \gamma) = Q \cos \beta + P \sin \beta.$$

Similarly

$$\sin a_0 = P_0 \cos \beta - Q_0 \sin \beta \qquad (6)$$
$$\cos a_0 = Q_0 \cos \beta + P_0 \sin \beta$$

where

$$P_0 = \frac{p_0}{R} \sin \beta \qquad (7)$$
$$Q_0 = (1 - P_0^2)^{1/2}$$

By an evident trigonometric transformation

$$\sin (\alpha - \alpha_0) = PQ_0 - P_0 Q \qquad (8)$$
$$\cos (\alpha - \alpha_0) = PP_0 + QQ_0$$

Substituting the values from equation (8) in equation (2)

$$x = (PP_0 + QQ_0) \, q - (PQ_0 - P_0 Q) \, d \qquad (9)$$
$$y = (PQ_0 - P_0 Q) \, q + (PP_0 + QQ_0 - 1) \, d$$

Equations (9) enable the coordinates of any desired point on the curve to be obtained, but the form remains an inconvenient one for the present purpose. If x_1, y_1 be the coordinates of a given point on the curve, r may be defined by the equation

$$r = \frac{x_1^2 + y_1^2}{2 y_1} \qquad (10)$$

where r is the radius of curvature of a circle which cuts the axis of the lathe at right angles at the origin and passes through the point x_1, y_1. If the generated curve were truly circular the value of r would be independent of the particular point x_1 y_1 selected. As has been mentioned, however, the generated curve is only approximately circular and r is a function of x_1 y_1.

Substitution from equations (9) may be made directly in equation (10). However, a more convenient form is obtained by a series development. Two auxiliary quantities are defined by the equations

$$\rho = \frac{Q_0 R}{2 \sin \beta}$$
$$\qquad (11)$$
$$n = q/d$$

From equations (1), (3), (7), and (11) one may write

$$P = P_0 + \frac{n \, d \, Q_0}{2 \rho} \qquad (12)$$

and; by substitution in equation (3), Q may be expressed as an infinite series and P_0 and Q_0.

By making these substitutions for P and Q in equations (9) and substituting these values of x and y for x_1 and y_1 in equation (10), a series development for r may be obtained of which the first few terms are given in equation (13).

$$r = \rho - \left(\frac{3}{4} + \frac{1}{4}\frac{P_0}{Q_0}n\right)d + \frac{1}{16}\left[1 - \left(\frac{P_0}{Q_0}\right)^3 n^2\right]\frac{d^2}{\rho}$$
$$- \frac{1}{64}\left(n\frac{P_0}{Q_0} - 1\right)\left[2\frac{P_0^2}{Q_0^3}n^3 + 2\frac{P_0}{Q_0}n + n^2 + 1\right]\frac{d^3}{\rho^2} \qquad (13)$$

The right-hand member is an implicit function of β as will be noted by reference to equations (7) and (11). In the applications which have been made this series is so rapidly convergent that it is necessary to retain only the first three terms and $\frac{P_0}{Q_0}$ changes so slowly for changes in r that the value of β corresponding to any value of r may be readily obtained. The angle β, as indicated in Figure 1, is considered positive, r and ρ of equation (16), for this case, are positive and a convex surface is produced. If the slope of the line DE changes sign, β, r and ρ are negative and a concave surface is produced.

IV. DIMENSIONS AND PERFORMANCE CHARACTERISTICS OF LINKAGE ADOPTED FOR CONSTRUCTION

The instrument which has been built enables tools 300 mm in diameter to be produced and permits the turning of either convex or concave surfaces with the radius of curvature greater than 500 mm. This range of curvatures is considered entirely satisfactory as more sharply curved surfaces may be readily and conveniently produced by mounting the tool on the end of a radius rod.

The constants which were adopted for the instrument are

$$\alpha_0 = 45°$$
$$R = 150\sqrt{2} \text{ mm}$$
$$d = 50 \text{ mm}$$

and the maximum value of n is 3. Reference to Figure 1 indicates that

$$p_0 = 150 \cot \beta + 150 \tag{14}$$

From equation (7)

$$P_0 = \frac{\cos \beta}{\sqrt{2}} + \frac{\sin \beta}{\sqrt{2}} = \sin (45 + \beta)$$
$$\dot{Q}_0 = \cos (45 + \beta) \tag{15}$$

and equation (13), with the last term omitted, becomes

$$r = \rho - \frac{3}{4}d - \frac{1}{4}nd \tan (45° + \beta) + \frac{1}{16} \frac{d^2}{\rho} - \frac{1}{16} \frac{n^2 d^2}{\rho} \tan^2 (45° + \beta) \tag{16}$$

As has been already mentioned, the surfaces which are generated by this apparatus are approximately but not strictly spherical. In equation (16) this is evidenced by the presence of n and n^2 in the right-hand member. In investigating this departure from sphericity several assumptions will be made in order to simplify the work without any important lessening of the accuracy.

1. The last two terms in equation (16) will be neglected.
2. It will be assumed that x of the coordinate system indicated in Figure 1 is equal to nd. Reference to the first of equations (2) will indicate the approximation involved.[1]
3. The sagitta of the approximately circular arc, KL, shown in Figure 2 will be assumed equal to $\frac{x^2}{2r}$.
4. In the quantity $(4\rho + 3d)$, which will occur, it will be assumed that ρ can be replaced by $r + 60$.

---

[1] The approximation of assumption 2 is applied successively to two nearly equal quantities, and only the difference between these two quantities is of importance in the following discussion. The approximation is therefore much closer than may at first appear. This is also true of the approximation of assumption 3.

If $n_1 d$ is the radius of the tool to be turned, the radius of curvature of the edge zone, as it has been defined, will be approximately given by the equation

$$r_1 = \rho - \frac{3}{4}d - \frac{1}{4}n_1 d \tan(45° + \beta)$$

The sagitta of a sperical surface, having this radius of curvature, at a distance nd from the center, will be

$$\frac{2n^2 d^2}{4\rho - 3d}\left[1 + \frac{n_1 d \tan(45° + \beta)}{4\rho - 3d}\right]$$

and, similarly, the sagitta of the actual surface, at the same distance from the center, will be

$$\frac{2n^2 d^2}{(4\rho - 3d)}\left[1 + \frac{nd \tan(45° + \beta)}{4\rho - 3d}\right]$$

The distance between these two surfaces, which will be/termed the departure from sphericity, for the zone nd will be

$$\frac{2 n^2 d^3 (n_1 - n)}{(4\rho - 3 d)^2} \tan(45° + \beta)$$

Applying assumption 4 one may write
Departure from sphericity

$$= \frac{2 (n_1 n^2 - n^3) d}{(4 r + 90)^2} \tan(45° + \beta) \tag{17}$$

For given values of r and n_1, this expression will assume a maximum value when $n = \frac{2}{3} n_1$ and, substituting this value

Maximum departure from sphericity

$$= \frac{8}{27} \frac{n^3_1 d^3}{(4 r + 90)^2} \tan(45° + \beta) \tag{18}$$

This expression gives the maximum distance, measured parallel to the axis of the tool, between the actual surface as turned and a spherical surface with a radius of curvature corresponding to the zone of radius $n_1 d$ and is a measure of the thickness of metal which must be removed to obtain a rigorously spherical surface. For a given radius of curvature it is evident that this departure varies as the cube of the diameter of the tool and inversely, approximately as the square of the radius of curvature. Consequently, within the working range of the instrument, the greatest values of the departure from sphericity will be obtained when the diameter of the tool is 300 mm ($n_1 = 3$) and when r equals $+500$ or -500 mm. For r equals $+500$ (convex surface) and r equals -500 (concave surface), β assumes the values $+6.7°$ and $-11.4°$, respectively, and tangent $(45° + \beta)$ assumes the values 1.27 and 0.55. This indicates that a concave surface generated by this attachment will, in general, be more nearly spherical than a convex surface although, as the absolute value of r increases, β approaches zero as a limit and the values of tan $(45° + \beta)$ approach equality for the two types of surfaces. For $n_0 = 3$ and $r = +500$ or -500, the maximum departures from sphericity are

	mm
For convex surface	0.3
For concave surface	.15

In considering these values of the departure from sphericity it should be remembered that these are the extreme values for the most unfavorable conditions and that the departures are very much less for the greater part of the working range of the instrument. For example, for tools 200 mm in diameter and r equals $+1,000$ or $-1,000$ mm the departures from sphericity are

	mm
For convex surface	0.02
For concave surface	.01

In equation (16) for all values of β and ρ lying within the range of the instrument, the third term of the right-hand member is the term which determines the sense in which the surface departs from true sphericity. This term does not change sign, and, accordingly, for a convex surface, the absolute value of the radius of curvature decreases from the center outward and for a concave surface the change is in the opposite direction. Consequently the marginal portions of a convex surface and the central portions of a concave surface are the more strongly curved.

In Table 1, the values of β corresponding to different values of r are tabulated. These values are determined for $n = 1.8$ and the values of β, as tabulated, may be used for tools 200 mm or less in diameter. For convex surfaces approaching 300 mm in diameter it is well to increase the desired value of r by 15 and to use this new value when entering the table.[2] Similarly, for a concave surface the absolute value of the desired r should be decreased by 10 before entering the table. For values of r greater than 10,000 mm, the value of β can be computed to a satisfactory approximation by the formula

$$\tan \beta = \frac{75}{r + 135} \qquad (19)$$

where r and β are positive for convex, negative for concave surfaces.

---

[2] As has been noted, Table 1 is computed for $n = 1.8$. For a tool approaching 300 mm in diameter it is desirable that the surface actually turned and the desired spherical surface coincide for a zone further from the center than that corresponding to $n = 1.8$. For a convex surface, this result will be attained if the value of r for $n = 1.8$ is made 15 mm greater than the nominal value desired for the complete surface. Similarly, for the concave surface, the value of r for $n = 1.8$ is less than the value corresponding to the complete surface.

TABLE 1.—*Values of β corresponding to r*

r	Convex surface β	Concave surface β
mm	°	°
500	6.66	−11.38
600	5.78	−9.04
700	5.10	−7.49
800	4.56	−6.39
900	4.13	−5.57
1,000	3.77	−4.93
2,000	2.01	−2.30
3,000	1.37	−1.50
4,000	1.04	−1.11
5,000	.84	−.88
6,000	.70	−.73
7,000	.60	−.63
8,000	.53	−.55
9,000	.47	−.48
10,000	.42	−.44

V. MECHANICAL DETAILS OF THE COMPLETED INSTRUMENT

The attachment has been built in such form that it replaces the usual compound slide rest on the lathe. No other alterations are necessary and the lathe may be used interchangeably for ordinary turning or the facing of spherical surfaces. In Figure 1 the constraints have been represented as pins against which the sides of the triangle bear. In the mechanical realization of this linkage the pins are necessarily replaced by pivoted slides of sufficient strength to oppose the forces on the cutting tool. The detailed design, as completed at the Bureau of Standards, is shown in Figure 2. The tool, indicated at L, is carried on a cross slide pivoted at A. The pivot at A corresponds to the similarly lettered constraint in Figure 1 and the line CD corresponds to the line CD of the triangle. The member which carries the tool is extended to the right of the slide and carries a sector FGB, pivoted at B. This sector is rotated about B and may be set in any desired position by means of the worm which is rotated by the handle at I. The angular setting of the sector can be read by an arc graduated in degrees and by a graduated drum which is mounted on the shaft of the worm. On the lower surface of the sector there is a male dovetailed member, indicated by the dotted lines parallel to the line DE of Figure 1. This dovetailed member corresponds to the side DE of the triangle of Figure 1 and it can be set in the required position for any desired radius of curvature by rotating the sector. The arc, carried by the sector, is graduated to permit the value of $β$ to be read directly. In Figure 2 the setting illustrated corresponds to $β=0$, CD and DE are parallel and the instrument will produce a plane surface. The dovetailed member which has been mentioned slides in a short dovetailed slide pivoted in the base at B. The small section in Figure 3 indicates the details of this arrangement. The pivots at A and B, corresponding to the constraints A and B, respectively, of Figure 1, are carried by the base of the attachment and remain invariant in position with respect to each other. It will be noted that the pivot about which the sector rotates is directly over the pivot at B when the main slide is at right

angles to the axis of the lathe and the point of the tool is on the axis
of the lathe.   Consequently, when in this position, the sector can
be rotated through any desired angle without causing any rotation
of the line DC about A.

As has been mentioned, this attachment is mounted on the carriage
of the lathe in place of the compound slide rest.   Reference to a table,
such as Table 1, gives the required value of β for the desired radius

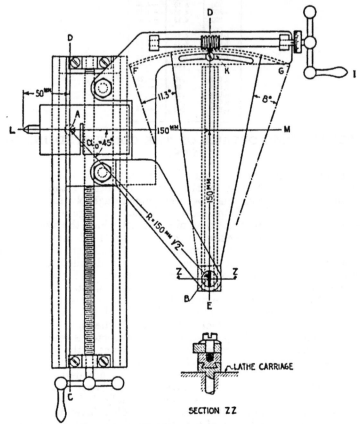

FIGURE 2.—*Mechanical realization of linkage*

and the setting is made by turning the handle I, after which the
sector is clamped in position by the nut at K.   The cut is then made
in the usual manner by turning the handle at I.   To advance the
tool for a second cut, the entire attachment is advanced toward the
work by traversing the carriage along the bed of the lathe.   The
operation of facing a spherical surface with this attachment is quite
as simple as facing a plane surface with a compound slide rest except
that no automatic cross-feed is provided.   This could be easily added
if it were considered essential.

One of these attachments has been built by the Mann Instrument Company, Cambridge, Mass. Although the details of the design shown in Figure 2 have been modified to correspond to their practice in instrument design, the fundamental dimensions and elements of design were retained unaltered Figures 3 and 4 give two views of the completed attachment mounted on a lathe in position for operation. In the two views the different details of construction to which reference has been made are clearly recognizable. The attachment has proved quite satisfactory for the production of surfaces for lens, tools, and similar purposes.

WASHINGTON, June 2, 1932.

WAVE LENGTHS AND ZEEMAN EFFECTS IN LANTHANUM SPECTRA

By William F. Meggers

ABSTRACT

The wave lengths corresponding to more than 1,500 lines photographed in the arc and spark spectra of lanthanum were measured relative to standards in the iron spectrum. The values extend from 2,142.81 A in the ultra-violet to 10,954.6 A in the infra-red. Data on the furnace spectra of 695 lanthanum lines studied by King and Carter in the interval 2,798 to 8,346 A are quoted. Measurements of Zeeman effects for 476 lines ranging in wave length from 2,791 to 7,483 A are also presented. Comparison of relative intensities and other characteristics of lines in the different sources permitted a sharp discrimination between three classes of lines; about 700 are ascribed to neutral atoms (constituting the La I spectrum), 800 originate with singly ionized atoms (La II spectrum), and 10 belong to doubly ionized atoms (La III spectrum). These data form a complete and accurate description of atomic lanthanum spectra, suitable for chemical indentification and for analysis of the spectral structures. A new infra-red sequence of La O bands is appended.

CONTENTS

I. INTRODUCTION

Lanthanum (La = 138.90; Z = 57) is the third element in the sixth period of the periodic classification, and is analogous to scandium and yttrium which appear in similar positions in the fourth and fifth periods. The emission spectra of scandium [1] and yttrium [2] have been described with satisfactory thoroughness and precision, but lanthanum data of the same quality have been lacking up to the present.

Attempts to analyze the structures of lanthanum spectra,[3] when based upon the published wave lengths and Zeeman effects, were only partially successful and it was evident that more extensive and accurate descriptive material was required before the structural analyses could be completed. Such data are now presented in this paper; they comprise wave-length measurements from 2,100 to 11,000 A, estimated relative intensities for 1,535 lines, and Zeeman effect observations for 476 lines in the interval 2,800 to 7,500 A.

---

[1] W. F. Meggers, B. S. Sci. Paper No. 549, vol. 22, p. 61, 1927.
[2] W. F. Meggers, B. S. Jour. Research, vol. 1 (RP12), p. 319, 1928.
[3] W. F. Meggers, La I, J. Wash. Acad. Sci., vol. 17, p. 25, 1927. W. F. Meggers, La II, J. Opt. Soc. Am., vol. 14, p. 191, 1927.

Analysis of the structures of lanthanum spectra and details concerning the hyperfine structure of lanthanum lines will be given in other papers.

The only remaining spectra of this type are those associated with actinium, the optical spectra of which have never been investigated.

II. EXPERIMENTAL

1. WAVE LENGTHS

The lanthanum salts used in this investigation consisted of $LaCl_2$ purified by Auer v. Welsbach, pure LaO_2 kindly supplied by Prof. B. S. Hopkins of the University of Illinois, and commercial $LaCl_2$ purchased from Eimer and Amend. The latter material contained a considerable amount of cerium, and its use was confined mainly to the observation of Zeeman effects.

Arc spectrograms were made with electric arcs between electrodes of graphite, silver, or copper, a small portion of lanthanum salt being placed on the lower electrode before striking the arc. The arc was fed by 5 to 6 amperes direct current from a circuit with 220 volts potential difference. Some of the spark spectrograms were made with the same metal electrodes which had been used previously for arc spectra, the salt fused on the rods in the arc sufficed to give well-developed spark spectra. Additional spark spectrograms were obtained with graphite electrodes upon which a solution of $LaCl_3$ was allowed to drip during the exposure. For excitation of spark spectra a 40,000-volt transformer consuming about a kilowatt was used, with condensers of $0.006H\mu f$ capacity in parallel with the spark in the secondary circuit. Comparison arc and spark spectra of graphite, silver, and copper were photographed adjacent to those of the electrodes plus lanthanum salt so that lines due to the electrodes or to atmospheric gases could be recognized and avoided during measurement. Finally, the arc spectrum of iron was recorded alongside each lanthanum spectrogram to supply the scale of standard wave lengths from which the values for lanthanum lines were derived.

The wave-length interval from 2,500 A in the ultra-violet to 11,000 A in the infra-red was investigated with concave gratings, while the shorter wave portion to 2,000 A was photographed with a quartz spectrograph. Since these spectrographs have already been described in connection with their use in the study of yttrium spectra[4] further details concerning them can be dispensed with here.

Most of the spectrograms were made on photographic plates of thin glass coated with Eastman 33 emulsion, the plates being sensitized with pinaverdol, pinacyanol, dicyanin, and neocyanin to photograph the longer wave portions. The longest wave length photographed in lanthanum spectra with neocyanin-stained plates was 9,737 A, but the infra-red limit was easily extended to 10,955 A during the past winter by employing Eastman "Q" type plates sensitized with xenocyanine.

2. ZEEMAN EFFECTS

Spectrograms for the study of lanthanum lines in a magnetic field were made with one of the stigmatic concave-grating spectographs

[4] W. F. Meggers, B.S. Jour. Research, vol. 1 (RP12), p. 320, 1928.

referred to above. The spectral range investigated extended from 2,500 A to 8,000 A, the first half of which was photographed in the second-order spectrum with a scale of 1.8 A per millimeter and the remainder in the first order. Photographic plates and procedure were the same as for wave-length measurements described above. The exposures ranged from 30 minutes in the violet to 5 hours in the red, the source being a spark between graphite electrodes upon which $LaCl_2$ solution was allowed to drip from a burette during the exposure.

The magnetic resolution was obtained with a water-cooled Weiss electromagnet operated with a speed-regulated motor-generator set supplying 150 amperes at 98 volts to the magnet windings. With a pole gap of 6 mm, the field strength was about 33,000 gausses. The actual field strength obtaining for each exposure was derived from the measured resolutions of standard lines. The following lines served as field standards:

Element	Wave length	Zeeman pattern
Ca	3,933.670	$\frac{(1)\ 3,5}{3}$
	3,968.475	$\frac{(2)\ 4}{3}$
Al	3,944.025	$\frac{(2)\ 4}{3}$
	3,961.537	$\frac{(1)\ 3,5}{3}$
Zn	4,680.140	$\frac{(0)\ 4}{2}$
	4,722.163	$\frac{(1)\ 3,4}{2}$
	4,810.534	$\frac{(0,1)\ 2,3,4}{2}$
Na	5,889.965	$\frac{(1)\ 3,5}{3}$
	5,895.932	$\frac{(2)\ 4}{3}$
K	7,664.94	$\frac{(1)\ 3,5}{3}$
	7,699.01	$\frac{(2)\ 4}{3}$

The Ca, Na, and K lines arose naturally from impurities in the La Cl$_2$ solution, and to these a 2-minute exposure of Al or Zn was added at the end of the La exposure. Thus either four or five lines were always present to reveal the intensity of the magnetic field and the average probable error of such a determination was about one-fourth per cent.

In making the Zeeman-effect spectrograms light from the source was projected onto the slit of the spectrograph by means of a fused quartz lens after passing through a quartz Wollaston prism. A double image was thus obtained which gave a complete separation of light polarized parallel and perpendicular to the lines of force, thus permitting the two polarizations to be photographed simultaneously without overlap.

III. RESULTS

Limited portions of the arc and spark spectra of La have already been described by different observers. The early observations up to 1910 are quoted by Kayser[5]; the only ones worth mentioning here are measurements in arc spectra by Rowland and Harrison (3,104.702 to 5,930.330 A), and by Wolff (2,610.428 to 5,762.040 A).[6]

A list of arc lines (2,610.43 to 6,774.52 A) and another of spark lines (2,216.20 to 6,643.10 A) was published by Exner and Haschek.[7] The arc spectrum of La was investigated in the region of longer wave lengths (5,501.349 to 9,078.99 A) by Kiess,[8] and in the ultra-violet (2,200 to 3,100 A) by Piña de Rubies.[9]

The above-mentioned descriptions refer mainly to arc spectra, they contain little or no information as to which lines belong to ionized atoms, and they cover their respective intervals with different scales of wave lengths and of intensities.

Altogether, their inadequacy as standard descriptions of the successive spectra characteristic of La atoms justifies the preparation and publication of the following results, appearing in Table 1. No mention is made in any of the earlier descriptions of La spectra of hyperfine structure; it was first noted by King[10] and studied by Meggers and Burns.[11] It now appears that a considerable number of the La lines are complex. (Indicated in Table 1 by the letter "c" following the wave length.) Since hyperfine structure of La lines has no particular importance in a general description, and is furthermore the subject of a special investigation, no further discussion of it will be made here; the results in Table 1, either for wave-length values or for Zeeman effects, apply to the center of gravity of a line whether it is narrow or appears widened due to partial resolution of the hyperfine components.

The probable errors of wave-length measurements reported in Table 1 are usually less than 0.01 A for the stronger lines between 3,100 and 6,000, but the errors for the remaining lines are somewhat larger since they were measured for the most part with smaller dispersion. Many lines appearing enhanced or only on the spark spectrograms are hazy and unsymmetrical; such lines are not susceptible of precise measurement and may be in error by several hundredths of an angstrom unit.

Any attempt to describe the atomic spectra of La must contend with the nuisance of bands due to molecular compounds. In an investigation of the band spectra of LaO, Meggers and Wheeler[12] measured more than 300 band heads in the ordinary arc spectrum and classified the bands in possibly nine systems. The bands appear with considerable intensity throughout almost the entire range from 4,300 to 9,800 A, causing difficulty and uncertainty in the detection

[5] H. Kayser, Handbuch der Spectroscopie, vol. 5, p. 655, S. Hirzel, Leipzig, 1910.
[6] A. Rowland and C. N. Harrison, Astrophys. J., vol. 7, p. 373, 1898; E. Wolff, Zeitschr. f. wis. Phot., vol. 3, p. 395, 1905.
[7] F. Exner and E. Haschek, Die Spektren der Elemente bei Normalem Druck, vol. 2, p. 114, 1911; vol. 3, p. 105; 1912. F. Deuticke, Wien.
[8] C. C. Kiess, B. S. Sci. Paper No. 421, vol. 17, p. 318, 1921.
[9] S. Piña de Rubies, Anales Soc. Esp. Fis. Quim., vol. 23, p. 444, 1925.
[10] A. S. King and E. Carter, Astrophys. J., vol. 65, p. 86, 1927.
[11] W. F. Meggers and K. Burns, J. Opt. Soc. Am., vol. 14, p. 449, 1927.
[12] W. F. Meggers and J. A. Wheeler, B. S. Jour. Research, vol. 6 (RP273), p. 239, 1931.

of the weaker atomic lines. Since the data for LaO band heads (3,565 to 9,150 A) have already been published, they will be omitted here, and no reference to bands will be found in Table 1 except when coincident or unresolved from lines belonging to the atomic spectra. Observations of the arc spectrum of La in the infra-red with xenocyanine plates revealed another long sequence of band heads extending from 9,111 to 9,729 A. These constitute the 0-2 sequence of System VII, and are presented here as Table 2.

The only published Zeeman effects of lanthanum lines are those observed by Rybar[13] in 1911; they extend over the range 2,791 A to 5,188 A and include about 215 lines. These observations are fairly satisfactory, and it was at first my intention only to supplement them with data for the longer waves. However, a trial exposure in the ultra-violet indicated that Rybar's results might be improved and extended by reobservation, and the desirability of having the magnetic resolutions of La lines strictly comparable over a wide range of spectrum finally persuaded me to extend measurements to the same ultra-violet limit. In column 5 of Table 1, the observed Zeeman effects are given for 460 La lines ranging in wave length from 2,798 A to 7,483 A. A few lines observed by Rybar, but not measurable on my spectrograms, are quoted and followed by the letter R. The patterns are presented in the standard notation for Zeeman effects; that is, the separations are expressed in decimal parts of a, the separation of a normal triplet, components polarized parallel to the magnetic field being inclosed in parentheses followed by the perpendicular components. In resolved complex patterns the strongest components are printed in bold-face type. For unresolved patterns a conscious effort was made to measure the strongest components when the type could be easily recognized and a symbol added to indicate the intensity distribution among the fused components. For this purpose, the notation used by Back[14] for distinguishing various types of intensity gradients was employed. The letters A or B after a Zeeman effect mean that the pattern is complex but unresolved, and A indicates that the maximum intensity for perpendicular components is at the edge of a group, while B signifies that it is in the middle. The distinction between strongest component inside or outside of the group is shown by A^1 and A^2, respectively. The interpretation of these Zeeman effects will be given in another paper dealing with the spectral series classification of La lines.

An important contribution to the description of the atomic spectra of La is found in the temperature classification of La lines by King and Carter.[15] From an examination of the furnace, arc, and spark spectra in the interval 2,800 A to 8,400 A data were obtained for a classification of 695 La lines according to the temperature required for initial appearance and the rate of change of line intensity as the temperature is increased. These lines were divided into five classes. Lines in Classes I and II appear at low temperature (2,000° C.), but those of Class I show a slower change from low to high temperature

---

[13] S. Rybar, Math. es Phys. Lapok, vols. 20-21, p. 198, 1911-12; Phys. Zeitschr. vol. 12, p. 889, 1911, a partial list.
[14] E. Back, Zeitschr. f. Phys., vol. 15, p. 212, 1923.
[15] A. S. King and E. Carter, Astrophys. J. 65, p. 86; 1927.

than those of Class II, and as a rule are less conspicuous in the arc. Lines of Class III are usually well developed at medium temperature (2,200° to 2,300° C.), while lines clearly associated with high temperatures (2,600° to 2,800° C.) are placed in Classes IV and V, those of Class V being absent or very faint in the furnace spectrum. A large number (230) of the lines thus classified are enhanced in the spark and more than half of these appear in the furnace. This indicates that La atoms are ionized with relative ease, which accounts for the great prominence of spark lines in the ordinary arc. In fact, a majority of the La atoms appear to be ionized in the 220-volt arc, and there can be no doubt that the best method of developing the spectrum of neutral atoms is by means of a temperature-controlled vacuum furnace which restricts ionization and also eliminates confusion with molecular spectra. A few lines not detected on my spectrograms were measured on furnace spectrograms by Professor Russell; they are marked H. N. R. in Table 1.

In the last column of Table 1 an attempt is made to assign each observed wave length to its proper atomic origin. This separation of lines into La I, La II, La III spectra is based primarily upon the relative intensities and other characteristics appearing in column 2. It is supported by the temperature classification and further description in column 3 and by Zeeman effects (column 5) describing combinations of spectral terms having even multiplicity for La I and La III lines, but odd multiplicity for La II lines. In the infra-red where data on spark intensities, temperature classes, and Zeeman effects are lacking, the lines are assumed to belong to the La I spectrum unless they are accounted for as combinations of the La II spectral terms. The significance of the symbols and abbreviations used in Table 1 is summarized as follows:

> h = hazy (n for King and Carter).
> l = shaded to long wave lengths.
> M = molecular spectrum, La O band head.
> p = part of band structure.
> d = double.
> c = complex.
> E = enhanced line.
> A = stronger in furnace than in arc.
> A^1 = strongest s-components of Zeeman effect inside.
> A^2 = strongest s-components of Zeeman effect outside.
> B = strongest s-components of Zeeman effect in center.
> w = wide.
> W = very wide.
> us = unsymmetrical.
> ur = unresolved.
> R = Zeeman effect by Rybar.
> H. N. R. = wave length by Henry Norris Russell.
> e = appearing only at negative electrode.

TABLE 1.—*Wave lengths and Zeeman effects in lanthanum spectra*

λ (air) I. A.	Intensity arc spark B. S.	Arc intensity and tempera-ture class K and C	ν (vac.) cm⁻¹	Zeeman effect	Spec-trum
10, 954. 6	3		9, 126. 09		II
10, 952. 0	1		9, 128. 25		I
10, 739. 66	5		9, 208. 73		I
10, 612. 56	10		9, 420. 21		I
10, 552. 41	6		9, 473. 91		I
10, 522. 09	10		9, 501. 21		I
10, 483. 0	2		36. 6		I
61. 69	15		56. 07		I
50. 82	20		66. 01		I
23. 4	1		9, 591. 2		I
10, 409. 55	3		9, 603. 93		I
10, 372. 4	1		38. 4		I
57. 70	20		52. 01		I
49. 08	40		60. 05		I
37. 20	3		71. 15		I
32. 40	2		75. 64		I
30. 3	1		77. 6		I
10, 318. 2	2		9, 689. 0		I
10, 294. 68	10d		9, 711. 09		I
85. 64	3		21. 52		I
81. 34	10		23. 69		I
78. 52	3		26. 37		I
74. 85	10		29. 84		I
34. 78	2		67. 93		I
23. 76	1		78. 46		I
19. 83	3		82. 22		I
10, 209. 85	2		9, 791. 78		I
10, 186. 5	2		9, 814. 2		·II
84. 60	20		16. 06		I
77. 74	6h		22. 67		I ·
54. 74	40		44. 92		I ·
43. 38	2		55. 95		I
41. 20	10		58. 06		I
30. 82	5		68. 17		I
10, 111. 9	2h		9, 886. 6		·I
10, 093. 54	1		9, 904. 62		·II
83. 96	2		14. 03		I
64. 77	6		30. 95		I
58. 79	2		34. 84		I
54. 82	2		42. 76		I
29. 74	2		67. 62		I
14. 45	4		82. 84		I
10, 005. 73	50		9, 991. 54		I
9, 988. 47	10		10, 008. 80		I
81. 24	6		016. 05		I
80. 38	10		016. 91		I
65. 70	3		031. 67		I
32. 72	2		064. 98		I
20. 82	150		077. 06		I
9, 911. 08	,3		086. 95		I
9, 893. 82	4		104. 55		II
81. 24	100		117. 42		I.
62. 60	3		136. 54		I
52. 58	6		146. 85		I
48. 70	4		150. 84		I
42. 0	2		157. 8		I
33. 30	3h		166. 74		I ·
05. 2	1h		195. 9		I
9, 804. 20	2		196. 92		I
9, 775. 09	8		227. 28		I
72. 24	20		230. 26		I
68. 82	3h		233. 85		I
37. 09	100		267. 20		I
13. 52	3		292. 11		I
9, 709. 45	10		10, 296. 42		I

TABLE 1.—*Wave lengths and Zeeman effects in lanthanum spectra*—Continued

λ (air) I. A.	Intensity arc spark B. S.	Arc intensity and temperature class K and C	ν (vac.) cm⁻¹	Zeeman effect	Spectrum
9,706.48	20		10,299.57		I
9,699.64	20		306.84		I
96.7	1		310.0		I
92.6	2		314.3		I
72.94	3		335.29		II
72.04	8		336.25		I
57.00	20		352.35		II
46.47	3		363.65		I
40.81	30		369.73		I
33.72	40		377.36		I
9,631.84	2		379.38		I
9,570.38	5		446.04		I
63.60	4		453.45		II
60.69	10		456.63		I
42.06	50		477.06		I
41.23	20		477.96		I
9,528.0	1A		492.5		I
9,485.15	15		539.91		I
84.2	1		541.0		I
76.98	3		548.99		I
74.45	5		551.81		I
67.25	2		559.84		I
61.82	60		565.89		I
57.62	2		570.59		I
41.7	1		588.4		I
38.30	100		592.23		I
15.64	3		617.71		I
9,412.65	100		621.08		I
9,396.2	1		637.4		I
90.56	4		646.07		I
77.71	3		660.66		I
76.10	3		662.49		I
72.57	30		661.5t		I
46.69	15		696.04		II
28.87	2		716.47		I
24.5	1		721.5		I
9,312.9	1h		724.5		I
9,293.3	2h		757.5		I
87.5	1h		764.2		I
60.42	3		795.69		II
54.70	10		802.36		I
50.06	20		807.78		I
26.63	30		835.22		I
9,219.64	10		843.44		I
9,172.88	5		896.71		I
72.39	10		899.30		I
57.13	10		917.46		I
51.62	6		924.03		I
46.75	2		929.85		II
43.78	5		933.40	`	I
42.24	6		935.24		I
27.5	1		952.9		II
19.18	20		962.89		I
09.25	2		974.84		I
9,101.10	2		984.67		II
9,096.71	3		10,989.97		II
79.10	50		11,011.29		I
58.63	2		036.17		I
56.53	5		038.73		I
46.97	2		060.39		I
25.05	4		077.23		I
16.80	2		087.37		II
9,008.26	6		097.88		I
8,977.39	2		136.04		I
8,970.07	3		11,145.13		I

TABLE 1.—*Wave lengths and Zeeman effects in lanthanum spectra*—Continued

λ (air) I. A.	Intensity arc spark B. S.	Arc intensity and temperature class K and C	ν (vac.) cm⁻¹	Zeeman effect	Spectrum
8, 965. 41	2		11, 150. 92		I
63. 53	10		153. 13		I
57. 74	50		160. 47		I
48. 89	2		171. 50		I
47. 95	1		172. 68		I
8, 917. 70	1h		210. 56		I
8, 891. 06	1		244. 17		I
84. 24	2		252. 80		I
79. 56	3		258. 73		I
75. 05	2		264. 45		I
71. 00	4		269. 60		I
67. 35	3		274. 24		I
39. 64	20		309. 57		I
25. 86	50		327. 23		I
21. 66	1h		332. 62		I
18. 96	20		336. 10		I
8, 810. 57	3		346. 89		II
8, 797. 6	2h		363. 6		I
81. 98	2		383. 83		II
72. 02	1		396. 76		I
67. 92	4		402. 07		I
60. 4	1		411. 9		I
48. 42	50		427. 50		I
34. 12	1		459. 33		I
20. 42	20		464. 19		I
8, 703. 13	5		486. 97		I
8, 674. 40	60		525. 01		I
72. 10	30		528. 07		I
50. 82	2		556. 43		II
24. 22	6		592. 07		I
8, 621. 55	2		595. 66		I
8, 590. 97	6		636. 94		I
58. 9	1		680. 5		I
45. 44	50		698. 94		I
43. 46	20		701. 65		I
29. 68	3		720. 55		I
14. 65	3		741. 24		II
13. 55	15		742. 76		I
8, 507. 37	10		751. 29		I
8, 484. 01	2		783. 65		II
76. 48	30		794. 12		I
67. 62	15		806. 45		I
8, 440. 06	3		845. 01		I
8, 379. 90	20		930. 18		I
58. 50	2h		960. 59		I
46. 60	100	12 III	977. 64		I
34. 44	3		11, 993. 11		I
24. 72	100	15 III	12, 009. 12		I
24. 59	H. N. R.	1 III A	009. 30		I
23. 35	1		011. 10		II
16. 05	10	2 IV A	021. 64		I
8, 302. 82	4		040. 80		I
8, 247. 46	60	10 III	121. 62		I
37. 90	3		135. 68		I
11. 65	2		174. 48		I
8, 203. 38	3		186. 75		I
8, 159. 05	10 M(?)		252. 96		II(?)
8, 086. 10	20+M4	15 III	363. 50		I
84. 53	3		365. 92		I
59. 5	3h 3h		404. 3		II
51. 38c	10 2	10 III	416. 82		I
8, 001. 91	4 1	4 III A	493. 58		I
7, 964. 86	5 2	6 III	551. 70		I
48. 30	H. N. R.	10 III	577. 85		I+M
7, 931. 18	1 —	10 III	12, 003. 00		I

TABLE 1.—*Wave lengths and Zeeman effects in lanthanum spectra*—Continued

λ (air) I. A.	Intensity arc spark B. S.		Arc intensity and temperature class K and C	ν (vac.) cm⁻¹	Zeeman effect	Spectrum
7,927.83	1	2		12,610.33		II
7,891.69	1	2		666.08		II
79.93	2	4		686.96		II
64.96	1	—		711.10		I
41.76	3	1	2 III A	748.74		I
7,838.83	1	1		753.50		II
7,740.54	(?)	2h		915.45		II+M
7,737.74	1h	(?)		12,920.12		I+M
7,664.38	4	2	? III	13,043.78		I
7,612.94	3	3		131.92		II
7,539.24	10	3	15 II	260.29		I+M
33.64	2	(?)		270.14		I
7,501.78	1	(?)		326.50		I
7,496.82	2	2	5 III A	331.76		I
89.15	(?)	1		346.96		II
83.48c	15	30	20 IV E	359.10	(0.00) 1.12	II
7,463.08c	5	2	10 III	395.61		I
7,382.73	6	2	5 III A	541.40		I
79.71			10 III A	546.94		I+M
45.36c	25	7	20 III A	610.30	(0.42) 1.18	I
44.42	H. N. R.		2 III A	612.03		I
40.08	—	2		620.06		II
34.18c	50	10	60 II	631.04	(0.00 w) 1.30 w	I
20.90	2	1		655.77		I
7,306.46	2h	(?)		679.01		I+M
7,297.09	—	2		698.63		II
85.83	1	(?)		721.50		I
82.36	100	150	100 III E	728.03	(0.00 w) 1.26	II
70.30	H. N. R.		5 II A	750.81		I
70.07	H. N. R.		10 III	751.24		I
66.13	2	5		758.70		II
63.68	2	(?)		763.34		I
62.83	2	(?)		764.95		I
50.35	1	—		788.59		I
19.92	15	2	15 II A	846.76		I
17.16	2	(?)		852.06		II
7,213.55	1	2		858.22		
7,161.25	40	10	60 III	960.20	(0.00) 0.91	I
58.11c	30	10	50 II A	966.32	(0.00) 1.27	I
51.92	H. N. R.		1 III A	978.42		
49.76	2	1	2 III	13,982.63		I
18.6	(?)	2		14,043.8		II
16.8	(?)	3		047.4		II
7,104.7	(?)	3h		071.3		II
7,096.18	2	(?)		084.24		I
92.40	H. N. R		2 I A	095.72		I
79.74	—	4		120.92		II
76.36	3	2	3 III	127.67		I
68.37	60	10	100 III	143.64	(0.69) (?)	I
66.24c	200	300	400 III E	147.90	(0.00) 0.70	II
62.4	1h	(?)		155.6		I
45.96c	200	30	300 II	188.62	(8.31, 0.65) 1.04, 1.47	I
32.07	25	5	15 III	216.65	(0.69) 1.27	I
23.67c	150	20	150 III	233.66	(0.00) 1.16	I+M
7,013.15	2	1		255.00		I
6,978.09	2	1		326.62		I
76.87	2	1	3 III	329.13		I
64.78	2	25	tr V E	345.76	(0.00) 0.81	II
58.11	30	100	20 IV E	367.76	(0.21) 0.96	II
54.54c	20	20	8 IV	375.14		II
52.52c	10	10	2 IV	379.31	(0.00) 1.10	II
35.03	50	8	50 III	415.58	(0.48) 1.14	I
25.27	80	15	100 III	435.90	(0.00w) 1.56 A²	I
18.33	8	2	8 III A	450.38		I
6,917.26	10	3	10 III A	14,452.61		I

TABLE 1.—*Wave lengths and Zeeman effects in lanthanum spectra*—Continued

λ (air) I. A.	Intensity arc spark B. S.		Arc intensity and temperature class K and O	ν (vac.) cm⁻¹	Zeeman effect	Spectrum
6,903.08	2	1	1 III A	14,482.30		I
6,902.06	3	8	1 III	484.40		I(?) II
6,899.63	2h	1	tr V	489.54		
98.41			tr V	492.10		I
59.03	4	5	tr V	575.31		II
37.91	15	15	3 IV(?) E	520.32		II
34.07	20	20	4 IV(?) E	528.54	(1.05) (?)	II
30.83	1	6	tr V E	535.49	(0.00) 0.98	II
23.80	40	10	15 II A	550.56	(0.00) 1.15	I
21.51	5d	2	3n V	655.47		I
19.14	1	—		660.57		I
16.29			1n V(?)	666.70		
13.68	5	50	1 V(?) E	672.32	(1.04) 1.29 B	II
08.88c	30	30	6 IV	682.66	(0.00h) 1.56h	II
6,801.38	1	5		698.85	(0.00) 1.10	II
6,796.73	4	1	4 II A	708.90		I
83.55	1	—	tr IV A	737.48		I
78.19	1	—	tr V	749.14		I
76.69	1	—	1 III A	752.40		I
74.28c	80	100	30 III E	757.65	(6.00w) 1.10	II
60.73	H. N. R.		1 III A	787.23		I
60.23	1	1		788.32		I
53.05	40	10	50 I A	804.04	(0.00W) 1.51 A'	I
50.47	1	1		809.70		II
48.12c	10	3	10 II A	814.86		I
41.20	2	1		830.07		I
37.29	1	—		838.67		I
32.80c	2	40	1 V E	848.57	(0.00) 0.99	II
18.68	2	60	tr V E	879.78	(0.69) 1.12 B	II
15.96	1	—	1 IV A(?)	885.80		I
14.08	5	80	2 V E	889.97	(0.00) 1.38	II
6,709.49	150	30	200 I	900.16	(0.00) 1.18	I
6,699.86	3	1	4 III A	921.57		I
99.26	2	2	2 III(?)	922.91		I
92.86c	20	4	30 I A	937.18	(0.00) 1.06	I
76.14	1	3		974.59		II
71.41c	15	40	6 IV(?) E	14,985.20	(0.68) 1.17	II
64.45	1	—		15,000.85		
61.40c	60	15	80 I A	007.72	(0.65) 1.23 B(?)	I
58.06	1	—		015.25		I
50.81c	80	20	100 I A	031.62	(0.00w) 0.88	I
45.15	7	3	8 IV	044.42		I
44.40c	30	5	40 I A	046.12		I
42.79	10	100	3 V E	049.77	(0.33) 0.82	II
36.53	3	5	1 V	063.96		II
31.20	3	3	3 V(?)	076.07		I(?)
28.4	1	1	1 IV A	082.4		I
19.10	—	2		103.63		II
16.59c	60	10	80 I	109.36	(0.52) 1.12 B	I
12.48	3	1		118.75		I
08.25c	40	5	40 II	128.43	(0.00) 1.01	I
07.7	1	—	3 IV	129.7		I+M
6,600.17c	25	5	50 II A	146.95	(0.35) 1.12	I
6,599.48			tr IV A	148.53		
93.45c	40	5	60 I	162.30	(0.00w) 1.30 A'	I
90.59	1	—		168.97		I
82.18	4	2	6 IV	188.35	(0.00h) 1.41	I
78.51c	200	100	400 I	196.82	(0.00) 1.00	I
70.96	2	(?)	2n III	214.28		II
68.54	1	(?)		219.89		I
65.45	40	(?)	40 I A	227.05		I
55.95			3 III A	249.12		I
55.11	2	(?)	1 V	251.07		I
54.18	2	(?)	1 V	253.23		II
6,551.78	1	(?)		15,258.82		I

Table 1.—*Wave lengths and Zeeman effects in lanthanum spectra*—Continued

λ (air) I. A.	Intensity arc spark B. S.		Arc intensity and temperature class K and C	ν (vac.) cm⁻¹	Zeeman effect	Spectrum
6,549.16	2	(?)	2n III(?)	15,264.92		I
43.17c	300	100	500 I	278.90	(0.33, 0.99) 0.00, 0.74, 1.41, 2.07	I
29.72c	4	4h	4 III(?)	310.37		II
26.99c	100	200	100 III E	316.77	(0.06, 0.22) 0.94 A²	II
23.86	2	1	2 III A(?)	324.12		
20.70	15	3	20 IV	331.55	(0.00h) 1.10h	I
19.30	3	1	4 IV	334.84		I
17.35	1	—		339.43		I
6,506.25	4	2	6 IV	365.59		I
6,498.19	30	250	7 IV(?) E	384.66	(0.00) 0.81	II
92.86	4	2	5 V	397.29		I
85.54c	20	5	20 II A	414.66	(0.00) 1.10	I
80.20	1	—		427.37		I
68.44c	10	3	8 II A	455.42		I
55.99	250	100	300 I	485.22	(0.00) 1.18	I
54.50c	150	50	200 I	488.80	(0.35, 1.11) 0.80, 1.54, 2.26	I
50.34	6	3	8 II A	496.78	(0.00) 1.29	I
49.94	3	—		499.74		I
48.25	4	2	10(?) II A	503.81		I
48.10	20	10	50(?) II A	504.17		I
46.62	15	200	5 V E	507.73	(0.00w) 1.43 A²	II
43.05	3	50h	1 V	516.32	(0.00w) 0.76 A¹	II
26.60	2	1		556.04		I
20.90	1	—		569.85		I+M
17.23	2	1	3 III(?) A	578.75		I+M
15.39	1	1		583.22		II
6,410.98	200	60	300 I	593.94	(0.00) 1.09	I
6,399.04	20	400	10 V(?) E	623.04	(0.00) 1.04	II
94.23c	400	100	600 I	634.79	(0.00 w) 1.02 A¹	I
90.48	100	200	80 III	643.96	(0.00) 1.03	II
80.48	1	1	1 V	664.48		I
75.50	2	—		780.72		I
75.11	2	—	2 V	681.68		I
74.08	3	30	1 V	684.21	(0.00 W) 1.67 A²	II
60.20c	30	10	30 II	718.44	(0.00) 0.89	I
58.12	20	30	5 IV	723.58	(0.99) 0.49, 1.52	II
56.38	3	1	4 III A	727.88		I
53.63	1	—		734.69		I
39.16	2	1	2 III A	770.61		I
37.88	2	3	1 V	773.79		II
33.74	3	2	2 IV(?)	784.10		I
33.24	2	1	2 III(?)	785.35		I+M
30.42	3	1	2 III A	792.38		I
25.90	100	20	150 I	803.67	(0.63, 0.99) 0.25, 0.63, 1.01, 1.38, 1.76	I
20.30c	100	200	80 III	817.44	(0.44) 0.87 B	II
18.26	5	2	12 III A	822.78		I
15.79	3	50	1 V	828.96	(0.95) 0.49, 1.43	II
10.91	8	200	4 V E	841.20	(0.27) 1.10 B	II
10.13	8	5	12 IV	843.16		I
08.87	2	?	2 III	846.32		I
08.21	2	?	3 III	847.98		I
07.25	1	20h	tr IV	850.40	(0.00) 0.51	II
6,305.46	10	10	5 IV	854.90		II
6,296.08c	100	300	50 IV E	878.52	(0.00 W) 1.20 A²	II
93.57c	60	10	80 II A	884.85	(0.00) 1.33	I
88.56	6	3	7 III A	897.50		I
87.73	7	3	8 III A	899.60		I
78.31	1	1		923.46		I
73.76	4	100	2n III(?)	935.01	(0.00) 1.04	II
66.00c	40	20	50 III	954.74	(0.00) 1.06	I
62.30	100	300	{30(?)/50(?)} III	964.17	(0.00 w) 1.10 A¹	II
49.92c	300	100	500 I	15,995.79	(0.00) 1.18	I
38.58	12	3	15 III A	16,024.86		I
34.76c	7	2	10 III A	029.54		I
6,236.17	8	3	12 IV	16,031.06		I

TABLE 1.—*Wave lengths and Zeeman effects in lanthanum spectra*—Continued

λ (air) I. A.	Intensity arc spark B. S.		Arc intensity and temperature class K and C	ν (vac.) cm⁻¹	Zeeman effect	Spectrum
6,234.85c	10	3	15 III A	16,024.45	(0.00) 1.26	I
33.51	10	3	15 III A	037.90		I
32.56	1	1		040.34		I
25.33	2h	1		058.97		I
24.24	1	1		061.78		I
19.46	2	1	2 V	074.13		I
18.19	5	2	7 III	077.41		I+M
14.35	2	1		087.34		
06.76	2	1		107.02		
6,203.51	3	501	2 V	115.46	(0.17) 1.14h	II
6188.09	5	1001	1 V	155.61	(0.00 W) 1.02 A¹	II
74.15	4	6	1 V	192.09	(0.00) 0.63	II
73.74			1 IV A	193.17		I
72.72	10	10	5 V	195.84	(0.00, 0.32, 0.63) 1.69 A²	II
67.69	2	1		209.05		I
65.69c	30	10	40 II	214.31	(0.88) 0.87, 1.42, 1.90	I
65.02	3	1	2 IV	216.07		I
54.55	1	—		243.66		
46.53	10	15	5 IV	264.85	(0.00 W) 1.69 A¹	II
45.29	3	2	3 III	268.13		I
42.98c	10	5	10 III	274.25	(0.00) 1.15	I
41.71	3	—		277.61		I
36.48	2	1	2 II A	291.49		I
34.39c	20	8	33 III	297.34	(1.34) 1.37	I
31.92	1	1		303.60		II
29.57c	30	50	15 IV(?)	309.85	(0.00 W) 0.99 A¹	II
27.04	8	5	12 III A	316.59		I
26.09	20	50	10 V(?)	319.12	(0.34) 0.92 B	II
23.75	2	1	2 IV	325.36		I
21.27			1 IV	331.97		I
20.34			1 IV	334.45		I II
11.71c	20	7	30 II	357.52	(0.00) 1.17	I
06.47c	40	10	60 II	366.19	(0.53) 1.34 B	I
07.26	4	2	12 II A	369.43		I
6,100.37	10	30	10 V	387.92	(0.00, 0.34) 0.84, 1.16	II
6,092.22	2	1	1 III A	408.85		I+M
88.00	2	1	1n IV	421.22		II
85.43	2	10	1n V	428.15	(0.00, 0.74) 1.20, 1.90	I
84.86	5	3	8 III	429.69		I
75.24			1 III A	455.71		I
74.01	3	5	1 III A	459.04	(0.21) 1.12	I II
72.04	3	2	4 III A	464.38		I
69.70c	20	6	30 III	473.44	(0.75) 0.48, 1.90	I
67.13	2	6	1 V(?)	477.71	(0.00) 1.49	II
61.42	2	2	2 V(?)	493.23		II(?)
46.07	2p(?)	2	1 IV(?)	535.10		II(?)
44.8	2	2	2 III A	538.6		I
41.6	2	2	2 III	547.3		I
38.57c	20	5	25 III A	555.64		I
37.98	1	2		557.25		II
35.12			2 V	565.11		
34.5	2h	2	3 V	566.8		I+M
32.28	5	2		572.63		I
31.46	5	2	3n IV(?)	575.15		I+M
25.09	2p(?)	1	1 IV(?)	592.68		I
17.16	3	2	1 III A	614.54		I
6,007.34c	50	10	50 III A	641.70	(0.00 W) 1.77 A²	I
5,992.35c	3	2	2 III A	683.33		I
91.98	1	4h		684.36		II
82.34	10	4	5 III A	711.25		I
75.75c	10	6	3 III A	729.68		I+M
73.82c	15	1201	5 V E	735.92	(0.00) 1.20	II
71.09	3	8		742.73	(0.00) 1.18	II
70.58	H. N. R.	2	1 III A	744.16		I
5,965.30	3	2	1 III	16,758.96		I

TABLE 1.—*Wave lengths and Zeeman effects in lanthanum spectra*—Continued

λ (air) I. A.	Intensity arc spark B. S.		Arc intensity and temperature class K and C	ν (vac.) cm⁻¹	Zeeman effect	Spectrum
5,962.59	4	2	1 III A	18,766.60		I
61.43	2	3		769.86		II
60.59	4	3	3 III A	772.23		I
57.90	—	4		779.80		II
48.30c	10	20		806.9	(0.52) 0.86 h	II+M
40.83c	3	1	1 III A	828.01		I
36.22c	15	20	12 V E	841.08	(0.00)'1.21	II
35.29c	20	5	15 II A	843.72	(0.00 W) 1.72 A²	I
30.68	100	20	} 400 I	856.81		I
30.61	200	40		857.01	(0.00) 0.87	I
28.48	5	4	4 III A	863.07		I
27.71	5	30		865.26	(0.00) 1.30	II
18.26	—	4		892.19		II
17.64c	20	6	15 II A	893.96	(0.27) 1.10 h	I
04.28	4	3	2 III A	932.20	(0.00) 1.25	I
01.95	5	40l	2 IV E	938.87	(0.00 W) 1.60 A²	II
5,900.75c	3	2	1 III A	942.31		I
5,894.84c	20	10	8 III A	980.30		I
92.66	3	4	1 V	963.58	(0.00) 1.54	II
85.23	—	1		16,986.99		II
80.63c	40	50	30 III E(?)	17,000.28	(0.06, 0.42) 0.54, 0.96, 1.38	II
77.96	3	2	1 III A	008.00		I
77.62	4	2	2 III A	008.98		I
74.72c	8	4	8 III A	017.38		I
74.00	5	6	3 IV	019.47	(0.00) 1.49	II
69.93	2	2	1 III A	031.27		I
63.70	30	80	20 V E	049.36	(0.00) 0.96	II
57.44	2	—		067.58		I
55.57c	20	8	15 II A	073.03	(0.00 W) 1.42 A²	I
52.26c	6	4	2 III A	082.69		I
48.95	3	20		092.36	(0.00) 0.54	II
48.37c	15	5	8 III A	094.05	(0.00) 0.90	I
45.02c	10	5	6 II A	103.85	(0.00) 1.18	I
39.77	3	2	2 II A	119.23		I
29.71c	20	10	10 III A	148.77	(0.00) 1.75	I
28.44	2	2		152.51		II
27.56c	8	3	3 III A	155.10	(0.50) 1.50(?)	I
23.82c	15	10	10 III A	166.11	(0.00 w) 0.96 h	I
21.98c	30	20	30 III	171.54	(0.00) 0.93	I
13.44	2	1	1 III A	196.76		I
08.63	4	8		211.00		II
08.31	40	60	15 IV(?)	211.95	(0.00) 0.75	II
06.56	3	8		217.14		II
05.77	80	120	50 III E	219.48	(0.00) 1.09	II
5,802.10c	2	1	1 III A	230.37		I
5,797.57	100	150	50 III E	243.84	(0.00) 1.24	II
91.32c	200	60	400 I	262.45	(0.00 W) 1.31 B	I
89.22c	150	40	250 I	268.71	(0.15) 1.21	I
81.02	2	3		293.20		II
79.91	2	4		296.52	(0.00) 1.38 h	II
69.97c	25	20	25 III A	326.32		I
69.32	80	30	50 I	328.27	(0.00) 1.01	I
69.06c	40	60	50 V E	329.05	(0.00) 1.09	II
61.53c	50	20	60 I	350.80	(0.94) 0.06, 0.68, 1.34	I
49.59	—	2		387.73		II
44.41c	60	20	80 III	403.41	(0.00 w) 1.24	I
42.93c	4	3	2 III A	407.90		I
40.65c	80	20	100 I	414.81	(0.47) 0.59 B	I
34.93c	6	4	5 III A	432.18	(0.00) 0.92	I
27.29	5	20	4 V E	455.43	(0.00) 1.45	II
20.01c	10	4	10 III A	477.65	(0.16, 0.48, 0.81) 1.93A²	I
14.55c			1 III A	494.35		I
14.01	4	3	5 III A	496.00	(0.62) (?)	I
12.30c	20	20	20 III E(?)	500.96	(0.00) 0.98	II
10.85	2p(?)	(?)	2 III	505.68		I
5,703.32	20	20	10 III	17,528.80	(0.00) 1.48	I II

TABLE 1.—*Wave lengths and Zeeman effects in lanthanum spectra*—Continued

λ (air) I. A.	Intensity arc spark B. S.		Arc intensity and temperature class K and O	ν (vac.) cm⁻¹	Zeeman effect	Spectrum
5,702.57			2 III	17,531.10		I
01.15			1 III	535.47		
5,700.25			3 III	538.24		Cu. (?)
5,699.32	3	2	.5 III	541.10		I
96.18	30	15	40 I	550.77	(0.00w) 0.80Λ¹	I
71.54	10	100	8 VE	627.02	(0.00) 0.79	II
69.97			1 IV	631.90		Ce (?)
61.34c	2	2	1 IIIA	658.78		I
57.71c	30	15	50 II	670.10	(0.29, 0.88) 0.73, 1.33, 1.94	I
56.54	2	2	1 IIIA	673.76		I
54.8c	20	10	3 IIIA	679.2		I+M
52.3	5	10h		687.0		II+M
48.24c	50	30	80 III	699.73	(00.0) 1.12	I
39.31c	8	5	5 IIIA	727.76		I
32.02	15	6	25 II	750.70	(0.00) 0.93	I
31.22	60	10	100 I	753.22	(0.00W) 1.45 Λ²	I
5,610.53	2	20		818.69		II
5,598.52	3	—		856.92	(0.00) 0.49	I
91.51	1	1		879.30		II
88.33c	100	20	80 II	889.48		I
70.37	3	2	5 IIA	947.16		I
68.45c	30	10	50 II	953.35	(0.44) 1.16	I
66.92	5	40	2 VE	958.28	(0.37) 1.24	II
65.70	10	5	20 II	962.22		I
65.43c	10	5	20 III	963.00	(0.00) 1.52	I
62.54	2	1	2 III(?)	17,972.42	(0.00) 1.13	I
47.56	2p(?)	3h		18,020.95	(0.28) 1.47	II
44.90	3	2	6 III	029.60	(0.00) 1.25	I
41.25c	15	10	20 III	041.47	(0.00) 1.25	I
35.66c	20	80	15 VE	059.69	(0.00W) 0.80 Λ¹.	II
32.17	4	10	3 III	071.08	(0.00W) 0.61h	I II
29.86	2	2	3 III	078.63		I
26.51			2 V	089.59		I
24.40			1 V	096.50		
17.34c	20	7	30 III	119.66	(0.00) 1.10	I
15.28c	5	4	10 III	126.42	(0.25) 1.42	I
07.33			2 III·	152.59		I
06.00c	20	5	40 II	156.97	(0.00w) 1.54h	I
03.80c	40	10	80 III	164.23	(0.70) 1.22 B	I
02.66	5	2	10 III	168.00		I
.02.24	3	2	4 III	169.38		I
5,501.34	200	70	300 I	172.35	(0.00) 0.80	I
5,498.70	2	1	2n III	181.08		I
93.45	10	20	20 VE	198.45	(0.00w) 1.57 Λ²	II
91.07	5	2	8 III	206.34		I
86.86	3	5	3 V	220.31		II
82.27c	20	40	80 VE	235.57	(0.00W) 1.89 Λ²	II
80.72	4	25	1 VE	240.72	(0.00) 1.03h	II
75.17c	10 .;	6	15 III	259.21	(0.00) 1.08	I
66.91	2	2	3 III	286.80		I
64.37	10	25	25 VE	295.30	(0.00, 0.49, 0.97) 0.00d, 0.49, 0.97	II
58.68c	10	50	20 VE	314.37	(0.00) 1.00	II
55.14	200	40	400 I	326.25	(0.00) 1.20	I
47.59	2	10		351.65	(0.22) 1.14	II
37.55	2	2	5 III	385.54	(0.00) 2.54	I
29.86c	6	5	15 III	411.57	(0.49) 1.04	I
23.82	4	4		432.08		II
22.10			3 IIIA	437.93		I
5,415.67	4	3	8 III	459.82	(0.36) 1.56w	I
5,390.63	(?)	1		545.56		I
81.91	15	100	} 15 VE	575.61	(0.00) 0.74	II
81.77	5	50		576.10		II
80.97	40	100	80 VE	578.86	(0.00) 0.77	II
80.00	3	1	5 III	582.21		I
5,377.08	20	200	2 VE	18,592.30	(0.00w) 0.82 Λ¹	II

TABLE 1.—*Wave lengths and Zeeman effects in lanthanum spectra*—Continued

λ (air) I. A.	Intensity arc spark B. S.		Arc intensity and temperature class K and C	ν (vac.) cm⁻¹	Zeeman effect	Spectrum
5,365.87	4	2	8 III	18,631.14		I
59.70	2	1	2 III A	652.59		I
57.85c	25	10	60 III	659.03	(0.00) 1.53	I
40.66	20	100	40 III E	719.09	(0.00) 1.07	II
33.42	1	2		744.50		II
30.64	1	2		754.27		I(?)
23.56	3	2	3 III	779.22		I
21.34	1	1		787.04		I
20.14	3	2	3 III	791.28		I
07.52	3	2	3 III	835.96		I
04.01c	20	5	30 III	848.43		I
03.54	50	100	100 III E	850.10	(0.06, 0.51) 0.57, 1.17, 1.78	II
02.62	30	150	40 V E	853.37	(0.00) 1.16	II
5,301.97c	80	200	200 III E	855.68	(0.00 w) 1.50 w	II
5,290.83	30	50	60 III E	895.38	(0.34) 0.82 B	II
87.45	1	1	1 V	907.46		I
79.11	6	40	2 V E	937.33	(0.00) 1.20	II
76.40	5	4	10 III	947.06	(0.00) 0.90	I
71.18	100	30	150 I	18,965.82	(0.00) 1.39 A²	I
59.38c	30	50	40 III E	19,008.37	(0.00 W) 0.46 A¹	II
57.83	4	3	3 III	013.98	(0.00) 1.07	I
57.28	—	2		015.96		II
53.45	100	30	100 I	029.83	(0.32, 0.69) 0.57, 1.03, 1.49, 1.95	I
40.81	4	3	2 III	075.72		I
39.54	4	3	4 III	080.35		I
34.27c	150	40	300 II	099.56	(0.00 W) 2.10 A²	I
26.20	2	40l	1 V E	129.05	(0.00 W) 0.32 W	II
22.48	—	3h		142.68	(0.00) 1.15	II
21.32	—	3h		146.93		II
17.83	—	10h		159.74		II
11.85c	150	40	300 II	181.72	(0.00 W) 1.73 A²	I
5,204.14	25	300	12 V E	210.14	(0.00) 0.86 A¹ (?)	II
5,191.50	1	3h		256.91	(0.00) 1.05	II
90.34	8	4		261.21		I
88.21	40	500	25 V E	269.12	(0.00) 1.09	II
83.91	10	10(?)	20 II	285.10		I
83.42c	200	400	400 III E	286.93	(0.00) 1.31	II
79.11	2	2	2 III	302.98		I
77.30c	150	50	300 II	309.72	(0.00) 1.07	I
73.83	10	25l	10 IV (?) E	322.67	(0.00) 1.06	II
72.89	5	20l	1 V E	326.19	(0.20) 1.03	II
68.95	2	2		340.92		I
67.79	20	10	20 III	345.26		I
67.28	4	10		347.17		II
64.03			1 V (?)	359.34		I
63.61	20	40	20 V (?) E	360.92	(0.74) 1.18 B	II
62.68	2	3		364.41		II
61.54	2	1		368.68		I
58.68	40	20	80 I	379.42	(0.19, 0.57) 1.05, 1.45, 1.85	I
57.43	15	150	15 V E	384.12	(0.00) 0.66	II
56.74	20	40	30 V E	386.71		II
52.31	1	1		403.38		I III
45.42	100	40	200 II	429.36	(0.00) 0.89	I
39.16	3	2		453.03		I
35.42	3	2	2 V	467.20		I
34.37	2	1		471.18		I
29.81	3	2		488.48		I
22.99	100	200	150 III E	514.43	(0.00) 1.16	II
20.87	10	4	10 III	522.51	(0.00) 0.92	I
14.55c	100	200	150 III (?) E	546.63	(0.00) 0.54	II
12.37	1	2		554.97		II
09.12	3	2	2 III A	567.40		I
07.54	—	6h		573.46	(0.00) 1.19	II
06.23c	100	40	150 II	578.48	(0.19) 0.62 A² (?)	I
5,103.11	3	2	2 III	19,590.45		I

TABLE 1.—*Wave lengths and Zeeman effects in lanthanum spectra*—Continued

λ (air) I. A.	Intensity arc spark B. S.		Arc intensity and temperature class K and C	ν (vac.) cm⁻¹	Zeeman effect	Spectrum
5, 096. 59	2	1	1 IV (?)	19, 615. 51		I
90. 56	2	201	tr V E	633. 74	(0.00 W) 1.33 A²	II
86. 71	—	3hl		653. 61		II
86. 22	2	1		655. 50		I
80. 21	10	40	1 V E	678. 76	(0.38) 0.98 w	II
79. 37	4	3	5 IV	642. 01		I
78. 92	3	2	2 V (?)	683. 76		I
72. 10	3	1	1 III A	710. 22		I
67. 90c	15	10	15 III	726. 56		I
66. 99	1	20h		730. 10	(0.00) 1.10 h	II
63. 76	2	3		742. 68		II
62. 91	10	20	4 V E	746. 00	(0.00, 0.66) 0.90, 1.56, 2.22	II
60. 85	2	3		754. 04		II
58. 56	—	1h		762. 97		II
56. 46c	60	20	80 II	771. 18	(1.20) 0.00, 0.79, 1.57	I
52. 10	1	1		788. 25		I
50. 57	60	20	80 II	794. 24	(0.80) 1.18 B	I
48. 04	4	301	1 V E	804. 16	(0 00) 1.16	II
46. 87c	30	15	60 III	808. 75	(0.00 w) 1.24	I
37. 60	2	2		845. 20		I
33. 24	2	1		862. 40		I
19. 50	10	5	8 III	916. 76		I
14. 45	5	30hl		936. 82		II
02. 12	10	40	3 V E	985. 96	(0.00) 1.09	II
5, 001. 78c	20	6	10 III A	937. 32		I
4, 999. 46c	100	200	200 III E	19, 996. 60	(0.00 w) 1.28 A²	II
96. 82	6	50	2 V E	·20, 007. 16		II
95. 95	1	1		010. 64		I
95. 17	1	1		013. 77		II
94. 64	2	1	1 III A	015. 90		I
93. 87	15	5	20 II	018. 98		I
91. 27c	40	80	20 IV E	029. 41	(0.00) 0.85	II
86. 82c	60	100	100 III E	047. 28	(0.00, 0.43) 0.57, 1.00, 1.43	II
84. 92	3	2	2 IV(?)	054. 92		I
84. 63	1	1		056. 09		I
83. 56	2	1		000. 40		I
77. 95	8	4	8 II A	083. 01		I
74. 20	—	4h		098. 14		II
70. 39	50	100	100 III E	113. 55	(0.00, 0.29) 0.83, 1.12 us	II
68. 59	4d	3	2 {III / V} A	120. 84		I
64. 84	3	2	4 III A	136. 03		I
57. 77	4	3		164. 75		I
56. 04	2	2	1 V	171. 79		II(?)
52. 06	10	40	5 V E	·188. 00	(0.00) 1.17	II
49. 76	50	20	200 I	197. 38	(0.00) 0.84	I
46. 47c	20	50	20 IV(?) E	210. 81	(0.00, 0.67) 1.05, 1.42	II
45. 84	3	2	5 III A	213. 39		I
35. 61	5	10	2 V E	255. 28		II
34. 83	40	100	20 V E	258. 48	(0.00) 0.97	II
25. 40	3	2	2 IV	297. 27		I(?)
21. 80	200	300	400 III E	312. 12	(0.00) 1.11	II
20. 98	200	300	400 III E	315. 50	(0.00) 1.00	II
16. 62	3	2	1 III(?) A	333. 52		I
11. 34	·4	10		355. 38		II
05. 13	4	3	4 III A	381. 15		I
04. 43	1	2h		384. 05		II
4, 901. 87	15	10	25 I	394. 70	(0.00) 0.94	I
4, 899. 92	150	200	300 III E	402. 82	(0.00) 0.80	II
94. 24	2	2	1 III	426. 49		I
91. 43	4	10		438. 23		II
87. 60	4	3	5 III	454. 24		I
86. 82	3	2	2 IV	457. 51		I
81. 94	1	1		477. 96		I
80. 22	1	10h		· 485. 17		II
4, 878. 86	10	10	15 III	20, 490. 89	(0.00) 1.10 R	I

TABLE 1.—*Wave lengths and Zeeman effects in lanthanum spectra*—Continued

λ (air) I. A.	Intensity arc spark B. S.		Arc intensity and temperature class K and C	ν (vac.) cm⁻¹	Zeeman effect	Spectrum
4,874.99	1	1		20,507.15		II
70.56	5	2	5 III	525.81		I
68.90	3	1	2 IV(?)	532.80		I
67.37	3	1		539.26		I
60.90	60	80	100 III E	566.59	(0.00) 1.19	II
59.18	2	5h		573.87		II
54.95c	8	6	8 III	591.80	(0.00) 1.21 R	I
60.81	20	10	20 I	609.37		I
50.58	20	30	5 V E	610.35	(0.00) 1.03	II
43.29	4	5	1 V	641.37		II
40.02	20	30	10 V E	655.32	(0.67) 1.30(?)	II
39.51	20	10	25 II	657.49	(0.00) 0.89	I
30.51	5	10	1 V E	695.98	(0.00) 0.76 R	II
26.87c	10	20	4 V E	711.59	(0.33) 1.08 R	II
24.05	80	100	80 III E	723.70	(0.00) 0.79	II
17.55	3	2		751.66		I
17.17	10	6	4 IV(?)	753.29		I
09.00c	60	100	80 V E	788.55	(0.00) 0.51	II
04.04c	50	80	50 V E	810.02	(0.95) 0.49, 1.43	II
4,800.24	9	6	8 III	826.49		I
4,799.99	8	5	8 III	827.57		I
96.67	10	25	4 V E	841.99		II
94.55	3	3	2 V E	851.20		II
92.46	1	1	tr III A	860.30		I
91.77	1	—		863.30		I
91.39	5	3	5 II	864.96		I
80.55c	2	2	1 V	912.27		II(?)
79.89c	4	3	4 II	915.16		I
75.14c	3	3	2 V	935.96		I(?)
70.43	10	8	15 II	956.63	(0.00) 0.93 R	I
67.80	1h	1h		968.19		I
66.89	60	30	100 I	20,972.19	(0.00) 0.99	I
59.71c	2	2	2 IV(?)	21,003.83		I
58.40	2	3		009.61		II
57.14c	3	2	2 IV	015.18		I
56.97			1 V	015.93		I
53.11c	2	1	1 IV	032.99		I
52.41c	3	2	3 III	036.09		I
50.41c	10	8	15 III	044.95	(0.00) 1.04 R	I
48.73	80	150	100 V E	052.39	(0.00) 1.00	II
43.06c	100	250	100 V E	077.47	(0.52 w) 0.95 B	II
40.27	100	120	150 III E	089.96	(0.00 w) 0.97 A¹	II
39.80	6	15	2 V E	092.06		II
33.82c	8	5	4 V(?)	118.70		I
30.73	2	3h		132.50		II
29.09	1	—	1 V	139.82		I(?)
28.41c	80	100	80 V(?) E	142.86	(0.00, 0.32) 1.31, 1.61, 1.91	II
24.42	20	40	6 V E	160.72	(0.65) 0.95 us	II
23.72	2	1	3 II	163.85		I
22.14	1	2h		170.94		II
19.93c	40	150	15 V E	180.85	(0.00) 1.06	II
17.58	10	50	4 V E	191.40	(0.50) 1.2	II
16.44c	40	80	25 V E	196.52	(0.00, 0.60) 0.92, 1.52, 2.12	II
14.14	4	2	5 I	206.86		I
12.92c	20	40	12 V E	212.35	(0.00, 0.47, 0.92) 1.83, 2.36 R	II
08.18c	8	8	8 III (?)	233.71	(0.00) 0.97 R	I
03.27c	30	150	20 V E	255.87	(0.00 W) 1.70 A¹	II
02.64	8	6	10 I	258.72		I
4,700.26c	8	8	8 III (?)	269.48		I
4,699.62c	20	50	20 V E	272.38		II
95.30	3	3		291.95		I
92.50c	50	200	50 V E	304.66	(0.00 W) 1.21 A¹	II
91.17c	25	50	30 V E	310.70	(0.00, 0.46) 0.46 us	II
88.65	8	40	2 V E	322.15	(0.00) 0.96	II
4,684.39	1	2h		21,341.54		II

TABLE 1.—*Wave lengths and Zeeman effects in lanthanum spectra*—Continued

λ (air) I. A.	Intensity arc spark B. S.		Arc intensity and temperature class K and C	ν (vac.) Cm⁻¹	Zeeman effect	Spectrum
4,682.12	4	5		21,351.89		II
73.53	—	1h		391.14		II
71.82c	30	200	40 V E	398.96	(0.00 w) 1.29h	II
68.91c	40	250	60 V E	412.30	(0.00 0.34, 0.68) 0.77, 1.12, 1.47, 1.81 us	II
63.76c	50	300	50 V E	435.94	(0.20) 0.99	II
62.51	100	200	200 III E	441.69	(0.00 W) 0.89 A²	II
60.70	8	8	8 III	450.02	(0.00) 0.92 R	I
55.49c	80	400	150 V E	474.02	(0.43) 1.20 B	II
53.90c	4	3		481.36		I
52.07	15	30hl	20 I	489.81		I II(?)
50.32	12	8	15 I	497.90	(0.20) 1.13 R	I
48.64	30	20	40 I	505.67	(0.43) ?	I
47.50c	25	100	8 V E	510.94	(0.00, 0.31, 0.61) 1.08 A²	II
46.33	10	10	12 III	516.36	(0.00) 0.89 R	I
45.28	40	100	40 V E	521.22	(0.60) 1.26 B	II
42.11c	8	5	5 III	531.28		I
41.40	—	2h		539.21		II
36.42	10	80	4 V E	562.35	(0.21) 1.00	II
34.95	2	25l		569.19	(0.00) 1.43	II
33.4	1	10hl		576.4		II
27.35	2	2		604.61		I
23.99	1	2h		620.31		II
19.87c	100	300	150 V E	639.59	(0.00) 0.70	II
15.06c	8	7	18 III	662.14	(1.31) 1.37 R	I
13.38	100	200	200 V E	670.03	(0.00, 0.33) 0.83, 1.16, 1.49	II
05.78	50	100	50 V E	705.79	(0.63) 0.89, 1.52	II
05.06	6	3	10 III	709.09		I
04.24	6	4	10 III	713.05		I
02.04c	10	10	20 III	723.43		I
01.65	—	3		725.27		II
4,600.59	1	5h		730.27		II
4,596.19	6	6	10 I	751.08		I
95.06	1	2h		756.42		II
89.89	5	5		780.93		I
87.14	—	2h		793.99		II
81.20c	10	10	12 III	822.25		I
80.05	50	150	100 V E	827.73	(0.00) 1.51	II
74.87c	150	200	300 III E	852.44	(0.49, 0.96) 0.49, 0.96, 1.46, 1.91	II
70.97	6	10	8 V E	871.08	(0.43) 1.00 R	II
70.02	60	50	250 I	875.63	(0.00) 1.34	I
67.90	50	40	200 I	885.78	(0.00) 1.24	I
64.85c	6	4	12 III	900.41	(0.00) 1.42 R	I
62.5	1	5h		911.7		II
59.28	30	100	50 V E	927.16	(0.00, 0.18) 1.62	II
58.46	100	200	250 III E	931.10	(0.00 W) 0.88 A¹	II
52.47	8	8	8 II A	959.96	(0.00) 0.57	I
50.76c	8	6	10 III A	968.21	(0.78) 0.38 R	I
50.16	4	3	5 III A	971.11		I
49.50	40	30	50 I	21,974.30	(0.00) 1.08	I
41.78c	10	8	15 III	22,011.65		I
40.71	3	10		016.84	(0.00) 1.27	II
38.87	1	8hl		025.76		II
37.57	2	2		032.07		I
30.54	4	15	2 V E	066.26		II
28.88	3	2		074.34		I
26.12	100	200	200 III E	087.81	(0.00 W) 1.01 A¹	II
25.31c	40	100	20 V E	091.76	(0.46) 1.19 B	II
22.37	200	400	500 III E	106.12	(0.00) 1.02	II
16.38	2	5hl		135.44		II
08.48	3	10		174.22	(0.00) 0.90	II
07.4	2	2		179.5		I
05.82	1	3hl		187.32		II
02.16	1	10hl		205.35	(0.00) 1.47	II
01.57	6	4	10 II A	208.26		I
4,500.21	30	30	40 II	22,214.97	(0.00 W) 1.10 A¹	I

TABLE 1.—*Wave lengths and Zeeman effects in lanthanum spectra*—Continued

λ (air) I. A.	Intensity arc spark B. S.		Arc intensity and temperature class K and C	ν (vac.) cm⁻¹	Zeeman effect	Spectrum
4,499.04c	10	10	10 III	22,220.75	(0.00) 1.02	I
98.76	2	10		222.13	(0.00) 1.14	II
97.00	2	2		230.83		II
94.71c	20	15	30 I	242.16	(0.51) 0.27, 0.58, 0.91 us	I
93.81	5	3	10 IA	246.61	(0.00) 0.72 R	I
93.11	15	10	25 I	250.08	(0.00) 1.16	I
91.76c	10	8	15 III	256.76	(0.00) 1.52	I
88.06	10	10	20 III	285.04		I
84.48	—	1h		292.90		II
81.21	3	25h1		309.16	(0.40) 0.54, 0.87	II
79.82c	6	5	15 II A	316.08		I
74.54	4	3	5 III A	342.42		I
74.03	2	10		344.96	(0.96, 0.66) 0.50, 1.17, 1.33	II
68.97	10	6	25 II	370.26		I
59.10	—	3		419.78		II
55.79	20	50	25 V E	436.43	(0.24) 1.41	II
55.21	3	2	10 II A	439.35		I
53.85			2 IV A	446.20		I
52.15	15	15	30 II	449.78	(0.00) 0.98	I
45.12	2	1	2 III A	490.79		I
43.94c	5	20h1	10 I	496.26	(0.00) 1.18 h	I II?
42.68	6	3	12 II	502.64		I
35.84	6	10	6 IV E	537.34	(0.86, 0.60, 1.19) 1.28, 1.86, 2.44	II
32.95	10(?)	20(?)		552.03		II+M
29.90c	200	400	500 III E	567.56	(0.86, 0.40) 0.51, 0.91, 1.31	II
27.52c	30	100	40 V E	579.69	(0.00w) 1.05	II
23.90	20	15	30 II	598.16	(0.00) 1.09	I
19.16	20	30	6 V E	622.41	(0.28) 1.04	II
17.14	2	2h	6n III	632.75		I II
13.45			2 III A	651.67		I
12.22	—	2h		657.98		II
11.21	1	25h1	2 V E	663.17	(0.00h) 1.13h	II
03.02c	2	2	8 III A;	705.33		I II?
4,402.64	5	4	15 III	707.29	(0.00) 0.86	I
4,397.04			2 IV A	736.21		I
96.79c			4 IV A	737.50		I
96.31			2 IV	739.98		I
93.52c	2	1	4 III	754.42		I
89.87	6	4	15 III	773.34	(0.00) 1.07	I
85.20c	10	40	20 V E	797.59	(0.00) 1.24	II
83.44	20(?)	100	25 V E	800.75	(0.86, 0.40, 0.80) 0.28, 0.68, 1.08, 1.48, 1.88	I }II+M
80.55	4	2	12 II A	821.79		I
78.10c	15	50	15 IV E	834.56	(0.30w) 0.95	II
64.06c	25	100	30 IV E	904.88	(0.00) 1.28	II
63.05	5	50l	4 V E	913.33	(00.0h) 1.32h	II
60.86	2	?	2 III A	924.84		I
60.49	2	2	2 III A	926.78		I
57.88	2	1	2 III A	940.51		I
56.18	—	1		949.46		II
54.79	20	?	25 III	956.79		I
54.40	60	200	80 IV E	22,958.84	(0.00) 1.06	II
40.72c	10	6	15 III	23,031.20	(0.00) 0.90	I
39.93	5	3	6 III A	035.39		I
37.78	2	10l		046.81	(0.00) 1.06h	II
34.96c	50	100	60 V E	061.80	(0.86, 0.29, 0.57) 0.69, 0.97, 1.26, 1.54	II
33.76c	300	500	600 III E	068.19	(0.17) 0.94	II
26.19	2	2		108.55		I
22.51c	60	100	130 III E	128.22	(0.00) 1.10	II
15.90c	10	30	12 V E	163.65	(0.86, 0.50, 1.00) 0.86, 1.36, 1.86, 2.36	II
11.73c	5	4	4 III A	186.05	(0.59) 0.96 R	I
4,306.00	6	5	6 II A	23,216.90	(0.14) 1.49	I

Table 1.—*Wave lengths and Zeeman effects in lanthanum spectra*—Continued

λ (air) I. A.	Intensity arc spark B. S.		Arc intensity and tempera-ture class K and C	ν (vac.) cm⁻¹	Zeeman effect	Spec-trum
4, 304. 11	1	10h1	3	23, 227. 10	(0.00h) 1.24h	II
00. 62	H. N. R.			245. 94		I
4, 300. 44	40	60	20 IV E	246. 92	(0.73, 1.46) 0.00, 0.72, 1.45, 2.16	II
4, 298. 05c	100	300	280 IV E	270. 67	(0.00) 1.48	II
91. 00	2	2		298. 06		I
89. 65	H. N. R.		1	305. 39		I
89. 01	H. N. R.		2	308. 87		I
86. 97c	80	300	150 V E	319. 96	(0.00) 1.08	II
80. 27	60	40	100 I	356. 46	(0.00) 1.14	I
75. 64	50	100	60 IV E	381. 75	(0.06, 0.35) 0.62, 1.17, 1.53	II
71. 14	4	3		406. 39		I
69. 50c	80	300	150 V E	415. 38	(0.00) 1.20	II
67. 74	2	2		425. 04		I
63. 59c	60	200	100 V E	447. 83	(0.00) 1.12	II
62. 35	10	5	15 II A	454. 66	(0.00w) 1.19	I
59. 51	—	2h		470. 30		II
56. 92c	6	5	6 III A	484. 58	(0.00) 1.29 R	I
56. 50	—	3		486. 89		II
52. 93	3	4		508. 61		II
49. 99c	20	100	25 V E	522. 87	(0.74) 0.50, 0.76, 1.02, 1.27, 1.53, 1.76	II
48. 32	—	2		532. 11		II
41. 20	1	15h1		571. 62		II
38. 59	4	(7)	10 III A	586. 13		I
38. 38c	200	400	400 III E	587. 30	(0.74) 0.56, 0.82, 1.06, 1.34, 1.60, 1.86	II
30. 95c	20	150	30 V E	628. 72	(0.80) 0.65, 0.85, 1.06, 1.26, 1.47, 1.67 ur	II
17. 56c	40	200	80 V E	703. 74	(0. 00) 1.22	II
16. 54	4	3		709. 47		I
10. 22	4	50h1		745. 06	(0.00h) 1.02h	II
07. 61	2	101		759. 79	(0.27) 0.68h	II
04. 03	40	100	100 V E	780. 02	(0.00) 1.48	II
4, 201. 50		6h		794. 34		II
4, 196. 55c	150	250	300 III E	822. 41	(0.32, 0.64) 0.51, 0.82, 1.13, 1.44	II
94. 36	4	30h		834. 85	(0.00h) (1.06h)	II
93. 34	1	5		840. 64	(0.25) 0.64	II
92. 72	H. N. R.		2	844. 17		I
92. 35c	40	100	80 V E	846. 27	(0.00) 1.06	II
87. 31	50	30	125 I	874. 98	(0.00w) 1.16 A²	I
80. 97	2	121		911. 18		II
77. 48	15	10	30 I	931. 15	(0.00) 1.50	I
72. 32	6	4	10 III A	960. 75		I
71. 13	5	4	8 III A	23, 967. 58		I
63. 31	5	4	8 III A	24, 012. 60		I
61. 94	2	8h		020. 51	(0.00h) 1.02	II
60. 26	20	10	30 I	030. 21	(0.00W) 1.65 w	I
57. 52	6	5	10 II A	046. 04		I
54. 59	—	2h		063. 00		II
52. 78	40	100	40 IV(?) E	073. 49	(0.43, 0.64) 0.27, 0.67, 1.06, 1.48	II
51. 98c	100	250	300 III E	078. 13	(0.29) 0.50, 0.77	II
50. 24	2	2		088. 22		I
48. 2	—	4h		100. 1		II
44. 36	4	2		122. 40		I
42. 92	5	(?)	12 III A	124. 96		I
43. 77	6	15		125. 83		II
41. 73c	80	200	{40} {50} IV E	137. 72	(0.00W) 1.50w	II
37. 91	—	2		160. 00		II
37. 05	20	10	40 I	165. 02	(0.00) 1.22	I
33. 33	—	6h1		186. 77	(0.00) 0.97	II
31. 50	1	10h1		191. 63	(0.00h) 0.83h	II
31. 74	—	5h		196. 08	(0.00h) 1.13h	II
4, 128. 28	200	400	600 III E	24, 246. 01	(0.00) 1.05	II

TABLE 1.—*Wave lengths and Zeeman effects in lanthanum spectra*—Continued

λ (air) I. A.	Intensity arc spark B. S.		Arc intensity and temperature class K and C	ν (vac.) cm⁻¹	Zeeman effect	Spectrum
4,117.67	8	5	20 III	24,278.75	(0.00) 0.93	I
15.35	—	1h		292.44		II
13.28	2	40l		304.66	(0.00h) 1.09h	II
09.80	10	6	20 I A	325.24	(0.00) 1.04	I
09.48	6	3	15 II A	327.14	(0.00) 1.26	I
04.87	30	20	60 I	354.46	(0.06, 0.25) 0.53, 0.69, 0.85 ur	I
4,101.01	—	3h		377.38		II
4,099.54c	40	150	60 V E	386.12	(0.06) 0.77	II
98.73	1	5		390.94		II
90.40	2	1		440.61		
89.61	25	12	50 I	445.33	(0.00) 0.93	I
86.72	200	300	400 III E	462.62	(0.32) 0.79 w	II
79.17	20	10	40 I	507.90		I
77.35c	200	300	400 III E	518.84	(0.00, 0.32) 0.57, 0.89, 1.26	II
76.71	10	40	30 IV E	522.68	(0.00, 0.31, 0.62) 0.79, 1.08, 1.38, 1.69	II
67.39c	60	100	125 IV E	578.88	(0.00w) 1.14 A²	II
65.58	15	6	30 II	589.82		
64.79	25	15	50 II	594.60	(0.00w) 1.03 A¹	I
60.33	30	20	60 II	621.61	(0.00w) 1.10 A¹	I
58.06	2	51		635.26	(0.45) 1.06, 1.47us	II
50.06c	50	200	60 V E	683.92	(0.00) 1.02	II
42.91	150	300	300 IV E	727.70	(0.00) 0.99	II
40.97	2	1		739.57		I
37.21	25	10	50 I	762.61	(0.00) 0.86	I
36.59c	7	15d	8 V E	766.41	(0.00) 1.54	II
31.68	100	300	300 III E	796.57	(0.00) 1.15	II
25.87	20	50	40 IV E	832.36	(0.00 W) 1.37 A²	II
23.58c	10	40	15 IV E	846.49	(0.00) 1.07	II
20.19	—	2h		867.44		II
15.39	25	15	50 I	897.17	(0.16, 0.31) 0.96, 1.10, 1.30, 1.48ur	I
07.64	1	7h		945.32	(0.00W) 1.01h	II
4,001.38	2	2		24,984.34		I
3,995.74c	200	400	600 III E	25,019.61	(0.30) 0.90w	II
94.50c	4	10		027.37	(0.57h) 1.25h	II
88.51c	300	500	800 III E	064.96	(0.00) 1.32	II
81.36	2	10l		109.97	(0.38) 0.98 h us	II
79.08	1	8l		124.35	(0.00) 1.26	II
63.04	1	5l		226.04	(0.00w) 0.82h	II
62.03	1	10l		232.47	(0.00) 1.10h	II
58.53	—	2		254.78		II
57.25	—	2		262.95		II
56.07	1	4		270.48	(0.00h) 1.29h	II
55.21	—	3h		275.98	(0.00) 1.15	II
53.67	10	5	40 II	285.83		I
53.36	—	2		287.81	(0.00) 1.06	II
51.43		3h		300.16		II
49.10c	400	600	1000 III E	315.09	(0.00) 1.13w	II
44.15	2	3		346.86		II
39.85	2	20l		374.52	(0.21) 1.27h	II
36.22	20	50	60 IV E	397.92	(0.06, 0.25, 0.48) 0.86, 1.10, 1.34, 1.58	II
32.53	3	10l		421.75	(0.44) 1.06 B	II
30.47	—	3		435.07	(0.00) 0.92	II
29.22c	100	300	500 III E	443.17	(0.00) 1.20	II
27.56	30	10	80 I	453.92	(0.00) 0.76	I
25.09	1	5		469.94		II
24.69	—	3		472.53	(0.00h) 0.72h	II
21.54c	70	200	300 III E	492.99	(0.00, 0.66) 0.52, 1.18, 1.84	II
16.05c	80	300	400 III E	528.73	(0.38) 0.51, 0.88	II
10.81	4	10l	15 IV (?) E	562.94	(0.00) 1.04	II
3,902.57	5	3	20 II	616.91		I
3,898.60	8	4	40 II	642.99		I
97.43	2	4		650.70		II
95.65			8 IV	662.42		Fe(?)
92.47	—	3		683.37	(0.00h) 1.4h	II
92.05	—	3		686.15		II
3,886.37	60	150	400 III E	25,723.69	(0.00W) 1.62 A²	II

TABLE 1.—*Wave lengths and Zeeman effects in lanthanum spectra*—Continued

λ (air) I. A.	Intensity arc spark R. S.		Arc intensity and temperature class K and C	ν (vac.) cm⁻¹	Zeeman effect	Spectrum
3,885.09	1	4		25,732.16		II
71.64	100	200	300 III (?) E	821.56	(0.00) 1.08	II
68.35	—	3h		843.52		II
64.49	6	100l		869.33	(0.00h) 1.10h	II
63.11	—	2		878.57		II
60.31	—	2		897.34		II
54.91	15	30		933.62	(0.00) 1.50	II
49.02	40	100	200 III (?) E	973.30	(0.00) 0.69	II
46.00	10	20	20 V (?) E	25,993.69	(0.36) 1.11 B	II
40.72	40	60	50 V (?) E	26,029.43	(0.00) 1.04	II
36.4	—	1		066.7		II
35.09	15	50	20 V E	087.64	(0.00) 1.49	II
17.24	1	8h		189.53	(0.00) 0.82h	II
16.25	2	10h		196.33	(0.00 h) 0.68	II
14.1	—	2		211.1		II
08.79	10	15		247.63	(0.00w) 1.18h	II
07.1	—	1		259.3		II
04.8	—	2h		275.2		II
3,801.0	—	1		301.4		II
3,798.19	—	2		320.88		II
94.78	150	400	600 III E	344.54	(0.00) 1.16	II
90.83	100	300	600 III E	371.99	(0.00) 1.00	II
84.81	8	15	20 V E	413.93	(0.17) 0.79	II
83.06	—	1		426.15		II
80.67	20	50 (?)	50 V E	442.85	(0.00) 1.54	II
80.53	5	50 (?)		443.84	(0.00h) 0.92h	II
73.12	10	150l		498.77	(0.00) 1.00h	II
68.98	1	3h		524.87		II
67.05	1	5h		538.46	(0.47) 1.24 B	II
66.58	1	3h		541.77		II
59.06	100	300	600 III E	594.72	(0.00) 1.25	II
53.04	—	2h		637.52		II
47.96	1	5l		673.63	(0.00w) 0.90h	II
44.85	—	2h		695.78		II
36.41	2	15l		756.08	(0.37) 1.20 B	II
35.85	6	10	20 IV E	760.09	(0.00w) 0.90d	II
35.09	—	1		765.53		II
31.42	1	8h		791.86		II
28.97	—	2h		802.46		II
25.05	10	20	40 IV E	837.67	(0.00, 0.38, 0.75) 0.79, 1.16, 1.53, 1.90	II
24.77			2 V	839.69		
20.75	—	2		868.69		II
17.99	1	2		888.63		II
15.33	20	50	80 IV E	906.44	(0.40, 0.72) 0.80, 1.12, 1.44, 1.79	II
14.87	10	40	60 V E	911.22	(0.00) 1.50	II
14.30	2	2		915.34		I
13.54	30	100	300 IV E	920.85	(0.00, 0.34, 0.68) 1.00, 1.35, 1.69, 2.03	II
10.61	—	2		942.11		II
05.81c	25	80	125 V E	977.01	(0.00) 1.51	II
04.54	10	4	40 II	26,986.25	(0.00) 1.08	I
3,701.81	4	40l	4 V E	27,006.15	(0.00W) 0.73A¹	II
3,699.57	4	2	12 III A	022.51		I
96.11	—	2h		047.80		II
95.2	—	2h		054.5		II
94.27	2	7h		061.27	(0.36) 0.88h	II
92.31	—	2h		075.64		II
78.24	—	2h		179.20		II
75.22	—	1		201.54		II
72.02	8	4	30 III	225.24		I
70.23	1	4h		238.52	(0.30) 1.06	II
69.27	—	3h		246.65		II
65.22	2	10l		275.75	(0.00d) 1.92d	II
62.08	15	30	50 IV E	299.14	(0.66) 1.16 B	II
59.40	1	3		328.60	(0.00) 1.0	II
3,658.04	—	1		27,329.29		II

TABLE 1.—*Wave lengths and Zeeman effects in lanthanum spectra*—Continued

λ (air) I. A.	Intensity arc spark B. S.		Arc intensity and temperature class K and C	ν (vac.) cm⁻¹	Zeeman effect	Spectrum
3, 652.62	—	1h		27, 369.84		II
50.19	25	80	125 V E	388.06	(0.48, 0.91) 0.29, 0.73, 1.13, 1.62	II
49.55	10	6	40 II	392.86	(0.00) 1.00	I
45.43	50	200	400 IV E	422.82	(0.00) 0.66	II
41.66	10	50l		452.21	(0.00) 1.21	II
41.53	20 (?)	(?)	100 II (+E)	453.19		I
41.10	—	2		456.43	(0.00) 1.01	II
39.25	1	3h		470.39		II
37.15c	10	40	50 V E	486.26	(0.00) 1.51	II
36.67	6	2	40 III	489.88		I
29.99	1	2hl		540.46		II
28.83	20	60	125 IV E	549.27	(0.00W) 1.68 A ²	II
21.77	2	4	4 V E	602.97		II
20.16	1	1		615.24		II
18.60	—	1		627.15		II
13.08	10	4	30 II	669.36	(0.00) 0.97	I
12.34	5	50	8 V E	675.02	(0.00) 1.11	II
11.09	1	2		684.60	(0.00) 1.57h	II
10.25	4	30 l		691.04	(0.00) 1.02	II
09.22	—	4		698.95	(0.00) 1.00	II
08.18	—	4		706.93	(0.00w) 1.08 A ²	II
06.42	—	4hl		720.45		II
3, 601.07	5	20bl	15 V E	761.63	(0.00) 1.18	II
3, 598.93	—,	3hl		778.14	(0.00) 1.15	II
96.65	—	4hl		795.75	(0.00) 1.22	II
93.29	—		2 V E	831.74		II
92.42	—	2h		836.48		II
90.66	—	1h		842.11		II
85.53	—	2		881.95	(0.00) 1.25	II
81.68	1	20hl		911.92	(0.00) 0.74	II
80.10	—	8h		924.24	(0.00) 1.52	II
78.89	—	5 h		933.68	(0.00) 1.44	II
76.56	—	2hl		951.88		II
74.43	20	6	50 II	27, 948.53	(0.00) 0.83	I
70.10	(?)	30hl		28, 002.38	(0.00) 1.08	II
57.26	3	8	20 IV (?) E	103.53		II
50.82	3	6	15 IV (?) E	154.49		II
36.37	—	3h		269.53		II
33.67	—	3h		291.13		II
30.67	4	8	20 V (?) E	315.17		II
26.77	—	2h		346.48		II
20.72	2	10chl		396.19	(0.00) 0.88	II
17.14c	5e	200	— V E	424.09	(0.72) 1.36	III
14.87	—	2h		442.45		II
14.07	6	3	20 II A	448.92		I
12.93	5	10	25 IV E	458.16	(0.53) 0.83, 1.10	II
10.00	6	15	25 IV E	481.91		II
3, 507.90	—	4hl		498.96		II
3, 493.97	—	2h		512.58		II
84.39	2	10l		691.25	(0.00) 1.33	II
80.61	3	2	8 III A	722.40		I
74.84	2	8l		770.10		II
66.46	—	1h		829.64		II
62.32	—	2h		874.13		II
61.18	10	3	25 III A	883.64		I
60.31	2	5l		890.90		II
53.17c	40	50	70 III E	950.63	(0.93) 0.37, 0.68, 1.00, 1.31, 1.63, 1.94.	II
52.18c	30	40	50 III E	958.94	(0.00) 0.95	II
51.12	—	3l		967.83		II
50.65	5	2	12 III A	28, 971.78		I
32.81	2	5		29, 122.34		II
27.57	2	8		166.85		II
21.9	3	5		198.1		II
22.44	1	2		210.57		II
3, 420.54	1	5h		29, 226.80		II

TABLE 1.—*Wave lengths and Zeeman effects in lanthanum spectra*—Continued

λ (air) I. A.	Intensity arc spark B. S.		Arc intensity and tempera- ture class K and C	ν (vac.) cm⁻¹	Zeeman effect	Spec- trum
3,411.76	2	20hl		29,302.01	(0.00 w) 1.48	II
07.00	1	8hl		342.95	(0.00) 1.50	II
3,404.53	10	3	15 III A	364.24		I
3,398.29	1	2h		418.15		II
97.77	4	40hl	4 V E	422.65	(0.24) 0.85	II
92.94	1	4h		464.54		II
90.40	1	4h		486.61		II
88.61	6	2	12 II A	502.18		I
81.42			15 II A	504.92		I
80.91	200	300	400 III E	569.36	(0.00, 0.37) 0.75, 1.11, 1.47	II
76.33	40	50	40 III E	609.49	(0.00) 0.88	II
74.89	1	3		632.12		II
68.36	3	—		679.54		I
64.88	2	—		710.24		I
62.04	7	3	12 III A	735.33		I
57.50	5	2	7 III A	775.54		I
51.89	—	3		825.37		II
49.82	2	—	3 III A	843.80		I
44.56	150	200	300 III E	890.73	(0.00) 0.53	II
42.23	10	5	20 II A	911.57	(0.00) 1.08	I
37.49	200	300	500 III E	29,954.05	(0.00w) 1.12 A¹	II
29.07	2	8		30,029.81		II
26.21	1	5		055.63		II
25.33	1	3		063.59		II
10.62	1	4		197.16		II
06.96	7	8	10 IV E	230.40	(0.00) 1.77 R	II
3,303.11c	100	150	200 III E	265.82	(1.00) 0.51, 1.51	II
3,296.72	—	5h		306.09		II
97.15	1	3		320.52		II
94.44	1	10		345.46		II
83.95	1	8h		442.39		II
77.83	—	4		499.23		II
67.31	1	3		597.43		II
65.67c	80	100	150 III E	612.79	(0.48, 0.80) 0.82, 1.17, 1.51, 1.86	II
63.98	2	5		628.64		II
56.60	2	1		698.05		I
53.41	2	10h		728.15	(0.00) 0.65	II
49.35	60	80	100 III E	766.54	(0.00, 0.50), 0.45, 0.94, 1.44	II
47.06	5	2	8 II A	788.24		I
45.13	100	150	200 III E	806.55	(0.00) 1.04	II
35.66	3	2	5 III A	896.71		I
26.03	1	2		30,988.93		II
24.71	—	1		31,001.62		II
17.12	2	8h		074.76		II
15.81	10	2	15 II A	087.42	(0.00) 1.13	I
12.56	1	5		118.86		II
09.13	2	6		152.12		II
08.13	1	6		161.83		II
05.75	3	4		184.96		II
3,204.55	1	3		196.64		II
3,194.70	—	2		292.83		II
93.02c	15d	25	{ 10(?) IV E / 15(?) IV E }	309.29	(0.00, 0.96) 0.61, 1.58, 2.54	II
91.39	1	10h		325.28		II
79.78	4	2	8 III A	439.65		I
75.99	8	2	15 II A	477.17		I
74.88	1	10hl		488.18		II
71.68c	8e	300	— V E	519.94	(0.37) 0.98, 1.76	II III
66.26	1	2		573.90		II
65.19	1	4		584.57		II
60.56	1	3		630.84		II
57.56	1	2		660.69		II
56.35	—	2		673.02		II
48.51	4	2	4 III A	751.89		I
45.7	1	2h		780.3		II
3,142.76	30	40	50 IV E	31,809.98		II

TABLE 1.—*Wave lengths and Zeeman effects in lanthanum spectra*—Continued

λ (air) I. A.	Intensity arc spark B. S.		Arc intensity and temperature class K and C	ν (vac.) cm-1	Zeeman effect	Spectrum
3,132.14	—	3		31,917.83		II
30.25	—	2		937.10		II
25.72	1	4hl		31,983.39		II
12.63	1	8h		32,117.88		II
09.42	8	3	12 II A	151.04		I
08.46	6	8	5 IV E	160.97		II
3,104.58	40	50	60 IV E	201.16	(0.00 W) 1.50 A °	II
3,095.02	8	2	10 II A	290.19		I
94.76	—	4h		303.34		II
88.53	—	4h		368.49		II
81.42	1	6h		443.18		II
75.51	—	4h		505.52		II
69.45	—	3		569.69		II
68.96	—	4		575.09		II
59.91	1	8		671.23		II
54.02	2	6		734.24		II
49.39	2	5		783.94		II
45.63	—	1		824.41		II
36.43	1	2		923.86		II
35.80	—	1		32,930.69		II
28.64	—	2		33,008.54		II
25.88	—	4(?)		039.85		II
22.26	1	5hl		078.22		II
18.95	1	6hl		114.48		II
10.78	4	3	2 IV A	204.34		I
07.32	2	5		242.54		II
04.69	—	5h		271.74		II
3,001.41	2	—		307.99		I
2,992.99	2	—		401.69		I
85.76	1	2		482.57		II
85.43	2	5		486.27		II
84.33	2	3		498.61		II
83.44	—	3		509.61		II
76.83	—	3		583.01		II
71.48	—	1		643.47		II
66.55	1	4		699.33		II
66.08	—	2		704.72		II
62.90	2	15		740.89		II
59.85	2	5		775.65		II
58.71	—	1		788.67		II
51.46	1	3		871.66		II
50.50	8	50	2 V E	882.68	(0.00) 1.04	II
48.82	—	1		901.99		II
43.56	1	6hl		33,952.57		II
39.61	—	3h		34,007.85		II
29.96	1	7		121.37		II
25.15	1	5h		176.31		II
23.90	2	20		190.92	(0.00) 1.23 (?)	II
13.60	—	2		311.78		II
10.05	—	1		353.63		II
09.65	2	—		358.36		I
05.53	1	4hl		407.07		II
2,904.62	1	—		417.85		I
2,899.80	1	4hl		475.06		II
97.76	2	5hl		499.32		II
93.06	10	60	4 V E	555.14	(0.00) 1.12	II
89.11	—	1		602.61		II
85.13	10	50	4 V E	650.35	(0.00) 1.04	II
83.35	—	1		671.74		II
80.65	8	40	3 V E	704.23	(0.00) 1.18	II
76.55	—	1		753.69		II
74.28	—	3		781.14		II
73.20	—	2		794.21		II
67.47	1	2h		863.74		II
2,862.96	5	15hl	1 V E	34,918.41		II

TABLE 1.—*Wave lengths and Zeeman effects in lanthanum spectra*—Continued

λ (air) I. A.	Intensity arc spark B. S.		Arc intensity and temperature class K and C	ν (vac.) cm⁻¹	Zeeman effect	Spectrum
2,862.37	3	6		34,925.85		II
59.76	3	5		34,957.73		II
55.90	10	50hl	1 V E	35,004.97	(0.00) 1.13	II
53.72	1	4h		031.71		II
49.51	1	2h		083.47		II
48.34	3	6		097.88		II
46.67	1	5		118.47		II
43.67	1	4		155.51		II
40.51	3	25hl		194.62		II
38.45	3	5l		220.16		II
32.53	1	5		293.77		II
25.82	—	1		377.57		II
25.51	1	5		381.45		II
21.03	1	5		437.64		II
19.73	1	2		453.97		II
18.40	1	3		470.70		II
17.46	2	—		482.54		I
15.36	3	6		509.00		II
13.72	2	5		529.70		II
13.06	1	3		538.16		II
09.35	1	2		584.96		II
08.39	40	150	10 V E	597.13	(0.00) 1.04	II
07.20	—	1		612.22		II
05.58	3	5		632.78		II
2,804.55	2	4		645.86		II
2,798.56	8	40hl	1 V E	722.15	(0.00) 1.08	II
96.40	3	5		749.75		II
94.03	4	1		780.07		I
91.51	8	25		812.37	(0.00) 1.04R	II
81.25	—	1		944.47		II
80.23	4	20		957.66		II
79.78	2	10		963.48		II
78.76	2	10		35,976.68		II
73.86	—	1		36,040.23		II
67.40	1	8		124.35		II
66.46	4	1		136.62		I
61.56	7	1		200.74		I
61.10	1	5		206.77		II
60.51	1	3		214.51		II
59.54	4	1		227.24		I
59.14	—	3		232.49		II
58.65	—	3		238.93		II
56.57	2	—		266.27		I
55.57	—	1		279.43		II
52.54	1	10		315.41		II
49.52	2	—		359.25		I
48.31	1	8		375.26		II
39.25	4	1		495.56		I
37.49	3	1		519.03		I
36.90	—	2		526.90		II
36.41	—	3		533.44		II
32.40	1	10		587.05		II
30.15	2	—		617.20		I
29.85	5	2		621.22		I
27.5	—	2		652.8		II
26.48	—	1		666.49		II
25.57	15	3		678.73		I
22.31	6	2		722.65		I
21.45	—	2		734.25		II
17.33	2	—		789.90		I
15.77	3	—		811.06		I
15.43	1	10hl		815.68		II
14.52	8	1		828.03		I
12.51	—	1		855.32		II
2,710.69	4	1		36,880.06		I

TABLE 1.—*Wave lengths and Zeeman effects in lanthanum spectra*—Continued

λ (air) I. A.	Intensity arc spark B. S.		Arc intensity and temperature class K and C	ν (vac.) cm⁻¹	Zeeman effect	Spectrum
2,709.92	—	3		36,890.54		II
07.07	3	1		929.38		I II
06.49	—	2		937.29		II
2,702.13	5	8		36,996.88		II
2,695.47	15	35		37,088.29		II
94.21	—	5		105.64		II
91.60	—	1		141.62		III
87.75	—	2		194.81		II
84.90	—	50hl		234.29		III
84.11	6	1		245.25		I
82.46	—	30hl		268.16		III
81.49	4	10		281.64		II
79.87	—	4h		304.17		II
77.77	2	—		333.43		I II
75.66	2	5		362.87		II
73.74	1	3		389.70		II
72.90	15	30		401.45		II
72.06	—	2		413.20		II
71.91	2	—		415.30		I
70.05	—	2		441.37		II
66.54	1	3		490.65		II
66.18	2	6		495.71		II
65.62	—	4		503.59		II
64.75	1	3		515.83		II
62.73	—	1h		544.29		II
61.66	1	3		559.38		II
61.36	1	4		563.61		II
53.48	2	—		675.16		I
51.60	—	300hl		701.87		III
49.61		1		730.18		II
47.36	1	4		762.25		II
47.13	2	—		765.53		I
44.70	—	1		800.23		II
42.27	—	1		834.96		II
40.15	—	1		865.37		II
39.00	1	5		881.87		II
36.66	—	1		915.48		II
31.94	2	8		983.47		II
31.52	1	4		37,969.54		II
20.01	2	7		38,156.42		II
17.29	—	1		196.07		II
16.32	2	7hl		210.23		II
13.09	1	4		257.46		II
10.34	50	150		297.76		II
04.18	—	1		388.35		II
02.87	—	1		407.66		II
01.79	1	5		423.60		II
00.86	—	2		437.34		II
2,600.33	1	4		445.18		II
2,596.32	1	3		504.55		II
96.06	6	20		508.11		II
92.87	—	3h		555.78		II
86.35	2	10		652.97		II
82.96	1	8		703.70		II
82.55	2	6		709.85		II
80.82	3	8hl		735.79		II
77.92	—	2		779.36		II
73.47	—	2h		846.42		II
66.09	3	10hl		38,968.13		II
61.84	2	20l		39,022.75		II
60.37	10	50		045.15		II
58.99	—	3		066.21		II
58.72	—	2		070.33		II
53.41	—	3h		151.57		II
2,552.60	—	7		39,164.00		II

Table 1.—*Wave lengths and Zeeman effects in lanthanum spectra*—Continued

λ (air) I. A.	Intensity arc spark B. S.		Arc intensity and temperature class K and C	ν (vac.) cm⁻¹	Zeeman effect	Spectrum
2,582.36	—	2		29,167.68		II
46.40	4	20hl		269.35		II
42.40	—	6		321.11		II
41.60	1	4		333.49		II
38.40	—	2		383.07		II
36.76	—	3		408.53		II
34.96	2	6		436.21		II
33.14	10	15		464.84		II
31.60	—	8		488.84		II
30.26	—	1		509.76		II
27.84	—	3		547.56		II
23.07	1	5hl		622.34		II
19.22	10	50		682.89		II
15.79	—	4		736.99		II
14.59	—	3		755.95		II
2,501.18	1	15hl		969.06		II
2,499.69	—	1		39,992.91		II
95.82	—	2		40,054.91		II
94.90	—	1		069.68		II
87.59	10	40		187.42		II
83.00	1	5hl		261.71		II
79.85	5?	10i		312.85		II
78.8	—	20hl		329.9		III
76.72	—	100hl		363.79		III
74.50	—:	3		400.00		II
72.44	1	10		433.65		II
71.90	15	20		442.49		II
71.06	—	5		456.23		II
70.55	—	3		464.58		II
68.11	—	1		504.58		II
56.15	—	2		668.69		II
55.36	—	10		706.28		II
54.30	—	1		732.48		II
52.73	1	8		756.55		II
51.59	—	2		777.50		II
45.56	—	10h		878.04		II
43.14	—	2h		918.53		II
42.80	—	3		924.22		II
39.08	—	2		986.63		II
38.42	—	10		40,997.82		II
38.02	2	20		41,004.45		II
37.14	1	10		019.26		II
36.42	1	15		031.37		II
31.40	—	6h		116.08		II
24.53	—	2h		232.58		II
21.61	—	5h		282.29		II
20.01	1	5hl		309.58		II
17.61	—	3h		350.59		II
12.08	—	2h		445.38		II
10.10	—	5hl		479.43		II
07.79	—	5hl		519.22		II
04.65	—	6		573.43		II
03.29	1	7		596.98		II
2,401.46	—	2h		628.06		II
2,399.64	3	20hl		660.22		II
96.70	—	3h		676.55		II
97.26	1	7hl		701.57		II
94.98	—	4		741.27		II
93.27	—	2h		771.10		II
89.84	—	3hl		831.04		II
88.96	—	2		846.45		II
86.26	—	2		893.80		II
84.28	—	3h		41,928.58		II
79.38	5s	200h		42,014.92		III
75.63	—	2h		081.24		II
2,370.47	—	2		42,172.84		II

TABLE 1.—*Wave lengths and Zeeman effects in lanthanum spectra*—Continued

λ (air) I. A.	Intensity arc spark B. S.		Arc intensity and temperature class K and C	ν (vac.) cm⁻¹	Zeeman effect	Spectrum
2,369.18	—	2		42,195.79		II
65.50	—	3		261.43		II
58.02	—	3h		395.48		II
56.10	—	1b		430.02		II
55.81	—	5h		435.24		II
53.40	—	2		478.70		II
53.03	—	1		485.43		II
51.93	—	1		505.24		II
49.86	—	2		560.79		II
41.83	—	4		688.18		II
28.75	4	20hl		42,928.29		II
22.78	—	3h		43,037.62		II
19.44	15	20		100.59		II
17.82	5	20hl		130.71		II
2,311.45	—	1		249.56		II
2,297.75	4e	200hl		507.40		III
93.47	—	2h		588.59		II
92.32	1	3		610.45		II
80.94	—	4h		828.01		II
76.06	—	1h		43,921.97		II
65.54	—	3		44,125.90		II
56.77	40	50		297.36		II
30.74	—	7		44,814.20		II
16.06	2e	100hl		45,110.63		III
07.08	—	1		294.56		II
2,202.76	—	1		383.38		II
2,195.91	—	4		524.94		II
90.67	—	1		633.82		II
87.87	30	40		45,692.22		II
63.66	5	20bl		46,203.12		II
61.36	—	4		252.58		II
2,142.81	2	20hl		46,642.94		II

TABLE 2.—*Band heads in the spectrum of lanthanum monoxide (La O)*

λ air I. A.	Intensity	ν vac cm⁻¹	System and ν', ν''
8,423.3	2	11,866.6	VII 14,14
8,994.5	3	11,114.9	VII 14,15
9,035.9	1	11,063.9	VII 15,16
9,073.4	1	11,018.2	VII 0, 2
9,111.5	2	10,972.1	VII 1, 3
9,150.1	2	925.9	VII 2, 4
9,188.8	4	879.8	VII 3, 5
9,228.15	4	833.4	VII 4, 6
9,267.8	4	787.1	VII 5, 7
9,307.8	4	740.7	VII 6, 8
9,348.2	3	694.3	VII 7, 9
9,388.8	3	648.1	VII 8,10
9,429.9	3	601.7	VII 9,11
9,471.4	3	555.2	VII 10,12
9,513.3	3	508.7	VII 11,13
9,555.5	3	462.3	VII 12,14
9,582.2	3	415.8	VII 13,15
9,641.1	3	369.4	VII 14,16
9,684.8	3	322.6	VII 15,17
9,729.1	3	10,275.6	VII 16,18

WASHINGTON, June 11, 1932.

RP469

NOTE ON THE FREEZING POINT OF "ISO-OCTANE" (2, 2, 4-TRIMETHYLPENTANE) [1]

By Johannes H. Bruun [2] and Mildred M. Hicks-Bruun

ABSTRACT

A high-grade sample of commercial iso-octane was purified by equilibrium melting. The freezing point of the pure hydrocarbon was found to be $-107.41°$ C. The purity of an iso-octane sample may be calculated from the equation: Mole per cent purity$=3.86_7 t_f + 514.8$ in which t_f is the initial freezing point of the sample in ° C.

"Iso-octane" (2, 2, 4-trimethylpentane) is used as the upper reference standard for antidetonation tests of motor fuels. As a

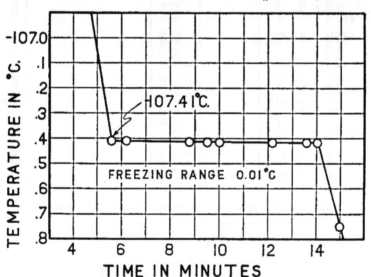

FIGURE 1.—*Time-temperature cooling curve of iso-octane*

criterion for the purity of commercial "iso-octane," it is desirable to have a reliable value for the freezing point of the pure hydrocarbon. The value ($-107.8°$ C.) reported in the literature [3] is apparently too low as many commercial samples were found to freeze at temperatures above this value.

[1] Financial assistance has been received from the research fund of the American Petroleum Institute. This work is part of Project No. 6, "The Separation, Identification, and Determination of the Constituents of Petroleum."
[2] Research Associate representing the American Petroleum Institute at the Bureau of Standards.
[3] G. M. Parks and H. M. Huffman, Ind. Eng. Chem., vol. 23, p. 1139, 1931.

A sample of high-grade commercial "iso-octane" was purified further by subjecting it to a number of fractionations by equilibrium melting in a centrifuge.[4] As a result of these fractionations four different fractions with freezing points ranging from $-107.7°$ to $-107.4°$ C. were obtained. The time-temperature cooling curve of the purest fraction of "iso-octane" was determined, and is shown in Figure 1.

From the value, $-107.41°$ C., found for the freezing point of pure iso-octane and from Parks and Huffman's value [5] (18.9 cal./g) for the heat of fusion, the purity of the iso-octane may be calculated from the laws of ideal solutions and is expressed by the equation: $P = 3.86_2 t_f + 514.8$ in which t_f is the initial freezing point of the sample in ° C., and P is the purity in mole per cent.

Temperatures were determined by means of a platinum-resistance thermometer calibrated at this bureau in accordance with the International Temperature Scale [6] as adopted in 1927.

WASHINGTON, May 4, 1932.

---

[4] For method see M. M. Hicks-Bruun and J. H. Bruun, B. S. Jour. Research, vol. 8, p. 527, 1932.
[5] See footnote 3, p. 269.
[6] B. S. Jour. Research, vol. 1, p. 635, 1928.

A TWIN-BOMB METHOD FOR THE ACCURATE DETERMINATION OF PRESSURE-VOLUME-TEMPERATURE DATA AND A SIMPLE METHOD FOR THE ACCURATE MEASUREMENT OF HIGH PRESSURES

By Edward W. Washburn

ABSTRACT

By filling one bomb, A, with the system under investigation and a twin bomb, B, with a reference substance and then adjusting the two pressures to exact equality (at a given temperature) with the aid of a pressure equalizer, the ratio of $\frac{PV_0}{T}$° per gram for the two systems can be accurately determined by weighing the two bombs. No pressure measurements are involved. If now the volume of bomb B is known, the value of $\frac{PV_0}{T}$ for the system under investigation can be computed to the accuracy with which $\frac{PV_0}{T}$ is known for the reference substance.

Furthermore, if a gas-filled bomb of volume V at a known temperature, T, is brought into pressure equilibrium with any system at the pressure P and the mass of the contained gas determined, the value of P can be computed with the accuracy to which $\frac{PV_0}{T}$ is known for the gas employed. In this way an ordinary balance and weights can be utilized as a laboratory tool for the accurate measurement of high pressures.

CONTENTS

I. INTRODUCTION

For practical purposes the task of accumulating accurate physical data concerning chemical substances and systems may be roughly divided into two categories. To the first category belong those pri-

mary measurements in which the measuring instruments employed are
more or less directly standardized in terms of the fundamental units
of science, for example in cgs (*e* or *m*) °K. units.[1]   Such measurements may be called primary or "absolute" measurements to distinguish them from the second category, which may be designated as
secondary or relative measurements.

This category comprises measurements in which the apparatus and
instruments employed are in part at least, standardized with the aid
of a substance or material for which accurate values are available
from primary measurements carried out as outlined above or which
is selected by convention as the reference substance.

Measurements in this class are fundamentally relative measurements, but they can be converted to absolute values through the
standardizing material employed, and if the accuracy of the relative
measurements is sufficient, the absolute values obtained in this way
should be as reliable as are the primary data for the standardizing
substance.   Examples of measurements belonging to the second
category are the determination of the viscosity of a liquid with a
viscosimeter standardized with water and the determination of the
heat of combustion of an organic substance with a bomb calorimetric
equipment standardized with benzoic acid.

In some cases the relative measurements are so much more precise
than any primary measurements that a conventional value is sometimes adopted for the standardizing substance.   This conventional
value may be purely arbitrary, as in the atomic weight table, or it
may be the best at-the-time available absolute value with zeros
assumed in all places following the last known figure, as in the measurement of current in so-called "international amperes" by means of a
silver coulometer.

The relative method frequently has the advantage of greater rapidity in the measurements combined with simplicity and relative in-
expensiveness in the equipment required, these advantages being
sometimes combined with a higher degree of accuracy than that obtainable in the absolute measurements.

So far as the writer is aware, no attempt has been made to determine directly and accurately the ratio of PV/T for two gases at high
pressures.[2]   The purpose of this paper is to outline a simple method
for doing this and to show that this method may be extended to the
determination of pressure-volume-temperature data for any system.

II. EQUIPMENT

The equipment to be employed consists of:
1. A pair of twin bombs as nearly identical as possible in all respects.
2. Inexpensive Bourdon gages.
3. A good thermometer.
4. A well-stirred constant-temperature bath, or baths, variable over
the temperature range desired.
5. A good balance and set of weights.
6. Means for obtaining the gases under the desired pressures.
No accurate pressure measuring equipment is required.

---

[1] Centimeter, gram, second (electrostatic or electromagnetic), °K.
[2] For very low pressures, below 1 atmosphere, an interesting and accurate relative method has been
described by Addingley and Whytlaw-Gray, Trans. Faraday Soc., vol. 24, p. 375, 1928.

III. EXPERIMENTAL PROCEDURE

The investigator first selects, on the basis of available data, a reference gas (or gases) for which satisfactory $P-V-T$ values are available over the pressure and temperature ranges within which he proposes to make his measurements. A typical experiment is carried out as follows: With the aid of an ordinary Bourdon gage one of the bombs is filled at about $t°$ C. with the reference gas, R, at a pressure slightly above that desired for the experiment. The second bomb is filled in the same way with the gas, E, under investigation. Both bombs are placed side by side in the constant-temperature bath regulated to the desired temperature, $t°$ C., and each one is then joined through capillary tubing (and a guard bomb if desired) to a pressure equalizer, for example, a large cylinder containing air (or other suitable fluid) at a pressure slightly below the pressure in the two bombs, and provided with a Bourdon gauge.

The valves of the two bombs are now opened simultaneously and when temperature equilibrium has been attained, the valves are closed, the capillary tubing is disconnected, and the bombs are removed from the bath, dried, and the difference in mass accurately determined. One of the bombs is then evacuated and the difference in mass again determined in the same way, or either of the two may be weighed against a closed dummy bomb of known mass. The ratio of $\frac{PV}{T}$ for unit masses of the two gases can now be calculated, as explained in section VII below.

IV. CONSTRUCTION OF THE TWIN BOMBS

The material used for constructing the bombs will be determined by the nature of the gas to be investigated and the pressure and temperature ranges to be covered.[3] The material selected and the wall thickness adopted should be such that the strain on the bomb will be well within the elastic limit. The thoroughly annealed bomb before calibration is first put through a number of cycles of compression and decompression (filling and emptying) with compressed gas, for the maximum pressure for which it is to be utilized at each temperature. It is now ready for calibration.

The calibration consists in determining the volume of one of the bombs and in determining, or reducing to a negligible amount, two small quantities both of which would be zero, if the bombs were identical twins. These two quantities are the difference in the masses and the difference in the volumes of the two bombs.

V. DETERMINATION AND EQUALIZATION OF THE VOLUMES OF THE TWIN BOMBS

With the aid of the constant-temperature bath and pressure equalizer described above, both bombs are filled with the same gas at the same moderate pressure (preferably a dense gas, such as CO_2,

---

[3] A discussion of materials suitable for containers for various conditions of temperature, pressure, and corrosive influences has been given by F. G. Keyes (Ind. Eng. Chem., vol. 23, p. 1378, 1931).

purity not essential) or with the same liquid, and the difference, $m_1 - m_2$, in the masses of the fluid, together with the mass, m_1, in one of the bombs is determined as in a regular experiment. The difference in the two volumes is given by the relation

$$\Delta V = V_1 - V_2 = \frac{(m_1 - m_2)V_1}{m_1} \tag{1}$$

To the larger of the two bombs there is now added, in the form of fine shot or wire for example, the right amount of a material, having negligible vapor pressure and negligible (or known) compressibility, to adjust the volumes to exact equality. There is some advantage in using for this purpose the material of the bomb itself.

After this adjustment has been made the now-much-smaller ΔV may be determined as a function of p and T over the proposed experimental range. If the construction and adjustment of the two bombs have been carefully carried out, this value of ΔV should be negligible.

The bombs having been adjusted to equality of volume, the volume of one of them is now determined as a function of pressure and temperature. For atmospheric pressure (zero pressure difference) this value will be known from the above determinations, if the density of the fluid used is known. The temperature coefficient can be calculated from the coefficient of thermal expansion of the bomb or the volume may be determined by calibration at different temperatures.

The small pressure coefficient can, for many purposes, be determined with the necessary accuracy by immersing all but the stem of the bomb in the liquid of a volumeter and reading the increases in volume of this liquid as successively increasing pressures are applied to the bomb. It may also be calculated from the elastic properties of the material composing the bomb. Formulas for this purpose have been given by F. G. Keyes.[4]

For the most accurate work, especially for extreme conditions of temperature and pressure, it may be necessary to check the volume calibration from time to time because of possible hysteresis effects.

VI. EQUALIZATION OF THE MASSES OF THE TWIN BOMBS

The evacuated bombs are now suspended from the arms of a sensitive balance and the masses are adjusted to equality. At the same time a closed dummy bomb can also be provided, if desired, and similarly adjusted to the same mass. This dummy may be employed as a counterpoise for determining the total mass of the gas in one of the bombs, as indicated in the procedure outlined in Section III above.

VII. COMPUTATION OF THE $\frac{PV_0}{T}$ RATIO

From the data obtained by the procedure described in Section III the value of $\pi_s = \frac{PV_0}{T}$ for unit mass of any gas at the pressure and temperature of the experiment is obtained from the relation

$$\frac{\pi_s}{\pi_R} = (1 - \Delta V/V_R)\ (1 + \Delta m/m_E) \tag{2}$$

---

[4] See footnote 3, p. 273.

in which τ_R is the value of $\frac{PV_0}{T}$ for the reference gas at the pressure and temperature of the experiment; V_R is the volume of the reference gas; $\Delta V, = V_R - V_E$, is the difference in the volumes of the two gases; m_E is the mass of the gas under investigation; and $\Delta m, = m_R - m_E$, is the difference in the masses of the two gases. After completing the experiment as described above, it may be repeated by interchanging the gases in the two bombs and ΔV may be eliminated from the two equations thus obtained.

Judging from the precision which should apparently be attainable in the measured quantities, the relative values obtained in this way should be more accurate than many of the absolute values at present available.

VIII. COMPUTATION OF $\frac{PV_0}{T}$ FOR THE REFERENCE GAS

In order to convert into absolute values the relative values obtained by the above procedure, it is necessary to compute the value of τ for the reference gas. If the data for this gas have been put into mathematical form by evaluating the parameters of an equation of state in which τ is given as a function of V and T, then the value of τ is readily calculable since both V and T are known with the necessary accuracy. If only tabulated values of $\frac{PV_0}{T}$ are available for various temperatures and for a series of pressures, then the experimental temperatures employed should include those available for the reference gas and the value of τ_R can be obtained by graphing values of $\frac{PV_0}{T}$ against $\frac{T}{V_0}$ in the experimental region and interpolating τ_R for the known value T/V_0. At the same time the value of P is obtained to the degree of accuracy corresponding to that of the $\frac{PV_0}{T}$ data of the reference gas.

IX. ACCURACY REQUIRED IN THE MASS DETERMINATIONS

For any specific case the accuracy required in the weighings for any given desired accuracy in τ_E/τ_R can be judged by inspection of equation (2).

In general, it may be said that an accuracy of 0.01 per cent in τ_E/τ_R should be practically always attainable, or stated in another way, the masses of the bombs will never need to be so great as to render difficult the attainment of the required accuracy in the weighing operations, and this accuracy will be obtainable with a comparatively inexpensive balance, except possibly in the cases of hydrogen and helium where a balance of high sensitivity might be needed. (See further, Sec. XIV below.)

X. ACCURACY REQUIRED IN THE VOLUME DETERMINATIONS

If the volumes of the two bombs are adjusted to substantial equality, a large percentage error is obviously allowable in both ΔV and V in determining the value of τ_B/τ_R. For interpolating the absolute value of τ_R, V_R must be known with an accuracy which varies with the slope of the $\pi - \dfrac{T}{V_0}$ curve for the reference gas in the experimental region. Thus, for a considerable region on both sides of the Boyle pressure,[5] large errors in V_R will have but little effect upon the result. In any case V_R can be determined with the required accuracy.

XI. SELECTION OF THE REFERENCE GAS

Since relative values of $\dfrac{PV_0}{T}$ can be determined with a high degree of accuracy and with comparatively simple equipment, absolute values should be available for one or for a few gases selected on the basis of their advantages as reference gases and these values should be known as accurately as possible and should cover a wide range of temperatures and pressures.

Among the gases which might be selected as suitable reference gases, air has the disadvantages of its oxidizing action at high temperatures and its lack of constancy in composition.[6] In spite of these disadvantages, however, dry, CO_2-free air is likely to be the favored reference gas for measurements of moderate accuracy, say ±0.1 per cent, because of its ready accessibility. For this reason reliable P-V-T data should be available for air (of known normal density) over wide ranges of temperature and pressure. Furthermore, if the investigator who proposes to use air as a reference gas determines the normal density of the sample of air employed, air would probably be suitable as a reference gas even for measurements of the highest accuracy.

A review of the various other possibilities leads to the conclusion that "atmospheric nitrogen",[6] methane, and carbon dioxide would be suitable as additional reference gases. These gases can be prepared in a high state of purity at reasonable cost. Carbon dioxide could be used for investigations confined to temperatures above, say, 40° C., while methane could be used for temperatures between, say, 200° and −80° C. "Atmospheric nitrogen" could be used at any temperature above, say, −140° C.

---

[5] That pressure at which for a given temperature the PV product is a minimum.
[6] See the extensive data on this question recently obtained by Moles (Gazz. Chim. Ital., vol. 56, p. 915, 1926). This investigator also found that after chemical removal of the oxygen from air, the residue, "atmospheric nitrogen," showed a much more nearly constant density, the maximum variation found being only 1 in 10,000. Apparently, therefore, "atmospheric nitrogen" would be an excellent reference gas for wide ranges of temperature and pressure.

XII. ADAPTATION OF THE METHOD TO LIQUIDS AND SOLIDS AND TO POLYPHASE SYSTEMS

It is obvious from the foregoing discussion that the method described in this paper might also be applied to the determination of P–V–T relations in many systems composed of or containing one or more solid or liquid phases. For this purpose a retaining or immersing liquid may be needed or preferred in the bomb containing the system under investigation and in that case the gas-filled bomb functions purely as a pressure gage.

XIII. THE USE OF A GAS-FILLED BOMB AS A PRESSURE GAGE

If a gas-filled bomb of known volume and temperature is brought into pressure equilibrium with a system at the pressure P, with the aid either of a pressure equalizer such as that described above or with the aid of a nul-point differential gage,[7] and the mass of gas in the bomb determined, then its pressure can be calculated as indicated above and the value thus obtained will be as accurate as are the available PV_0/T data for the gas employed. In this way a balance and weights can be utilized as a laboratory tool for the accurate measurement of high pressures.

The accuracy required in V and T will be determined by the accuracy wanted in P or by the accuracy possessed by $\dfrac{PV_0}{T}$, whichever happens to be the determining quantity.

The method might also find some application for calibrating an electrical or mechanical pressure gage, in case a dead-weight pressure gage is not available.

XIV. DETERMINATION OF THE MASS OF THE GAS AFTER REMOVAL FROM THE BOMB

The range of pressure over which a given pair of twin bombs can be employed is limited on the high pressure side by the elastic limit of the material of which the bomb is composed and on the low-pressure side by the mass of contained gas which can be determined with the required accuracy by weighing the bomb. If, therefore, a given gas is to be investigated over a very wide range of pressures, a set of bombs would be required. For many gases this necessity can be avoided and a single pair of heavy-walled bombs can be employed in all parts of the pressure range by arranging to remove the gas from the bomb for the purpose of weighing it. Thus CO_2 could be weighed after condensation or after absorption in ascarite (NaOH-asbestos mixture), H_2O and certain alcohols after absorption in Dehydrite $(Mg(ClO_4)_2 . 3H_2O)$, a combustible gas after passage through a combustion furnace followed by absorption of CO_2 and/or H_2O as above, etc. These methods are somewhat more time-consuming than direct weighing, but a high degree of accuracy is obtainable and certain other obvious advantages are secured.

[7] For example, the sensitive differential gage described by Osborn, Stimson, and Flock, B. S. Jour. Research, vol. 5, p. 430, 1930. The differential gage is required whenever direct contact between the system and the fluid of a pressure equalizer is undesirable.

XV. CONCLUSION

The method outlined above will probably find its chief application in chemical laboratories and industrial laboratories which do not have available accurate dead-weight pressure gages, but in which the need occasionally arises of obtaining P-V-T- data for gases and vapors and their mixtures. The recent publication of several papers giving rather rough data of this character obtained with Bourdon gages illustrates a growing need for reliable P-V-T data for a considerable number of gases and gas mixtures for which no information is at present available. It is hoped that the method described in this paper may be found useful to the occasional investigator who finds it necessary to determine such data for himself.

WASHINGTON, December 28, 1931.

O

U. S. DEPARTMENT OF COMMERCE
ROY D. CHAPIN, Secretary

BUREAU OF STANDARDS
LYMAN J. BRIGGS, Acting Director

BUREAU OF STANDARDS
JOURNAL OF RESEARCH

September, 1932

Vol. 9, No. 3

UNITED STATES
GOVERNMENT PRINTING OFFICE
WASHINGTON : 1932

For sale by the Superintendent of Documents, Washington, D. C.   -   -   Price 25 cents; $2.50 per year on subscription

A NEW DETERMINATION OF THE ATOMIC WEIGHT OF OSMIUM [1]

By Raleigh Gilchrist

ABSTRACT

A method for the preparation of pure osmium is described. The value of the atomic weight of osmium is calculated from the average percentage of osmium found in carefully prepared samples of ammonium chloroosmate and of ammonium bromoosmate. The value obtained from the ratio $(NH_4)_2OsCl_6$: Os is 191.53, and from the ratio $(NH_4)_2OsBr_6$: Os is 191.57, based upon the values for nitrogen, hydrogen, chlorine, and bromine given in the International Table of Atomic Weights for 1932. The weighted average value is 191.55.

The densities of ammonium chloroosmate and of ammonium bromoosmate at 25° C. were found to be 2.93 g/cm³ and 4.09 g/cm³, respectively.

CONTENTS

I. INTRODUCTION

Osmium was first identified as a new element by Tennant,[2] who discovered it in the portion of crude platinum which is not attacked by aqua regia. Osmium occurs as a constituent of a natural alloy with iridium known as osmiridium or iridosmine. The name osmium, which was chosen for the element, was taken from the Greek, οσμή, a smell, suggestive of the pronounced odor of the volatile tetroxide, the formation of which is the most outstanding property of the metal.

The earliest attempt to determine the atomic weight of osmium was made by Berzelius,[3] who analyzed potassium chloroosmate, K_2OsCl_6, and obtained the value 198.94 from the ratio Os:2 KCl.

Seubert,[4] 60 years later, undertook to make careful analyses of potassium chloroosmate and of ammonium chloroosmate, $(NH_4)_2 OsCl_6$,

[1] Submitted as partial fulfillment of the requirements for the degree of doctor of philosophy, The Johns Hopkins University, June 1922.
[2] Smithson Tennant, Phil. Trans., vol. 94, p. 411, 1804.
[3] J. J. Berzelius, Ann. Physik (Pogg.), vol. 13, p. 530, 1828.
[4] K. Seubert, Ber., vol. 21, p. 1839, 1888; Ann., vol. 261, p. 258, 1891.

279

and it is upon his experiments that the value of the atomic weight was based which has appeared in the International Table of Atomic Weights for many years. Seubert used the ratios $(NH_4)_2OsCl_6:Os$; $6AgCl:(NH_4)_2OsCl_6$; $K_2OsCl_6:Os$; $K_2OsCl_6:2$ KCl; and 4 $AgCl:K_2OsCl_6$. The values obtained for the atomic weight range from 190.27 to 192.22 when recalculated on the basis of the International Table of Atomic Weights for 1932. The value calculated from Seubert's experiments on the osmium content of ammonium chloroosmate is 191.24.

In the last 41 years only one published account of new work on the atoic mweight of osmium has appeared, that of Seybold[4], describing the determination of the osmium content of ammonium chloroosmate. The value calculated from a set of three experiments is 191.09, while that from five other experiments is 189.33.

II. PREPARATION OF PURE OSMIUM

1. EXTRACTION OF OSMIUM FROM CRUDE MATERIAL

The osmium used in the experiments reported in this paper was obtained from two sources. That used in ammonium chloroosmate, Series VI, was obtained from one of the platinum companies, while that used in the other experiments was recovered from the osmiridium residues from grain platinum.

The crude material was fused with sodium hydroxide and sodium nitrate in a gold dish. The aqueous extract of this melt was acidified with nitric acid and distilled. The osmium tetroxide thus obtained was absorbed in a 10 per cent solution of sodium hydroxide and again distilled as just described. The remaining platinum metals,[5] under these conditions, do not appear in the distillate. The osmium was recovered by electrolysis from the alkaline solution, into which the tetroxide had been distilled the second time, using platinum electrodes. The osmium separated at the cathode as a black, loosely adherent deposit which, according to Moraht and Wischin,[6] is the dioxide. The precipitated material was washed as free from alkali as possible, dried, and ignited to metal in hydrogen. Spectrographic examination of the metal thus obtained showed that it was free from the other platinum metals, but that it did contain traces of sodium and of iron.

2. PURIFICATION OF THE OSMIUM

In considering possible reactions feasible for the preparation of pure osmium, it appeared highly desirable to employ volatile reagents. The reaction between osmium tetroxide and hydrochloric acid had been studied by Milbauer,[7] who observed that osmium tetroxide was decomposed by concentrated hydrochloric acid at room temperature with the evolution of chlorine. Ruff and Mugdan,[8] however, using acid of specific gravity 1.124, stated that the reaction reported by Milbauer did not take place. A few years later, Remy[9] made a careful study of the reaction at room temperature and concluded that the reaction depends considerably upon the concentration of the

[4] F. Seybold, Inaugural-Dissertation, 1912, Friederich-Alexanders Universität, Erlangen.
[5] R. Gilchrist, B. S. Jour. Research, vol. 6, p. 421, 1931.
[6] H. Moraht and C. Wischin, Z. anorg. Chem., vol. 3, p. 153, 1893.
[7] J. Milbauer, J. prakt. Chem. (2), vol. 96, p. 187, 1917.
[8] O. Ruff and S. Mugdan, J. prakt. Chem. (2), vol. 98, p. 143, 1918.
[9] H. Remy, J. prakt. Chem. (2), vol. 101, p. 341, 1921.

hydrochloric acid. He found that with hydrochloric acid of specific gravity greater than 1.160 the osmium tetroxide was decomposed with appreciable velocity, with the evolution of chlorine. He further found that the osmium was converted to the quadrivalent state and not to the bivalent, as asserted by Milbauer. Krauss and Wilken[11] later substantiated Remy's conclusion regarding the quadrivalent state.

In a recent study by Crowell[12] of the reaction between octavalent osmium and hydrobromic acid at 100° C., it was found that the products formed are quadrivalent osmium and bromine and that an equilibrium is reached. It was also found that in concentrated hydrobromic acid (about 8 normal) the reaction goes to practical completion in the direction of the formation of quadrivalent osmium and bromine, while in dilute acid solutions (about 0.1 normal) the reaction goes to completion in the reverse direction.

In 1920, previous to the appearance of Remy's paper, the author observed that reaction occurred when a mixture of osmium tetroxide and 20 per cent[13] hydrochloric acid (constant-boiling acid) was heated to incipient boiling and that the presence of a small quantity of ethyl alcohol assisted in the decomposition of the tetroxide. After about three hours the odor of osmium tetroxide disappeared and the solution, which had gradually turned deep brown in color, became reddish yellow and transparent. This solution, when evaporated to sirupy consistency, appeared to suffer no decomposition. Ammonium chloride, added to an acid solution of the sirupy residue, precipitated the brick red ammonium chloroosmate, $(NH_4)_2OsCl_6$. Ignition of the precipitate in hydrogen produced metallic osmium in the spongy form. It was further observed that the reaction between osmium tetroxide and 20 per cent hydrobromic acid proceeded with greater ease and rapidity than that between osmium tetroxide and hydrochloric acid. Considerable bromine was produced. Enough alcohol was added just to destroy the bromine vapor in the refluxing flask. The resulting solution was deep brown in color and likewise appeared to suffer no decomposition on evaporation to a sirup. Ammonium bromide precipitated the deep brownish black ammonium bromo-osmate, $(NH_4)_2OsBr_6$, from an acid solution of the sirupy residue.

The osmium metal obtained by reduction of the product of electrolysis previously described was converted in turn into osmium tetroxide, chloroosmic acid, ammonium chloroosmate, and osmium sponge as will be described in detail in connection with the preparation of the compounds for analysis. This cycle of operations was repeated twice, using purified reagents. The resulting metal was found on spectrographic examination to be free from impurities and was used in the final preparation of the compound ammonium chloroosmate.

III. PREPARATION OF REAGENTS

Hydrochloric acid.—Chemically pure hydrochloric acid of commerce (specific gravity 1.18) was diluted with an equal volume of distilled water and distilled three times from a ground-glass stoppered Pyrex flask whose side arm extended a considerable distance into the tube of a condenser. In each case the first and last fractions were discarded.

---

11 F. Krauss and D. Wilken, Z. anorg. allgem. Chem., vol. 137, p. 349, 1924.
12 W. R. Crowell, J. Am. Chem. Soc., vol. 54, p. 1324, 1932.
13 Here and elsewhere in this paper the percentage of acid is percentage by weight.

Bromine and hydrobromic acid.—Chemically pure bromine of commerce was purified by the procedure described by Baxter and Grover.[14] Hydrobromic acid was prepared by reducing bromine with hydrogen in the presence of a catalyst consisting of pumice impregnated with platinum. The hydrogen was produced by electrolysis in a cell described later. The first portion of the hydrobromic acid was discarded. The remainder was dissolved in water and distilled once.

Ammonium chloride.—Ammonium hydroxide, which had been freshly prepared by saturating distilled water with ammonia from a cylinder of liquid ammonia, was gently heated and the ammonia conducted into the purified hydrochloric acid until the acid was neutralized. A slight excess of acid was then added to the solution of ammonium chloride.

Ammonium bromide.—Ammonium bromide was prepared in the same manner as the ammonium chloride.

Alcohol.—Ethyl alcohol of commerce (95 per cent) was allowed to remain in contact with lime for several days, after which it was distilled three times. The first and last portions from each distillation were discarded.

Water.—In the preliminary work, the water used was purified by distilling from an alkaline permanganate solution and then from a

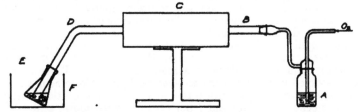

FIGURE 1.—*Apparatus for the preparation of osmium tetroxide*

dilute sulphuric acid solution, using a Pyrex distilling flask and a block-tin condenser. The preparation reported as ammonium chloroosmate, Series VI, was likewise made with water thus prepared. In the other three preparations reported, the water used was that which was regularly supplied to the chemical laboratories at the Bureau of Standards from a still by a pipe of block tin. This water gave no test for halides and had a specific conductance of about 1×10^{-6} reciprocal ohm at room temperature.

IV. PREPARATION OF AMMONIUM CHLOROOSMATE

Preparation VI.—Osmium, which had been prepared from the material obtained from one of the platinum companies as described in Section II, was converted to osmium tetroxide by heating it in a current of oxygen. Figure 1 shows the apparatus used. A bottle, *A*, containing water, served to indicate the flow of oxygen. A tube, *B*, of combustion glass, having a diameter of about 2.5 cm, was bent at an angle and the end inserted into a Pyrex flask which contained purified constant-boiling hydrochloric acid. The flask and contents were cooled by immersion in crushed ice in order to prevent too great a loss of osmium tetroxide. The combustion tube was heated by a

14 G. P. Baxter and F. L. Grover, J. Am. Chem. Soc., vol. 37, p. 1029, 1915.

12-inch Hoskins electric furnace of the split type. When heated to a temperature of from 220° to 230° C. the osmium, contained in porcelain boats, absorbed oxygen rapidly and reached a glowing temperature. This rapid absorption of oxygen was repeatedly observed and probably indicated the formation of osmium dioxide. As the temperature was raised the oxidized osmium was converted to tetroxide which condensed to a pale yellowish mass, as well as to white crystals, in the cooler portion of the combustion tube. Gentle warming of the tube with a flame loosened the solidified tetroxide, while a jet of hot hydrochloric acid solution from a wash bottle completely removed all tetroxide from the tube as it was lifted from the flask.

The contents of the flask, consisting of osmium tetroxide and hydrochloric acid, were quickly transferred [15] to a refluxing flask to which a water-cooled condenser was attached by a ground joint. A U tube containing a solution of sodium hydroxide was attached to the top of the condenser to prevent the escape of any tetroxide vapor. A quantity of purified constant-boiling hydrochloric acid, four times the amount necessary to form chloroosmic acid, H_2OsCl_6, was added to the refluxing flask. From 3 to 5 ml of purified alcohol was added, and the mixture gently heated for one hour. During this time the color changed from faint yellow to deep brown. As heating was continued, droplets of osmium tetroxide gradually ceased to appear on the walls of the flask and the solution finally became reddish yellow in color and transparent. The total time of heating was about three hours, during the last hour of which the solution was kept at incipient boiling. The solution of chloroosmic acid which had been formed was evaporated to a sirup on the steam bath. On cooling, the sirupy solution solidified to a mass of crystals. These crystals were dissolved in purified constant-boiling hydrochloric acid and the resulting solution diluted with water so as to form a solution containing 4 per cent of osmium and 7 per cent of hydrochloric acid. This solution was filtered first through hardened paper, in a Gooch crucible, and then through blue ribbon paper (S. and S. No. 589).

A 15 per cent solution of purified ammonium chloride, in slight excess, was added slowly to the constantly stirred solution of chloroosmic acid at room temperature. The precipitate of ammonium chloroösmate, brick red in color, was caught on a hardened filter in a Gooch crucible and washed with a 15 per cent solution of purified ammonium chloride until the wash waters were colorless. The ammonium chloroosmate thus prepared was recrystallized by making a saturated solution of the compound in a solution containing 7 per cent of hydrochloric acid at about 90° C. The hot saturated solution was quickly filtered through a hardened filter and cooled to about 5° C. while being stirred. The supernatant liquid was decanted and the crystals of ammonium chloroosmate stirred with several portions of 7 per cent hydrochloric acid. The crystals were then caught on a hardened filter, in a Gooch crucible, drained by suction, washed with a dilute solution of hydrochloric acid, and finally washed with purified alcohol. The crystals were drained free from alcohol, spread on a watch glass and dried in a partially evacuated desiccator containing phosphorus pentoxide. The dried salt was thoroughly mixed by grinding in an agate mortar and kept in a desiccator until used.

---

[15] It is, perhaps, needless to mention that an operation of this kind should be conducted with care and that it is preferable to wear a mask to prevent the vapor of the tetroxide from attacking the eyes, nose, and throat of the operator.

Preparation VII.—Osmium, which had been pre red from osmiridium residues from grain platinum as described irpSection II, was converted into chloroosmic acid. A solution of the chloroosmic acid was made which contained 1.5 per cent of osmium and 1.7 per cent of hydrochloric acid. This solution was slowly added, at room temperature, to a dilute solution (approximately 2 per cent) of purified ammonium chloride which was constantly stirred, according to one of the procedures used by Archibald [16] in the preparation of ammonium chloroplatinate. The precipitate of ammonium chloroosmate was not recrystallized, but was washed, dried, and mixed as described under Preparation VI.

V. PREPARATION OF AMMONIUM BROMOOSMATE

Preparation VI.—The ammonium bromoosmate was made from the osmium sponge [17] resulting from the analyses of ammonium chloroosmate. The source of the osmium was the osmiridium residues from grain platinum. This osmium was first converted into tetroxide as previously described, then into bromoosmic acid, ammonium bromoosmate, and again into metal using purified reagents. The metal prepared by this cycle of reactions was again converted into bromoosmic acid. The resulting solution of bromoosmic acid was evaporated to a sirup on the steam bath, then evaporated twice more with 20 per cent hydrobromic acid. The crystalline mass was dissolved in 20 per cent hydrobromic acid and diluted with water to form a solution containing 1.5 per cent of osmium in 7 per cent of hydrobromic acid. This solution was filtered twice through filter paper (S. and S. No. 589, blue ribbon). A 2 per cent solution of purified ammonium bromide, in slight excess, was added slowly to the constantly stirred solution of bromoosmic acid at room temperature. The precipitate of ammonium bromoosmate, deep brownish black in color, was caught on a hardened filter and washed with a 15 per cent solution of ammonium bromide until the wash waters were colorless. The compound thus prepared was recrystallized by making a saturated solution of it in a solution containing 15 per cent of hydrobromic acid at about 90° C. The hot saturated solution was quickly filtered through a hardened filter and cooled to about 10° C. while being stirred. The supernatant liquid was decanted and the crystals washed with 7 per cent hydrobromic acid. The crystals were then caught on a hardened filter, in a Gooch crucible, drained by suction, washed with a dilute solution of hydrobromic acid and finally washed with purified alcohol. The crystals were drained free from alcohol and dried over hosphorus pentoxide. The dried salt was then thoroughly mixed by grinding it in an agate mortar and kept in a desiccator over phosphorus pentoxide until used.

Preparation VII.—This preparation was made from another portion of the same sponge as Preparation VI and in the same manner except that the compound was recrystallized from 10 per cent hydrobromic acid.

---

[16] E. H. Archibald, Proc. Roy. Soc., Edinburgh, vol. 29, p. 721, 1908-9.
[17] A portion of this sponge was used by Meggers to measure the spectral lines of osmium. W. F. Meggers, Arc Spectra of the Platinum Metals, B. S. Sci. Papers No. 499, Jan. 23, 1925.

VI. DETERMINATION OF THE OSMIUM CONTENT OF AMMONIUM CHLOROOSMATE AND OF AMMONIUM BROMOOSMATE

1. APPARATUS

The apparatus used in determining the osmium content of the preparations just described is shown in Figure 2. It consisted of an electrolytic cell for the generation of hydrogen; two series of scrubbing and drying towers; and a reaction tube, electrically heated.

The electrolytic cell was equipped with platinum electrodes and contained a dilute solution of potassium hydroxide. An electrically heated porcelain combustion tube, containing asbestos and quartz both impregnated with palladium, was inserted in the line to convert to water any oxygen which might have diffused into the hydrogen chamber of the cell. The series of five towers for the purification of the hydrogen contained silver sulphate, sulphuric acid which had previously been heated to heavy fumes, soda-lime, sulphuric acid, and phosphorus pentoxide which had been sublimed into the tower. Glass beads were used in all liquid towers, and several layers of beads were used in the soda-lime tower.

The nitrogen was the commercial gas produced by the fractionation of liquid air. The method used to remove the small amount of oxygen present was based upon the experiments of Badger, [18] who found that a solution of ammonium hydroxide (1 volume of ammonium hydroxide, specific gravity 0.90, diluted with 1 volume of water), when saturated with ammonium chloride and kept in contact with metallic copper, absorbed oxygen quantitatively. In order to insure the complete removal of oxygen a tube 6 feet in length, containing copper and the ammoniacal solution, was placed between the wash bottle and the first tower, which also contained the same solution. The second tower in this series contained 50 per cent sulphuric acid to absorb ammonia. The last five towers were similar to the set of towers described for the purification of hydrogen.

The reaction tube of combustion glass was fitted to the purification train by a carefully ground joint. Two U tubes, containing glass beads and sulphuric acid, sealed the exit of the reaction tube. The Hoskins electric furnace which was previously mentioned was used to heat the reaction tube. At a few places in the apparatus where it was not possible to make sealed-glass connections, the ends of the rigid tubes were held flush by rubber tubing. A heavy coating of De Khotinsky cement was then applied so that the rubber tubing was completely covered. In order to make certain that no oxygen reached the reaction chamber, either from the hydrogen or from the nitrogen, these gases were alternately passed through the heated tube in the presence of spongy metallic osmium. A tube containing phosphorous pentoxide, attached directly to the reaction tube, indicated no formation of water.

2. BALANCES, WEIGHTS, AND WEIGHING

The experiments on ammonium chloroosmate, Series VI, were made at The Johns Hopkins University. The balance used in these experiments was a new analytical balance known as Ainsworth QA.

The experiments on ammonium chloroosmate, Series VII, and on ammonium bromoosmate, Series VI and VII, were made at the

---

18 W. L. Badger, J. Ind. Eng. Chem., vol. 12, p. 161, 1920.

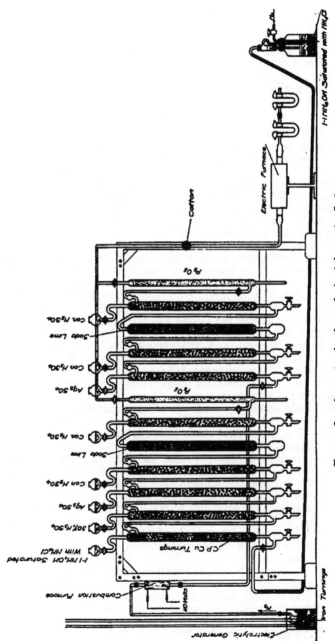

Figure 2.—*Apparatus for the analysis of the osmium salts*

Bureau of Standards. The balance used in these experiments was made by Ruprecht,[19] of Vienna, and was designed to carry a maximum load of 1 kg on each pan. This balance was inclosed in a case of blackened copper which served as a radiation screen. The weights used were carefully calibrated by the Bureau of Standards just previous to use.

Two glass weighing tubes of nearly the same size and weight, each fitted with a ground-glass stopper, were used. The boats were of porcelain. Weighing was done by the method of substitution, the vessels being allowed to remain in the balance case for three hours previous to weighing.

3. ANALYSIS

One is led to infer that Seubert[20] dried his compounds only over phosphorous pentoxide in a desiccator. Seybold,[21] cognizant of Seubert's procedure, in addition heated his preparations at 105° C., most likely in air so far as one is able to tell from his description.

Archibald,[22] in the analysis of ammonium chloroplatinate and of ammonium bromoplatinate, found that the purest samples which he was able to pre re began to decompose slightly at a temperature above 185° C. and that all hydrochloric acid and hydrobromic acid seemed to be driven off below 150° C.

With ammonium chloroosmate and ammonium bromoosmate it was found that slight decomposition occurred, in an atmosphere of dry purified nitrogen, if the temperature was above 170° C. At temperatures up to 150° C. no evidence of decomposition was observed. In experiment No. 1, ammonium chloroosmate, Series VII, the sample was heated for three 2-hour periods at 156° to 160° C. There was no change in weight during the second and third heating periods. When the temperature was raised to 166° to 170°, however, a faint white sublimate was formed and the sample continually decreased in weight. A solution of the sublimate in water gave a precipitate with silver nitrate. The weight recorded for the sample was that obtained at 156° to 160° C. These same temperature limits were also observed with ammonium bromoosmate.

Experiments previously made on other preparations, in which the drying was done at about 175° C., gave values for the atomic weight about 0.5 of a unit higher than those reported here. These values must be in error for the reason stated above and were rejected.

In the experiments reported here, the samples taken for analysis were dried to constant weight in a current of dry purified nitrogen. The samples in Series VI, ammonium chloroosmate, were dried from 3 to 5 hours at 140° to 145° C. Those in Series VII, ammonium chloroosmate, with the exception of No. 1 discussed above, were dried from 4 to 7 hours at 145° to 150° C. The samples in Series VI, ammonium bromoosmate, were dried from 10 to 14 hours at 150°, while in Series VII the first sample was dried for 14 hours at 145° and the second for 12 hours at 150° C. During the heating of the sample in experiment No. 2, ammonium chloroosmate, Series VII, a tube containing phosphorous pentoxide showed no increase in weight after the first heating period of 3 hours, although the sample was heated during two additional periods of 2 hours each.

[19] W. A. Noyes, Bulletin of the Bureau of Standards, vol. 4, p. 179, 1907.
[20] See footnote 4, p. 279.
[21] See footnote 5, p. 280.
[22] See footnote 16, p. 284.

The decomposition of the compounds was accomplished as follows: The porcelain boat with its contents was placed in the reaction tube. The tube was swept with dry purified nitrogen for one hour. The nitrogen was displaced by pure dry hydrogen for one and one-half hours, at the end of which time the temperature of the tube was gradually raised. When the decomposition appeared to be complete, the temperature was increased to 700° to 725° C. and maintained at this point for two hours. While at this temperature the hydrogen was displaced by dry purified nitrogen for one hour. The boat, now containing metallic osmium, was allowed to cool to room temperature in nitrogen, the furnace having been removed. The boat was placed in the glass-stoppered weighing tube and allowed to remain in the balance case for three hours before being weighed.

This procedure for the very gradual decomposition of the osmium compounds had been found, in previous experiments, to give a metal sponge which remained constant in weight upon reignition in hydrogen. No mechanical loss of material could be detected during the decomposition. With the exception of four instances where the metal sponges were returned to the cleaned reaction tube for an additional heating period of three hours in hydrogen at 700° to 725° C., the weights recorded in this paper were those obtained after the above treatment. The metal sponges in experiments Nos. 1 and 2, ammonium chloroosmate, Series VII, lost 0.00091 and 0.00051 g, respectively, on reheating in hydrogen. In these two cases the time of the first reduction had been cut somewhat short owing to the late hour of the night, so that all traces of volatile matter had apparently not been completely eliminated. The weight of the metal sponge in experiment No. 2, ammonium bromoosmate, Series VI, remained unchanged on reheating, while that in experiment No. 2, ammonium bromoosmate, Series VII, lost 0.00006 g. The weights recorded for these four sponges were those after the second heating period.

It had been repeatedly observed that metallic osmium, which had been ignited and cooled in hydrogen, was rapidly attacked by the air with the formation of osmium tetroxide. When the reaction tube was opened, the surface of the metal sponge usually glowed and a distinct odor of the tetroxide was noticeable. When the atmosphere of hydrogen was displaced as described in the experiments reported here, no odor of osmium tetroxide was detected either on opening the reaction tube or several days later when the boat was finally removed from the weighing tube. The limiting quantity of osmium tetroxide which von Wartenberg[23] could detect by the sense of smell was 2×10^{-5} mg/cm.

4. DENSITIES OF THE SUBSTANCES WEIGHED

In order to reduce the weights obtained to vacuum, the densities of the two osmium compounds and of the resulting osmium sponge were determined at 25° C. by displacement of toluene which had been dried and redistilled.

Two determinations of the density of ammonium chloroosmate were made, using 6.2303 and 8.7552 g of the compound. The values found were 2.941 and 2.908 g/cm³. The average value was 2.93 g/cm³.

Two determinations of the density of ammonium bromoosmate were made, using 10.2161 and 10.5552 g of the compound. The

---

23 H. von Wartenberg, Ann., vol. 440, p. 97, 1924.

values found were 4.102 and 4.084 g/cm³. The average value was 4.09 g/cm³.

One determination of the density of the osmium sponge was made, using 6.7478 g. The value found was 19.13 g/cm³.

A correction of +0.00026, +0.00014, and −0.00009 g was applied to each gram of ammonium chloroosmate, ammonium bromoosmate, and osmium sponge, respectively.

5. RESULTS

The results of the determinations are given in Tables 1 and 2.

TABLE 1.—*Results of the analysis of ammonium chloroosmate*

	Experiment No.	Weight of $(NH_4)_2OsCl_6$ in vacuum	Weight of osmium in vacuum	Percentage of osmium	Atomic weight
$(NH_4)_2OsCl_6$, Series VI	1	*g* 3.81131	*g* 1.65758	43.491	
	2	3.46016	1.50505	43.496	
	3	1.11090	.48320	43.497	
				43.495	191.53
$(NH_4)_2OsCl_6$, Series VII	1	7.80602	3.39547	43.499	
	2	7.21775	3.13949	43.496	
	3	7.04888	3.06644	43.502	
	4	7.54170	3.27946	43.484	
				43.495	191.53

TABLE 2.—*Results of the analysis of ammonium bromoosmate*

	Experiment No.	Weight of $(NH_4)_2OsBr_6$ in vacuum	Weight of osmium in vacuum	Percentage of osmium	Atomic weight
$(NH_4)_2OsBr_6$, Series VI	1	*g* 5.85596	*g* 1.58647	27.091	
	2	5.59080	1.51471	27.093	
	3	3.91634	1.06117	27.082	
				27.089	191.55
$(NH_4)_2OSBr_6$, Series VII	1	4.98331	1.35016	27.093	
	2	4.53546	1.22899	27.097	
				27.095	191.61

The average percentage of osmium in ammonium chloroosmate, combining the results of the two series, was 43.495. The average percentage of osmium in ammonium bromoosmate, again combining the results of the two series, was 27.091. Using the values

$$\begin{aligned} N &\quad 14.008 \\ H &\quad 1.0078 \\ Cl &\quad 35.457 \\ Br &\quad 79.916 \end{aligned}$$

the ratio $(NH_4)_2OsCl_6$: Os yields the value 191.53 for the atomic weight while the ratio $(NH_4)_2OsBr_6$: Os yields 191.57. The weighted average value is 191.55.

VII. DISCUSSION OF RESULTS

It is recognized that complex compounds of the platinum metals, of the type considered here, sometimes undergo a change of composition through hydrolysis. Such is known to be the case with potassium chloroplatinate when formed under the conditions which obtain in the usual analytical separation of potassium from sodium. The change of composition comes about through the partial replacement of halogen by the hydroxyl group. The effect of such a change would be to raise the percentage of osmium and hence the apparent atomic weight. Except by accidental compensation the effect would be to produce a divergence in the values of the atomic weight calculated from similar ratios on analogous compounds. The probability that such change occurred in the compounds examined here is held to be highly unlikely, since the compounds were precipitated and recrystallized from solutions containing varying quantities of acid, and the value of the atomic weight as calculated from one compound agrees so closely with that calculated from the other analogous compound.

No impurities were detected in the metallic osmium used to prepare the compounds for analysis. The compounds were made by using only volatile reagents.

In the present work the samples of ammonium chloroosmate and of ammonium bromoosmate were dried to constant weight at temperatures of 140° to 150° C. A loss of weight was always observed in the first drying period when the salts, which had merely been desiccated over phosphorus pentoxide, were heated at 140° to 150° C. No evidence of decomposition was detected below 170° C. The probability that the compounds so treated still retained significant amounts of volatile impurities is held to be unlikely because of the close agreement of the values of the atomic weight calculated from the two ratios. The net effect of such impurities would be, of course, to lower the percentage of osmium and hence the apparent atomic weight, but a divergence in the ratios would result unless the relative amounts of the impurities in the two compounds happened to be such as to compensate for it.

The decomposition of the compounds in hydrogen was accomplished so gradually that the possibility of mechanical loss appears improbable.

The formation of only volatile products upon decomposition and the constancy of weight of the resulting metallic osmium, when heated in hydrogen at an elevated temperature, make it highly improbable that any impurity contaminated the final metal. The removal of hydrogen at an elevated temperature by nitrogen undoubtedly prevented the attack of the metal sponge by atmospheric oxygen during the time necessary to make the weighings.

The identical agreement of the average determinations of the osmium content of the two preparations of ammonium chloroosmate, when different quantities of the substances were taken and when different balances were used, makes it improbable that significant errors were introduced in weighing.

VIII. ACKNOWLEDGMENTS

The author wishes to express his gratitude for the help and encouragement received from Prof. J. C. W. Frazer, of The Johns Hopkins University, under whose supervision the research was conducted, and for the many valuable suggestions received from Edward Wichers.

WASHINGTON, June 8, 1932.

RP472

"MOISTURE EXPANSION" OF CERAMIC WHITE WARE

By R. F. Geller and A. S. Creamer

ABSTRACT

This report supplements the reports of a number of investigations of "moisture expansion." The relative susceptibility of various materials, individually and as consitituents of bodies, was studied, as was also the effect of the temperature to which the specimens had been heated. The relative susceptibility to moisture expansion of the principal ingredients of ceramic bodies appears to be approximately as follows: Fused feldspar, 2.4; calcined clay, 0.40; unfused feldspar, 0.20; calcined flint, 0.06. The expansion caused by reaction with moisture may be decreased at temperatures as low as 120° C. The removal of the moisture is relatively sluggish at 120° C., proceeds faster as the temperature is raised, and apparently may be completed at 250° C. The marked difference is susceptibility to moisture expansion of fused and unfused feldspar has an important effect on ceramic bodies. The data indicate that at temperatures below approximately "cone 8" and absorptions above 10 per cent the ratio of fused to unfused feldspar is the predominating factor determining the resistance to moisture expansion while at higher temperatures and lesser absorptions the relative absorption is the predominating factor. The data obtained offer no evidence regarding the nature of the reaction involved.

CONTENTS

I. INTRODUCTION

This report supplements the reports of many investigations [1] which have been made during the past six years on the general subject of moisture expansion or expansion due to the action of water. It was undertaken primarily to determine which of the usual constituents of white-ware bodies are most susceptible to moisture expansion.

---

[1] G. E. Merritt and C. G. Peters, J. Am. Ceram. Soc., vol. 9 (5), p. 332, 1926; H. G. Schurecht, vol. 11 (5), p. 271, 1928, and vol. 12 (2), p. 118, 1929; H. G. Schurecht and G. R. Pole, vol. 12 (9), p. 596, 1929; R. G. Mills, vol. 13 (12), p. 903, 1930; H. H. Holscher, vol. 14 (3), p. 207, 1931; H. G. Schurecht and G. R. Pole, vol. 14 (4), p. 313, 1931, and B. S. Jour. Research, vol. 3 (RP98), August, 1929; O. S. U. Eng. Exp. Sta. Cir. Nos. 18, J. Otis Everhart, and 22, H. H. Holscher.

II. MATERIALS AND METHODS

1. COMMERCIAL BODIES

Dinner plates of the same quality as supplied regularly to the trade, and representing the product of 16 manufacturers of earthenware, were available for the study of commercial bodies. The specimens for water-absorption determinations were irregularly shaped pieces broken from the plates and tested without removing the glaze, using three specimens of each brand. The determinations were made by the so-called 5-hour boiling test in accordance with the method prescribed in Federal specification No. 243a for vitrified chinaware. Indirect moisture-expansion measurements were made by means of the interferometer according to the method described by Schurecht.[2] These indirect measurements are based on the assumption that the difference in total thermal expansion obtained at 400° C. (or at a higher temperature) in the first and second tests of specimens which have had the opportunity of reacting with moisture, either during prolonged storage or in an autoclave, is a measure of the moisture expansion.

2. EXPERIMENTAL BODIES

The compositions of the experimental bodies, together with the heat treatment received, are given in Table 1 and in part A, Table 2. The potter's flint and the feldspar (No. 3)[3] were composed of particles varying in size from 10μ to 80μ, the separation from coarser and finer particles having been made by means of a No. 325 United States standard sieve and an air elutriator.[4] The clay (Georgia kaolin) was not sized in the elutriator because it "balled" in the apparatus and made the separation of any particular range of particle sizes impracticable.

The mixtures were made into specimens 6 inches long and one-half inch in diameter. The extrusion method was used to form the specimens of bodies containing clay, while the flint-feldspar series, containing gum tragacanth as a binder, were tamped. The specimens which had been heated as described in section B, Table 1, were dried at $115° \pm 5°$ C., weighed, measured, and autoclaved at 150 lbs./in.² steam pressure (183° C.) for one and one-half hours while submerged in water. They were then removed from the autoclave, reweighed and remeasured before and after drying (constant in weight to 0.01 g) at $115° \pm 5°$ C. and at $270° \pm 5°$ C. to determine changes in weight and length due to these drying treatments. The absorption was taken as the percentage gain in weight during the autoclave treatment.[5] The changes in length resulting from the autoclave treatment and subsequent drying were measured by means of a comparator using a graduated invar bar as reference.[6] Reference points in the specimens were obtained by cementing capillary Pyrex glass tubes in small holes approximately 12 cm apart, using a hydraulic high alumina cement. The precision in these length measurements was found to be approximately ± 0.01 per cent of the length measured. Room temper-

[2] H. G. Schurecht, J. Am. Cer. Soc., vol. 11 (5), p. 271, 1928.
[3] H. Insley, J. Am. Cer. Soc., vol. 10 (9), p. 651, 1927.
[4] J. C. Pearson and W. H. Sligh, B. S. Tech. Paper No. 48.
[5] This method of determining water absorption is believed to be fully as effective as the 5-hour boiling method because it has been shown by the authors that a 10-hour autoclave treatment at 150 lbs./in.² steam pressure is as effective as 240 hours boiling. See B. S. Tech. News Bull. No. 167, March, 1931.
[6] J. C. Pearson, J. Am. Concrete Inst., 1921.

atures were noted at all readings, but did not vary sufficiently to justify correcting the measurements for this factor.

TABLE 1.—*Experimental bodies and heat treatments*

A. BODIES

[Three specimens of each mixture for each heat treatment]

Series A			Series B			Series C			Series D			
Body No.	Flint	No. 3 Feld-spar	Body No.	Clay	No. 3 Feld-spar	Body No.	Clay	Flint	Body No.	Clay	Flint	No. 3 Feld-spar
	Per cent	*Per cent*		*Per cent*	*Per cent*		*Per cent*	*Per cent*		*Per cent*	*Per cent*	*Per cent*
A1....	80	20	B1...	80	20	C1...	80	20	D1...	20	40	40
A2....	60	40	B2...	60	40	C2...	60	40	D2...	40	30	30
A3....	50	50	B3...	50	50	C3...	50	50	D3...	50	25	25
A4....	40	60	B4...	40	60	C4...	40	60	D4...	60	20	20
A5....	20	80	B5...	20	80	C5...	20	80	D5...	80	10	10

B. HEAT TREATMENT

Series A (flint-feldspar)

Series B (clay-feldspar)

Fired to cones:	02	2	4	6	8	10	12	14	16	18

Series C (clay-flint)

Series D (clay-flint-feldspar)

3. INDIVIDUAL MATERIALS

The individual raw materials included: (*a*) Feldspars Nos. 3 and 18; (*b*) pulverized quartz obtained by crushing a sample of standard Ottawa sand; (*c*) washed kaolin from Georgia containing about 4 per cent of impurities, such as mica and quartz; (*d*) two samples of lepidolite;[7] (*e*) Cornwall stone as supplied to the pottery trade; and (*f*) Lemoor china clay. Three samples of the china clay were available: (*a*) The original clay; (*b*) a granular separate containing particles larger than 1μ in diameter; and (*c*) a "colloidal" separate containing particles less than 1μ in diameter.[8] With the exception of the feldspar and the Lemoor china clay used in the determination of the effect of particle size, each of the individual materials tested as powders was heated at approximately 1,225° C.[9] in a small gas-fired furnace, cooled rapidly, and crushed to pass a No. 120 United States standard sieve. It is realized that differences in particle size (or ratio of surface to unit weight) introduce an unmeasured variable which may lessen appreciably the significance of the values. For this reason the materials were ground with a mortar and pestle and sieved repeatedly as the grinding progressed so as to produce a minimum of fines.

[7] The following chemical compositions were obtained from the concerns furnishing the lepidolite:

	SiO_2	Al_2O_3	Fe_2O_3	MnO	K_2O	Li_2O	Na_2O	F	H_2O
	Per cent	*Per cent*	*Per cent*	*Per cent*	*Per cent*	*Per cent*	*Per cent*	*Per cent*	*Per cent*
Sample A............	49	27	0.3	0.5	8-10	4-5	2	....	1.5
Sample B............	63	24	.05	.35	6.8	2.9	1.5	1.5	.2

[8] The fractions were obtained by K. Langenbeck during the course of an investigation at the Bureau of Standards, the results of which have not been published.
[9] This temperature was chosen because it lies between the temperatures used by earthenware and vitrified chinaware manufacturers to mature their ware.

The determination of the relative degree of reaction of individual materials with moisture was based on weight change, assuming that expansion (or volume change) would be proportional to weight change. The crushed samples were dried to constant weight [10] at $115° \pm 5°$ C., exposed to water vapor at 150 lbs./in.^{2} steam pressure for five hours (precautions being taken to prevent condensation of moisture on the samples), weighed immediately on removal from the autoclave and again after drying at $115° \pm 5°$ C. and at $250° \pm 10°$ C.

TABLE 2.—*Effect of autoclave treatment on individual materials*

A. 1 BY 1 BY 5 INCH BARS (THREE SPECIMENS OF EACH BODY) AUTOCLAVED FOR THREE HOURS AT 150 LBS./IN.^{2} STEAM PRESSURE (183° C.)

	Temperature of heating			
	Cone 5 to 6 (approximately 1,180° C.)		Cone 8 to 9 (approximately 1,230° C.)	
	Expansion	Absorption	Expansion	Absorption
	Per cent	*Per cent*	*Per cent*	*Per cent*
50 per cent feldspar No. 18—50 per cent flint	0.03	6.0	0.00	1.4
50 per cent feldspar No. 3—50 per cent flint	.15	15.1	.03	3.9
Florida kaolin [1]	.04	14.8	.02	13.3
Georgia kaolin [1]	.83	20.7	.02	17.8

B. POWDERED SPECIMENS HEATED AT 1,225° C., RECRUSHED AND AUTOCLAVED FOR FIVE HOURS AT 150 LBS./IN.^{2} STEAM PRESSURE

		Weight gain during autoclave treatment	Percentage of water absorbed during autoclave treatment which was removed by drying at—	
			115° ± 5° C.	250° ± 10° C.
		Per cent	*Per cent*	*Per cent*
Feldspar No. 3	1	2.4	32	84
	2	2.4	30	95
	3	2.5	30	93
	4	2.5		
Quartz	1	.06	67	92
	2	.06	69	86
	3	.05	17	81
	4	.05	15	78
Kaolin	1	.46	64	100
	2	.34	37	91
Lepidolite	A 1	1.7	35	84
	A 2	1.6	33	83
	B 1	1.3	27	78
	B 2	2.0	21	77
Cornwall stone	1	1.4	23	76
	2	1.4	22	74
	3	1.3	20	73

[1] Taken from Table 7 of H. H. Holscher's paper. (See footnote 12, p. 299.)

[10] In this phase of the work the specimens were considered to have reached constant weight when the weight difference after 2-day intervals of drying did not exceed 0.5 mg.

III. RESULTS

1. COMMERCIAL BODIES

(a) MOISTURE EXPANSION OF UNAUTOCLAVED SPECIMENS

As a preliminary study, specimens of five brands (Nos. 1, 5, 6, 14, and 16) were taken from glazed plates which had been stored in the laboratory for several months. The moisture expansion was found to vary from 0.03 to 0.06 per cent for four brands which had water absorptions of from 9.4 to 12.5 per cent; the other brand (No. 1) showed no moisture expansion and the water absorption was 2.0 per cent. Typical curves are given in Figure 1 and the values in Table 3.

Since the moisture expansion tests of unautoclaved glazed plates were made on specimens of the body immediately after being taken from the plates, it is indicated that the expansion noted had taken place regardless of the supposedly impervious coating of glaze.[11] The data, therefore, justify the conclusion that glazed earthenware may be subject to crazing caused by penetration of moisture through the glazed and the resultant moisture expansion of the body.

(b) MOISTURE EXPANSION OF AUTOCLAVED SPECIMENS

One specimen of each of the 16 brands was autoclaved at 150 lbs./in.^{2} steam pressure for one hour, the pieces being immersed. Typical results are shown by the curves in Figure 1. For any one specimen the expansions during the second and third heatings were practically identical. The curves in Figure 2 show also that the contraction during the cooling, following the first heating of an autoclaved specimen, is practically along the same curve as the expansion during the second heating. The data indicate that the first heating and the first cooling are sufficient to give the desired information. Pertinent data obtained in this phase of the investigation are summarized in Table 3 and the calculated moisture expansions are plotted against percentage water absorption in Figure 3. Although the bodies are all of one type and probably had been matured at about the same temperature, there are undoubtedly sufficient differences in composition and structure to affect the porosity-expansion relation. As shown by data given in this paper, feldspar alone may be an important factor.

TABLE 3.—*Effect of moisture on specimens either stored in laboratory air or autoclaved*

Brand No.	Water absorption	Moisture expansion after storage [1]	Moisture expansion after autoclave [2]	Brand No.	Water absorption	Moisture expansion after storage [1]	Moisture expansion after autoclave [2]
	Per cent	Per cent	Per cent		Per cent	Per cent	Per cent
1	2.	0	0.05	9	11		0.12
2	7.5		.07	10	11		.11
3	9.		.07	11	11		.14
4	9.5		.08	12	11		.09
5	9.5	0.03	.11	13	12		.08
6	10	[3] .04	.10	14	12	0.06	.12
7	10		.10	15	12		.09
8	10		.08	16	12	.03	.09

[1] Difference in total expansion at 400° C. between first and second heating of unautoclaved specimen which had been taken from a whole plate stored in the laboratory under normal atmospheric conditions for several months.

[2] Difference in total expansion at 400° C. between first and second heating of specimen autoclaved for one hour in steam at 150 lbs./in.^{2} pressure.

[3] Especially selected and tested for unbroken glaze before specimen was taken from plate.

[11] It was shown by Schurecht (footnote 1, p. 291) that glazes may undergo moisture expansion. This is further evidence that they are not impervious to moisture.

A number of observations were made by holding autoclaved specimens at several temperatures (maintaining the temperatures as nearly constant as possible) and observing length changes. Seventeen tests

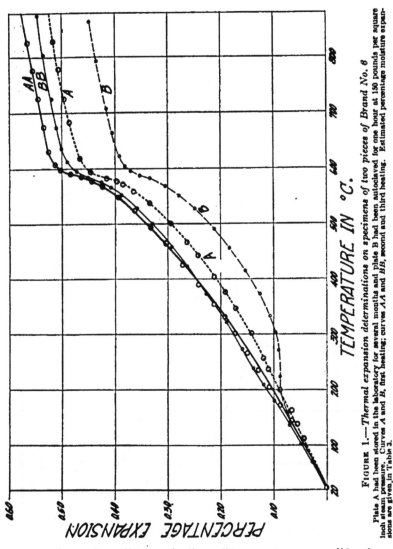

FIGURE 1.—*Thermal expansion determinations on specimens of two pieces of Brand No. 6*

Plate A had been stored in the laboratory for several months and plate B had been autoclaved for one hour at 150 pounds per square inch steam pressure. Curves A and B, first heating; curves AA and BB, second and third heating. Estimated percentage moisture expansions are given in Table 3.

were made on three different bodies. These tests were modifications of the following, the results of which are typical and are discussed in detail:

1. A specimen of brand No. 5 was "autoclaved" for two hours at 150 lbs./in.² steam pressure while immersed in water. Length changes

were then observed while the specimen was heated in a furnace to approximately 118° C. and held for 5½ hours, reheated to approximately 180° C. and held for 1½ hours and again heated to 246° C. and held for 1½ hours and finally heated to 600° C. The specimen was permitted to cool to room temperature after each heat treatment.

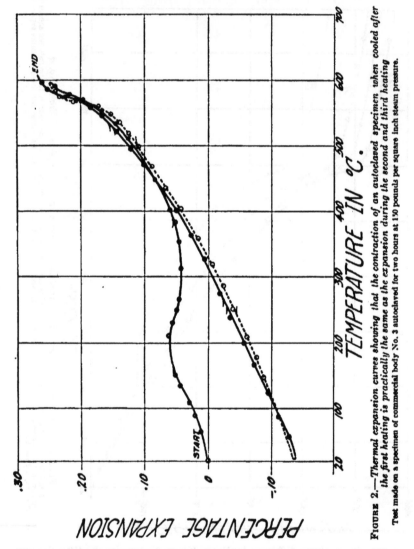

FIGURE 2.—*Thermal expansion curves showing that the contraction of an autoclaved specimen when cooled after the first heating is practically the same as the expansion during the second and third heating*

Test made on a specimen of commercial body No. 3 autoclaved for two hours at 130 pounds per square inch steam pressure.

The results obtained are shown by the curves in Figure 4. The specimens were held at each temperature until further length changes were negligible.

2. Length changes of a second specimen of brand No. 5 were observed while the specimen was held at approximately 245° C. for

five hours. As is evident from the curve in Figure 5, the rate of contraction had decreased considerably at the end of the 5-hour period, but conditions had not reached equilibrium. The specimen, after cooling to room temperature, was again heated to approximately 245° C. and held for four and one-half hours, after which the temperature was raised to 325° C.

3. Two specimens of brand No. 15 were autoclaved for two hours at 150 lbs./in.[2] steam pressure while immersed in water. Length

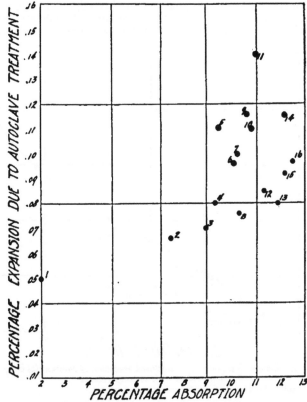

FIGURE 3.—*Water absorption and difference in total thermal expansion at 400° C. (during first and second tests) of autoclaved specimens of 16 brands of commercial whiteware (data in Table 3)*

The difference in total expansion at 400° is considered as moisture expansion.

changes of one specimen were then observed while the specimen was heated to approximately 260° C. and held for two hours (*A*, fig. 6), after which the temperature was raised to 380° C. Length changes of the second specimen were observed while heating it to 680° C. at a constant rate of temperature increase (*B*, fig. 6) permitting it to cool to room temperature, and again heating to 680° C. at a constant rate of 3° C. per minute.

The information given in Figures 4, 5, and 6 indicates that the reactions induced by heating autoclaved specimens, and evidenced by contraction, are relatively sluggish at 120° C., proceed faster as the temperature is raised, and are markedly accelerated at temperatures above 230° C.

2. EXPERIMENTAL BODIES

(a) EFFECT OF COMPOSITION, HEAT TREATMENT AND ABSORPTION ON MOISTURE EXPANSION

As originally outlined this phase of the investigation included the making and testing of 435 specimens representing 20 different bodies, each of which was to be heated at several temperatures as specified

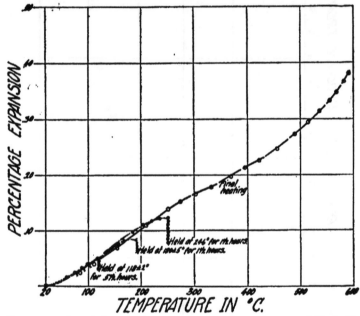

FIGURE 4.—*Length changes of an autoclaved specimen of commercial body No. 5 when held at various temperatures*
All tests made on the same specimen after one autoclave treatment.

in Table 1. Three specimens were prepared for each heat treatment of each body.[13] Because of the extreme weakness of many of the specimens, particularly the *A* series (flint feldspar), and the specimens of the *B* series (clay feldspar) heated at the lower temperatures, only 215 were available for the entire series of tests. Had this been anticipated, fewer bodies and more specimens of each would have

---

[13] It has been shown by previous investigators (footnote 1, p. 291) that the relative porosity is an important factor in determining the extent of the reaction of ceramic bodies with moisture. Obviously the relative moisture expansions of two bodies varying in composition are not comparable unless the porosity is the same both in extent and in nature. Not only is the nature of the pore space difficult of determination (ratio of fine to coarse pores, etc.), but it would be practically impossible to develop the same porosity in two bodies of the whiteware type, varying in composition, without using a different heat treatment for each. This introduces a third variable which, it was anticipated, would be a significant factor. It was this practical difficulty of changing only one variable at a time that prompted the study of such a large number of specimens in the hope that the data could be studied as group averages which might indicate the effect to be expected by changing either composition, porosity, or heat treatment.

been made. However, such values as could be determined are presented in Table 4. This table gives the average percentage moisture expansion and water absorbed for each body during the autoclave treatment, but only the group averages of each series of bodies for the determinations which were made after the drying treatments.

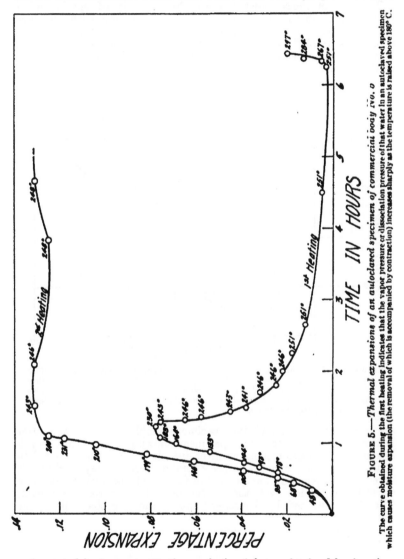

FIGURE 5.—*Thermal expansions of an autoclaved specimen of commercial body No. 0*

The curve obtained during the first heating indicates that the vapor pressure or dissociation pressure of that water in an autoclaved specimen which causes moisture expansion (the removal of which is accompanied by contraction) increases sharply as the temperature is raised above 180° C.

As stated in footnote 12, the analysis of data, obtained by heating a number of different ceramic bodies to a series of temperatures and subjecting them to moisture in an autoclave at elevated temperatures, is made difficult by the fact that one factor can not be changed without changing others. In studying the data (Table 4) it should be remembered that only one clay, one feldspar, and one flint were investigated. Also, little significance can be attached to certain irregular absorption-temperature relations because the bodies were heated by the draw-trial method, which is comparatively rapid and not suitable for a study of absorption-temperature relations.

The length measurements show that autoclaved specimens of earthenware which have expanded due to reaction with moisture apparently return to their original length when held for a reasonably short time at 250° to 270° C. in air and under atmospheric pressure. However, the most noticeable development is the unique behavior of the C (or clay-flint) series of bodies. Although the compositions

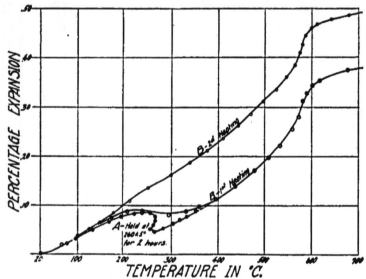

FIGURE 6.—*These thermal expansion curves for commercial body No. 15 (together with the curves given in figs. 4 and 5) show that subjection to a temperature of approximately 250° C. for a relatively short time is sufficient to overcome the effect of autoclave treatment on length change, while prolonged treatment at lower temperatures is not*

are very refractory, and the specimens highly porous, even after heating to cone 18, the only specimens showing a significant length change, or moisture expansion, after autoclaving are those heated to cone 8. This indicates that a body composed of the clay used in this investigation (Georgia kaolin) and of flint would be very resistant to moisture expansion even though the porosity was relatively high. The data obtained by Holscher [13] would indicate that this might not be true for all clays.

[13] J. Am. Cer. Soc., vol. 14 (3), p. 207, 1931.

Table 4.—Percentage change in weight and length after autoclave treatment and also after subsequent drying

Body	Heat treatment (approximate °C.) — Cone																			
	1,100 — 02		1,125 — 2		1,170 — 4		1,190 — 6		1,215 — 8		1,250 — 10		1,300 — 12		1,380 — 14		1,440 — 16		1,490 — 18	
	Change in— Weight (Per cent)	Length (Per cent)	Weight	Length	Weight	Length	Weight	Length	Weight	Length	Weight	Length	Weight	Length	Weight	Length	Weight	Length	Weight	Length
Flint-feldspar:																				
A1	25.0°	0.10°	21.5°		29.5°	0.04°	28.0	0.06°	21.5°	0.05°										
A2	25.0	.05	20.0	.13	27.5°	.07°	27.5	.07	17.0	.08										
A3	20.5	.09	20.0	.14			29.0°	.05	15.0°	.14°	7.5	.06°	0.0°	.00°	0.0	0.00				
A4	20.0	.10°	13.0	.15	15.0°	.18	9.0	.18	8.0	.12										
A5	23.5°	.10°	22.0	.11	15.0°	.14°	16.0	.03°												
Average after autoclave [2]	22.8	.09	20.0	.13	11.5	.17	19.5		5.0	.09	3.5	.04	3.0	.09						
Average after first drying [3]	-.0	.03	.2	.05	.3	.07		.07	.3	.05	.1	.04	.1	.02	0.0	0.00				
Average after second drying [4]	-.2	.00	.2	.08	-.2	.01		.01	-.1	-.01	.2	.00	.1	.00		0.00				
Clay-feldspar:																				
B1							19.5	.11	7.0°	.12	6.5	.12°	7.0°	.08°						
B2					12.0	.17	11.0	.20	4.5°		4.0°	.05°	4.0	.10						
B3					9.5	.18	9.0	.18	5.0°	.14°	4.0°	.05°								
B4					10.0°	.18	16.0	.13	3.0°	.01°	0.0	.00								
B5					15.0°	.14°	16.0													
Average after autoclave [2]					11.5	.17	14.0	.16	5.0	.09	3.5	.04	3.0	.09						
Average after first drying [3]					.3	.07	.2	.07	.3	.05	.1	.04	.1	.02						
Average after second drying [4]					-.2	.01	-.1	.00	-.1	-.01	.2	.00	.1	.00						

Clay-flint:																							
C1	—	—	—	—	22.0	.08	17.5	.02	16.0	.00	13.5	.02	12.0	0.01	8.5	0.00
C2	—	—	—	—	22.0	.07	21.0	.00	20.0	.00	17.8	.00	17.0*	.00*	13.5*	.00*
C3	—	—	—	—	22.8	.04	21.5*	.00*	19.5	.00	18.9	.00	18.5	.01	14.5	.00
C4	—	—	—	—	22.6	.00			20.5		20.0		20.0	.03*	16.5	.00
C5	—	—	—	—									22.5*	.00	23.5	.00
Average after autoclave [2]					22.5		19.0	.01	19.0	.00	23.5		18.0	.01	16.5	.00
Average after first drying [3]					−.1		−.1	.01	.0	.00	−.1		−.1	.00	.0	.00
Average after second drying [4]					−.2		−.2	.00	−.1	.00	−.1		−.1	.00	−.1	.00
Clay-flint-feldspar:																
D1	19.5	.17	20.0*	.27*	16.0	.13	11.0	.07	8.0	.06	1.0	.09	−.3	.08	−.1	.00
D2	17.0	.11	17.0*	.21*	14.0	.14	11.5*	.08	8.5	.04	5.5	.09	.2	.03	.2	.00
D3	17.0	.17	15.5	.15	13.0	.12	11.5	.10	7.0	.05	5.5	.10*	.1	.02	.0	.00
D4	16.5*	.13*	16.0*	.15*	13.6	.13	12.0	.06	8.5	.05	6.0	.02				
D5	23.0	.00			13.5	.06	12.0	.05	9.0		6.5	.04				
Average after autoclave [2]	18.5	.13	17.0	.19	14.0	.12	11.5	.08	7.0	.03	5.0	.03	.2	.02	.1	.00
Average after first drying [3]	.1	.00	.1	.08	.0	.04	−.1	.02	.2	.03	−.0	.03	−.0	.00	.0	.00
Average after second drying [4]	−.3	.00	−.2	.02	−.2	−.02	−.2	−.02	−.1	.01	−.2	.01	−.1	−.1	.0	.00

[1] The negative values for averages 3 and 4 indicate that the specimens were lighter, or shorter, than after the preliminary drying (115° ±5° C.) before autoclave treatment.
[2] The individual values in the table of which these values are the average, represent determinations on from 1 to 3 specimens of each body and each heat treatment after autoclaving for 1½ hours at 150 lbs./in.² steam pressure (183° C.). The asterisk (*) identifies values representing determinations on 1 specimen only.
[3] Average difference in weight and length, from weight and length, after drying at 115° ±5° C.
[4] Average difference in weight and length, from weight and length before autoclaving, after drying at 270° ±5° C.

A point of interest concerning the B series is the progressively decreasing resistance to expansion which was shown by the bodies fired to cones 02, 2, and 4 even though the percentage absorption also decreased. (Table 4 and fig. 7.) This led to the supposition that crystalline or unaltered feldspar is less susceptible to moisture expansion than feldspar glass. Therefore, a specimen of raw feldspar No. 3 and one of the same sample fused at 1,225° C., both ground to pass a No. 325 United States standard sieve (<80μ), were subjected to identical autoclave treatment. Also, specimens of body B4 heated to cones 02, 6, and 10 were examined petrographically.[14] It was found that the body heated to cone 02 was composed of two phases—unmelted feldspar (index of refraction = 1.52) and dehydrated clay

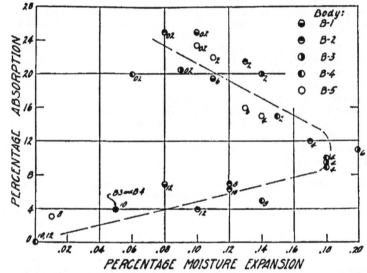

FIGURE 7.—*Relation between water absorption and moisture expansion of the clay-feldspar series of bodies (B, Table 4)*

The numbers indicate the temperatures, in terms of cones, to which the specimens were heated.

(index of refraction = 1.53 to 1.54). The specimen fired to cone 6 contained three phases—melted feldspar (index of refraction = 1.49), unmelted feldspar (index of refraction = 1.52), and dehydrated clay (index of refraction = 1.53 to 1.54). The cone 10 specimen contained the same phases as that fired to cone 6, but much more of the feldspar had melted. The absorption values obtained for raw and fused feldspar, which are given in Table 5 and supported by the other values for fused feldspar in Table 2 when considered in connection with the phase compositions of the B bodies examined may be accepted as supporting the supposition. Therefore, commercial earthenware, which is matured at about cone 6 and has an absorption averaging 10 per cent, may have a structure peculiarly susceptible to moisture expansion.

---

[14] Examinations made by R. H. Ewell.

3. INDIVIDUAL MATERIALS

(a) RELATIVE REACTION TO MOISTURE

Holscher [15] found that a body composed of feldspar, clay, and flint showed greater length changes after autoclave treatment than the pure clay although the absorptions were lower. This indicates that the other constituents of the bodies he investigated, namely, the feldspar or flint, may be even more reactive to moisture than the clay. Schurecht [16] investigated a series of bodies in which ball clay, feldspar and flint were varied between wide limits. He found that a mixture of 40 per cent feldspar and 60 per cent flint showed the greatest percentage length change in the series (0.188 per cent), while 60 per cent feldspar and 40 per cent flint showed only 0.100 per cent. The absorptions, however, were 13.4 per cent for the former and 5.0 per cent for the latter mixture. Whether the relatively high moisture expansion of the former mixture was induced by a reaction between the flint and feldspar, as Schurecht surmised, or because the presence of the flint produced sufficient porosity to permit the water vapors to react with a greater portion of the body had not been proven. It is worthy of note that the three bodies of Schurecht's series which showed unusually low expansions, considering their high absorptions, concontained no feldspar. The data indicate feldspar to be a material highly reactive to water vapor. This indication was strengthened by the results of preliminary tests given in part A of Table 2. The values given in part B of Table 2 are for individual powdered materials heated, recrushed and autoclaved as described in Part II.

TABLE 5.—*Effect of particle size on percentage gain in weight during autoclave treatment (samples exposed to vapor only)*

Material	Particle size	Gain in weight		
		Sample 1	Sample 2	Sample 3
Feldspar No. 3: [1]	μ	*Per cent*	*Per cent*	*Per cent*
Fused	<80	2.8		
Raw	<80	.18	0.22	
	80–10	.27	.28	0.28
	<10	1.7	1.2	1.2
Lemoor china clay [2]	Original.	2.3	2.2	
	>1	1.7	1.7	
	<1	3.2	3.3	

[1] Autoclaved at 500 lbs/in.² steam pressure for 1 hour (250° C.).
[2] Autoclaved at 320 to 350 lbs./in.² steam pressure for 16 hours (225°±5° C.).

The relative reactions of individual materials with moisture (Table 2) lead to conclusions similar to those indicated by the results of the much longer series of tests with experimental bodies. The weight increase of the feldspar glass investigated was the greatest, that of dehydrated clay was considerably less, and that of the calcined flint almost negligible. This is in accord with the observed fact that the bodies containing no feldspar (series C, Table 4) showed the least expansion. No lepidolite or Cornwall stone having been used in the experimental bodies, the data obtained with them are of general interest only. However, they are of possible value, when compared with those for feldspar, to those concerns using lepidolite or Cornwall stone in their bodies and glazes.

(b) EFFECT OF PARTICLE SIZE

The values obtained by autoclave treatment of fractions of one feldspar and of one clay, the fractions differing in the range of particle sizes comprised, are given in Table 5. They may be taken as further evidence that reactions with moisture in an autoclave will increase with increase in the surface exposed per unit mass.. This is to be expected since there is no breaking down of the structures and the reaction is confined to the original surfaces or must proceed from these surfaces.

(c) PETROGRAPHIC EXAMINATIONS

These petrographic examinations and the following report were made by Dr. H. Insley.

Georgia clay (heated to 1,225° C., recrushed to pass a No. 120 sieve, and autoclaved at 760 to 800 lbs./in.² steam pressure or at approxmiately 275° C., for 22 hours). Mean index of refraction of grains about 1.57. No difference between autoclaved and unautoclaved samples.

Cornwall stone (same treatment as Georgia kaolin). No apparent difference between autoclaved and unautoclaved samples.

Mica and quartz (same treatment as Georgia kaolin). No perceptible effect of autoclaving.

Lepidolite (same treatment as Georgia kaolin). Some etching marks on surfaces of grains, but otherwise indistinguishable from lepidolite which has been heated but not autoclaved.

Raw feldspar No. 3 (<10 μ; autoclaved at 800 lbs./in.² steam pressure or 275° C., for 16 hours). Identical with untreated material.

Raw feldspar No. 3 (10 to 80 μ; same treatment as other raw feldspar). Identical with untreated feldspar.

Feldspar No. 3 (heated at 1,225° C.; recrushed to pass a No. 120 sieve and autoclaved for 24 hours at 700 to 1,000 lbs./in.² steam pressure or approximately 265° to 290° C.). The feldspar is fused. Index of grains is slightly but unmistakably higher than the unautoclaved sample. (Unautoclaved=1.493, autoclaved=1.496.) Autoclaved sample shows a slight etching of the surface of the grains. The unautoclaved sample shows an extremely narrow layer on each grain somewhat higher in index than the body of the grain. The higher index of the autoclaved sample may be due to the annealing action during the autoclave treatment.

4. NATURE OF REACTION TERMED MOISTURE EXPANSION

While the results obtained in this investigation throw further light on the factors affecting the resistance of ceramic whiteware to moisture expansion, they offer no evidence regarding the nature of the reaction involved. Any of the observed facts can apply equally well, whether this expansion is caused by adsorption, absorption, or solid solutions. It is possible that the nature of the reaction could be proven using the method described by Washburn [17] for studying the dissociation equilibrium of autoclaved material at different temperatures. This method has been used by Schwarz [18] and Huttig [19] in a study of silica and its hydrates.

[17] J. Franklin Inst., p. 753, June, 1922.
[18] Ztschr. Elektrochem., vol. 32, p. 415, 1926.
[19] Ztschr. anor. Chem., vol. 121, p. 243, 1922.

IV. CONCLUSIONS

The most important conclusions justified by the data obtained in this investigation may be summarized briefly as follows:

1. Fused feldspar is the most susceptible to moisture expansion of the materials investigated.

2. The increasing content of fused feldspar and the decreasing absorption of whiteware bodies heated to progressively higher temperatures are counteracting influences which apparently result in least resistance to moisture expansion for those bodies matured at cone 6 to 8 and having an absorption of about 10 per cent.

3. Glazed earthenware may be subject to crazing caused by penetration of moisture through the glaze and the resultant moisture expansion of the body.

4. Reactions induced by heating autoclaved specimens and evidenced by contraction are relatively sluggish at 120° C., proceed faster as the temperature is raised, and are markedly accelerated at temperatures above 230° C.

5. Autoclaved specimens of whiteware which have expanded due to reaction with moisture apparently return to their original length when cooled to room temperature after holding for a reasonably short time at 250° to 270° C. in air and under atmospheric pressure.

WASHINGTON, June 21, 1932.

INFRA-RED ARC SPECTRA PHOTOGRAPHED WITH XENOCYANINE

By W. F. Meggers and C. C. Kiess

ABSTRACT

A new sensitizing dye, xenocyanine, has made possible the preparation, by the Eastman Kodak Co., of plates highly sensitive to infra-red light ranging in wave length from 8,000 to 11,000 A. With such pla es the infra-red arc spectra of about 50 elements have been observed at the Bureau of Standards. New wave-length measurements have been made for Ti, Fe, Co, Ni, and Zr, and are presented in this paper. Many of these new lines have been accounted for as combinations between previously known terms of the neutral atoms. In the case of Ti a new term, a^3D, has been found. Both Ti and Fe are rich in strong lines in the region investigated and are recommended as suitable comparison spectra for the measurement of wave lengths in the infra-red.

CONTENTS

I. INTRODUCTION

More than a decade has passed since the authors [1] published several papers on the infra-red emission spectra of various metallic arcs. Use was made of photographic plates bathed in solutions of dicyanin which was the best photographic sensitizer then known for the infrared. Dicyanin exhibits its maximum sensitivity in the red at wave length 7,000 A, but it is characterized by a broad band which made it possible to record, with prolonged exposures, the strongest lines between 9,000 and 10,000 A. It appears now that the practical limit of dicyanin as employed by us is about 9,800 A, and nearly all of the lines observed with apparently greater wave length have been explained as Lyman ghosts produced by the diffraction grating when very long exposures are used.

A new infra-red sensitizer, called neocyanine,[2] was discovered in 1926; its sensitizing action is a maximum at 8,100 A and it has been used extensively for the photography of spectra. However, its sensitizing band is much narrower than that of dicyanin, the action falling off very rapidly beyond 9,000 A, so that no appreciable advance was made in the study of infra-red spectra.

---

[1] W. F. Meggers, B. S. Sci. Papers, vol. 14 (S312), p. 371, 1918. W. F. Meggers and C. C. Kiess, B. S., Fe . Papers, vol. 14 (S324), p. 637, 1918. C. C. Kiess and W. F. Meggers, B. S. Sci. Papers, vol. 16 (S372) p. 51, 1920.
[2] M. L. Dundon, A. L. Schoen, and R. M. Briggs, J. Opt. Soc. Am. and Rev. Sci. Inst., vol. 12, p. 397, 1926.

The recent discovery, in the Eastman Kodak research laboratories, of a new infra-red sensitizer, xenocyanine, is now recognized as beginning a new era in infra-red spectroscopy. Photographic emulsions containing xenocyanine have two remarkable properties—(1) a very broad sensitizing band which ranges from about 8,000 to 11,000 A with a maximum at 9,600 A, (2) extraordinary speed, which is estimated at one hundred to one thousand times that of the best materials heretofore available for the interval 9,000 to 10,000 A. At the Bureau of Standards more than 50 spectra have now been explored in the infra-red with these new plates, and in nearly all cases many new lines have been discovered. In this paper results are presented for titanium, iron, cobalt, nickel, and zirconium.

II. APPARATUS AND METHODS

The light source in each case was a direct-current arc carrying 6 to 8 amperes with an applied potential of 220 volts. Rods of iron, cobalt, and nickel served as arc electrodes, while in the cases of titanium and zirconium, pieces of metal were used in a copper arc.

The spectrograms were made with a concave grating, mounted stigmatically, as described elsewhere,[3] the dispersion being 10.4 A/mm in the first-order spectrum. The infra-red spectra were photographed in the first order with exposures of 20 to 60 minutes and adjacent portions of the plate were exposed for 1 minute to the second-order spectrum of iron for the purpose of wave-length measurements. For titanium and iron the stronger lines were photographed also with the grating ruled with 20,000 lines per inch, giving a dispersion of 3.6 A/mm.

The xenocyanine plates were hypersensitized in a dilute ammonia bath just before use in the spectrograph, and they were developed in Eastman Kodak X-ray developer at about 18° C. immediately after exposure.

Wave-length determinations in the infra-red spectra involved careful measurement of the position of the lines relative to lines in the blue and green portions of the iron spectrum recorded in juxtaposition. The values of the iron lines were those adopted as secondary standards by the International Astronomical Union,[4] their apparent values for the interpolation of wave lengths corresponding to infra-red lines being assumed to be exactly double their actual values.

III. RESULTS

1. TITANIUM

In Table 1 are presented the new wave lengths measured in the arc spectrum of Ti and extending it 1,000 A beyond the longest line previously recorded. The most complete list of Ti ɪ lines published to date is that compiled by Prof. H. N. Russell[5] from various sources and giving the classifications of nearly 1,400 lines as combinations between terms of the singlet, triplet, and quintet systems. Our list overlaps Russell's by nearly 1,500 A, and a comparison of the two well illustrates the advantage of the new photographic materials over

[3] W. F. Meggers and K. Burns, B. S. Sci. Papers (No. 441), 18, p. 191, 1922.
[4] Trans. Internat. Astron. Union, vol. 3, p. 77, 1928.
[5] Astrophys. J., vol. 66, p. 347, 1927.

those formerly used. Not only are the intensity estimates of the lines greatly increased but many faint lines, predictable from term combinations and heretofore unobserved, are now recorded.

In addition to wave length and intensity estimates, Table 1 records for each line its vacuum wave number and its term combination. Most of the classifications are based on the terms given by Russell which we have expressed in the current standard notation.[6] A key for translating Russell's notation into that used at present is to be found in a recent paper by Miss Moore.[7]

As shown in Russell's analysis of Ti I a group of metastable terms is to be expected arising from the electron configuration $3d^4$. Of this group of terms we have found a^3D which actually lies higher in the energy diagram than the lowest odd terms ($z^5D°$ and $z^5F°$) of the middle group with which it combines. The values of the components of a^3D relative to $a^5F_2 = 0$ are given in Table 2.

TABLE 1.—*Wave lengths in the infra-red spectrum of titanium*

λ_{air} I. A.	Intensity	ν_{vac} cm^{-1}	Term combination
10, 774. 82	1	9, 278. 36	$a^5F_3-z^5G_4^\circ$
10, 732. 84	1	9, 314. 65	$a^5F_2-z^5G_3^\circ$
10, 731. 11	0	9, 316. 15	$z^3D_3^\circ-f^3F_4$
10, 726. 27	4	9, 320. 35	$a^5F_1-z^5G_2^\circ$
10, 689. 33	1	9, 352. 56	$z^3D_3^\circ-f^3F_3$
10, 676. 90	2	9, 363. 44	$a^5F_1-z^5G_1^\circ$
10, 661. 50	8	9, 376. 98	$a^5F_3-z^5G_4^\circ$
10, 607. 71	1	9, 424. 52	$a^5F_2-z^5G_3^\circ$
10, 584. 57	12	9, 445. 12	$a^5F_3-z^5G_3^\circ$
10, 552. 87	1	9, 473. 50	$a^3H_5-y^3G_4^\circ$
10, 496. 08	15	9, 524. 76	$a^5F_4-z^5G_5^\circ$
10, 459. 98	3	9, 557. 63	$a^3H_6-y^3G_5^\circ$
10, 396. 78	20	9, 615. 72	$a^5F_2-z^5G_2^\circ$
10, 145. 40	5	9, 853. 98	
10, 120. 87	7	9, 877. 87	$a^3D_3-z^3D_3^\circ$
10, 066. 47	8	9, 931. 25	$a^3D_2-z^3D_1^\circ$
10, 059. 87	12	9, 937. 77	$b^3F_2-z^3G_3^\circ$
10, 057. 69	25	9, 939. 92	$a^3D_2-z^3D_2^\circ$
10, 050. 11	5	9, 947. 41	
10, 048. 78	12	9, 948. 73	$b^3F_2-z^3G_1^\circ$
10, 034. 45	15	9, 962. 94	$b^3F_4-z^3G_5^\circ$
10, 011. 72	15	9, 985. 56	$a^3D_1-z^3D_1^\circ$
10, 003. 02	25	9, 994. 24	$a^3D_2-z^3D_2^\circ$
9, 997. 94	15	9, 999. 32	$a^3G_3-y^3F_2^\circ$
9, 981. 16	5	10, 016. 13	
9, 948. 98	8	10, 048. 53	$a^3D_1-z^3D_2^\circ$
9, 941. 33	8	10, 056. 26	$a^3D_2-z^3D_3^\circ$
9, 930. 30	1	10, 067. 44	$y^1G_4^\circ-e^1G_4$
9, 927. 35	20	10, 070. 42	$a^3G_4-y^3F_3^\circ$
9, 923. 25	2	10, 074. 58	$\left\{\begin{array}{l} a^3D_3-y^3G_4^\circ \\ a^3D_3-y^3G_3^\circ \\ y^3G_3^\circ-e^1F_4 \end{array}\right.$

[6] Phys. Rev., vol. 33, p. 900, 1929.   [7] Astrophys. J., vol. 75, p. 235, 1932.

TABLE 1.—*Wave lengths in the infra-red spectrum of titanium*—Continued

λ_{air} I. A.	Intensity	ν_{vac} cm⁻¹	Term combination
9, 879. 41	3	10, 119. 29	$a^3G_3-y^3F_2^o$
9, 832. 15	25	10, 167. 93	$a^3G_5-y^3F_4^o$
9, 813. 45	5	10, 187. 30	$z^3D_2^o-a^3D_3$
9, 787. 67	50	10, 214. 14	$a^5F_1-z^5F_2^o$
9, 783. 59	20	10, 218. 40	$a^5F_2-z^5F_1^o$
9, 783. 30	40	10, 218. 70	$a^5F_4-z^5F_3^o$
9, 770. 28	40	10, 232. 32	$a^3F_3-z^5F_2^o$
9, 768. 22	5	10, 234. 47	$z^3D_3^o-a^3D_2$
9, 746. 86	15	10, 256. 91	$z^5D_3^o-a^5D_4$
9, 743. 60	50	10, 260. 34	$a^5F_1-z^5F_1^o$
9, 737. 77	5	10, 266. 48	$z^5D_2^o-a^3D_1$
9, 728. 36	60	10, 276. 41	$a^5F_3-z^5F_3^o$
9, 718. 96	25	10, 286. 35	$a^1G_4-z^1F_3^o$
9, 717. 00	10	10, 288. 42	$z^3D_3^o-a^5D_3$
9, 715. 51	3	10, 290. 00	$z^5D_1^o-a^5D_0$
9, 705. 64	80	10, 300. 47	$a^5F_3-z^5F_4^o$
9, 702. 86	3	10, 303. 42	$z^5D_3^o-a^5D_2$
9, 690. 62	2	10, 316. 43	$a^3D_2-z^5P_2^o$
9, 688. 86	30	10, 318. 31	$a^5F_1-z^5F_2^o$
9, 678. 98	3	10, 328. 85	$z^5D_2^o-a^5D_1$
9, 675. 55	90	10, 332. 50	$a^5F_4-z^5F_4^o$
9, 663. 19	3	10, 345. 72	$z^5D_1^o-a^5D_2$
9, 661. 42	10	10, 347. 61	$a^3D_3-z^5P_3^o$
9, 652. 38	1	10, 357. 30	$z^5D_3^o-a^5D_3$
9, 651. 66	1	10, 358. 07	$z^5D_3^o-a^5D_4$
9, 647. 91	1	10, 362. 10	
9, 647. 40	50	10, 362. 65	$a^5F_2-z^5F_3^o$
9, 638. 28	100	10, 372. 45	$a^5F_5-z^5F_4^o$
9, 606. 77	3	10, 406. 48	$y^5F_3^o-f^5F_4$
9, 602. 38	1	10, 411. 23	$y^5D_3^o-e^5F_4$
9, 599. 53	50	10, 414. 32	$a^5F_3-z^5F_2^o$
9, 590. 15	3	10, 424. 51	$y^5D_4^o-e^5F_5$
9, 588. 77	4	10, 426. 01	$y^5F_4^o-f^5F_3$
9, 570. 08	4	10, 446. 37	$y^5F_3^o-f^5F_2$
9, 550. 11	2	10, 468. 21	$y^5F_2^o-f^5F_1$
9, 546. 07	50	10, 472. 64	$a^5F_4-z^5F_3^o$
9, 511. 80	8	10, 510. 37	$y^5F_3^o-f^5F_2$
9, 511. 55	10	10, 510. 65	$y^5F_1^o-f^5F_1$
9, 510. 81	12	10, 511. 47	$y^5F_3^o-f^5F_3$
9, 508. 49	20	10, 514. 03	$y^5F_4^o-f^5F_4$
9, 506. 04	25	10, 516. 76	$y^5F_4^o-f^5F_5$
9, 473. 51	2	10, 552. 86	$y^5F_1^o-f^5F_2$
9, 453. 22	3	10, 575. 51	$y^5F_2^o-f^5F_3$
9, 431. 77	3	10, 599. 56	$y^5F_3^o-f^5F_4$
9, 409. 69	2	10, 624. 44	$y^5F_4^o-f^5F_5$
9, 312. 48	4	10, 735. 34	$x^3F_2^o-e^3F_2$
9, 305. 04	1	10, 743. 92	$y^3F_4^o-e^3G_4$
9, 285. 04	5	10, 767. 06	$x^3F_3^o-e^3F_3$
9, 257. 62	7	10, 798. 95	$x^3F_4^o-e^3F_4$
9, 246. 14	10	10, 812. 36	$y^3F_5^o-e^3G_5$

TABLE 1.—*Wave lengths in the infra-red spectrum of titanium*—Continued

λ_{air} I. A.	Intensity	ν_{vac} cm^{-1}	Term combination
9,170.38	2	10,901.68	$v^3D_{\frac{1}{2}}-g^3F_4$
9,167.53	8	10,905.07	$y^3F_{\frac{1}{2}}-e^3G_4$
9,135.92	2	10,942.80	$c^3P_1-w^3D_{\frac{2}{1}}$
9,123.14	5	10,958.13	$y^3F_{\frac{1}{2}}-e^3G_3$
9,090.70	25	10,997.24	$a^5P_2-z^5S_2$
9,087.67	2	11,000.90	$c^3P_1-w^3D_{\frac{2}{1}}$
9,086.94	2	11,001.79	$a^3P_1-y^3F_2$
9,027.32	15	11,074.45	$a^5P_2-z^5S_2$
9,023.65	2	11,078.95	$a^3P_1-y^3F_2$
8,989.44	12	11,121.11	$a^5P_1-z^5S_2$
8,985.80	2	11,125.62	$a^3P_1-y^3F_2$
8,863.09	3	11,279.64	$z^3G_{\frac{1}{2}}-f^3G_3$
8,821.14	12	11,333.29	$a^5P_2-y^3D_3$
8,819.39	8	11,335.54	$a^3P_2-z^3D_1$
8,794.40	8	11,367.75	$z^1H_{\frac{1}{2}}-e^1G_4$
8,778.66	30	11,388.13	$a^5P_2-z^3P_2$
8,766.64	75	11,403.75	$a^3P_2-z^3D_{\frac{3}{2}}$
8,761.44	15	11,410.52	$a^5P_2-y^3D_3$
8,737.31	7	11,442.03	$z^3G_{\frac{1}{2}}-f^3G_4$
8,734.70	75	11,445.45	$a^3P_1-z^3D_1$
8,725.76	6	11,457.18	$a^5P_1-y^3D_2$
8,719.56	30	11,465.33	$a^5P_2-z^3P_2$
8,692.34	100	11,501.23	$a^3P_0-z^3D_1$
8,682.99	125	11,513.61	$a^3P_1-z^3D_{\frac{2}{3}}$
8,675.38	150	11,523.71	$a^3P_2-z^3D_{\frac{3}{2}}$
8,641.47	40	11,568.93	
8,636.38	18	11,575.75	
8,629.33	18	11,585.20	
8,618.44	20	11,599.85	$b^3P_1-w^3D_1$
8,618.14	15	11,600.25	
8,612.91	7	11,607.29	$a^3P_2-y^3D_1$
8,600.98	25	11,623.39	$\left\{\begin{array}{l} b^3P_2-w^3D_{\frac{2}{3}} \\ a^3P_1-y^3D_0 \end{array}\right.$
8,598.18	60	11,627.18	$b^1G_4-z^3G_{\frac{3}{2}}$
8,578.40	15	11,653.99	$a^3P_1-y^3D_1$
8,569.72	50	11,665.79	$b^3P_0-w^3D_1$
8,565.45	25	11,671.61	$a^3P_2-y^3D_{\frac{3}{2}}$
8,550.54	25	11,691.96	$a^5P_2-y^5D_{\frac{3}{2}}$
8,548.07	100	11,695.34	$a^3G_3-x^3F_2$
8,539.36	60	11,707.26	$b^3P_1-w^3D_{\frac{2}{3}}$
8,531.36	15	11,718.25	$a^3P_1-y^3D_{\frac{2}{3}}$
8,526.36	8	11,725.11	
8,525.99	8	11,725.63	
8,518.37	100	11,736.11	$a^3G_4-z^3F_{\frac{3}{2}}$
8,518.05	60	11,736.55	$z^5F_{\frac{1}{2}}-a^5D_4$
8,496.03	60	11,766.97	$\left\{\begin{array}{l} b^3P_2-w^3D_{\frac{2}{3}} \\ a^1F_3-v^1F_3 \end{array}\right.$
8,495.51	15	11,767.69	
8,494.42	30	11,769.20	$a^3P_2-y^3D_{\frac{3}{2}}$
8,483.16	25	11,784.83	$a^3G_3-x^3F_{\frac{3}{2}}$
8,468.46	100	11,805.28	$a^3G_5-z^3F_{\frac{3}{2}}$
8,467.15	75	11,807.11	$z^5F_{\frac{1}{2}}-a^5D_3$

TABLE 1.—*Wave lengths in the infra-red spectrum of titanium*—Continued

λ_{air} I. A.	Intensity	ν_{vac} cm^{-1}	Term combination
8, 460. 96	7	11, 815. 74	
8, 457. 10	40	11, 821. 14	$a^3P_2-y^3D_2^\circ$
8, 450. 89	75	11, 829. 74	$a^3H_5-z^3G_5^\circ$
8, 442. 98	20	11, 840. 91	$b^3P_2-z^3D_3^\circ$
8, 438. 93	75	11, 846. 60	$a^3H_4-z^3G_4^\circ$
8, 435. 68	300	11, 851. 15	$a^5F_5-z^5D_4^\circ$
8, 434. 98	300	11, 852. 14	$a^5F_4-z^5D_3^\circ$
8, 426. 50	200	11, 864. 07	$a^5F_3-z^5D_2^\circ$
8, 424. 41	50	11, 867. 01	$z^5F_2^\circ-a^5D_3$
8, 423. 10	20	11, 868. 86	$a^1G_4-z^3F_3^\circ$
8, 418. 70	10	11, 875. 06	
8, 417. 54	25	11, 876. 70	$z^5F_1^\circ-a^5D_4$
8, 416. 97	60	11, 877. 50	$a^3H_4-z^3G_3^\circ$
8, 412. 36	150	11, 884. 01	$a^5F_2-z^5D_1^\circ$
8, 407. 87	4	11, 890. 37	
8, 402. 54	5	11, 897. 90	$a^3H_5-z^3G_5^\circ$
8, 396. 93	90	11, 905. 84	$a^5F_1-z^5D_2^\circ$
8, 389. 48	25	11, 916. 42	$z^5F_3^\circ-a^5D_1$
8, 382. 82	90	11, 925. 89	$a^5F_1-z^5D_1^\circ$
8, 382. 54	100	11, 926. 29	$a^5F_3-z^5D_3^\circ$
8, 377. 90	100	11, 932. 90	$a^5F_2-z^5D_2^\circ$

Table 2.—*The term a^5D*

J	ν	$\Delta\nu$
4	28, 952. 10	69. 66
3	28, 882. 44	53. 93
2	28, 828. 51	36. 89
1	28, 791. 62	18. 76
0	28, 772. 86	

2. IRON

The arc spectrum of iron is of exceptional interest because it is the main source of internationally adopted secondary standards of wave length. Furthermore, it is the spectrum in which the largest number of lines have been classified on the basis of modern theories of spectral structure, more than 2,400 lines now being accounted for as combinations of established atomic energy levels characteristic of the neutral atoms of iron.

Extensive analyses of the arc sprectrum of iron have been published by Burns and Walters [8] and by Catalán,[9] the former being based on new measurements of wave lengths in the vacuum arc, while the latter uses all available data for the arc in air and presents the most complete list of energy levels. Since our observations were also made at atmospheric pressure, we are quoting term combinations from the latter paper in Table 3 below. The results for the infra-red arc spectrum of iron appearing in Table 3 include measured wave lengths, estimated relative intensities, computed wave numbers in

[8] K. Burns and F. M. Walters, Jr., Pub. Allegheny Obs., vol. 6, No. 11, 1930.
[9] M. A. Catalán, Anales Soc. Esp. Fis. y Quim., vol. 28, p. 1239, 1930.

vacuum, and spectral term combinations. In the range covered by these photographic observations lines were observed radiometrically by Randall and Barker.[10] Without disparaging the excellent quality of the radiometric results as such, a comparison, nevertheless, shows the enormous advantage of the photographic method for recording details and improving precision of the measurements.

TABLE 3.—*Wave lengths in the infra-red spectrum of iron*

λ_{air} I. A.	Intensity	ν_{vac} cm^{-1}	Term combination
10,863.4	1	9,202.68	$b^3D_2'-a^3F_4$
10,817.9	1	9,241.38	
10,783.0	1	9,271.36	
10,532.12	10	9,492.17	
10,469.55	20	9,548.89	
10,452.60	5	9,564.38	
10,423.55	3	9,591.04	
10,395.75	8	9,616.68	$a^5P_3-a^3F_4'$
10,353.7	2h	9,655.77	$d^3D_1'-\beta^3D_2$
10,348.16	4h	9,660.91	
10,345.2	1.	9,663.70	
10,340.77	4	9,667.81	$a^5P_3-a^3F_3'$
10,218.36	3	9,783.63	$c^3F_4'-20C_5$
10,216.42	100	9,785.48	$b^3D_2'-a^3F_4$
10,195.11	2	9,805.94	$a^3G_5-a^3F_2'$
10,187.4	1	9,813.35	
10,167.4	1	9,832.64	$a^5P_3-a^5F_2'$
10,145.64	80	9,853.75	$b^3D_2'-a^3F_3$
10,142.82	2	9,856.49	$c^3F_3'-29C_3$
10,115.18	1	9,883.42	
10,113.86	2	9,884.71	
10,071.88	2	9,925.91	
10,065.09	60	9,932.61	$b^3D_1'-a^3F_2$
10,057.64	3	9,939.97	$c^3F_4'-24C_3$
9,996.85	1h	10,000.41	
9,980.55	2h	10,016.74	
9,977.52	1	10,019.78	$c^3F_3'-35C_2$
9,944.13	3h	10,053.43	
9,917.93	1p?	10,079.99	
9,897.69	1p?	10,100.59	
9,889.11	40	10,109.36	$c^3F_4'-30C_5$
9,881.4	1	10,117.2	
9,868.09	3	10,130.90	$c^3F_4'-46C_3$
9,861.83	30	10,137.33	$c^3F_3'-39C_3$
9,839.38	1	10,160.46	
9,834.04	3h	10,165.97	$c^3F_4'-21C_{5,4}$
9,811.36	2	10,189.47	
9,800.42	20	10,200.85	$c^3F_2'-49C_3$
9,794.91	1	10,206.59	
9,786.62	2	10,215.24	$b^3F_3'-a^5F_4$
9,783.96	3	10,218.01	$c^3F_3'-22C_5$
9,763.91	15	10,238.99	$c^3F_4'-33C_4$
9,763.34	15	10,239.59	$c^3F_1'-56C_2$
9,753.15	10	10,250.29	$b^3D_2'-a^3F_2$
9,747.24	2	10,256.50	

[10] H. M. Randall and E. F. Barker, Astrophys. J., vol. 49, p. 42, 1919.

TABLE 3.—*Wave lengths in the infra-red spectrum of iron*—Continued

λ_{air} I. A.	Intensity	ν_{vac} cm^{-1}	Term combination
9, 738. 73	200	10, 265. 47	$c^5F_5'-23C_5$
9, 699. 70	6h	10, 306. 77	
9, 693. 69	1	10, 313. 16	$a^5S_2'-48C_2$
9, 683. 57	1	10, 323. 94	$a^5S_2'-49C_2$
9, 676. 42	1	10, 331. 57	
9, 673. 16	1h	10, 335. 05	
9, 666. 59	2	10, 342. 07	
9, 658. 94	3	10, 350. 26	$c^5F_3'-47C_{1,2}$
9, 657. 30	4	10, 352. 03	$c^5F_2'-56C_1$
9, 653. 18	20	10, 356. 44	$b^3D_2'-a^5F_2$
9, 637. 55	2	10, 373. 25	
9, 634. 22	5	10, 376. 82	$c^5F_3'-49C_3$
9, 626. 60	30h	10, 385. 00	$c^5F_4'-39C_3$
9, 602. 07	2	10, 411. 57	
9, 569. 95	40h	10, 446. 51	$c^5F_3'-30C_3$
9, 556. 56	1	10, 461. 14	
9, 550. 90	2	10, 467. 35	$c^5D_2'-29C_3$
9, 529. 31	4h	10, 491. 06	$b^3G_5'-\beta^5F_5$
9, 527. 7	1	10, 492. 8	$c^5F_3'-54C_4$
9, 513. 21	10h	10, 508. 82	{ $c^5F_4'-45C_5$ $b^3G_5'-\beta^5F_5$
9, 462. 97	2	10, 564. 61	$c^5D_3'-24C_3$
9, 454. 24	4h	10, 574. 37	$c^5F_1'-61C_2$
9, 452. 45	2	10, 576. 37	$c^5F_4'-33C_4$
9, 443. 98	10h	10, 585. 85	$c^5F_2'-60C_2$
9, 437. 91	2	10, 592. 66	$b^3F_2'-a^5F_2$
9, 430. 07	4	10, 601. 47	
9, 414. 14	20h	10, 619. 41	$c^5F_3'-58C_2$
9, 410. 1	1h	10, 624. 0	
9, 401. 09	10h	10, 634. 15	$c^5F_4'-50C_4$
9, 394. 71	3h	10, 641. 37	
9, 388. 28	3h	10, 648. 66	$c^5D_2'-36C_1$
9, 382. 83	3h	10, 654. 84	
9, 372. 84	6	10, 666. 20	{ $b^3F_2-a^3F_1'$ $c^3D_1'-43C_2$
9, 362. 36	4	10, 678. 13	$a^1P_1-a^1P_2'$
9, 359. 37	3	10, 681. 55	$b^3F_4-a^3D_3'$
9, 350. 52	10	10, 691. 66	$b^3F_4'-a^5F_4$
9, 343. 40	3	10, 699. 80	$c^5F_4'-a^5D_3$
9, 333. 94	2	10, 710. 65	$c^5F_4'-38C_4$
9, 318. 13	3	10, 728. 82	$c^5D_2'-29C_2$
9, 307. 94	2	10, 740. 57	$c^5F_4'-54C_4$
9, 294. 66	2	10, 755. 02	$c^5F_4'-\beta^6D_4$
9, 259. 05	15	10, 797. 29	$c^5D_4'-21C_{3,4}$
9, 258. 40	20	10, 798. 04	$b^3F_3'-a^5F_4$
9, 246. 54	2	10, 811. 89	$b^3F_2-a^3D_3'$
9, 242. 32	2	10, 816. 82	$b^3F_1'-a^5D_1$
9, 233. 2	1	10, 827. 5	$b^3G_3'-a^3G_3$
9, 217. 54	5h	10, 845. 91	$c^5F_3'-45C_5$
9, 214. 45	6	10, 849. 54	$c^5D_4'-22C_5$
9, 210. 02	6	10, 854. 76	$b^3P_1-b^5D_2'$
9, 199. 52	2h	10, 867. 15	

TABLE 3.—*Wave lengths in the infra-red spectrum of iron*—Continued

λ_{air} I. A.	Intensity	ν_{vac} cm^{-1}	Term combination
9, 178. 57	1h	10, 891. 96	$c^3D_2'-35C_3$
9, 173. 46	4h	10, 898. 02	$b^4F_5-a^3D_1'$
9, 166. 44	3h	10, 908. 74	$c^3D_1'-24C_2$
9, 156. 9	2	10, 917. 7	$c^3D_2'-46C_2$
9, 155. 9	1	10, 918. 9	
9, 147. 91	5h	10, 928. 46	$c^3F_3'-\beta^5D_2$
9, 146. 11	3	10, 930. 61	$b^3F_5-a^3F_3'$
9, 121. 1	1	10, 960. 6	
9, 118. 87	20	10, 963. 26	$b^3P_2-b^3D_2'$
9, 117. 10	2	10, 965. 39	$c^3D_1'-a^3F_2$
9, 116. 14	2	10, 966. 55	
9, 103. 64	1	10, 981. 60	$b^3F_4'-a^3D_4$
9, 100. 50	5h	10, 985. 40	$c^3D_4'-26C_3$
9, 089. 40	30	10, 998. 81	$b^3G_5-a^5G_6'$
9, 088. 22	40	11, 000. 24	$b^3P_1-a^3P_2'$
9, 084. 2	1	11, 005. 1	$b^3F_3'-a^3D_2$
9, 080. 48	3h	11, 009. 62	{ $c^3D_2'-39C_3$ $c^3F_1'-60C_2$
9, 079. 65	8	11, 010. 62	$b^3F_1'-a^3F_2$
9, 070. 42	2	11, 021. 82	{ $b^3F_1'-a^3D_2$ $c^3F_1'-a^3D_1$
9, 062. 24	2	11, 031. 77	$c^3F_3'-\beta^5D_3$
9, 052. 6	1	11, 043. 5	$b^3G_5'-a^3G_4$
9, 044. 63	1	11, 053. 25	
9, 036. 9	1	11, 062. 7	$c^3D_2'-a^3D_2$
9, 030. 67	1	11, 070. 34	$b^3P_1-b^3D_1'$
9, 024. 47	15	11, 077. 94	$c^3D_4'-30C_3$
9, 019. 84	2	11, 083. 63	$c^3F_1'-\beta^5D_1$
9, 013. 90	1	11, 090. 93	$a^3P_2-a^3P_2'$
9, 012. 10	30	11, 093. 15	$c^3F_1'-\beta^5D_4$
9, 010. 55	2	11, 095. 06	$b^3F_2-a^3F_3'$
9, 008. 37	2	11, 097. 84	$c^3D_3'-52C_{2,3}$
9, 006. 72	1	11, 099. 78	$c^3D_3'-53C_4$
8, 999. 54	100	11, 108. 63	$b^3P_2-a^3P_2'$
8, 995. 20	1p?	11, 113. 99	
8, 984. 87	3	11, 126. 77	$c^3F_1'-\beta^5D_0$
8, 975. 36	15	11, 138. 56	{ $b^3G_5-a^5G_4'$ $c^3D_2'-a^5F_2$
8, 946. 25	1	11, 174. 80	$b^3P_1-b^5D_0'$
8, 945. 15	20	11, 176. 17	$c^3F_4-\beta^5D_3$
8, 943. 00	3	11, 178. 86	{ $b^3P_2-b^5D_1'$ $c^3D_1'-46C_3$
8, 929. 04	5	11, 196. 34	$c^3F_3'-\beta^5D_1$
8, 919. 95	10	11, 207. 75	{ $c^3D_1-33C_4$ $c^3F_1'-\beta^3D_2$
8, 916. 26	1	11, 212. 39	$a^3F_2-a^7P_2'$
8, 876. 13	2	11, 263. 08	
8, 868. 42	3	11, 272. 97	$b^3G_5-a^5G_3'$
8, 866. 92	150	11, 274. 78	$b^3F_1'-a^3F_4$
8, 863. 64	1p?	11, 278. 95	

TABLE 3.—*Wave lengths in the infra-red spectrum of iron*—Continued

λ_{air} I. A.	Intensity	ν_{vac} cm^{-1}	Term combination
8, 840. 82	5	11, 300. 40	$c^3D_1'-61C_2$
8, 838. 36	30	11, 311. 21	$b^3P_0-a^3P_1$
8, 824. 18	200	11, 329. 39	$a^3P_2-a^3P_2'$
8, 814. 5	2	11, 341. 8	
8, 812. 0	1p?	11, 345. 0	
8, 809. 1	2	11, 348. 8	
8, 808. 3	4h	11, 349. 8	
8, 804. 56	10	11, 354. 64	$a^3P_2-a^3P_1'$
8, 801. 98	1p?	11, 357. 96	
8, 796. 42	2	11, 365. 14	$c^3D_2'-54C_4$
8, 793. 38	120	11, 369. 07	$b^3F_2'-a^3F_3$
8, 790. 62	10h	11, 372. 64	$c^4D_2'-60C_2$
8, 784. 44	5	11, 380. 64	$c^3D_2'-\beta^3D_4$
8, 764. 02	100	11, 407. 16	$b^3F_2'-a^3F_2$
8, 757. 16	50	11, 416. 09	$b^3P_1-a^3P_1'$
8, 747. 32	2	11, 428. 94	$b^3G_4-a^3G_4'$
8, 729. 1	2	11, 452. 8	
8, 713. 19	10	11, 473. 70	$b^3G_5-a^3G_5'$
8, 710. 29	20h	11, 477. 53	$c^3D_4-45C_3$

3. COBALT

About 1,200 lines of the cobalt arc have been classified by Catalán and Bechert,[11] but only 10 of these have wave lengths exceeding 8,648 A. Our results for cobalt are shown in Table 4, the construction of which is identical with that of the preceding wave-length tables. The line classifications have all been derived from Catalán's term tables without any effort to extend these. A considerable number of lines, including some very strong ones, remain unclassified, indicating that the term analysis for cobalt is less complete than for iron. The arc spectrum of cobalt has also been explored radiometrically in the infra-red by Randall and Barker.[12] In the interval 9,000 to 11,000 A the wave lengths and relative intensities were measured for 10 lines, all but two of which are identifiable with the strongest ones in our list. Here again the superior advantages of photographic over bolometric methods of studying complex spectra are strikingly exemplified.

[11] M. A. Catalán and K. Bechert, Zeit. f. Phys., vol. 32, p. 336, 1925. M. A. Catalán, Zeit. f. Phys., vol. 47, p. 89, 1927.
[12] H. M. Randall and E. F. Barker, Astrophys. J., vol. 49, p. 54, 1919.

TABLE 4.—*Wave lengths in the infra-red spectrum of cobalt*

λ_{air} I. A.	Intensity	ν_{vac} cm^{-1}	Term combination
11, 091. 94	2	9, 013. 09	$b^2D_2'-a^4F_4$
10, 805. 97	1	9, 251. 61	$c^4F_4'-[^4F_5]$
10, 681. 80	4	9, 359. 15	
10, 660. 15	5	9, 378. 16	
10, 584. 99	2	9, 444. 75	
10, 521. 36	2	9, 501. 87	$b^2D_2-b^4D_2'$
10, 471. 96	10	9, 546. 69	$b^2F_2-a^4F_3$
10, 447. 40	3	9, 569. 14	$b_3-[^4F_5]$
10, 442. 10	15	9, 574. 00	
10, 398. 37	4	9, 614. 26	$c^4G_6'-[H_7]$
10, 382. 22	30	9, 629. 26	$b^2D_2-a^2F_3'$
10, 374. 65	1	9, 636. 24	$c^4F_3-[F_3]$
10, 369. 12	2	9, 641. 38	$c^4G_6'-[^4F_5]$?
10, 367. 95	2	9, 642. 47	$c^4F_4'-[D_3]$
10, 364. 43	1	9, 645. 74	$c^4G_6'-[H_6]$
10, 354. 45	60	9, 655. 04	$b^2F_4-a^4F_4$
10, 348. 00	1	9, 661. 1	$c^4F_2'-[F_4,D_2]$
10, 335. 40	4	9, 672. 84	
10, 332. 65	3	9, 675. 41	$c^4G_4'-[^4G_5,^4F_5]$
10, 327. 32	2	9, 680. 40	
10, 276. 80	5h	9, 727. 99	$(m^2F_3')-[F_3]$
10, 265. 4	1h	9, 738. 8	
10, 244. 64	1	9, 758. 52	
10, 239. 1	1h	9, 763. 8	$^2G_4-^4G_4'$
10, 222. 12	1	9, 780. 03	$a^2D_3-^4D_1'$
10, 195. 3	1	9, 805. 75	
10, 172. 83	20h	9, 827. 41	$c^4F_3'-[^4D_4]$
10, 167. 58	200	9, 832. 49	$b^2D_3'-a^2F_2$
10, 154. 90	2	9, 844. 77	
10, 152. 93	4	9, 846. 68	
10, 131. 38	2	9, 867. 63	$b^4P_1-^4D_2'$
10, 128. 05	150	9, 870. 87	$b^2D_2'-a^2F_3$
10, 111. 1	1h	9, 887. 4	
10, 105. 40	2	9, 892. 09	
10, 092. 0	2h	9, 906. 1	$c^4G_4'-[F_4]$
10, 078. 62	100	9, 919. 34	
10, 052. 96	8	9, 944. 60	$c^4F_5'-[^4F_5]$
10, 048. 80	3	9, 948. 71	$c^4F_3'-[H_5]$
10, 046. 31	150	9, 951. 18	$b^2D_3-a^2F_4'$
10, 031. 44	5	9, 965. 93	$b^4P_3-^4D_3'$
10, 021. 48	4h	9, 975. 83	
10, 019. 05	30h	9, 978. 25	$c^4F_5'-[^4G_6,^4F_5]$
10, 007. 78	3	9, 989. 49	
9, 999. 7	2h	9, 997. 6	$b^2D_2-b^4D_2'$?
9, 952. 2	3h	10, 045. 3	$b^2F_3'-a^4F_2$
9, 940. 69	2	10, 056. 91	$(a^2P_2')-[F_3,D_2]$
9, 918. 1	1	10, 079. 8	$b^4P_3-^4D_4'$
9, 912. 73	10	10, 085. 27	
9, 909. 52	1h	10, 088. 54	
9, 890. 92	30	10, 107. 51	$b^2D_3-b^4D_4'$

Table 4.—*Wave lengths in the infra-red spectrum of cobalt*—Continued

λ_{air} I. A.	Intensity	ν_{vac} cm⁻¹	Term combination
9, 859. 90	1	10, 139. 31	
9, 852. 5	1	10, 146. 9	
9, 847. 7	2	10, 151. 9	$c^4F_4'-[F_3]$
9, 823. 52	4h	10, 176. 87	$c^4F_4'-[F_4,D_2]$
9, 798. 37	2h	10, 202. 99	$c^4F_4'-[F_3,D_2]$
9, 785. 39	40	10, 216. 52	
9, 769. 0	1h	10, 233. 7	
9, 764. 53	5h	10, 238. 34	
9, 746. 02	100	10, 257. 79	
9, 738. 46	2	10, 265. 78	
9, 735. 53	2	10, 268. 84	
9, 729. 54	3	10, 275. 16	
9, 696. 60	5	10, 310. 06	$b^2D_1-a^3D_3'$
9, 694. 0	2	10, 312. 8	
9, 678. 21	10h	10, 329. 66	
9, 670. 20	2	10, 338. 21	
9, 659. 94	3	10, 349. 20	
9, 638. 21	1	10, 372. 53	
9, 629. 83	3h	10, 381. 56	
9, 626. 72	1	10, 384. 91	
9, 618. 32	2	10, 393. 98	
9, 613. 46	4	10, 399. 23	
9, 606. 52	5	10, 406. 75	
9, 597. 90	200	10, 416. 09	$b^2F_3'-a^2F_3$
9, 592. 3	2h	10, 422. 2	
9, 585. 28	2	10, 429. 80	
9, 580. 63	3	10, 434. 87	
9, 569. 00	5h	10, 447. 55	$c^4G_5'-[F_4]$
9, 548. 66	4	10, 469. 80	
9, 544. 52	300	10, 474. 34	$b^2F_4'-a^2F_4$
9, 527. 17	10h	10, 493. 42	
9, 517. 33	1	10, 504. 27	
9, 513. 42	1	10, 508. 59	
9, 500. 01	1	10, 523. 42	
9, 482. 75	1	10, 542. 57	
9, 470. 74	2	10, 555. 94	$b^4P_5-^6D_3'$
9, 454. 23	3h	10, 574. 38	
9, 442. 34	4h	10, 587. 69	
9, 435. 70	3	10, 595. 14	$c^4G_3'-[^4F_3]$
9, 428. 8	1	10, 602. 9	
9, 425. 8	1	10, 606. 3	
9, 422. 60	3h	10, 609. 88	
9, 406. 12	4h	10, 628. 46	
9, 395. 11	2	10, 640. 92	
9, 356. 98	200	10, 684. 28	
9, 351. 06	3	10, 691. 04	
9, 347. 88	2h	10, 694. 68	
9, 344. 93	20	10, 698. 06	
9, 340. 54	3h	10, 703. 08	
9, 319. 53	2	10, 727. 21	

TABLE 4.—*Ware lengths in the infra-red spectrum of cobalt*—Continued

λ_air I. A.	Intensity	ν_vac cm⁻¹	Term combination
9, 280. 42	5	10, 772. 42	$b^2F_4'-a^4F_5$
9, 269. 90	1	10, 784. 64	
9, 262. 5	2	10, 793. 3	$b^2D_2-b^4F_3'$
9, 258. 18	4	10, 798. 30	$c^4F_4'-[F_4]$
9, 245. 60	1	10, 812. 99	$b^2D_3-a^2G_4'$
9, 233. 64	1	10, 827. 00	$a^2P_1-a^4F_3'$
9, 217. 80	2h	10, 845. 60	
9, 207. 96	1	10, 857. 19	
9, 204. 11	5	10, 861. 73	$b^2D_3-a^2F_3'$
9, 197. 29	1	10, 869. 79	$c^4F_4'-[F_4D_3]$
9, 185. 95	2	10, 883. 21	
9, 181. 75	5	10, 888. 18	
9, 177. 93	20	10, 892. 72	
9, 165. 52	2	10, 907. 47	
9, 150. 65	2h	10, 925. 19	
9, 133. 24	6	10, 946. 02	$c^4F_4'-[^4F_4]$
9, 130. 50	2	10, 949. 30	
9, 111. 64	1	10, 971. 96	
9, 095. 37	50	10, 991. 58	
9, 071. 35	4h	11, 020. 69	
9, 057. 29	2h	11, 037. 80	
9, 052. 44	2	11, 043. 72	$b^2D_2-b^4F_3'$
9. 040. 0	1	11, 058. 9	
9, 037. 87	50	11, 061. 52	
8, 997. 3	1	11, 111. 4	
8, 986. 51	3	11, 124. 74	
8, 972. 89	7	11, 141. 62	
8, 958. 37	15	11, 159. 68	$b^4F_3'-a^4F_4$
8, 953. 72	2	11, 165. 48	
8, 939. 14	5	11, 183. 69	
8, 926. 21	50	11, 199. 89	$b^2D_1-a^2D_2'$
8, 904. 63	30	11, 227. 03	
8, 892. 49	1	11, 242. 36	$c^4D_2'-[F_3]$
8, 888. 70	8h	11, 247. 15	
8, 886. 28	2	11, 250. 22	
8, 878. 28	3	11, 260. 35	
8, 872. 59	1	11, 267. 58	$c^4D_2'-[F_4D_3]$
8, 870. 70	8	11, 269. 98	
8, 856. 56	3	11, 287. 97	
8, 850. 70	30	11, 295. 44	$b^2F_4'-a^2F_3$
8, 837. 90	4h	11, 311. 80	
8, 835. 21	20	11, 315. 25	$b^4F_4'-a^4F_6$
8, 819. 10	100	11, 335. 92	$c^4G_6'-x_6$
8, 779. 20	3	11, 387. 44	
8, 774. 71	2	11, 393. 26	$c^4D_4'-[^4D_4,^4P_3]$
8, 772. 04	2	11, 396. 73	
8, 766. 55	4h	11, 403. 87	$c^4D_4'-[^4D_4]$
8, 759. 58	3	11, 412. 94	
8, 750. 13	60	11, 425. 27	$c^4F_4'-[F_3D_2]$

TABLE 4.—*Wave lengths in the infra-red spectrum of cobalt*—Continued

λ$_{air}$ I. A.	Intensity	ν$_{vac}$ cm^{-1}	Term combination
8, 745. 56	8	11, 431. 24	$b^4G_2'-a^4F_4$
8, 744. 37	10	11, 432. 79	
8, 733. 27	40	11, 447. 32	
8, 728. 5	2	11, 453. 6	
8, 722. 12	2	11, 461. 96	
8, 678. 65	3	11, 519. 37	
8, 675. 02	20	11, 524. 19	$a^2D_2'-a^4F_2$
8, 673. 02	2	11, 526. 85	
8, 661. 06	80	11, 542. 76	$b^2D_2-a^2D_3'$
8, 658. 14	3	11, 546. 66	$b^2D_2-b^4F_4'$
8, 655. 73	3h	11, 549. 85	
8, 648. 79	4	11, 559. 14	$a^2P_3-a^4D_2'$

4. NICKEL

All of the known and classified lines (1,071) of Ni I were published by Russell in 1929,[11] but only a dozen lines had been observed photographically beyond 8,700 A. The new results obtained with xenocyanine plates appear in Table 5, in which the combinations have all been derived from Russell's term tables. The infra-red, as well as other portions of the Ni I spectrum, appears to be less complex than Co I, and relatively simple compared with Fe I. Almost all of the new lines have been accounted for, and the analysis of the Ni I spectrum may be regarded as very nearly completed.

The infra-red arc spectrum of nickel was investigated radiometrically by Randall and Barker,[12] and in the range 9,520 to 10,980 A six lines coincide, within the errors of measurement, with the strongest we have recorded photographically. Comparison of our intensity estimates with their galvanometer deflections indicates that the photographic sensitivity of xenocyanine plates for wave length 11,000 A is approximately 5 per cent that for wave lengths below 10,300 A.

[11] H. N. Russell, Phys. Rev., vol. 34. p. 821, 1929.
[12] H. M. Randall and E. F. Barker, Astrophys. J., vol. 49, p. 55, 1919.

TABLE 5.—*Wave lengths in the infra-red spectrum of nickel*

λ_{air} I. A.	Intensity	ν_{vac} cm^{-1}	Term combination
10, 979. 87	5	9, 105. 09	$y^3D_3-e^3D_3$
10, 891. 25	2	9, 179. 17	$y^3F_3-e^3D_2$
10, 762. 24	2	9, 289. 20	$y^3D_3-e^3D_1$
10, 530. 53	20	9, 493. 60	$y^3F_3-e^3D_2$
10, 378. 62	100	9, 632. 56	$y^3F_3-e^3D_3$
10, 330. 23	50	9, 677. 68	$y^3F_3-e^3D_3$
10, 321. 10	5	9, 686. 23	$w^3F_3-g^3F_3$
10, 302. 61	50	9, 703. 62	$y^3D_3-e^3D_1$
10, 295. 05	5	9, 710. 75	
10, 226. 15	4	9, 776. 18	$w^3D_3-g^3F_3$
10, 193. 25	100	9, 807. 73	$z^1P_1-e^3D_2$
10, 145. 37	20	9, 854. 01	$y^3D_1-e^3D_2$
10, 061. 29	10	9, 936. 36	$y^3G_3-g^3F_2$
10, 048. 60	10	9, 948. 91	$y^3D_1-e^3D_1$
9, 898. 90	40	10, 099. 36	$y^3D_1-e^1D_2$
9, 842. 04	2	10, 157. 71	
9, 710. 1	1	10, 295. 74	$y^3D_1-f^1F_3$
9, 689. 35	3	10, 317. 78	$z^1D_3-g^3F_3$
9, 520. 06	100	10, 501. 26	$y^3F_3-e^3D_1$
9, 447. 29	5	10, 582. 15	$z^3F_3-g^3F_4$
9, 396. 57	20	10, 639. 26	
9, 385. 62	10	10, 651. 68	$y^3F_3-e^1D_2$
9, 258. 47	5	10, 797. 96	
9, 196. 18	2	10, 871. 10	$a^1G_4-y^3F_3$
9, 106. 40	30	10, 978. 28	$w^3F_4-g^3F_4$
9, 085. 25	3	11, 003. 84	$z^3G_3-e^3D_2$
9, 078. 70	10	11, 011. 77	$w^3F_3-f^1F_3$
9, 058. 56	6	11, 036. 25	
9, 005. 14	5	11, 101. 72	$w^3D_3-f^1F_3$
8, 982. 35	2	11, 129. 89	$z^1P_1-e^3D_1$
8, 968. 20	30	11, 147. 45	$y^3G_3-g^3F_4$
8, 965. 94	50	11, 150. 26	$y^3D_1-e^1D_2$
8, 954. 65	2	11, 164. 32	$z^1D_2-e^3D_2$
8, 877. 07	10	11, 261. 89	$y^3G_3-f^1F_3$
8, 862. 59	100	11, 280. 29	$z^1P_1-e^3D_2$
8, 824. 2	1	11, 329. 4	
8, 809. 47	30	11, 348. 30	$z^1D_3-e^3D_3$
8, 770. 68	10	11, 398. 50	$a^1G_4-y^3D_3$
8, 716. 58	3	11, 469. 24	$w^3D_3-g^3F_4$
8, 706. 05	1	11, 483. 12	
8, 702. 49	6	11, 487. 81	$a^1G_4-z^1G_4$
8, 688. 64	3	11, 506. 12	
8, 680. 2	1	11, 517. 3	
8, 661. 85	2	11, 541. 71	
8, 637. 04	15	11, 574. 86	$z^1F_3-e^3D_2$
8, 606. 45	10	11, 616. 00	$z^3D_3-g^3F_4$
8, 598. 84	2	11, 626. 29	$z^3G_3-e^3D_2$
8, 586. 2	1	11, 643. 40	$z^1D_2-f^1F_3$
8, 580. 04	2	11, 651. 76	$y^3G_3-h^3F_4$

5. ZIRCONIUM

About a year ago Kiess and Kiess [14] published an analysis of Zr I based on a list of approximately 1,600 wave lengths measured at the Bureau of Standards. To this list, which terminated at 9,276 A, we are now able to add the lines given in Table 6. All of the strong lines and many of the fainter ones are accounted for as combinations between the terms given in the paper cited above: No new terms have been revealed by these supplementary observations, although several metastable terms arising from the configuration $4d^4$ are still to be found.

A conspicuous feature of the infra-red Zr spectrum is a group of bands, shaded toward longer wave lengths, lying between 9,300 and 9,500 A. No regularities have been detected among these bands that will relate them to the systems recently described by Miss Lowater.[15] The wave lengths, estimated intensities, and wave numbers of the heads of these bands are given in Table 7.

TABLE 6.—*Wave lengths in the infra-red spectrum of zirconium*

λ_{air} I. A.	Intensity	ν_{vac} cm^{-1}	Term combination
10, 738. 94	1	9, 309. 36	$a^3D_1-y^3F_2^a$
10, 696. 67	1	9, 346. 14	$a^3G_3-z^3G_3^a$
10, 654. 11	1	9, 383. 48	$a^3G_4-z^3G_3^a$
10, 551. 46	1	9, 474. 77	$a^3D_1-y^3F_2^a$
10, 515. 86	1	9, 506. 84	$b^3F_4-z^3G_3^a$
10, 433. 73	1	9, 581. 67	
10, 328. 41	1	9, 679. 38	$a^3D_1-y^3D_2^a$
10, 242. 85	2	9, 760. 24	$a^5F_3-z^5G_3^a$
10, 210. 44	10	9, 791. 21	$a^3G_5-z^3G_4^a$
10, 199. 42	4	9, 801. 80	$b^3F_4-z^3G_3^a$
10, 105. 30	1	9, 893. 08	$b^3F_2-z^3G_3^a$
10, 084. 70	12	9, 913. 30	$a^5F_1-z^5G_2^a$
10, 045. 15	3	9, 952. 33	$a^5F_3-z^5G_3^a$
10, 028. 68	2	9, 968. 67	$c^3P_2-z^3D_3^a$
10, 016. 97	2	9, 980. 33	
9, 990. 44	1	10, 006. 83	
9, 958. 58	2	10, 038. 84	
9, 928. 53	1	10, 069. 22	
9, 909. 76	3	10, 088. 30	
9, 822. 30	20	10, 178. 12	$a^5F_1-z^5G_3^a$
9, 820. 42	3	10, 180. 07	$a^5F_4-z^5G_4^a$
9, 812. 85	10	10, 187. 93	$b^3F_3-z^3G_3^a$
9, 792. 74	6	10, 208. 85	$b^3F_3-z^3G_3^a$
9, 780. 40	18	10, 221. 73	$b^3F_4-z^3G_3^a$
9, 773. 30	0	10, 229. 15	
9, 666. 82	1	10, 341. 83	
9, 547. 26	25	10, 471. 34	$a^5F_3-z^5G_3^a$
9, 493. 47	1	10, 530. 67	$c^3P_2-y^3P_2^a$
9, 483. 35	3	10, 541. 91	
9, 441. 26	1	10, 588. 90	

14 B. S. Jour. Research (R P296), vol. 6, p. 621, 1931.
15 Proc. Phys. Soc. (London), vol. 44, p. 51, 1932.

TABLE 6.—*Wave lengths in the infra-red spectrum of zirconium*—Continued

λair I. A.	Intensity	νvac cm⁻¹	Term combination
9, 438. 35	1	10, 592. 17	
9, 419. 36	1	10, 613. 53	
9, 405. 00	2	10, 629. 73	$c^3P_1-y^3P_2$
9, 358. 32	2	10, 682. 74	
9, 352. 43	1	10, 689. 48	
9, 318. 19	4	10, 728. 76	
9, 290. 30	0	10, 760. 96	
9, 276. 89	25	10, 776. 52	$a^5F_4-z^5G_5$
9, 251. 17	5	10, 806. 48	$a^3G_3-y^3F_2$
9, 242. 65	5	10, 816. 44	$\begin{cases} c^3P_1-w^3F_2 \\ a^3G_3-y^3D_2 \end{cases}$
9, 229. 33	3	10, 832. 05	
9, 182. 30	0	10, 887. 53	
9, 171. 50	3	10, 900. 35	$a^3G_4-y^3D_3$
9, 140. 40	0	10, 937. 44	
9, 139. 36	10	10, 938. 69	$a^3H_4-y^3G_4$
9, 134. 23	1	10, 944. 83	
9, 099. 90	2	10, 986. 12	$\begin{cases} c^3F_2-z^3F_2 \\ a^3G_3-y^3D_3 \end{cases}$
9, 069. 41	15	11, 023. 05	$a^3H_5-y^3G_4$
9, 015. 16	20	11, 089. 39	$a^5F_5-z^5G_5$
9, 011. 34	2	11, 094. 09	$a^3G_4-y^3F_3$
8, 975. 78	0	11, 138. 05	$c^3F_4-z^3F_4$

TABLE 7.—*Infra-red zirconium bands*

λair I. A.	Intensity	νvac cm⁻¹	λair I. A.	Intensity	νvac cm⁻¹
9, 463. 55	2	10, 541. 69	9, 397. 21	2	10, 638. 54
9, 476. 65	1	10, 549. 36	9, 387. 26	2	10, 649. 82
9, 453. 55	1	10, 575. 14	9, 370. 74	3	10, 668. 59
9, 443. 93	1	10, 585. 91	9, 360. 34	3	10, 680. 45
9, 438. 24	1	10, 592. 29	9, 358. 33	1	10, 682. 74
9, 431. 75	2	10, 599. 58	9, 356. 12	3	10, 685. 26
9, 412. 03	2	10, 621. 79	9, 343. 19	4	10, 700. 05
9, 408. 92	1	10, 625. 30	9, 329. 93	5	10, 715. 26
9, 402. 84	2	10, 632. 17	9, 315. 87	5	10, 731. 43
9, 401. 09	2	10, 634. 15	9, 299. 56	3	10, 750. 25

IV. CONCLUSION

The data presented in the tables of this paper furnish ample proof of the effectiveness of xenocyanine in bringing into the domain of precise measurement a region of the spectrum heretofore accessible only with radiometric devices. Their application to the investigation of the structure of near-infra-red band spectra, a problem in which great efforts are being made to secure the advantages of increased resolving power, is obvious. In astrophysics the identification of unknown lines in the spectra of the sun and stars is advanced by each extension of our knowledge of the emission spectra of the elements. Reference to the Revision of Rowland's Preliminary Table of Solar

Spectrum Wave-Lengths [16] will show that several of the unidentified infra-red solar lines are due to the five elements whose spectra are described above. Both Ti and Fe are rich in strong lines in the region from 9,000 to nearly 11,000 A and may be used to supply comparison spectra for observations made with dispersing systems other than concave gratings. However, the values of the Fe wave lengths to be used as standards of reference may be calculated from the terms determined by interference methods more accurately than the observed values given in Table 3.

As stated above, we have completed a survey of the infra-red spectra of about 50 elements. We shall present the new wave lengths in subsequent papers in this Journal as soon as the data have been compiled for publication.

WASHINGTON, June 21, 1932.

---

[16] Carnegie Institution of Washington, Publ. No. 396, 1928.

RP474

TENSILE PROPERTIES OF CAST NICKEL-CHROMIUM-IRON ALLOYS AND OF SOME ALLOY STEELS AT ELEVATED TEMPERATURES

By William Kahlbaum[1] and Louis Jordan

ABSTRACT

The tensile properties as measured in "short-time" tests were determined for a medium-manganese steel at 900° F.; for a series of cast nickel-chromium-iron alloys containing about 0.5 per cent carbon, 35 per cent chromium and from 10 to 45 per cent nickel, at a temperature of 1,550° F.; and for three tungsten-chromium-vanadium steels and four molybdenum-chromium-vanadium steels at temperatures of 850° and 1,000° F.

CONTENTS

I. INTRODUCTION

The most recently published report by the Bureau of Standards on the so-called short-time tensile properties (that is, the strength under relatively rapid loading) of steels at elevated temperatures[2] gave data on several groups of low-alloy steels. Among these were tungsten-chromium-vanadium steels with and without the addition of silicon or aluminum. Tests were also reported on wrought chromium-nickel-iron alloys containing approximately 10 per cent chromium and from 25 to 60 per cent nickel. The present report deals with the tensile properties, as determined by short-time tests, of some additional similar steels and alloys.

II. MATERIALS

The materials tested consisted of (1) a steel in some respects similar to the 0.45 per cent carbon "boiler drum" steel included in the preceding paper,[3] but containing in the present case 1.08 per cent manganese; (2) a series of five cast nickel-chromium-iron alloys, all containing about 35 per cent chromium and from 10 to 45 per cent of nickel; (3) a series of three tungsten-chromium-vanadium steels and (4) a series of four molybdenum-chromium-vanadium steels. The chemical compositions of these materials are given in Table 1.

---

[1] Research associate, The Midvale Co., Philadelphia, Pa.
[2] William Kahlbaum, R. L. Dowdell, and W. A. Tucker, The Tensile Properties of Alloy Steels at Elevated Temperatures as Determined by the "Short-time" Method, B. S. Jour. Research, vol. 6 (RP 270), p. 199, 1930.
[3] See footnote, 2.

TABLE 1.—*Chemical composition*

Type of alloy	Designation	C	Mn	P	S	Si	Cr	Ni	V	W	Mo
		Per cent	Per cent	Per cent	Per cent	Per cent	Per cent	Per cent	Per cent	Per cent	Per cent
Carbon steel (medium manganese).	8/1819	0.40	1.08	0.037	0.034	0.23					
Nickel - chromium - iron (cast).	E E1522	.57	.59			1.04	36.2	10.2			
	E E1523	.50	.65			1.07	35.2	21.0			
	E E1524	.54	.63			1.13	56.9	30.1			
	E E1525	.47	.71			1.68	34.5	38.4			
	E E1526	.36	.53			1.15	33.5	45.2			
Tungsten - chromium - vanadium steels.	11 F-1/159	.29	.51			.58	2.28		0.36	1.70	
	11 F-1/167	.26	.86			.69	2.25		.33	2.26	
	11 F-1/168	.32	1.02			.17	1.63		.26	2.17	
Molybdenum - chromium - vanadium steels.	11 F-1/166	.20	.45			.55	1.55		.26		0.57
	11 F-1/165	.29	.88			.18	1.47		.22		.52
	11 F-1/177	.31	1.38			.49	2.31		.29		.55
	11 F-1/178	.21	2.23			1.33	1.28		.17		1.00

III. METHOD OF TESTING

The equipment and the method used for the short-time tension tests have been described and illustrated previously.[4][5] All of the tests were made with a hydraulic testing machine, and measurements of strain were made with a Tuckerman optical strain gage.[6]

As in one previous work, a thermocouple mounted in the fillet of the 0.505-inch diameter test bar was used for measuring the temperature of the specimen. The temperature gradient from the center to either end of the gage length of the test bar was approximately 13° F. (7° C.).

Proportional limits were determined by plotting the differences between observed and calculated strain and taking a strain of 1×10^{-5} as indicating departure from a straight line.[7]

IV. RESULTS AND DISCUSSION

1. MEDIUM-MANGANESE STEEL

The data from tests of the 0.4 per cent carbon medium manganese (1.08 per cent) steel at room temperature and at 900° F. are given in Table 2. The point of chief interest in these results lies in the comparison they afford with similar tests made in previous work [8] on a 0.4 per cent carbon steel (boiler-drum steel) containing only a normal proportion of manganese, namely, 0.55 per cent. These previous tests of the lower manganese steel showed that it maintained a proportional limit of approximately 15,000 lbs./in.² from 600° to 750° F., but that its proportional limit decreased noticeably, to 11,000 or 12,000 lbs./in.², at a temperature of 800° F. The higher manganese steel had a proportional limit of 15,000 to 16,000 lbs./in.² at 900° F., fully 150° F. higher than the temperature at which the lower-manganese steel had an equivalent proportional limit. The higher manganese steel did not, however, show a corresponding superiority in tensile strength over the lower manganese steel at the elevated temperature.

[4] H. J. French, Methods of Test in Relation to Flow in Steels at Various Temperatures, Proc. Am. Soc. Test. Materials, vol. 26, Pt. II, p. 7, 1926, also Eng. News Record, vol. 97, p. 22.
[5] H. J. French, H. C. Cross, and A. A. Peterson, Creep in Five Steels at Different Temperatures, B. S. Tech. Papers, No. 362, 1928.
[6] L. B. Tuckerman, Optical Strain Gages and Extensometers, Proc. Am. Soc. Test. Materials, vol. 23, Pt. II, p. 602, 1923.
[7] L. B. Tuckerman, Discussion of paper by R. L. Templin, Proc. Am. Soc. Test. Materials, vol. 29, Pt. II, p. 538, 1929.
[8] See footnote 2, p. 327.

TABLE 2.—*Tensile properties of a medium manganese steel at elevated temperatures as determined by short-time tests*

Designation [1]	Temperature of test	Proportional limit	Tensile strength	Modulus of elasticity	Elongation in 2 inches	Reduction of area	Heat treatment prior to testing
	°F.	Lbs./in.²	Lbs./in.²	Lbs./in.² ×10⁴	Per cent	Per cent	
S/1919	70	[2] 32,000	86,000	....	27.5	53.9	1,600° F., 12 hours, air cooled; 1,200° F., 10 hours, cooled slowly.
	900	16,000	59,500	21.7	25.5	61.3	
	900	15,000	61,000	20.7	25.0	60.6	

[1] See Table 1 for chemical composition.
[2] Determined by the Midvale Co. A strain of 5×10⁻⁴ taken as indicating departure from a straight line.

2. NICKEL-CHROMIUM-IRON ALLOYS

The results of short-time tensile tests at 1,550° F. of cast nickel-chromium-iron alloys are given in Table 3. These alloys contained from 0.4 to 0.6 per cent carbon and about 1 per cent silicon and 35 per cent chromium. The nickel content of EE1522 was approximately 10 per cent and this increased to about 45 per cent in the last alloy of the series, EE1526.

This series showed no significant differences in the proportional limits of the various alloys at 1,550° F. There was, however, a progressive increase in the ductility of the alloy as the nickel content increased from 10 to 45 per cent, and above 30 per cent nickel there was a noticeable decrease in tensile strength.

It is in this case also of interest to recall the results of short-time tensile tests given in a previous report [9] for nickel-chromium-iron alloys containing 10 per cent chromium and either 35 or 57 per cent nickel. Only a very general correlation is possible as the earlier tests were made with maximum testing temperatures of only 1,360° F. as compared with 1,550° F. for the more recent tests; the alloys tested in the earlier work were rolled material and in the later work were "as cast"; and also the high-nickel alloy of the earlier work contained over 3 per cent tungsten.

TABLE 3.—*Tensile properties of cast nickel-chromium-iron alloys at 1,550° F. as determined by short-time tests*

Designation [1]	Proportional limit	Tensile strength	Modulus of elasticity	Elongation in 2 inches	Reduction of area
	Lbs./in.²	Lbs./in.²	Lbs./in.² ×10⁶	Per cent	Per cent
E E 1522	7,000	49,000	11.6	2.5	3.4
	5,000	41,500	9.4	2.5	3.1
E E 1523	5,000	49,000	14.4	3.0	4.2
		54,500		4.0	5.0
E E 1524	6,000	49,000	13.2	4.0	7.3
	7,000	55,000	13.1	5.0	6.5
E E 1525	6,000	44,000	14.4	8.5	13.3
	6,000	41,500	12.7	8.5	16.2
E E 1526	6,000	33,500	11.0	15.5	34.0
		39,000	....	14.0	23.7

[1] See Table 1 for chemical compositions.
[9] See footnote 7, p. 327.

Nevertheless, it is to be noted that the tensile strengths of the cast alloys of the present series, containing 35 per cent chromium and either 10, 20, or 30 per cent nickel, were of the order of 50,000 lbs./in.² and the proportional limits of the order of 5,000 to 7,000 lbs./in.² at 1,550° F. as compared with tensile strengths of only 35,000 to 40,000 lbs./in.² and proportional limits of 5,000 to 6,000 lbs./in.² at 1,360° F. for rolled alloys containing 35 per cent nickel and 10 per cent chromium.

Increased proportional limits and tensile strengths in alloys of this type at temperatures of 1,300° to 1,500° F. appear to result from an increased chromium content rather than an increased nickel content. With this increase in strength, there is also noticeably less ductility as measured by elongation and reduction of area.

3. TUNGSTEN-CHROMIUM-VANADIUM STEEL

The chief differences in the chemical compositions of the three tungsten-chromium-vanadium steels (Table 1) were that the first of the series (HF-1/169) had a tungsten content of only 1.70 per cent, which was somewhat below the normal composition of the series; the second of the series (HF-1/167) may be considered of the normal composition, namely, about 2¼ per cent each of chromium and tungsten; the last steel of the group was low in chromium (1.63 per cent) but also had a manganese content (1.02 per cent) distinctly higher than the normal for the group.

All of these steels were tested after normalizing from 1,700° F., hardening, and tempering at 1,200° F. The hardening treatment consisted of air quenching from a temperature 50° F. above the upper critical temperature of the individual steel. The exact temperatures are shown in Table 4.

The results of the short-time high-temperature tests of these steels (Table 4) did not show any very striking differences in the behavior of the three steels. On the whole, the steel containing 1 per cent manganese (HF-1/168), with the lower chromium and the normal tungsten content, appeared to be superior with respect to the maintenance of a high proportional limit at elevated temperatures.

TABLE 4.—*Tensile properties of tungsten-chromium-vanadium steels at elevated temperatures as determined by short-time tests*

Designation [1]	Temperature of test	Proportional limit	Tensile strength	Modulus of elasticity	Elongation in 2 inches	Reduction of area	Hardening heat treatment [2]
	°F.	Lbs./in.²	Lbs./in.²	Lbs./in.² ×10⁶	Per cent	Per cent	
HF-1/169.........	70	[3] 92,000	128,000	....	20.2	62.6	}1,515° F., 30 minutes, air cooled.
	850	32,000	86,000	25.2	22.0	64.7	
	1,000	26,000	71,500	22.1	23.5	72.9	
HF-1/167.........	70	[3] 107,000	139,500	....	19.2	59.9	}1,510° F., 30 minutes, air cooled.
	850	30,000	98,500	23.6	21.5	62.3	
	1,000	16,000	77,500	22.1	23.0	70.1	
HF-1/168.........	70	[3] 109,000	136,500	....	18.2	55.7	}1,480° F., 30 minutes, air cooled.
	850	43,000	99,500	24.3	19.0	54.9	
	1,000	24,000	84,500	19.9	22.5	65.6	

[1] See Table 1 for chemical compositions.

[2] All specimens were normalized before the hardening treatment and tempered after the hardening treatment indicated in the last column of this table. In all cases the normalizing treatment consisted of heating at 1,700° F. for 30 minutes and cooling in air; the tempering treatment, of heating at 1,200° F. for 30 minutes and cooling in air.

[3] Determined by The Midvale Co. A strain of 5×10⁻⁵ taken as indicating departure from a straight line.

4. MOLYBDENUM-CHROMIUM-VANADIUM STEELS

The molybdenum-chromium-vanadium steels are listed in both Table 1 and Table 5 in the order of increasing manganese content. The second steel of the series differed from the first in that it had slightly higher carbon (0.29 per cent) and manganese (0.88 per cent) contents than the first steel of the series (0.20 per cent carbon, 0.45 per cent manganese). This difference in composition apparently accounted for a distinctly higher proportional limit and tensile strength in the steel of higher carbon and manganese contents at temperatures of 850° and 1,000° F. Further increase in the proportion of manganese, and also of chromium, in the third steel of the series (HF-1/177) resulted in a still higher proportional limit at 850° F., but produced no increase in either proportional limit or tensile strength at 1,000° F. The fourth and last steel of this series had the highest manganese content of all (2.23 per cent), about double the molybdenum content (1.00 per cent) of the other steels, and a high silicon content (1.33 per cent). This steel had tensile strengths, at the high temperatures, equivalent to or higher than the other steels of the series. Its proportional limit at 850° F. was also high. Its proportional limit at 1,000° F., however, failed to show an equivalent increase and was only of the order of the proportional limit of the first steel of this series, the low-carbon, low-manganese composition, at the corresponding temperature. In previous work [10] it has been observed that 1.25 per cent silicon had a very marked effect in lowering the proportional limit of a tungsten-chromium-vanadium steel at about 1,000° F.

TABLE 5.—*Tensile properties of molybdenum-chromium-vanadium steels at elevated temperatures as determined by short-time tests*

Designation [1]	Temperature of test	Proportional limit	Tensile strength	Modulus of elasticity	Elongation in 2 inches	Reduction of area	Hardening heat treatment [2]
	°F.	Lbs./in²	Lbs./in²	Lbs./in² ×10⁶	Per cent	Per cent	
HF-1/166	70	[3] 60,000	104,000	--------	24.7	66.2	
	850	25,000	83,000	24.7	23.5	67.5	1,490° F., 30 minutes, air cooled.
	1,000	19,000	71,000	22.6	24.0	71.6	
HF-1/165	70	[3] 80,000	118,000	--------	20.7	65.0	
	850	32,000	96,000	24.0	21.5	67.5	1,460° F., 30 minutes, air cooled.
	1,000	26,000	81,000	21.7	23.5	74.5	
HF-1/177	70	[3] 75,000	112,000	--------	23.0	63.3	
	850	37,000	97,000	24.8	20.5	60.1	1,460° F., 30 minutes, air cooled.
	1,000	25,000	71,500	20.2	25.0	73.3	
HF-1/178	70	[3] 74,000	124,000	--------	21.5	54.4	
	850	36,000	104,000	23.3	23.5	63.5	1,460° F., 30 minutes, air cooled.
	1,000	20,000	78,500	18.9	27.0	77.4	

[1] See Table 1 for chemical compositions.
[2] All specimens were normalized before the hardening treatment and tempered after the hardening treatment indicated in the last column of this table. In all cases the normalizing treatment consisted of heating at 1,700° F. for 30 minutes and cooling in air; the tempering treatment, of heating at 1,200° F. for 30 minutes and cooling in air.
[3] Determined by The Midvale Co. A strain of 5 x 10⁻⁴ taken as indicating departure from a straight line.

[10] See footnote 2, p. 327.

V. SUMMARY

1. Short-time tensile tests at elevated temperatures were made of a medium-manganese steel; of a series of cast nickel-chromium-iron alloys containing about 0.50 per cent carbon, 35 per cent chromium, and nickel from 10 to 45 per cent; and of two series of low-alloy steels, namely, tungsten-chromium-vanadium and molybdenum-chromium-vanadium steels, all normalized, hardened, and tempered at 1,200 ° F.

2. The medium-manganese steel (1.08 per cent) had a proportional limit of 15,000 to 16,000 lbs./in.² at 900° F. This is fully 150° F. higher than the temperature at which a 0.55 per cent manganese boiler-drum steel of the same carbon content (0.4 per cent carbon) possessed the same proportional limit.

3. The nickel-chromium alloys showed no marked change in proportional limit at 1,550° F. over the range of composition studied. With nickel contents over 30 per cent tensile strengths of the alloys decreased noticeably and the ductility increased.

4. A medium-manganese content (1.02 per cent) in one of the steels of the tungsten-chromium-vanadium series was accompanied by proportional limits as high as or higher than in the other steels of the series.

5. Increased manganese contents (up to 1.4 per cent) in the molybdenum-chromium-vanadium steels resulted in higher proportional limits at 850° and 1,000° F. A steel having a still higher manganese content (2.23 per cent) accompanied by high silicon (1.33 per cent) and 1.00 per cent molybdenum had a high proportional limit at 850° F., but a comparatively low proportional limit at a temperature of 1,000° F. Similarly, low proportional limits at 1,000° F. were observed in previous work with tungsten-chromium-vanadium steels having a high silicon content (1.25 per cent Si).

WASHINGTON, July 6, 1932.

THE COMPARISON OF HIGH VOLTAGE X-RAY GENERATORS

By Lauriston S. Taylor and K. L. Tucker [1]

ABSTRACT

It is shown that when measurements are made in the customary manner, the X-ray emission of a tube operated on two mechanical rectifiers may differ by ±20 per cent, although the electrical indications are the same. This also applies to quality evaluations whether of the full absorption curve, half-value layer or effective wave-length type.

Determinations of percentage depth dose do not show any such marked differences since this is a comparatively insensitive indicator. It is seen, moreover, that there is little gain in percentage depth dose in going from 160 to 200 kv peak; the only change of significance being in the actual 10 cm depth dose.

It is shown on the other hand, when the potential of three generators including constant potential is measured in "effective kilovolts," that for the same effective voltage the outputs for a given filter are all very nearly the same, both as regards intensity and quality.

CONTENTS

I. INTRODUCTION

In the technical and clinical use of high voltage X rays, a wide variety of generators have come into common use. To save strain on the X-ray tube, unidirectional voltage, obtained by mechanical or by thermionic rectifiers, is usually applied. Since the rectification characteristics of such generators differ widely, there has been considerable confusion as to their relative effectiveness in producing the desired therapeutic or technical results.

Experience in this laboratory has led us to the conclusion that, due largely to the lack of properly controlled experimental conditions, comparisons between X-ray generators which have been made in the past, are of questionable soundness. For example, we have found that a different X-ray emission, as expressed in terms of the ionization produced in air, is obtained from a given tube and generator when either the aerial system or the tube inclosure is changed.

It was decided, therefore, to make a careful comparison of several typical X-ray generators having as few variables as possible and yet under as nearly clinical operating conditions as obtainable in a labora-

---

[1] Research associate, Radiological Research Institute. This work was started by K. L. Tucker, who, by reason of illness, was obliged to withdraw from active participation before the experimental work was complete.

tory. For the most part, the various quantities were measured in
the same units and in the same manner as in the clinic so that the
results may be readily interpreted in commonly used terms.

The physical equality of X-ray beams produced by different
generators, so far as concerns therapeutic application, has been shown
to be based on 10 or more factors.[2] Included in these are tube
current, tube voltage (peak), filter, quality (H. V. L. or λ_e),[3] Röntgens
per minute delivered by tube,[4] and percentage depth dose.[5]

II. APPARATUS

Three generators, *A*, *B*, and *C*, were chosen as typical. *A* and *B*
were rated to deliver full wave 220 to 230 kv (peak) at 30 ma (milliamperes). *C* was rated to deliver 200 kv (peak) at 10 ma. *A* was a
mechanical rectifier having a single high-tension transformer and
rectifying approximately 30° of the cycle. *B* was a mechanical
rectifier having a divided secondary high-tension transformer and
rectifying approximately 20° of the cycle. *C* was a valve tube and
condenser ripple potential generator (so-called constant potential)
having a ripplage[6] of only about 1.5–2 per cent per milliampere,
and hence an X-ray emission not differing appreciably from strictly
constant potential.

Any one of the three generators could be connected to the same
overhead system without changing its capacitance. As will be shown,
this is essential. The X-ray tube was of the Coolidge type, having
thin walls. Since tubes of the same design vary slightly, a single
tube was left in position for the six months during which the measurements were made. It was inclosed in a rectangular ¼-inch lead box,
4 by 4 by 7 feet, having forced ventilation. Previous experience
indicated that X-ray tubes operate more smoothly in unconfined space
than in some of the conventional tube drums. For example, it was
found that for a tube operating at 200 kv (peak) a spacing of about 12
inches from center to ground wall produced unsteadiness, whereas an
18-inch spacing appeared sufficient to avoid such difficulty. Accordingly, we used the 24-inch spacing and had no difficulty with tube
unsteadiness at voltages up to 230 kv (peak).

Control of the tube with generators *A* and *B* was observed by
means of a d. c. milliammeter in the high-tension circuit and a wide-

[2] E. A. Pohle, Am. J. Roent., vol. 18, p. 55, 1927.
[3] The half-value layer (H. V. L.) in copper or aluminum is a measure of the "penetrating power" of an X-ray beam. It is defined as the thickness of copper (or aluminum) which interposed in an X-ray beam reduces its intensity to one-half its initial value (as measured in terms of air ionization).
The effective wave length of an X-ray beam is the wave length of the homogeneous radiation having the same absorption coefficient in copper (or aluminum) as the heterogeneous radiation in question. (See discussion of various methods of expressing X-ray qualities in a paper by L. S. Taylor, B. S. Jour. Research, vol. 5 (RP212), p. 517, 1930.)
[4] The "Röntgen" (r), is defined as the quantity of radiation which, when the secondary electrons are fully utilized and the wall effect of the chamber is avoided, produces in 1 cm³ of atmospheric air at 0° C., and 76 cm mercury pressure such a degree of conductivity that one electrostatic unit charge is measured at saturation current.
[5] The "percentage depth dose" is the ratio of the X-ray intensity as measured at a given depth in a body of homogeneous material, to the intensity as measured at the irradiated surface. It must be recognized that these intensities are measured in terms of Röntgens per minute and hence do not give true indications of the energy absorbed in a given volume element.
[6] Up to the present the term "constant potential" has been used carelessly in describing the potential supplied by kenotron or other valve tube rectification. We will use a more accurate designation of voltage which are actually not constant but fluctuate about a certain average value. Thus by a "ripple quantity" (potential or current) is meant a simple periodic quantity

$$y = V_0 + V_1 \sin(\omega t + \alpha_1) + V_2 \sin(2\omega t + \alpha_2) + \ldots$$

in which the constant term (V₀) is so large that all values of the quantity are positive (or negative). The amount of ripple ("ripplage" or "ripplance") in a ripple quantity is the ratio of the difference between the maximum and minimum values of the quantity to the average value.

scale voltmeter connected directly across the transformer primary.
Duplication of results over a period of months indicated that this
control method was sufficient.   In the case of the generator C ("constant potential") the control was effected by means of a shielded
high-tension voltmeter [7] connected directly across the tube leads.

Peak voltage was measured with a sphere gap in all cases.   Average
voltages were measured by means of the high-tension voltmeter, which
in the case of generator C gave readings very slightly less than those
of the sphere gap, depending upon the tube current.   Since the rectification characteristics of generators A and B vary with load, this was
kept constant at 4 ma (average) throughout the entire work.

The rectifying switch of generator B was fixed in a permanent
position on the synchronous motor shaft, so measurements made with

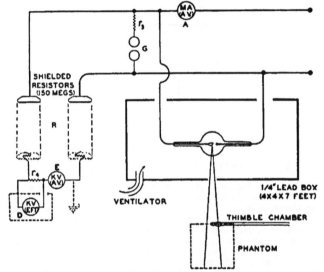

FIGURE 1.—*Diagram of apparatus showing current and voltage measuring
systems*

B were arbitrarily chosen as a basis in comparing with A.   The switch
on A was so arranged that its phase position could be readily changed.
The positions of the switches were at the points set by the agents of
the respective manufacturers—presumably best suited for use at
200 kv (peak).

Ionization measurements were made with a calibrated thimble
ionization chamber arranged according to the diagram in Figure 1.
The distance from focal spot to center of the chamber was 104 cm.
The ionization chamber system was tested for leakage by covering
the thimble chamber cap with lead and then exposing it to the X-ray
beam.   No leakage was detectable in an exposure time double the
longest used in the measurements.

---

[7] L. S. Taylor, B. S. Jour. Research, vol. 5 (RP217), p. 609, 1930.

For depth dose measurements a cubical wax phantom about 35 cm on a side was used. Wax was selected instead of water largely for convenience since its difference from water has been shown [8] to be insignificant. The wax was carved out so that the chamber in the surface position was half submerged leaving no air pockets between it and the wax. The focal spot-surface distance was likewise 104 cm. A beam area of 10 cm diameter (78 cm²) at the position of the chamber was chosen in order to work with a clearly defined field receiving little stem radiation. Table 1 from Grebe and Nitzge shows that within wide limits variations of depth dose with quality are independent of the field area.

TABLE 1.—*Relationship between radiation quality and percentage depth dose for two different field areas*

A, field area 400 cm²; B, field area 50 cm²

H. V. L.	Per cent depth dose		Ratio
	A	B	
mm Cu			A/B
2.00	52	30	1.73
1.6	51	30	1.70
1.0	50	29	1.72
.8	48	28	1.72
.6	47	27	1.74
.4	40	24	1.67
.2	28	16	1.75
.1	16	9	1.78
.05	7	4	1.75

III. INTENSITY MEASUREMENTS IN AIR

The curves in Figure 2 give for generators A and B the beam intensity measured in air (in arbitrary units) as a function of the copper filtration in the beam. Each curve is for a constant peak voltage as measured with a sphere gap. It will be noticed that, for the same tube current and 200 kv (peak) generator B gives about 20 per cent more radiation than A while at 180 kv (peak) both are about the same. At 160 kv (peak) A has a greater X-ray emission than B and from 140 kv (peak) down both are essentially the same.

Figure 3 gives a similar set of curves for generator C where the voltages are expressed in kilovolts average. The curve for 150 kv (average), replotted as the broken line on Figure 2, approximates very closely the curve for 200 kv (peak) on generator B. Similarly, it will be found for the conditions used here that the other approximate intensity equivalents given in Table 2 are obtainable for radiation filtered through 0.5 mm of copper.

Thus to obtain an X-ray intensity of 178 units from any of the three generators, through 0.5 mm copper filter, would require that A be operated at 200 kv (peak), B at 189 kv (peak), and C at 142 kv (average). This is, of course, for the same tube current in all cases. Similarly, an intensity of 80 units is obtained at 138 kv (peak), 140 kv (peak), and 99 kv (average), respectively.

[8] Grebe and Nitzge, Strahlen (Sonderbände), vol. 14, 1930.

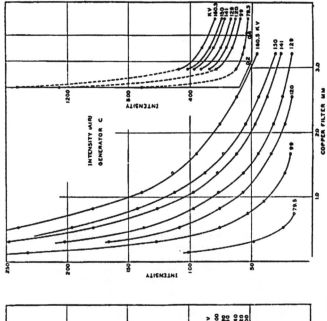

FIGURE 3.—*Beam intensities for generator C measured in air*

FIGURE 2.—*Beam intensities for generators A and B measured in air*

Broken curve is for 150 kv (average) on generator C, taken from Figure 3.

TABLE 2.—*Potentials of generators A, B, and C required to produce equal X-ray intensities or qualities*

Intensity	λ_e	H. V. L.	Absorption curve	A kv (peak)	B kv (peak)	C kv (average)
	A.	*mmCu*				
215				[1]237	200	152
			X	(215)	200	(155-160)
178				200	189	142
			X	200	(190)	(150)
	0.192	0.775		200	195	161
				200	195	159
151				173	180	132
			X	(180)	180	(135-140)
	.199	.68		178	180	147
				177	180	146
122				154	160	122
			X	(150-155)	160	(125-130)
	.206	.60		153	160	124
				154	160	126
80				138	140	99
			X	(133-140)	140	(105-110)
	.218	.525		132	140	112
				134	140	110

[1] Extrapolated value, probably too high.

To obtain some quantitative idea of the effect of shortening the aerial system, a section about 12 feet long was disconnected. This reduced the capacitance of the high-tension system by about a third (from $158\mu\mu f$ to $110\mu\mu f$) and caused a noticeable decrease of the X-ray emission at the higher voltages and a slight increase at lower voltages. It is thus shown clearly that had two different aerial systems been used comparisons between the two machines would have been grossly misleading, favoring one machine at higher voltages and the other at lower voltages.

IV. QUALITY MEASUREMENTS IN AIR

It is believed that at present the most nearly adequate method of expressing the quality of an X-ray beam is to give its full absorption curve in copper or aluminum.[9] This is best done by plotting the logarithm of the intensity (or the per cent transmission) as a function of the filtration. It is not possible by this method to express a radiation quality as a single numerical magnitude.

For two radiation qualities to be equivalent, the curvatures of their respective absorption curves must be coincident. Where two curves do not exactly coincide, the difference in quality must be estimated. Wilsey has shown, however, that actual or estimated matching of absorption curves permits the most accurate reproduction of a given X-ray quality.

One of the advantages of giving a full absorption curve is that all other quality expressions, such as the half-value-layer or effective wave length, may be obtained from it. The slope of such a curve at any point gives the effective absorption coefficient of the radiation corresponding to the particular filter for which the point was chosen, and from this in turn may be obtained the true effective wave length of the beam.[10]

[9] E. A. Pohle, and C. S. Wright, Radiology, vol 14, p. 17, 1930. R. B. Wilsey, Radiology, vol. 17, p. 700, 1931.
[10] L. S. Taylor, B. S. Jour. Research, vol. 5, p. 517, 1930. This corresponds to Mutscheller's "average wave length."

Figures 4, 5, and 6 give the copper absorption curves for generators *A*, *B*, and *C* respectively. These curves are from the same data used for Figures 2 and 3. It is significant to note that the 200 kv (peak) curve for generator *B* indicates a generally more penetrating radiation than for generator *A* at the same voltage. At 180 kv (peak) the qualities are about the same, while at 150 kv (peak) the reverse obtains. From 140 kv (peak) down, the qualities are again roughly the same. In other words, as shown in Table 2, the qualities of the radiations bear a similar relationship to one another as do the intensities so that, for a given peak voltage, and a given filtration, if the radiation

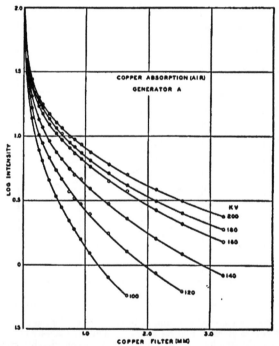

FIGURE 4.—*Semilogarithmic copper absorption curve for generator A*

emission of *A* is greater than *B* the penetration of *A* is likewise greater than *B*. It is unsafe to draw any generalization from these results, but it may be noted that such is roughly the case for all conditions thus far encountered in this work.

In Figure 6, giving the copper absorption curves for generator *C,* the broken lines are transposed from Figures 4 and 5. It will be noted that the 200-kv (peak) curve for generator *B* (upper broken curve) corresponds very nearly in slope to the 150-kv (average) curve for generator *C*. (It will be recalled that the corresponding intensity curves also nearly coincide.) Similarly, it is found possible to match

the other absorption curves for the different generators and we find
that Table 2 above for intensity equivalents holds approximately true
for quality equivalents, also.

V. DEPTH DOSE MEASUREMENTS

Measurements of the percentage depth dose were made over the
whole range of radiation qualities used. Figures 7, 8, and 9 give,
respectively, the surface and 10-cm depth intensities for generators
A, *B*, and *C*. Comparison of the curves for the two mechanical rec-

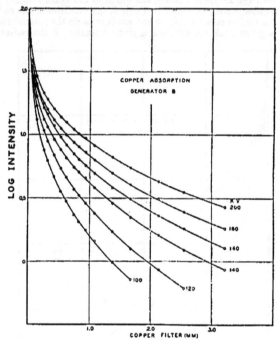

Figure 5.—*Semilogarithmic copper absorption curve for generator B*

tifiers shows the same general similarities as were evidenced by the air
intensity curves given in Figures 2 and 3. Thus the 200-kv curve for
generator *B* (broken line curve in fig. 7) shows a considerably greater
surface dose than the 200-kv curve for generator *A*.

Again the intensities measured at 10 cm depth for the two machines
are related in the same manner as the air intensities.

The broken line curves in Figure 8 are intensities for generator *B*
when used with the shortened aerial system. As noted for the air
intensity measurements, the long aerial system has a greater X-ray
emission at 200 and 180 kv and a smaller emission at 160 kv than the
short aerial system. Below 160 kv there appears to be no significant
difference between generators.

In comparing generator C with A and B, it is found (fig. 9) that 160.5 kv (average) on C produces slightly greater air and surface intensities than 200 kv (peak) on generator B while 150 kv (average) on C is about equivalent to 200 kv (peak) on generator A. Also 140 kv (average) on C is seen to be equivalent to 180 kv (peak) on either A or B.

Quality measurements as here carried out, with the thimble chamber at the phantom surface and at 10 cm depth, have no real

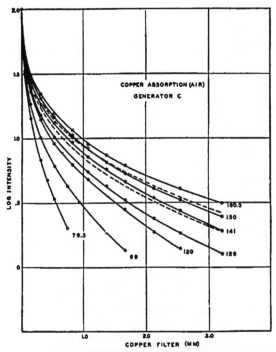

FIGURE 6.—*Semilogarithmic copper absorption curve for generator C*
Upper broken line curve for generator B at 200 kv. Lower broken line curve for generator A at 180 kv.

significance in relating the radiation quality in a phantom to that in air. This is because radiation under such conditions contains a major proportion of very soft scattered X rays that introduce a large and unknown wall correction into the measurements. Since it is impossible to reproduce such radiations entering a standard chamber, the thimble chamber can not be calibrated for equivalent radiation qualities.

Such curves have significance, however, in comparing radiation qualities measured under identical experimental conditions, and as such have been used to compare depth dose qualities.

Figures 10, 11, and 12 give the copper absorption (corresponding to the intensity curves of figs. 7, 8, and 9) measured at the surface and 10 cm depth for generators *A*, *B*, and *C*, respectively. On the assumption that similar absorption curves, obtained under identical conditions, imply equivalent radiations, these curves show precisely the same relation between the 10 cm depth radiation qualities for the

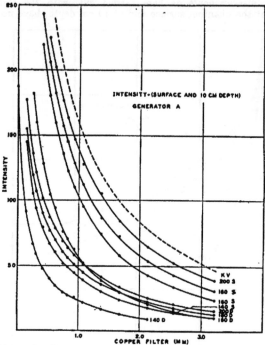

INTENSITY-(SURFACE AND 10 CM DEPTH)

GENERATOR A

FIGURE 7.—*Beam intensities for generator A measured at phantom surface (S) and at 10 cm phantom depth (D)*
Broken curve is for 200 kv on generator *B* measured at surface.

three generators as was indicated by the absorption curves measured in air.

Percentage depth doses for all the conditions here used may be obtained directly from the curves in Figures 7, 8, and 9. It happens that the percentage depth dose changes but very slowly with increasing hardness of radiation after one or two tenths of a millimeter copper filtration. It also changes but slightly with increase of voltage above 160 kv (peak). Consequently, the change in percentage depth dose is an insensitive indicator of radiation equalities. It can not, however, be neglected; since, if two radiations, having otherwise similar properties, should differ materially in percentage depth dose, there would be no justification for saying that they were equivalent.[11]

---
[11] It should be noted that a percentage depth dose is the ratio between two measurements and hence its error may be considerably larger than the individual errors of the original measurements.

The change of depth dose with filtration for the three generators is given in Table 3. It is seen that, while the accuracy is none too good, the percentage depth dose with generators *A* and *B* are about the same at equal voltages. The depth doses at 160 kv appear to be slightly greater than at higher voltages which is probably unreasonable but may be due to small cumulative errors. In comparing generator *C* with the mechanical rectifiers, it is found that the depth doses at 160

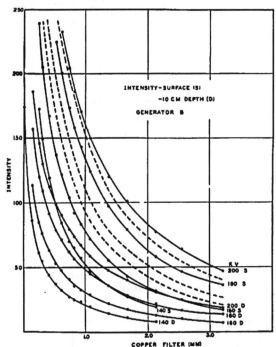

FIGURE 8.—*Beam intensities for generator B measured at phantom surface (S) and at 10 cm phantom depth (D)*

Broken line curves (top to bottom) for surface intensities for same generator except with shortened aerial operating at 200, 180, and 160 kv (peak).

kv (average) for *C* slightly exceed those at 200 kv (peak) for *A* and *B* and likewise the depth doses at 140 kv (average) for *C* are slightly greater than those at 180 kv (peak) for *A* and *B*. At 160 kv (peak) generators *A* and *B* give a slightly greater depth dose than *C* operating at 120 kv (average). The importance of these depth-dose measurements rests in their agreement and no inference should be drawn from the small differences found.

TABLE 3.—*Percentage depth doses in wax phantom (field 10 cm diameter) for different copper filtrations*

	kv (peak)		kv (average)	kv (peak)		kv (average)	kv (peak)		kv (average)
	200	200	160.5	180	180	140	160	160	120
Filter									
					Machine				
	A	B	C	A	B	C	A	B	C
mm Cu									
0	18	18	......	17	16	......	16	16	......
0.14	35	34	34.8	34	31	34.2	34	33	29.3
0.304			36.9			37.5		0	35.9
0.415	37	38	......	37	35	......	37	38	......
0.6	38	38	38.3	37	37	38.4	37	39	37.7
0.8	38	39	......	38	37	......	39	39	......
1.0	38	39	40.0	38	37	38.9	40	39	......
1.5	39	40	......	39	38	......	40	40	38.7
2.0	39	40	41.3	40	38	40.4	40	39	......
2.5	38	39	......	38	37	......	40	40	......
3.0	39	39	......	40	38	......	......		40.0

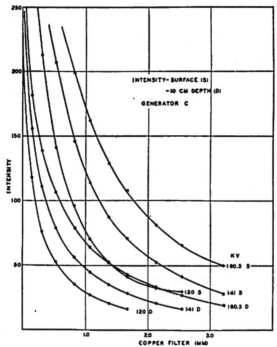

FIGURE 9.—*Beam intensity for generator C measured at phantom surface (S) and at 10 cm phantom depth (D)*

VI. HALF-VALUE LAYERS AND TRUE EFFECTIVE WAVE LENGTHS

The results above have shown that the copper absorption curves for the three generators give a fairly reasonable comparison of the resulting tube emission as regards quantity, quality, and depth dose. From the curves given, effective wave lengths or half-value layers in copper for any desired beam of radiation may be obtained directly. In practice, radiation qualities have been variously expressed by either

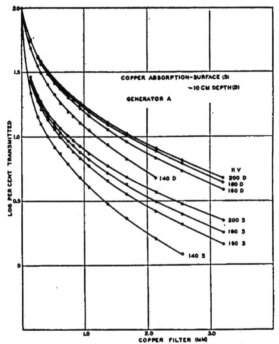

FIGURE 10.—*Semilogarithmic copper absorption curves for radiation measured at surface (S) and 10 cm depth (D) on phantom for generator A*

of these two methods. Hence, for facilitating comparisons of the radiation output of generators, values of the copper half-value layer (H. V. L.) and true effective wave length (λ_e), covering all the radiations used in this study, have been plotted in Figures 13 and 14. It should be emphasized that the values for mechanical rectifiers are likely to vary somewhat between machines and that the values here given hold strictly only for our generators. They should, however, serve as a fairly close guide.

Values of H. V. L. and λ_e also depend upon the thickness of the tube wall, this probably accounting for the differences in H. V. L. for "constant potential" reported by several workers. Probably a more

serious source of difference between workers lies in their method of
voltage measurement and in the amount of ripplage present in their
generator.   When using generator *C* the ripplage was found to be
about 8 per cent and to vary slightly with voltage.   Voltage measurements were made by means of a shielded high resistance "voltmeter
multiplier" in series with a d. c. microammeter.[12]   Thus the potentials
are expressed in average kilovolts whereas the potentials used by other
workers in measuring H. V. L. were usually, if not always, in peak
kilovolts.   To show how various observers agree, their results are
plotted in Figure 15.   There appears to be no systematic difference

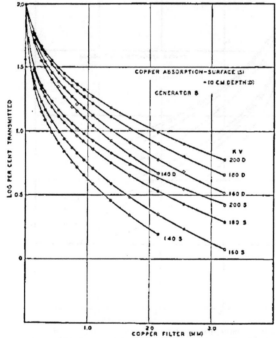

FIGURE 11.—*Semilogarithmic copper absorption curves for radiation
measured at surface (S) and 10 cm depth (D) on phantom for
generator B*

except that Holthusen's are consistently higher than all others.   Our
measurements fall between the others over a large portion of the
range covered.

The "quality" comparisons of the three generators, made from the
full absorption curves, are borne out very well by the H. V. L. and
λ_e curves of Figures 13 and 14.   For example, an X-ray beam, filtered
with 0.5 mm copper and having an H. V. L. of 0.75 mm Cu, may be
obtained with 200 kv on generator *A*, 195 kv on *B*, and 160 kv on *C*.

[12] See footnote, 7 p. 335.

Again a beam filtered with 0.25 mm copper and having an H. V. L. of 0.50 mm copper may be obtained with 210 kv on *A*, 202 kv on *B*, and 150 kv on *C*. We find a similar relationship from the effective wavelength curves. Thus a beam filtered with 0.5 Cu and having a value of $\lambda_e = 0.195$ Angstrom may be obtained with 200 kv on *A*, 196 kv on *B*, and 160 kv on *C*. Again a 0.25 mm copper filtered beam having $\lambda_e = 0.235$ Angstrom is obtained by 212 kv on *A*, 203 kv on *B*, and 150 kv on *C*. The agreement of these with the results obtained from the H. V. L. curves is probably within the experimental error.

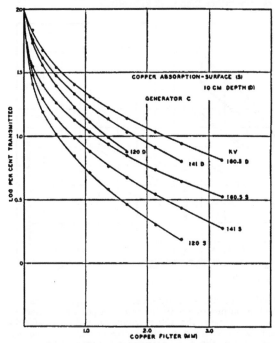

FIGURE 12.—*Semilogarithmic copper absorption curves for radiation measured at surface (S) and 10 cm depth (D) on phantom for generator C*

VII. VOLTAGE MEASUREMENTS

A partial explanation, at least, for the variation in X-ray emission under apparently identical conditions, is found to lie in the method of voltage measurement. The emission of an X-ray tube operating on a mechanically rectified generator depends upon the wave form of the voltage and the voltage-space current characteristic of the tube. If the tube current saturation is reached at a comparatively low voltage, then the voltage wave form of the generator is the predominant factor in determining the tube output. The wave form in turn depends upon current load drawn from the generator, the voltage, and the setting of the rectifying switches. It happens that, in general, no single setting of the rectifiers will suffice to yield the maximum output for all current and voltage combinations. In practice a single rectifier setting is used for all operating conditions and as a consequence the optimum output is realized only for a narrow range of conditions.

In measuring the voltage with a sphere gap the peak of the wave is the quantity determined, regardless of the wave form. Thus while the peak voltage of two waves may be the same, if one is a sharp-topped wave form while the other is broad-topped, it is obvious that the tube output for the latter will exceed that of the former. This condition may be largely met if the generator output voltage be measured in *effective* rather than peak kilovolts.

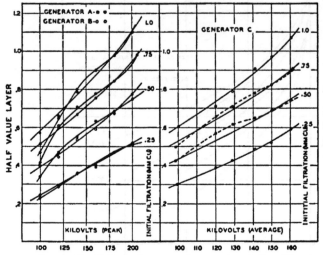

FIGURE 13.—*Half-value layer curves for generators A, B, and C*

This was readily accomplished by means of the high voltage voltmeter described in conjunction with the constant potential generator. As used up to the present, the 150 megohm noninductive shielded resistor [13] was in series with a d. c. microammeter and thus measured *average* voltage. When the d. c. instrument is replaced by an a. c. microammeter the potential is measured in *effective* kilovolts. In this work, the latter instrument not being readily available, we used a Kelvin multicellular electrostatic voltmeter to measure the potential drop across 75,000 ohms placed in series with the main high resistance. The meter then read the voltage across the line.

In the case of generator *C* the effective and average voltages are nearly identical. However, for mechanically rectified potentials,

[13] These have been described in paper referred to in footnote 7, p. 335. It was found that the separate units making up the final resistor had a very slight negative reactance at 1,000 cycles.

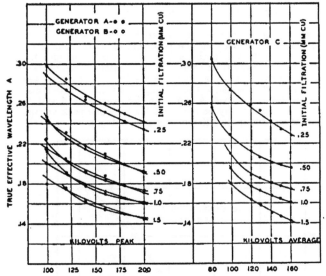

FIGURE 14.—*Effective wave length curves for generators A, B, and C*

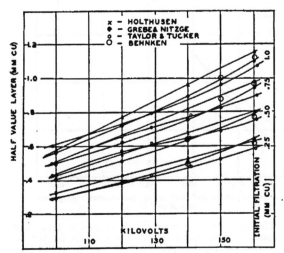

FIGURE 15.—*Comparison of half-value layer measurements by different observers*

the effective voltage is considerably higher than the average. Figure 16 shows the relation between the peak and effective voltage for generators *A* and *B* used in this study. In order to produce the same total load on the rectifier as used in the previous experiments—thus having nearly the same wave form—the tube current was adjusted so that when added to that through the voltmeter (0–1 ma) the total was 4 ma. Curves *A* and *B* are for generators *A* and *B*, respectively, whence it may be seen that for the same peak voltage the two systems have quite different effective voltages.

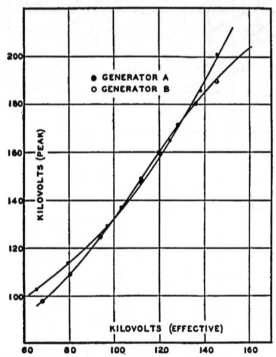

FIGURE 16.—*Effective and peak voltage measurements on generators A and B*

That the effective voltage is more closely related to tube output than peak voltage may be readily seen. For example, for the same 200 kv peak, generator *B* had a greater output; and, as seen in Figure 16, the effective voltage of *B* exceeded that of *A*. Similarly, at 180 kv (peak) the outputs were approximately equal and it is likewise found that the effective voltages were about the same, and so on.

To compare the output of the three generators, when the potential is measured in effective kilovolts, the curve in Figure 17 shows the relationship between the beam intensity (filtered through 0.525 mm.

of copper) and the effective kilovoltage. Curve C is for generator C ("constant potential") and the other points are as indicated. It is seen at once that the output at a given voltage is of the same order of magnitude for all three generators. Since the radiation quality was found to be roughly proportional to the intensity for a given tube current, it follows that the effective voltage also presents a fairly close indication of the quality. These results will be discussed in greater detail in a later paper.

VIII. CONCLUSION

It is believed that any value to be ascribed to this study lies in showing physical similarities, and not differences, between high voltage X-ray generators. Care has been taken to avoid any comparisons of the biological effectiveness per se of the radiations used. One of the outstanding biological problems is how to administer a desired

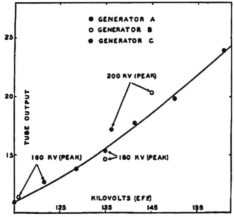

FIGURE 17.—*Plot showing tube output as a function of effective voltage for generators A, B, and C*

dose of radiation within the body without at the same time producing a dangerous skin erythema or destroying intervening tissue. It has been shown here how equivalent depth dose may be obtained with different generators and a wide range of qualities and intensities. The clinical application of this wide range of radiations depends on two factors which must be decided by the clinician, not by the physicist. The first, is the relationship between quality and the erythema dose. The second is the economics of administering radiation,[14] for obviously it would not be economical to use highly filtered low-voltage radiation for deep therapy when there is available a much greater intensity of less filtered radiation—assuming the biological effect to be the same.

---

14 G. Failla, Dissert. L'Univ. de Paris, Serie A, No. 1776, 1922.

It is hoped that the results of the present study will enable the clinician to better compare his technique with that of others using different generators and making his measurements by different methods.

This study has been made possible through the support and the generous cooperation of the Radiological Research Institute and the X-ray manufacturers of this country, for which we express our appreciation. In addition, we thank G. Singer and C. F. Stoneburner, of this laboratory, for their assistance, particulary with the effective voltage measurements.

WASHINGTON, June 6, 1932.

THE STRUCTURE OF THE CHROMIC ACID PLATING BATH; THE THEORY OF CHROMIUM DEPOSITION

By Charles Kasper

ABSTRACT

The structure of solutions that are of importance to the theory of chromium deposition from the chromic acid bath was investigated by cryoscopic and conductivity measurements, and absorption spectra. It was shown that the first step in the reduction of chromic acid is the formation of chromium dichromate, $Cr_2(Cr_2O_7)_3$, a strong electrolyte which forms negative molecular ions. This compound is found not to exist in the "green" form. The next product of reduction is the basic chromium chromate, $Cr(OH)_2 \cdot Cr(OH)CrO_4$. This compound is a colloid, which may exist in relatively acid solutions. If sulphate is present, it forms chromic sulphate only the green form of which exists in chromic acid solutions.

If the basic colloid does not have its electrophoretic velocity reduced, it coats the cathode and prevents further reduction of chromic acid. The beneficial action of the sulphate is caused by the fact that it lowers that velocity by adsorption. The sulphate reaches the cathode film by being transported as a nonreactive positive molecular ion, $[Cr_4O(SO_4)_4(H_2O)_x]^{++}$.

The above theory was confirmed by employing it to correlate facts and principles of chromium plating.

CONTENTS

I. INTRODUCTION

Chromium is electrodeposited commercially from baths containing chromic acid (2 to 4 M), a small amount of sulphate (0.02 to 0.05 N), and a variable concentration of trivalent chromium (derived from the partial reduction of the chromic acid). No entirely satisfactory explanation has been offered for the mechanism of this process, and in particular for the function of the sulphate.

The research naturally divided itself into two parts—first, the structure of the initial solution and of any system to which it may give rise on electrolytic reduction, and, second, the deportment of such systems in an electric field.

By the structure of a solution is meant the type and extent of the linkages existing between the various species present. Two types of such bonds are now recognized (23),[1] although modern research indicates that an indefinite number may exist (29).

Unless otherwise stated, all strong electrolytes [2] will be regarded as completely ionized at all concentrations. This characteristic is associated with systems in which only the ionic type of linkage exists.

The structure of the solutions was studied by measuring their freezing point lowerings, conductivities, and absorption spectra. From a consideration of published data on the systems of interest and admittedly similar ones, and from new experimental evidence, a theory of the electrodeposition of chromium has been developed. This theory correlates and explains practically all known facts concerning the process and also permits valid predictions to be made.

II. MATERIALS AND SOLUTIONS

A special lot of pure chromic acid was employed which had the composition (on a dry basis) shown in Table 1. The methods of analysis were the same as those employed by Moore and Blum (26).

TABLE 1.—*Composition of chromic acid*

Material	Per cent
Chromic acid (CrO_3)	99.75
Trivalent chromium (Cr_2O_3)	.15
Sulphate (SO_4)	.01
Alkali ($Na_2O + K_2O$)	.01
Insoluble	.01
Total	99.93

A number of methods were employed to prepare chromic acid solutions containing trivalent chromium and practically no other substances. Reduction with alcohol gave solutions from which the resultant acetic acid could not be readily removed. By dissolving pure Cr_2O_3 (produced by igniting CrO_3) in strong chromic acid, solutions with compositions corresponding to degrees of reduction up to 20 per cent [3] were obtainable. Dissolving of basic chromate, calcined at 200° C., in chromic acid yielded the same result. The more highly reduced solutions were prepared by the addition of concentrated hydrogen peroxide. These will be described and discussed in a subsequent section.

The solutions for the cryoscopic measurements were prepared, as customary, on a weight basis (moles per 1,000 g of water). Those for the measurement of conductivity and light absorption were prepared

[1] The numbers in the text here and throughout the text refer to the bibliography at the end of the paper.
[2] In this paper the term electrolyte is used in the same sense as ionogen; that is, a substance which when dissolved in a specified medium produces a conducting solution (which latter is frequently called the "electrolyte").
[3] The relative content of trivalent chromium in these solutions may be conveniently expressed in terms of the total chromium content. For example, a solution corresponding to $Cr_2(CrO_4)_3$, chromium dichromate, has one-fourth of its chromium content in the trivalent state, and, hence, represents a solution that is 25 per cent reduced. Similarly, the basic-chromate, $(Cr(OH)_2)_2 \cdot Cr(OH)CrO_4$ represents a reduction of about 67 per cent.

on a volume basis (moles per liter of solution). The stated concentrations of free chromic acid are based on the assumption that the trivalent chromium is present as the dichromate, $Cr_2(Cr_2O_7)_3$. Sulphate was introduced as dilute sulphuric acid, which was added at room temperature, and, therefore, produced no appreciable heating.

III. STRUCTURE OF SOLUTIONS

1. PURE CHROMIC ACID

The published data analyzed by Abegg (7) show conclusively that chromic acid is a strong acid, which at moderate concentrations exists principally as dichromic acid, $H_2Cr_2O_7$, of which both hydrogens are strongly dissociated. In more concentrated solutions trichromic acid, $H_2Cr_3O_{10}$, and possibly tetrachromic acid, $H_2Cr_4O_{13}$, exist. In very dilute solutions the monochromate ions, $HCrO_4^-$, and CrO_4^{--}, are the principal forms. The content of CrO_4^{--} ions at any finite acid concentration is extremely low.

2. CHROMIUM DICHROMATE

As it appeared probable from chemical evidence that the first step in the reduction of chromic acid results in the formation of chromium dichromate, the structure of partly reduced solutions was first investigated.

(a) PREVIOUS WORK

Some investigators (13), (16), and (19) have considered chromium dichromate as a colloid, while others (22) have regarded it as a very weak electrolyte. In all cases the evidence is unsatisfactory. The principal arguments were based upon ultramicroscopic observations. In order that a colloid be detectable by this means it must possess a certain size, a certain refractive index, and a certain transmittancy. It is difficult to see how scattered (not reflected) light could be observed in such highly absorbing solutions as these, if the dispersoid is not a metal. Some solutions like those previously investigated were examined with the ultramicroscope. From these experiments the conclusion was reached that the variable results (and hence the variable interpretations) were caused by dust. No great weight should be attached to ultramicroscopic observations on these solutions.

In addition to obtaining negative results with the ultramicroscope, Ollard (22) cited the qualitative evidence of the migration of the trivalent chromium toward the cathode in these solutions. This observation does not prove that chromium dichromate is a very weak electrolyte. It might be a strong electrolyte or a lyophilic colloid (a micelle).

(b) CRYOSCOPIC EVIDENCE

The measurements were made with the usual type of apparatus; in the 1F series a platinum resistance thermometer was used, and in the 2F series a Beckmann. The results are given in Table 2.

TABLE 2.—*Freezing-point depressions*

Solution No.	Moles per 1,000 g H_2O					Depression
	Cr^{VI}	Cr^{III}	$Cr_2(Cr_2O_7)_3$	Free CrO_3	H_2SO_4	
						°C.
1FI	4.340	1.106	0.554	1.016		13.72
1FII	4.340	1.106	.554	1.016	0.04	13.97
1FIII	4.340	1.106	.554	1.016	.20	14.47
1FIV	1.016			1.016		2.98
2FI	1.084	.278	.139	.249		1.72
2FII	1.084	.278	.139	.249	.06	2.17
2FIII	.249			.249		.66

In the designation of a given solution, for example 2FI, the Arabic numeral (2) refers to the apparatus employed, the letter (F) denotes the type of measurement (in this case freezing point), and the Roman numeral (I) refers to the number of the solution in that series. These three symbols define the composition of the solution. Sometimes it was found necessary to heat the solutions for one hour at 100° C.; such solutions are designated with an additional symbol, B.

The data relating to solutions 1FI and 1FIV show that the depression produced by 0.554 mole of chromium dichromate is 13.72° − 2.98° = 10.74°, or 19.40° per mole. If this salt were completely ionized into chromic and dichromate ions, and "perfect" in the Arrhenius sense, there would be a molal depression of 9.3°. The solute is a "superperfect" one, and most certainly is not a colloid or a very weak electrolyte. This point is shown even more conclusively by the data for the more dilute solutions. In the same manner the molal depression calculated from the data for the 0.139 *M* chromium dichromate is 7.6°; the value for violet chromic sulphate (7) (a comparable solute) at the same dilution is 4.15°.

Heating the above solutions for one hour at 100° C. caused no change in the depressions. As practically all inorganic chromic compounds exhibit changes in properties (from the violet to the green form[4]) when they are heated, this observation is important. This evidence is indicative that the "green" form of chromium dichromate did not exist in these solutions. In this respect this salt is similar to chromic nitrate and perchlorate. The absence of a change in solutions containing sulphate shows that the state of the sulphate is the same in the heated as in the unheated solutions.

(c) CONDUCTIVITY MEASUREMENTS

A solution that was 1.086 *M* in Cr^{VI} and 0.277 *M* in Cr^{III}, and which was, therefore, 0.255 *M* in free CrO_3, had at 25° C. a conductivity of 0.666 mho–cm. This is about 15 per cent higher than the interpolated value of 0.579 of Moore and Blum (26) for 0.255 *M* CrO_3. The increase in conductivity caused by 0.138 *M* chromium dichromate is 0.086 mho–cm, which corresponds to a molal conductance of $\frac{1,000 \times 0.086}{0.138} = 623$ mho–cm. The data in Abegg (7) show that the molal conductance of violet chromic sulphate at the same

---

[4] For example, the violet chloride $[Cr(H_2O)_6]Cl_3$, is transformed by heat into the green compound $[Cr(H_2O)_4Cl]Cl_2$ and $[Cr(H_2O)_4Cl_2]Cl$.

dilution is about 800 mho–cm. The difference is of the order expected in view of the greater mobility of the sulphate ion ($\mu_\infty = 136$) as compared with the dichromate ion ($\mu_\infty = 50$).

The cryoscopic and conductimetric measurements that we have made indicate that chromium dichromate is similar in structure to violet chromic sulphate; that is, it is an electrolyte that does not form molecules or molecular ions that include the negative radical. The spectrophotometric measurements to be described indicate that, while this conclusion is in the main correct, true molecular ions do exist.

(d) ABSORPTION SPECTRA

The cells for these measurements were made entirely of glass and were usually 1 mm thick. Thin cells were necessary on account of the high absorbencies of most of the solutions. The measurements were limited to wave lengths from 500 to 700 mμ. The 1P series were measured on a Koenig-Martens visual spectrophotometer.[5] The 2P series were executed on a Bausch & Lomb visual spectrophotometer. Although the accuracy of our measurements with the latter instrument is not quite as high as with the former, the results are sufficiently precise for our purposes.

In expressing the results the nomenclature of the Optical Society of America (12) is followed as closely as possible. The transmittancy, T, is equal to I/I_o, where I is the light intensity through a cell filled with the solution, and I_o that through a similar cell filled with the solvent. The absorbency, A, is equal to $-\log_{10} T$. Also, A is equal to $\Sigma(a_i c_i)1$, where 1 is the length in centimeters of the path traversed, c_i is the concentration of species i in moles per cubic centimeter, and a_i is the molal absorbency of i. If the values of a_i are independent of those of c_i, the system is said to obey Beer's law.

All measurements were made at about 25° C. The composition of the solutions upon which spectrophotometric measurements were made is given in Table 3.

TABLE 3.—*Composition of solutions for spectrophotometric measurements*

Designation No.	Moles per liter							
	Cr^{VI}	Cr^{III}	Free CrO_3	Al^{III}	Y^{III}	$Cr(SO_4)_{3/2}$	$Cr(NO_3)_3$	H_2SO_4
1PI	1.086	0.277	0.254					
1PII	1.086	.277	.254					
1PIII	1.086	.277	.254					0.01
1PIV	1.086		1.086					.10
1PV	.254		.254					
1PVI							0.544	
1PVII	1.086		.254		0.277			
2PI						0.066		
2PII	1.086		.254	0.277				
2PIII	.0225	.00519	.00635					
2PIV	.518	.277						
2PV	.255	.185						

[5] These measurements were made by Mabel E. Brown, of the colorimetry section.

132919—32——6

The first five solutions (and the corresponding heated ones) were so selected that the following salient facts could be determined: First, the absorbency of $Cr_2(Cr_2O_7)_3$ as such; second, the apparent absorbency of Cr^{III}; third, the possible existence of the "green"

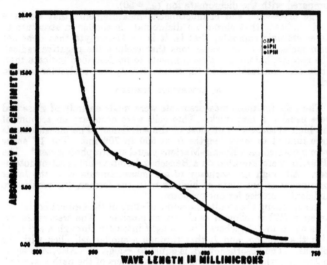

FIGURE 1.—*Effect of sulphuric acid on absorbency of solutions containing free chromic acid (0.254 M) and chromium dichromate (0.139 M)*

	Sulphuric acid
	M
1PI	0
1PII	.01
1PIII	.10

form of chromium dichromate; fourth, the state of the sulphate in cold solutions; and fifth, the state of the sulphate in heated solutions. The data are reported in Table 4 and are plotted in Figures 1, 2, and 3.

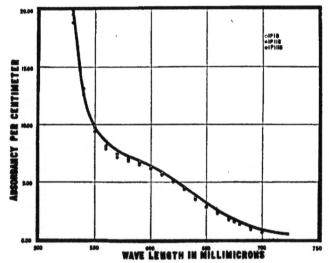

FIGURE 2.—*Effect on absorbency produced by heating solutions of Figure 1 to 100° C. for one hour*

(Curve same as in fig. 1)

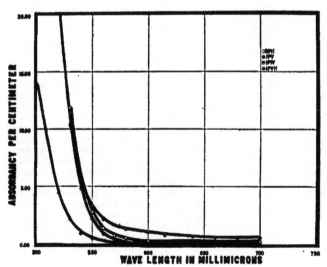

FIGURE 3.—*Effect of trivalent ions on absorbency of chromic acid*

	CrO₃	Addition	M
1PIV	1.086		
1PV	.254		
1PII	.254	Al⁺⁺⁺	0.277
1PVII	.254	Y⁺⁺⁺	.277

TABLE 4.—*Absorbency per centimeter at 25° C. of chromic acid solutions containing trivalent chromium*

λ in mμ	1PI	1PII	1PIII	1PIB	1PIIB	1PIIIB	1PIV	1PV
500								14.0
10								8.24
20							21.0	4.42
30	19.6	20.0	19.6	19.6	19.6	18.9	10.4	2.13
40	13.3	13.6	12.6	13.1	13.1	12.6	4.84	.88
550	9.91	10.3	9.87	9.79	9.67	9.36	2.15	.43
60	8.24			8.18	8.07	7.83	.84	
70	7.57	7.83	7.45	7.50	7.38	7.13	.36	.11
80	7.10			7.08	6.95	6.78	.16	
90	6.80	6.97	6.78	6.76	6.60	6.48	.10	
600	6.40			6.33	6.22	6.13	.04	.03
10	5.78	5.97	5.80	5.78	5.62	5.58		
20	5.20			5.12	5.02	5.00		
30	4.51	4.60	4.49	4.47	4.32	4.33		
40	3.83			3.63	3.55	3.55		
650	3.03	3.18	3.10	2.90	2.79	2.70	.02	.02
60				2.33	2.26	2.28		
70				1.71	1.72	1.74		
75	1.89	1.86	1.76	1.51	1.51	1.56		
700	.67	.87	.87	.90	.62	.59	.03	.00

In Figures 1 and 2 it will be noted that there are only small differences due to the sulphate; this point will be discussed in a later section. No significant difference was found on heating the solutions such as would indicate the formation of a green form of the dichromate. The process would result in an increased absorption in the red by the heated solutions.

The absorbency of chromium dichromate was found to be an unsatisfactory mode of approach. If chromium dichromate were a simple violet salt, subtraction of curve 1PIV from 1PI (which differs from 1PIV by 0.277 M Cr^{III}) should give values approximately one-half of those of *1PVI* (0.554 M Cr(NO₃)₃). However, it will be seen in Figure 4 that this is far from being true. This discrepancy is made even worse by correcting for the difference in hydrogen ion concentration, which would change the concentration of Cr₂O₁₀⁻. Bjerrum (3) has shown that the absorbency of the violet chromic ion is practically independent of the concentration and of the charge and size of the anion (Cl⁻, NO₃⁻, and SO₄⁻⁻). This is to be interpreted as indicating that the water "molecules" of the molecular ion, Cr(H₂O)₆⁺⁺⁺, protect the optical electrons from excessive coupling with external fields. Therefore, the above discrepancy can not be explained on the basis of a simple deformation of the chromic ion.

Hantzsch and Garrett (5) have shown that the absorbency of the chromate ion is greatly affected by the size and charge of the cation. Data on the dichromates of the alkali metals indicate slight deformation, but this possiblity can not be ignored when salts of high valence types are involved. To obtain the desired information regarding deformation of the dichromate ion, the absorbencies of solutions

2PII and 1PVII were measured. In each there is exactly the same content of hexavalent chromium as in 1PIV, but 2PII contains aluminum and 1PVII yttrium, each in a concentration equivalent to that of trivalent chromium in 1PI. The data are reported in Table 5, and are plotted in Figure 3. Table 5 also includes solutions that will be discussed later in this paper.

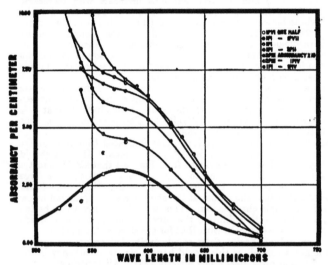

FIGURE 4.—*Absorbencies of various chromium solutions*

TABLE 5.—*Absorbency per centimeter of various chromium solutions at 25°C.*

λ in mμ	2PI	2PII	2PIII [1]	1PVI	1PVII
500	0.67			1.95	
10					
20				3.05	19.6
30		11.9	12.00		10.3
40	.97	5.68	6.64	4.56	5.4
550		2.73			3.16
60		1.32	4.75	5.99	2.12
80	1.31	.50	4.50	6.35	1.30
600		.23	4.10	5.56	1.04
20	1.31		3.19	4.09	.78
40			2.26	2.63	.67
60	.92		1.23	1.41	.62
80			.79	.89	.60
700	.41	.17	.38	.41	.58

[1] Absorbency ×10.

It is well known that in ionic systems deformation effects can be related to the size and charge of the "companionate" ions. Only relative sizes of the bare (not hydrated) ions are involved. The available data indicate that no simple analysis is possible, but it is legitimate to employ the conception for interpolative purposes.

Crystal structure data (28) indicate that the radii of the trivalent ions of Al, Cr, and Y are, respectively, 0.55, 0.70, and 0.90 A°. So, from a knowledge of the deformation produced by aluminum and yttrium, we can estimate that to be expected from chromium. The data plotted in Figure 3 show that the deformation is, as demanded by the above discrepancy, greater with the increasing charge (H^+ (1PIV) to Al^{+++}) and for trivalent ions increases with increasing radius. The deformation caused by Cr^{III}, may be assumed to be intermediate between those of Al and Y (that is, between the curves 1PI to 2PII and 1PI to 1PVI). Evidently this deformation by Cr^{III} is insufficient to account for the difference in question; that is, between 1PI and one-half 1PVII. Some other mode of interaction than the ionic is evidently effective; that is, the molecular. This is not unusual with chromic compounds (29).

It appears probable, therefore, that chromium dichromate exists as a Werner coordination complex. Its exact nature was not investigated, but the character of the absorption bands indicate that it is $[(H_2O)_2Cr(Cr_2O_7)_2]^-$ or $[Cr(Cr_2O_7)_3]^{---}$. The possibility of a true undissociated molecule is not excluded.

The cryoscopic and conductivity data indicated that this, unlike most coordination complexes, is highly dissociated. If so, it should not obey Beer's law. It is evident from Figure 4 that 2PIII (a forty-fold dilution of 1PI examined through a 40 mm cell) shows a large decrease in absorbency. On formally subtracting 1PIV from 2PIII it will be noted that the values at the shorter wave lengths fall below those of the simple chromic ion. This correction is known to be too large, and with further dilution the curve would, no doubt, coincide at the longer wave lengths with that for the simple chromic ion.

The absorption spectra show conclusively that chromium dichromate is not a colloid, as this would have a strong absorbency in the red and would obey Beer's law strictly (34). All experiments were consistent with the interpretation made, and inconsistent with any other that was entertained. No variations in absorbency occurred on standing for long periods.

3. HIGHLY REDUCED SOLUTIONS

Ollard (22) reported that by treating chromic acid solutions with concentrated hydrogen peroxide, a reduction up to 50 per cent ($Cr(OH)CrO_4$) could be effected, but that beyond that point a precipitate of $Cr(OH)_2Cr(OH)CrO_4$, (9) (14) (67 per cent reduction) was produced. In our experiments it was found that a reduction as high as 60 per cent could be thus effected without immediate precipitation. On standing one or more days the above precipitate appeared in which the ratio of the trivalent to hexavalent chromium was 2 to 1 and the filtrate corresponded to about 37 per cent reduction. The precipitation occurred with all solutions more highly reduced than 37 per cent. The rate of precipitation varied directly as the degree of reduction; it was accelerated by heating. These observations indicate that the system is a mixture of a sol, whose dispersoid is $Cr(OH)_2Cr(OH)CrO_4$, and a true solution of an electrolyte, chromium dichromate. The reason that solutions with 37 per cent reduction are stable is that then the hydrogen ion concentration is so high that the particles of the basic sol become sufficiently electropositive to remain in suspension. This hypothesis was confirmed by precipitating the basic chromate out of partially reduced solutions (42 per cent) with sodium sulphate.

Liebreich (9) and Müller (14) prepared colloidal solutions of the basic chromate by digesting the precipitate with water for a long time. The method employed in the present research was to take an unstable, highly reduced solution and dilute it strongly. Sols prepared in this manner exhibited a strong Tyndall beam and a dichroic appearance. They were quite murky and hence had fairly large particles, though none were visible with the ultramicroscope. As previously noted this instrument is not well adapted for these solutions.

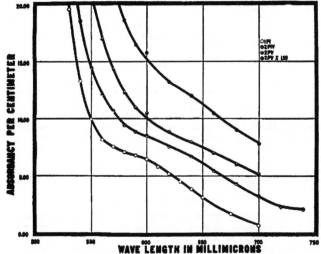

FIGURE 5.—*Absorbencies of solutions with equal contents of trivalent chromium and decreasing contents of hexavalent chromium*

The absorbency data for the highly reduced solutions, 2PIV and 2PV (prepared in the cold), are recorded in Table 6 and shown in Figure 5.

TABLE 6.—*Absorbency per centimeter of highly reduced solutions*

λ in mμ	2PIV	2PV	2PV×1.5
540	18.5		
50	14.4		
60	12.1	17.0	
70	10.7	14.3	21.5
80	9.43	12.4	18.6
90	8.89	11.0	16.5
600	8.51	10.5	15.8
20	7.56	8.86	13.3
40	6.68	7.98	12.0
60	5.42	7.00	10.5
80	4.30	6.00	9.00
700	3.20	5.16	7.80
20	2.37		
40	2.07		

The data in the last column have been recalculated to correspond to the same concentration of trivalent chromium as the other two. It will be noted that an increase in reduction causes an increased absorption over the entire range. The increase at the shorter wave length is probably caused in part by the deformation of the chromate radical. The increase in absorption in the red could be caused either by a complex (green) electrolyte, or a colloid, or both.

An effort to settle the question by freezing point measurements was only partially successful in that the data could be interpreted equally well in terms of a weak electrolyte or a mixture of a strong electrolyte and a colloid. Like the other solutions, these also coagulate, indicating that the latter interpretation is the correct one.

Electrometric methods of titration have a greater range of applicability than the conductimetric. Solutions with a constant content of Cr^{VI} (0.184 M) and a variable content of Cr^{III} were prepared and their pH values were determined with a modified glass electrode. In effect, therefore, chromic acid was titrated with chromium hydroxide. The results are given in Table 7.

TABLE 7.—*pH of chromic acid (0.184 M) with additions of trivalent chromium*

Cr^{III} M	0.000	0.033	0.058	0.061	0.065	0.079	0.118	0.123	0.126	0.129	0.133	0.137
pH	1.08	1.20	1.26	1.31	1.32	1.47	2.22	2.30	2.42	2.54	2.65	2.82

A plot of these data does not indicate a sharp break of any kind. More highly reduced solutions (corresponding to 60 per cent reduction) had a pH of about 3.5. According to the above interpretation of the structure of this system, we may have large amounts of the basic colloid at a pH as low as 3. This point is important for the theory of chromium deposition.

4. STATE OF THE SULPHATE

The above recorded cryoscopic and spectrophotometric measurements included some data on solutions containing chromic acid, chromium dichromate, and sulphuric acid. As noted before, the cryoscopic measurements indicated that there was no difference in the state of the sulphate in the heated and unheated solution. This behavior would be impossible with the violet chromic sulphate, so the green salt must be present.

The spectrophotometric measurements confirm this conclusion. It will be noted in Figures 1 and 2 that the absorbencies of the solutions with the largest content of sulphate are lower at the shorter, and (usually) higher at the longer, wave lengths than those with no sulphate, whether the solutions are heated or not. This is to be expected if green chromic sulphate replaces some chromium dichromate. The difference can be calculated [6] from the data on 1PI and 2PI, which latter is a solution of green chromic sulphate that was

---

[6] As this is rather involved, a sample calculation, that at 540μμ is given. According to Table 1 the absorbence per centimeter of the chromium dichromate in 1PI is 13.90−0.88=12.4. 1PIII contains in addition 0.1 M of sulphate. If the complex is $[Cr_2O(SO_4)_4(H_2O)_x]^{++}$, then according to Tables 3 and 5 the expected absorbency of the green chromic sulphate will be $0.97 \times \frac{3}{2} \cong 1.5$;

There will be a loss in absorbency corresponding to $\frac{1}{2.77} \times 12.4 \cong 4.1$. So the change in absorbency will be

4.1−1.5−1.3=1.3. The minus 1.3 denotes the increase due to the formation of free chromic acid.

heated for one hour at 100° C.   The effects, though small, are quite definite and beyond experimental error; the interpretation explains why only a small change is found.   The results are given in Table 8.

TABLE 8.—*Change in absorbency produced by addition of sulphate to solutions containing Cr^{VI} and Cr^{III}*

λ in mμ	540	570	580	590	620	650	700
Ac [1]	1.3		0.37		0.06	−0.51	−0.40
A_f	.68	0.12		0.02			−.23
A_{fB}	.50		.40		.12	.05	.04

[1] A_c is the calculated change in absorbency, A_f is that found for the unheated series, and A_{fB} that for the heated series.

The green complexes of chromic sulphate possess very remarkable properties.   Some are true colloids, but most of them are electrolytes. The colloidal form is obtained only with drastic treatment.   Part of the sulphate in the ordinary green forms is combined with the trivalent chromium to form a positive molecular ion $[Cr_2O(SO_4)_4(H_2O)_x]^{++}$. In the green chloride we have $[CrCl(H_2O)_5]^{++}$ and $[CrCl_2(H_2O)_4]^{+}$. These complexes are remarkable in that apparently there is absolutely no dissociation of the coordinated groups.   Electron pair bonds are undoubtedly effective (29).   These positive molecular ions furnish an ideal means of conveying the sulphate radical into the cathode film.

The pH of pure violet chromic sulphate solutions is about 2.   It was noted previously that the maximum pH found in solutions containing colloidal basic chromium chromate was 3.5.   These data indicate that, barring complex formation (which, however, is not favored at a intermediate pH like 3), free sulphate ions can exist in these colloidal solutions.   This point is also important for the theory of chromium deposition.

In pure solutions the violet form is the more stable at room temperature and below; the green is more stable above room temperature. The rate of conversion of the green into the violet is low, especially in the presence of acid, which acts as a negative catalyst and may even shift the quasiequilibrium.   Such a shift is recorded in this paper for the sulphate in chromic acid solutions; a similar case has been worked out by Bjerrum (2) for violet chromic chloride in hydrochloric acid solutions.   The fairly high temperatures employed in chromium deposition probably assist in this shift of equilibrium.

5. CONCLUSIONS

1. Chromium dichromate is a strong electrolyte and forms true molecular ions.

2. The "green" form of chromium dichromate does not exist in the solutions investigated.

3. More highly reduced solutions than the dichromate contain, in addition to that compound, a sol of the basic chromium chromate $Cr(OH)_2Cr(OH)CrO_4$.

4. In partially reduced solutions with an excess of chromic acid, the sulphate exists principally as the green chromic sulphate.

IV. THE THEORY OF CHROMIUM DEPOSITION

1. SURVEY OF THEORIES

The principal theories of chromium deposition that have been advanced will be briefly considered, including that proposed by the author in this investigation.

1. Sargent (6) recognized a number of features that are extremely important; that the cathode film is only slightly acid at high current densities, that reduction of chromic acid at high current densities must occur through the medium of the chromic-chromous couple (see p. 369), that sulphate is necessary for metal deposition, and that pure chromic acid can be reduced at low but not at high current densities. The latter feature he attributed to a colloidal layer which he designated as nonpermeable, though he tacitly assumed it to be permeable to hydrogen ions. The mechanism of the action of the sulphate was not clearly set forth.

2. Liebreich (9) (10) measured polarization curves by means of which he endeavored to show the successive stages through which chromic acid must pass before being reduced to the metallic state. The existence of this "step-wise" reduction is the principal feature of Liebreich's theory. The action of the sulphate was not explained. He demonstrated that the cathode film is less acid than the solution, and that in it there probably is an electropositive colloid, $Cr(OH)_3$-$Cr(OH)CrO_4$.

3. Fink (11) advanced a theory in which the sulphate is designated as a catalyst, but he gave no clear conception of its action. He assumed that the hydrogen that is always evolved in chromium deposition protects the metal from oxidation.

4. E. Müller (14) recognized that pure chromic acid is not reducible at high current densities. He attributed this fact to the formation of a colloidal layer in the slightly acid cathode film as a result of the migration of an electropositive sol. The idea that the cathode layer is permeable to hydrogen ions was expressly set forth. His first explanation of the action of the sulphate was that it precipitated and destroyed the cathode layer, and thus made possible the reduction of chromic acid. Later (20), (21), (24), and (33) he abandoned this interpretation in favor of one which states that the bisulphate ion, HSO_4^-, can, on account of its relatively small size, pass through interstices in the colloidal layer, thus permitting the existence of a higher hydrogen ion concentration in the cathode film. This condition is less favorable for the formation of the basic colloid. It is important to note that he found it necessary to assume that the bisulphate ion is effective at a pH of 3 or 4, and that its attraction toward the cathode is secondary. Müller demonstrated that the basic dispersoid, $Cr(OH)_2Cr(OH)CrO_4$, is electropositive. This theory is somewhat similar to that developed in this paper.

5. Haring (13), (17), and later Piersol (27) suggested that the action of the sulphate is in accordance with the following equation:

$$Cr_2(Cr_2O_7)_3 + 3H_2SO_4 \rightarrow Cr_2(SO_4)_3 + 3H_2Cr_2O_7$$

| (colloid) | (strong electrolyte) | (strong electrolyte) | (strong electrolyte) |

Chromium dichromate is not a colloid, and a pure metathesis (like the one represented) between a slightly soluble (or ionizable) salt and a strong electrolyte to produce two strong electrolytes in dilute solution is thermodynamically impossible.

6. Ollard's (22) theory denies that chromic acid can be reduced directly, and states that all reduction must proceed through the medium of the chromic-chromous couple. The action of the sulphate was explained by a reaction similar to that proposed by Haring and Piersol except that chromium dichromate was considered to be a very weak electrolyte.

7. In this research a theory of chromium deposition has been developed, which involves some features that have been expressed in previous theories, but that have been modified in the light of additional information and of a different viewpoint.

Deposition of bright chromium depends upon maintaining in the cathode film a definite hydrogen ion concentration. This must be sufficiently low to permit metal deposition and yet sufficiently high to prevent hydrolysis and thus yield oxide-free deposits.

At low current densities direct reduction of hexavalent to trivalent chromium undoubtedly occurs. At high current densities, which require cathode potentials sufficient to liberate hydrogen gas, the conditions are vastly different. On account of the tremendous migration velocity of the hydrogen ion in the cathode film, the latter becomes relatively alkaline. If the hydrogen ion concentration drops to 10^{-3}, a basic electropositive dispersoid of $Cr(OH)_2 \cdot Cr(OH)CrO_4$ exists. In the absence of the sulphate the dispersoid migrates toward the cathode and covers it with a colloidal layer which prevents reduction of everything except hydrogen ion, to which it is permeable.

The action of the sulphate is to decrease the electrophoretic velocity of the colloid by adsorption; which causes a reduction of the positive charge and prevents the formation of a dense, adherent colloidal layer. Other ions can then reach the cathode and be reduced. It is improbable, on account of the repulsive action of the cathode toward negative ions, that the simple sulphate ion can approach the cathode sufficiently close to be very effective. Some other means must be found for bringing it to the cathode.

It is well known that ordinary green chromic sulphate forms nondissociable, positive molecular ions like $[Cr_4O(SO_4)_4(H_2O)_n]^{++}$. This ion can, on account of its nonreactiveness, convey the sulphate to the cathode surface, and when the complex ion is discharged and reduced, the sulphate is set free to wander toward the anode. Before leaving the vicinity of the cathode, it may be adsorbed by the positive dispersoid and may thus limit the formation of the colloidal layer. While reduction at high current densities is made possible by the presence of sulphate, the actual reduction occurs mainly through the chromic-chromous couple.

2. POSSIBLE REDUCTION PROCESSES

It will be profitable to examine the principal reactions in terms of their standard electrode potentials (4). It will be seen that the conditions that are required for chromium deposition are those that inhibit certain undesirable reactions.

TABLE 9.—*Standard potentials of possible reactions in chromium deposition*

No.	Reaction	Standard potential
		Volts
1	1/2 $Cr_2O_7^-$ +7H⁺ +3ê = Cr^{+++} +7/2 H_2O	1.3
2	1/2 $Cr_2O_7^-$ +7H⁺ +4ê = Cr^{++} +7/2 H_2O	.9
3	1/2 $Cr_2O_7^-$ +7H⁺ +6ê = Cr +7/2 H_2O	.4
4	H⁺ + ê = 1/2 H_2	.0
5	Cr^{+++} + ê = Cr^{++}	−.4
6	Cr^{++} +2ê = Cr	−.5
7	Cr^{+++} +3ê = Cr	−.5

ê denotes an electron.

According to reaction 3, it should be theoretically possible to arrange a cell having a chromium-coated metal as cathode and a hydrogen electrode as anode, and deposit chromium without the use of external electrical energy. In fact, there would be a gain of electrical energy that could be employed to generate some of the hydrogen used. The reason that this is impracticable is because reaction 1 is the more probable under any condition that is favorable for reaction 3. If instead of chromic acid a chromic salt were employed electrical energy would be required in order to effect chromium deposition.

3. REACTIONS IN THE CATHODE FILM

During deposition at appreciable current densities the solution close to the cathode, the cathode film, may differ greatly in composition from the body of the solution. However, a steady state is set up between the factors causing depletion and restoration of the various species in the film. The former include, first, the rates of discharge of the various ions present, and, second, the forces that exist between the ions and the cathode. The restorative factors include, first, migration of the ions toward the cathode, second, diffusion processes and, third, convection currents. The electrical forces very close to the cathode are relatively large and hence there is a large concentration gradient in the cathode film. The processes that actually occur at the cathode under specified conditions depend upon the composition and structure of the cathode film.

There are three imperative reasons why the cathode film must possess an intermediate pH (3 or 4) before deposition of metallic chromium can occur. First, the difference in free energy between the standard chromium-chromic ion potential and the standard hydrogen-hydrogen ion potential is 0.5 volt faraday per equivalent. This demands, first, that the hydrogen ion concentration be low for chromium to be deposited, although the permissible hydrogen ion concentration will be conditioned by the hydrogen overvoltage. Second, it must not be too low because the chromic ion is easily hydrolyzed and, if the film becomes too basic, oxides or basic compounds will be deposited. Third, in the reduction of chromic acid large amounts of oxygen are converted to oxygen ions or hydroxyl ions, which must be neutralized by hydrogen ions to prevent the formation of oxides and hydroxides.

If pure chromic acid is electrolyzed at such low current densities that the cathode potential is not sufficient to cause hydrogen evolution, the cathode film will have a higher hydrogen ion concentration than the solution. Under these conditions true undissociated molecules of chromic acid can exist close to the cathode and may be reduced directly by the electrons to the trivalent state. The formation of metallic chromium is possible but improbable with the acidic conditions obtaining.

If the cathode potential and current density are raised so that hydrogen is discharged the conditions are vastly different. The hydrogen ion concentration in the cathode film tends to drop very rapidly on account of the tremendous migration velocity of the hydrogen ion within the film. The decrease in hydrogen ion concentration is favorable to the deposition of chromium, especially as large amounts of trivalent chromium must have been formed before this condition could arise. Reactions, 1, 2, and 3 show that the direct reduction of chromic acid will be tremendously affected by a variation in the hydrogen ion concentration. This leads to the conclusion that with increasing current density the chromic-chromous couple becomes the dominant process in effecting the reduction of chromic acid. This couple consists of having chromic ion reduced at the cathode to the chromous ion, which on escaping into the body of the solution reduces chromic acid to trivalent chromium. There is no justification for the suggestion that cathodic reduction is effected by atomic hydrogen. Such a process is in direct conflict with thermodynamic principles.

It has been shown above that when this system attains a hydrogen ion concentration of 10^{-3}, a sol of the basic chromium chromate, $Cr(OH)_3 \cdot Cr(OH)CrO_4$, exists. This sol and any other form of the colloid are probably permeable to hydrogen ions, in view of the fact that in migration and diffusion the radius of the bare hydrogen ion is effective and not that of the hydrated or hydronium ion. The chain theory (30) of the abnormally high migration velocity of the hydrogen ion indicates that although the colloid may permit hydrogen ions to pass through, it will hamper their motion. By this action the portions closer to the cathode will possess a still lower hydrogen ion concentration.

The amount of the colloid formed and the particle size will vary inversely, whereas the electrophoretic velocity will vary directly as the hydrogen ion concentration. The Brownian movement of the dispersoid of the sol, and hence the permeability of the sol, is greatest at the highest hydrogen ion concentration consistent with its existence. However, the high electrophoretic velocity under these conditions favors the formation of the colloidal layer, which would tend to exclude all ions but hydrogen and thus prevent reduction. As shown elsewhere in this paper the amount of colloid varies very rapidly with the hydrogen ion concentration, whereas the electrophoretic velocity does not vary in the range probably involved. No change in conditions is to be expected by varying the hydrogen ion concentration in the cathode film by means of the current density. It appears from this analysis that the formation of the objectionable colloidal layer is a precipitative effect and that any explanations of the beneficial action of the sulphate on the basis of a similar effect is illogical.

4. ACTION OF NEGATIVE IONS, SUCH AS SULPHATE

Müller's first theory that the sulphate coagulates the colloid and thus destroys the colloid layer is untenable because the resultant precipitate would probably hinder the reduction as completely as the original sol. He abandoned this view in favor of one which states that the particles of the sol tend to arrange themselves in a more or less regular manner on the surface of the cathode. In consequence, there will be spaces with a certain average size through which relatively small negative ions may pass. This condition permits the existence of a higher hydrogen ion concentration at the cathode surface and is less favorable to the formation of the colloid.

A number of criticisms have been leveled against the conception of colloids and colloidal layers in chromium deposition. It is important to note that the sol is only a part of the cathode film; though it may under certain conditions constitute almost exclusively a definite section of it; that is, the colloidal layer. The colloidal layer and the sol are stable because the cathode film is low in hydrogen ions. The supposed difficulty of the discharged hydrogen breaking up the colloidal layer and destroying the film structure does not exist because the cathodic processes discussed are microscopic and not macroscopic. The rates of the reactions involved are undoubtedly much faster than bubble formation; furthermore, bubble formation would not involve the destruction of the film.

If electrophoretic migration is important in determining the reduction processes, adsorption is also. Since the sol is electropositive, the adsorbable ions must be negative. Theories of the structure of the cathode film (31) indicate that close to the cathode there is a paucity of negative ions, because they are repelled by the cathode. This is the most important region, hence means must be found for bringing negative ions into this forbidden territory.

The question naturally arises as to why the dichromate ion, $Cr_2O_7^{--}$, is ineffective (chromate ion, CrO_4^{--}, does not exist at a pH of 3 or 4). Being a negative divalent ion, it could be adsorbed by the sol and thus alter the electrophoretic velocity. It is admitted that this effect does exist in the outer fringes of the film, but apparently not close to the cathode proper. Only high-energy negative ions can penetrate the repulsive barrier and be adsorbed by the electropositive sol. It has been demonstrated above that chromium dichromate does not exist in the green form and hence does not form positive molecular ions like $[Cr(Cr_2O_7)(H_2O)_4]^+$, which on account of their nonreactive nature would be able to penetrate the film. Even if it were so conveyed, the dichromate would be reduced at the cathode and hence become ineffective. An effective addition agent must be nonreducible. The nonexistence of the "green" form of the dichromate favors, however, the formation of oxide-free deposits.

 The action with the sulphate is vastly different. In consequence of its small size and high charge it will be readily adsorbed by the electropositive sol. It readily forms green complexes with trivalent chromium in which the sulphate is in an extremely nonreactive, positive molecular ion, possibly $[Cr_4O(SO_4)_4(H_2O)_x]^{++}$. This ion is capable of passing through the "forbidden" territory with little change. On reaching the cathode surface the ion is reduced so that the chromium is no longer trivalent and the sulphate ion is set free. While still in the cathode film it may either first, reform the green complex (which is improbable at the pH prevailing); second, exist in the free state (a possibility already demonstrated in this paper); or, third, it may be adsorbed by the particles of the sol. This latter action will lower the electrophoretic velocity, and render the formation of a dense collodial layer improbable. The action of the chloride would be similar. The positive molecular ions that would then be effective are undoubtedly $[CrCl(H_2O)_5]^{++}$ and $[CrCl_2(H_2O)_4]^{+}$.

Evidence of the relative adsorption of ions by basic chromium chromate was obtained in migration experiments reported in Table 10. A suspensoid of the basic chromate, $Cr(OH)_3Cr(OH)CrO_4$, was prepared by grinding a precipitate in a colloid mill until the particles were of the order of 1 μ. The migration apparatus was of the type used by Abramson and Freundlich (16). Electrode effects were rendered negligible by having nonpolarizable electrodes. The electro-endosmose effect was eliminated in the manner advocated by Smoluchowski (8). Relative values only are reported, as absolute measurements were impracticable with the apparatus employed. Either sodium or potassium salts were used.

TABLE 10.—*Relative electrophoretic velocities of basic chromate in the presence of added ions at 25° C.*

Concentration in equivalent per ml ×10⁻⁷	SO--	CrO--	Cr₂O--	NO₃-	ClO₄-	H₂PO₄-	Cl-	OH-
0	+0.133	+0.133	+0.182	+0.133	+0.133	+0.133	+0.170	+0.173
1	+.122	+.130						
4								+.089
5	+.067	+.040						
8								+.050
10			+.100				+.106	
12								−.067
16								−.100
20			+.095					
40			+.076					
50	+.010	−.092					+.048	
100							+.009	
150			+.073					
200			+.050					−.204
500	±.000	−.127		+.095	+.130	+.099	±.000	
1,000	−.038	−.130		+.084	+.127	+.077		

It is obvious that three ions are outstanding in their adsorption; namely, the sulphate, chromate, and hydroxide. The effects of the latter two ions are undoubtedly specific, since we are dealing with a basic chromate. At a pH of 3 or 4 neither chromate nor hydroxide ions can exist in significant concentrations, hence they are unimportant in the cathode film. The dichromate and the chloride are adsorbed to about the same extent, but only the chloride is an effective addition agent. This difference must be due to the fact that the green form of chromic chloride exists, whereas that of chromic dichromate does not. The perchlorate ion is much less adsorbed than the nitrate or phosphate.

pH measurements with the glass electrode were made on solutions with hydroxide additions. It was established that the isoelectric point occurred at a pH of about 6. The maximum electropositive velocity was attained at a pH of 4.5 and remained constant with a further decrease in pH. The maximum electronegative velocity was attained at 7 and did not change with a further increase in pH.

5. CONCLUSIONS

(a) The processes of reduction at low and high current density are different.

(b) Metal deposition from the chromic acid bath depends on maintaining a definite pH in the cathode film.

(c) At high current densities pure chromic acid can not be reduced owing to the formation of a colloidal layer.

(d) The action of the sulphate is to lower the electrophoretic velocity of the particles of the sol and thus render the formation of a dense adherent layer improbable.

(e) The sulphate is effective because it is a negative ion that is nonreducible and strongly adsorbable, and that forms positive molecular ions with trivalent chromium.

V. APPLICATIONS OF THE PROPOSED THEORY

The utility of the proposed theory was tested by explaining the principal known facts of chromium deposition, and by making predictions that were verified experimentally. The data used in these considerations were derived principally from the curves of Haring and Barrows (17) and of Farber and Blum (25), whose experimental conditions were clearly defined. These data refer chiefly to the 2.5 M solutions of chromic acid.

1. The sulphate ion is the most effective addition agent tried. This is to be expected in view of the fact it forms positive molecular ions with trivalent chromium and that it is highly adsorbed by the basic chromate.

2. The effect of the sulphate is independent of the form in which it is added. This follows from the theory which involves only the sulphate ion or its complex with trivalent chromium.

3. A low concentration of sulphate is effective. This is in accordance with adsorption phenomena and with the migration experiments recorded in this paper.

4. The curve for the sulphate concentration versus the cathode efficiency passes through a maximum. The increase in the cathode efficiency up to the maximum represents the favorable effect of the sulphate in preventing the formation of a dense colloidal layer on the cathode. Beyond the maximum the adsorption is so great as to produce excessively large colloidal particles with a sluggish Brownian movement. This feature decreases the permeability and hence the efficiency. At the isoelectric point a colloidal layer is formed through precipitation. The deposits produced with excess sulphate are usually nonadherent, owing probably to the precipitated colloid. The beneficial action of the sulphate occurs on the positive side of the isoelectric point.[7]

5. At higher current densities the curve for the sulphate concentration versus the cathode efficiency becomes flatter. This is because at the higher current densities more colloid is formed and the effect of a given increase in sulphate content is proportionally less.

6. At higher temperatures the sulphate content is less critical, as was shown by Willink (32). This is because the sulphate is then not so strongly adsorbed, which tends to make the curve of current

---

[7] Compare Frölich (15).

efficiency versus sulphate concentration broad and similar to that obtained with the fluoride at lower temperatures. A higher concentration of sulphate is probably necessary in order to realize the maximum efficiency.

7. The curve for fluoride concentration versus cathode efficiency has a long flat maximum, in contrast to the sharp maximum of the sulphate curve. The adsorption of the divalent sulphate ion will be much greater than that of monovalent ions, such as fluoride and chloride. No electrophoretic measurements were made on the effect of the fluoride, but it is probably similar to that of the chloride. On this basis the difference in behavior can be explained.

8. The acetate ion is ineffective. Acetates and other organic anions do not form green chromic complexes.

9. The nitrate ion is ineffective. Carvath and Curry (1) reported that it produces black deposits with low efficiencies. The nitrate does not form green complexes under the conditions that obtain in these solutions and is not readily adsorbed by the sol.

10. According to Müller phosphates are completely ineffective. While the dihydrogen phosphate ion is adsorbed at least as much as the nitrate, chromium phosphate is very insoluble at the pH of the cathode film.

11. Perchlorate is ineffective according to Müller's polarization curves. As this is a stable radical it would appear ideal according to the theories of Haring, Ollard, and Piersol. However, the green form of chromium perchlorate does not exist, and the perchlorate ion is only slightly adsorbed; hence, according to the theory presented in this paper, it is ineffective. This was confirmed by numerous experiments. In all cases a brown powder or no deposit was obtained and the metal efficiencies were very low (compare (33)). The solutions contained 2.5 M CrO_3 and from 0.1 to 1 M $HClO_4$, the current densities ranged from 10 to 50 amp./dm², and the temperature was 45° C.

12. Pure chromic acid can be reduced to trivalent chromium at low but not at high current densities. This is because at low current densities the chromium compounds have access to the cathode, while at high current densities the semipermeable colloidal layer prevents reduction.

13. As the concentration of chromic acid is increased, the cathode efficiency is decreased. In view of the fact that the chromic ion is less readily reduced and slower than the hydrogen ion, the hydrogen ion concentration in the cathode film will be the most sensitive factor dependent on the rate of supply. With a greater concentration of acid in the body of the solution, the hydrogen ion concentration in the film will be greater, and hence the efficiency will be less.

14. The cathode efficiency does not change greatly if the acidity is decreased by neutralizing part of the chromic acid with alkali. This conclusion (Farber and Blum) is in apparent contradiction with No. 13. The two cases are not strictly comparable. In this case the slower and less reducible positive ion (sodium) would permit a larger concentration of dichromate ions to exist in the cathode film, and cause an even lower hydrogen ion concentration than would be expected on the basis of lowering the free acid. Apparently what occurs is that this condition favors the formation of excessive amounts of the colloid, which

counteracts the increase in efficiency due to lowering the hydrogen ion concentration.

15. At 25° C. the bright plating range is decreased by an increase in the trivalent chromium content (18). This behavior is readily explained by the theory. It was also predicted that this effect could be counteracted by an addition of sulphate. The trivalent chromium increases the cathode efficiency, but also increases the colloid formation at the cathode, especially at high current densities. By adding sulphate this effect should be eliminated. This was confirmed by experiments with 2.5 M CrO_3 in which the presence of about N trivalent chromium required an increase of sulphate from 0.05 N to 0.15 N to widen the plating range.

16. The cathode efficiency increases with an increase in current density. This is because at the higher current density the hydrogen evolution is greater (absolutely but not relatively) and hence the cathode film becomes less acid, and a higher potential is required for hydrogen evolution, which factor favors a higher efficiency for the reduction of chromic acid.

17. A still further increase in current density produces "burnt" deposits. These contain oxides derived from the hydrolysis of the chromic ion when the hydrogen ion concentration in the cathode film is too low.

18. The cathode efficiency decreases with an increase in temperature. The hydrogen overvoltage decreases with increase in temperature, and hence hydrogen evolution is increased. This effect partially results from the more rapid convection with increasing temperature, which brings a larger supply of chromic acid into the cathode film and increases the hydrogen ion concentration.

19. Although many of the factors that give rise to the widened plating range at higher temperatures are beyond the theory presented, there is one important factor that can be accounted for. The shift of the plating range at higher temperatures to higher current densities is due to the fact that then a higher current density is necessary to yield the intermediate hydrogen ion concentration necessary for chronium deposition.

20. The initial cathode efficiency is highest on metals with the highest hydrogen overvoltage. This feature is strictly beyond the theory, but is consistent with it.

A number of other features of the process have been analyzed by means of the theory and explained. Only those that are regarded as the most outstanding have been reported.

VI. ACKNOWLEDGMENTS

Appreciation is extended to W. Blum for his suggestion of the problem, his ever available assistance, and his able counsel. M. R. Thompson kindly made the pH measurements that disclosed important relationships.

VII. BIBLIOGRAPHY

1. H. R. Carveth and B. E. Curry, J. phys. chem., vol. 9, p. 353, 1905.
2. N. Bjerrum, Z. physik. chem., vol. 59, p. 596, 1907.
3. N. Bjerrum, Z. anorg. chem., vol. 63, p. 140, 1909.
4. R. Abegg, F. Auerbach and R. Luther, Abhandlungen der Deutschen Bunsengesellschaft No. 5 Halle, 1911.
5. A. Hantzsch and P. Garrett, Z. physik. chem., vol. 84, p. 321, 1913.

6. G. J. Sargent, Trans. Am. Electrochem. Soc., vol. 37, p. 479, 1920.
7. R. Abegg, Handbuch der Anorganischen Chemie, Band IV, 1921.
8. M. Smoluchowski in Gratz, Handbuch der Elektrizität, Band II, p. 366, 1921;
9. E. Liebreich, Z. Elektrochem., vol. 27, pp. 94, 452, 1921.
10. E. Liebreich, Z. Elektrochem., vol. 29, p. 208, 1923.
11. C. G. Fink, United States Patent, 1581188, 1925.
12. Opt. Soc. Am., vol. 10, p. 169, 1925.
13. H. E. Haring, Chem. Met. Eng., vol. 32, p. 692, 1925.
14. E. Müller, Z. Elektrochem., vol. 32, p. 399, 1926.
15. P. K. Frölich, Trans. Am. Electrochem. Soc., vol. 69, p. 395, 1926.
16. H. Freundlich and H. A. Abramson, Z. physik. chem., vol. 126, p. 25, 1927.
17. H. E. Haring and W. P. Barrows, B. S. Tech. Paper No. 346, 1927.
18. R. Schneidewind, Univ. Michigan Eng. Res. Bull., No. 10, 1928.
19. D. T. Ewing, J. O. Hardesty, and T. H. Kao, Bull. No. 19, Michigan Eng. Exper. Station, 1928.
20. E. Müller and P. Eckwall, Z. f. Elektrochem., vol. 35, p. 84, 1929.
21. E. Müller and I. Stscherbakow, Z. f. Elektrochem., vol. 35, p. 222, 1929.
22. E. A. Ollard, Electroplaters' and Depositors' Tech. Soc., vol. 3, p. 5, 1929.
23. S. Freed and C. Kasper, J. Am. Chem. Soc., vol. 52, p. 2632, 1930.
24. E. Müller and O. Essin, Z. Elektrochem., vol. 36, p. 2, 1930.
25. H. L. Farber and W. Blum, B. S. Jour. Res., vol. 4, p. 27, 1930.
26. H. R. Moore and W. Blum, B. S. Jour. Res., vol. 5, p. 255, 1930.
27. R. J. Piersol, Metal Cleaning and Finishing, vol. 3, p. 207, 1931.
28. R. W. Wyckoff, Structure of Crystals, 2d ed., Chem. Catalogue Co., 1931.
29. L. Pauling, J. Am. Chem. Soc., vol. 53, p. 1396, 1931.
30. M. Huggins, J. Am. Chem. Soc., vol. 53, p. 3190, 1931.
31. R. W. Gurney, Proc. Roy. Soc., vol. 134A, p. 137, 1931.
32. A. Willink, Trans. Electrochem. Soc., preprint 61-5, 1932.
33. E. Müller, Z. Elektrochem., vol. 38, p. 205, 1932.
34. B. Lange, Z. physik. chem., vol. A 159, p. 277, 1932.

WASHINGTON, July 6, 1932.

RP477

PHYSICAL PROPERTIES AND WEATHERING CHARACTERISTICS OF SLATE

By D. W. Kessler and W. H. Sligh

ABSTRACT

Tests on 343 samples of slate from the various districts gave the following average values: Modulus of rupture, 11,700 lbs./in.^{2}; modulus of elasticity in flexure, 13,500,000 lbs./in.^{2}; toughness, 0.192; abrasive hardness, (H_a) 7.6; absorption, 0.27 per cent by weight; porosity, 0.88 per cent; bulk density, 2.771; weight per cubic foot, 172.9 pounds. The strength, elasticity, and toughness values given above were obtained on oven-dried specimens tested in the strongest grain direction. Strength determinations on specimens that had been soaked in water for several days showed considerably lower values. The average ratio of the modulus of rupture of soaked specimens to that of dry specimens was 0.69.

The examination of a considerable number of slate shingles which had been in service for periods varying from 12 to 130 years indicated that slate deteriorates mainly from a combination of chemical and physical causes. Slates containing both pyrite and calcite in appreciable amounts are subject to decay due to the conversion of a part of the calcite to gypsum. The increase in molecular volume causes scaling of the surface. Slate shingles exposed on the roof decay more rapidly on the downward surface, which is probably due to the leaching and concentration of gypsum. A similar type of decay can be produced by alternately soaking and drying the slate. The slates which are more subject to this type of weathering can be disintegrated by 40 or 50 cycles of this treatment.

Frost may cause deterioration, but the rate of this action is very slow. Aside from chipping off the scales near the edges already loosened in the trimming process, frost plays a very minor part in slate weathering. However, the freezing of water between shingles which are nailed down tightly may cause breakage.

CONTENTS

I. INTRODUCTION

This paper embraces the results of laboratory studies on the commercial slates of this country and is a continuation of the general investigation relating to various types of stone.

The samples were from nine different States, and consist, mainly, of the materials now found on the market. Fifty samples in the form of large blocks and a large number in the form of shingles were collected especially for this work. Results are also given for numerous samples submitted for various Government buildings. Studies of slate weathering were made of shingles taken from buildings representing exposures of from 12 to 130 years.

While the scope of the paper is limited mainly to physical properties, a brief description of the general characteristics and uses of slate is included as an aid in the interpretation of the results.

The cooperation of the National Slate Association, Bangor Slate Association, Pennsylvania Slate Institute, The Structural Slate Co., and many individual producers was of great assistance and is gratefully acknowledged. The Albion Vein Slate Co. supplied samples from each bed in its quarry at a depth from 190 to 240 feet. The tests on these samples indicate the variations in a section across this quarry and may also be applicable to other quarries on the same "vein."

II. GENERAL DESCRIPTION OF SLATE

1. ORIGIN

All slates considered herein are assumed to have been formed from very fine argillaceous silts, originally deposited under water in horizontal beds. These beds were compacted by vertical pressures due to subsequent deposits of other sediments. Finally, horizontal pressures distorted the original beds into folds, some of which are now standing in a vertical position, while others are overturned and are resting in a horizontal position. The dynamic action which caused the folding of the beds also caused further consolidation and, by the aid of heat, a process of recrystallization took place which resulted in a marked change in the mineral constituents. The predominating mineral in slate is found to be mica in the form of fine flakes. These flakes are arranged in parallel order, with the flat faces roughly parallel to the axes of the folds. As the original silts were formed from decomposing igneous rocks, the chemical composition of slate bears some relation to that of the parent rock. The geologic age of the slates ranges from Cambrian to Silurian.

2. DISTRIBUTION

Although slate deposits are known to exist in 17 States, it is now being produced in only 5 or 6. The latest statistics indicate that 95 per cent of the roofing slate is supplied by Vermont, Pennsylvania, and Virginia. A large portion of the slate from Maine is used in electrical equipment. The Vermont-New York region supplies red, green, and purple slates for roofing, electrical, and structural purposes, as well as some dark and gray varieties for roofing only. Pennsylvania also produces a large amount of structural and blackboard slate, as well as a considerable quantity for electrical purposes.

3. CLASSIFICATION

Slate is broadly classified under the headings of mica, clay, and igneous slates. A mica slate is one formed from clay sediments, but the process of metamorphism is so complete that the clay content is very small. A clay slate is one in which clay still exists in appreciable quantities, while an igneous slate is one resulting from the metamorphism of a ledge of igneous rock. All materials included in this report are classified as mica slates. The slates of this country have sometimes been subdivided into fading and unfading varieties; but due to the misconceptions arising from such terms and difficulties involved in defining such terms, this phase of the subject will be discussed under a separate heading.

4. CLEAVAGE

The distinguishing characteristic of slate is its fissility. Most types of rock may be split more easily in one direction than in another, but in no other type is this property so prominent as in slate. The plane of splitting is commonly called the cleavage plane. Cleavage is evidently due to the parallel arrangement of the mica flakes and the elongation of other mineral constituents in the same general direction. As may be expected, the flattening of the mineral particles occurred on the sides exposed to the greatest pressure and the elongation in the direction of least resistance. The horizontal thrust which caused the folding of the original beds also caused the parallel arrangement of minerals; hence the cleavage is found to make various angles with the original direction of bedding, depending on the part of the fold being considered. At either the apex or trough of a fold the cleavage is commonly perpendicular, while on the sides of the folds it is roughly parallel to the bedding. Not all deposits of slate are as simple as the case just described, and it can not be definitely claimed that horizontal pressures are necessary for the formation of slate. Probably the only necessary conditions are pressures from two opposite directions on a clay stratum with a limited degree of freedom to flow or yield in some other direction. The direction of yielding will determine the cleavage.

5. GRAIN

A slab of slate having its broad faces parallel to the cleavage can be broken transversely in one direction more readily than at right angles to this direction. This direction of easiest breaking is called the "grain," and is caused by the arrangement of the minerals. This feature is assumed to be due to a greater elongation of the mineral in the grain direction than at 90°.

The grain or "sculp" of slate is advantageous in quarrying processes. Mill blocks, or pieces taken from the quarry in convenient size for handling, commonly have the largest faces in the cleavage planes, the next largest in the grain direction, and the least across grain. Likewise in the finished product, when in the form of rectangular slabs, the largest, next largest, and smallest faces will bear the same relation to the cleavage, grain, and cross-grain directions.

There are various ways of determining the grain direction in slate. Slate workers usually resort to the hammer and ascertain the easiest breaking direction in a slab. A means proposed by Jannetaz [1]

---

[1] Jannetaz, E., Memoire sur les clivages des roches et sur leur reproduction. Soc. Geol. France, Bul. 3d ser., vol. 12, p. 211, 1884.

depends on an assumed difference in thermal conductivity parallel and perpendicular to the grain. This means has not proven very satisfactory as applied to the slates of this country. A method used with considerable success in this study is as follows: A square slab of slate is laid on an asbestos mat with a hole in the center of the latter about 1 inch in diameter. The flame of a Bunsen burner is applied directly to the slate surface through the hole. In a few seconds the slate will usually break in the grain direction.

6. TEXTURE

The term "texture" is commonly used by geologists, and mineralogists in describing the state of aggregation of mineral components in rocks. This term in the slate trade has come into a somewhat more restricted use in describing the character of the split surface. A slate which presents a rough cleavage surface when split in the usual way is said to have more "texture" than one that splits evenly. Within recent years architects have often shown a preference for what they call "textural roofs." By using the rougher splitting slates in various sizes and thicknesses they obtain what is designated as "roof texture."

7. COLORS

The prevailing color of slates now being produced in Maine, Pennsylvania, and Virginia is dark gray with slight tinges of blue, green, or occasionally brown. Generally the slates from Maine and Virginia present a somewhat darker appearance than the Pennsylvania slates, although the Peach Bottom slates from Maryland and certain beds from Pennsylvania are quite dark. That known as "gray bed" from Pennsylvania, representing a minor part of the total production of that slate, is gray with a tinge of green. The production of red, green, and purple slate is now mainly limited to the Vermont-New York region. A purple slate was once worked in Maryland. Black, red, and green slate deposits occur in Arkansas, and green slates are now being worked to some extent in Georgia and Tennessee.

8. COLOR PERMANENCE

The apparent fading which occurs in some of the dark-gray slates seems to be due to an efflorescence of calcium sulphate which partially conceals the true color. During rains, when a slate roof is wet, such fading effects are not apparent, although the same roof may show considerable fading when dry. The source of the calcium sulphate will be discussed under the heading "Weathering processes."

Some of the dark-gray varieties change to a rusty color after a period of exposure. This appears to be due to decomposition of the sulphides of iron in the slate. The first change is probably the formation of sulphate of iron, which is finally oxidized on the exposed surface.

Some of the green varieties of slate also change to buff or rusty brown after a few months of exposure. Such slates have been called "sea green" or "weathering green" slates. This change has been ascribed to the conversion of finely divided particles of iron carbonate to the oxide.

The red, most of the purple, and some of the green varieties undergo no appreciable change of color during long periods of exposure.

As no satisfactory laboratory test for color permanence has been developed, it has been the practice to rely on past experience with the various deposits. Generally those slates which are low in carbonates undergo the least color changes.

9. RIBBONS

Bands of different color, texture, and composition in slate are commonly called "ribbons." The most common ribbons are darker in color than the main part of the slate and vary in width from a fraction of an inch up to several inches. Ribbons follow the original bedding planes of the clay deposits and represent periods of accumulated coarse particles in the sediments. The dark color is due to carbonaceous matter. Such ribbons are usually more absorptive and contain more calcite than the clear part of the slate. Fresh slate containing these ribbons usually shows no appreciable difference in strength whether broken through a ribbon or through the clear part. On account of having coarser particles of quartz the ribbon offers more resistance to cutting or grinding than other parts of the material. The cleavage surface is more uneven and is often deflected somewhat at the ribbon.

Fading effects as mentioned in section 8 above are more common along ribbons, and generally this part is less durable than the clear portion when exposed to the weather.

III. FACTORS AFFECTING SERVICE

1. SERVICE CONDITIONS

Since the most extensive use of slate is for roofing purposes, the foremost point of concern is durability. A roofing material is obviously exposed to the most severe weather conditions. Roof temperatures may often reach 140° F., and the exposure offers no protection against the lowest winter temperatures. Moisture, ice, snow, and hail all contribute to the severe conditions. Wind storms deflect the rafters, and where slates are nailed down too tightly may strain them to the breaking point. Besides the normal weather conditions, slate is exposed to flue gases and the leaching from deposits of soot. Evidently the durability of slate in such exposure depends upon the resistance of the mineral component; hence a study of the composition is important. Strength, elasticity, and density are also important characteristics.

Slate also finds extensive use in the form of heavier slabs for steps, floor tile, window sills, toilet stalls, tubs, sinks, etc. In such uses strength and density are important, but probably the most important property is resistance to abrasion. Where exposed to severe wear, as in floors or steps, the presence of hard veins is undesirable because they cause uneven wear.

For blackboards the material has few competitors. Color and texture are the main considerations. In this use of the material, physical tests afford little information of value.

Electrical slate is limited mainly to the products of those deposits which have high insulating values. A slate of high absorption is not apt to prove satisfactory for this purpose unless it is thoroughly dried and treated with some preparation to prevent the entrance of

moisture. Strength and toughness are important properties, but the
final test is that of electrical resistance, and this test it of little value
unless applied to the entire panel. The producers of switchboards
should have appliances for conducting such tests.

2. FABRICATION PROCESSES

Certain processes used in quarrying and working slate into the
finished product may have a bearing on its service. The use of
explosives in quarrying is quite general. Considerable waste material
results, due to shattering, and it seems inevitable that some of the
finished product will contain portions of the injured material.

In splitting slabs from quarry blocks the practice is to separate the
block into two equal parts by a middle split. The two halves are
then divided into halves and so on until the desired thicknesses are
obtained. In roofing slate, where no further finishing of the split
face is required, there is necessarily a considerable variation in thickness. Slabs which are to be planed to definite thicknesses, as for
structural or electrical purposes, must be split to a thickness considerably greater than the finished piece. Slates with curved cleavage are not suitable for such purposes.

The splitting process is varied somewhat according to the nature
of the cleavage. Slates having an easy cleavage are split with a
long-handled wedge driven into the block on one of the long edges
and used as a lever. Slate blocks somewhat more reluctant to split
are separated with one wedge, which is driven in the middle of one
end and used as a lever. In the most difficult cases the driving of
two or more wedges to start the crack is required, and the final separation is made by forcing a long thin blade, shaped like a spatula, down
from one end. In all of these methods there is an unavoidable strain
produced which is transverse to the slab and greatest in the weakest
direction. Thin roofing slates occasionally find their way into
the stacks of marketable material which have partially opened transverse cracks running the long way of the shingles. The presence of
such cracks can usually be detected by supporting the shingle on the
knuckles of one hand and striking it with a light tool or pencil. If
free from cracks it should ring clearly.

Slate shingles are usually trimmed to size, after being split to the
desired thickness, by shearing with heavy steel blades. Although
this would seem to be a severe treatment for so brittle a material, no
very appreciable injury appears to be caused by shearing. However,
it probably accounts for the fact that small fragments of slate are
often found on the ground near slate roofs. That the slight fracturing
near the sheared edge does not affect the durability of the slate is
indicated by the fact that decay is more rapid on other parts.

Nail holes are more frequently punched than drilled. Machine
punching is usually considered to be more satisfactory than hand
punching, although the process is about the same in both. The
punch is driven from the side that is to be laid downward. Where the
point comes through there is enough chipping around the hole to
produce a recess that serves the purpose of countersinking for the
nail head. Observations on old shingles seem to indicate that punching is more injurious than the shearing, as decay often begins around
the hole before it does on other parts. Some trimming and punching

)e done by the roofer. As this is done under less favorable
ons than at the plant, it is more apt to prove injurious. Fur-
re, slate is less brittle when fresh from the quarry than after
become seasoned.

IV. COMPOSITION OF SLATE

1. MINERAL COMPOSITION

prevailing mineral in the commercial slates of this country is
mica, which accounts for 30 to 40 per cent of its bulk. Next
er is usually quartz or carbonate of lime, although chlorite,
or carbon is occasionally found to be next in abundance.
magnetite, and rutile are usually present, and occasionally
ematite, rhodochrosite, zirocon, tourmaline, pyrrhotite, apatite,
te, and feldspar. These minerals occur as small scales, lenses,
stals, the greatest dimension of which is generally a small
n of a millimeter.

2. CHEMICAL COMPOSITION

·cal analyses made in connection with this study have been
d mainly to weathered slates to determine what changes have
ed. Chemical analyses of various slates have been compiled
e in United States Geological Survey Bulletin No. 586. The
ons shown for slates from different regions are as follows:

	Vermont	New York		Virginia
	Per cent	*Per cent*	*Per cent*	*Per cent*
....	59 to 68....	56 to 68....	55 to 65....	54 to 62.
....	14 to 19....	10 to 13....	15 to 22....	17 to 25.
....	0.8 to 5.2....	1.5 to 5.6....	1.4 to 4.5....	7.0 to 7.8.
....	2.5 to 6.8....	1.2 to 3.3....	2.3 to 9.0....	
....	0.3 to 2.2....	0.1 to 5.1....	0.2 to 4.2....	0.4 to 1.9.
....	2.2 to 3.4....	3.2 to 6.4....	1.5 to 3.8....	1.5 to 3.9.
....	3.5 to 5.5....	2.8 to 4.4....	1.1 to 3.7....	
....	1.1 to 1.9....	0.3 to 0.8...·....	0.5 to 3.5....	
....	0.1 to 3.0....	0 to 7.4....	1.6 to 3.7....	0.2 to 2.0.

chemical composition of various minerals occurring in slate
n in textbooks on mineralogy are as follows:

Muscovite............................ $H_2KAl_3(SiO_4)_3$.
iotite............................ $(HK)_2(Mg, Fe)_2(Al, Fe)_3(SiO_4)_3$.
uarts............................ SiO_2.
alcite............................ $CaCO_3$.
·hlorite (clinochlore)............ $H_8(Mg, Fe)_5Al Si_3O_{18}$.
rite............................ FeS_2.
utile............................ TiO_2.
Iagnetite............................ Fe_3O_4.
ourmaline............................ R_9SiO_5, $R = Al, K, Mn, Ca, Li$.
aolinite............................ $H_4Al_2Si_2O_9$.
iderite............................ $FeCO_3$.

to the complex nature of some of these minerals, it is a very
t, if not impossible, task to determine the true composition
chemical analysis. Referring to the molecular formulas given
for the predominating minerals in slate, it is seen that silica
may be ascribed to muscovite, biotite, quartz, chlorite, tour-

maline, or kaolinite; iron oxides (Fe_2O_3 and FeO) to biotite, chlorite, magnetite, or siderite; alumina (Al_2O_3) to muscovite, biotite, chlorite, tourmaline, or kaolinite; lime (CaO) to calcite or tourmaline; magnesia to biotite or chlorite; potassia (K_2O) to muscovite, biotite, or tourmaline; soda (Na_2O) to muscovite; carbon dioxide to calcite or siderite; sulphur (S) to pyrite; titanium (TiO_2) to rutile. Water (H_2O) determined below 110° C. is considered to be present in the form of moisture within the pores, while water determined above 110° C., may form a part of the minerals muscovite, biotite, kaolinite, or one of the chlorite group. Sulphur anhydride (SO_3) is seldom present in appreciable quantities in fresh slate, but in weathered slates it indicates the presence of gypsum ($CaSO2_4H_2O$).

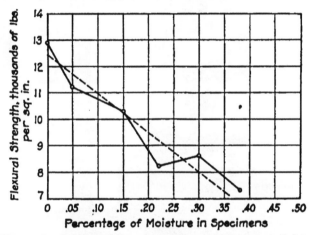

FIGURE 2.—*Relation of flexural strength to moisture content of slate*
[Each point on the curve is the average of six strength tests.

V. PHYSICAL TESTS

1. STRENGTH

The flexural strength of slate is commonly determined instead of compressive, tensile, or shearing strength because of facility in preparing the specimens. In this work a specimen 12 inches long, 4 inches wide, and of thickness varying from one-eighth to one-half inch was generally used. These were tested on the apparatus shown in Figure 1. The pieces were supported on knife-edges of the rocker type spaced 10 inches and loaded at the middle through a third knife-edge. Loads were added in 10-pound increments. The strength was computed by means of the modulus of rupture formula $R = \dfrac{Wl}{bd^2}$ in which W = breaking load, l = length of span, b = width, and d = thickness, the load being in pounds and dimensions in inches.

The strength was usually determined "across the gr in"; that is, the long dimension of the specimen was parallel to the grain direction, and hence the fracture was transverse to this direction. Several tests were also made "parallel to the grain" in which the fractural occur-

B. S. Journal of Research, RP477

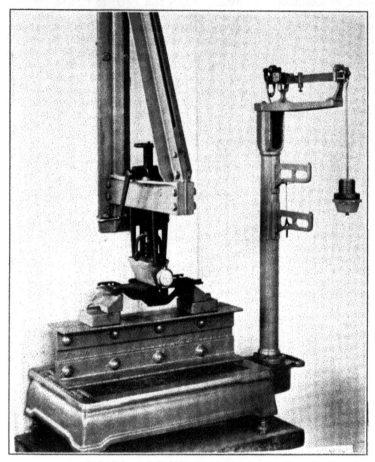

FIGURE 1.—*Apparatus used for determining the flexural strength and elasticity of slate*

red in the grain direction. Results of strength tests on slate in the dry condition are given in Table 1.

The effect of absorbed moisture on the strength was studied by tests on several samples after 14 days' soaking. The pronounced reduction in strength is shown in the last column of Table 2. Figure 2 shows strength determinations on several specimens of one slate tested with various amounts of absorbed water. This shows the importance of drying before the strength test if truly comparative results are desired.

TABLE 1.—*Modulus of rupture—specimens dry*

MAINE SLATES

Serial No.	Number of samples	Number of tests	Modulus of rupture, thousands of lbs./in.²			Source of sample	Variety
			Maximum	Minimum	Average		
1.........	1	⁴ 3	10.9	10.0	10.6	Portland Monson Co......	Blue black.
2.........	2	⁴ 7	12.1	10.2	11.0	Monson Maine Co.........	Do.
	1	⁴ 3	9.0	7.6	8.3	.....do.....................	Do.
3.........	1	⁴ 6	15.3	9.8	13.3	Rising & Nelson Slate Co.	Do.

VERMONT-NEW YORK SLATES

4.........	4	¹ 11	15.5	5.3	9.8	Norton Bros...............	Sea green.
	2	⁴ 5	10.3	5.9	7.6	.....do....................	Do.
5.........	2	⁴ 4	14.9	12.0	13.9	.....do....................	Silver gray.
	1	⁴ 3	12.2	9.5	11.1	.....do....................	Do.
6.........	2	⁴ 4	14.6	8.4	12.0	.....do....................	Gray and black.
	1	⁴ 3	12.9	9.5	11.3	.....do....................	Do.
7.........	2	⁴ 4	10.2	7.2	8.6	.....do....................	Green and purple.
	1	⁴ 3	8.7	7.2	8.2	.....do....................	Do.
8.........	2	⁴ 4	12.7	10.4	10.8	.....do....................	Unfading green.
	1	⁴ 3	10.3	8.8	9.7	.....do....................	Do.
9.........	1	⁴ 1			10.7	.....do....................	Dark purple.
10.........	1	⁴ 2	12.2	11.6	11.9	Wells, Vt.............	Green and purple.
13.........	1	⁴ 5	11.1	7.7	9.2	J. D. Emack...........	Colonial gray.
14.........	1	⁴ 3	12.3	11.1	11.8	Hydeville, Vt.............	Green and purple.
	1	⁴ 3	9.4	9.0	9.2	.....do....................	Do.
	1	⁴ 3	8.4	7.5	7.8	.....do....................	Purple.
15.........	1	⁴ 2	8.2	7.5	7.85	.....do....................	Do.
16.........	1	⁴ 3	13.3	8.6	10.9	.....do....................	Green.
	1	⁴ 3	9.3	5.6	7.0	.....do....................	Do.
17.........	1	⁴ 2	11.0	8.7	10.2	Fair Haven.............	Green and purple.
	1	⁴ 3	9.9	6.2	8.4	.....do....................	Do.
18.........	1	⁴ 3	12.6	12.0	12.3	.....do....................	Unfading green.
	1	⁴ 3	8.3	7.3	7.8	.....do....................	Do.
19.........	1	⁴ 3	10.5	9.8	10.0	.....do....................	Green and purple.
	1	⁴ 3	10.1	8.5	9.5	.....do....................	
20.........	1	⁴ 3	9.5	8.0	9.0	Granville..........	"Arabian red."
	1	⁴ 3	9.7	6.1	8.0	.....do....................	Do.
21.........	1	⁴ 3	15.3	15.2	15.2	Poultney..........	Semiweathering gray.
	1	⁴ 3	14.2	9.0	12.2	.....do....................	Do.
22.........	1	⁴ 2	11.5	11.4	11.4	Rupert, Vt..........	Mottled purple.
		⁴ 3	14.2	9.5	11.1	.....do....................	Do.

PENNSYLVANIA SLATES

HARD VEIN REGION

23.........	11	¹ 44	16.5	10.0	13.8	Chapman................	Smooth texture.
	1	⁴ 3	6.0	4.9	5.4	.....do....................	Do.
24.........	5	¹ 19	16.5	9.0	13.4	.....do....................	Rough texture.
25.........	1	⁴ 6	12.7	8.4	10.8	Belfast................	Smooth texture.

See footnotes at end of table.

TABLE 1.—*Modulus of rupture—specimens dry*—Continued

PENNSYLVANIA SLATES—Continued

BANGOR REGION

Serial No.	Number of samples	Number of tests	Modulus of rupture, thousands of lbs./in.²			Source of sample	Variety
			Maximum	Minimum	Average		
26...... {	3	18	16.2	9.6	13.7	Structural Slate Co......	Bangor clear.
	3	15	10.8	6.1	7.8	.....do.....	Do.
27...... {	3	18	14.8	8.2	11.5	.....do.....	Bangor ribbon.
	3	16	9.0	5.5	7.3	.....do.....	Do.
28......	3	17	16.0	7.9	12.2	Vendor Slate Co......	Bangor clear.
29......	2	16	12.6	11.5	12.1	.....do.....	Bangor ribbon.
30......	2	14	16.0	12.9	14.1	Wallace & Gale......	Bangor clear.
31......	1	14	14.6	12.3	13.4	Griffith Consumers Co....	Do.
32......	1	12	14.3	13.9	14.1	Starrett Equipment Co....	Do.
33...... {	6	18	13.9	9.9	12.2	N. Bangor Slate Co......	Do.
	1	3	8.3	7.3	7.8	.....do.....	Do.
34......	1	14	11.2	10.4	10.8	.....do.....	Bangor ribbon.
35......	1	14	15.2	12.0	13.8	Bangor Quarry Co......	Bangor clear.
36...... {	2	18	13.8	11.0	12.6	.....do.....	Bangor ribbon.
	1	1	......	......	8.1	.....do.....	
37......	1	14	12.2	11.1	11.7	Old Bangor Quarries......	Bangor clear.
38......	2	10	14.7	10.6	12.9	.....do.....	Bangor ribbon.
39......	1	14	12.5	10.4	11.2	Bangor Ideal Quarry......	Bangor clear.
40......	1	14	12.1	8.8	10.5	.....do.....	Bangor ribbon.
41......	1	14	15.8	13.8	14.6	Columbia Bangor Quarry.	Bangor clear.
42......	1	14	16.1	12.6	13.8	.....do.....	Bangor ribbon.
43......	1	14	13.1	9.7	11.7	E. Bangor Consolidated....	Bangor clear.
44......	1	14	12.4	10.9	11.5	.....do.....	Bangor ribbon.

PEN ARGYL REGION

Serial No.	Number of samples	Number of tests	Maximum	Minimum	Average	Source of sample	Variety
45......	8	31	14.3	8.8	11.7	Jackson Bangor......	Deep bed clear.
46...... {	6	18	12.9	8.6	11.1	.....do.....	Albion clear.
	1	3	11.8	8.4	10.5	.....do.....	Do.
47......	1	14	11.9	10.8	11.6	.....do.....	Albion black bed.
48......	1	14	12.0	10.0	11.0	.....do.....	Albion ribbon.
49...... {	2	17	13.0	8.5	10.9	Parson Bros......	Clear.
	1	5	11.7	6.4	8.9	.....do.....	Do.
50......	2	12	14.1	9.6	11.1	.....do.....	Deep bed clear.
51......	1	4	11.1	7.7	10.1	.....do.....	Ribbon.
52...... {	54	212	15.9	6.8	12.0	Albion Vein Slate Co......	Blue gray.
	38	134	12.6	4.6	7.4	.....do.....	Do.
52a......	6	22	15.9	9.1	12.7	.....do.....	Blue gray, 300-foot level.
53...... {	3	10	14.0	10.8	12.9	Albion Vein Slate Co......	Gray Bed.
	1	4	8.6	7.8	8.1	.....do.....	Do.
54...... {	2	8	12.1	9.7	11.0	.....do.....	Ribbon.
	1	4	7.9	6.8	7.2	.....do.....	Do.
55......	1	2	11.8	11.1	11.4	Keenan Structural Slate Co.	Clear.
56......	4	20	11.5	8.8	10.3	Doney Slate Co......	Do.
57......	3	11	11.7	9.1	10.7	Colonial Slate Co......	Do.
58...... {	4	12	11.1	9.0	10.4	Structural Slate Co......	Do.
	3	9	9.5	8.2	8.5	.....do.....	Do.
59...... {	4	12	12.1	9.0	10.6	.....do.....	Ribbon.
	5	16	10.3	7.1	8.0	.....do.....	Do.
60......	1	4	13.4	11.2	12.2	.....do.....	Gray.
61......	2	3	12.8	9.9	11.0	Vendor Slate Co......	Clear.
62......	2	10	14.2	11.3	12.7	Stephens-Jackson......	Gray.
63......	1	5	14.3	12.9	13.7	.....do.....	Blue gray.
64......	2	6	12.1	10.6	11.4	Belmont Slate Co......	Do.

See footnotes at end of table.

TABLE 1.—*Modulus of rupture—specimens dry*—Continued

PENNSYLVANIA SLATES—Continued

WIND GAP REGION

Serial No.	Number of samples	Number of tests	Modulus of rupture, thousands of lbs./in.[1]			Source of sample	Variety
			Maximum	Minimum	Average		
65	1	[1] 2	11.4	11.3	11.35	Structural Slate Co	Clear.
	5	[1] 14	7.7	5.6	7.2	.....do	Do.
66	1	[1] 3	12.3	10.4	11.3	.....do	Ribbon.
	1	[1] 5	8.8	4.8	7.1	.....do	Do.
67	1	[1] 4	11.0	9.1	10.0	Phoenix Slate Co	Rough texture.
68	2	[1] 12	13.0	10.4	11.0	Imperial Slate Blackboard Co.	Clear.
	1	[1] 3	7.7	6.2	6.8	.....do	Do.

SLATINGTON REGION

70	2	[1] 4	12.3	11.6	12.4	Franklin Big Bed	Clear.
69	1	[1] 3	10.9	9.8	10.3	Structural Slate Co	Do.
	3	[1] 10	9.2	4.6	6.8	.....do	Do.
71	1	[1] 2	13.0	11.3	12.15	Vendor Slate Co	Rough texture.
72	5	[1] 17	14.4	9.4	12.1	Washington Big Bed	Clear.
73	2	[1] 5	12.5	9.2	11.7	Slatington Slate Co	Do.
74	2	[1] 7	11.1	9.7	10.4	Kern Slate Co	Rough texture.
75	1	[1] 4	9.7	7.5	8.5	Shenton Slate Co	Clear.

VIRGINIA

76	7	[1] 22	14.2	8.1	10.2	Arvonia	Blue black.
	1	[1] 3	13.3	11.3	12.5	.....do	Do.
77	7	[1] 19	12.9	7.3	10.6	Ore Bank	Do.
	1	[1] 2	8.2	6.4	7.3	.....do	Do.
78	1	[1] 3	9.4	5.8	7.5	Eamont	Greenish gray.
	1	[1] 2	2.9	2.8	2.85	.....do	Do.

MARYLAND

79	1	[1] 3	9.4	5.8	7.5	Ijamsville	Purple.
	1	[2] 2	2.9	2.8	2.85	.....do	Do.
80	1	[3] 3	4.1	3.4	3.4	.....do	Green.
81	2	[1] 3	15.4	15.2	15.3	Peach Bottom	Blue black.

TENNESSEE

82	1	[1] 2	10.6	9.2	9.9	Tellico Plains	Green.
	1	[1] 3	10.0	8.5	9.0	.....do	Do.

GEORGIA

83	1	[1] 3	14.5	7.9	11.0	Fairmont	Green.
	1	[1] 3	8.2	5.7	6.7	.....do	Do.

ARKANSAS

84	1	[1] 2	6.5	3.4	4.95	Missouri Mountain	Black.
85	3	[1] 13	11.4	5.2	8.1	.....do	Green.
86	2	[1] 5	7.6	3.7	5.4	.....do	Red.

[1] Specimens broken across the grain.   [2] Specimens broken along the grain.

The remarkably high flexural strength of slate is probably due to its compactness and the reinforcing effect of the mica particles. The difference in strength between the two grain directions indicates the effect of the elongation of the mica flakes in the grain direction. The lubricating action of water in its effect on strength of stone has been emphasized by Julien [2] and by Baldwin-Wiseman and Griffith.[3]

Besides affording a ready means of comparing various slates, the flexural strength is of interest in connection with most of the uses of the material. When slate shingles are nailed down too tightly they may be strained to the breaking point due to wind pressures deflecting the rafters. Breakage of shingles on roofs due to workmen walking on the slate or placing ladders on the roof is an important consideration. The flexural strength is a factor in determining the thickness of slabs for various uses. There appears to be no general relation between the strength and durability of slates from different districts.

TABLE 2.—*Results of modulus of rupture tests—specimens wet*

Serial No. [1]	Number of specimens	Number of tests	Modulus of rupture, thousands of pounds/in.²			Ratio' wet strength to dry strength
			Maximum	Minimum	Average	
2	1	³ 3	8.0	7.2	7.7	0.93
14	1	³ 3	7.4	7.0	7.2	.61
15	1	³ 3	5.2	3.9	4.7	.60
16	1	³ 3	10.9	8.2	9.1	.83
17	1	³ 3	5.9	5.7	5.8	.57
18	1	³ 3	6.3	4.4	5.1	.42
19	1	³ 3	7.2	6.0	6.5	.65
20	1	³ 3	7.4	6.8	7.1	.89
21	1	³ 3	11.2	9.7	10.6	.89
22	1	³ 3	9.6	6.6	8.1	.73
29	1	³ 4	8.8	7.9	8.4	.69
52	38	³ 141	12.3	4.9	7.7	.66
53	1	³ 4	11.0	9.1	9.8	.74
54	1	³ 4	9.1	7.8	8.3	.74
62	1	³ 5	9.2	6.9	8.1	.66
63	1	³ 5	9.3	8.2	8.6	.63
76	1	³ 3	10.5	9.8	10.0	.80
81	1	³ 3	10.1	9.4	9.8	.64
83	1	³ 3	13.5	9.9	12.3	1.12

[1] Identification of serial numbers given in Table 1.
² Specimens broken along the grain.
³ Specimens broken across the grain.

The average modulus of rupture and extreme values for slate from various regions obtained by tests on dry specimens broken across the grain are as follows:

District	Number of tests	Average	Maximum	Minimum
Maine	21	11,700	15,300	9,800
Vermont-New York	68	10,600	15,600	5,300
Hard Vein, Pa	73	13,600	16,600	8,400
Bangor, Pa	156	12,500	16,300	7,900
Pen Argyl, Pa	431	11,500	15,900	6,800
Wind Gap, Pa	21	10,900	13,000	8,800
Slatington, Pa	35	11,800	14,400	9,200
Virginia	62	10,500	14,200	6,700

² Building Stones, J. Franklin Inst., April, May, and June, 1899.
³ Minutes of the Proc. Inst. of Civ. Eng., vol. 179, p. 290, 1909.

2. ELASTICITY

The modulus of elasticity in flexure is easily determined in connection with the modulus of rupture test by making deflection measurements at various loads. Determinations in connection with this work were made by suspending the deflectometer shown in Figure 1 from rods laid across the specimen above the supporting knife-edges. The gage used indicated deflections of 0.0001 inch. Usually deflection readings were recorded for each 20 pounds increment of load. The modulus of elasticity, E, was computed by means of the formula

$$E = \frac{W' \, l^3}{4\Delta b d^3}$$

in which Δ is the deflection corresponding to some chosen load W', and l, b, and d are the same dimensions used in the modulus of rupture test.

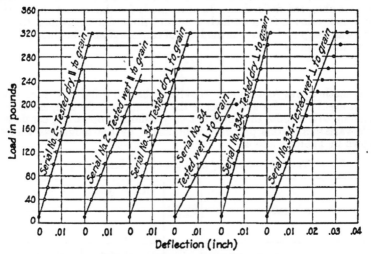

FIGURE 3.—*Stress-strain curves for slate in flexure*

Figure 3 shows a few stress-strain relations for slate specimens in flexure. In general there is a small deviation from a straight line for loads above half of the breaking load, but no definite yield point is in evidence.

Elasticity determinations have been made in both grain directions on dried specimens and across the grain on several wet specimens. The modulus values are usually somewhat lower for tests on wet specimens, but show no appreciable difference for tests made in different grain directions on dry specimens. The average value of E obtained was 13,500,000, with a maximum of 18,000,000. There appears to be less variation in the E values for slates from different regions than in other properties. The average E and extreme values determined across the grain on dry specimens for the main producing districts are:

District	Number of tests	Average	Maximum	Minimum
Maine	18	15,100,000	18,000,000	13,100,000
Vermont-New York	63	14,500,000	17,600,000	12,000,000
Hard Vein, Pa	45	15,300,000	16,700,000	13,000,000
Bangor, Pa	129	13,500,000	16,300,000	10,200,000
Pen Argyl, Pa	259	12,900,000	15,900,000	9,800,000
Slatington, Pa	29	13,400,009	15,200,000	11,300,000
Wind Gap, Pa	17	12,400,000	13,800,000	9,800,000
Virginia	56	14,700,000	16,400,000	12,400,000

3. TOUGHNESS

Merriman's test [4] for toughness with slight modifications has been employed in this work. This consists in determing the maximum deflection for a given span and thickness of specimen when loaded in the middle. Merriman used a 22-inch span and recorded the highest observed deflection. In tests made for this report the maximum deflection Δ for a 16-inch span and three-sixteenths-inch thickness was computed from the modulus of rupture R and elastic modulus E by the formula

$$\Delta = \frac{227\ R}{E}$$

This method was found more expedient than to grind the specimens to an exact thickness and observe the maximum deflection, since the grinding to a definite thickness is a very tedious process. A considerable variation in this property is indicated by the toughness values shown in Tables 3 and 4.

TABLE 3.—*Results of toughness tests—specimens dry*

Serial No.[1]	Number of samples	Maximum deflection of 16-inch span and 3/16 inch thickness			
		Number of tests	Maximum	Minimum	Average
			Inch	*Inch*	*Inch*
1	1	3	0.142	0.125	0.135
2	2	7	.193	.152	.174
	1	3	.139	.125	.131
3	1	6	.245	.159	.200
4	4	11	.212	.093	.143
	2	5	.164	.111	.130
5	2	4	.217	.212	.214
	1	3	.182	.144	.167
6	2	4	.226	.153	.193
	1	3	.202	.146	.174
7	2	4	.161	.119	.135
	1	3	.140	.116	.132
8	2	4	.194	.159	.172
	1	3	.153	.125	.142
9	1	1			.187
10	1	2	.182	.180	.181
13	1	2	.171	.131	.151

See footnotes at end of table.

[4] Strength and Weathering Qualities of Roofing Slates, Trans. Am. Soc. of Civ. Engrs., No. 551, vol. 27 September, 1892.

TABLE 3.—*Results of toughness tests—specimens dry*—Continued

Serial No.[1]	Number of samples	Maximum deflection of 16-inch span and 3/16 inch thickness			
		Number of tests	Maximum	Minimum	Average
			Inch	*Inch*	*Inch*
14	1	‡3	0.197	0.173	0.184
	1	‡3	.167	.155	.161
15	1	‡3	.117	.108	.111
	1	‡2	.130	.121	.126
16	1	‡3	.197	.125	.160
	1	‡3	.155	.105	.134
17	1	‡3	.180	.141	.166
	1	‡3	.163	.115	.142
18	1	‡3	.213	.204	.201
	1	‡3	.134	.117	.125
19	1	‡3	.181	.166	.172
	1	‡3	.169	.111	.159
20	1	‡3	.136	.113	.128
	1	‡3	.154	.066	.126
21	1	‡3	.230	.226	.228
	1	‡3	.208	.147	.184
22	1	‡2	.186	.182	.184
	1	‡3	.226	.155	.182
23	9	‡25	.249	.153	.208
24	4	‡10	.232	.154	.198
25	1	‡3	.189	.166	.175
26	3	‡8	.247	.173	.222
	5	‡14	.173	.099	.130
27	3	‡8	.233	.141	.187
	3	‡6	.166	.002	.122
28	3	‡5	.271	.190	.219
29	1	‡2	.199	.190	.194
30	2	‡4	.254	.237	.246
31	1	‡4	.238	.209	.225
32	1	‡2	.246	.227	.236
33	6	‡18	.223	.168	.197
	1	‡3	.148	.127	.140
34	1	‡4	.206	.184	.193
35	1	‡4	.240	.190	.216
36	2	‡8	.219	.189	.203
	1	‡1	....	....	.130
37	1	‡4	.209	.186	.196
38	2	‡7	.222	.172	.204
39	1	‡4	.233	.184	.206
40	1	‡4	.235	.191	.207
41	1	‡4	.231	.196	.212
42	1	‡4	.232	.185	.208
43	1	‡4	.236	.175	.207
44	1	‡4	.238	.198	.220
45	7	‡22	.244	.157	.212
46	11	‡33	.241	.160	.206
47	1	‡3	.184	.137	.166
	1	‡4	.218	.195	.210
48	1	‡4	.211	.189	.200
49	1	‡5	.205	.115	.160
50	2	‡12	.254	.170	.210
51	1	‡4	.195	.139	.179
52	47	‡104	.248	.127	.207
	1	‡2	.148	.138	.143
53	1	‡2	.250	.202	.226
54	2	‡6	.207	.174	.197
55	1	‡2	.202	.188	.195
56	3	‡10	.221	.177	.198
57	3	‡6	.211	.182	.199
58	4	‡12	.206	.161	.190
	3	‡8	.208	.139	.157
59	4	‡12	.220	.168	.193
60	5	‡16	.179	.122	.152
	1	‡2	.204	.187	.196

See footnotes at end of table.

TABLE 3.—*Results of toughness tests—specimens dry*—Continued

Serial No.[1]	Number of samples	Maximum deflection of 16-inch span and ¾₆ inch thickness		
		Number of tests	Maximum	Minimum	Average
			Inch	*Inch*	*Inch*
61	2	[3]3	0.211	0.191	0.201
62	2	[3]6	.222	.183	.196
63	1	[3]3	.219	.208	.215
64	2	[3]6	.222	.191	.208
65	1	[3]2	.195	.186	.190
	5	[3]15	.140	.112	.132
66	1	[3]3	.207	.182	.194
	1	[3]3	.143	.103	.119
67	1	[3]4	.211	.188	.200
68	2	[3]8	.227	.151	.197
	1	[3]3	.149	.123	.134
	1	[3]3	.180	.159	.169
69	3	[3]9	.163	.102	.132
70	2	[3]2	.229	.199	.214
71	1	[3]2	.224	.196	.210
72	4	[3]11	.226	.160	.190
	1	[3]3	.155	.146	.150
73	2	[3]5	.229	.151	.204
74	2	[3]7	.223	.178	.196
75	1	[3]4	.193	.151	.168
76	7	[3]22	.200	.130	.162
	1	[3]3	.189	.153	.175
77	6	[3]16	.198	.142	.170
78	1	[3]3	.130	.110	.123
	1	[3]2	.125	.110	.118
	1	[3]3	.158	.132	.144
79	1	[3]2	.103	.076	.090
80	1	[3]2	.102	.092	.097
81	1	[3]3	.198	.189	.193
	1	[3]3	.137	.009	.123
82	1	[3]2	.185	.162	.174
	1	[3]3	.183	.149	.166
83	1	[3]3	.180	.104	.140
	1	[3]3	.136	.096	.113
84	1	[3]1	----------	----------	.164
85	3	[3]12	.160	.101	.137
86	2	[3]5	.230	.084	.130

[1] Identification of serial numbers given in Table 1.
[2] Specimens broken across the grain.
[3] Specimens broken along the grain.

The average and extreme values from tests across the grain on dry specimens for various regions are:

District	Number of tests	Δ Average	Δ Maximum	Δ Minimum
Maine	18	0.178	0.245	0.125
Vermont-New York	63	.163	.230	.093
Hard Vein, Pa	45	.204	.249	.153
Bangor, Pa	125	.205	.271	.141
Pen Argyl, Pa	263	.205	.250	.127
Wind Gap, Pa	17	.196	.227	.151
Slatington, Pa	29	.197	.229	.151
Virginia	56	.164	.200	.106

TABLE 4.—*Results of toughness tests—specimens wet*

Serial No.[1]	Number of samples	Maximum deflection of 16-inch span and $\frac{3}{16}$ inch thickness			
		Number of tests	Maximum	Minimum	Average
			Inch	*Inch*	*Inch*
2	1	[2]2	0.126	0.120	0.124
14	1	[2]3	.136	.131	.134
15	1	[2]3	.069	.065	.069
16	1	[2]3	.180	.131	.161
17	1	[2]3	.134	.126	.130
18	1	[2]3	.127	.090	.105
19	1	[2]3	.150	.118	.131
20	1	[2]3	.132	.121	.125
21	1	[2]3	.183	.165	.176
22	1	[2]3	.172	.113	.143
52	33	[2]68	.202	.114	.168
53	1	[2]2	.192	.169	.180
54	1	[2]2	.185	.172	.178
62	2	[2]4	.176	.132	.152
63	1	[2]2	.184	.170	.177
76	1	[2]3	.151	.140	.146
81	1	[2]3	.180	.167	.172
83	1	[2]3	.176	.127	.160

[1] Identification of serial numbers given in Table 1.
[2] Specimens broken along the grain.
[3] Specimens broken across the grain.

This test is of value in showing the flexibility of the material. Other properties being equal, a slate of high toughness is less apt to break under a given strain than one of lower toughness. This property is sometimes used as a specification requirement.

4. ABRASIVE HARDNESS

The apparatus used in determining the abrasive hardness values given in Table 5 is described in detail in the Proceedings of the American Society for Testing Materials, volume 28, part 2, 1928, pages 855–867. The result is expressed as a reciprocal of the volume abraded under controlled conditions in a given period of time. The specimens, which were 2 inches square and 1 inch thick, were weighed and subjected to the abrading process, after which they were weighed again. By means of the apparent specific gravity of the material the weight loss is converted to volume, since this gives a more definite basis of comparison. The reduction formula is

$$H_a = \frac{10\,(2,000 + W_a)G}{2,000\,W_a}$$

in which W_s is the weight of the specimen, G the apparent specific gravity, and W_a the loss of weight by abrasion. The constant load on the specimen during the grinding is 2,000 G which is augmented by the variable weight of the specimen. In applying the correction for the weight W_s the average of the original and final weights is used. The reciprocal relation is used in order to give values in conformity with the usual conceptions of hardness; that is, low values for soft materials and higher values to harder materials. The factor 10 in the numerator is used arbitrarily to avoid values less than unity for soft materials.

TABLE 5.—*Results of abrasive hardness tests*

Serial No.[1]	Number of samples	Hardness values (H_a)			
		Number of tests	Maximum	Minimum	Average
1	1	3	9.4	8.8	9.1
2	1	5	9.8	9.3	9.6
4	1	5	8.0	7.6	7.7
5	1	5	7.8	7.2	7.5
6	1	5	7.2	7.0	7.1
7	1	6	15.2	7.4	11.7
8	1	6	7.6	7.2	7.5
14	1	5	8.5	7.6	8.0
15	1	5	9.0	6.9	8.3
16	1	6	8.1	7.0	7.6
17	1	6	9.2	8.8	9.0
18	1	6	8.7	8.1	8.5
19	1	4	8.9	7.4	7.9
20	1	3	12.2	11.0	11.8
21	1	5	7.9	7.6	7.7
22	1	5	7.3	6.7	6.9
23	1	6	9.3	8.1	8.7
26	3	9	7.2	6.8	6.9
55	1	5	7.2	5.8	6.5
58	2	12	7.2	6.3	6.8
59	2	15	7.2	6.2	6.6
65	4	12	6.4	5.6	6.0
66	2	6	7.2	6.5	6.9
69	2	6	5.8	5.5	5.6
76	1	5	12.2	9.9	10.9
81	1	3	5.7	5.5	5.6
83	1	5	8.9	7.9	8.4

[1] Identification of serial numbers given in Table 1.
[2] All specimens contained ribbons.
[3] Nine specimens contained ribbons.

A comparison of the H_a values in Table 5 with the service records of various slates shows considerable evidence that hardness is directly proportional to durability. However, the very low H_a value for one slate, which has so thoroughly established its good weathering quality by long service, can be cited to prove that no general relation between hardness and durability exists.

The range in hardness values for slates from different districts was not found to be as great as for other materials of a given type, as marble, limestone, etc. The average and extreme H_a values for various regions were as follows:

District	Number of tests	H_a average	H_a maximum	H_a minimum
Maine	8	9.4	9.8	8.8
Vermont-New York	72	8.3	15.2	6.7
Hard Vein, Pa	9	8.2	9.3	7.0
Bangor, Pa	6	6.8	7.0	6.8
Pen Argyl, Pa	32	6.7	7.2	5.8
Wind Gap, Pa	18	6.3	7.2	5.6
Slatington, Pa	6	5.6	5.8	5.5
Virginia	5	10.9	12.2	9.9

The dark "ribbons" in the slates from Pennsylvania are usually more resistant to abrasion than the parts of the slate free from ribbons. According to Dale [5] and Behre,[6] this is due to the presence of coarser quartz particles in the ribbons.

The abrasive hardness of the material is of interest in cases where it is to be used in floors, steps, sills, sinks, table tops, etc., but probably has no significance in connection with roofing slate.

5. ABSORPTION

The absorption test is frequently used in roofing-slate specifications, and there appears to be a relation between absorption values and weathering qualities. The values given in Table 6 were determined on slabs from 4 to 6 inches square and from three-sixteenths to one-fourth inch thick. These were dried at 100° C. for 24 hours for the dry weight, and then completely immersed in water at room temperatures for 48 hours. Absorption results are given in percentage by weight, which were determined by dividing the weight of water absorbed by the weight of the dry specimen. The weight ratio is commonly used in specification tests, but the volume ratio is a more definite means of comparison. The errors arising from the use of the weight ratios are not as great for slate as for most other materials, since the variation in "bulk density" from one slate to another is comparatively small.

TABLE 6.—*Results of absorption tests*

Serial No.[a]	Number of samples	Absorption by weight			
		Number of tests	Maximum	Minimum	Average
			Per cent	*Per cent*	*Per cent*
1	1	9	0.08	0.05	0.06
2	2	7	.06	.02	.03
3	1	3	.04	.03	.04
4	4	26	.20	.06	.12
5	2	8	.11	.07	.09
6	2	8	.29	.08	.11
7	2	17	.23	.09	.13
8	2	8	.20	.11	.12
9	1	2	.17	.17	.17
10	1	2	.30	.29	.30
11	1	1	-----------	-----------	.25
12	1	1	-----------	-----------	.42
13	1	4	.15	.09	.11
14	1	6	.15	.12	.14
15	1	12	.18	.09	.12

[a] Identification of serial numbers given in Table 1.

[5] U. S. Geological Survey Bulletin No. 586.

[6] Slate in Northampton County, Pa., Bul. M9, Topographic and Geologic Survey of Pennsylvania, 1927.

TABLE 6.—*Results of absorption tests*—Continued

Serial No.	Number of samples	Absorption by weight			
		Number of tests	Maximum	Minimum	Average
			Per cent	*Per cent*	*Per cent*
16	1	9	0.18	0.14	0.16
17	1	6	.12	.11	.11
18	1	6	.14	.10	.12
19	1	6	.14	.12	.13
20	1	12	.15	.07	.12
21	1	5	.06	.05	.06
22	1	4	.13	.12	.13
23	10	43	.37	.10	.15
24	5	15	.20	.12	.15
25	1	4	.26	.21	.24
26	5	54	.35	.14	.18
27	3	34	.37	.11	.17
28	1	3	.35	.33	.34
29	1	4	.67	.56	.60
30	1	3	.27	.25	.26
31	1	4	.40	.29	.33
33	4	24	.43	.16	.26
34	1	4	.45	.43	.44
35	1	4	.30	.28	.29
36	2	12	.31	.23	.26
37	1	4	.44	.33	.38
38	2	10	.44	.24	.32
39	1	4	.34	.31	.32
40	1	4	.42	.31	.35
41	1	4	.23	.21	.22
42	1	4	.32	.28	.30
43	1	4	.33	.25	.30
44	1	4	.40	.35	.38
45	6	27	.38	.24	.30
46	10	57	.39	.11	.24
47	1	8	.50	.45	.47
48	1	7	.33	.29	.31
49	3	22	.46	.26	.36
50	2	13	.32	.26	.28
51	1	8	.32	.29	.30
52	53	244	.48	.19	.31
52a	6	34	.32	.19	.26
53	3	16	.21	.17	.19
54	2	12	.50	.27	.41
55	1	2	.32	.21	.26
56	4	28	.57	.26	.44
57	3	11	.56	.43	.50
58	5	61	.32	.20	.23
59	6	81	.34	.21	.25
60	1	4	.21	.19	.20
61	2	8	.49	.19	.44
62	2	12	.14	.10	.12
63	1	6	.21	.18	.20
65	4	35	.47	.28	.36
66	3	33	.43	.23	.32
67	1	8	.46	.42	.45
68	2	15	.58	.51	.54
69	3	40	.35	.21	.26
70	1	2	.20	.17	.18
71	4	25	.42	.19	.31
72	2	6	.45	.35	.41
73	2	14	.36	.30	.34
74	1	8	.93	.89	.91
75	3	21	.04	.00	.01
76	1	2	.03	.02	.02
77	1	13	.18	.13	.15
78	1	14	1.63	1.23	1.41
80	1	6	1.31	1.01	1.12
80	1	6	.21	.18	.20
81	1	4	.28	.25	.27
82	1	12	.13	.08	.10
83	1	4	.59	.54	.57
84	3	14	.55	.33	.40
85	2	6	1.54	1.09	1.27

A study of the relations between absorption and certain other properties of slate has been made by computing the correlation coefficients. A high degree of correlation would be expected to exist between absorption and porosity, but the correlation coefficient obtained for the results of these determinations was $r = 0.70$. Two factors may be expected to influence this relation—first, the rate of absorption and, second, mineral composition. A better correlation between these properties might be obtained by determining the absorption values for a longer period of immersion and using the volume ratio instead of the weight ratio. The loss of strength due to soaking in water suggests that there may be a relation between the amount of water absorbed and the strength loss; however, the computation of the correlation coefficient for these values gave $r = 0.25$, which indicates a very small relationship. In the acid tests, to be described below, the decay is caused by converting the particles of calcite to gypsum. Since the measurements are based on the depth below the surface to which the decay occurs, one might expect a relation to exist between the absorption values and the depth of softening for slates of similar composition. A study of such relations was made from 61 determinations on the slate from one quarry which gave $r = -0.18$. The negative value indicates a reciprocal relation, or that the more absorptive slates are more resistant to the acid action. The reason for this is believed to be due to a variation in the calcite content of the various samples rather than to differences in absorption.

The average absorption and the range (48 hours' immersion) in percentage by weight for various regions were as follows:

District	Number of tests	Absorption		
		Average	Maximum	Minimum
		Per cent	*Per cent*	*Per cent*
Maine	27	0.05	0.08	0.02
Vermont-New York	154	.13	.42	.06
Hard Vein, Pa	77	.16	.37	.10
Bangor, Pa	250	.28	1.24	.11
Pen Argyl, Pa.	648	.30	.75	.11
Wind Gap, Pa	91	.38	.58	.23
Slatington, Pa	81	.29	.45	.17
Virginia	61	.06	.31	.00

6. POROSITY

The porosity of a material is usually expressed as a percentage of the void space to the total volume. Table 7 gives the results of several porosity determinations on slate. The values were deduced from the results of true and apparent specific gravity determinations as follows:

$$P = \frac{100 \ (T\text{-}A)}{T}$$

where T and A are respectively the true and apparent specific-gravity values.

TABLE 7.—*True specific gravity, apparent specific gravity, porosity, and weight per cubic foot*

Serial No.[1]	True specific gravity					Apparent specific gravity					Porosity per cent by volume	Weight per cubic foot (dry)
	Number of samples	Number of tests	Maximum	Minimum	Average	Number of samples	Number of tests	Maximum	Minimum	Average		
1	1	5	2.833	2.816	2.823	1	9	2.809	2.806	2.807	0.57	175.2
2	1	3	2.831	2.826	2.829	1	6	2.822	2.820	2.821	.28	176.0
3						1	6	2.809	2.796	2.804	......	174.8
4	3	11	2.803	2.768	2.787	3	18	2.776	2.758	2.767	.74	172.6
5	1	3	2.782	2.778	2.780	1	6	2.773	2.770	2.772	.29	173.0
6	1	3	2.776	2.767	2.772	1	6	2.767	2.763	2.764	.29	172.4
7	1	3	2.812	2.806	2.809	1	6	2.803	2.790	2.796	.46	174.5
8	1	3	2.788	2.781	2.786	1	6	2.770	2.768	2.769	.61	172.8
14	1	5	2.814	2.810	2.812	1	6	2.803	2.791	2.798	.50	174.6
15	1	3	2.828	2.825	2.827	1	6	2.812	2.810	2.811	.57	175.4
16	1	3	2.806	2.805	2.806	1	6	2.786	2.782	2.784	.78	173.7
17	1	3	2.821	2.819	2.820	1	6	2.811	2.806	2.810	.35	175.4
18	1	3	2.808	2.802	2.804	1	6	2.795	2.780	2.788	.57	174.0
19	1	3	2.809	2.808	2.809	1	6	2.792	2.788	2.790	.68	174.1
20	1	3	2.805	2.802	2.803	1	6	2.795	2.794	2.795	.29	174.4
21	1	3	2.775	2.772	2.774	1	5	2.765	2.764	2.764	.36	172.5
22	1	3	2.810	2.805	2.808	1	4	2.791	2.774	2.783	.89	173.7
23						2	4	2.770	2.755	2.764	......	172.4
24						1	2	2.769	2.766	2.768	......	172.7
26	4	16	2.795	2.774	2.785	5	47	2.775	2.754	2.763	.79	172.4
27	2	6	2.775	2.769	2.772	2	23	2.771	2.751	2.758	.51	172.1
28						1	2	2.773	2.771	2.772	......	173.0
29						1	2	2.780	2.778	2.779	......	173.4
30						2	6	2.780	2.762	2.772	......	173.0
31						1	4	2.764	2.749	2.758	......	172.1
32						1	2	2.776	2.776	2.776	......	173.2
33	1	4	2.810	2.792	2.802	5	17	2.802	2.749	2.770	.50	172.9
36						1	5	2.791	2.774	2.783	......	173.7
45						3	10	2.769	2.760	2.564	......	172.5
46	1	5	2.781	2.760	2.771	5	15	2.784	2.739	2.763	.29	172.4
49	1	5	2.789	2.779	2.785	1	15	2.751	2.738	2.743	1.51	171.2
50						1	6	2.767	2.761	2.764	......	172.5
52						2	9	2.771	2.753	2.764	......	172.5
55						1	2	2.747	2.741	2.744	......	171.3
58	5	20	2.800	2.775	2.786	5	61	2.769	2.750	2.761	.84	172.3
59	6	22	2.808	2.775	2.791	6	64	2.780	2.753	2.768	.80	172.8
61						2	3	2.760	2.753	2.756	......	172.0
65	5	22	2.788	2.765	2.780	5	48	2.754	2.739	2.748	1.15	171.5
66	2	10	2.812	2.789	2.802	2	21	2.787	2.736	2.753	1.72	171.8
68						1	9	2.743	2.738	2.740	......	171.0
69	3	11	2.799	2.772	2.785	3	34	2.771	2.753	2.760	.94	172.2
70						2	6	2.773	2.770	2.771	......	172.9
71						1	2	2.761	2.761	2.761	......	172.3
72	1	4	2.783	2.778	2.780	2	9	2.760	2.742	2.752	1.15	171.8
73						1	4	2.750	2.750	2.750	......	171.6
76	1	3	2.802	2.789	2.795	5	17	2.796	2.777	2.788	.14	174.0
77						6	24	2.811	2.774	2.798	......	174.6
78	1	10	2.908	2.890	2.898	1	5	2.874	2.868	2.871	.93	179.2
79	1	4	2.883	2.869	2.878	1	7	2.757	2.742	2.754	4.31	171.8
80	1	5	2.854	2.839	2.848	1	3	2.760	2.743	2.758	3.16	172.1
81	1	3	2.885	2.887	2.888	1	6	2.903	2.878	2.888	......	180.2
82	1	3	2.813	2.806	2.810	1	4	2.778	2.777	2.778	1.14	173.4
83	1	3	2.832	2.821	2.827	1	6	2.814	2.809	2.811	.57	175.4
84						1	4	2.694	2.689	2.692	......	168.2
85						1	3	2.745	2.744	2.744	......	171.2
86						1	3	2.727	2.718	2.723	......	170.2

[1] Identification of serial numbers given in Table 1.

Porosity values are of interest in studying the weathering qualities of a material, but, in general, the determination is too difficult and tedious for specification purposes. A sufficiently accurate indication of porosity for comparative purposes is obtained by the absorption test. Evidently the porosity value for any material should be equal to or greater than the volume absorption value. In the specific-gravity determinations the true value was determined on one specimen of the sample and the apparent on another. The porosity values are so small that a slight error in either determination or a slight variation of the material will lead to erroneous porosity results.

Most of the porosity values found were between 0.3 and 1 per cent, although one value was 0.14 per cent and another over 4 per cent. The latter was obtained on a sample taken from the surface of the ground and hence is not representative of material below the weathered zone.

7. SPECIFIC GRAVITY

The "true specific gravity" of a porous material may be considered as the ratio of the weight of a certain volume of material to the weight of the same volume of water at the same temperature, assuming that the material has been compressed until the pores are entirely filled. It is not feasible to eliminate the pores by compression, but the material can be pulverized until the component particles are separated. Determinations given in Table 7 were made on material pulverized and passed through a 100-mesh sieve. Various types of specific-gravity apparatus were tried, but the most concordant results were obtained with the LeChatelier flask. As this apparatus requires a much larger sample than other types and the process is less tedious, it was employed for the tests. By using considerable care in removing the air from the powders and correcting for temperature differences between the initial and final volume readings it was found possible to obtain quite consistent results. A 95 per cent ethyl alcohol was used as the liquid, since there was no difficulty in immersing all of the powdered material in it, while other liquids had a tendency to float some of the fine powder.

The apparent specific gravity of a porous material is the ratio of weight of a certain volume of the material in its natural state to the weight of the same volume of water at the same temperature. The results given in Table 7 were determined from dry weights of regular-shaped specimens and the corresponding volumes obtained by weighing the saturated specimens immersed in water. The reduction formula was

$$A = \frac{w_1 d}{w_2 - w_3}$$

where w_1 = dry weight of specimen, d = apparent density of the water at the temperature of observation, w_2 = weight of the specimen after immersion in water for 48 hours, and w_3 = weight of saturated specimen suspended in water. It is necessary to use soaked specimens for determining the weight suspended in water to prevent errors due to absorption during the operation.

8. WEIGHTS PER CUBIC FOOT

Having determined the apparent specific gravity or "bulk density" of a material, it is only necessary to multiply this value by the weight of a cubic foot of water in order to deduce the weight per cubic foot of the sample. The results in the last column of Table 7 were obtained by multiplying the average apparent specific-gravity value for each sample by 62.35. The average and extreme weights per cubic foot for slate from various districts were as follows:

Weight per cubic foot

District	Number of tests	Average	Maximum	Minimum
		Pounds	*Pounds*	*Pounds*
Maine	21	175.3	176.1	174.3
Vermont-New York	93	173.7	175.5	172.1
Hard Vein, Pa	16	172.7	173.2	171.9
Bangor, Pa	108	172.6	174.8	171.5
Pen Argyl, Pa	185	172.4	174.0	170.9
Wind Gap, Pa	78	171.5	173.9	170.7
Slatington, Pa	51	172.2	173.0	171.8
Virginia	52	174.8	179.3	173.1

The above values, considered in conjunction with service records of the slates, afford considerable evidence that a correlation exists between bulk density and durability. However, as will be shown later, the durability of slate is influenced to such a large extent by the presence of small amounts of certain minerals, it would not be logical to base judgment upon this determination alone.

VI. WEATHERING CHARACTERISTICS

1. WEATHERING PROCESSES

In the study of slate weathering, about 60 samples of old shingles were collected which had been exposed to the elements for various periods of time up to 131 years. Where an appreciable amount of decay was in evidence the decayed portion was subjected to a partial chemical analysis to determine if any alteration in composition of the less stable minerals had occurred. Specimens were also subjected to strength, toughness, and absorption tests to determine the ph sical alterations. A summary of these results is shown in Table No y8.

TABLE 8.—*Summary of tests on weathered slate*

Producing region	Number of samples examined	Average weathering period	Average modulus of rupture		Average toughness values		Average absorption by weight		Indicated change per year due to weathering[1]		
			Fresh slate	Weathered slate	Fresh slate	Weathered slate	Fresh slate	Weathered slate	Modulus of rupture	Toughness	Absorption
		Years	*Lbs./in.²*	*Lbs./in.²*	*Inch*	*Inch*	*Per cent*	*Per cent*	*Lbs./in.²*	*Inch*	*Per cent*
Maine	4	46	11,700	10,100	0.178	0.168	0.05	0.06	−35	−0.0002	+0.0002
Vermont-New York	3	40	10,000	8,700	.163	.160	.13	.23	−47	−.0007	+0025
Hard Vein, Pa	12	54	13,600	9,000	.204	.168	.16	.85	−85	−.0006	+0128
Soft Vein, Pa[1]	29	24	11,700	9,800	.201	.189	.31	.68	−100	−.0005	+0154
Virginia	9	81	10,500	9,600	.164	.144	.06	.15	−11	−.0002	+0011

[1] Slates from Bangor and Pen Argyl region.

A large range in the resistance of the various slates was evident from the examination of weathered samples. The chemical analysis revealed, in most cases, a considerable amount of gypsum in the weathered portion, while the fresh slates showed almost none. It seemed probable, therefore, that some of the calcite in the slate was converted to gypsum in the weathering process. Two means were considered possible for this change—first, the action of sulphur acids from the air on the calcite, and, second, an interaction between the sulphide minerals and calcite. Sulphides of iron are usually present in slate.

If gypsum could be formed from an interaction of the constituent minerals of slate, it was assumed that the weathering process could be simulated in the laboratory by merely soaking and drying the material several times. The following experiments were made for that purpose: Three samples of slate were selected which were known to have poor weathering qualities. Determinations of SO_3 were made on each in the original condition and after the samples had been subjected to 30 cycles of soaking and drying. The following results were obtained:

		SO_3
		Per cent
Sample No. 1:		
(a)	In original condition	0. 01
(b)	After 30 cycles of soaking and drying	. 50
Sample No. 2:		
(a)	In original condition	. 01
(b)	After 30 cycles of soaking and drying	. 50
Sample No. 3:		
(a)	In original condition	. 01
(b)	After 30 cycles of soaking and drying	. 23

A portion of each of these three samples, after the soaking and drying process, was powdered and leached with distilled water. The leach was then filtered and evaporated. An abundance of crystals was obtained which was examined with the petrographic microscope and proven to be gypsum. In these experiments it was considered that the sulphur-acids of the air could play no appreciable part and that the gypsum was due practically entirely to the reaction between the calcite and sulphide minerals, the oxygen necessary for completing the SO_3 radicle being drawn from the air.

It was desirable to study the reaction between iron sulphide and calcite alone as well as in the presence of carbon. For this purpose four mixtures were made, as follows: (1) Equal parts of calcite and pyrite; (2) equal parts of calcite and marcasite; (3) same as (1) with a small addition of carbon; (4) same as (2), with a small addition of carbon. The four mixtures were placed in open beakers, covered with distilled water, and allowed to stand for seven hours at room temperatures. The beakers were then set in a drying oven at 110° C. for 17 hours. The 7 hours' soaking and 17 hours' drying constituted a cycle which was repeated one hundred times. The water-soluble matter was then extracted and determined as follows:

Sample No.	$CaSO_4$
	Per cent
1	7. 0
2	2. 3
3	14. 9
4	6. 0

These experiments indicate that a chemical reaction between pyrite and calcite or between marcasite and calcite occurs under such conditions and results in the formation of gypsum. They also indicated that the presence of carbon accelerates the rate of reaction.

Since this chemical change has been found to occur in slates containing appreciable amounts of calcite and iron sulphides, it remains to be considered how this may produce decay. The gypsum molecule requires about twice as much space as the calcite molecule, hence the conversion to gypsum of a calcite crystal, which is embedded in another material, must produce stresses in the surrounding material. Slates containing considerable amounts of both pyrite and calcite, when subjected to several cycles of soaking and drying, disintegrated almost completely. Slates of the same kind in actual exposure on the roof of a building are found to decay more rapidly on the unexposed parts. This seems to indicate that in service the solubility of gypsum comes into consideration. As the rain water leaches downward through the slate some of the gypsum is carried in solution and deposited at a lower level. By this process the gypsum is concentrated on the lower sides of shingles and the recrystallization within the pores causes internal stresses. The analyses made on weathered shingles showed, in most cases, a concentration of gypsum in the decayed parts.

Soaking and drying tests have been made on numerous samples of slates which were known to have poor weathering qualities by drying at 110° C. for 17 hours and soaking 7 hours. Under such conditions some of the samples showed signs of decay in 30 cycles and an advanced state of disintegration in less than 50 cycles. Some producers contended that the drying temperatures were too high in these tests and out of proportion to those of service conditions. For this reason a series of tests was made in which the drying temperature was 50° C. This temperature was assumed to be about equivalent to roof temperatures on hot summer days. The results are given in Table 9, and show that the deterioration occurs under such conditions but at a much slower rate.

TABLE 9.—*Soaking and drying tests with absorption and weight determinations after 200 cycles* [1]

Serial No.[2]	Specimen No.	Average absorption by weight of specimens		Average change in weight of specimens	Final appearance of specimens
		Original condition	After 200 cycles		
		Per cent	*Per cent*	*Per cent*	
1	1	0.06	0.08	−0.05	No apparent change.
	2				Do.
	3				Do.
2	1	.02	.04	−.05	Do.
	2				Do.
	3				Do.
4	1	.09	.14	−.06	Do.
	2				Do.
	3				Do.
5	1	.08	.13	−.05	Do.
	2				Do.
	3				Do.
6	1	.08	.12	−.03	Do.
	2				Do.
	3				Do.
7	1	.14	.16	−.03	Do.
	2				Do.
	3				Do.

[1] In the first 100 cycles each cycle consisted of 17 hours drying at 50° C., followed by 7 hours immersion in water at room temperatures. In the second 100 cycles each cycle consisted of 24 hours drying at 50° C. and 24 hours immersion in water at room temperatures.
[2] For identification of serial numbers see Table 1.

TABLE 9.—*Soaking and drying tests with absorption and weight determinations after 200 cycles*—Continued

Serial No.	Specimen No.	Average absorption by weight of specimens		Average change in weight of specimens	Final appearance of specimens
		Original condition	After 200 cycles		
		Per cent	*Per cent*	*Per cent*	
8	1 2 3	0.11	0.16	+0.04	No apparent change. Do. Do.
14	1 2 3	.14	.15	−.01	Do. Do. Do.
15	1 2 3	.12	.12	−.04	Do. Do. Do.
16	1 2 3	.16	.22	+.05	Do. Do. Do.
17	1 2 3	.11	.12	−.03	Do. Do. Do.
18	1 2 3	.12	.15	−.03	Do. Do. Hair cracks on 2 edges.
19	1 2 3	.13	.19	−.04	No apparent change. Do. Do.
20	1 2 3	.10	.11	.00	Do. Do. Do.
21	1 2 3	.06	.07	−.02	Do. Do. Do.
22	1 2 3	.13	.18	−.05	Do. Do. Hair cracks on 4 edges.
23	1 2 3	.15	.32	−.04	No apparent change. Do. Do.
26	1 2 3	.17	.27	−.06	Hair cracks on 1 edge. Cracks along 2 edges. No apparent change.
27	1 ₂2 ₃3	.14	.19	−.04	Hair cracks on 1 edge. No apparent change. Hair cracks on 2 edges.
49	1 2 3	.39	.60	−.02	Hair cracks on 1 edge. No apparent change. Do.
58	1 2 3	.22	.33	−.02	Do. Do. Do.
59	1 ₂2 3	.27	1.54	−.02	Ribbon badly cracked No apparent change. Hair cracks on 4 edges.
65	1 2 3	.36	.70	+.03	Hair cracks on 3 edges. Hair cracks on 4 edges. No apparent change.
66	1 ₂2 ₃3	.32	.77	−.50	Ribbon disintegrated. Do. Hair cracks on 4 edges.
69	1 2 3	.26	.64	.00	Do. Do. No apparent change.
76	1 2 3	.01	.02	−.04	Do. Do. Do.
81	1 2 3	.20	.25	+.02	Do. Do. Do.
82	1 2 3	.27	.32	−.07	Do. Do. Do.
83	1 2 3	.10	.18	−.04	Do. Do. Do.

¹ Specimens contained a dark ribbon. Those not so marked are clear slate.

In order to determine if a relation exists between the results obtained in the soaking and drying tests and the resu ts of actual weathering it is necessary to have some definite measure of the deterioration.   It is fair to assume that a decrease in strength or an increase in absorption is indicative of deterioration.   In Table 9 is shown the computed rate of "strength loss" and "absorption increase" per year for weathered slate from five producing districts.   These rates are based on the average strength and absorption results obtained on the fresh material as compared to the results obtained on slate shingles after exposure to the weather.   It was found that these yearly rates of change were approximately one-twentieth part of the corresponding changes obtained in 200 cycles of the soaking and drying test when the specimens were dried at 50° C.   Figure 4 shows the "absorption increases"

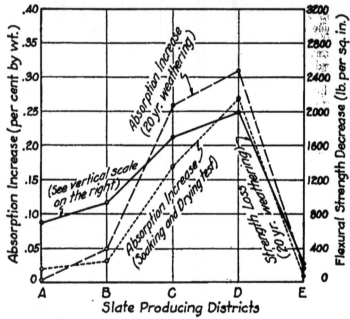

FIGURE 4.—*Relation between strength loss for 20 years weathering, absorption increase for 20 years weathering and absorption increase in the soaking and drying test*

Plotted points are the respective averages of all specimens from five districts as follows: *A*, Maine; *B*, Vermont-New York; *C*, Hard Vein, Pa.; *D*, Soft Vein, Pa., *E*, Virginia.

obtained in the soaking and drying tests plotted with the computed "absorption increases" and "strength losses" for 20 years weathering. These curves indicate that the soaking and dry;ng test gives results similar to those of actual weathering.

Thermal expansion and contraction of the component minerals has often been mentioned as a possible cause of deterioration.   This was studied by means of a series of tests on samples from five districts. The samples were heated in an oven for 17 hours and laid out in the laboratory for 7 hours.   Two of the samples were disintegrated by 63 cycles but the others were continued in the test to 554 cycles.

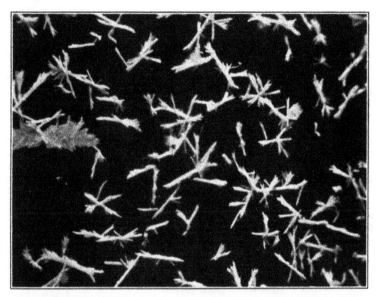

FIGURE 5.—*Photomicrograph of calcium sulphate crystals obtained by evaporating the leach from a sample of Pen Argyl (Pa.) slate which had been heated and cooled 310 cycles*

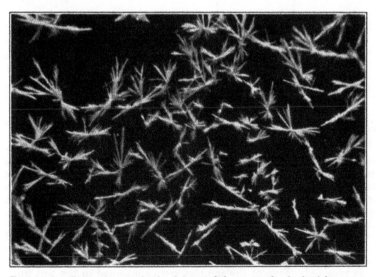

FIGURE 6.—*Photomicrograph of calcium sulphate crystals obtained by evaporating the leach from a sample of Bangor (Pa.) ribbon slate that had been heated and cooled 63 cycles*

The results were judged by means of inspection and by absorption tests after 25, 50, and 554 cycles. The detailed results are given in Table 10. The s ates which had previously shown a high resistance to the soaking and drying test showed little deterioration in the heating and cooling procedure. As those samples which decayed in this test presented about the same appearance as samples which decayed in the soaking and drying test, they were examined for gypsum and it was found in considerable amounts. Figures 5 and 6 are photomicrographs of the crystals obtained by leaching serial Nos. 58 and 27, respectively. Hence, it is probable that decay was caused mainly by the formation of gypsum. Most of the samples showed considerable resistance to the heating and cooling process, especially those of low calcite content, which seems to indicate that the effect of temperature changes on slate is slight unless it results in such chemical alterations as described above. The tests also indicate that considerable amounts of moisture are not necessary to bring about the reaction. Furthermore, they supply a c.ue to the cause of decay sometimes noted on interior slate installations.

TABLE 10.—*Effect of repeated heating and cooling on slates from five different regions*

Serial No.[1]	Observed condition of specimens				Per cent absorption by weight after 48-hour immersion			
	After 20 cycles	After 63 cycles	After 100 cycles	After 554 cycles	Original condition	After 25 cycles	After 50 cycles	After 554 cycles
2	No visible effect....	No visible effect....	No visible effect....	No visible effect....	0.04	0.04	0.02	0.04
15	....do.............	....do.............	....do.............	....do.............	.15	.16	.15	.16
23	....do.............	....do.............	....do.............	Swelling at ribbons.[2]	.12	.12	.10	.40
26	....do.............	....do.............	....do.............	No visible effect....	.32	.24	.23	.48
27	Swelling at ribbon..	Badly decayed......			.45	1.76	(3)	
58	No visible effect....	No visible effect....	....do.............	Considerable decay.	.37	.40	2.88	(3)
59	Swelling at ribbon..	Badly decayed......			.46	1.20	(3)	

[1] Identification of serial numbers given in Table 1.
[2] On 1 specimen the ribbons were crumbling at the surface.
[3] Specimens were too badly decayed for absorption determination. 1 specimen of No. 58 was in an advanced state of decay at 310 cycles.

Frost action is considered to be one of the common causes of stone weathering. Table 11 gives the results of freezing tests on 18 samples from the various districts. The samples were slabs 4 inches square and approximately one-half inch thick. They were soaked 14 days before freezing was started and then thawed each time by immersion in water at about 20° C. for 30 minutes. The temperature of the freezing chamber was maintained at approximately −10° C. and two freezings were made each day. In these tests only one sample failed and this occurred after 1,374 cycles. This sample was originally taken from the surface of the ground and showed signs of weathering when received. Five of the samples were continued in the test until 2,436 cycles were reached and the others were discontinued after 1,743 cycles. The results were judged by inspection and by absorption tests.

TABLE 11.—*Results of freezing tests*

Serial No.[1]	Specimen No.	Number of cycles[2]	Condition[3]	Average absorption by weight			Remarks
				A[4]	B[5]	C[6]	
				Per cent	Per cent	Per cent	
1	1 2 3 4 5 6	1,743	b b b d c c	0.15	0.08	0.06	Sharp corners chipped off in some places.
4	1 2 3 4 5 6	2,436	b b b b c c	.29	.22	.16	Corners chipped off in places and some cleavage cracks.
26	1 2 3 4 5 6	1,743	c c c c c c	.39	.33	.15	Corners chipped slightly and some scaling on cleavage faces.
27	1 2 3 4 5 6	1,743	c c c c c c	.34	.29	.14	Do.
33	1 2 3 4 5 6	2,436	d d e d c d	.46	.38	.19	Considerable scaling on cleavage faces; more in evidence at ribbons.
46	1 2 3 4 5 6	2,436	c e c d c d	.36	.28	.15	Effects mainly confined to scaling on cleavage faces.
58	1 2 3 4 5 6	1,743	c c c c c b	.43	.38	.24	Some scaling on cleavage faces.
59	1 2 3 4 5 6	1,743	d e c c c c	.46	.43	.25	Considerable scaling on cleavage faces; more pronounced at ribbons.
65	1 2 3 4 5	1,743	e c c e c	.69	.63	.35	Considerable scaling on cleavage faces and corners chipped.

[1] Identification of serial numbers given in Table 1.
[2] A cycle consisted of 1 freezing and 1 thawing as outlined on p. 405.
[3] The final state of the specimens was estimated by inspection and expressed in 8 stages as follows: "a" signifies no decay, "h" advanced state of decay. The intervening letters represent about equal stages of decay between "a" and "h."
[4] Percentage of water absorbed during the freezing and thawing process.
[5] Absorption obtained by drying the specimens after the freezing procedure and then immersing them for 48 hours.
[6] Absorption of samples in original condition.

TABLE 11.—*Results of freezing tests*—Continued

Serial No.[1]	Specimen No.	Number of cycles [2]	Condition [3]	Average absorption by weight			Remarks
				A [4]	B [5]	C [6]	
				Per cent	*Per cent*	*Per cent*	
66	1 2 3 4 5 6	1,743	c c c c d d	0.65	0.56	0.32	Surface scaling on cleavage faces; corners chipped.
69	1 2 3 4 5 6	1,743	b d d d c c d	.54	.46	.26	Some specimens scaling considerably.
72	1 2 3 4 5 6	2,436	e e d d c c	.73	.64	.36	Considerable surface scaling.
78	1 2 3 4 5 6	2,436	a a a a a a	.61	.45	.15	No definite sign of decay.
79	1 2 3	1,374 1,469 1,374	h h h			1.41	Cracked and crumbled.
80	1 2	1,743	d d	1.49	1.28	1.12	Surface scaling and some cleavage cracks.

See footnotes on p. 406.

While all of the samples showed an increase in absorption after repeated freezing and thawing, and while surface scaling was observed in most cases, it can not be said that any sample (except one mentioned above) had reached an advanced state of decay. In view of the large number of cycles of freezing it may be concluded that the action of frost on slate is not a very serious cause of decay. This is especially true of those slates which are subject to the chemical changes discussed above, since that action is several times as severe. However, it should be considered that when decay has started from any cause then frost action can proceed at a more rapid rate.

2. WEATHERING TEST PROCEDURE

The soaking and drying test requires a month or more to carry the weathering process to a definite visible stage. A more rapid acceptance test is often desirable. Several procedures, such as the change in strength or absorption, have been used to measure the effects of the weathering test at some intermediate stage of the process. The strength criterion has not proven very satisfactory, because one can not determine the original strength of the actual specimens used in the weathering test. Determinations on separate specimens of the sample may prove very misleading, due to the natural variation of the material. Figure 7 shows the average strength losses for a large number of specimens after 10, 20, and 30 cycles of soaking and drying. In this series of tests the specimens were dried

17 hours at 110° C. and soaked 7 hours at about 20°. This shows that the strength loss up to 20 cycles is slight and unless a large number of specimens are averaged the effect might not be discerned. Absorption tests can be made on the same specimens before and after the soaking and drying process, and such tests afford a more reliable criterion than strength determinations. An indirect means which offers considerable promise consists in determining the gypsum content of the sample before and after the weathering process, and using the gypsum increase as a measure of weathering.

A test procedure which has been used to some extent for judging the weathering qualities of slate consists of soaking the specimens for seven days in a 1 per cent solution of sulphuric acid. The depth of softening is then determined by gaging the thickness of the specimens at several points, scraping off the softened layer at these points and gaging again. By using a dull blade for the scraping and standardizing the conditions of the process it is possible to obtain fairly

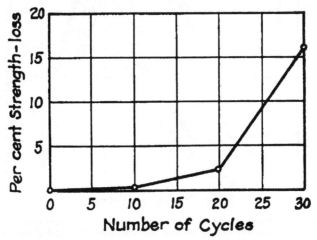

FIGURE 7.—*Average strength-loss for 16 samples of slate from 1 district after 10, 20, and 30 cycles of soaking and drying*

consistent results. This test converts the calcite near the surface to gypsum and causes decay similar to that obtained in the soaking and drying test. Slates which have a low calcite content are not appreciably affected by this test. The absorption of the slate may have a marked effect. Evidently this test would lead to erroneous conclusions if applied to slates containing considerable calcite, but very little sulphide of iron. All of the slates now being produced in this country contain considerable amounts of pyrite so the acid test is applicable to our present commercial materials.

Table 12 gives the results of such tests on several samples representing the various producing districts. Figure 8 shows the relations obtained by plotting the average acid resistances of slates from the various districts with absorption increases due to actual weathering as well as absorption increases noted in the soaking and drying test. This test indicates no appreciable difference between the durability of slates from districts *A*, *B*, and *E*, all of which are generally accepted to be durable materials. It does show a greater difference between districts C and D than is indicated by either the soaking and drying test or by the examination of weathered samples. The test is advantageous in cases where the results must be had in a short period of time. It seems to be fairly reliable in distinguishing between the good and poor grades within certain districts but its general application to all types of slates is not justified by the results of this investigation.

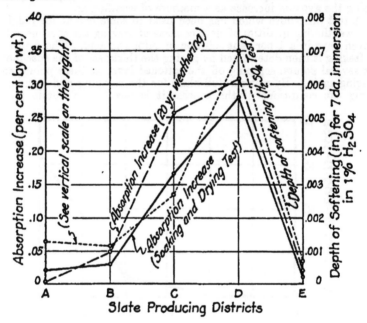

FIGURE 8.—*Relation between absorption-increase for 20 years weathering, absorption increase during soaking and drying test and depth of softening in sulphuric acid test*

Plotted points are the respective averages of all the specimens from five districts as follows: A, Maine; B, Vermont-New York; C, Hard Vein, Pa.; D, Soft Vein, Pa.; E. Virginia.

TABLE 12.—*Sulphuric acid tests with depth of softening measurements after seven days in 1 per cent H_2SO_4*

Serial No.[1]	Number of samples	Depth of softening			
		Number of tests	Maximum	Minimum	Average
			Inch	Inch	Inch
1	1	8	0.002	0.000	0.001
2	2	16	.003	.000	.002
3	1	4	.001	.000	.001
4	1	4	.001	.000	.001
5	1	8	.002	.000	.001
6	1	8	.003	.001	.002
7	1	8	.001	.000	.001
8	1	8	.003	.000	.002
13	1	6	.002	.001	.002
14	1	8	.002	.001	.001

[1] Identification of serial numbers given in Table 1.

TABLE 12.—*Sulphuric acid tests with depth of softening measurements after seven days in 1 per cent H_2SO_4—Continued*

Serial No.	Number of samples	Depth of softening			
		Number of tests	Maximum	Minimum	Average
			Inch	*Inch*	*Inch*
15	1	8	0.001	0.000	0.001
16	1	8	.002	.000	.001
17	1	8	.001	.000	.001
18	1	8	.001	.000	.000
19	1	8	.002	.001	.001
20	1	8	.001	.000	.001
21	1	8	.003	.000	.001
23	4	26	.005	.000	.002
24	2	12	.006	.001	.003
25	1	6	.007	.001	.004
29	2	14	.024	.008	.015
34	1	6	.018	.013	.016
36	1	3	.012	.010	.011
39	1	4	.017	.012	.015
40	1	6	.016	.007	.012
42	1	6	.016	.009	.013
45	1	9	.007	.001	.003
47	1	9	.024	.010	.016
49	1	9	.011	.006	.009
51	1	6	.004	.001	.002
52	40	150	.017	.001	.006
52a	2	4	.011	.002	.004
53	1	3	.003	.001	.002
54	2	10	.010	.001	.005
56	1	6	.004	.003	.004
57	3	15	.007	.001	.005
60	1	6	.003	.001	.002
61	1	6	.015	.002	.007
62	2	16	.005	.001	.003
63	1	8	.012	.007	.010
67	1	6	.017	.004	.010
72	1	2	.010	.007	.008
73	1	6	.014	.007	.012
74	1	6	.005	.002	.003
75	1	3	.020	.017	.018
76	3	20	.002	.000	.000
81	1	8	.002	.001	.002
82	1	8	.002	.001	.001
83	1	8	.002	.001	.001
85	2	24	.005	.000	.002

VII. CONCLUSIONS

1. The flexural strength of slate is much higher than other types of rock used for structural purposes. Although most slates are stronger across the grain than along the grain, this ratio varies greatly with different deposits. The effect of moisture in decreasing the strength is more marked than for other common types of rock.

2. The elastic modulus of slate is high in comparison with other types of rocks. The high flexural strength of the material enables it to be used in relatively thin slabs which have considerable flexibility.

3. Although the porosity determinations in some cases were of rather uncertain precision, they seem to indicate that the void spaces in slate are usually less than 2 per cent of the volume.

4. The bulk density (apparent specific gravity) of slate varies between the limits 2.74 and 2.89, the average being 2.78.

5. The abrasive hardness of slate is about equal to that of the harder oolitic limestones in present-day use in this country.

6. Slate is probably affected to some extent by frost and atmospheric acids, but the major weathering action is due to, first, chemical interaction between certain mineral constituents and second, internal stresses resulting from the newly-formed compounds.

7. The weather resistance is decreased by appreciable amounts of calcite and pyrite existing together. The presence of free carbon with these two minerals lessens weather resistance.

8. Tests on several samples of weathered slate shingles indicate that the early stages of decay are accompanied by an increase in absorption and decrease in strength. These facts may be found of value in deciding upon the use of shingles which have been exposed to the weather for a few years at the plant or the reuse of shingles removed from a building.

9. The most reliable weathering test for slate appears to be the soaking and drying test. For the slates now being produced in this country the sulphuric acid test gives a sufficiently accurate indication of weathering qualities for specification purposes. For slates containing a considerable amount of calcite but no pyrite the sulphuric acid test would probably lead to erroneous conclusions.

10. The possibility of injury to slate due to diurnal temperature changes alone is remote. The chemical transformations occurring in slates, containing considerable amounts of calcite and pyrite, proceed even in dry weather, but at a much slower rate than when more moisture is present. This fact probably accounts for the decay sometimes noted in cases of interior slate construction.

11. Color changes in slate are of two kinds: First, those due to concealment of the true color by an efflorescence giving the effect of fading, and, second, oxidation of ferrous minerals resulting in a more permanent color.

12. The so-called fading of dark-colored slate is confined mainly to those varieties containing considerable amounts of calcite and pyrite. This phenomenon accompanies the weathering process and is believed to be a result of the same chemical changes which produce decay. Color changes due to the oxidation of ferrous minerals are confined mainly to certain varieties of green slate and results in a change from green to brown. Occasionally dark-colored slates take on a brown color after exposure to the weather. This appears to be due to a decomposition of the pyrite and oxidation of the iron compounds.

WASHINGTON, June 21, 1932.

RADIATION FROM CÆSIUM AND OTHER METALS BOMBARDED BY SLOW ELECTRONS

By C. Boeckner

ABSTRACT

The absolute intensity and intensity wave-length distribution of the radiation from Cs, Cr, Ni, Mo., and W is measured and found to be similar for all five metals. Control measurements are also made allowing the effect of surface contamination to be estimated. It is found that the radiation from contaminated surfaces is usually much more intense than from clean metals. It is pointed out that the similarity between the radiation from cæsium and other metals makes it probable that the radiation is analogous to the continuous X rays and is not due to the excitation of electron levels characteristic of the metal.

CONTENTS

I. INTRODUCTION

It has been observed in this laboratory that a small metal electrode (Langmuir probe), drawing a heavy electron current in a gas discharge, emits radiation. The properties of this radiation, dependence upon the metal used as electrode, the energy of bombarding electrons, etc., have been set forth in two previous papers.[1]

The principal facts may be summarized as follows: The spectrum of radiation is continuous and of almost constant intensity throughout the visible and ultra-violet. Exceptions are silver and copper which possess intense selective emission bands. The absolute intensity of the radiation for most metals is about the same. For electron energies of 2 or 3 volts the spectrum has a short wave-length limit, the frequency of which is connected with the work function of the metal and the electron energy by the familiar Einstein photo-electric equation. The intensity of the radiation is of the order expected for the "bremsstrahlung" produced by electrons with energies of the order used, which suggests that the radiation is similar in orgin to the continuous X rays.

The present work is a continuation of the studies described in the first two papers in the earlier work. The absolute intensity measurements were few and the results discordant. It was therefore decided to make more extensive absolute measurements and at the same time look for unknown factors, such as surface contamination, which might influence the results.

---

[1] F. L. Mohler and C. Boeckner, B. S. Jour. Research, vol. 6 (RP297), p. 673, 1931; vol. 7 (RP371), p. 751, 1931.

It was also decided to measure the radiation from several metals not studied previously, namely, Cs, Cr, Mo, and Ni. The radiation from cæsium was considered of particular interest because of the exceptional properties of the metal and because a cæsium electrode surface in a cæsium discharge could be kept very clean.[2]

II. EXPERIMENTAL

1. METHOD

For information concerning the technique, the original papers should be referred to. Some details, however, will be mentioned.

The intensity measurements were made by the usual photographic methods, using an incandescent tungsten strip at various temperatures as a comparison source. An E2 Hilger quartz spectrograph was used to photograph the spectrum of the radiation and quartz fluorite achromats were used to focus the image of the electrodes on the slit.

A low-voltage hot cathode discharge in cæsium vapor was used as a source of electron current to the probe. The discharge tube was placed in a furnace at about 170° C. to maintain the required Cs vapor pressure (several thousandths of a millimeter).

The electrodes were usually in the form of cylinders 2 to 4 mm in diameter. The sides were insulated by glass tubing and the flat polished ends exposed to the discharge.

2. CONTROL MEASUREMENTS

A number of control measurements were made in order to locate possible sources of contamination of the probe surfaces. In the earlier work the probes had been placed in the negative glow of the discharge, near the barium oxide coated platinum cathode. It was suspected, however, that barium or oxygen evaporated from the cathode might contaminate the probe surface. Measurements were therefore made in the negative glow, using a tungsten cathode, and also in the positive column some 15 cm from the cathode.

In most of the work the probes were outgassed at 400° C. and the metal surface cleaned by several hours' bombardment with 100-volt Cs ions. To ascertain the effect of better outgassing and a possibly cleaner surface, a tungsten electrode was used in the form of a wire 0.4 mm in diameter and 1 cm long. The wire could be heated to incandescence and outgassed by current passed through it from insulated supporting leads.

3. MEASUREMENT OF THE CÆSIUM RADIATION

A specially designed tube was required for the measurement of the radiation from a cæsium electrode. A tungsten wire, 2 mm in diameter and several centimeters long, was sealed into the glass wall of a discharge tube, the inner end being flush with the glass wall. The other end projected from the tube and through the walls of the furnace into the air. Upon blowing air against the end of the wire the end in the discharge was cooled sufficiently to condense a bright layer of cæsium upon it. If the electron current to the probe was too large

---

[2] A cæsium vapor discharge was found to be most suitable as a source of current to the electrodes due to the relative scarcity of lines in its arc spectrum. Scattered light from the discharge is a serious difficulty in the type of work described here.

the cæsium evaporated; if too small it would spread over the surrounding glass wall. A current and voltage could easily be found for which the cæsium just covered the end of the wire. The cæsium could be evaporated and deposited many times during the observations, thus insuring a clean surface. The diameter of the cæsium

ergs/micron.ampere

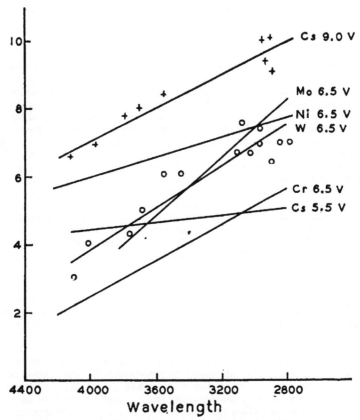

FIGURE 1.—*Absolute intensities of radiation from metals*

Ordinates, intensities in ergs per ampere in a micron wave-length range. "o" tungsten 6.5 volts; "+" cæsium 9 volts.

disk, a knowledge of which was necessary in order to determine the current density, was evaluated from the size of the image on the spectrograms. It may be mentioned that a monatomic adsorbed cæsium layer present on the other metals studied does not, of course, constitute a difficulty in the case of a cæsium electrode.

III. RESULTS

1. ABSOLUTE INTENSITIES

The curves of Figure 1 give the results of the intensity measurements. The intensities are given in ergs in a micron wave-length range emitted by a square centimeter of the electrode surface and for a current density of 1 ampere per square centimeter.[3] The measurements are given together in one figure in order to illustrate the similarity between the metals. It will be seen from the following paragraph that the differences between them are not much greater than the experimental uncertainty.

2. EXPERIMENTAL UNCERTAINTY

A series of measurements on tungsten electrodes will illustrate the influence of various factors on the intensity and the magnitude of the experimental uncertainty. An electrode near the cathode yielded 4.5 ergs per ampere at 3,700 A for electron energies of 6.5 volts; in the positive column, 15 cm from the cathode, 4.1 ergs were obtained; the filament electrode, which could be cleaned and outgassed by heating to incandescence, gave 5.6 ergs. The first two electrodes were outgassed at 400° C. and cleaned by ion bombardment. The agreement is fair and shows that the latter method of outgassing and cleaning is satisfactory and that contamination by material evaporated from the cathode is not serious. Fluctuations in check measurements were usually of the magnitude illustrated above. As causes of the variations there may be mentioned, in addition to surface contamination and photometric difficulties, the uncertainties in the current collecting area of electrodes due to variation in fit between the glass insulators and the metal cylinder.

3. RADIATION FROM CONTAMINATED SURFACES

Before cleaning by ion bombardment the electrodes were sometimes visibly stained. The radiation from such surfaces was usually very intense; after a few seconds ion bombardment the intensity would drop to a much lower value. Changes could usually be detected, however, after an hour's cleaning by ion bombardment.[4] It was suspected in some cases that a layer of tungsten oxide evaporated from the tungsten cathode might be responsible for the intense radiation. Measurements were, therefore, made with a layer of tungsten oxide evaporated upon the electrode from a near-by oxidized tungsten filament. The radiation was found to be ten times more intense than that from a metal, but to have about the same intensity distribution between 4,500 and 2,700 A. Measurements made at low voltage indicated a work function of about 2.4 volts, somewhat higher than that for a metal in cæsium vapor. The surface layer was, however, soluble in water and was probably a compound of cæsium and tungsten oxide formed during evaporation in the presence of cæsium vapor. A grayish white layer of oxide, formed during evaporation in the absence of cæsium, emitted radiation two or three times stronger than the metallic radiation.

---

[3] The variation of intensity with angle is supposed to be similar to that of a black body.
[4] Published data on sputtering indicates that the ion current densities used here remove of the order of an atomic layer in a second.

The measurements of the radiation from aluminum previously reported were from a contaminating layer which seems to form upon an aluminum surface in a cæsium discharge. The peculiar minimum in the voltage intensity relation reported in the earlier paper is characteristic of this layer and not of aluminum.

IV. DISCUSSION

The fact that the radiation from cæsium is similar to that from other metals is of some interest in connection with the theory of its origin. One possibility is that the radiation is analogous to the continuous X-ray spectrum and is produced by the deceleration of the bombarding electrons in the fields of the metal atoms. Another suggestion supposes that the primary electron ejects a free electron from the metal and that another free electron then falls into the vacated level with emission of radiation. The latter explanation would predict that, for sufficiently great energy of the bombarding electrons, the spectrum would possess a fixed short wave-length limit, the frequency of which corresponds to the energy spread of the free electron levels. For most metals the width of these levels is of the order of 10 volts and the limit would lie in the far ultra-violet. For cæsium, however, this width, according to most writers,[5] is near 2 volts and the cæsium spectrum should have a high frequency limit in the red or yellow. Since this is not observed one may conclude that the "bremsstrahlung" explanation is probably the correct one.

This view is suggested by the observations of O. W. Richardson and his coworkers [6] that the total intensity of soft X rays produced by 300 to 6,000 volt electrons is very nearly independent of the metal from which the target is made. The extreme variation between 14 metals was less than twofold. This fact is analogous to the similarity of the radiative properties of metals in the low-voltage region described in the present paper, and suggests that the phenomena have a common origin.[7]

The observed intensity may be compared with that predicted by quantum theory for the continuous radiation generated by slow electrons impinging on protons.[8] The radiation emitted in ergs in frequency range $d\nu$ is

$$J(\nu) = g32 \; \pi^2 \; Z^2 \; e^6 \; d\nu/3\sqrt{3}c^3 m^2 v^2 \tag{1}$$

The intensity is given in ergs per proton for a current of 1 electron passing through a square centimeter in a second. Z is the charge on the proton; e, m, and v are the charge, mass, and velocity of the electron; g is a numerical factor equal to unity classically, and also in the quantum theory for the case of slow electrons.

It is of interest to compute the intensity radiated by 6-volt electrons, for example, falling on a monatomic layer of protons (0.25×10^{16} on a square centimeter). In the units of Figure 1, one obtains at 3,000 Å roughly 80 ergs, about ten times greater than is observed.

[5] Ig. Tamm, Phys. Rev., vol. 39, p. 170, 1932.
[6] O. W. Richardson and F. S. Robertson, Proc. Roy. Soc., vol. 124, p. 188, 1929.
[7] The similarity is surprising when one considers the extreme difference in the properties of tungsten and cæsium, for example: the atomic volume of cæsium is seven times that of tungsten; the spread of energies of the free electrons in tungsten is perhaps three or four times that of cæsium; tungsten has the highest melting point of the metals, while cæsium was a liquid under the conditions of the measurements.
[8] A. W. Maue, Ann. der Physik, vol. 13, p. 161, 1932.

When it is considered that many layers of charged centers contribute to the radiation in the experimental case of electrons striking a metal, the discrepancy becomes even greater.[9] "Z" was made unity in the computation because the field of the singly charged nucleus was thought to represent more nearly the fields actually encountered by the bombarding electrons.

Leo Neledsky [10] has computed the intensity to be expected when the nucleus is surrounded by its shielding electrons. Using a rough model of such a system of charges, he obtained values much lower than those given by equation (1) and in better agreement with the observed magnitudes. The classical explanation is of course that the external electrons prevent the bombarding electrons from approaching near the nucleus where most of the radiation is emitted.

WASHINGTON, June 20, 1932.

---

[9] The agreement with the theoretical intensity for thick targets reported in an earlier paper was due to the use of the Thomson-Whiddington formula for the depth of penetration of electrons into metals. This relation can not be applied to slow electrons.
[10] Leo Neledsky, Phys. Rev., vol. 39, p. 552, 1932. (Abstract.)

A SIMPLIFIED PRECISION FORMULA FOR THE INDUC-TANCE OF A HELIX WITH CORRECTIONS FOR THE LEAD-IN WIRES

By Chester Snow

ABSTRACT

A precision formula is here given for the self inductance of a single-layer helix wound with ordinary round wire. This is neither more nor less accurate than that previously published, from which it has been derived by evaluating certain correction terms and replacing them by approximation formulas which are much simpler to compute.

CONTENTS

I. INTRODUCTION

The formula derived in this paper is a modification, without loss of precision, of the expression for the inductance of a single-layer helix of round wire, given in Bureau of Standards Scientific Paper No. 537 (vol. 21, p. 431, 1926–27). The notation has been changed, and the correction terms transformed so that the present formula is more simple from the point of view of the computer.

The diameter of the wire is d; the mean diameter of the solenoid is D; and its length l is the axial distance from the center of the wire at the begininng of the first turn to the center of the wire at the end of the Nth turn, N being the total turns so that the pitch of the winding is l/N. The modulus k of the complete ellipitic integrals K and E is given by

$$k^2 = \frac{D^2}{l^2+D^2}$$

The principal term L_s is the current-sheet formula of Lorenz.

$$L_s = \frac{4\pi N^2}{3}\sqrt{l^2+D^2}\left[K-E+\frac{D^2}{l^2}(E-k)\right] \tag{1}$$

As one of the correction terms there appears M, the mutual inductance between the two end-circles of the solenoid, which is computed by

$$M = 4\pi D\left[\frac{K-E}{k}-\frac{k}{2}K\right] \tag{2}$$

The inductance L of an actual helix is (to a precision which neglects terms of the order of $\left(\frac{l}{ND}\right)^3\log\frac{l}{ND}$)

419

$$L = L_s + \pi D \left[2N \left(ln \frac{l}{Nd} - 0.89473 \right) + \frac{1}{3} ln \frac{N\pi D}{l} \right] + lP \left(\frac{l}{D} \right)$$
$$- \frac{1}{6}M - \frac{2}{\pi} \sqrt{l^2 + D^2} (E-k) \left[2 \pm \left(\frac{N\pi d}{2l} \right)^2 \right]$$
$$- \frac{l\pi}{2} \left[1 - \frac{l}{D} sin^{-1} \frac{D}{\sqrt{l^2 + D^2}} \right]$$

(3)

where

$$P(\eta) = \frac{1}{4\eta^2} + 2ln4\eta \text{ when } \eta \geqq 1$$
$$= 3\eta - \eta ln\eta \text{ when } \eta \leqq 1$$

(4)

The ambiguous sign is plus for the "natural" and minus for the uniform distribution of current over the section of the wires. A current density inversely proportional to the distance from the axis of the solenoid is called the natural distribution.

The formula for L here given is obtained by change of notation and by transformation of the function $A_2(k)$ in formula (114) of the paper cited. (This formula contains a misprint in the term $\frac{1}{3} log \frac{\alpha}{p}$ which should be $\frac{1}{3} log \frac{a}{p}$). In the present notation this formula is

$$L = L_s + \pi D \left[2N \left(ln \frac{l}{Nd} - 0.89473 \right) + \frac{1}{3} ln \frac{N\pi D}{l} \right]$$
$$+ \pi D A_2(k) \mp \frac{2}{\pi} \sqrt{l^2 + D^2} (E-k) \left(\frac{N\pi d}{2l} \right)^2$$

(A)

The function $A_2(k)$ is defined by formula (89)

$$A_2(k) = \frac{4}{\pi^2}[B_0(k) - B_1(k) + B_2] + 0.66267 - \frac{1}{3}ln\pi$$

The numerical quantity B_2 was defined in (90) as a series which has been found to be -0.60835. Hence

$$\pi D A_2(k) = D \left[\frac{4}{\pi}(B_0(k) - B_1(k)) + 0.11 \right]$$

(5)

The term 0.11 in the parenthesis will be omitted since 0.1 D will be considered negligible in this small correction term. As most solenoids to be used as precision standards would have an inductance greater than 10 millihenries (10^7 cm), it is evident that with a diameter of say 30 cm, 0.1 D would be 3 cm, which is about 3 parts in 10,000,000. Hence, in the transformations of $B_0(k)$ and $B_1(k)$ which follow, terms of the order of 0.1 may be considered negligible.

II. TRANSFORMATION OF $B_0(k)$

On page 507 of the paper quoted, B_0 is given by

$$B_0(k) = 1 - \frac{E(k)}{k} + \frac{\sqrt{1-k^2}}{k} \int_k^1 \frac{K(x)dx}{x\sqrt{1-x^2}}$$

(6)

If we let

$$\eta = \frac{\sqrt{1-k^2}}{k} \quad \text{and} \quad x^2 = \frac{1}{1+\lambda^2} \quad \text{and}$$

$$P(\eta) \equiv \frac{4}{\pi}\int_k^1 \frac{K(x)dx}{x\sqrt{1-x^2}} = \frac{4}{\pi}\int_0^\eta \frac{d\lambda}{\sqrt{1+\lambda^2}} K\left(\frac{1}{\sqrt{1+\lambda^2}}\right)$$

$$= \frac{4}{\pi}\int_0^{\frac{\pi}{2}} d\theta \int_0^{\frac{\pi}{2}} \frac{d\lambda}{\sqrt{\lambda^2+sm^2\theta}} = \frac{4}{\pi}\int_0^{\frac{\pi}{2}} \log\left[\frac{\eta+\sqrt{\eta^2+\sin^2\theta}}{\sin\theta}\right]d\theta \quad . \tag{7}$$

then (6) gives

$$\frac{4D}{\pi} B_0(k) = -\frac{4}{\pi}\sqrt{l^2+D^2}[E(k)-k]+lP\left(\frac{l}{d}\right) \tag{8}$$

We may evaluate $P(\eta)$ first for the case where $\eta \geq 1$; that is, where $l \geq D$. Using the formula

$$\int_0^{\frac{\pi}{2}} \log \sin\theta\, d\theta = -\frac{\pi}{2}\log 2$$

we may write (7) in the form

$$P(\eta) = 2\log 2 + \frac{4}{\pi}\int_0^{\frac{\pi}{2}} \log(\eta+\sqrt{\eta^2+\sin^2\theta})\,d\theta$$

$$= 2\log 2\eta + \frac{4}{\pi}\int_0^{\frac{\pi}{2}} \log\left(1+\sqrt{1+\frac{\sin^2\theta}{\eta^2}}\right)d\theta$$

or since (for $\eta \geq 1$)

$$\log\left(1+\sqrt{1+\frac{\sin^2\theta}{\eta^2}}\right) = \log 2 - \frac{1}{2\sqrt{\pi}}\sum_{n=1}^{\infty} \frac{(-1)^n}{n}\frac{\Gamma\left(n+\frac{1}{2}\right)}{\Gamma(n+1)}\frac{\sin^{2n}\theta}{\eta^{2n}}$$

$$P(\eta) = 2\log 4\eta + \frac{1}{4\eta^2} - \sum_{n=2}^{\infty} \frac{(-1)^n}{n}\left[\frac{1.3.5\ldots(2n-1)}{2.4.6\ldots 2n}\right]^2\frac{1}{\eta^{2n}} \text{ if } \eta \geq 1 \tag{9}$$

Since the sum of this alternating series is numerically less than the value of its first term $\frac{9}{128\eta^4}$, which is never greater than about $\frac{1}{14}$ if $\eta \geq 1$, it is sufficient to take

$$P(\eta) = 2\log 4\eta + \frac{1}{4\eta^2} \qquad \text{when } \eta \geq 1 \tag{10}$$

When $\eta = 1$ this becomes $2\log 4 + \frac{1}{4} = 3.02$ while the exact expression (9) becomes 2.98. We may, therefore, take $P(1) = 3$. On the other hand, when $0 \leq \eta \leq 1$, the approximation

$$P(\eta) = 3\eta + \eta\log\frac{1}{\eta} \qquad \text{when } 0 \leq \eta \leq 1 \tag{10}'$$

will be in error by not more than 0.1. This may be seen from the fact that (10)′ is correct at $\eta=0$ and $\eta=1$, while the error between 0 and 1 is

$$Y(\eta)\equiv 3\eta+\eta\,\log\frac{1}{\eta}-P(\eta)$$

which has its greatest numerical value when

$$\eta=\frac{\sqrt{1-k^2}}{k}=0.34904$$

corresponding to $k^2=0.8914$. This is the root of the equation

$$Y'(\eta)=0=2+\log\frac{1}{\eta}-P'(\eta)=2+\frac{1}{2}\,\log\frac{k^2}{1-k^2}-\frac{4k}{\pi}K_{(k)}$$

To estimate this greatest error, we may obtain an upper and lower limit to $P\,(\eta)$ as follows:
Write

$$P(\eta)=\log 4+\frac{1}{\pi}\int_0^\pi\log\,(\eta+\sqrt{1+\eta^2-\cos^2\theta})^2 d\theta$$

$$=\log 4+\frac{1}{\pi}\int_0^\pi\log\,(1+2\eta^2-\cos^2\theta+2\eta\sqrt{1+\eta^2-\cos^2\theta})\,d\theta$$

Now

$$2\eta\sqrt{1+\eta^2-\cos^2\theta}=2\eta\sqrt{1+\eta^2}-\frac{\eta}{\sqrt{1+\eta^2}}\cos^2\theta-Z\,(\eta,\,\theta)$$

where

$$Z\,(\eta,\,\theta)=\frac{\eta\sqrt{1+\eta^2}}{\sqrt{\pi}}\sum_{s=2}^{\infty}\frac{\Gamma\left(s-\frac{1}{2}\right)}{\Gamma(s+1)}\,k^{2s}\cos^{2s}\theta$$

where

$$k^2=\frac{1}{1+\eta^2}$$

The function Z is never negative. It vanishes when $\theta=\frac{\pi}{2}$ and has its greatest value when $\theta=0$ and $\theta=\pi$ which is

$$Z\,(\eta,\,\theta)=\frac{\eta(\sqrt{1+\eta^2}-\eta)^2}{\sqrt{1+\eta^2}}$$

Hence $P(\eta)$ may be written in the form

$$P(\eta)=\frac{1}{\pi}\int_s^\pi\log 4\left\{(\sqrt{1+\eta^2}+\eta)^2-Z(\eta,\,\theta)-\left(\frac{\sqrt{1+\eta^2}+\eta}{\sqrt{1+\eta^2}}\right)\cos^2\theta\right\}d\theta$$

If in this integral we replace the variable function $Z(\eta,\,\theta)$ by its smallest value, zero, we get a function $F(\theta)$, which is an upper limit for $P(\eta)$ for all values of η.

$$F(\eta)\equiv\frac{1}{\pi}\int_s^\pi\log 4\left\{(\sqrt{1+\eta^2}+\eta)^2-\left(\frac{\sqrt{1+\eta^2}+\eta}{\sqrt{1+\eta^2}}\right)\cos^2\theta\right\}d\theta$$

If we replace $Z(\eta,\theta)$ by its greatest value, we get a function $f(\eta)$ which is a lower limit to $P(\eta)$ for all values of (η) where

$$f(\eta) \equiv \frac{1}{\pi} \int_\delta^\pi \log 4 \left\{ (\sqrt{1+\eta^2}+\eta)^2 - \frac{\eta(\sqrt{1+\eta^2}-\eta)^2}{\sqrt{1+\eta^2}} - \left(\frac{\sqrt{1+\eta^2}+\eta}{\sqrt{1+\eta^2}} \right) \cos^2\theta \right\} d\theta$$

That is $f(\theta) < P(\eta) < F(\eta)$ if $\eta > 0$.

Both of these functions may be evaluated by means of the definite integral formula

$$\frac{1}{\pi} \int_0^\pi \log 4 \; (a^2 - b^2 \cos^2\theta) \; d\theta = 2 \log \; (a + \sqrt{a^2 - b^2}) \text{ if } a \geqq b$$

which gives

$$F(\eta) = 2 \log \; (\sqrt{1+\eta^2} + \eta) \left(1 + \sqrt{\frac{\eta}{\sqrt{1+\eta^2}}} \right)$$

$$f(\eta) = 2 \log \left(2\eta + \sqrt{1 + 4\eta^2 + \frac{\eta}{\sqrt{1+\eta^2}}} \right)$$

This gives

$$F(0) = 0, \; F(1) = 2.984$$

$$f(0) = 0, \; f(1) = 2.958$$

At the point of maximum error in the range $0 < \eta < 1$, which is at $\eta_1 = 0.349$, we find

$$\left. \begin{array}{l} F(\eta_1) = 1.581 \\ f(\eta_1) = 1.431 \end{array} \right\} \text{whence } P(\eta_1) = 1.506 \; \pm 0.075$$

and

$$3 \; \eta_1 + \eta_1 \log \frac{1}{\eta_1} = 1.415$$

which shows that the maximum error made by using (10)' for $P(\eta)$ is approximately 0.1 in the range $0 \leqq \eta \leqq 1$.

III. TRANSFORMATION OF $B_1(k)$

The function $B_1(k)$ is defined in the paper quoted, on page 453, by

$$kB_1(k) = \frac{\pi^2}{12} E(k) - \int_0^{\frac{\pi}{2}} \theta^2 \sqrt{1 - k^2 \sin^2 \theta} \, d\theta$$

$$- \left(1 - \frac{k^2}{2} \right) \left[\frac{\pi^2}{12} K(k) - \int_0^{\frac{\pi}{2}} \frac{\theta^2 d\theta}{\sqrt{1 - k^2 \sin^2 \theta}} \right]$$

Now

$$\int_0^{\frac{\pi}{2}} \theta^2 \sqrt{1 - k^2 \sin^2 \theta} \, d\theta = \frac{\pi^2}{4} E(k) - \int_0^{\frac{\pi}{2}} \phi(\pi - \phi) \sqrt{1 - k^2 \cos^2 \phi} \, d\phi$$

and

$$\int_0^{\frac{\pi}{2}} \frac{\theta^2 d\theta}{\sqrt{1 - k^2 \sin^2 \theta}} = \frac{\pi^2}{4} K(k) - \int_0^{\frac{\pi}{2}} \frac{\phi(\pi - \phi)}{\sqrt{1 - k^2 \cos^2 \phi}} \, d\phi$$

Hence

$$\frac{4D}{\pi} B_1(k) = \frac{4\pi D}{6} \left[\frac{K(k) - E(k)}{k} - \frac{kK(k)}{2} \right] + DU(k)$$

where

$$U(k) \equiv -\frac{\pi k}{2} \int_0^{\frac{\pi}{2}} \frac{4\phi(\pi-\phi)}{\pi^2} \frac{\cos 2\phi}{\sqrt{1-k^2\cos^2\phi}} \, d\phi$$

The factor of the integrand $y(\phi) = \frac{4}{\pi^2}\phi (\pi-\phi)$ is represented by a parabola which vanishes when $\phi=0$ and has the maximum value unity when $\phi=\frac{\pi}{2}$, where it also has a horizontal slope. It is therefore very approximately the same as $\sin \phi$. Replacing it by $\sin \phi$ gives the sufficient approximation

$$U(k) = -\frac{\pi k}{2} \int_0^{\frac{\pi}{2}} \frac{\cos 2\phi \sin \phi}{\sqrt{1-k^2\cos^2\phi}} \, d\phi = \frac{\pi\sqrt{1-k^2}}{2k}\left[1-\frac{\sqrt{1-k^2}}{k}\sin^{-1}k\right]$$

$$= \frac{\pi l}{2D}\left[1-\frac{l}{D}\sin^{-1}\frac{D}{\sqrt{l^2+D^2}}\right]$$

Hence

$$\frac{4D}{\pi} B_1(k) = \frac{M}{6} + \frac{l}{2}\left[1-\frac{l}{D}\sin^{-1}\frac{D}{\sqrt{l^2+D^2}}\right] \tag{11}$$

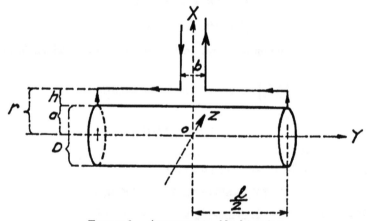

FIGURE 1.—*Arrangements of lead-in wires*

where M is defined by equation (2) and is obviously the mutual inductance between the two end circles of the solenoid. By use of equations (10) or (10)' and (11), the equation (A) is transformed into formula (3) of this paper.

IV. CORRECTIONS DUE TO LEAD-IN WIRES

The inductance of the helix and lead-in wires together is

$$L' = L + L_1 + 2M_M$$

where L is that of the helix given by formula (3), L_1 is the self-inductance of the lead-in wires and M_M the mutual inductance between these wires and the helix. In computing M_M, the helix may be considered a current sheet, and only its axial component of current contributes to M_{hl}. If the leads are disposed as shown in Figure 1, it is evident that these leads which are perpendicular to the axis of the cylinder have no mutual inductance with the cylinder. The latter carries unit current in the $y=$ direction so that the current density is $\frac{1}{2\pi a}=\frac{1}{\pi D}$. Consequently, the term M_{hl} is given by

$$M_{hl}=2\int_0^{b/2}A(r,\,y)dy-2\int_0^{l/2}A(r,\,y)dy \qquad (12)$$

where

$$A\,(r,\,y)=\frac{1}{2\pi}\int_{-l/2}^{l/2}dy'\int_0^{2\pi}\frac{d\,\phi}{\sqrt{(y-y')^2+a^2+r^2-2\,a\,r\cos\phi}}$$

$$=\frac{2}{\pi}\int_{-l/2}^{l/2}dy'\int_0^{\pi/2}\frac{d\,\theta}{\sqrt{(y-y')^2+(a+r)^2-4\,a\,r\sin^2\theta}}$$

$$=\frac{2}{\pi}\int_0^{\frac{l}{2}+y}dz\int_0^{\pi/2}\frac{d\,\theta}{\sqrt{z^2+(a+r)^2-4\,a\,r\sin^2\theta}}$$

$$+\frac{2}{\pi}\int_0^{\frac{l}{2}-y}dz\int_0^{\pi/2}\frac{d\,\theta}{\sqrt{z^2+(a+r)^2-4\,a\,r\sin^2\theta}}$$

or

$$A\,(r,\,y)=\frac{2}{\pi}\int_{k_y}^1\frac{d\,x}{x\sqrt{1-x^2}}K\left(\frac{2\sqrt{ar}}{a+r}x\right)+\frac{2}{\pi}\int_{k_{-y}}^1\frac{d\,x}{x\sqrt{1-x^2}}K\left(\frac{2\sqrt{ar}}{a+r}x\right) \qquad (13)$$

where

$$\frac{1}{k_y^2}=1+\left(\frac{\frac{l}{2}+y}{a+r}\right)^2 \qquad (14)$$

As a special case of this

$$A\,(a,\,o)=\frac{4}{\pi}\int_{k_o}^1\frac{K\,(x)\,d\,x}{x\sqrt{1-x^2}}=P\,(\eta_0)\text{ by }(7) \qquad (15)$$

Where

$$k_0^2=\frac{1}{1+\left(\frac{l}{4\,a}\right)^2}=\frac{1}{1+\left(\frac{l}{2\,D}\right)^2}\text{ and }\eta_0=\frac{l}{2\,D} \qquad (16)$$

Now since b and $r-a=h$ are both small, we may write

$$2\int_0^{b/2}A(r,y)dy=2A(a,0)\int_0^{b/2}dy=bA(a,0)=bP(\eta_0) \qquad (17)$$

In the first integral of (12) we may write

$$A(ry)=A(a+h,y)=A(a,y)+h\left[\frac{\partial A(r,y)}{\partial r}\right]_{r=a}$$

Computing $\left[\dfrac{\partial A(ry)}{\partial r}\right]_{r=a}$ from (13), since $\left[\dfrac{\partial}{\partial r}\dfrac{2\sqrt{ar}}{a+r}\right]_{r=a}=0$,

one obtains after placing $r=a$

$$\frac{\partial k_y}{\partial r}=\frac{k_y(1-k^2{}_y)}{D}\quad\text{and}\quad\frac{\partial k_{-y}}{\partial r}=\frac{k_{-y}(1-k^2{}_{-y})}{D}$$

so that

$$-2h\int_0^{l/2}\left[\frac{\partial A(r,y)}{\partial r}\right]_{r=a}dy=\frac{4h}{\pi D}\left[\int_0^{l/2}\sqrt{1-k^2{}_y}\,K(k_y)dy+\int_0^{l/2}\sqrt{1-k^2{}_{-y}}\,K(k_{-y})dy\right]$$

where

$$k^2{}_y=\frac{1}{1+\left(\dfrac{l/2+y}{D}\right)^2}$$

Changing the variable of integration from y to k_y gives

$$-2h\int_0^{l/2}\left[\frac{\partial A(ry)}{\partial r}\right]_{r=a}dy=\frac{4h}{\eta}\left[-\int_{k_0}^{k}\frac{K(x)dx}{x^2}+\int_{k_0}^{1}\frac{K(x)dx}{x^2}\right]=\frac{4h}{\pi}\int_k^1\frac{K(x)dx}{x^2}$$

$$=\frac{4h}{\pi k}\Big[E(k)-k\Big]\quad\text{where }k^2=\frac{D^2}{l^2+D^2}\qquad(18)$$

Similarly

$$-2\int_0^{l/2}A(a,y)dy=-\frac{4}{\pi}\left[\int_0^{l/2}dy\int_{k_y}^1\frac{K(x)dx}{x\sqrt{1-x^2}}+\int_0^{l/2}dy\int_{k_y}^1\frac{K(x)dx}{x\sqrt{1-x^2}}\right]$$

$$=-\frac{4D}{\pi}\int_k^1\frac{dz}{z^2\sqrt{1-z^2}}\int_z^1\frac{K(x)dx}{x\sqrt{1-x^2}}=-\frac{4D}{\pi}B_0(k)\quad(19)$$

Adding (17), (18), and (19) gives by (12)

$$M_{hl}=\frac{4}{\pi}\left[-DB_0(k)+\frac{h(E(k)-k)}{k}\right]+bP(\eta_0)\qquad(20)$$

where

$$k^2=\frac{1}{1+\dfrac{l^2}{D^2}}\quad\text{and}\quad k_0{}^2=\frac{1}{1+\left(\dfrac{l}{2D}\right)^2}$$

By use of (8) this may be put in the general form

$$M_{hl}=\left(1+\frac{h}{D}\right)\sqrt{l^2+D^2}\cdot\frac{4}{\pi}\Big[E-k\Big]-lP\left(\frac{l}{D}\right)+bP\left(\frac{l}{2D}\right)\qquad(21)$$

where $P(x)$ is defined by (4).

Washington, June 17, 1932.

REGISTER STUDIES IN OFFSET LITHOGRAPHY

By C. G. Weber and R. M. Cobb [1]

ABSTRACT

A serious economic waste in offset lithography results from lack of knowledge of the optimum printing properties of lithographic papers. The most important losses result from register difficulties, and studies of register have comprised a large part of the research on lithographic papers now being carried on at the Bureau of Standards in cooperation with the Lithographic Technical Foundation. Information on factors influencing register was obtained by making experimental printings in a commercial plant under routine operating conditions. The paper samples used were prepared by cooperating manufacturers, knowledge of the history of manufacture of the papers was supplemented by complete laboratory analyses, and their response to offset printing in controlled atmosphere was observed to find the influence of paper characteristics on register of prints. Most satisfactory register was obtained with the papers that had received the least drastic processing of fibers in manufacture. Internal sizing in machine-finish and coated papers appeared essential, but the amount used, within normal limits, was not important. Thorough conditioning of paper to equilibrium with pressroom atmosphere before the first printing was required to obtain register on subsequent printings, and the longest seasoning period practicable between printings was found desirable. Closeness of register was also influenced by variations in pressure, ink, and water used in printing and by uniformity of plates in respect to thickness. An accurate rule of special design for measuring prints and a sword type of hygroscope for determining the hygrometric state of paper were found invaluable in the plant studies.

CONTENTS

I. INTRODUCTION

The Bureau of Standards is cooperating with the Lithographic Technical Foundation in a study of the offset lithographic process. The work was undertaken at the request of the foundation. This technical organization of the domestic lithographic industry comprises over 400 members, including lithographic concerns, paper manufacturers and dealers, ink and varnish manufacturers, press and machinery manufacturers, metal lithographers, litho finishers, and photo-composing machine manufacturers. The research is being

---

[1] Research associate, representing the Lithographic Technical Foundation.

carried on with the advice and active cooperation of an advisory committee from the foundation under the chairmanship of R. F. Reed, director of the department of lithographic research, University of Cincinnati. The study was requested because the industry suffers an enormous annual loss resulting from lack of scientific knowledge concerning the characteristics of paper required for optimum results in offset printing. The principal limiting factor in production is apparently the failure of paper to meet the requirements of the modern high-speed offset press, and it has been conservatively estimated that not over 75 per cent of the theoretical production of the average press is attained for that reason.

Preliminary investigational work was previously done by the bureau[2] for the purpose of finding the principal difficulties in offset lithography for which paper is wholly or partly responsible. A survey was made of 31 lithographic plants, and the results served as a basis for planning the present study.   Misregister was the most serious difficulty found; therefore, a study of the register of prints was undertaken first, and this article deals with some results already obtained.

A series of practical printing tests were made in the Buffalo, N. Y., plant of the American Lithographic Co. under routine operating conditions to isolate and study the important factors influencing the closeness of register. An air-conditioned pressroom was selected in order to eliminate variables due to changes in atmospheric humidity. By using specially prepared papers, of which the history of preparation as regards composition, degree and kind of sizing, degree of hydration of fibers, and fiber length were known, information relative to the influence of paper characteristics on closeness of register was developed.   Printing identical papers on a press with controlled variations of such factors as pressure, water, and ink gave information on the effects of press variables.   The papers used in experimental printings were subjected to complete laboratory tests of physical properties, composition, water penetration, and behavior under tension at different humidities to obtain information for correlation with results developed relative to the response of the papers to the printing tests.

II. BEHAVIOR OF PAPER ON THE OFFSET PRESS

The offset lithographic process is indirect in that the print is transferred from the plate to the paper by means of a rubber blanket. Three cylinders of the same dimensions run one against another in fixed positions with a gear on the end of each engaged with a similar gear on the next. The cylinders are driven through these gears, providing the same speed of rotation for all. They rest one against another with fixed pressure on narrow metal bands, known as bearers, at the ends of the cylinders. The plate with a positive image is clamped around the first or plate cylinder and this bears against the second or blanket cylinder which is covered with a thin rubber blanket. The blanket cylinder bears against a third cylinder known as the impression cylinder. In operation, the plate after being progressively moistened and inked, prints on the rubber blanket which in turn deposits the print on the paper pressed against it by the impression cylinder.

---

[2] The preliminary work was done by F. H. Thurber, formerly an associate scientist at the Bureau of Standards.

The dimensions of paper may change during passage through the press or between printings, and the nature and extent of such changes determine, largely, the closeness of register in subsequent printings. Changes in length in the short or around-the-cylinder direction are relatively unimportant because any reasonable change in this dimension can be compensated for by press adjustments. However, changes in dimension in respect to the length along the back edge (long direction) always result in register difficulties. In passage through the press, paper does one of three things in respect to length along the back edge before the instant of taking the print. It shortens or "pulls in," remains unchanged, or lengthens by "fanning-out" under stress. If the paper has stretched before taking the impression and recovers afterward with resultant contraction of the printed design, the resulting print will be shorter than the plate image and, if after taking the print the paper expands, the print will be longer than the image.

Register will be obtained regardless of changes in the paper, if at the instant of taking the succeeding print the paper responds in such a manner as to bring the preceding print back to exactly blanket image length. Misregister will occur if the preceding print is not brought to image length. Hence, paper can change in length as measured and the print stay in register; while paper may show no change in length as measured and yet be out of register. A consideration of the mechanics involved indicates that there is much more chance of a short first print stretching or "fanning-out" to plate image length, to obtain register as it takes the next print, than there is of a long print being shortened or "pulled in" to plate image length as it takes the next print. Therefore, papers that expand after taking a print, that is, "print longer than the plate image," are particularly prone to give misregister.

III. DESCRIPTION AND PROPERTIES OF THE EXPERIMENTAL PAPERS

The papers used in the experimental printings were studied in the laboratory in order to obtain complete information for correlation with their response to the printing processes. Through the cooperation of p er manufacturers, four papers of the same composition with different degrees of beating, the same papers with surface sizing, three coated papers differing as to the degree of sizing in the raw stock (paper before coating), and three extra strong machine-finish papers differing as to degree of internal sizing were made available for the studies. The properties of the papers as indicated by the test data for them are given in Table 1. All were chemical wood papers.

Since mechanical deformation of paper is a factor in obtaining register, laboratory studies were made of the elongation of various samples of lithographic papers under tension. Eleven papers were conditioned and tested in atmospheres of 30, 45, 65, and 75 per cent relative humidity. Tensions of from 9 to 455 g were applied to specimens 15 mm wide and the elongation for each increment of tension was observed by means of a horizontal microscope. The results did not indicate any significant differences in elongation between the various samples tested which included machine-finish litho, supercalendered litho, coated litho, and surface-sized paper.

The greatest differences in the elongation of the different papers occurred at 75 per cent relative humidity and there the widest difference was less than 0.1 per cent for the cross direction. Differences for the machine direction were much smaller. Such differences in expansion are not significant; however, the relation of relative humidity to expansion was significant. The minimum variation between the expansions of the different papers occurred at 45 per cent relative humidity and changes in expansion for changes in relative humidity were of smaller magnitude in the vicinity of 45 per cent relative humidity than at other humidities used. These results indicate definite advantages in the selection of an atmosphere of approximately 45 per cent relative humidity for lithographic plants.

TABLE 1.—*Test data of experimental papers*

Sample No.	Weight, 25 by 40 inches, 500 sheets	Thickness	Bursting strength[3]	Folding endurance[1]		Tensile properties[2]				Water resistance (Dry-indicator method)	Ash	Opacity
				Machine direction	Cross direction	Breaking load		Elongation at rupture				
						Machine	Cross	Machine	Cross			
	Pounds	*Inch*	*Points*	*Double folds*	*Double folds*	*kg*	*kg*	*Per cent*	*Per cent*	*Seconds*	*Per cent*	*Per cent*
1	78.1	0.0040	24.3	52	15	6.9	3.3	1.5	3.2	53.7	9.9	96.2
2	74.3	.0042	21.1	18	11	5.5	3.0	1.6	3.7	31.9	11.7	96.2
3	73.8	.0044	21.4	25	10	5.9	3.0	1.1	2.6	44.1	9.9	95.2
4	69.2	.0040	20.7	14	11	5.2	2.9	1.3	3.9	31.5	11.5	94.4
5	79.8	.0048	33.0	62	29	8.5	4.1	1.7	3.7	48.4	9.5	95.2
6	82.0	.0049	25.9	12	15	7.6	3.9	1.5	3.8	63.3	12.1	96.1
7	77.7	.0047	27.9	46	17	8.0	3.8	1.3	3.7	43.4	9.1	94.4
8	76.6	.0047	25.1	14	10	7.1	3.8	1.5	4.2	55.3	10.5	95.6
9	70.2	.0032	21.7	55	10	5.5	2.6	2.3	3.1	19.1	35.7	95.1
10	75.3	.0036	22.8	48	16	5.9	2.8	2.1	3.6	23.1	33.6	95.5
11	76.2	.0037	21.2	38	12	5.4	2.6	2.1	3.4	26.3	35.2	96.4
13	56.3	.0033	20.0	28	12	5.9	2.8	1.5	3.3	12.1	7.2	87.7
14	54.4	.0034	19.6	24	10	5.5	2.5	1.6	3.4	19.3	8.0	88.4
15	53.9	.0032	20.1	26	11	5.5	2.5	1.5	4.0	25.9	8.0	88.9

[1] For test specimen, 15 mm wide and 90 mm between jaws.
[2] For test specimen, 15 mm wide and 100 mm between jaws.
[3] Bursting pressure in pounds per square inch through a circular orifice 1.2 inches in diameter.

Description of special papers

Sample No.	Description
1	Machine-finish litho, light beating, light jordanning.
2	Machine-finish litho, light beating, heavy jordanning.
3	Machine-finish litho, normal beating, light jordanning.
4	Machine-finish litho, normal beating, heavy jordanning.
5	No. 1 surface sized with starch.
6	No. 2 surface sized with starch.
7	No. 3 surface sized with starch.
8	No. 4 surface sized with starch.
9	Coated litho, no sizing in raw stock.
10	Coated litho, one-half normal sizing in raw stock.
11	Coated litho, normal sizing in raw stock.
13	Extra strong machine-finish litho with no sizing.
14	Extra strong machine-finish litho with one-half normal sizing.
15	Extra strong machine-finish litho with normal sizing.
12	Commercial machine-finish litho with which special papers were printed.

NOTE.—The papers were cut 38 by 52 inches with the long dimension in the machine direction of the paper.

IV. TECHNIQUE IN EXPERIMENTAL PRINTINGS

In the experimental printings to determine the relation between paper characteristics and closeness of register, the samples of experimental papers were inserted in the center of the piles and printed with papers being used for regular commercial multicolor printings requiring very close register. To determine the effects of press variables, one press was set aside for experimental work. This press was made ready with the plate at exactly bearer height and with a minimum pressure between the blanket and plate. The printing pressure was varied within closely controlled limits, while printing with a dry plate, with water alone, and with ink and no water to find the extent of effect of each of those factors on register.

Changes in paper dimensions were determined by means of a rule designed for measuring the displacement of reference marks on large sheets of paper. (Fig. 1.) It is equipped with two magnifying glasses, each with a small glass reticule having a vertical cross hair mounted directly underneath. One magnifier may be placed at any point within the range of the scale (24 to 64 inches) and readily set so that its cross hair will coincide with any division on the scale. The other magnifier (fig. 2) is set over the second reference point by means of micrometer adjustment. The distance between the reference points may then be read to 0.001 inch.

Data relative to dimension changes of papers were determined by measuring distances between register marks on the plate, distances between reference marks on the paper before printing and after each printing, and the distances between register prints after each printing. Length of plate image in the around-the-cylinder direction was measured on the press with a thin steel tape.

Since all changes in moisture content of paper are accompanied by dimension changes, a very careful check was maintained on the hygrometric state of the papers, especially in respect to the surrounding atmosphere. The sword type of paper hygroscope [3] was invaluable as a tool for determining quickly whether or not paper was in equilibrium with the pressroom atmosphere. The instrument resembles a sword, the blade of which contains a hygroscopic element. Expansion and contraction of the element actuate the pointer of an indicating device mounted on the handle of the instrument. The hygroscope is first set for the pressroom atmosphere by waving it in the air until the pointer comes to rest and then turning the dial to the position where the zero reference mark is directly under the pointer. The blade is then inserted between the sheets of the pile of paper to be tested, and left until the pointer comes to rest. Movement of the pointer from the zero reference mark indicates that the paper is not in equilibrium with the surrounding atmosphere.

V. RESULTS OF STUDIES OF FACTORS INFLUENCING REGISTER

The results of the printings in respect to permanent dimension changes (figs. 3 and 4), the relation of length of first print to plate image length with subsequent printings (fig. 7), and effects of press variables (figs. 5 and 6), are presented graphically.

---

[3] Robert F. Reed, The Paper Hygroscope, Sales Bull. No. 1. Research Series No. 6, Lithographic Technical Foundation, 220 East Forty-second Street, New York, N. Y.

A number of important factors were found to affect register in lithographic printing when constant atmospheric conditions were maintained and the response of papers of known history to a number of the factors was studied. Some study of mechanical factors was

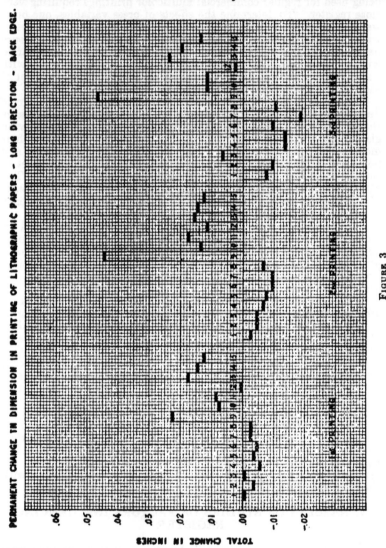

necessary in order to eliminate the variables not influenced by paper characteristics.

1. ATMOSPHERIC

The moisture content of the commercial papers as received at the plant was found to vary from 2.8 to 6.2 per cent while that required for hygrometric equilibrium with the press room atmosphere was

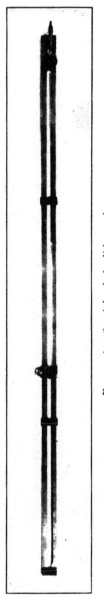

FIGURE 1.—*Special rule for lithography*

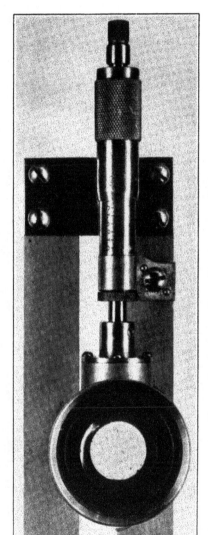

FIGURE 2.—*Rule magnifier with micrometer adjustment*

approximately 5.0 per cent.   When cases of paper having a moisture content more than 1 per cent below room equilibrium were opened and exposed to room atmosphere in the pile, the exposed edges and top of the pile absorbed moisture and waves and wrinkles quickly developed from expansion of those portions.   Excessive conditioning

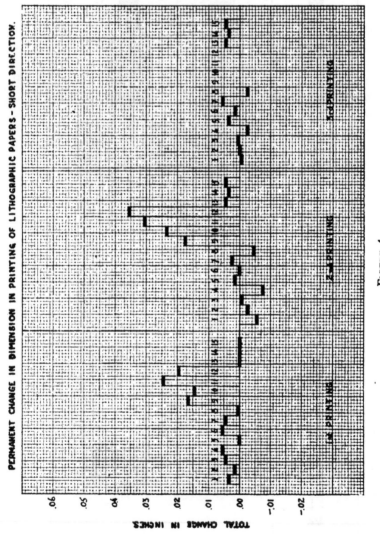

FIGURE 4

time was required for removal of such wrinkles and when they were not removed, misregister resulted as a result of the paper distortion. The benefits of controlled atmosphere in the press room are, of course, lost if the paper is not thoroughly conditioned before the first printing. It is important to remember that papers differ as to time required to thoroughly condition them.   Certain papers, such as moisture-resistant bond paper or paper badly out of equilibrium with press room atmosphere, require extra conditioning time.

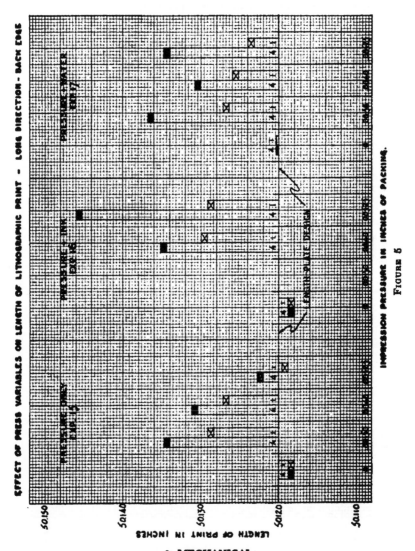

FIGURE 5

2. MECHANICAL

Thickness of plates and the stress applied in clamping them on the cylinder both influence the effective image length.   The length is increased due to curvature in bending the plate around the cylinder and the amount of increase varies with the plate thickness, although no mathematical formula seems to apply.   Increase in length of plate due to mechanical stress in clamping it varies inversely with its thickness. All plates measured were from 0.050 to 0.085 inch longer in 35 inches in the around-the-cylinder direction, when clamped on the cylinder than when flat. From these considerations, thickness standardization for plates appears very important.

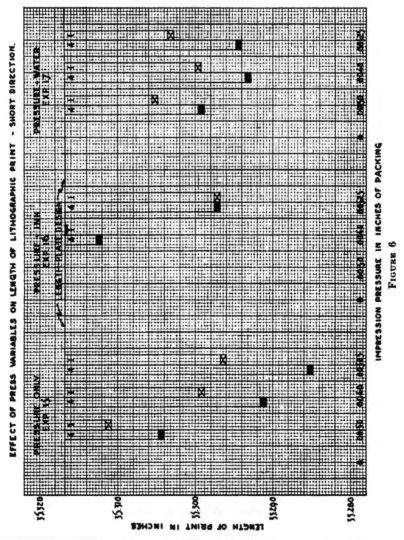

FIGURE 6

Studies were not made to determine the effects of packing under the plate and blanket. In the studies of other factors, the plate was built up to bearer height and the blanket overpacked 0.0025 to 0.003 inch, although it is common practice to build both plate and blanket above bearer heights. The bearers were 18 inches in diameter. With neither ink nor water on the plate, increase in pressure between the blanket and paper decreased slightly the excess in length of print over plate image along the back. In the short direction, the print was shortened. With ink alone on the plate, the effect of pressure was less along the back edge, but was little changed in the short direction.

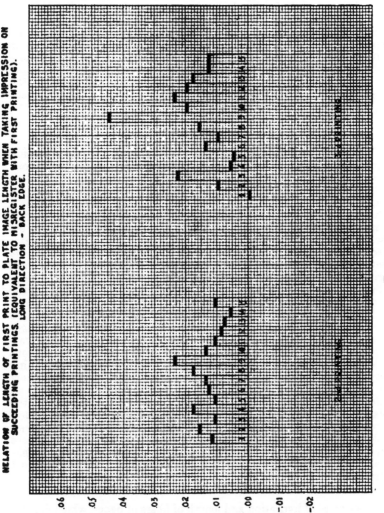

FIGURE 7

With water alone on the plate, the effects of increased pressure were less in both directions.

A print may be lengthened in the short or around-the-cylinder direction by removing packing from beneath the plate and placing it under the blanket. The increased length is caused by slippage introduced by increasing the circumference and thereby the peripheral speed of the blanket, and decreasing the peripheral speed of the plate by decreasing its circumference. This slippage is probably taken up largely by blanket creep, but it results in an increase in the length of print corresponding to the difference between the effective cylinder circumferences. Contrary to rather common belief, the increase in plate image length due to change in curvature by removing packing was too slight to require consideration. However, the difference in peripheral speed between plate and blanket introduced by transferring 0.003 inch packing from the plate to the blanket was calculated to be 0.0191 inch in 35 inches. The length of print in the around-the-cylinder direction may also be increased by pasting a sheet of paper on the back cylinder. By this method the peripheral speed of the paper is increased, introducing between the blanket and paper a creeping or slippage which results in a longer print.

3. WATER AND INK

The water used on the press had two effects on the paper passing through the press. The immediate effect, due to absorption of moisture, was an expansion of about 0.005 inch in 50 inches. When conditioned until equilibrium with the atmosphere was again reached, the paper shrank to about 0.005 inch less than its original length, a phenomenon that has not been satisfactorily explained. Ordinarily, paper left in piles between printings is in a condition somewhere between the two extremes at the second printing. The amount of water used is apparently of little importance in respect to action of the paper during passage through the press. During a normal first printing, the moisture content increased about 0.1 per cent. Doubling the amount of water used had no serious immediate effect on the expansion and contraction of the paper.

Ink affects register to a degree contingent on the extent of the area covered. When printing large solid ink areas, the prints were longer than the plate image along the back edge, possibly a result of lubricating action of the ink in permitting the blanket to fan-out without taking the paper with it. Second prints on first prints containing large solid ink areas on which the ink was not entirely dry were short in the around-the-cylinder direction. This may have been due to an effect of the moist ink on the friction between the paper and the blanket, or it may have been due to increased elasticity of the paper, permitting more temporary stretch in the cross direction of the sheet. Ink tackiness was not studied, but is apparently important. These effects of water, ink, and pressure on length of print are shown graphically in Figure 5 (long direction, back-edge) and Figure 6 (short direction).

4. PAPER PROPERTIES

The degree of hydration and length of fibers of paper, which depend on the beating [4] and jordanning [5] the fibers have been given during manufacture, is extremely important. In studying the influence of

---

[4] Beating is the term applied to mechanical treatment given to paper-making materials, suspended in water, to prepare them for forming a sheet on the paper machine. Beating separates, brushes, and frays-out the fibers and causes them to absorb water by a process known as "hydration."
[5] Jordanning is a refining process that usually follows beating to complete the preparation of the materials for forming a paper of the desired character. In the Jordan the fibers are freed from lumps and cut to the desired length.

these factors on register, seven of the papers having the same composition, but with different beating, jordanning, and sizing treatments were inserted in the center of a pile of paper on a nine single-color run which required hairline register, and not separated from the pile until the nine printings were finished. The length of the first color (yellow) impression on each paper was measured and compared with those of the regular run. The sample with the least drastic beating and jordanning in its manufacture had the shortest first-color impression along the back edge and was the best as regards register, while the sample with the most drastic treatment in those respects was poorest. The sample with the least beating and jordanning had the longest first-color impression, hence was best in repsect to register in the around-the-cylinder direction, while the sample having the most drastic treatment was again poorest.

Samples of the eight papers (Nos. 1 to 8, Table 1) were also inserted in the center of a pile of commercial paper and printed six colors, two colors each printing. In these tests, as in the preceding, the sample that had the least drastic beating and jordanning was best in respect to register, and misregister increased with increased beating and jordanning. (Figs. 3 and 4.) These results indicate the desirability of manufacturing papers for lithography with the least beating and jordanning necessary to obtain satisfactory strength and printing surface.

The degree of internal sizing and the presence of surface sizing are factors. Surface sizing decreases slightly the dimension changes during printing, but did not have marked effect on the samples available. Increasing the internal (beater) sizing of coated papers reduced the tendency to stretch along the back edge, but was not as effective in the short direction. (Nos. 9, 10, 11, figs. 3, 4, 7.) The effects on register were slight. Paper with one-half normal sizing remained in register best in both directions. With extra strong machine-finished papers, increased degree of internal sizing improved the register on the back edge, but did not improve it in the short direction. (Nos. 13, 14, 15, figs. 3, 4, and 7.) One-half normal sizing was practically as effective as normal sizing. It appears that while internal sizing is desirable for good register, the amount used, within normal limits, was not a very important factor.

5. INCIDENTAL HANDLING

The time that paper is allowed to season between printings will influence results in subsequent printings. In a conditioned plant, it was found that the longest seasoning period practicable is desirable since the seasoning period permits an even distribution of moisture picked up in passing through the press and allows the paper to recover from stresses set up in printing.

Register can be controlled within certain limits by set-up of the feeders. By certain adjustments, a "dip" may be introduced in the sheet that will materially shorten long prints already on the sheet, and at the same time lengthen to somewhat smaller extent the current print. In one case noted, a red color found to be printing 48.866 inches along the back edge was 0.028 inch longer than a succeeding black. By pressing down on the sheet with a 1-inch roller 10 inches inside the edge, while the paper was being fed to the press, the current

black print was lengthened 0.004 inch along the back edge and the
paper was drawn in to shorten permanently the preceding red print
· 0.020 inch. The resulting misregister was therefore reduced to
0.004 inch. Such methods can be employed satisfactorily only on
the last printing.

VI. SUMMARY

In the plants having controlled atmosphere, thorough conditioning
of paper is essential if the full benefits of atmospheric control are to
be obtained. The moisture content of paper to be used in such
plants should be adjusted at the paper mill to approach equilibrium
with the pressroom conditions in order to facilitate conditioning.
Paper badly out of equilibrium should be hung immediately after
unwrapping. The hygroscopic condition of paper with reference to
pressroom atmosphere in which it is unwrapped can be quickly
determined with the sword-type hygrometer. The longest seasoning period practicable between printings is desirable for uniform
results, and there appear to be definite advantages in the selection
of 45 per cent relative humidity for lithographic plants. For the
plant without controlled atmosphere, every precaution should be
taken to protect paper from atmospheric changes after starting the
first printing. The paper should be protected between printings by
covering with waterproof wrappers.

Thickness standardization of plates was found important to
insure uniformity of plate stretch in clamping plates in place. With
neither ink nor water on the plate, changes in printing pressure
affected register only slightly and the effects of pressure variations
were even less with ink alone and with water alone. Prints may be
lengthened in the around-the-cylinder direction by transferring
packing from the plate to the blanket or adding packing to the impression cylinder. The amount of water used did not appear important as doubling the amount used had no serious immediate
effects. The effect of ink was contingent on the extent of the area
covered, the presence of large solid ink areas resulting in increased
register troubles.

Paper characteristics are important. The best register was
obtained on the papers that had received the least drastic beating
and jordanning in manufacture. Internal sizing was desirable but
the amount used, within normal limits, was relatively unimportant.
Approximately one-half normal sizing gave best results. The
presence of surface sizing had no marked effect on the papers studied.

The closeness with which the imprint length on the paper agrees
with the length of the plate design on the first color will indicate
the closeness of register that may be expected with subsequent
printings. By measuring the first sheets printed on the first color,
information may be obtained at the start of any multicolor "job"
that will indicate whether register trouble will be encountered in
succeeding printings. A first print on the paper longer than the plate
design along the back edge is always indicative of register troubles.
The special rule described should be of great value to the lithographer
for use here in forestalling trouble.

VII. ACKNOWLEDGMENT

Acknowledgment is made of the invaluable assistance in this study of the advisory committee from the Lithographic Technical Foundation; of the Lowe Paper Co., who made the services of Miss R. M. Cobb available; of the American Lithographic Co. in the experimental printings; and of the following paper manufacturers who furnished the experimental papers: Dill & Collins Co., Rex Paper Co., West Virginia Pulp & Paper Co.

WASHINGTON, July 12, 1932.

RP481

CREEP AT ELEVATED TEMPERATURES IN CHROMIUM-VANADIUM STEELS CONTAINING TUNGSTEN OR MOLYBDENUM

By William Kahlbaum[1] and Louis Jordan

ABSTRACT

Determinations of creep at temperatures between 750° and 1,100° F. were made on two tungsten-chromium-vanadium and a molybdenum-chromium-vanadium steel. These steels were tested as tempered after mechanical working (rolling) and are compared with steels of similar compositions which had been oil quenched and tempered.

CONTENTS

I. INTRODUCTION

This paper reports the results of tests of creep in tension made on 3 alloy steels; namely, 2 tungsten-chromium-vanadium steels and 1 molybdenum-chromium-vanadium steel. The chemical compositions of these steels, their mechanical properties at room temperature, and the heat treatments to which they were subjected prior to testing are all listed in Table 1. There are also included in Table 1 three additional steels (a tungsten-chromium-vanadium steel, a chromium-molybdenum, and a molybdenum-chromium-vanadium steel) which had been tested previously for creep.[2] These steels so closely resemble the steels tested in the present work as to make desirable a comparison of the results of creep tests of the two groups.

The two tungsten-chromium-vanadium steels of the present report differ chiefly in carbon content (0.50, and 0.38 per cent). One of the steels previously tested (EE1139) very closely duplicated the two tungsten-chromium-vanadium steels of this report, especially the higher carbon steel. The heat treatments, however, differed in the two cases. The two steels recently tested were simply tempered at a rather high temperature after rolling, while the similar steel of the earlier work was oil quenched and tempered.

The third steel with which this report is primarily concerned is a molybdenum-chromium-vanadium steel (SE208) containing about 1.2 per cent manganese. This steel is compared with two of the steels previously tested, of which one (E1549) was of a very similar composition, also with high manganese (0.95 per cent) but without vanadium; the other was also similar to steel SE208 of the presen report, except that this comparison steel (E1490) was of higher carbon and lower manganese content.

[1] Research associate, representing The Midvale Co., Philadelphia, Pa.
[2] H. J. French, William Kahlbaum, and A. A. Peterson, Flow Characteristics of Special Fe-Ni-Cr Alloys and Some Steels at Elevated Temperatures, B. S. Jour. Research, vol. 5 (RP192), p. 125, 1930.

TABLE 1.—*Compositions, mechanical properties, and heat treatment of alloy steels tested for "creep" in tension*

Designation	Chemical composition							Mechanical properties [1]				Mechanical and heat treatment
	C	Mn	Si	Cr	V	W	Mo	Tensile strength	Proportional limit [2]	Elongation in 2 inches	Reduction area	
	Per cent	*Per cent*	*Per cent*	*Per cent*	*Per cent*	*Per cent*	*Per cent*	*Lbs./in.²*	*Lbs./in.²*	*Per cent*	*Per cent*	
EE1546	0.50	0.62	0.76	2.35	0.27	1.66	------	183,500	148,000 ------	13.2	38.5	Rolled; 1,275° F., 1 hour; cooled slowly.
EE1554	.28	.49	.38	2.24	.20	1.92	------	167,500	138,000 ------	16.0	47.8	Rolled; 1,275° F., 1 hour; cooled slowly.
SF308	.34	1.18	.17	1.60	.21	------	0.47	171,500	146,000 ------	16.0	56.2	Rolled; 1,650° F., 30 minutes; air cooled; 1,200° F., 30 minutes; air cooled.
EE1139 [3]	.46	.37	.19	2.2	.25	1.65	------	231,500	207,000 (y. p.) [4]	7.8	22.8	Rolled; 1,600° F., 2 hours; cooled slowly; 1,740° F., 1 hour, oil; 1,000° F., 1 hour; air cooled.
E1549 [3]	.53	.95	.23	1.07	.27	Ni .15	.44	143,000	124,000 (y. p.) [4]	[5] 18.7	[5] 56.0	Forged; 1,550° F., 1 hour, oil; 1,200° F., 4 hours, cooled slowly.
E1400 [3]	.75	.52	.81	1.45	.27	.15	.54	174,000	------	[5] 17.5	[5] 16.7	Forged; 1,850° F., 1 hour, oil; 1,100° F., 4 hours, cooled slowly.

[1] Determined at room temperature.
[2] Determined on the basis of a strain of 8X10⁻⁴ indicating departure from a straight line.
[3] Steels previously tested; included for convenience of comparison.
[4] Yield point by drop of beam.
[5] Determined on tensile specimens with a section ⅜ by 0.1565 inch.

II. METHOD OF TESTING

The method employed for determining "creep" was substantially the same as previously described,[3] except that all of the horizontal furnaces and loading machines were replaced by vertical units, as shown in Figure 1. As a result the frictional losses of the mechanical lever loading system were materially reduced.

Figure 1 shows the method of loading the specimen in the vertical furnace. The specimen was 0.250 inch in diameter within the 2-inch gage length. The top adapter was suspended from a spherical seat.

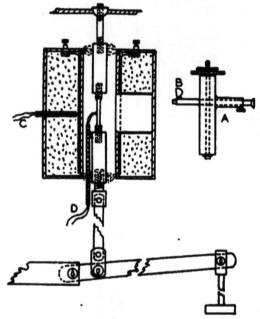

FIGURE 1.—*Diagram of vertical test units for creep
tests at elevated temperatures*

A, traveling microscope; B, lamp; C, furnace control thermocouple;
D, specimen temperature thermocouple.

The temperature of the specimen was measured by the thermocouple *D*, the hot junction of which was tightly wired to the surface of the specimen at the middle of the gage length. An automatic temperature control which regulated the temperature of the furnace was actuated by the thermocouple *C*, the hot junction of which was placed close to the furnace windings. During operation the top and bottom openings of the furnace around the adapters were tightly packed with asbestos to minimize convection currents within the furnace. A narrow rectangular opening through one side of the furnace (covered except when actually observing the gage marks) permitted measurements of the extension of the specimen during testing.

The gage length of the specimen was defined by cutting a shallow groove close to the fillet at each end of the reduced section of the specimen and securing tightly in this groove a loop of 32 B. & S. gage (0.008 inch diameter) platinum wire. The coefficient of expansion of platinum being less than that of steel, the wire did not loosen as the temperature rose.

The expansion was observed by means of a traveling microscope of long focus (*A*, fig. 1) whose motion was governed by a screw of 0.5 mm pitch and a disk carrying 100 divisions. The value of one division on the disk was, therefore, 0.005 mm (2×10^{-4} inch).

The accuracy of the readings depended considerably on the illumination, and therefore the illumination was kept as constant as possible throughout the period of test. As shown in Figure 1, the microscope carried a small lamp, *B*, so arranged that the rays were reflected from the wire at nearly normal incidence. Through the telescope the illuminated spot appeared pointed at each end, and upon the points the cross hair of the traveling microscope was set.

Observations were made over a period of from 400 to 1,000 hours, usually at intervals of 24 hours. In the case of specimens carrying the lower loads at any temperature, there was frequently insufficient creep in 24 hours to be detected. The total creep over erio s of 400 to 1,000 hours permitted the detection of creep rates ofpthe drder of 10^{-6} inches per inch per hour.

A survey of the temperature uniformity of the specimen showed that the temperature indicated by the thermocouple, *D*, at the surface of the middle of the gage length differed from the temperature at the center of the specimen at the midpoint of the gage length by only 1° F. at 800° F. and by 3° F. at 1,200 °F. Furthermore, the maximum variation in temperature along the length of the test bar (the temperature measurements being made at the ends of the specimen in the shoulder of the bar just outside of the reduced section) was in all cases greatest between the bottom end and the middle of the gage length. This difference amounted to 23° F. (13° C.) at 800° F. (426° C.) and to 18° F. (10° C.) at 1,200° F. (648° C.). The temperature difference between the middle of the gage length and the top end of the bar was not detectable at 800° F. (426° C.) and was only 9° F. (5° C.) at 1,200° F. (648° C.).

III. RESULTS

The time-extension curves for the three steels, showing on each curve the stress in lbs./in.² of original cross section and, in parentheses, the identification number of the specimen, are given in Figures 2, 3, and 4. The two tungsten-chromium-vanadium steels were tested at 750°, 850°, 950°, 1,050°, and 1,100° F.; the single molybdenum-chromium-vanadium steel at 650°, 850°, and 1,050° F.

In Figures 5, 6, and 7 are plotted for the several testing temperatures the stress and the initial flow; that is, the relatively rapid deformation at the time of, or immediately after, loading. Also, there are plotted the stress and the average rates of flow, calculated from the so-called second stage of the time-extension curves; that is, that portion of rather uniform rate of extension following the initial flow.

Figures 8, 9, and 10 show the stress which, at any selected temperature, results in 0.1, 1, 2, or 5 per cent secondary elongation, or creep, in 1,000 hours. These figures also show in some cases the stress resulting in 0.1, 1, or 2 per cent initial elongation preceding the stage of secondary elongation.

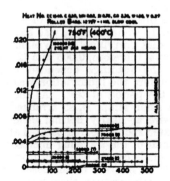

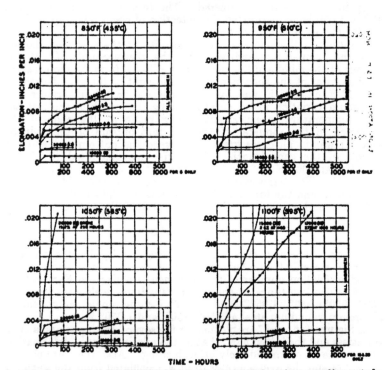

FIGURE 2.—*Time-elongation curves of tungsten-chromium-vanadium steel*
EE1546 under different loads and at different temperatures

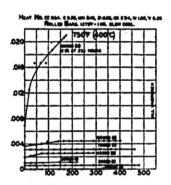

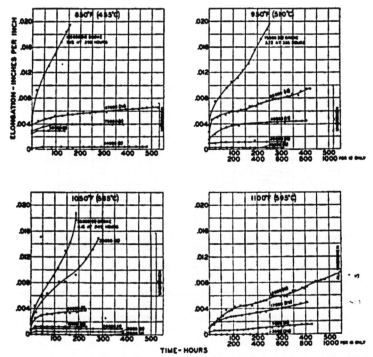

FIGURE 3.—*Time-elongation curves of tungsten-chromium-vanadium steel EE1554 under different loads and at different temperatures*

IV. DISCUSSION AND SUMMARY

The discussion of the results of the present creep tests and correlation of these results with previously tested steels of similar compositions is most conveniently based on Figure 11. In this figure the top and bottom of a solid black rectangle indicate, respectively, the stress producing 1 per cent and 0.1 per cent initial deformation. When there is given only a small solid triangle, the base of the triangle indicates the stress producing 0.1 per cent deformation and there is no value for 1 per cent deformation. Similarly, the open (light) rec-

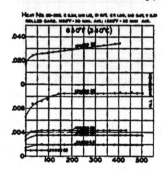

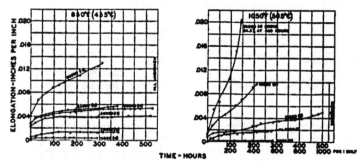

FIGURE 4.—*Time-elongation curves of molybdenum-chromium-vanadium steel SE208 under different loads and at different temperatures*

tangles indicate secondary flow in 1,000 hours of 1 per cent and 0.1 per cent.

A comparison of the two tungsten-chromium-vanadium steels of the present tests (EE1546 and EE1554) shows that appreciably higher stresses are sustained by the steel of lower carbon content (EE1554) at 860°, 950°, and 1,000° F. before 0.1 or 1 per cent creep results than is the case with the higher carbon steel. Also there is for the lower carbon steel a wider range at 950° and 1,000° F. between the stresses producing 0.1 and 1 per cent creep than for the higher carbon steel.

The previously tested tungsten-chromium-vanadium steel (EE1139) was oil hardened and tempered before testing, while steel EE1546 of very similar composition was simply tempered after rolling. From Figure 11 it appears that at the higher testing temperatures, 950°

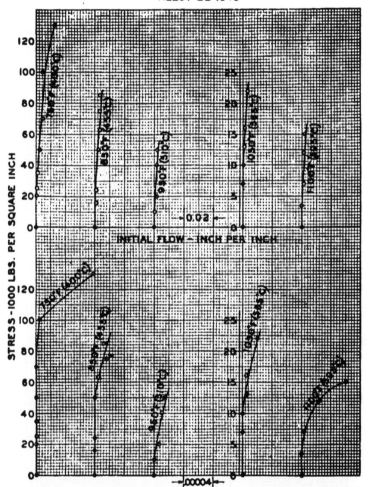

FIGURE 5.—*Flow data for tungsten-chromium-vanadium steel EE1548 at different temperatures*

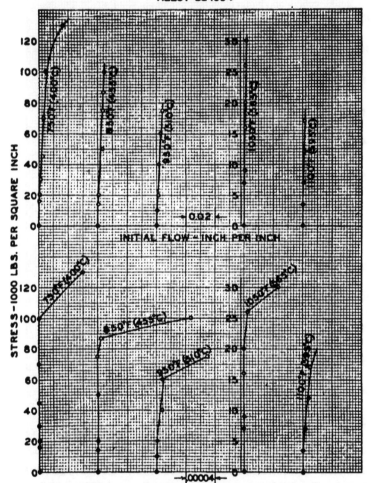

FIGURE 6.—*Flow data for tungsten-chromium-vanadium steel EE1554 at different temperatures*

HEAT NO. SE-208. C 0.34, MN 1.18, SI 0.17, CR 1.60, MO 0.47, V 0.21
ROLLED BARS. 1650°F-30 MIN. AIR; 1200°F -30 MIN. AIR.

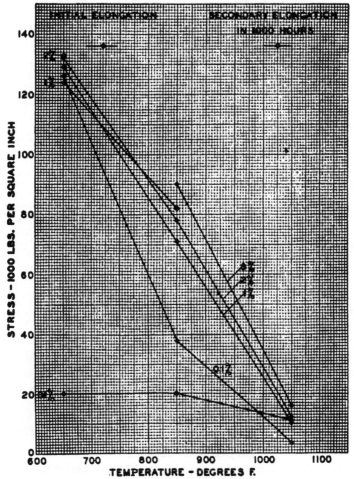

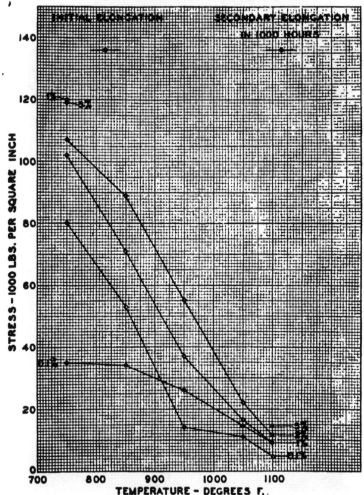

FIGURE 8.—*Flow chart for tungsten-chromium-vanadium steel EE1546 at elevated temperatures*

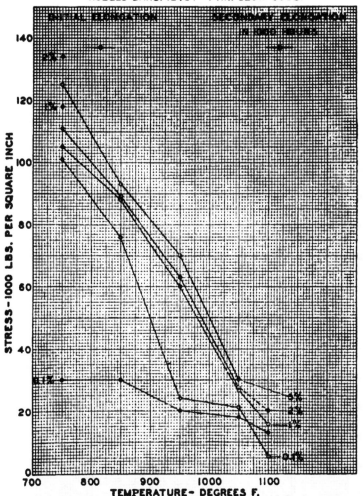

FIGURE 9.—*Flow chart for tungsten-chromium-vanadium steel EE1554 at elevated temperatures*

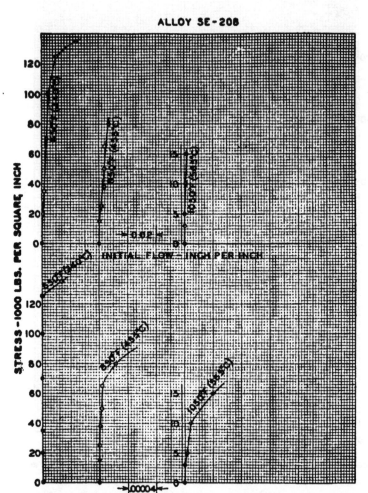

FIGURE 10.—*Flow chart for molybdenum-chromium-vanadium steel SE 208 at elevated temperatures*

132919—32——12

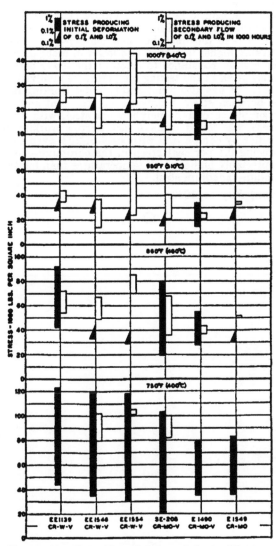

FIGURE 11.—*Comparison of steels on the basis of the stresses producing both initial deformation and secondary flow (secondary flow is over period of 1,000 hours)*

and 1,000° F., the stress producing 0.1 per cent creep was appreciably higher in the quenched and tempered steel. However, this hardened and tempered steel showed 1 per cent creep at stresses very much lower than the corresponding rolled-and-tempered steel (EE1554) at 860°, 950°, and 1,000° F.

The other comparison to be made in Figure 11 is between the two molybdenum-chromium-vanadium steels, SE208 (normalized and tempered) and E1490 (oil quenched and tempered), the latter steel being considerably higher in carbon and lower in manganese than the former. The greater resistance to creep of the lower carbon higher manganese steel at all of the testing temperatures involved is obvious from Figure 11. It has also a noticeably wider working range; that is, the range between stress producing 0.1 per cent and 1 per cent creep. The simple chromium-molybdenum steel, with about 1 per cent manganese, was found to have such narrow working ranges of stress as to suffer by comparison with steel SE208, even through the stresses producing 0.1 per cent creep in the chromium-molybdenum steel were appreciably higher at all temperatures than for the molybdenum-chromium-vanadium steel.

WASHINGTON, July 13, 1932.

O

U. S. DEPARTMENT OF COMMERCE
ROY D. CHAPIN, Secretary

BUREAU OF STANDARDS
LYMAN J. BRIGGS, Acting Director

BUREAU OF STANDARDS
JOURNAL OF RESEARCH

October, 1932
Vol. 9, No. 4

UNITED STATES
GOVERNMENT PRINTING OFFICE
WASHINGTON: 1932

For sale by the Superintendent of Documents, Washington, D. C. · · Price 25 cents; $2.50 per year on subscription

THE SYNTHESIS, PURIFICATION, AND CERTAIN PHYSICAL CONSTANTS OF THE NORMAL HYDROCARBONS FROM PENTANE TO DODECANE, OF n-AMYL BROMIDE AND OF n-NONYL BROMIDE[1]

By B. J. Mair

ABSTRACT

The normal paraffin hydrocarbons from pentane to dodecane, n-amyl bromide and n-nonyl bromide have been synthesized and prepared in very pure condition. The boiling points, freezing points, and refractive indices (N_D^{20}) of these compounds have been measured and the values compared with those obtained by other investigators on hydrocarbons isolated from petroleum. Two crystalline modifications of n-nonyl bromide freezing at −29.06° and −30.71° C., respectively, were discovered.

CONTENTS

I. INTRODUCTION

An investigation on the electrical properties of insulating liquids in progress at this bureau necessitated the preparation of some very pure hydrocarbons. The normal paraffin hydrocarbons were selected because the resistivities of liquids in this series are among the highest known, and because an interpretation of the electrical properties of liquids in terms of molecular structure should be more readily accomplished with their simple molecules than with molecules of more complex structure. All the members of this series from pentane to dodecane were prepared. An examination of the literature when this work was begun showed that many of the physical properties of these substances were not well established. Recently, however, Shepard, Henne, and Midgley (1)[2] isolated these hydrocarbons in a high degree

---

[1] This work is part of an investigation on the electrical properties of insulating liquids which is being carried on in cooperation with the Utilities Research Commission (Inc.), Chicago, Ill.

[2] The figures in parentheses here and throughout the text relate to the bibliography at the end of this paper.

of purity from petroleum, and have published values for their physical constants which, in general, are in satisfactory agreement with those obtained in this investigation. Since, however, the hydrocarbons obtained in this work were prepared synthetically, a brief description of the methods and a comparison of the values obtained for the physical constants with those of Shepard, Henne, and Midgley (1) seems desirable.

In addition to the hydrocarbons, two of the intermediates in their preparation, n-amyl bromide and n-nonyl bromide, were obtained in a pure condition and certain of their physical constants measured. Nonyl bromide in particular proved interesting since it was found to exist in two crystalline modifications, a fact not previously recorded.

II. METHODS

The methods used in the purification, in the determination of purity, and in the measurement of the physical constants were similar for all the hydrocarbons and are referred to frequently in discussing the individual substances. It is therefore advisable to describe these methods briefly at this point before proceeding to a more detailed description of the individual syntheses and purifications.

1. PURIFICATION

The physical methods employed were fractional distillation and fractional crystallization. Apparatus and methods for efficiently pursuing these processes have been developed in this bureau, and were used with only minor modifications. The liquids were distilled through a 10-plate bubble cap fractionating column (2, 3) fitted with a variable reflux regulator of the type described by Marshall (4). The reflux ratio during the final distillation was always greater than 10 : 1. The distillation range was read on a thermometer extending into the vapor from the top of the column. The method employed in fractional crystallization (5) was to transfer the frozen material to a vacuum funnel, stir it to a mush, and draw off the mother liquor with suction. For the liquids whose freezing points were higher than that of decane, melting took place so slowly that a vacuum funnel was unnecessary. In the case of decane, the crystals were separated from the mother liquor in a centrifuge designed for use at low temperatures by Hicks-Bruun and Bruun (6).

2. FREEZING POINT DETERMINATIONS

The freezing point determinations were made with a glass-incased, 25.5 ohm (at the ice point), potential terminal, platinum resistance thermometer of the strain-free type, using as accessories a Mueller thermometer bridge and commutator. The thermometer was calibrated by the heat division of this bureau.

The sample was contained in a double-walled Pyrex tube. By varying the air pressure in the annular space of this double walled tube, any desired rate of cooling could be obtained. This vessel was about 11 inches in length and was filled to a depth of 8½ inches with the liquid in question. With the thermometer in place, the liquid extended 4 inches above the resistance coil. The liquid could be stirred vigorously by means of a spiral of nickel chromium wire which surrounded the thermometer. After the sample (about 70 ml) had been introduced and the thermometer and stirrer adjusted in place, a thin brass tube (as a safety measure) was fitted over the vessel, which was then immersed in liquid air. Freezing took place from the walls toward the center. Stirring was possible until the sample was about half frozen. A change in the slope of the cooling curve was noticed (particularly if the sample was impure) as soon as stirring stopped. This may be explained on the supposition that equilibrium conditions were no longer maintained, and the impurities crystallized simultaneously with the pure substance. The freezing point of any liquid was taken as the temperature at which the cooling curve (figs. 2 to 11) became approximately horizontal. The principal criterion of purity was the slope of this approximately horizontal portion of the freezing point curve.

Considerable precision was attained with these arrangements. Duplicate experiments were made with hexane, octane, decane, and dodecane, and in no instance was a variation greater than 0.002° C. in the freezing point obtained.

An examination was made of the thermometric technique in order to determine how accurately the measured temperatures agreed with the international scale of temperature. The question of conduction through the leads was studied by varying the height of the liquid above the resistance coil. Determinations of the freezing point of hexane in which this height was varied from 5 to 2 inches agreed within 0.001° C., and indicated that with the apparatus in question this effect was negligible. For the thermometer regularly used, the values of the constants given by the heat division of this bureau permitted the calculation of the temperature with an accuracy of 0.02° C. In order to obtain a further test on the accuracy of this thermometer and accessories, freezing-point determinations were made on two samples, using the equipment of the low-temperature laboratory of this bureau, with the assistance of R. B. Scott. The values −56.82° and −95.36° C. obtained on samples of octane and hexane with the equipment of the low-temperature laboratory compare well with the values −56.82° and −95.37°C., which had been obtained on these samples with the regular equipment. Thus with two different thermometers, bridges, and operators the maximum discrepancy was 0.01° C. It seems reasonable to conclude that the thermometric measurements do not differ from the international scale by more than ±0.02° C.

3. BOILING-POINT DETERMINATIONS

The boiling points were determined with the aid of a platinum resistance thermometer in the apparatus shown in Figure 1. Vapor given off from the boiling liquid in the flask had, first, to pass through liquid in the bubble cap before reaching the thermometer, thus preventing superheating. The outer jacket was electrically heated sufficiently to permit the liquids with high boiling points to distill over, but was always kept somewhat below the boiling temperature. About half of the 50 ml sample was distilled over during a boiling-point determination. The temperature generally rose less than 0.01° C. except during the distillation of the first few milliliters when the thermometer was warming up. Barometric pressures within 0.02

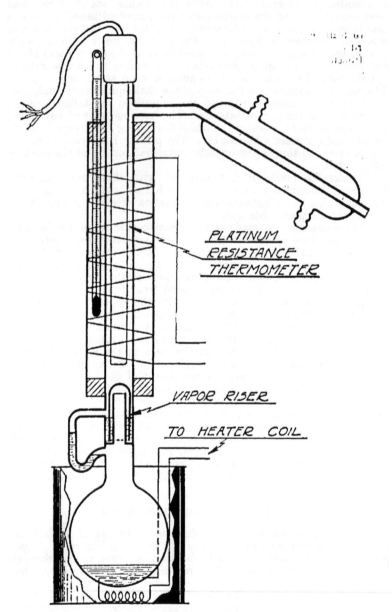

FIGURE 1.—*Boiling-point apparatus*

mm were obtained immediately before and after the boiling-point reading from the aeronautic-instruments section of this bureau. The boiling points of the hydrocarbons were corrected to 760 mm with the aid of the values given by Shepard, Midgley, and Henne (1) for $\frac{dt_B}{dp}$, the rate of change of boiling point with change of pressure.

For *n*-amyl bromide several boiling-point determinations in the vicinity of 760 mm were used to compute $\frac{dt_B}{dp}$. Duplicate boiling-point de ermina ions were made in most cases and agreed within 0.01° C. t t

4. REFRACTIVE INDEX DETERMINATIONS

Refractive indices were measured by L. W. Tilton, of the optical instruments section of this bureau, with three different Abbe refractometers. Each was calibrated with distilled water and with several prismatic standards of refractivity whose indices had been previously measured on a spectrometer by the method of minimum deviation (7). The observations on these liquids were made at water jacket temperatures of 25° C. in a room the temperature of which was between 23° and 24° C. On each instrument two series of measurements were made in which the liquids were sampled in the direct and also in the reverse order of their optical densities. Each of these samplings consisted of. two successive applications of liquid to the refractometer block by means of a clean glass rod, and the instruments were set and read twice after each application.

The precision of these refractive index readings is characterized by probable errors of a few units in the fifth decimal place. With instruments so calibrated an accuracy of $\pm 5 \times 10^{-5}$ is attainable when using solid samples. With liquids, however, the accuracy may be affected by contamination of the small samples when they come in contact with the cement which holds the refractometer block in its water jacket. Also, the high volatility of some of the samples of lower index causes them to cool appreciably and, perhaps, to maintain a temperature lower than that indicated by the thermometer which is immersed in the jacketing stream. In accord with this supposition the results on the different refractometers do not agree as well for the liquids having the lower indices. In no case, however, does the index as determined by any one refractometer depart from the mean recorded in Table 1 by more than $\pm 12 \times 10^{-5}$, the average of such departures being $\pm 3 \times 10^{-5}$.

III. SYNTHESIS AND PURIFICATION

The synthesis and purification of the hydrocarbons and some of their intermediate products are described in the following paragraphs.

1. *n*-PENTANE

The Sharples Solvent Corporation, of Philadelphia, kindly supplied a gallon of amyl alcohol for this preparation. This material was stated by them to be a mixture of 2-pentanol and 3-pentanol with not more than 1 per cent of tertiary amyl alcohol and possibly a trace of diamylene. Fractional distillation yielded a large middle fraction boiling between 116.4° and 118.4° C., which was undoubtedly a mixture of 2 and 3 pentanol. Both of these substances on dehydration and subsequent hydrogenation yield *n*-pentane. The dehydration was performed with the use of sulphuric acid exactly as described in Organic Syntheses (8). The pentenes thus obtained were fractionated and the portion which distilled between 35.8° and 36.4° C. was retained. The pentenes were then hydrogenated by passage through a reaction tube filled with nickel catalyst maintained at 140° C. Practically complete reduction took place in one passage. The re-

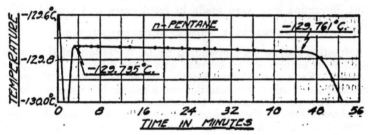

FIGURE 2.—*n-pentane cooling curve*

sulting pentane was shaken three times with concentrated sulphuric acid, then washed twice with a 10 per cent solution of sodium carbonate, and finally with water. After drying with calcium chloride, the pentane was fractionally distilled three times. About 750 ml boiling over a range of 0.03° C. was collected as pure material. An examination of the freezing point curve (fig. 2) indicates the high purity of the sample. The material froze between −129.735° and −129.761° C., a range of 0.026° C.

2. *n*-HEXANE

The starting material in this synthesis was *n*-propyl alcohol. It was converted into the bromide by the hydrobromic-sulphuric acid

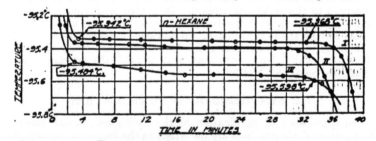

FIGURE 3.—*n-hexane cooling curves*

method (9). The bromide was then converted into hexane by treatment with sodium shot, using the method employed by Faillebin (10). The last traces of *n*-propyl bromide were removed by refluxing the crude hexane with liquid sodium potassium alloy for four hours. The hexane was then shaken with cold concentrated sulphuric acid until the acid layer was colorless, then shaken with a 10 per cent solution of sodium carbonate, then with water, and finally dried with anhydrous calcium chloride. This material was fractionally distilled thrice, and 1,500 ml which distilled within 0.04° C., was obtained. It was collected in two nearly equal fractions which showed a pronounced difference in freezing behavior, although distilling within 0.04° C. (Fig. 3.) The second fraction (Curve *II*) gave a more nearly horizontal curve and froze 0.12° C. higher than the first fraction. (Curve *III*.) A 400-ml portion of the second fraction was stirred with 175 ml of chlorosulphonic acid at 45° C. for seven hours. (This reagent has been reported as efficacious in the removal of branch chain hydrocarbons (11, 12).) After washing thrice with a 10 per cent solution of sodium hydroxide and once with water, the sample was dried with calcium chloride and again distilled. The freezing range of this sample as shown in Curve *I* (−95.342° to −95.368° C.) was only 0.026° C., and indicated that the hexane was of high purity.

3. n-HEPTANE

Several liters of heptane, the original source of which was the Jeffrey pine, was furnished by Graham Edgar, of the Ethyl Gasoline Corporation. This material was of very high purity, as is illustrated

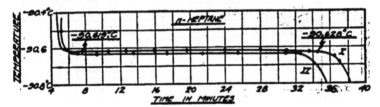

FIGURE 4.—*n-heptane cooling curves*

by its freezing behavior. (Fig. 4, Curve *II*.) Vigorous treatment with chlorosulphonic acid and subsequent distillation raised the freezing point 0.008° C. Curve *I* illustrates the freezing behavior of the treated sample. This sample froze from −90.619° to −90.628° C., a range of only 0.009° C., indicative of exceptionally high purity. The same value, −90.62° C., was obtained on a sample of heptane isolated from petroleum by Hicks-Bruun and Bruun (6).

4. n-OCTANE

The starting material in this synthesis was n-butyl alcohol. It was converted into the bromide by the hydrobromic-sulphuric acid method (9). The bromide was then converted into octane in a manner similar to that employed by Lewis, Hendricks, and Yohe (13). The procedure was as follows: Sodium in 73 g portions was converted to a finely divided shot by melting and shaking under xylene in a 1-liter 3-necked Pyrex flask. The xylene was decanted and the sodium washed several times with anhydrous ether. It was then covered with 300 ml of anhydrous ether and the flask fitted with a dropping funnel, stirrer with mercury seal, and reflux condenser with calcium chloride tube. The flask was placed in an ice-water bath, the stirrer started, and 384 g of n-butyl bromide run in over a period of three hours. The excess sodium was destroyed by running water into the mixture. The octane layer was drawn off, and the ether and octane separated by fractional distillation. The impure octane was then refluxed with a 10 per cent solution of sodium hydroxide in alcohol for several hours to destroy any unconverted butyl bromide. Water was then added, and the octane separated from the aqueous layer. It was shaken with cold concentrated sulphuric acid until the acid layer remained colorless; then shaken with a 10 per cent solution of sodium carbonate, then with water, and finally dried over calcium chloride. In this way 3,200 ml of crude material was prepared. It was fractionally distilled four times. At the end of this fractionation, 1,800 ml of material distilling within a range of 0.04° C. was obtained. This material was collected in two equal fractions which had the same freezing point. However, judging from the slope of the freezing-point curve (fig. 5, Curve *III*), it was not yet pure. A sample further purified by three fractional crystallizations froze as shown in Curve *II*. This material, although freezing at a higher temperature, still showed almost as great a freezing range.

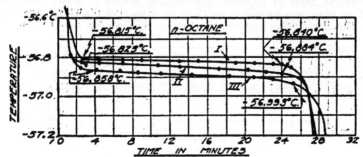

FIGURE 5.—*n-octane cooling curves*

Another sample was stirred with chlorosulphonic acid at 45° C. for seven hours and this treatment repeated with fresh chlorosulphonic acid three times. The octane was separated from the chlorosulphonic acid, washed with a 10 per cent solution of sodium hydroxide, then with water, dried over calcium chloride, and again fractionally distilled. As a result, 270 ml of material was obtained, freezing between −56.815° and −56.840° C., a range of 0.025° C., which indicates its high purity.

5. n-NONANE AND n-NONYL-BROMIDE

The starting materials in this synthesis were *n*-heptyl bromide and ethylene oxide obtained from Eastman Kodak Co. and from Carbide & Carbon Chemicals Corporation. From these *n*-nonyl alcohol was prepared according to the following reactions:

$$CH_3 (CH_2)_5 CH_2 Br + Mg \longrightarrow CH_3 (CH_2)_5 CH_2 Mg Br$$

$$CH_3 (CH_2)_5 CH_2 Mg Br + CH_2 - CH_2 \longrightarrow$$
$$\underset{O}{\diagdown \diagup}$$

$$CH_3 (CH_2)_7 CH_2 OMg Br \xrightarrow{H_2O} CH_3 (CH_2)_7 CH_2 OH$$

This method is exactly analogous to the preparation of *n*-hexyl alcohol from *n*-butyl bromide and ethylene oxide described in Organic Syntheses (14). The *n*-nonyl alcohol was fractionally distilled under a pressure of 83 mm. From the original 1,000 g of *n*-heptyl bromide, 545 ml of *n*-nonyl alcohol distilling between 143.7° and 144° C. was obtained. This purified alcohol was converted into *n*-nonyl bromide by the hydrobromic-sulphuric acid method (9). The resulting product was fractionally distilled at 83 mm and the material distilling between 144.55° and 144.60° C. used in the synthesis of *n*-nonane.

The *n*-nonyl bromide was converted in the usual manner into nonyl magnesium bromide, which was then decomposed with ice and the nonane separated by steam distillation. It was then shaken in turn with concentrated sulphuric acid, 10 per cent sodium carbonate solution, dried over calcium chloride, and fractionally distilled once. A yield of 210 ml of *n*-nonane distilling within a range of 0.03° C. was obtained. The freezing behavior is shown in Figure 6. The freezing range from −53.700° to −53.754° C. was 0.054° C., which is somewhat greater than that obtained for the other hydrocarbons, indicating a less pure material. The freezing point, −53.70° C., is

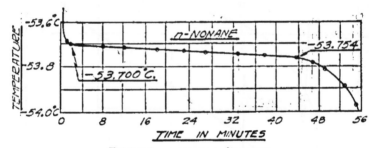

FIGURE 6.—*n-nonane cooling curve*

lower than the value −53.68° C. given by Shepard, Henne, and Midgley (1), and the value −53.65° C. obtained on a sample of very pure nonane isolated by White and Rose (15) from petroleum.

The behavior of *n*-nonyl bromide during its freezing point measurements indicated clearly that two crystalline modifications of this substance exist. Curve *I* (fig. 7) is a time temperature cooling curve obtained in the usual manner with stirring. The temperature, after remaining constant at −30.71° C. for about seven minutes, rose abruptly to −29.06° C., and again remained approximately constant for a short time. Simultaneously with this rapid temperature change, stirring became much easier. If the liquid was not stirred, crystallization occurred at −30.71° C. without the appearance of the β-modification. Curves *II* and *III* are heating curves, Curve *II* showing the behavior of a sample which had been frozen with stirring and in which the β-phase had made its appearance, while Curve *III* shows the behavior of a sample frozen without stirring in which the β-phase had not appeared. These curves show a difference in melting behavior and afford confirmation of the existence of two modifications. The freezing point of the α-modification is −30.71° C. The freezing point of the β-modification can not be stated so definitely, since it is possible that the approximately flat portion of the Curve (*I*) is caused by an equality between the loss of heat to the surroundings and the gain in heat caused by the transformation. The freezing point of the β-modification, however, can not be below −29.06° C., and judging from the heating Curve (*III*), it seems improbable that it can be appreciably higher. Moreover, in another time-temperature cooling experiment, a horizontal portion of the curve was obtained at −29.06° C. Thus it seems probable that −29.06° C. represents the true freezing temperature of the β-modification.

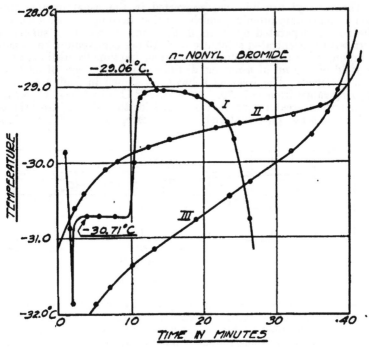

FIGURE 7.—*n-nonyl bromide*

I, cooling curve, *II* and *III* heating curves

6. n-DECANE AND n-AMYL BROMIDE

The starting material in this synthesis was Eastman's *n*-propyl alcohol. It was converted to *n*-propyl bromide by the hydrobromic-sulphuric acid method (9). The resulting bromide was fractionally distilled once. The bromide was then converted to *n*-amyl alcohol by the use of ethylene oxide in a manner exactly analogous to that of the preparation of *n*-hexyl alcohol from *n*-butyl bromide and ethylene oxide, described in Organic Synthesis (14). The *n*-amyl alcohol was fractionally distilled twice and 1,600 ml of material distilling between 137.82° and 137.92° C. obtained. This was converted to the bromide by the hydrobromic-sulphuric acid method (9) and 1,150 ml of the bromide distilling between 129.40° and 129.50° C. obtained. The freezing range of this material (fig. 8) was only 0.014° C. and

indicated it to be of high purity. This material was used to synthe-
size decane as follows: Pea-sized sodium pellets were added through
a reflux condenser to 100 ml portions of amyl bromide in a 500 ml
flask at such a rate that the liquid, which was first heated to boiling,
continued to boil gently. After no more sodium could be added
(owing to accumulation of solid), the reaction mixture was cooled,
treated with water, and the hydrocarbon layer separated and dried
with calcium chloride. This process was repeated three times before

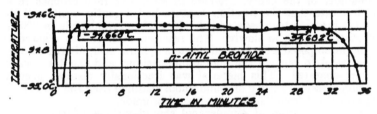

FIGURE 8.—*n-amyl bromide cooling curve*

the amyl bromide was practically completely converted. Finally
the decane was refluxed over sodium potassium alloy for six hours to
remove the last traces of amyl bromide. The decane was decanted
from the alloy and shaken in turn with several portions of cold concentrated sulphuric acid until the acid remained colorless, then shaken
with 10 per cent sodium carbonate solution, and finally with water.
It was then fractionally distilled once. Practically all the sample of
505 ml distilled between 173.37° and 173.47° C. It was collected
in two nearly equal fractions, one between 173.37° and 173.43° C.,
the other between 173.43° and 173 47° C. The freezing points of
these two fractions were −29.88° and −29.76° C., respectively.

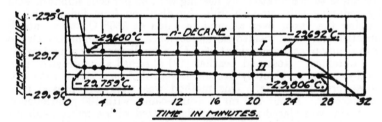

FIGURE 9.—*n-decane cooling curves*

The cooling curve for the sample which gave −29.76° C. as the
freezing point is shown in Figure 9 (Curve *II*). Shortly after this
sample was obtained Bruun and Hicks-Bruun (16) obtained from
petroleum by distillation and by crystallization, with the use of a
centrifuge, a decane sample of exceptional purity freezing at
−29.68° C. By fractional crystallization with the same centrifuge,
the freezing point of the synthetic sample was raised to the identical
value −29.68° C. The freezing range from −29.680° to −29.692° C.
(Curve *I*, fig. 9) indicates very high purity.

7. n-UNDECANE

The starting materials in this synthesis were Eastman's best grade of n-heptyl bromide and n-butyraldehyde from which 4-undecanol was synthesized as follows: A 50 ml portion of a mixture of 537 g of n-heptyl bromide in 900 ml of anhydrous ether was run onto 75 g of magnesium in a 5-liter 3-necked flask, fitted with reflux condenser, stirrer with mercury seal, and dropping funnel. A crystal of iodine was added and the reaction started by warming slightly. The remainder of the heptyl bromide in ether was then dropped in at such a rate that the reaction mixture boiled gently. After all the heptyl bromide had been added, 216 g of butyraldehyde in 200 ml of ether was run in slowly. The reaction mixture was allowed to cool, decomposed with ice and dilute sulphuric acid, filtered through glass wool to remove the magnesium residue, and the 4-undecanol layer separated. This material was fractionally distilled once at 93 mm pressure; 1,230 g of 4-undecanol distilling from 155° to 157° C. was obtained from 3,000 g of n-heptyl bromide.

The dehydration of 4-undecanol should yield a mixture of two undecylenes, depending on the manner in which water splits off. Thus

$$C_2H_5 \cdot CH \cdot CH \cdot CH_2 \cdot C_6H_{13} \rightarrow C_2H_5 \cdot CH = CH \cdot CH_2 \cdot C_6H_{13}$$
$$\boxed{H\ \ OH}$$

or

$$C_2H_5 \cdot CH_2 \cdot CH \cdot CH \cdot C_6H_{13} \rightarrow C_2H_5 \cdot CH_2 \cdot CH = CH \cdot C_6H_{13}$$
$$\boxed{OH\ \ H}$$

Both of these undecylenes should yield n-undecane on hydrogenation.

However, the dehydration of 4-undecanol proved to be difficult. Attempts to dehydrate it by heating with 60 per cent sulphuric acid, by refluxing with zinc chloride, by distilling from iodine, and by passing the vapor over aluminium oxide at 350° C., were entirely unsatisfactory and yielded no undecylenes. By running the alcohol into sirupy phosphoric acid at 250° C., a small quantity of undecylene was obtained. Ross and Leather (17) obtained an undecylene from methyl nonyl carbinol by heating the alcohol on a steam bath with phosphorus pentoxide. This method was tried and proved more satisfactory. About 200 ml of 4-undecanol, with 100 g phosphorus pentoxide was placed in a 500-ml flask fitted with a reflux condenser and calcium chloride tube. The flask was then heated for eight hours on a steam bath. The resulting pasty material was extracted with ether and the extract fractionally distilled. In addition to the undecylenes distilling from 191° to 193° C., a considerable quantity of material (almost half) distilled in the range from 222° to 223° C. This material gave the phenylhydrazine test for a ketone. It was probably propyl heptyl ketone and resulted from the oxidizing action of phosphorus pentoxide. This ketone was reduced to 4-undecanol by mixing with an equal volume of amyl alcohol, and adding an excess of sodium in small pieces to the boiling mixture. It was then washed with water, dried, and separated from the amyl alcohol by fractional distillation. It distilled at atmospheric pressure from 228° to 229° C. The 4-undecanol thus obtained was treated with phosphorus pentoxide as before and more undecylene obtained. Altogether 800 ml of undecylene was obtained from 1,230 g of 4-undecanol.

Attempts to hydrogenate the undecylene to undecane by passage over a nickel catalyst at 200° C. were not satisfactory. The first small portion which was collected was almost completely hydrogenated, but the $c_{atalyst}$ soon became inactive, presumably owing to poisoning by impurities in the undecylene. The undecylene was, however, satisfactorily hydrogenated by passage over a copper chromite catalyst (18) at 200° C., this catalyst being less readily poisoned than nickel. The undecylenes remaining unconverted were removed by shaking with cold concentrated sulphuric acid. The undecane was washed with a 10 per cent solution of sodium carbonate and dried with calcium chloride. Approximately 700 ml of crude undecane was obtained. This material was quite impure. After four fractional distillations 350 ml was obtained distilling over a range of 0.5° C. Freezing point measurements showed that this material was still quite impure. Fractional crystallization was used for final purification and the progress of purification followed with the platinum resistance thermometer. After many fractional crystallizations 75 ml of

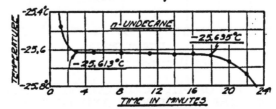

FIGURE 10.—*n-undecane cooling curve*

very pure material was obtained, 75 ml only slightly less pure, and 100 ml of fairly pure material. The freezing range of the purest sample, from −25.613° to −25.635° C., was only 0.022° C. and is indicative of high purity. (Fig 10.)

8. *n*-DODECANE

The starting materials in this synthesis were *n*-butyl bromide and ethylene oxide from which hexyl alcohol was synthesized as described in Organic Syntheses (14). The hexyl alcohol resulting was fractionally distilled and the portion distilling between 156° and 157° C. converted into the bromide by the hydrobromic-sulphuric acid method (9). Practically all of this hexyl bromide distilled between 155.20° and 155.32° C.

Dodecane was prepared from the hexyl bromide by the use of sodium in exactly the same manner as decane from amyl bromide. Finally the dodecane was refluxed over sodium potassium alloy for six hours to remove the last traces of hexyl bromide. The dodecane was then decanted from the alloy and shaken with several portions of cold concentrated sulphuric acid until the acid remained colorless, then shaken with a 10 per cent solution of sodium carbonate and finally with water. Two fractional distillations at atmospheric pressure yielded 570 ml of dodecane with a distilling range of 0.2°. It had a slightly acrid odor, and evidently contained unsaturated compounds caused by cracking at the high temperature necessary to distill at atmospheric pressure. Its freezing behavior indicated that it was not pure. (Fig. 11, Curve *II*.) After many crystallizations, a 75 ml sample with the freezing behavior shown in Curve *I* was obtained. This material froze from $-9.609°$ to $-9.669°$, a range of $0.06°$ C., indicative of a fairly pure material, through probably not as pure as some of the other hydrocarbons obtained.

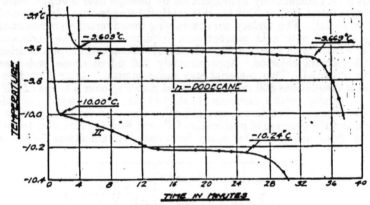

FIGURE 11.—*n-dodecane cooling curves*

IV. COMPARISON OF PHYSICAL CONSTANTS WITH THOSE OF SHEPARD, HENNE, AND MIDGLEY

Physical constants of all the hydrocarbons described in this paper are available in the literature. In general, the criteria of purity were not as strict as those employed in this investigation and the methods of determining the constants, in particular the methods of determining temperature, were not sufficiently refined to yield results of high accuracy. In consequence, the freezing points reported in the literature differ widely. The values obtained on samples isolated from petroleum at this bureau, to which reference has already been made (6, 15, 16) and the values obtained by Shepard, Henne, and Midgley (1) are notable exceptions to this statement. Only Shepard, Henne, and Midgley (1) have recorded values for the entire series from pentane to dodecane. Their values, moreover, were obtained on samples isolated from petroleum, so that a comparison with values on samples obtained in an entirely different manner; that is, by synthesis, is valuable in establishing the true value of the constants. For these reasons only the values of Shepard, Henne, and Midgley (1) are recorded for comparison in Table 1.

TABLE 1.—*Physical constants of normal hydrocarbons*

Substance	Boiling point at 760 mm [1]		Freezing point in air		Refractive Index N$_D^{25}$	
	Author	Shepard, Henne, and Midgley	Author	Shepard, Henne, and Midgley	Author	Shepard Henne, and Midgley
	°C.	°C.	°C.	°C.	°C.	°C.
Pentane	36.06	36.00	−129.73	−129.93	1.35470	1.35495
Hexane	68.70	68.71	−95.34	−95.39	1.37224	1.37230
Heptane	98.38	98.38	−90.62	−90.65 −90.62 [3]	1.38510	1.38553
Octane	125.59	125.59	−56.82	−56.90	1.39509	1.39534
Nonane	150.72	150.71	−53.70	−53.68 −53.65 [3]	1.40318	1.40340
Decane	174.02	174.06	−29.68	−29.76 −29.68 [3]	1.40961	1.40986
Undecane [2]			−25.61	−25.65	1.41495	1.41516
Dodecane [2]			−9.61	−9.73	1.41952	1.41967
Amyl bromide	129.58		−94.67		1.44199	
α-Nonyl bromide			−30.71		1.45221	
β-Nonyl bromide			−29.06		1.45221	

[1] The values for *dt/dp* for the hydrocarbons were taken from the paper by Shepard, Henne, and Midgley (1) while that for amyl bromide was determined and found equal to 0.048°/mm.

[2] The boiling point of undecane and dodecane and *n*-nonyl bromide were not determined owing to the probability of some thermal decomposition.

[3] Values obtained on samples isolated from petroleum by Hicks-Bruun and Bruun (6) (heptane), White and Rose (15) (nonane), and Bruun and Hicks-Bruun (16) (decane).

The agreement between these two sets of values of physical constants, although somewhat less satisfactory than is to be desired, is nevertheless much superior to that generally obtained by investigators on the physical constants of organic liquids. The greatest discrepancy occurs with pentane in which there is a difference of 0.06° in the boiling point and 0.2° in the freezing point. The agreement on the boiling points from hexane to nonane is excellent. The freezing point agreement is not so satisfactory. With the exception of nonane (which was not quite pure), the freezing point values obtained in this work are all higher than those obtained by Shepard, Henne, and Midgley (1). It is probable that the higher values should be considered the more reliable since impurities, except with one type of solid solution, lower the freezing point. Dodecane, as shown by the slope of the freezing point curve, was not quite pure, and it seems possible that the true freezing point may be still higher than the value recorded here, by from 0.05° to 0.1° C. Unfortunately, sufficient dodecane was not available to continue the fractional crystallization and test this point. The other hydrocarbons were purer than nonane and dodecane and their freezing points are probably correct within 0.02° to 0.05° C.

The values recorded here for the refractive indices are lower than those given by Shepard, Henne, and Midgley (1). The average value of the differences is 22 × 10^{-5}.

Refractive indices are not a sensitive test for small amounts of impurities, particularly if the refractive index of the impurity is close to that of the pure substance. Thus 1 per cent of heptane in 99 per cent of octane with refractive indices of 1.38510 and 1.39509, respectively, would give a liquid with refractive index of 1.39499, differing by only 10 × 10^{-5} from that of pure octane. A calculation based on the differences between the freezing point values given by Shepard, Henne, and Midgley (1), and those recorded here, shows that the difference in purity is generally much less than 1 mole per cent. Since the impurities most likely to be present in both cases are isomers, the refractive indices of which do not differ greatly from those of the normal hydrocarbons, it is improbable that the average value for the difference, 22×10^{-5}, can be accounted for by impurities, but instead is attributable to differences in the instruments or methods of measurement.

V. BIBLIOGRAPHY

1. A. F. Shepard, A. L. Henne, and T. Midgley, jr., J. Am. Chem. Soc., vol. 53, p. 1948, 1931.
2. J. H. Bruun, Ind. Eng. Chem., Anal. ed., vol. 1, p. 212, 1929.
3. J. H. Bruun and S. E. Schicktanz, B. S. Jour. Research, vol. 7, p. 851, 1931.
4. M. J. Marshall, Ind. Eng. Chem., vol. 20, p. 1379, 1928.
5. M. M. Hicks: B. S. Jour. Research, vol. 2, p. 484, 1929.
6. M. M. Hicks-Bruun and J. H. Bruun, B. S. Jour. Research, vol. 8, p. 525, 1932.
7. L. W. Tilton and A. Q. Tool, B. S. Jour. Research, vol. 3, p. 622, 1929.
8. Organic Syntheses, vol. 7, p. 77. J. Wiley & Sons, 1927.
9. Organic Syntheses, vol. 1, p. 5. J. Wiley & Sons, 1921.
10. M. Faillebin, Bull. Soc. Chim. (4), vol. 35, p. 160, 1924.
11. O. Aschan, Ber., vol. 31, p. 1801, 1898.
12. A. F. Shepard and A. L. Henne, Ind. Eng. Chem., vol. 22, p. 356, 1930.
13. H. F. Lewis, Robert Hendricks, and G. R. Yohe, J. Am. Chem. Soc., vol. 50, p. 1993, 1928.
14. Organic Syntheses, vol. 6, p. 54: J. Wiley & Sons, 1926.
15. J. D. White and F. W. Rose, jr., B. S. Jour. Research, vol. 7, p. 907, 1931.
16. J. H. Bruun and M. M. Hicks-Bruun, B. S. Jour. Research, vol. 8, p. 583, 1932.
17. R. Ross and J. P. Leather, J. Soc. Chem. Ind., vol. 21, p. 676, 1902.
18. H. Adkins and R. Connor, J. Am. Chem. Soc., vol. 53, p. 1092, 1931.

WASHINGTON, March 11, 1932.

RP483

A STUDY OF SOME CERAMIC BODIES OF LOW ABSORPTION MATURING AT TEMPERATURES BELOW 1,000° C.

By R. F. Geller and D. N. Evans

ABSTRACT

This paper contains a report of an investigation involving the development and testing of a number of ceramic bodies of the whiteware type containing synthetic flux and maturing below 1,000° C. The bodies compare favorably in transverse breaking strength with earthenware, are nearly white in color and vary in absorption from 1 to 5 per cent. The resistance to mechanical abrasion is higher than that of nonceramic flooring materials, such as marble. Glazed specimens resisted crazing satisfactorily in laboratory tests.

CONTENTS

I. INTRODUCTION

The idea of using artificially prepared fluxing agents in ceramic bodies to produce structures, having desirable strength and porosity, at temperatures below those required by the feldspathic fluxes is not new.[1] According to the Encyclopædia Britannica this idea had its inception not later than 1581, when a factory was established at Florence under the patronage of Francesco de Medici to produce white porcelain by introducing common glass into the body. The manufacture was continued at Pisa, and similar porcelain is attributed to Candiana near Padua. Another factory, founded at St. Cloud,

---

[1] Encyclopædia Britannica, vol. 18, 14th ed., p. 351, which refers to "Pottery and Porcelain, A Handbook for Collectors." Vol. III, E. Hannover (Ed. B. Lockham). A. B. Searle, Encyclopædia of Ceramic Industries, vol. 2, p. 377.

had grown to considerable size by the end of the seventeenth century. Searle states that porcelains, heated at low temperatures and made translucent by the aid of especially prepared glassy mixtures of frit were first produced at St. Cloud in 1727 and at Vincennes in 1740. These porcelains are now known as Pate Tendre or soft paste and are matured at temperatures above 1,000° C.[2] The same idea has been applied in more recent times to the manufacture of a type of ware named belleek, presumably after Belleek in County Fermanagh, Ireland.

The last few years have witnessed a renewed interest in the use of artificially prepared fluxes. With the development of methods for utilizing the electric current as a means for producing controlled temperatures, there naturally arises the question of the possibility of adapting this source of heat to the maturing of ceramic whiteware. It has been proven practicable to heat some classes of whiteware by passing the electric current through nonmetallic resistors. However, it would be an added economy and convenience if bodies of satisfactory qualities could be matured at, or below, 1,000° C. and consequently in furnaces of the metallic resistor type. The information contained in the literature [3] is lacking in specific data, particularly with reference to bodies produced at temperatures below 1,000° C. Moreover the field covered by the patent literature referred to is in general that of earthenware.

It was therefore the purpose of this investigation to determine the properties of a limited number of bodies having low absorptions and maturing below 1,000° C., and to obtain results serving as a guide to future industrial developments.

---

[2] Temperatures are expressed in degrees Centigrade throughout this report.
[3] German Patent No. 182107 (Mar. 19, 1907); British Patents Nos. 253184 (June 9, 1926), 302519 (Dec. 20, 1928), 318188 (Aug. 30, 1929), and 323379 (Jan. 2, 1930); U. S. Patents Nos. 1749642 (Mar. 4, 1930), Re. 17656 (May 6, 1930) and 1819686 (Aug. 18, 1931). Also, P. F. Collins, J. Am. Cer. Soc., vol. 11 (9), p. 706, 1928; and vol. 15 (1), p. 17, 1932.

TABLE 1.—*Compositions of fluxes, given in molecular proportions*

Flux No.	Na$_2$O	MgO	SrO	ZnO	CaO	Al$_2$O$_3$	H$_2$O$_3$	Sb$_2$O$_3$	P$_2$O$_5$	SiO$_2$	ZrO$_2$	F	Remarks
	(RO=1.0)						(R$_2$O$_3$=1.3)			(RO$_2$=1.7)			
1	1.0						1.3			1.7			Fuses readily, pours easily; glass colorless.
2	.8	0.2					1.3			1.7			Same as No. 1.
3	.6	.4					1.3			1.7			Fuses more readily than Nos. 1 and 2, pours easily; glass colorless.
4	.8	.2				.1	1.2			1.7			Tendency to foam during melting, melts and pours less readily than No. 1; glass clear light yellow (light blue when melted in platinum).
5	.8	.2				.2	1.1			1.7			Tendency to foam greater than No. 4; otherwise similar.
6	.95		.05				1.3			1.7			Fuses readily, pours easily; glass clear to very light yellow.
7	.95		.05			.2	1.1			1.7			Tendency to foam, melts and pours similar to No. 5. Color clear to light yellow.
8	.9		.1			.2	1.1			1.7			Melts and pours similar to No. 4; glass clear yellow (clear light blue when melted in platinum).
9	.8			.2			1.3			1.7			Tendency to foam but melts and pours readily; glass colorless.
10	.5			.5			1.3			1.7			Melts and pours less readily than No. 9, otherwise similar.
11	.2			.8			1.3			1.7			Melts fairly readily to a white viscous glass; sample melted in platinum shows small amounts of doubly refracting crystals.
12				1.0		.3	1.0			1.7			More difficult to melt but less opaque than No. 11. Too viscous to pour. Melt consisted of nearly pure glass of index 1.566.
13				1.0		.3	1.0			1.7			More refractory than No. 12; no satisfactory melts obtained of either Nos. 12, 13, or 14.
14				1.0			1.0	0.3		1.7			Did not fuse completely. Evidence of the reduction of zinc, also volatilization of ZnO or Sb$_2$O$_3$, or both.
15	.9		.05	.05			1.3			1.7			Fuses and pours readily, glass colorless.
16	.9	.05		.05		.2	1.1			1.7			Same as No. 15, except for tendency to foam during melting.
17	.8	.2					1.3			1.7			Same as No. 15, except that sample melted in platinum was a clear light yellow.
18	.85				.15		1.3			1.2	0.5		Same as No. 15, except amber in color; sample melted in platinum was clear greenish yellow.
19	1.0						1.3			1.2	.3		Less fusible than No. 15, but pours satisfactorily; glass clear light yellow.
20	1.0					.3	1.0			1.4		0.3	Tendency to foam, fuses fairly readily; glass clear light yellow.
21	.6	.4				.05	1.25			1.4		.3	Melts and pours readily; glass clear amber.
22	.5			.5		.05	1.25			1.4		.3	Foams during melting; fuses with difficulty; glass clear with bluish tinge.
23	.8		.1			.05	1.2			1.4		.3	Fuses and pours satisfactorily although more viscous than Nos. 19 to 22, inclusive; glass clear with bluish tinge.
24					.2		1.2		0.1	1.7			Very refractory and unsatisfactory as a frit; quenched mass composed principally of glass of index 1.515 to 1.522 and 4 to 5 per cent of very fine crystals.
25	1.0					.3	1.0			1.7			Fuses fairly readily, pours easily, no foaming; glass clear with tinge of yellow.
26	1.0					.6	.7			1.7			Considerable foaming during fusion, more refractory than No. 25 and does not pour readily; glass clear greenish yellow.
27	1.0					.8	.5			1.7			Could not be melted satisfactorily.
28	1.0					.45	.85			1.7			Tendency to foam, easier to fuse and pour than No. 26; glass clear with yellow tinge.

¹ Flux No. 1 is of the same composition as flux No. 24 in Auxiliary Fluxes in Ceramic Bodies, by P. F. Collins, J. Am. Cer. Soc. vol. 15 (1), p. 17, January, 1932.

II. MATERIALS

The following materials were used in preparing the artificial fluxes (Table 1): Borax, boracic acid, hydrous magnesium carbonate, zinc oxide, zirconium oxide, fluorspar, cryolite, antimony oxide, potters' flint, and kaolin from Georgia, all of commercial grade; also, sodium nitrate, strontium nitrate, strontium carbonate, and calcium phosphate, all of reagent quality.

The bodies (Table 2) were composed of two or more of the following materials in addition to one of the fluxes: Kaolin from Georgia, ball clay from Tennessee, feldspar containing approximately 50 per cent plagioclase feldspar, 40 per cent potash feldspar and 10 per cent quartz, and artificial "mullite" which was an electric furnace product containing approximately 65 per cent mullite, 12 per cent corundum, and 23 per cent glass. A commercial "soda-lime-silica" bottle glass, ground to pass a No. 100 sieve,[4] was substituted as the flux in one body having the type composition of the "BM Series."

All of the materials used in preparing the glaze (Table 2) were of commercial grade as regularly supplied to the pottery trade.

TABLE 2.—*Compositions of bodies and glaze used in the final series of tests*

A. BODIES [1]

| | Series— | | | |
	B	K	BM	KM
	Per cent	Per cent	Per cent	Per cent
Flux	25	25	25	25
Feldspar	35	35	35	35
Ball clay	40	20	35	17½
Kaolin	---	20	---	17½
Mullite	---	---	5	5

B. GLAZE

Fritt		Glaze [2]	
	Per cent		Per cent
Borax	10.3	Fritt	54.4
Boric acid	27.8	Whiting	4.8
White lead	24.3	White lead	16.5
Flint	37.6	Kaolin	24.3

[1] Body compositions are identified by a number followed by the series designation. Thus, 10 BM denotes a body of Series BM in which flux No. 10 was used.
[2] The coefficient of linear thermal expansion (calculated by means of factors in "The Influence of Chemical Composition on the Physical Properties of Glazes," by F. P. Hall, J. Am. Cer. Soc., vol. 13 (3), p.182, 1930) is 5.9×10⁻⁴.

III. METHODS

The fluxes were melted in clay crucibles using a gas-fired pot furnace and, when molten, were poured into water, dried, and dry ground to pass a No. 100 sieve. The crucibles held from 1½ to 2 kg of the molten flux and a different crucible was used for each flux.

Relative solubility of the fluxes was determined on specimens obtained by dry sieving a portion of each flux through No. 200 and

---

[4] Sieves manufactured to meet the requirements of the United States Standard Sieve Series were used throughout this investigation.

No. 270 sieves, using 10 g of that portion which passed the No. 200 and was retained on the No. 270. The specimen was placed in a glass container, together with 200 ml of distilled water, agitated for 48 hours, the liquid removed by filtering, and the residue dried and weighed.

Thermal expansions and "softening temperatures" of the fluxes were determined by the interferometer method [5] on samples of flux which were remelted in platinum, cooled rapidly, and then annealed by heating to their annealing temperatures at an average rate of 2° per minute and holding them at these temperatures for one hour and cooling slowly with the furnace. The "softening temperature" was taken arbitrarily as that temperature at which contraction due to softening began to exceed the expansion due to temperature increase.

Indices of refraction of the fluxes were determined by the Becke method.[6]

The bodies were prepared by wet grinding the constituents in a ball mill for two hours, passing them over a magnetic separator and through a No. 100 sieve, and partly drying on plaster. This procedure, and the compositions used, were established after preliminary trials in which various compositions, modifications of wet and dry grinding, addition of flux before and after grinding, and filter pressing were tried.

Specimens for measuring transverse strength, linear shrinkage, and absorption were prepared by kneading the partly dried bodies and extruding them as rods one-half inch in diameter which were cut into 6-inch lengths. The rods were heated to about 750° C. overnight in an electrically heated furnace and matured the next day. Transverse strength is expressed as the modulus of rupture and was calculated from values obtained by breaking the rods across a 5-inch span. Percentage water absorption (which is proportional to porosity) was calculated from the weights of the specimens after having been dried at $115° \pm 5°$ C. and after having been saturated by immersing in water and "autoclaving" for five hours at a steam pressure of 150 lbs./in.[2]. Linear shrinkage of the rods was determined by measuring the distance, after the heating, between two gage marks which had been pressed 10 cm apart into the freshly extruded rods.

Additional extruded and dried rods were later reground, moistened with about 10 per cent water, passed through a No. 35 sieve, and used in making tile (mold 2¾₂ inches square) and plates (mold 3 inches in diameter) by the "dry-press" method. Total pressure used for both tile and plates was 12,000 pounds. The tile and plates were heated to about 750° C. overnight, in an electrically heated furnace, and matured the next day. Linear shrinkage, absorption, and relative resistance to abrasion of the dry pressed tile were determined on unglazed tile. Linear shrinkage was calculated from the dimensions of the steel mold and of the finished tile. Absorption was determined by the same method used for the extruded rods. Resistance to abrasion, a purely relative value expressed as the H_a value, was obtained by using the apparatus and method of calculation devised by D. W. Kessler [7] for wear tests of flooring materials. Glazed tile and glazed

[5] C. G. Peters and C. H. Cragoe, B. S. Sci. Paper No. 393.
[6] E. S. Larson, U. S. Geological Survey Bulletin No. 679, p. 14, 1921.
[7] A. S. T. M. Proc., vol. 28, Pt. II, p. 855, 1928.

plates were "autoclaved" in water for one and one-half hours at 150 lbs./in.[2] steam pressure and the number of specimens crazed as a result of this treatment was noted.   The "translucency" of a glazed plate was estimated by comparing the intensity of the light it transmitted with that transmitted by an arbitrarily chosen plate of medium "translucency," when such plates were placed over a 2½-inch diameter opening in a box which contained a 75-watt electric light, the filament of which was as near to the opening as possible.   The glazed plates were between 3½ and 4 mm in thickness.

TABLE 3.—*Some physical properties of the fluxes*

Flux No.	Solubility		Softening temperature±10° C. Annealed[2]	Index of refraction	Coefficient of linear thermal expansion.[3] Annealed[2]	Flux No.	Solubility		Softening temperature±10° C. Annealed[2]	Index of refraction	Coefficient of linear thermal expansion.[3] Annealed[2]
	A[1]	B[1]					A[1]	B[1]			
			°C.		×10⁻⁶				°C.		×10⁻⁶
1	26	28	535	1.524	11.8	16	7	8	545	1.525	11.0
2	21	22	540	1.514	11.0	17	17	17	530	1.586	10.1
3	11	11	545	1.516	9.4	18	25	28	525	1.538	11.2
4	16	18	540	1.515	10.4	19	12	11	515	1.526	11.7
5	8	8	540	1.523	10.6	20	21	21	490	1.504	12.3
6	15	15	540	1.527	11.2	21	17	17	500	1.506	10.0
7	7	10	545	1.517	11.4	22	16	17	510	1.515	12.1
8	5	6	540	1.522	11.2	23	15	16	480	1.519	9.1
9	24	24	530	1.527	10.9	25	10	11	525	1.524	11.5
10	15	15	525	1.526	8.0	26	1	2	535	1.516	11.6
11	5	6	540	1.529	5.4	28	......	2	545	1.515	11.1
15	14	13	535	1.524	10.7						

[1] A and B are duplicate determinations calculated on the basis of using 50 g of flux per liter of water.
[2] Fluxes Nos. 20, 21, 22, and 23 were annealed at 475° C., all others at 515° C.
[3] Average coefficient for the temperature range from room temperature to the lower critical temperature. For further information on the "critical temperatures" noted in studies of the thermal expansion of glass see the reference given in Footnote 5. Values to be multiplied by 10⁻⁶.

IV. RESULTS

1. FLUXES

A brief description of the melting behavior of the fluxes and their appearance after quenching is given in Table 1.   Flux No. 1 is of the same composition as flux No. 24 in the series investigated by Collins[3] and was selected as having a melting point and water solubility justifying its use as a basis on which to build the series of fluxes for this investigation.   In outlining the series it is evident that no attempt was made to exhaust the field of possibilities.

Six of the twenty-eight fluxes (Nos. 3, 4, 7, 10, 11, and 28) were selected for intensive study.   This choice was based on the observed melting behavior, the physical properties given in Table 3, and the unreported results obtained in a preliminary study.   The compositions of the six selected fluxes as determined by calculations based on the batch materials, and by chemical analysis, are given in Table 4. The physical properties determined are presented in Table 5.

Although no specific tests were made of abrasives bonded with the fluxes investigated, it is believed, from observation, that at least some

[3] J. Am. Cer. Soc., vol. 15 (1), p. 17, January, 1932.

of the fluxes have possibilities as bonds for certain classes of manufactured abrasive wheels.

TABLE 4.—*Chemical composition* [1] *of fluxes Nos. 3, 4, 7, 10, 11, and 28*

Flux No. and method	SiO_2	Al_2O_3	CaO	MgO	Na_2O	B_2O_3	ZnO	SrO	Loss on ignition
	Per cent	Per cent	Per cent	Per cent	Per cent	Per cent	Per cent	Per cent	Per cent
3A [2]	37.6	0.42	0.09	5.1	16.7	41.0			0.31
3C [3]	41.5			6.5	15.1	36.8			
4A	38.8	3.9	.08	3.1	19.5	32.9			.22
4C	40.3	4.0		3.2	19.5	33.0			
7A	35.0	7.4		.13	24.1	32.3		1.8	.32
7C	38.8	7.7			22.4	29.1		1.9	
10A	37.8	.42	.05	.03	12.2	35.3	14.9		.23
10C	38.6				11.7	34.3	15.4		
11A	38.5	.53	.02	.02	4.8	33.7	23.6		.34
11C	37.8				4.6	33.5	24.1		
28A	36.6	17.4	.07	.12	23.6	22.6			.20
28C	38.0	17.0			23.0	22.0			

[1] The chemical analyses were made by J. F. Klekotka and are based on samples dried at 110° C.
[2] Compositions determined by chemical analysis (A).
[3] Calculated compositions based on batch materials (C).

2. BODIES

(a) PRELIMINARY STUDY

The results of this study are not presented in detail, but may be summarized briefly as follows:

Either a dry grinding of all of the constituents, or the addition of the flux after the other constituents of the body have been wet ground, produced coarse textured bodies.

Wet grinding four hours or more introduced so much dissolved flux into the slip that both the slip and the body were unworkable.

Sufficient flux was dissolved during two hours wet grinding to make filter pressing difficult.

Filter cakes contained sufficient water to carry flux to the surface, resulting in the formation of a hard, discolored "skin." (It is suggested that excess water could be removed from these bodies most efficiently by means of spray driers or drum driers.)

No method was found for preparing bodies which could be "jiggered" successfully. Such faulty specimens as were made contained sufficient water to carry dissolved flux to the surface during the drying.

Casting slips made with bodies containing fluxes 3 and 10 worked well on a laboratory scale and thin-walled pieces could be cast without difficulty. Fluxes Nos. 4, 7, 8, 11, and 28 were fairly satisfactory, while others, notably Nos. 1, 16, and 20, produced slips entirely unsuited for casting. However, rapid disintegration of molds was observed with all of the body slips because of the crystallization of dissolved flux carried into them by water from the slip. It was noted that pH values of the slips when removed from the ball mills could not be used as indicators of casting properties in so far as the particular bodies studied are concerned. Slips of bodies containing flux No. 10 showed the lowest pH values determined (approximately 6.0), while the slips containing flux No. 3 showed the highest (approximately 8.5).   McDowell[9] states the pH value at which maximum dispersion is attained is different for each clay and for each electrolyte.

TABLE 5.—*Some physical properties of the fluxes used in the final series of tests as determined by calculation* [1]

Flux No.	Tensile strength	Modulus of elasticity	Coefficient of thermal expansion [2]	
			Calculated	Observed [3]
	kg/mm^2	kg/mm^2		
3	31.7	6,400	8.6	9.4
4	31.8	6,100	10.3	10.4
7	28.9	5,700	11.6	11.4
10	26.8	7,200	8.6	8.0
11	26.4	7,600	6.6	5.4
28	30.6	6,500	11.9	11.1

[1] Calculations based on the information contained in the reference given in the footnote to Table 2.
[2] Values to be multiplied by 10^{-6}.
[3] Values taken from Table 3.

No difficulty was experienced in making satisfactory specimens of tile and plates by the dry-press process.

Specimens made with 75 per cent flint and 25 per cent flux, when heated to 1,000° C., were translucent, bluish white in color, and of good texture, but the percentage tridymite developed was very high and the specimens were correspondingly sensitive to temperature change in the region of the α-β_1-β_2 tridymite transformations (117° and 163° C.).[10]   The average coefficient of expansion for the temperature interval 100° to 200° C. was 41×10^{-6}.

Specimens made with various ratios of feldspar and flux formed highly translucent bodies, but the indicated maturing range did not exceed 10°.

A large majority of the trials indicated that additions of kaolin at the expense of ball clay darkened the color of the product.

Care must be observed in heating to allow ample time between 650° and 750° C. for complete combustion of organic matter before there is appreciable fusion of the flux and consequent sealing of the pores.

Holding at the maximum temperature for not less than one and one-half hours appears to permit the flux to distribute itself more uniformly throughout the body and produce a "smoother" appearing texture.

Specimens in which a commercial bottle glass had been used as a flux, and which had been heated at 990° ± 5° C., had an average absorption of 11.5 per cent.   This indicates that commercial bottle glass would probably be unsuitable as a flux for bodies designed to mature below 1,000° C., and which are designed to have relatively low absorptions.

(b) PROPERTIES OF EXTRUDED BARS

The six fluxes selected for more intensive study were used in four series of bodies (Table 2) which have been labeled B, BM, K, and KM.

[9] J. Am. Cer. Soc., vol. 10 (4), p. 225, April, 1927.
[10] C. N. Fenner, Am. J. Sci., vol. 36, p. 331, 1913.

Although it was found that specimens could not be formed successfully from the plastic bodies by jiggering, it was thought advisable to obtain approximate results for their maturing range and other properties by means of conventional tests on extruded bars. Results were obtained for the B and BM series and are presented in Table 6. The results obtained give no ground for significant exceptions to the following statements: All of the rods developed maximum strength and shrinkage and minimum absorption when heated to 975° C. Heating to 1,000° C. was harmful. The minimum absorption was lower and the maturing range was shorter for bodies containing 5 per cent mullite than for bodies containing no mullite. Bodies made with a flux containing zinc (Nos. 60 and 61) showed a higher shrinkage and higher strength but not a lower absorption than the average for the series. The maturing range, whether or not it was desired to develop bodies of the lowest possible or of intermediate absorptions, did not exceed approximately 25° C.

The average values for modulus of rupture of the rods heated to 975° C. (Table 6) are very nearly equal to similar values which are found in the literature [11] for "semivitreous" bodies heated to cone 9 (about 1,250° C.) and which vary from 4,700 to 6,600 lbs./in.^{2}; but they are much lower than the values for vitreous bodies heated to cones 11 and 12 (about 1,285° and 1,310° C., respectively) which vary from 8,100 to 9,900 lbs./in.^{2}.

TABLE 6.—*Some properties of extruded bars (¼ inch diameter by 6 inches long) heated for one-half hour at the temperatures indicated*

Body No.	900° C.			925° C.			950° C.			975° C.			1,000° C.		
	S	R	A	S	R	A	S	R	A	S	R	A	S	R	A
3B	1.4	3,000	8.0	2.1	3,400	6.5	2.2	3,500	5.0	3.0	4,000	3.5	3.1	3,500	2.0
4B	1.4	3,800	9.5	2.1	3,600	7.0	2.9	3,900	5.5	3.6	4,700	.5	3.1	4,000	4.5
7B	1.6	3,600	9.5	2.1	3,300	9.5	2.6	3,500	6.0	3.0	4,500	2.5	2.3	3,300	8.0
10B	1.7	3,200	10.0	3.1	3,700	7.5	3.8	4,500	5.0	4.9	4,300	3.5	3.3	3,500	7.0
11B	2.0	3,200	14.0	2.9	3,600	8.0	4.3	4,400	6.0	4.1	4,800	4.5	4.3	4,500	4.5
28B	1.4	3,300	12.5	2.5	3,800	9.0	3.3	4,400	7.0	3.5	4,600	6.0	3.5	4,300	5.0
3BM	1.6	3,600	9.0	2.0	4,100	6.0	2.2	3,900	5.0	3.0	4,600	.0	3.0	4,000	1.0
4BM	1.4	3,400	9.5	2.2	3,800	6.5	2.5	3,700	4.5	3.4	4,300	.5	2.5	3,400	4.0
7BM	1.6	3,800	8.0	2.0	3,800	6.0	2.4	4,000	3.5	2.8	4,600	.5	2.0	3,200	4.5
10BM	1.8	3,200	10.5	2.6	3,200	8.0	4.1	4,000	5.0	5.7	4,900	2.5	2.5	4,100	5.0
11BM	1.6	3,400	16.0	3.4	3,600	8.0	4.3	4,700	5.5	5.5	4,800	4.5	5.0	4,400	4.0
28BM	1.2	4,000	14.5	2.2	3,400	10.0	2.8	4,400	8.0	5.3	4,400	6.0	3.5	4,200	5.0

S = linear shrinkage, in per cent.
R = modulus of rupture, in pounds per square inch.
A = water absorption, percentage by weight.

(c) PROPERTIES OF DRY PRESSED TILE

(1) *Resistance to abrasion.*—The values obtained are given in Table 7. Apparently the compositions of the experimental bodies had no significant effect on their relative resistance to abrasion.

For comparison, three specimens of one brand of commercial white vitrified floor tile were tested and the average H_a value for nine determinations was 78. The following average H_a values are calculated from data in the report by Kessler:[12] Average of 45 marbles,

[11] E. E. Pressler and W. L. Shearer, B. S. Tech. Paper No. 310, 1926. R. F. Geller and A. S. Creamer, J. Am. Cer. Soc., vol. 14 (1), p. 30, January, 1931.
[12] See footnote 7, p. 477.

18; average of 15 limestones, 9; average of 14 slates, 8; average of 4 sandstones, 5.

(2) *Absorption.*—The absorption percentages reported in Table 7 are the lowest which it was possible to obtain without developing a vesicular or "overfired" structure. The group of bodies containing flux No. 3 has the lowest average absorption of the six groups differentiated by means of the flux. The BM series of bodies containing mullite but no kaolin has the lowest average absorption. Although there is little difference in the average absorption values given in Table 7 for the various series and groups, it should be remembered that those bodies containing flux No. 11 or No. 28 require a maturing temperature 30° C. higher than is required by the other bodies.

TABLE 7.—*Some properties of "dry pressed" tile*

Flux No.	Heat treatment		Resistance to abrasion;[1] series—				Absorption;[2] series—				Linear shrinkage;[3] series—			
	Maximum temperature, ±12° C.	Time at maximum temperature	B	BM	K	KM	B	BM	K	KM	B	BM	K	KM
	°C.	*Hours*		H_a value [4]			*Per cent*	*Per cent*	*Per cent*	*Per cent*	*Per cent*	*Per cent*	*Per cent*	*Per cent*
3	960	2–3	44	45	42	43	2.0	1.0	2.5	1.3	4.5	4.0	5.0	4.5
4	960	2–3	42	41	40	44	3.5	2.5	3.5	3.0	4.5	4.0	4.0	3.5
7	960	2–3	41	40	41	39	1.0	1.5	3.0	3.0	3.0	2.5	3.5	3.0
10	960	2–3	45	45	40	40	3.5	2.5	4.5	4.5	4.5	4.0	4.0	4.0
11	990	3	52	39	45	49	2.5	2.0	4.0	3.0	6.0	5.0	5.5	5.5
28	990	3	51	53	42	44	3.0	1.5	5.0	3.0	4.5	5.0	4.5	4.5

	Standard deviation		Maximum individual deviation
	Maximum	Minimum	
H_a value	7	1	10
Absorption ....per cent..	1.3	.3	2.5
Shrinkage ....do....	.8	.2	1.7

[1] Bodies 3, 4, 7, and 10, values are average of 6 specimens from 2 heatings. Bodies 11 and 28, values are average of 3 specimens from 1 heating.

[2] Bodies 3, 4, 7, and 10, values are average of 9 specimens from 3 heatings. Bodies 11 and 28, values are average of 6 specimens from 2 heatings.

[3] Number of specimens used same as for absorption. These values are the total linear shrinkage during drying and heating.

[4] Calculated by means of the formula $H_a = \dfrac{10\,(2,000 + W_a)\,G}{2,000\,W_a}$ (see footnote 7, p. 477) where:

W_a = weight of material lost by abrasion.
W_i = original weight of specimen.
G = apparent density.

The absorptions of 180 specimens representing 17 brands of commercial tableware were determined. The lowest average absorption for specimens of any one brand was 0.2 per cent, the highest 9.5 per cent, and the average for all brands was 7.2 per cent. The deviations for the experimental bodies (Table 7) may be compared with the following deviations for the commercial bodies: The maximum standard deviation, in per cent, for the specimens of a single brand was 2.1, the minimum 0.2; the maximum deviation of an individual specimen from the mean of a single brand was 5.0 per cent.

(3) *Linear shrinkage.*—The values for total percentage linear shrinkage during drying and heating are given in Table 7. The low values for the bodies containing flux No. 7 are an interesting development in view of the fact that these bodies are also among the least absorbent.

(4) *Color.*—None of the bodies are white. Within each series the colors vary according to the flux, as follows: Flux No. 28 produced the darkest body, then Nos. 7, 4, 3, 10, and 11. When compared with color standards [13] the B and BM series may be described as varying from cartridge buff, Plate XXX 19″ Y-O-Yf, toward white while the K and KM series vary from tilluel buff, Plate XL 17‴' O-Yf, toward white. Before making the comparisons the eyes of the investigator were tested for sensitivity to color in the colorimetry section, and found to be normal.

(5) *Resistance to crazing.*—The relative resistance to crazing is indicated by the values in Table 8. The BM series of bodies appears to have a resistance appreciably higher than the others and it will be noted that this series (Table 7) has the lowest average absorption. The total number of crazed specimens in the three groups containing flux Nos. 3, 7, or 28 is significantly lower than in the other three. No explanation for their apparently higher resistance was found.

(6) *Thermal expansion.*—The coefficient of thermal expansion of each of the bodies of the B and BM series was calculated from tests on one specimen of each body. (Table 9.) Their coefficients do not differ significantly from those of 16 commercial whiteware bodies (one test on each body) which varied from 6.7 to 7.5×10^{-6}.

TABLE 8.—*Relative translucency and resistance to crazing*

Flux No.	Tile [1]					Plates [2]					Relative translucency average applies to each series
	Maximum temperature $\pm10°$ C.	Number of specimens crazed (3 tested); series—				Maximum temperature $\pm10°$ C.	Number of specimens crazed (2 tested); series—				
		B	BM	K	KM		B	BM	K	KM	
3	960	1	0	0	0	945	0	0	1	0	"Bright".
4	960	2	0	2	0	950	1	0	2	2	Do.
7	960	0	0	1	1	940	0	0	0	0	Medium.
10	960	3	0	3	3	955	0	0	2	1	Between 28 and 11.
11	990	0	1	3	2	970	1	0	2	1	Nearly opaque.
28	900	0	0	1	0	980	0	0	1	1	Below medium.

[1] Bisque heated 3 hours at the maximum temperature. Glaze matured 1 hour at $950° \pm 5°$.
[2] Bisque heated from ½ to 1½ hours at the maximum temperature. Glaze matured at $950° \pm 5°$ – time 1 to 1½ hours.

(d) PROPERTIES OF GLAZED PLATES

(1) *Resistance to crazing.*—Values indicating relative resistance to crazing are given in Table 7. The resistance of the BM series was highest and the average resistance of the "ball clay bodies" (B and BM) was higher than that of the "kaolin bodies" (K and KM), which is in agreement with the results of the tests on glazed tile. Results of the tests on tile and plates may not be directly comparable because the former were glazed on one side only while the latter were completely coated.

---

[13] Robert Ridgway, Color Standards and Color Nomenclature, printed by A. Hoen & Co., Baltimore, Md.

(2) *Translucency.*—The values for translucency, which are purely relative, were determined for three specimens of each body. As a means for comparison, plates of several brands of domestic vitrified hotel china (thickness 4.3 to 5.1 mm) were tested also. Their translucency varied from the arbitrarily selected "medium" to "bright." The light transmitted by the least translucent specimens (containing flux No. 11) was approximately of the same intensity as that transmitted by a specimen of commercial earthenware of 4 per cent water absorption.

The translucency values for the experimental bodies are given in Table 8 and, as indicated, they varied according to the flux. Other variations in body composition did not affect the light transmission sufficiently to be detected by the test method. That the bodies containing flux No. 11 are least translucent is explained probably by the fact that the flux itself is nearly opaque. (Table 1.)

TABLE 9.—*Coefficients of thermal expansion of bodies of the B and BM series* [1]

Body No.	Coefficient of expansion [2]	Body No.	Coefficient of expansion [2]
3B	7.6	3BM	7.4
4B	7.9	4BM	7.8
7B	8.6	7BM	9.1
10B	7.3	10BM	7.5
11B	6.5	11BM	6.4
28B	8.4	28BM	8.7

[1] The temperature range is from room temperature to a temperature below the point at which the body shows a marked increase in rate of expansion (approximately from 25° to 400° C.). The temperature at which the rate of expansion increases is nearly the same as the lower critical temperature of the flux used.
[2] Values to be multiplied by 10^{-4}.

V. SUMMARY

The results obtained in this investigation indicate that the production of ceramic bodies having water absorptions lower than the average for commercial earthenware, and maturing below 1,000° C., would require careful supervision and control of all of the manufacturing processes.

It would probably be necessary to wet-grind the flux with the other ingredients of the body to obtain a sufficiently fine-grained and uniform product. This would be done at the cost of introducing dissolved flux in the "slip."

No difficulty was encountered in forming tile and plates approximately 2 inches square and 3 inches in diameter, respectively, by the dry-press process in a laboratory press.

The particular body compositions used could not be matured on what would usually be considered a short heating schedule, since it was necessary to allow sufficient time for all organic matter to oxidize and volatile products to escape before the flux softened sufficiently to seal the pores of the body.

Comparatively small variations in the chemical composition of the flux or of the body in which it is used may have a disproportionate effect on the properties of the product. For example, the use of kaolin in place of 50 per cent of the ball clay decreased the resistance to abrasion of 9 out of 12 bodies (Table 7); the average absorption of the bodies containing kaolin was about 50 per cent higher than the average for the others (Table 7) and their resistance to crazing when glazed was considerably lower (Table 8). Also, the substitution of kaolin for ball clay produced bodies slightly darker in color. The results, therefore, indicate that it would be advantageous to use all ball clay, preferably "washed" and low in organic matter, rather than a mixture of ball clay and kaolin.

The introduction of 5 per cent of "mullite" at the expense of the clay appeared to have some desirable effects. (Tables 7 and 8.)

Within any one series of bodies the flux appeared to be the predominating factor in determining the relative color and translucency of bodies. (Table 8.) Effects on the other properties are not so pronounced.

It is believed that at least some of the fluxes have possibilities as bonds for certain classes of manufactured abrasive wheels.

VI. CONCLUSIONS

In general, it may be stated that ceramic bodies comparable in transverse breaking strength to earthenware, nearly white in color and varying in water absorption from 1 to 5 per cent, have been made in the laboratory by the dry-press process, of commercially available materials, and by maturing below 1,000° C. The bodies may be expected to resist mechanical abrasion better than nonceramic flooring materials, such as marble, and glazed specimens resisted crazing satisfactorily in laboratory tests. There are no data available on which to base conclusions regarding the cost of manufacture under commercial conditions.

WASHINGTON, May 9, 1932.

RP484

THE DETERMINATION OF MAGNESIA IN PHOSPHATE ROCK

By James I. Hoffman

ABSTRACT

A method for the determination of magnesia in phosphate rock is presented. Calcium is separated as sulphate in alcoholic solution, and magnesium is determined in the filtrate by precipitating it as magnesium ammonium phosphate. Citric acid is added to prevent interference by iron, aluminum, and titanium. In this procedure part of the manganese always accompanies the magnesium. This is determined in the ignited residue by the periodate method. The data show that arsenic does not interfere and that the method is also applicable if larger amounts of magnesium are present than are usually found in phosphate rock.

CONTENTS

I. INTRODUCTION

Practically all samples of phosphate rock contain less than 0.5 per cent of magnesia. To determine this a procedure must be used in which large amounts of calcium and moderate amounts of iron, aluminum, titanium, manganese, and fluorine are eliminated. Calcium can be separated by precipitating it as oxalate in an oxalic acid solution,[1] or as sulphate in an alcoholic solution. If the former procedure is used, the precipitation with oxalate must be repeated. Precipitating calcium as sulphate is more suitable in this case because the separation of small amounts of magnesium is so nearly complete in a single precipitation that the operation need not be repeated. Iron, aluminum, titanium, and part of the manganese are eliminated if magnesium is precipitated as magnesium ammonium phosphate in a solution containing ammonium citrate, and a correction for any manganese which is precipitated can be applied to the finally weighed residue. Fluorine is eliminated at the start by evaporating with sulphuric acid. Arsenic, if it is not previously eliminated, is volatilized during the evaporation of the alcohol after the calcium sulphate is separated.

The following method was found to yield satisfactory results when applied to synthetic mixtures and to the Bureau of Standards

---

[1] G. E. F. Lundell and J. I. Hoffman, The Analysis of Phosphate Rock, J. Assoc. Official Agric. Chem. vol. 8, p. 184, 1924.

standard sample of phosphate rock, No. 56, which has the following percentage composition: P_2O_5, 31.3; Fe_2O_3, 3.3; Al_2O_3, 3.1; CaO, 44.8; SiO_2, 6.7; TiO_2, 0.15; MnO, 0.25; MgO, 0.32; F, 3.6. This composition is typical of phosphate rocks excepting that in most cases smaller percentages of the oxides of manganese and magnesium are present.

II. METHOD FOR THE DETERMINATION OF MAGNESIA

Transfer 2 g of phosphate rock, ground to pass a No. 100 sieve and dried for one hour at 105° C., to a 250 ml beaker, cover, add 15 ml of diluted hydrochloric acid $(2+1)$[2] and 5 ml of nitric acid, and boil gently for 10 to 15 minutes. Remove the beaker from the source of heat, add 6 ml of diluted sulphuric acid $(1+1)$, remove the cover, and evaporate until fumes of sulphuric acid appear. Cool slightly, wash down the inside surface of the beaker with a jet of water and again evaporate until fumes of sulphuric acid appear. Cool, add 10 ml of water, stir thoroughly, and digest on the steam bath for 10 to 15 minutes. Remove from the steam bath, add 100 ml of 95 per cent alcohol,[3] stir so that the calcium sulphate is well dispersed throughout the liquid, and then allow to stand for 30 minutes or longer. Filter by means of suction through a tight plug of filter paper pulp, using a Gooch crucible, carbon funnel, or Büchner funnel, and wash five times with 5 ml portions of 95 per cent alcohol containing 1 ml of sulphuric acid per 100 ml.

Evaporate the alcoholic filtrate as far as possible on the steam bath. Transfer the solution to a 300 ml Erlenmeyer flask, dilute to 75 to 100 ml, and add 2 g of citric acid and 15 ml of a 25 per cent solution of diammonium phosphate, $(NH_4)_2HPO_4$. Add ammonium hydroxide until the solution is alkaline to litmus and then add 10 ml in excess. Add 5 to 10 glass beads, tightly stopper the flask, and shake on a shaking machine for at least one hour. Allow to stand in a cool place for four hours or preferably overnight. Filter through a tight paper containing a little paper pulp, and wash with diluted ammonium hydroxide $(5+95)$, containing 50 g of diammonium phosphate per liter, until the precipitate and paper are free from iron and aluminum. Pass 25 ml of hot diluted hydrochloric acid $(5+95)$ through the paper into the flask, transfer the solution to a 150 ml beaker, and wash the paper and flask thoroughly with more of the diluted acid. To the solution in a volume of 50 to 75 ml and containing no glass beads, add ½ ml of a 25 per cent solution of diammonium phosphate, cool, and then add ammonium hydroxide slowly and with stirring until the solution is alkaline to litmus. Stir for a few minutes, then add 3 to 4 ml of ammonium hydroxide and allow to stand for four hours or overnight. Transfer the precipitate to a small filter and wash with diluted ammonium hydroxide $(5+95)$. Transfer the paper and precipitate to a platinum crucible, ignite slowly at a temperature below 900° C. until the carbon is burned (preferably in a muffle furnace with pyrometric control), and then at about 1,100° C. for one to two hours. Cool and weigh.

---

[2] This denotes 2 volumes of hydrochloric acid, specific gravity 1.18, mixed with 1 volume of water. This system of designating the diluted acids or ammonium hydroxide is used throughout this paper. If no dilution is specified, the concentrated reagent is intended.

[3] In most of this work 95 per cent ethyl alcohol was used. Mixtures of 100 parts of 95 per cent ethyl alcohol with from 5 to 25 parts of methyl alcohol proved equally satisfactory.

The residue consists of $Mg_2P_2O_7$ and possibly $Mn_2P_2O_7$ and $Ca_3(PO_4)_2$. If the alcoholic filtrate was clear, the tricalcium phosphate, $Ca_3(PO_4)_2$, will not exceed 0.3 mg and can be neglected unless very accurate results are desired.[4] The correction for manganese is made as follows: Dissolve the residue in 10 ml of diluted sulphuric acid (1+9), transfer the solution to a 300 ml Erlenmeyer flask, and add 50 ml of diluted nitric acid (1+3), 2 ml of sirupy phosphoric acid, specific gravity, 1.7, and 0.2 g of potassium periodate, KIO_4. Boil for 15 to 20 minutes, cool, and dilute to a convenient volume. In another flask containing the same amounts of the reagents treated in a similar way, match the color by adding a standard solution of potassium permanganate. From the volume of the solution of permanganate required, calculate the weight of $Mn_2P_2O_7$ in the residue. Subtract this weight from the total weight, and regard the difference as $Mg_2P_2O_7$, which contains 36.21 per cent of MgO.

Table 1 gives results obtained by this method for magnesia in phosphate rock and in mixtures approximating the composition of phosphate rock. The standard value for magnesia (0.32 per cent) in the Bureau of Standards standard sample of phosphate rock, No. 56, was obtained by twice precipitating the calcium as oxalate in dilute oxalic acid solution [5] and determining the magnesia in the combined filtrates by precipitating as magnesium ammonium phosphate in the presence of ammonium citrate and then igniting to magnesium pyrophosphate. Corrections were made for calcium and manganese.

TABLE 1.—*Results obtained for magnesia in phosphate rock and in synthetic mixtures*

Analysis No.	Material analyzed	MgO present or added	MgO found	MgO present or added	MgO found	Error
		g	*g*	*Per cent*	*Per cent*	*Per cent*
1	0.12 g Fe_2O_3+0.12 g Al_2O_3+0.1 g CaF_2+1.65 g $Ca_3(PO_4)_2$	0.0008	0.0007	0.04	0.035	−0.005
2	do	.0008	.0008	.04	.04	.000
3	do	.0032	.0031	.16	.155	−.005
4	do	.0079	.0078	.395	.40	+.005
5	do	.1140	.1130	5.70	5.65	−.05
6	0.12 g Fe_2O_3+0.12 g Al_2O_3+0.1 g CaF_2+.0015 g MnO +1.65 g $Ca_3(PO_4)_2$	.0032	.0034	.16	.17	+.01
7	do	.0032	.0030	.16	.15	−.01
8	0.12 g Fe_2O_3+0.12 g Al_2O_3+0.1 g CaF_2+0.0045 g MnO +1.65 g $Ca_3(PO_4)_2$	.0079	.0079	.395	.395	.000
9	2.0 g Bureau of Standards standard sample of phosphate rock No. 56.			.32	.32	.00
10	do			.32	.33	+.01
11	do			.32	.31	−.01
12	2.0 g Bureau of Standards standard sample of phosphate rock No. 56+0.0092 g MgO			.78	.79	+.01
13	2.0 g Bureau of Standards standard sample of phosphate rock No. 56+0.0175 g As_2O_3			.32	.31	−.01

[4] If a correction for calcium is to be made, dissolve the residue in 10 ml of diluted sulphuric acid (5+95), evaporate to about 5 ml, and then add 25 ml of 95 per cent alcohol. Allow to stand for one to two hours, filter on a small paper, and wash with alcohol. Dry the paper in the funnel and dissolve the precipitate of calcium sulphate in 20 ml of hot diluted hydrochloric acid (1+99). Add a few crystals of ammonium oxalate, heat, render the solution faintly ammoniacal, and allow to stand for one-half to one hour. Filter, ignite, and weigh as CaO. Calculate to tricalcium phosphate, $Ca_3(PO_4)_2$, and subtract from the apparent weight of $Mg_2P_2O_7$.

If calcium is determined in this way, the alcoholic filtrate is used for the determination of manganese. Before applying the periodate oxidation, this filtrate must be evaporated and heated until fumes of sulphuric acid appear, and then treated with nitric acid to make sure that reducing substances are absent.

[5] See The Analysis of Phosphate Rock, J. Assoc. Official Agric. Chem., vol. 8, p. 203; 1924.

The magnesium that was used in making up the mixtures was added in the form of a standard solution of magnesium chloride. The iron and aluminum were added as solutions of their chlorides. The Bureau of Standards standard sample of fluorspar No. 79 was used as the source of the CaF_2 added. The percentage composition of this sample is: CaF_2, 94.8; $CaCO_3$, 2.17; SiO_2, 1.9; Zn, 0.35; Pb, 0.23; S, 0.13; Fe_2O_3, 0.15; Al_2O_3, 0.02; MgO, 0.13. The tricalcium phosphate that was used contained 0.3 mg of MgO in 1.65 g. Proper corrections were made for the magnesia thus introduced.

The calcium sulphate obtained by precipitation with alcohol in analysis No. 5 was dissolved in hot hydrochloric acid and then evaporated with 6 ml of diluted sulphuric acid (1 + 1). The precipitation by alcohol was repeated and the magnesia determined in the filtrate as before. The amount of MgO found was 0.0011 g. This indicates that if large amounts of magnesium are present, some is retained by the calcium sulphate. Less than 0.005 per cent of MgO was retained by the calcium sulphate in the case of phosphate rock No. 56 (analyses Nos. 9 to 13).

III. DISCUSSION OF THE METHOD

1. DECOMPOSITION OF THE SAMPLE

A 2 g sample each of Tennessee brown rock (MgO — 0.02 per cent), Tennessee blue rock (MgO — 0.36 per cent), Florida pebble rock (MgO — 0.25 per cent), Florida hard rock (MgO — 0.05 per cent), Wyoming phosphate rock (MgO — 0.08 per cent) and Idaho phosphate rock (MgO — 0.31 per cent)[6] was decomposed by treatment with hydrochloric and nitric acids as described in the preceding section. The following test was made to determine the amount of magnesia in the insoluble residue: After adding 50 ml of water and digesting, the insoluble matter was collected on a filter, washed, and ignited in platinum. The ignited residue was brought into solution by treating with hydrofluoric and sulphuric acids and then evaporating until fumes of sulphuric acid were copiously evolved. The contents of the crucible were transferred to a small beaker by means of 25 to 30 ml of 95 per cent alcohol, and the calcium sulphate which separated was removed by filtration. Magnesium was determined in the filtrate by the method above outlined for the determination of magnesia in phosphate rock. An average of less than 0.005 per cent of magnesia, MgO, was found in the insoluble matter, and in no case did the percentage exceed 0.01.

2. SEPARATION OF CALCIUM SULPHATE BY MEANS OF ALCOHOL

The results in the table show that the separation of calcium sulphate by means of alcohol is very satisfactory for the small amounts of magnesium present in phosphate rock and that it is applicable even if much larger amounts are involved. The amount of sulphuric acid specified is the minimum that should be used, because if the sulphates become dry and form a cake during the evaporation with sulphuric acid, it is difficult to extract all of the magnesium with

---

[6] These representative samples were furnished through the courtesy of K. D. Jacob, of the United States Department of Agriculture.

alcohol. There is less danger of calcium sulphate running through the filter if a little sulphuric acid is added to the alcohol used for washing and if the filter is not allowed to run dry. The presence of considerable amounts of organic matter, such as are found in some samples of phosphate rock, does not adversely affect the separation.

3. PRECIPITATING MAGNESIUM AS MAGNESIUM AMMONIUM PHOSPHATE

In order to avoid tedious separations, it is desirable to precipitate the magnesium in the presence of the iron and aluminum. About 2 g of citric acid must be added to hold in solution the amounts of these elements that may be expected in phosphate rock. In such solutions it is difficult to precipitate small amounts of magnesium ammonium phosphate. In some cases complete precipitation is not obtained even if the solution is vigorously stirred or shaken mechanically, but if the solution is shaken in a flask containing a few glass beads, complete precipitation can readily be had. The precipitate thus obtained is very finely divided, which necessitates the use of a very tight filter. The most satisfactory filtering medium consists of a tight paper containing some fine paper pulp. Glass beads are not used in the second precipitation.

If small amounts of magnesium, such as are likely to be encountered in phosphate rock, are precipitated as magnesium ammonium phosphate in the presence of ammonium citrate without agitation, a few hours are sometimes required before the crystals begin to separate. These transparent crystals adhere to the beaker and stirring rod and are slightly soluble in the ordinary wash solutions. If they are broken in transferring them to the paper, more will dissolve in the wash water than if they remain undisturbed.

If manganese is present in a solution in which magnesium ammonium phosphate is precipitated, it is quantitatively precipitated under the usual conditions such as are outlined for the last precipitation, but if citric acid is added, as in the first precipitation, less than one-third of the manganese accompanies the magnesium when the two are present in the proportions usually found in phosphate rock. This quantity is not constant, and it is necessary to determine the manganese in the ignited residue that is finally obtained. If it is known that the phosphate rock contains less than 0.02 per cent of manganese oxide (MnO), the correction may be neglected.

IV. ACKNOWLEDGMENT

Acknowledgment is made to Dr. G. E. F. Lundell for technical advice.

WASHINGTON, July 6, 1932.

RP485

COLLISIONS OF THE FIRST AND SECOND KIND IN THE POSITIVE COLUMN OF A CÆSIUM DISCHARGE

By F. L. Mohler

ABSTRACT

The power input minus the power lost by recombination at the tube walls measures the number of quanta dissipated by inelastic collisions at the resonance potential, 1.42 volts. For low-pressure low-current conditions this is equal to the number of collisions $C_{1\,2}$ and the effective area for a collision is

$$q_{1\,2} = 5.4 \pm .6 \times 10^{-15} \text{ cm}^2$$

At higher pressures and currents collisions of the second kind give energy back to the electrons and the number of quanta dissipated is $C_{1\,2} - C_{2\,1}$ where $C_{2\,1}$ is the number of collisions of the second kind. This gives a basis for computing the number of excited atoms n_2. The value of n_2 can also be measured spectroscopically by analysing light from an incandescent lamp passed through the discharge and finding the lamp temperature T_L at which absorption lines match the background. T_L is nearly the electron temperature, confirming the evidence that collisions of the first and second kind become nearly equal in number.

The measured radiation intensity is equated to the radiation at a temperature T_L over a spectrum range $\Delta\lambda$ to find $\Delta\lambda$. This effective width of the resonance lines ranges from 0.6 to 19 A between 0.006 and 0.35 mm pressure. At constant pressure the radiation is proportional to n_2 and the ratio gives the depth of the layer of vapor from which the radiation comes; 0.003 to 0.001 mm in the above pressure range.

CONTENTS

I. INTRODUCTION

In a recent paper I have reported measurements of power input and power losses in the positive column of a cæsium discharge.[1] The experimental material included radiation measurements and measurements of probe characteristics for a wide range of currents and pressures in tubes 1.8 cm in diameter. The present paper uses these measurements as a basis for a study of the atomic processes involved in the dissipation of power. It was found that in general

---

[1] F. L. Mohler, Power Input and Dissipation in the Positive Column of a Cæsium Discharge, B. S. Jour. Research, vol. 9 (RP455), p. 25, July, 1932. Referred to as RP455 in the text.

the radiation is an important part of the power loss and that most of the radiation is in the first doublet of the principal series at 8,521 and 8,944 A. The excitation potential is 1.42 volts, and it is known from critical potential experiments[2] that this excitation gives the predominant type of inelastic collision in cæsium. This justifies the approximation that the power input is expended in inelastic collisions at the first resonance potential and in ionization, and that other excitation processes are negligible. This view of the power loss has been proposed and critically examined for the case of the mercury positive column by Killian.[3]

I will assume that the power lost by inelastic collisions at the first resonance potential is equal to the power input minus the power lost by recombination of ions at the walls; $P-W$ per centimeter of the column in the notation of RP455. For low pressure low current conditions this is practically equal to the radiation loss and is entirely justified; but, at the other extreme, my assumption becomes rather arbitrary. If ionization by collision with excited atoms is important, then part of the loss W is at the expense of inelastic collisions. If recombination in free space is important, then W does not account for the full ionization loss. The energy per collision or per quantum is $eV_r = 2.26 \times 10^{-19}$ joules, where V_r is the excitation potential; and the number of quanta dissipated is $(P-W)/2.26 \times 10^{-19}$ per centimeter of the column per second.

The number of inelastic collisions per centimeter of the column per second, $C_{1\,2}$, can be expressed in terms of the effective collision area $q_{1\,2}$ of an atom for such a collision.

$$C_{1\,2} = q_{1\,2}N_e n_1 v_e f(v)$$

where N_e is the number of electrons per cm of the column, n_1 is the number of atoms per cubic centimeter (accurately the number in the normal state), v_e is the electron velocity and $f(v)$ is the fraction of electron-collisions with kinetic energy greater than eV_r. The electrons have a Maxwell distribution of energy, here expressed in terms of V_0, where

$$eV_0 = kT = 2/3 \text{ mean energy}$$

Introducing for $f(v)$ the integral over energies between $V_r e$ and infinity

$$C_{1\,2} = q_{1\,2}N_e n_1 v_e \left(\frac{V_R}{V_0} + 1\right) \exp\left(-\frac{V_R}{V_0}\right) \tag{1}$$

The probability of an inelastic collision is a function of V which rises abruptly from zero at V_R to a maximum very close to V_R and then decreases less abruptly over a range of several volts.[4] $q_{1\,2}$ is the integrated effect of all electrons of a Maxwell distribution which exceed V_R and so it is a function of V_0 which presumably increases with decreasing V_0.

From the above remarks it seems to follow that the number of collisions is equal to the number of quanta dissipated.

$$C_{1\,2} = (P-W)/2.26 \times 10^{-19} \tag{2}$$

[2] Foote, Rognley, and Mohler, Phys. Rev., vol. 13, p. 59, 1919.
[3] Killian, Phys. Rev., vol. 35, p. 1238, 1930.
[4] Darrow, Electrical Phenomena in Gases, Williams, Wilkins Co., 1932.

It will be shown that this identity is not justified except for low pressures and currents. Electron energy expended in producing excited atoms may be given back to the electrons by collisions of the second kind so that in general the power dissipated is less than $C_{1\,2} \times 2.26 \times 10^{-19}$ watts. In general,

$$C_{1\,2} - C_{2\,1} = (P-W)/2.26 \times 10^{-19} \tag{3}$$

where $C_{2\,1}$ is the number of collisions of the second kind per centimeter of the column per second.

II. EXPERIMENTAL PROCEDURE

The type of discharge tube was illustrated in RP455 (fig. 1). It was a termionic discharge in a tube 1.8 cm in diameter. Commonly there were two small probes near the axis and 12 cm apart to measure the voltage gradient, electron energy and electron concentration; and a disk electrode against the wall to measure the flow of ions to the wall. The cæsium was in a side tube, maintained at a lower temperature than the body of the tube. Vapor pressures were computed from the temperature by the relation [5]

$$\log p \ (mm) = -\frac{3{,}966}{T} + 7.1650$$

The temperature measured by a thermocouple in contact with the tube wall in the region of the positive column was taken as the gas temperature in computing the number of atoms, n_1.

The discharge is characterized by exceptionally low electron speeds and high electron concentration. This makes the precision low, and an added difficulty is the electrical leakage over the cæsiated glass surfaces. The high electron concentration necessitates the use of a small probe surface. The leakage was greatly reduced by avoiding a close fit between the insulator and the probe but this leaves the effective collecting area uncertain. Two sizes of probe were used to cover the range of electron currents obtained; platinum wires 2 mm long and 0.4 mm in diameter, and wires of this diameter cut off flush with the insulator.

Figure 1 (taken from RP455) shows curves of the log of the probe current versus voltage relative to the anode for the two small probes 12 cm apart. The experimental points negative to the space potential (the intersection of the two lines) fall on a straight line showing that the electrons have a Maxwell distribution. The electron energy V_0 is given by the relation

$$V_0 = (V_2 - V_1) \log e/(\log I_1 - \log I_2)$$

The experimental uncertainty is about 0.02 volt and, as V_0 may have values as low as 0.2 volt, the percentage error is high. Recognized sources of error tend to make V_0 too high. The value of V_0 at the tube wall is nearly the same as at the axis.

The random current density at the center of the tube, i_0, is given by the current at the space potential divided by the collecting area.

---

[5] Rowe, Phil. Mag., vol. 3, p. 534, 1927.

For the 2 mm probes the measured area 0.0257 cm² was used and the effective area of the wire, cut off flush with the insulator, was obtained by comparison with the 2 mm probe. This area 0.002 cm² is greater than the end of the wire because the edge is not completely insulated

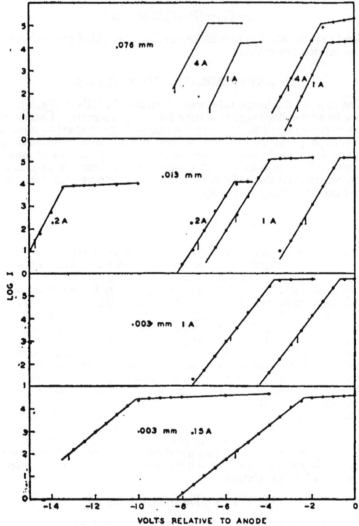

FIGURE 1.—*Probe characteristics for two small probes 12 cm apart at the axis of the positive column*

from the discharge. The number of electrons per cubic centimeter is given by the relation

$$n_0 = 2.51 \times 10^{19} i_0 / v_e$$

The space current decreases from the tube axis to the walls and the variation of current density across the diameter of the tube is roughly parabolic.[6] Two rather inaccurate experiments give a basis for estimating the total number of electrons per unit length of the tube from n_0. Comparison of the currents to a probe at the axis and a fine wire extending the full length of a radius gave the mean current density along the radius as $0.815\ i_0$. Assuming a parabolic distribution, the mean current density over the cross section is $0.72\ i_0$. In the second experiment the currents to similar probes at the axis and the walls were compared, and the value $0.325\ i_0$ for current density at the wall leads to a mean current density across the tube of $0.66\ i_0$ on the basis of a parabolic distribution. These experiments showed no systematic variation of distribution over a considerable range of pressure and current. The theoretical distribution for the case of low pressure and ionization proportional to the electron concentration gives a mean current density of $0.7\ i_0$.[7] I will use this value as it is consistent with the experimental values though the conditions do not conform to the postulates of the theory. Since v_s is constant

$$N_s = 0.7 \pi r^2 n_0$$

Where r is the tube radius.

III. COLLISIONS OF THE FIRST AND SECOND KIND

Assuming that the number of inelastic collisions is equal to the number of quanta dissipated, neglecting collisions of the second kind, equations 1 and 2 give

$$q_{1\,2} = (P-W) \Big/ \left[2.26 \times 10^{-19} N_s n_1 v_s \left(\frac{1.42}{V_0} + 1 \right) \exp\left(-\frac{1.42}{V_0} \right) \right] \quad (4)$$

Values of $q_{1\,2}$, computed on this basis from measurements at low pressures, are plotted in Figure 2 as a function of the discharge current. For pressures 0.003 and 0.0016 mm, there is a range of current, 0.5 amp and less, in which the value does not seem to depend on the current. There were 12 measurements in this range (some points in fig. 2 represent means of several measurements) which give a mean value

$$q_{1\,2} = 5.4 \pm .6 \times 10^{-16} \text{cm}^2.$$

Any dependence of $q_{1\,2}$ on V_0 was concealed by the large experimental error; but, in view of the comparatively small range of values of V_0, this value can be used to give at least a rough estimate of the number of inelastic collisions under all conditions.

Brode's [8] measurements of the scattering of electrons by cæsium atoms give a value for the effective scattering area for 1.42-volt electrons of 56×10^{-15}cm². There are no direct measurements of $q_{1\,2}$. Loveridge [9] has obtained values for sodium and potassium based on absolute measurements of the radiation excited by beam of electrons of uniform speed. His maximum value is about 1 per cent of

[6] Killian, Phys. Rev., vol. 35, p. 1238, 1930.
[7] Tonks and Langmuir, Phys. Rev., vol. 34, p. 876, 1929.
[8] R. Brode, Phys. Rev., vol. 34 p. 673, 1929.
[9] Loveridge, Univ. of Cal., Thesis, 1931.

the scattering area found by Brode. The experiment is extremely difficult and the uncertainty even greater than for the method used here. My value for cæsium is 10 per cent of the scattering area and this is an average value and not a maximum.

An examination of equation (4) shows the correlation between the curves of Figure 2 and other variables. At constant pressure the important variables are N_e and the exponential factor involving V_0. A curve of V_0 as a function of current is included in Figure 4. N_e increases roughly as the square of the current and up to 0.5 amp at 0.003 mm the decrease in V_0 compensates for the increase in N_e to keep the value of $q_{1\,2}$ constant. With higher currents, V_0 is constant and the increase in N_e gives a decrease in $q_{1\,2}$ that is entirely incompatible with the assumption that it measures a collision area. This is the case at all currents for higher pressures.

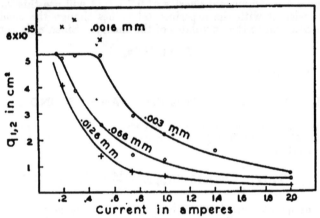

FIGURE 2.—*Values of $q_{1\,2}$ computed by equation (1) assuming that $C_{1\,2}$ is the number of quanta dissipated*

The difficulty is removed by the recognition that collisions of the second kind become increasingly important at higher currents. Figure 3 shows the data at 0.003 mm presented in accord with this idea. $C_{1\,2}$ is the number of collisions of the first kind as given by equation 1, and $C_{1\,2} - C_{2\,1}$ gives the number of quanta dissipated as given by equation 3. The two curves are nearly coincident up to 0.5 amp and above this point rapidly diverge with $C_{1\,2}$ nearly proportional to N_e. At high pressure it increases somewhat faster than N_e over the entire range of current. The number of collisions of the second kind, $C_{2\,1}$, is given by the difference between the two curves and it rises from a negligible quantity at 0.5 amp to 0.6 $C_{1\,2}$ at 1 amp, 0.8 $C_{1\,2}$ at 2 amps and 0.9 $C_{1\,2}$ at 3 amps. The predominant process at high currents is a reversible interchange of energy between the atoms and the electrons without dissipation of power.

The number of collisions of the second kind is expressed by an equation analogous to equation 1

$$C_{2\,1} = q_{2\,1} N_e n_2 v_e \qquad (5)$$

where $q_{2\,1}$ is the effective area of the excited atoms and n_2 is the number of atoms per cubic centimeter in the excited state. $q_{2\,1}$ can be computed from $q_{1\,2}$ by the principle of detailed balance which states that in thermal equilibrium between cæsium atoms and electrons $C_{2\,1}$ must equal $C_{1\,2}$. Equations (1) and (5) are applicable to the equilibrium state at a temperature $T = V_0/11,600$. In equilibrium n_2 is given by Boltzmann's equation

$$n_2 = \frac{g_2}{g_1}\, n_1 \exp\,(-V_R/V_0) = 3n_1 \exp\,(-1.42/V_0) \tag{6}$$

where g_2 and g_1 are the weights of the $2P_{1\,2}$ and $1S_1$ states, respectively. Equating (1) and (5) and eliminating n_2 by equation (6) gives

$$q_{2\,1} = \frac{g_1}{g_2}\, q_{1\,2} \left(\frac{V_R}{V_0} + 1\right) \tag{7}$$

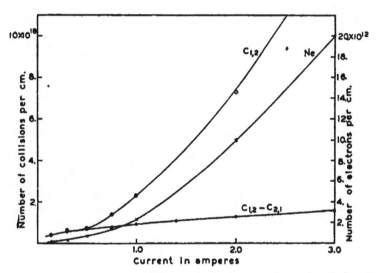

FIGURE 3.—*The number of collisions of the first kind and the number dissipated per cm of the column (left scale); the number of electrons per cm (right scale); from measurements at 0.003 mm pressure*

This is not the usual form of the Klein and Rosseland relation.[10] It is the equation applicable to all electrons of the Maxwell distribution and not to the particular speeds corresponding to V_0 and $V_R + V_0$.

Equation (5) now becomes

$$C_{2\,1} = \frac{1}{3}\, q_{1\,2} \left(\frac{V_R}{V_0} + 1\right) N_e n_2 v_e \tag{8}$$

This equation and equation (1) can be used to evaluate n_2 from the measured value of $C_{1\,2} - C_{2\,1}$. I will first give some direct experimental evidence as to the concentration of excited atoms.

[10] Klein and Rosseland, Zeits. für Phys., vol. 4, p. 46, 1921.

IV. THE NUMBER OF EXCITED ATOMS

A direct measure of the number of excited atoms can be obtained by a spectroscopic method which has been applied to measure flame temperatures. Light from an incandescent lamp is passed through the discharge and analyzed by a spectrograph. The resulting spectrum will show the cæsium absorption lines either as dark or bright lines, depending on whether the lamp temperature is above or below a critical temperature T_L, here called the reversal temperature. Hedwig Kohn [11] has shown that T_L, for a line emitted in a transition between states 2 and 1, differing in energy by $E_{2\,1}$, is related to the relative number of atoms in the two states by an equation identical in form to Boltzmann's equation

$$n_2/n_1 = (g_2/g_1) \exp\left(-E_{2\,1}/kT_L\right) \qquad (9)$$

where the g's are the weights of the two states. The only assumption is that n_2/n_1 has a definite value along the light path. In equilibrium all lines reverse at the same temperature, but in general they do not. The strength of absorption, the length of path, and the resolution of he spectrograph influence the accuracy of the setting but not the value of T_L.

Light from a tungsten strip lamp passed diametrically through the discharge tube and was focused near the axis. The image was focused by a second lens on the slit of a 3-prism Steinheil spectrograph. The brightness temperature of the image in the tube was found by measuring the apparent temperature with the tube in place and with the tube removed by means of an optical pyrometer.

Photographs of the spectra show the reversals of the $2P_{1\,2}$ doublet clearly with pressures above 0.01 mm and at higher pressures T_L for $3P_{1\,2}$ was measurable. With pressures above 0.08 mm the temperature falls in the range attainable with a strip lamp. The range was extended by the device of keeping the discharge on for only a fifth or a tenth of the exposure time. If S is the lamp temperature

$$\frac{1}{S} - \frac{1}{T_L} = \frac{\log R}{7,320}$$

where R is the ratio of total exposure to the time the discharge is on. The method was checked in the range where T_L could be obtained by the lamp, but theoretically it is open to objection. The discharge lowers the vapor density so that n_1 is not the same as with the discharge off.

At the higher pressures the line at 8,521 A showed a sharp self-reversal which did not disappear at T_L. The resolution of 8,944 was less and it disappeared over a range of temperature of 50° or 100°. The failure of the method with high resolution and the range of uncertainty with low resolution comes from the variation of n_2/n_1 across a diameter.

In Table 1 and Figure 4 reversal temperatures are expressed in terms of electron volts ($V_L = T_L/11,600$) for comparison with values of V_0 measured electrically. The experimental error in V_0 is considerably larger than the inherent uncertainty in V_L. In equilibrium $V_L = V_0$

11 Kohn, Phys. Zeits., vol. 29, p. 49, 1928.

and under actual discharge conditions V_L must be somewhat less than V_0 and the difference should decrease with increasing current and pressure. The observed differences are in the right direction, but the variation with current and pressure is not systematic. All differences fall within the range of the estimated error in V_0 so there is little basis for computing the departure from equilibrium. Reversals could not be measured at the lowest pressures where a large difference is to be expected. The small difference between values of T_L for 2P and 3P is real and shows that the number of atoms in the 3P state is further from the equilibrium value than the 2P state.

The results confirm the prediction that at moderate pressures and currents there is nearly an equilibrium between the electrons and the atoms in the first excited state. Or accepting this view they give a desirable check on the electrical method of measuring electron temperatures. Values from curves of Figure 4 (the upper one where there are two) will be taken for V_0 in the following computations.

TABLE 1.—*Electron temperature and reversal temperature in electron volts and the number of excited atoms*

Pressure	Current	V_0	V_L (2P)	V_L (3P)	$\frac{n_1}{n'_2}$ (2P)	$\frac{n_1}{n'_2}$ (3P)
mm	*Amp.*	*Volts*				
0.35	0.5	0.185	0.168	0.161	0.73	0.27
	1	.200	.180	.174	.91	.50
	2	.217	.195	.186	.96	.45
.16	.2	.175	.179	.163		
	.5	.183	.184	.171	.79	.22
	1.0	.205	.192	.180	.93	.33
	2	.220	.206	.192	.97	.37
.076	.2	.200	.189	.174		
	.5	.205	.195	.180	.81	.20
	1.0	.218	.202	.186	.93	.27
	2	.228	.211	.199	.97	.45
.032	.5	.232	.216	....	.76	....
	1	.232	.218		.92	....
	2	.234	.224		.96	....
	4	.266	.236		.97	....
.0126	.2	.257	.228			....
	.5	.257	.246		.61	....
	1	.257	.249		.87	....
	2	.257	.251		.95	....
	4	.291	.263		.96	....
.0056	.2	.298	.270			....
	.5	.283	.276			....
	1.0	.283	.288		.72	....
	2	.283	.283		.91	....

The actual values for the number of excited atoms can best be computed from the value of $C_{12} - C_{21}$ with C_{12} and C_{21} as given by equations 1 and 8.

$$C_{12} - C_{21} = q_{12} N_e v_e \left(\frac{V_R}{V_0} + 1\right)\left[n_1 \exp\left(-\frac{V_R}{V_0}\right) - \frac{1}{3}n_2 \right]$$

$$n_2 = n'_2 - 3(C_{12} - C_{21})/\left[q_{12} N_e v_e \left(\frac{V_R}{V_0} + 1\right)\right] \qquad (10)$$

where

$$n'_2 = 3n_1 \exp\left(-\frac{V_R}{V_0}\right)$$

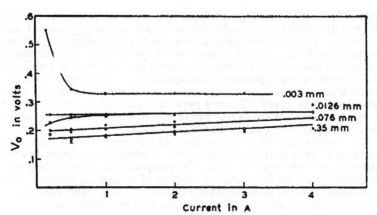

FIGURE 4.—*Measurements of electron energy V_0 shown by dots, V_L from reversal temperature of $1S-2P_{12}$ shown by circles, V_L from reversal temperature of $1S-3P_{12}$ shown by crosses*

Equation 10 gives the number of excited atoms in terms of the number that would exist in equilibrium at the electron temperature minus a term that becomes very small for high pressures and currents. Thus the fact that experimental uncertainties in the term on the

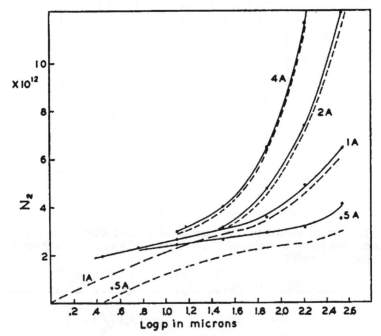

FIGURE 5.—*The number of excited atoms in the $2P_{12}$ states*

Full lines are the number that would exist in equilibrium at the electron temperature; broken lines give the actual number as estimated by equation 10.

right may become large under these conditions does not seriously affect the conclusions.

In Figure 5 the solid lines are values of n'_2 computed from V_0 as given in Figure 4. The broken lines are values of n_2 as given by equation 10. The values of n'_2 increase slowly with pressure, the equilibrium concentration n'_2/n_1 decreases. For 0.5 amp it ranges from 3.4×10^{-2} at 0.003 mm to 7×10^{-4} at 0.35 mm. Table 1 includes values of the ratio n_2/n'_2 which is a measure of the approach to equilibrium and Figure 6 shows this ratio as a function of the current. Dots are computed by equation 10 and are fairly accu-

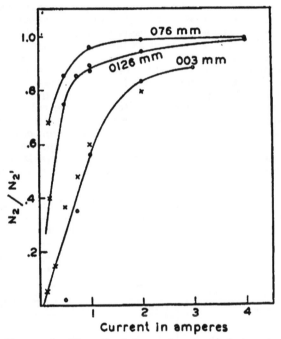

FIGURE 6.—*The ratio of the number of excited atoms to the equilibrium number measuring the approach to equilibrium as a function of current*

Dots are based on equation 10 while crosses are computed from the assumption that n_2 is proportional to the radiation intensity.

rate as the ratio approaches 1. The values become indeterminate at low current because of arbitrary assumptions in the derivation of 10. It will be shown in the following section that n_2 is roughly proportional to the intensity of the radiation, and the crosses in Figure 6 have been computed on this basis from a single high-current value of n_2 as given by equation 10. The rapid decrease in the ratio at low current comes in large part from the increase in n'_2 Table 1 includes some values of the ratio n_3/n'_3 for the $3P_{1\,2}$ state. This is derived from the tabulated values of n_2/n'_2 and the difference in T_L for the two doublets.

V. RADIATION INTENSITY AND REVERSAL TEMPERATURE

The luminous cæsium vapor is transparent except at the absorption lines and over the width of the lines it is completely opaque. The cæsium radiation may be considered as the radiation of an incandescent body at the temperature T_L which has zero emissivity except at the resonance lines. (The contribution of other lines is negligible.) The radiation per cm^2 can be expressed in the form of Wien's law:

$$J = C_1 \lambda^{-5} \exp\left(-V_R/V_L\right) \Delta\lambda$$

where λ is the mean value for the two lines and $\Delta\lambda$ is the sum of the widths of the lines. Table 2 gives values of $\Delta\lambda$ computed from measured values of T_L and of the total radiation as given in RP455. This effective width is about the same as the apparent width of the lines on the spectrograms though the dispersion was too small for an accurate comparison. It is nearly constant at constant pressure, so the increase in J with current comes from the increase in T_L. Since Wien's law and Boltzmann's law have the same form, the radiation is proportional to the number of excited atoms per cubic centimeter.

At constant current J remains nearly independent of pressure. The decrease in T_L with increasing pressure compensates for the increase in $\Delta\lambda$. The radiation does not remain proportional to n_2 for, because of the increasing opacity, the radiation comes from a smaller depth of the vapor at higher pressure. This depth can be estimated as given below.

TABLE 2.—*Effective line width and effective depth of radiating vapor*

	0.35 mm	0.16 mm	0.076 mm	0.0125 mm	0.0056 mm
0.2 ampere		8.5 A	5.6 A	1.25 A	0.55 A
0.5 ampere	20 A	9.35 A	5.64 A	1.48 A	.603 A
1.0 ampere	18.6 A	9.45 A	5.5 A	1.37 A	.625 A
2.0 amperes	17.8 A	9.0 A	5.86 A	1.52 A	.725 A
$\Delta\lambda$ mean	18.8	9.05 A	5.65 A	1.40 A	.626
Δr in cm	1.22×10^{-3}	1.42×10^{-3}	1.62×10^{-3}	2.28×10^{-3}	3.1×10^{-3}
$\dfrac{\Delta\lambda\times10^{-12}}{\Delta r \times n_1}$	2.6	2.3	2.5	2.8	1.9

The rate of radiation of atoms in the $2P_{1\,2}$ states is very closely that of a classical oscillator of the frequency of the cæsium lines [12]

$$A = \frac{8\pi^2 e^2}{3\ c\ m\lambda^2} = 2.8\times10^7 \text{ per second}$$

The radiation from n_2 atoms is then $2.8\times10^7\ n_2$ quanta per second. The measured radiation per square centimeter, divided by this number, gives the volume or the depth from which the radiation comes. Values of this effective depth, Δr, are included in Table 2 and range from 1 to 3×10^{-3} cm. The ratio $\Delta\lambda/\Delta r$ should be proportional to the number of atoms per cubic centimeter. The last row of the table shows that this proportionality holds except perhaps for the lowest pressure.

[12] Handbuch der Phys., vol. 20.

VI. DURATION OF EXCITED STATES

While the radiation life of an excited atom is $1/A = 3.56 \times 10^{-8}$ seconds, the duration of excitation within the discharge tube may be much longer than this. The radiation will be absorbed and reemitted in a distance comparable with Δr computed in the preceding section. If this experimental quantity is taken as the mean free path of a quantum of resonance radiation, then the time of diffusion of resonance radiation out of the cylindrical tube will be—

$$\tau_r = 0.75 \ (r/\Delta r)^2 \times 3.56 \times 10^{-8} = 0.015 \text{ to } 0.0025 \text{ second} \qquad (11)$$

using extreme values of Δr. This can only be right in order of magnitude because of the approximations used in the derivation of Δr.

The actual duration of the excited state in the discharge, τ, will be shorter than this because of collisions of the second kind. This time can be computed by equating the number of excited atoms to the rate of production times the duration—

$$n_2 = C_{12} \ \tau/\text{vol} \qquad (12)$$

The time between collisions τ_{21} is given by the relation—

$$1/\tau_{21} = q_{21} \ N_e v_e/\text{vol} \qquad (13)$$

Some values are given in Table 3 and it is seen that except at low currents and pressures τ and τ_{21} are nearly equal and by equation 13 inversely proportional to N_e which has a large range of values. The extreme value of $\tau = 3.8 \times 10^{-8}$ is nearly equal to the radiation life which means that in the 4 amp discharge the duration of an excited state is reduced to half its normal value.

TABLE 3.—*The duration of excited states and the time between collisions*

Pressure	Current	τ	τ_{21}	τ_0
mm 0.003	*Amperes* 2	0.88×10^{-8}	0.70×10^{-8}	3.3×10^{-8}
	1	1.84	2.65	3.7
	.5	3.24	12	4.4
	.15	4.95	90	5.2
.0126	2	.25	.264	5.0
	1	.77	.88	5.9
	.5	1.90	3.0	5.9
.076	4	.038		
	2	.098	.101	3.3
	1	.28	.30	4.0
.35	2	.094	.098	2.5
	1	.24	.26	2.9

The duration of an excited state in the absence of collisions of the second kind τ_0 is given by—

$$\frac{1}{\tau_0} = \frac{1}{\tau} - \frac{1}{\tau_{21}} \qquad (14)$$

It can be shown that $\tau_{21}/\tau = n'_2/n_2$ so that the small quantity $1/\tau_0$ can be evaluated by equation (14) in spite of experimental errors in τ

and τ_{21}. The values τ_0 which remain of the magnitude 5×10^{-6} seconds for all conditions are much smaller than the diffusion times as given by equation 11. Possibly collisions of the first kind with excited atoms are a more important factor than diffusion. However, the diffusion time and τ_0 are quantities of secondary importance in the discharge and, as the theory has been limited to first-order approximations, conclusions as to second-order effects are scarcely legitimate.

VII. KILLIAN'S MEASUREMENTS ON THE MERCURY POSITIVE COLUMN [13]

Killian has published a detaile dstudy of electrical conditions in a mercury discharge in a tube 6.2 cm in diameter with a constant current of 5 amps at three pressures. The precision is presumably much higher than that obtained with cæsium. The dissipation of power apart from recombination is ascribed by Killian to inelastic collisions at 4.9 volts. In Table 4 I present the essential data and the results of computations such as have been made for cæsium. The effective collision area q_{12}, computed by equation 1, is far from constant and I assume that the low-pressure value is correct and that collisions of the second kind are important at the higher pressures. In mercury there are three low-excitation potentials at 4.66, 4.86, and 5.43 volts, giving the states 2^3P_0, 2^3P_1, and 2^3P_2 of weights 1, 3, and 5. The assumption that the three states are in their equilibrium ratio leads to values of the number of excited atoms which seem unreasonably high and there are other good reasons to doubt that 2^3P_2 is populated in proportion to its weight. If most of the atoms are in 2 P_0 and 2 P_1, populated in the ratio of 1 to 3—

$$q_{21}=\frac{1}{4}\ q_{12}\left(\frac{4.9}{V_0}+1\right)\tag{15}$$

TABLE 4.—*Collisions of the first and second kind in mercury*

	0.0002 mm	0.001 mm	0.0058 mm
Atoms/cm³	0.61×10¹³	3.1×10¹⁴	16.1×10¹⁴
N_0	1.11×10¹⁷	2.18×10¹⁷	4.87×10¹⁷
V_0	3.27 volts	2.37 volts	1.71 volts
C_1 ʳ$-C_2$ ʟ	0.31×10¹⁶	0.96×10¹⁶	1.71×10¹⁶
q_{12}	0.68×10⁻¹⁵ cm²	(0.355×10⁻¹⁵)	(0.13×10⁻¹⁵)
q_{21}		0.52×10⁻¹⁵ cm²	0.66×10⁻¹⁵ cm²
Excited atoms per cm³, n_2		0.50×10¹³	2.48×10¹¹
Normal atoms, n_1		2.6×10¹³	13.6×10¹⁴
Duration, τ		0.97×10⁻⁴ seconds	0.905×10⁻⁴ seconds
Time between collisions, τ_{21}		2.56×10⁻⁴	1.14×10⁻⁴
Duration, τ_0		1.56×10⁻⁴	4.35×10⁻⁴

(Inclusion of 2^3P_2 changes ¼ to ⅙.) With this rather arbitrary assumption, the tabulated values for the number and duration of excited atoms are derived. n_2/n_1 is so high that n_1 is appreciably less than the total number of atoms. At the highest pressure the duration of excited states is nearly equal to the time between collisions of the second kind and the concentration n_2 has reached nearly its equilibrium value.

[13] Killian, Phys. Rev., vol. 35, p. 1238, 1930.

The computed value of τ_0, the duration in the absence of collisions, can be compared with direct measurements of the duration of 2^3P_0 and 2^3P_1 atoms. Couilette [14] has measured the diffusion time of 2^3P_0 atoms between concentric spheres 3.17 cm apart and finds the time 2.5×10^{-4} seconds at 0.0013 mm and 5×10^{-4} seconds at 0.0051 mm. The conditions are roughly comparable with Killian's conditions and the diffusion times are seen to be almost equal to the two values of τ_0. Measurements of the diffusion time of resonance radiation which give the duration of 2^3P_1 lead to values of the order of one-twentieth of this.[15]

The agreement between the computed values of τ_0 and the observed diffusion time of 2^3P_0 atoms implies a predominance of 2^3P_0 atoms in contradiction to my assumptions. Actually the number of atoms in 2^3P_0 must be more than a third of n_2 and the values for n_2, τ and τ_0 will be somewhat less than in Table 4. Apart from this minor complication, the results are consistent with the view that collisions of the second kind play an important part in the mercury discharge, even with low pressures and current densities.[16]

VIII. MEASUREMENTS OF KOPFERMANN AND LADENBURG ON THE NEON POSITIVE COLUMN [17]

Studies of the anomalous dispersion of excited neon give detailed information as to the concentration of different excited states under a wide range of conditions. Measurements were made on the transitions between the lower states s_3, s_4, s_5, and the various p states (Paschen notation). The lower states are analogous to the 2^3P states of mercury but in inverted order. The concentrations of these states approach constant values in the equilibrium ratios with a current density of about 0.1 amp/cm^3 for a pressure of 1 mm. Only at very low currents is there a marked preponderance of the metastable states as compared with the radiating state. With increasing pressure up to 9 mm the equilibrium is attained at lower currents in accord with the results with cæsium, but unlike cæsium the number of excited atoms decreases with increasing pressure. Only with the highest current density, 1.4 amp/cm^2, is the number of atoms in the higher states near the equilibrium value. The equilibrium value for the p states at 1 mm pressure corresponds to an electron energy $V_0 = 2$ volts while the s states correspond to values between 1.6 and 1.85 volts. Measurements of electron energies in the neon positive column[18] give values within this range for comparable conditions of pressure and current.

[14] Coulliette, Phys. Rev., vol. 32, p. 636, 1928.

[15] Webb and Messenger, Phys. Rev., vol. 33, p. 319, 1929.

[16] Latysceff and Leipunksy, Zeits. f. Phys., vol. 65, p. 111, 1930, give direct measurements of collisions of the second kind in mercury. The demonstration of the effect is very convincing, but their estimate of the absolute value of the probability is based on an assumed value of the number of excited atoms which is certainly far too high. Their value of the probability is much lower than that deduced from Killian's measurements. Published estimates of q_1, $_2$ cover a very wide range of values and are, in general, much less than I deduce. Killian's experimental data can not be wrong in order of magnitude and the assumptions I have used tend to give a value of q_1 $_2$; that is, too low rather than too high.

[17] Kopfermann and Ladenburg, Zeits. f. Phys., vol. 48, pp. 15, 26, 51, and 192, 1928. Zeits. f. Phys., vol. 65, p. 167, 1930.

[18] Seeliger and Hichert, Ann. der Phys., vols. 5–11, p. 817, 1931.

IX. CONCLUSIONS

In a cæsium positive column with pressures above 0.006 mm there are simple relations between the electron temperature T_e, the number of excited atoms, n_2, and the radiation intensity. The number of excited atoms in the first excited state is nearly equal to the number which would exist in equilibrium at a temperature T_e. With increasing current T_e and n_2 slowly increase and the radiation intensity remains proportional to n_2. The electron concentration and the number of collisions of the first kind increase greatly, but the number of collisions of the second kind increase at an equal rate and the only primary effect is to reduce the life of the excited states. With increasing pressure, T_e and the concentration of excited atoms decrease while n_2 gradually increases. The radiation remains nearly constant.

The studies of Kopfermann and Ladenburg of anomalous dispersion in the neon positive column and my deductions from Killian's electrical measurements in the mercury positive column indicate that the relation between the number of excited atoms and T_e are qualitatively similar for neon, mercury, and cæsium. These atoms are extremely different in structure and one can conclude that the relation is quite general. While the approach to equilibrium concentration occurs at low current for the first excited state (or close group of states) it requires high currents to approach equilibrium in the second states. Thus there is no reason to expect a distribution of intensity in the line spectrum (in a series of lines, for instance) that is simply related to the electron temperature.

WASHINGTON, June 13, 1932.

RP486

THE AREAS AND TENSILE PROPERTIES OF DEFORMED CONCRETE-REINFORCEMENT BARS

By A. H. Stang, L. R. Sweetman, and C. Gough

ABSTRACT

In order to compute the yield point and tensile strength of a deformed reinforcement bar from tensile tests, the cross-section area must be known. Four practical methods of measuring this area have been studied. The determination of the weight and length of a bar (assuming a density of 0.2833 lb./in.^{3}) and the measurement of the volume of liquid displaced by a bar of known length gave results about ten times as consistent as those obtained with a micrometer or a planimeter. The first method is considered the most practical.

Specifications for deformed bars permit the tests of bars which have been machined to a cylindrical cross section. Data are lacking as to the comparative tensile properties of machined and unmachined specimens. Tensile tests were therefore made on bars having the original lugs, bars from which the lugs had been filed and bars which had been machined to a cylindrical cross section. The results of these tests showed that the yield point and tensile strength were increased slightly by filing and more by machining, but that the differences were too small to warrant the cost of machining the specimens.

CONTENTS

I. SCOPE AND PURPOSE OF THE TESTS

Specifications [1] for deformed concrete-reinforcement bars list limiting values of the stress (lbs./in.^{2}) at the yield point and the tensile strength of the bars. When reinforcing bars are tested the load in pounds at the yield point and the maximum are observed. The stresses are then computed by dividing the load by the original cross-sectional area of the bar and compared with the values given in the specification.

---

[1] United States Government master specification for bars, reinforcement, concrete, Federal Specifications Board Specification No. 350a. American Society for Testing Materials standard specifications for billet-steel concrete reinforcement bars, A. S. T. M. designation: A-15-13 and standard specifications for rail-steel concrete reinforcement bars, A. S. T. M. designation: A-16-14; A. S. T. M. standards, 1930, Pt. I, Metals, pp. 131 and 135, respectively.

The specifications are drawn up for the purpose of insuring that the bars shall have adequate strength, that they shall be made of a suitable quality of material, and, in addition, that the purchaser shall not be required to pay for an undue amount of overweight in delivery. The insuring of adequate strength does not require the specification of the "yield point" or "tensile strength" of the material in the bars. This purpose would be equally well, if not better, served by making use of the principles of specifications already applied to manila and wire rope, specifying for each nominal size a "minimum yield load" and a "minimum tensile load." That, in fact, is the form in which the designer will use the specified values, regardless of the particular manner in which they are expressed in the specifications.

II. THE SPECIMENS

Deformed bars used in this country vary in size from ¼ to 1¼ inches. One-half inch and 1-inch bars, both round and square, were selected for these tests in which four types of deformation and three grades of steel were used. (See Table 1.) The rail steel bars were donated by the Buffalo Steel Co. The other bars were purchased from commercial stock.

Twenty-four bars were used in the tests. Beginning at one end of a bar, 12 test specimens were cut in sequence, A1, B1, C1, D1; A2, B2, C2, D2; A3, B3, C3, D3. The A specimens were 12 inches long, the others 20 inches. They were used as follows:

Type A, for area and density determination.
Type B, for tensile test, as received.
Type C, for tensile test after the lugs had been filed, by hand, from the middle 9-inch length.

TABLE 1.—*The bars, description and density*

Type of deformation	Grade of steel	Size	Shape	Density
		Inch		*Lbs./in.³*
Corrugated	Intermediate, billet	½	Round	0.2825
		½	Square	.2826
		1	Round	.2827
		1	Square	.2829
Diamond	....do	½	Round	.2834
		½	Square	.2825
		1	Round	.2823
		1	Square	.2827
	Structural, billet	½	Round	.2827
		½	Square	.2826
		1	Round	.2825
		1	Square	.2826
Havemeyer	Intermediate, billet	½	Round	.2823
		½	Square	.2822
		1	Round	.2822
		1	Square	.2827
	Hard, billet	½	Round	.2824
		½	Square	.2827
		1	Round	.2825
		1	Square	.2827
Rail	Hard, rail	½	Round	.2820
		½	Square	.2823
		1	Round	.2824
		1	Square	.2825

Type D, for tensile test after having been machined for the middle 9-inch length to the largest cylindrical section possible with removal of all scale.

III. DETERMINATION OF AREA

The area determinations by the different methods were made only on the type A specimens of the rail steel bars. These specimens had relatively small deformations and were more regular in cross section than some of the other bars.

1. WEIGHT-LENGTH METHOD

By this method the area generally is calculated from the weight of a measured length of the bar on the assumption that a steel bar 1 square inch in section and 1 foot long weighs 3.400 pounds, or 0.2833 lb./in.^{3} This value for the density is given in all American handbooks. To check the accuracy of this value, the density of one specimen of each type of deformation, grade of steel, and size was determined by the capacity and density section of the bureau. These densities, given in Table 1, are all within one-half of 1 per cent of the nominal density of 0.2833 lb./in.^{3} The average density, 0.2825 lb./in.^{3}, is less than 0.3 per cent from the nominal value.

The weight-length method is obviously an indirect volume measurement and the areas obtained represent average rather than minimum cross sections.

For this investigation each of three observers measured the length of the bars with a steel scale graduated to 0.01 inch and weighed them on an equal arm balance to the nearest 0.001 pound. Areas were then calculated by the following formula

$$A = \frac{W}{\rho L}$$

where
    A is the area, square inch.
    W is the weight, pound.
    L is the length, inch.
    ρ is the density, pounds per cubic inch.
The areas calculated by this method are given in Table 2. This table also lists the deviation of each observation from the average.

2. IMMERSION METHOD

For this method the specimens were immersed in denatured alcohol contained in a cylindrical glass graduate and the areas given in Table 2 were computed by the formula

$$A = \frac{V_1 - V}{16.39 L}$$

where
    A is the cross-sectional area of the specimen, square inch.
    V is the graduate reading before the specimen is immersed, cubic centimeter.
    V_1 is the graduate reading after the specimen is immersed, cubic centimeter, and
    L is the length of the specimen.
The length, L, was determined in the same way as for the weight-length method. This method has been recommended by Scheirer.[2]

---

[2] Accurate Method for Determining Actual Areas for Deformed Steel Reinforcing Bars, M. K. Scheirer, Concrete, vol. 33, p. 24, September, 1928.

TABLE 2.—*Cross-sectional area of deformed bars*

[Deformed rail steel concrete-reinforcement bars, type A specimens]

Shape	Speci-men No.	Observer	Weight-length method		Immersion method		Micrometer method		Planimeter method	
			Area	Deviation	Area	Deviation	Area	Deviation	Area	Deviation
			Square inch	*Per cent*	*Square inch*	*Per cent*	*Square inch*	*Per cent*	*Square inch*	*Per cent*
Round..	1	1	0.1904	0.11	0.1898	0.05	0.1901	2.87	0.183	1.67
		2	.1899	.16	.1898	.05	.1855	.38	.180	.00
		3	.1903	.05	.1900	.05	.1787	3.30	.177	1.67
		Average	.1902		.1899		.1848		.1800	
.do....	2	1	.1907	.16	.1911	.00	.1909	1.49	.190	1.77
		2	.1902	.11	.1910	.05	.1847	1.81	.183	1.98
		3	.1904	.00	.1911	.00	.1886	.27	.187	.16
		Average	.1904		.1911		.1881		.1867	
.do....	3	1	.1886	.05	.1890	.21	.1901	2.04	.183	1.10
		2	.1883	.11	.1895	.05	.1862	.00	.177	2.31
		3	.1885	.00	.1896	.11	.1825	2.04	.183	1.10
		Average	.1885		.1894		.1863		.1810	
Square..	1	1	.2559	.04	.2560	.12	.2549	.12	.250	1.92
		2	.2559	.04	.2568	.20	.2539	.27	.243	.94
		3	.2561	.04	.2561	.08	.2540	.12	.243	.94
		Average	.2560		.2563		.2546		.2453	
.do....	2	1	.2555	.16	.2559	.16	.2549	.12	.250	.93
		2	.2546	.20	.2554	.04	.2539	.27	.243	1.90
		3	.2552	.04	.2551	.16	.2540	.12	.250	.93
		Average	.2551		.2555		.2546		.2477	
.do....	3	1	.2541	.04	.2553	.04	.2529	.28	.250	.00
		2	.2539	.04	.2550	.08	.2539	.12	.250	.00
		3	.2541	.04	.2553	.04	.2539	.12	.250	.00
		Average	.2540		.2552		.2536		.2500	
Round..	1	1	.764	.04	.767	.00	.754	.40	.748	.04
		2	.764	.04	.767	.00	.757	.00	.737	.85
		3	.765	.09	.767	.00	.760	.40	.750	.90
		Average	.7643		.7670		.7570		.7438	
.do.	2	1	.765	.00	.768	.04	.776	.82	.757	.17
		2	.765	.00	.768	.04	.762	1.00	.753	.36
		3	.765	.00	.769	.09	.771	.17	.757	.17
		Average	.7650		.7683		.7697		.7557	
.do...	3	1	.766	.00	.768	.13	.767	.86	.760	.30
		2	.766	.00	.769	.00	.760	.04	.753	.62
		3	.766	.00	.770	.13	.754	.83	.760	.30
		Average	.7660		.7690		.7603		.7577	
Square..	1	1	.979	.07	.979	.03	.967	.21	.970	.45
		2	.978	.03	.978	.07	.961	.41	.967	.13
		3	.978	.03	.979	.03	.967	.21	.960	.59
		Average	.9783		.9787		.9650		.9657	
.do...	2	1	.975	.07	.975	.03	.963	.13	.957	.28
		2	.974	.03	.975	.03	.959	.55	.943	1.18
		3	.974	.03	.974	.07	.971	.69	.963	.91
		Average	.9743		.9747		.9643		.9543	
do.	3	1	.981	.00	.981	.07	.973	.28	.960	.34
		2	.981	.00	.981	.07	.967	.34	.960	.34
		3	.981	.00	.983	.13	.971	.07	.953	.49
		Average	.9810		.9817		.9703		.9577	
		Grand average		.05		.07		.64		.77

The bars were held by a small copper wire, the immersed volume of which was less than $1/20$ cm^3. Alcohol was used rather than water because, due to its lower surface tension, the level of the liquid in the graduate could be more accurately determined. It also wets the steel bar so that air bubbles are not trapped on the surface as with water. The calibrated graduate used with the ½-inch bars was graduated to 0.5 ml and that used with the 1-inch bars to 1.0 ml. On both readings were estimated to 1/10 ml.

3. MICROMETER METHOD

The minimum areas were calculated from measurements with a micrometer caliper of the sides of the square bar or the diameter of round bars. Two measurements to the nearest 0.001 inch were taken on each side of two sides (or of two diametral planes perpendicular to each other) and the four measurements were averaged. Correction was made for the rounded corners of the square bars, the radii of these corners being measured with radius gages. The areas are given in Table 2.

4. PLANIMETER METHOD

The smooth end of a bar was impressed on a paper and the area of the impression measured with a polar planimeter. It is customary in some laboratories to ink the end of the bar, and force it, using a testing machine, against the paper which may be backed up with blotting paper or some other relatively soft substance. Preliminary studies with round and square machined specimens showed that this procedure gave areas smaller than the area computed from the dimensions. This difference was probably due to the soft backing material.

The method finally adopted was as follows: The end of the bar was finished smooth, care being taken to avoid rounding the corners. This surface was inked with an inking pad, the surplus ink around the edges wiped off, and several impressions were made on buff detail paper. The weight of the specimen, 12 inches long, supplied the impressing force. The second or third impression was usually better than the first due to the smaller amount of ink on the end of the bar. This method resulted in clear cut perimeters.

The areas were measured with a Coradi polar planimeter with an arm length such that each division of the vernier amounted to 0.01 in.^{2}. The mean of three measurements by each observer is given in Table 2.

5. COMPARISON OF THE DIFFERENT METHODS

The deviations given in Table 2 are measures of the variation in the areas as determined by different observers and by different methods. If the average of the deviations for one method is smaller than for another method, it is probable that the first method will give more consistent values for different observers than the second. On this basis, the weight-length method is best and then come, in order, the immersion method, the micrometer method, and finally the planimeter method.

This rating is supported by a consideration of the observational errors involved. If three sets of observations made by the weight-length method are not in error more than plus or minus 0.01 inch in

12 inches as regards length and not more than plus or minus 0.001
pound in 0.668 pound as regards weight (⅝-inch round bar), then the
average departure of an individual determination from the mean of
three should never exceed 0.2 per cent. For the immersion method,
with the same uncertainty as regards length and a possible error of
plus or minus 0.1 ml in 38.7 ml in measuring the displacement, the
average departure should not exceed 0.3 per cent; while in the planimeter method with an uncertainty of 0.01 in.^{2} in the measurement of
an area of only 0.2 in.^{2}, the average departure of three determinations
from the mean may reach 4.3 per cent. While these percentages are
limiting values (not the most probable average deviation) they should
conform approximately in relative magnitude with the grand averages of the observed deviations in Table 2. Considering the methods
in the order discussed, the computed relative magnitudes of the
deviations for the three methods are 1:1.5:21, while the relative
magnitudes of the corresponding observed deviations are 1:1.4:15.
A similar analysis of the micrometer method was not attempted.
Deformed bars do not have simple regular sections, and the cross
sectional area may vary from point to point along the bar, so that
the average deviation depends more upon the location of the measuring stations on the bar than upon the precision of the micrometer
readings.

Current specifications permit the deliveries of reinforcement bars
one-half inch in diameter or larger which deviate not more than plus
or minus 7.5 per cent from the nominal area. If the nominal area is
used in the determination of yield point, tensile strength and elongation from the values observed during a tensile test, errors of this
order of magnitude may be introduced in the results. If an accuracy
of this order of magnitude is sufficient to insure that the material is
of suitable quality without an undue percentage of rejections of
borderline material, it would be simplest and most economical to
require that yield points and tensile strengths for specification purposes be calculated on the basis of the nominal areas, which are used
by the engineer in his design calculations.

If an accuracy of this order is not considered sufficient to insure
that the material is of suitable quality without undue rejections of
borderline material, the weight-length method of area measurement
is undoubtedly to be preferred to any of the others. It requires no
special apparatus, the measurements are easy to make, and the errors
(maximum plus or minus 0.2 per cent) are much below those tolerated
(plus or minus 1 per cent) in testing machines.

IV. THE TENSILE TESTS

1. METHOD OF TESTING

The ½-inch bars were tested in a testing machine having
50,000 pounds capacity and the 1-inch bars in a testing machine
having a capacity of 100,000 pounds. The specimens were held
with wedge grips, well lubricated with grease and graphite on the
sliding surfaces. The rate of separation of the heads of both machines
under no load, was 0.4 inch per minute. The yield point was determined by the drop of beam. All specimens had definite drop of
beam yield points.

2. CALCULATION OF RESULTS

The yield point and tensilve strength values were calculated from the test loads for the different types of tensile specimens by dividing by the area found as follows:

Type B specimens tested as received.—The areas were determined by the weight-length method.

Type C specimens with lugs filed off.—The specimens were weighed and their length measured. The fractional length with lugs left on was then determined from a count of the total number of original lugs and of the lugs left on. The weight of this length was determined from the calculated weight of an equal length of the corresponding A specimens. The difference between the total weight of the C specimen and the weight of the fractional length with lugs left on gave the weight of the fractional length from which the lugs had been filed. The area was then calculated by the weight-length method.

Type D specimens, machined to a cylindrical section.—The diameter was measured to the nearest 0.001 inch with a micrometer caliper and the area calculated from this diameter.

The elongation in 8 inches was measured with dividers and a steel scale to 0.01 inch. For the few specimens which broke outside of the middle half of the gage length, the elongation was computed according to the method outlined in A. S. T. M. tentative specification, serial designation: E8-27T.[3]

3. AVERAGE RESULTS

The yield-point and tensile-strength values for the three specimens of each type from each bar were very consistent. The elongation values varied more. The greatest variation of a single value from the mean for each of the 72 groups of three specimens was less than 3½ per cent for the yield point values and less than 2½ per cent for the tensile strengths (except for one group). For 54 of the groups the maximum deviation of the elongation values was less than 5 per cent. The average results are given in Table 3, each value for yield point, tensile strength and elongation being the average for the three similar specimens from each bar.   •

---

[3] Tentative Methods of Tension Testing of Metallic Materials, Proc. A. S. T. M., vol. 27, Pt. I, p. 1078.

TABLE 3.—Average results of tensile tests of concrete reinforcement bars

Type of deformation	Grade of steel	Size of bar (inch)	Shape	Yield point Type B as received (Lbs./in.²)	Yield point Type C lugs filed off (Lbs./in.²)	Yield point Type D cylindrical section (Lbs./in.²)	Ratio of yield points C/B	Ratio of yield points D/B	Tensile strength Type B as received (Lbs./in.²)	Tensile strength Type C lugs filed off (Lbs./in.²)	Tensile strength Type D cylindrical section (Lbs./in.²)	Ratio of tensile strengths C/B	Ratio of tensile strengths D/B	Elongation in 8 inches Type B as received (P. ct.)	Elongation in 8 inches Type C lugs filed off (P. ct.)	Elongation in 8 inches Type D cylindrical section (P. ct.)	Ratio of elongation C/B	Ratio of elongation D/B
Corrugated	Intermediate billet	½	Round	48,300	49,300	49,300	1.02	1.02	72,200	73,100	73,900	1.01	1.02	23.7	23.3	24.6	0.98	1.04
		½	Square	43,700	44,300	44,900	1.02	1.03	73,300	75,300	75,500	1.03	1.04	21.5	21.8	22.3	1.11	1.06
		1	Round	48,500	48,800	51,500	1.01	1.06	82,600	82,800	84,200	1.00	1.02	23.7	22.6	22.3	.95	.94
		1	Square	39,600	40,200	40,200	1.02	1.02	70,100	71,100	73,000	1.01	1.04	26.7	27.6	26.9	1.03	1.01
Diamond	Intermediate billet	½	Round	46,100	47,100	47,500	1.02	1.03	70,400	71,500	72,900	1.02	1.04	24.7	24.5	22.2	.99	.90
		½	Square	55,200	55,100	56,300	1.02	1.03	83,300	84,500	85,800	1.01	1.03	22.4	21.7	19.5	.97	.87
		1	Round	50,300	50,300	53,100	1.02	1.06	72,200	72,600	73,100	1.01	1.02	24.3	23.8	21.2	1.00	1.00
		1	Square	39,700	40,100	42,700	1.01	1.06	68,700	69,600	70,300	1.01	1.02	28.2	28.2	28.4	.97	.97
Havemeyer	Structural billet	½	Round	41,700	42,300	43,600	1.01	1.05	57,500	57,000	59,500	1.00	1.04	29.9	27.9	28.1	.93	.94
		½	Square	41,900	43,700	42,100	1.04	1.05	59,300	60,700	61,300	1.01	1.01	31.8	27.0	27.2	1.20	1.21
		1	Round	38,500	38,100	38,600	1.02	1.01	58,500	58,800	59,200	1.01	1.02	31.9	31.7	31.5	1.01	1.01
		1	Square	38,400	39,100	40,500	1.01	1.06	60,600	61,000	61,700	1.01	1.03	31.9	31.7	30.5	.99	.96
Havemeyer	Intermediate billet	½	Round	52,400	52,500	52,300	1.00	1.00	77,000	77,800	78,200	1.01	1.02	22.8	22.0	21.9	.97	.98
		½	Square	48,400	50,100	49,000	1.03	.99	70,900	71,100	72,100	1.01	1.01	26.1	24.2	24.2	.97	.93
		1	Round	47,600	47,600	48,200	1.00	1.01	77,700	78,400	78,600	1.01	1.01	26.4	24.2	25.3	1.02	.99
		1	Square	43,500	45,100	46,500	1.04	1.06	73,300	75,100	75,200	1.02	1.02	26.4	26.6	25.9	1.01	.98
Rail	Hard billet	½	Round	53,200	53,000	52,300	1.00	.99	80,000	81,600	82,000	1.01	1.04	20.1	21.6	21.5	1.07	1.07
		½	Square	52,400	53,100	53,700	1.01	1.02	89,200	89,900	91,600	1.00	1.02	16.9	16.5	17.4	.98	.98
		1	Round	51,800	51,600	51,000	1.00	.99	88,200	88,200	88,200	1.01	1.02	22.5	21.2	22.1	.94	1.03
		1	Square	50,300	50,700	51,200	1.01	1.02	86,000	86,500	87,300	1.01	1.02	21.3	20.7	21.9	.97	.98
Rail	Hard rail	½	Round	59,100	59,500	60,800	1.01	1.03	92,300	92,700	95,600	1.00	1.04	15.3	16.3	17.0	1.07	1.11
		½	Square	66,200	65,900	66,900	1.00	1.01	100,100	100,300	102,200	1.00	1.02	13.8	15.9	17.6	1.15	1.28
		1	Round	57,200	57,300	58,900	1.00	1.03	98,100	98,700	99,900	1.01	1.02	18.6	19.5	18.3	1.03	.98
		1	Square	48,800	49,000	49,000	1.00	1.01	90,400	81,100	82,200	1.01	1.02	23.2	21.8	22.8	1.00	.98

4. THE EFFECT OF FILING OFF THE LUGS, ON THE TENSILE STRENGTH

The average maximum load in pounds of the type C specimens with lugs filed off was less in all cases than the average maximum load in pounds of the type B specimens which were tested as received. The areas based upon weight-length measurements were also less,·but Table 4 which lists the decrease in maximum load and area of the type C specimens with respect to the type B specimens shows that the decrease in area was in all cases greater than the decrease in maximum load. The lugs were therefore effective in resisting tensile forces to some extent. An example will be used to show how the relative effectiveness of the area of the lugs in tension as given in Table 4 has been calculated.

Example.—½-inch round bars, Havemeyer type of deformation, intermediate grade of billet steel:

Type of specimen	Average area	Average tensile load	Average tensile strength
B, as received............	*Square inch* 0.1935	*Pounds* 14,910	*Lbs./in.* [2] 77,000
C, lugs filed off............	.1880	14,630	77,800
B-C......................	.0055	280	....

Tensile strength of lug area, $\dfrac{280}{0.0055} = 50,900 \ lbs./in.$ [2]

$$\frac{\text{Tensile strength of lug area}}{\text{Tensile strength for Type C}} = \frac{50,900}{77,800} = 0.65$$

It may, therefore, be concluded that the lug area of these bars was 65 per cent effective in resisting tensile stresses. Similar values for all bars are given in Table 4. Since these values depend on small differences between relatively large numbers, the results are somewhat variable. They show, however, that the lug area of the Havemeyer and rail steel bars were more than 50 per cent effective in resisting tensile forces. These lugs were parallel to the axis of the bars. The corrugated and diamond deformations were not parallel to the axis of the bars and the lug area was less effective in resisting tensile forces than for the other types. Since, however, the lug areas in the Havemeyer bars were always much larger than in the other types of bars (see Table 4) the effectiveness of the total area in resisting tensile load was practically the same for all types of bars as shown in C/B ratios of Table 3.

TABLE 4.—*Effect of filing off lugs on the maximum load and on the area*

Type of deformation	Grade of steel	Nominal size	Shape	Decrease in maximum load	Decrease in area	Relative effectiveness of area of lugs in tension
		Inch		*Per cent*	*Per cent*	*Per cent*
Corrugated	Intermediate	½	Round	0.9	2.1	44
		½	Square	1.0	3.7	27
		1	Round	.6	.9	73
		1	Square	.7	2.0	33
Diamond	do	½	Round	.4	1.8	20
		½	Square	.9	2.4	39
		1	Round	.9	1.4	63
		1	Square	1.1	2.5	44
Havemeyer	Structural	½	Round	2.6	2.7	95
		½	Square	6.1	8.3	72
		1	Round	1.7	2.7	63
		1	Square	4.4	5.0	88
	Intermediate	½	Round	1.9	2.8	65
		½	Square	5.3	6.2	86
		1	Round	2.4	2.7	92
		1	Square	4.3	6.4	67
	Hard	½	Round	2.1	2.9	71
		½	Square	5.4	6.1	88
		1	Round	3.1	3.2	96
		1	Square	3.6	4.3	85
Rail	do	½	Round	1.8	2.2	80
		½	Square	2.3	2.4	93
		1	Round	1.0	1.5	63
		1	Square	1.1	2.0	54

5. COMPARISON OF TENSILE PROPERTIES OF UNMACHINED AND OF MACHINED BARS

Table 3 gives the average results of the tensile tests. The tensile strength of type D specimens machined to cylindrical cross section was in all cases greater than that of the type C specimens, with lugs filed off. The increase in average strength of the type D specimens over the type B specimens was, however, in all cases less than 5 per cent.

The same general conclusions can be drawn from the yield point values although in two instances the average yield point for the type D specimens was 1 per cent below the value for the type B specimens.

The effect of type of specimens on the elongation was erratic. On the average, the elongation was not much affected by filing off the lugs or machining the bar. The average elongation in 8 inches was 23.7 per cent for the bars tested as received, 23.9 per cent for the bars from which the lugs had been filed and 23.7 per cent for the bars which were machined to a cylindrical cross section.

Table 3 shows that the tensile properties are not influenced by the size or shape of the bar, or by the type of deformation.

The small difference in the tensile properties of machined and unmachined specimens leads to the suggestion that the use of the expensive machined specimen be eliminated from the specifications. If the use of the original sections of deformed bars were made mandatory for tensile tests, one point of controversy would be eliminated.

The results listed in Table 3 show only two border-line cases in which specimens which failed to comply with the requirements of the specification as to yield point or tensile strength when tested as received, did comply when machined. These are the 1-inch square corrugated and diamond bars. For the intermediate grade of steel, the specified minimum yield point and tensile strength values are 40,000 and 70,000 lb./in.², respectively.

6. DETERMINING WHETHER A SPECIMEN CONFORMS TO THE REQUIREMENTS OF THE SPECIFICATION

By chance, the tensile strength of the 1-inch square rail steel specimen B–3, tested as received, affords an example of the difficulties of determining whether a specimen conforms to the requirements of the specification, when no method of area determination has been specified. Table 5 gives data as to the breaking load of this bar, of the area as determined by different methods (average results from Table 2 have been used), and of the tensile strength, when computed with regard to the different areas.

In this case, the values given in Table 5 show that the tensile strength computed by the nominal area, weight-length, and immersion methods, did not conform to the requirements of the specification. The tensile strength, computed from areas determined by the micrometer and planimeter methods, did comply with the specified requirement. In border-line cases such as this, it is apparent that much confusion results and honest differences of opinion arise when the method of area determination is not specified. The necessity of specifying some method of area determination is obvious.

TABLE 5.—*The tensile strength of rail steel specimen B–3, 1 inch square, tested as received*

[Breaking load, 77,970 pounds]

Method of determining area	Area	Tensile strength
	Square inch	*Lbs./in.²*
Nominal area	1.000	77,970
Weight-length	.9780	79,700
Immersion	.9767	79,800
Micrometer	.9703	80,300
Planimeter	.9577	81,400
Specified minimum		80,000

V. CONCLUSIONS

1. Area determinations by different observers of deformed concrete-reinforcement bars, one-half and 1 inch in size, both round and square in section were made by four different methods. The conclusions reached as a result of this study are as follows:

(a) If specifications for concrete-reinforcing bars specify the yield point and tensile strength in pounds per square inch of the material in the bars, the method of determining the area to be used in calculating them should be definitely specified.

(b) The nominal areas which are used by the engineer in his design calculations should be used in computing yield point and tensile strength values from the test results if the accuracy of the nominal area is considered sufficient. Since current specifications permit the deliveries of reinforcement bars one-half inch in diameter or larger which deviate not more than plus or minus 7½ per cent from the nominal area, errors of this order of magnitude may be introduced in the results.

(c) If accuracy of the order of plus or minus 7.5 per cent in the determination of yield point, tensile strength, and elongation of concrete reinforcement bars is not sufficient to insure that the material is of suitable quality, without an undue percentage of rejections of border-line material, the specifications should require the use of the weight-length method.

(d) The immersion method under ordinary laboratory conditions gives no greater accuracy than the weight-length method, and is much less easy to apply.

(e) The micrometer method applied to deformed bars gives much less accurate results than either the weight-length or the immersion method.

(f) The planimeter method is costly and time consuming. With the usual planimeters and within the usual range of sizes it gives much less accurate results than either the weight-length or the immersion method.

2. Tensile tests were made of deformed concrete-reinforcement bars, of three grades of steel, four types of deformation, one-half and 1 inch in size, both round and square in section. Each bar was tested (1) in the original condition, (2) after the lugs had been filed by hand from the middle 9-inch length, and (3) after the bars had been machined to a cylindrical cross section for the middle 9-inch length. The conclusions reached as a result of the tensile test are as follows:

(a) Removing the lugs from deformed concrete-reinforcement bars decreased slightly the load (pounds) carried by the bars at the yield point and at failure. The percentage decrease in load, however, was in no case as large as the percentage decrease in the average area as determined by the weight-length method, so that the yield point (pounds per square inch) increased from 0 to 4 per cent and the tensile strength (pounds per square inch) from 0 to 3 per cent, with no significant difference between the different types of bars.

(b) There was, however, a marked difference in the manner in which the different types of lugs contributed to the strength of the bar. The area of the Havemeyer lugs was approximately 50 per cent effective in resisting tensile stresses, while the areas of the corrugated and diamond type lugs were less effective.

(c) Machining deformed concrete-reinforcement bars to a cylindrical cross section increased the tensile strength (pounds per square inch) over the tensile strength of similar bars tested with the lugs on. The increase was, however, in no case as large as 5 per cent.

(d) The yield point (pounds per square inch) of deformed concrete-reinforcement bars machined to a cylindrical section was in general greater than the yield point of similar bars not machined. The maximum increase was 8 per cent.

(e) The effect of machining deformed concrete-reinforcement bars on the elongation was small and erratic. Sometimes the machined bars showed higher elongation and sometimes lower. The maximum difference for the elongation of unmachined and machined bars was less than 5 per cent in 8 inches. In the specimens tested, the difference was in no case sufficient to cause rejection of the material in the unmachined condition.

(f) These differences seem to be altogether too small to warrant the cost of machining concrete-reinforcement bars when testing under specifications.

WASHINGTON, August 1, 1932.

A CALORIMETRIC METHOD FOR DETERMINING THE INTRINSIC ENERGY OF A GAS AS A FUNCTION OF THE PRESSURE

By Edward W. Washburn

ABSTRACT

If a known mass of gas compressed in a bomb at a known pressure be immersed in a calorimeter and allowed to expand slowly to atmospheric pressure, the cooling effect can be compensated by electrical heating. Since the external work is accurately known, a precision discussion shows that the method should yield an accurate value ($\pm < 1$ cal. for a liter bomb) for the accompanying change, ΔU, in the internal or intrinsic energy of the gas. In order to obtain reliable results, the energy effects associated with the changes in stress on the walls of the bomb must be corrected for or reduced to a negligible amount by proper bomb design. From the standpoint of simplicity, rapidity, and accuracy the method is probably superior to any other means of determing this quantity.

CONTENTS

I. NOMENCLATURE

B Barometric pressure.
C_p Heat capacity at constant pressure.
j Joule.
M Molecular weight.
m Mass.
p Pressure.
Q Quantity of heat.
R Gas constant per mole.
T Absolute temperature.
t Centigrade temperature.
U Intrinsic energy.
v Volume.
W Work.
α Linear coefficient of thermal expansion.
μ Joule-Thomson coefficient.

II. INTRODUCTION

For any system whose thermodynamic condition is determined solely by temperature and pressure, the first law of thermodynamics assumes the form .

$$dU = dQ - pdv \tag{1}$$

in which U is the so-called "total," "internal," or "intrinsic" energy of the system, Q is the heat added to it, p is the pressure upon it, and v is its volume.

For any elastic system the above relation when combined with the second law yields the familiar thermodynamic equation

$$-\left(\frac{\partial U}{\partial p}\right)_T = p\left(\frac{\partial v}{\partial p}\right)_T + T\left(\frac{\partial v}{\partial T}\right)_p \tag{2}$$

in which T is the absolute temperature.

The integral of this equation, $U_T = f(p)$ for a gaseous system at the temperature T, is the quantity sought. Aside from its general thermodynamic importance this quantity is of practical interest in connection with bomb calorimetry where a knowledge of its value for oxygen and for mixtures of oxygen with carbon dioxide is needed in order to evaluate the changes in the heat content of these gases with the pressure.

III. METHODS OF EVALUATING $U_T = f(p)$

1. THE p-v-T METHOD

By operating upon an appropriate equation of state, the right-hand member of equation (2) can be obtained as a pressure function with numerical coefficients and the equation can then be integrated. In practice this method suffers from the disadvantage that there are very few gases for which p-v-T data are known with sufficient accuracy, and the acquisition of such data is a difficult and tedious undertaking.

In the absence of more exact information, it is however, convenient to remember that as a first approximation for moderate pressures the following equation, obtained by operating upon the Beattie-Bridgeman equation of state, [1] may be employed

$$-\Delta U\Big]_0^p = \left(\frac{A_0}{RT} + \frac{3c}{T^3}\right)p \tag{3}$$

in which A_0 and c are constants appearing in the equation of state.

2. THE ADIABATIC JOULE-THOMSON METHOD

In this method $\left[\dfrac{\partial(pv)}{\partial p}\right]_T$ is evaluated as a pressure function with the aid of measured pv values at the temperature T or, failing these, with the aid of an equation of state, and $\left(\dfrac{\partial U}{\partial p}\right)_T$ is obtained from the thermodynamic relation

$$-\left(\frac{\partial U}{\partial p}\right)_T = \mu C_p + \left[\frac{\partial(pv)}{\partial p}\right]_T \tag{4}$$

[1] J. A. Beattie, Phys. Rev., vol. 32, p. 699, 1928.

in which μ is the Joule-Thomson coefficient and C_p the heat capacity at constant pressure. Both μ and C_p are pressure functions.

This method also suffers from the disadvantage that reliable μ values are available only for two or three gases and the technic of obtaining them is exceedingly difficult. From the standpoint of attainable accuracy, however, the method is perhaps superior to the p–v–T method, especially at temperatures remote from room temperature.

3. THE ISOTHERMAL JOULE-THOMSON METHOD

This method differs from the preceding one only in that the product, $\mu\,C_p$, is determined calorimetrically by supplying measured electrical energy at such a rate as to compensate for the temperature drop which occurs in the classical Joule-Thomson experiment. The experimental technic of this "isothermal porous-plug experiment" has only recently been developed to the point where the method gives promise of yielding valuable and reliable results.[2]

4. THE DIRECT CALORIMETRIC METHOD

Since ΔU is an energy quantity, the method of measuring it directly with the aid of a calorimeter would at first sight appear to be the most convenient and accurate one to employ. With the exceptions noted below, however, attempts to employ this method have failed to give results of sufficient accuracy. The chief difficulty appears to arise from the energy effects associated with the changes in stress on the walls of the containing vessel.

One simple method of carrying out the calorimetric determination is to determine

$$Q = \int_p^B \left(\frac{\partial U}{\partial p}\right)_T dp + W = \Delta U \rbrack_B^p + B\Delta v \qquad (5)$$

where $B\Delta v$ is the work done against the atmosphere when the gas contained in a bomb is allowed to expand isothermally from the initial pressure p to the barometric pressure B, or in general to any low pressure P such that for all lower pressures the integral $\int_p^0 \left(\frac{\partial U}{\partial p}\right)_T dp$ is calculable or is negligibly small.

In this form the method has been used by Douglas[3] to determine the total energy change accompanying the evaporation of a liquefied gas, and by Bennewitz and Andreewa[4] to determine U as a function of p for gases in the neighborhood of the critical point. In both of these investigations the value of ΔU was a comparatively large quantity, and great refinement of calorimetric technic was not required in order to obtain a fair percentage accuracy. For temperatures far above the critical temperature, however, ΔU is a small quantity, calorimetric technic of high precision is required and the contribution of the containing vessel to the total heat effect can not be ignored.

[2] See (ª) F. G. Keyes and S. C. Collins, Proc. Nat. Acad. Sci., vol. 18, p. 328, 1932; (ᵇ) A. Eucken, K. Clusius, and W. Berger, Z. tech. Physick, vol. 13, p. 267, 1932.
[3] Douglas, Phil. Mag., vol. 18, p. 159, 1909.
[4] Bennewitz and Andreewa, Z. physik. Chem., vol. 142A, p. 37, 1929.

IV. THE CALORIMETRIC TECHNIC

The gas to be investigated, compressed into a suitable bomb, is immersed in a constant temperature bath at the temperature T. When temperature equilibrium has been established, the pressure p in the bomb is measured, the valve is closed, the bomb is removed from the bath and weighed to determine the mass of the contained gas. The bomb is then transferred to the calorimeter and its outlet connected to a long coil of metal tubing which is likewise immersed in the calorimeter. When equilibrium has been established in the calorimeter, observations are started and the time-temperature fore-period accurately established. The valve of the bomb is now opened gradually and the gas escapes through the long coil, which is so designed as to bring the gas to the calorimeter temperature and to barometric pressure before it escapes into the atmosphere. Coincident with this operation and throughout its course, measured electrical energy is added to the calorimeter at such a rate as to maintain the temperature constant. When the pressure in the bomb has fallen to atmospheric, the electrical heat is cut off and the calorimetric after-period is accurately determined. The heat, Q, of the process is now calculable as is also the work done against the atmosphere, and from equation (5) the quantity $\Delta U]_p^a$ is then obtained.

The contribution of the bomb itself to the observed heat effect can, for a properly designed bomb, be computed as explained below or it may be eliminated by standardizing the operation with a gas for which accurate ΔU values are available.

V. THE ENERGY CONTENT OF THE BOMB AS A FUNCTION OF THE PRESSURE

The bomb itself is an elastic system and its intrinsic energy changes with the tension on its walls. Unlike the gas, however, it may not return promptly to its original energy state on release of the tension, but instead may exhibit a lag.

For this reason it is desirable (1) to design a bomb for which the change of intrinsic energy with pressure shall be as small as possible, and (2) to suitably anneal the bomb and to age it by subjecting it to a number of cycles of extension and contraction by repeated filling and emptying with compressed gas.

Since the bomb is itself an elastic system the following thermodynamic relation analogous to equation (2) is applicable

$$\left(\frac{\partial U}{\partial p_b}\right)_T = p_b\left(\frac{\partial v_i}{\partial p_b}\right)_T - B\left(\frac{\partial v_e}{\partial p_b}\right)_T + T\left(\frac{\partial v_i}{\partial T}\right)_{p_b} \tag{6}$$

in which p_b is the internal pressure, v_i the internal volume, B the constant external barometric pressure, and v_e the exterior or bulk volume. Assuming the approximations

$$\left(\frac{\partial v_i}{\partial p_b}\right)_T = \left(\frac{\partial v_e}{\partial p_b}\right)_T \text{ and } \frac{1}{v_i}\left(\frac{\partial v_i}{\partial T}\right)_{p_b} = 3\alpha$$

both independent of the pressure, and integrating gives

$$\Delta U]_B^{p_b} = \frac{(p_b - B)^2}{2}\left(\frac{\partial v_i}{\partial p_b}\right)_T + 3\alpha v_i T(p_b - B) \tag{7}$$

in which α is the linear coefficient of thermal expansion of the material composing the bomb.

For a cylindrical steel bomb with hemispherical ends, a wall thickness of 1.65 mm and a capacity of about 1 l, $\left(\dfrac{\partial v_e}{\partial p_b}\right)_{300°K.}$ was found to be constant and equal to 39×10^{-6} l/atm., for stresses well within the elastic limit.

For $p_b = 50$ atm. and $B = 1$ atm. we have therefore at 300°K.

$$\Delta U]_1^{50} = \frac{49^2}{2} \times 39 \times 10^{-6} + 3 \times 11 \times 10^{-6} \times 1 \times 300 \times 49 \text{ l-atm.} \quad (8)$$

$$= 4.74 + 49.16 = 53.9 \text{ j} \quad (9)$$

If the contained gas is air at 50 atm. it will give $-\Delta U]_1^{50} = 560$ j approximately. The contribution of the bomb therefore amounts to about 10 per cent of intrinsic energy change of the gas and, if the bomb is slow in returning to its original energy state, its contribution can not in practice be calculated with sufficient accuracy.

Now the first term in equation (9) is a work term, and the quantity $\left(\dfrac{\partial v_e}{\partial p_b}\right)_T$ can be directly measured with sufficient accuracy by immersing the filled bomb in a constant-temperature volumeter and releasing the pressure. For a steel bomb no appreciable lag is observed in this quantity. The lag therefore affects only the second term which is the larger of the two. This term can be materially reduced by constructing the bomb of a material having a low coefficient of thermal expansion. Thus if a suitable invar steel is selected for this purpose, equation (9) for the same conditions becomes

$$\Delta U]_1^{50} = 4.74 + 4.92 = 9.7 \text{ j} \quad (10)$$

The total bomb contribution will now be only 1.7 per cent of the gas effect and that part of it which is affected by lag will be less than 1 per cent and the uncertainty arising from the lag will be a negligible quantity in comparison with the other errors of measurement.

VI. EFFECTS OF ADSORPTION

In all experimental work involving p–v–T or energy relations of gases the influences of adsorbed gas on the walls of the containing vessel must be considered. These influences are usually measured and eliminated by a series of experiments with different areas of exposed surface in contact with the gas. In the present instance such experiments can be readily made by placing wire or foil (of the same material as the bomb) inside the bomb in order to increase the area of metal surface exposed. Since, however, in the present method the pressure of the gas does not fall below atmospheric during the experiment, the adsorbed gas which leaves the walls of the bomb is only the small and probably negligible amount which becomes adsorbed when the pressure is raised from 1 atmosphere to the initial pressure p_b.

VII. THE TEMPERATURE COEFFICIENT OF U_T

For correcting the calorimetric determinations to a common temperature or for reducing a measured value at one temperature to a neighboring other temperature, it is convenient to remember that the small temperature coefficient can be approximately obtained by differentiating equation (3) which gives

$$\frac{100}{\Delta U}\frac{d(\Delta U)}{dt} = \frac{-100}{T}\left[1 + 6c\ T^3\left(\frac{A_0}{RT} + \frac{3c}{T^3}\right)\right] \quad (11)$$

Thus for O_2 at 25° C. this equation gives 0.4 per cent per degree.

VIII. PRECISION DISCUSSION

In equation (5) the work done by the expanding gas will be

$$W = B\Delta v = Bv - Bvi = \frac{m}{M}RT\ (1 + \beta_T B) - Bv_i \quad (12)$$

and equation (5) may be written

$$Q = m\left[f(p) + \frac{RT(1 + \beta_T B)}{M}\right] - Bv_i \quad (13)$$

in which m is the mass of gas contained in the bomb of volume v_i at the pressure p and temperature T, R is the gas constant, M the molecular weight of the gas, β_T is a small coefficient expressing the deviation from the perfect gas law at atmospheric pressure and T°K, B is the prevailing barometric pressure, and $f(p)$ is the value of ΔU_s^2 for 1 g of the gas at T°K.

Using the value of $\Delta U]_s^1 = f(p)$ for air, one can compute the precision required in each measured quantity in order to obtain a desired precision in the quantity ΔU for this gas. If we take $t = 25$° C., $v_i = 1$ liter and assume an allowable error of 0.25 cal. from each measured quantity, we obtain the results shown in the following table.

TABLE 1.—*Error in each measured quantity which produces an error of 0.25 cal. in the quantity $\Delta U]^1_p$*

	$\Delta U]^{1\ atm}_{p\ atm}$ (air at 25° C.), cal.	22.0	50.0	89.2	139
1	p .... atm..	20	30	40	50
	δp .... do....	0.22	0.14	0.11	0.09
2	t_i .... l..	1	1	1	1
	δt_i .... cm³..	10	10	10	10
3	m .... g..	24	36	48	60
	δm .... mg..	11	11	11	11
4	B .... mm Hg..	760	760	760	760
	δB .... do....	8.2	8.4	8.6	9.1
5	Electrical energy equivalent .... cal..	485	757	1,048	1,348
	δt .... ° C..	0.0036	0.0036	0.0036	0.0036
	$\delta t'$ .... do....	.0001	.0001	.0001	.0001
6	Stirring energy .... cal..	10	10	10	10
	100 δ (S. E.)/(S. E.) .... per cent..	2.5	2.5	2.5	2.5
7	T .... ° K..	298	298	298	298
	δT .... do....	.16	.11	.06	.06
8	δr .... atm.⁻¹..	.00037	.00037	.00037	.00037
	$\delta\delta r$ .... do....	.0005	.0003	.0003	.0002

Of the items shown in the table, Nos. 2, 4, 7, and 8 can be eliminated as sources of error since they can obviously be easily measured with much more than the required accuracy.

Item 1.—For the pressure measurements a good calibrated Bourdon gage could be employed for pressures up to 50 atm. At higher pressures a more accurate method of measuring the pressure would be required.[5]

Item 3.—In the mass determination, somewhat more than the required accuracy can be easily obtained with a liter bomb and with a balance sensitive to a few milligrams under loads up to 2 kg.

Item 5.—In determining the electrical energy required to equalize the cooling effect of the expanding gas no significant errors will be made in measuring the actual energy input employed. The total error from this source will be determined by the accuracy of the temperature control during the experiment and the accuracy in determining the small difference between the initial and final equilibrium temperatures. The value of δt shown in the table represents the permissible integrated average temperature difference between the calorimeter and the jacket during the experiment, assuming a reaction time of 10 minutes and an over-all heat-transfer coefficient of 7.2 cal. min.⁻¹ ° C.⁻¹. The value $\delta t'$ represents the accuracy with which the difference between the initial and final temperatures must be known with a calorimeter having a water equivalent of 2,500 g.

Item 6.—An efficient stirrer will supply energy to the calorimeter at the rate of about 1 cal. per minute. For an allowable error of not more than 0.25 cal. from this source it is obvious that throughout the period of the experiment (30 minutes) the stirring power should be constant within ± 2.5 per cent.

From the foregoing discussion it appears that it should be possible without much difficulty, to determine the quantity $\Delta U]^1_p$ for 1 liter of

---

[5] The necessity of employing a pressure gauge can be dispensed with, if desired, by the use of twin bombs. See Washburn, B. S. Jour. Research, vol. 9, p. 271, 1932.

gas with an error of less than 1 cal. A forthcoming paper by Rossini and Frandsen describes some experimental results obtained by the method.

In conclusion, it may be noted that by combining the calorimetric value for $\left(\dfrac{\partial U}{\partial p}\right)_T$ with the calorimetric value for $\mu\, C_p$, both $p\left(\dfrac{\partial v}{\partial p}\right)_T$ and $T\left(\dfrac{\partial v}{\partial T}\right)_p$ can be directly obtained, if the volume is known, and this is probably the most accurate method of determining these two coefficients.

WASHINGTON, August 3, 1932.

RP488

THE PHOTOGRAPHIC EMULSION; VARIABLES IN SENSITIZATION BY DYES

By B. H. Carroll and Donald Hubbard

ABSTRACT

These experiments were designed to test the effect of independent variables in the emulsion on the relative spectral sensitization by a given dye in a given emulsion. Four typical dyes representing four series were used in combination with emulsions of different types. Relative spectral sensitivity increased slowly with the concentration of dye. In agreement with Sheppard, it was found that the relative spectral sensitivity was little affected by the formation of sensitivity nuclei (from allyl thiocarbamide, for example) which greatly increased the absolute sensitivity to any wave length. Increased alkalinity increased relative spectral sensitivity by an extent which depended on the dye. Increased silver ion concentration generally increases sensitization by any dye, but there are differences between individual dyes which may be explained on the hypothesis that spectral sensitization depends on adsorption of the ion of the dye by the oppositely charged ion of the silver halide lattice; changes in adsorption of basic dyes may be sufficient to counteract the general trend at sufficient excess of silver.

CONTENTS

I. INTRODUCTION

The spectral sensitivity of a given photographic material in either absolute or relative terms is generally recognized to depend both on the dye or dyes, and on the emulsion. The many variables in the emulsion which influence the sensitization by the dye may be divided into those which are characteristic of the particular emulsion, such as the grain size and proportion of iodide, and those which may readily be varied in a given dye-emulsion combination, such as the silver and hydrogen ion concentrations. This paper will be primarily concerned with the latter group of variables. While it is not possible to draw a sharp distinction between these variables and those characteristic of the emulsion and dye, a study of the more general variables of sensitization will make it easier to distinguish characteristics of individual emulsions by comparing them under constant conditions. The process of hypersensitization (1) (2) [1] must obviously be connected with variation of conditions in a given dye-emulsion combination. This will be treated in full in a separate communication; a preliminary report (3) has already been made, using some of the data in this paper.

---

[1] Numbers in parentheses here and throughout the text refer to the list of references at the end of the paper.

The experimental methods used in this investigation, including emulsion making, sensitometry, and determination of bromide ion concentration, have been described for the most part in previous communications (4) (5). Emulsion formulas followed the types described in one of these references (4) and will be designated by the same letters; full details are available to any interested parties. A measure of the added spectral sensitivity conferred by the dyes was obtained with the sector wheel sensitometer, inserting an appropriate filter between the light source and plate in addition to the Davis-Gibson filter used for correction of the incandescent source to sunlight quality. A Wratten "Minus Blue" (No. 12) filter was most commonly used, as it gives approximately the total sensitivity to all wave lengths longer than those absorbed by the silver bromide; absorption by the filter is not over 15 per cent in the region of sensitization by any of the dyes except pinaflavol. This general method does not determine the relative spectral sensitivity, but it is the most accurate means available for comparison of the sensitization produced by a given dye under varying conditions.

Four dyes were used in the investigation, each representing a different type. Erythrosin, which is the only acid dye in common use as a sensitizer, was included because it was important to have an acid dye for comparison with the basic sensitizers. It is soluble in water in concentrations much higher than those used in emulsions and ionizes into the colorless positively charged sodium ion, and the negatively charged ion of the acid tetraiodofluorescein. The other three dyes are basic, and ionize into colorless iodide or chloride ions, and positively charged ions of the corresponding complex nitrogen bases. They represent three distinct types—pinacyanol is a carbocyanine (6); pinaverdol an isocyanine (6); and pinaflavol, a newer type which has not been assigned a general name (7). The first two are very insoluble in water, being completely extracted from it by solvents such as chloroform, and their suspensions are readily flocculated by electrolytes, especially halides, while pinaflavol does not show these colloidal characteristics and apparently forms true solutions in water at low concentrations. However, even the insoluble dyes may be considered as colloidal electrolytes, as they appear to be highly dissociated; the iodide ion in any of these dyes may be titrated electrometrically with silver nitrate giving values corresponding closely to the calculated molecular weight.

The basic dyes were commercial products, used without further purification; the pinacyanol was the chloride of its base, the others iodides. The erythrosin used in these experiments had been prepared from an old sample of a Kahlbaum product by precipitation of the acid and recrystallization of the sodium salt made from it. It was recently found to be low in iodine, the acid containing 26.5 per cent instead of 60.75 per cent theoretical (8); apparently considerable loss had occurred on storage, with resulting formation of fluorescein. Photographic comparison with a known sample of pure erythrosin received from the color laboratory, Bureau of Chemistry and Soils, showed that it produced its maximum sensitization in exactly the same spectral region as the pure dye, but in about half the proper intensity, and that there was faint sensitization at shorter wave lengths corresponding to that produced by fluorescein. Fortunately this contamination by the parent acid dye does not affect our conclusions.

The dyes were normally added to the emulsions just before coating, in the form of dilute solutions; the alcoholic stock solutions of the basic dyes were diluted further with water to avoid coagulation of gelatin on contact with the alcohol. In all cases, the dye was added to the liquid emulsion before coating; none of the following experiments deal with sensitization by bathing finished plates.

II. CONCENTRATION OF DYES

The concentration of a given dye in an emulsion is reported to have an optimum value (9), but no quantitative data are available. Eder believed that the sensitivity fell off after passing through the maxi-

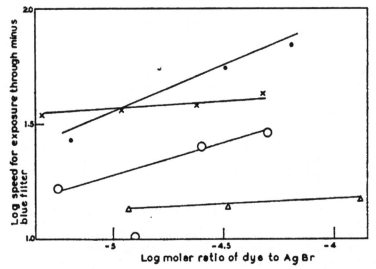

FIGURE 1.—*Effect of concentration of dye on spectral sensitization*

O, Pinacyanol.   . Pinaflavol.
X, Pinaverdol.   Δ Erythrosin.

mum because of filter action by the dye. With any of the basic sensitizers used in these experiments the upper limit seems to be set rather by the tendency of the dye to cause fog.

The data illustrated in Figure 1 indicate that within the range of concentrations tested, which covers the normal values for these dyes, the sensitization by the dye increases slowly with its concentration. The emulsions used in these experiments were of the neutral (type C, reference (4)) type with 4.0 mol per cent AgI, and were coated at a bromide ion concentration 0.9 to 1.1×10^{-4} and pH 7.1 to 7.3. Speeds were in all cases measured by exposure through the Minus Blue filter.[2] Speed numbers were somewhat dependent on development time, and were, therefore, compared by interpolating the value cor-

---

[2] Absorption by this filter was not taken into account in calculating the speed numbers, since relative values only were involved.

responding to $\gamma = 1.0$ from plots of speed against γ for each case. The contrast of the emulsions was not appreciably affected by the concentration of dye. The data are plotted on a logarithmic scale for compactness; concentrations have been reduced to the common basis of the molar ratio of dye to silver bromide.[3]

The slopes of the curves for pinaflavol and pinacyanol sensitization are larger than those for the other two dyes. This does not correspond to any classification of dyes; correlation with the adsorption of the dyes on silver bromide may be possible when data are available.

III. AFTER-RIPENING

After-ripening an emulsion, either with the sensitizers naturally occurring in gelatin or with known materials, such as allyl thiocarbamide, can produce a large increase in sensitivity, which is not associated with change in grain size, and which can be quite definitely ascribed to the formation of sensitivity nuclei. The relation of this effect, which we may call nuclear sensitization, to spectral sensitization by dyes, is of obvious importance. Sheppard (10) has briefly described experiments on this subject. An emulsion was made up with inert gelatin, and portions of it were sensitized with erythrosin, pinacyanol and a green sensitizer not specifically named. Another emulsion was made up with active gelatin so as to give the same mean grain size and the same size-distribution, and portions sensitized in the same way. It was found that for each dye, the ratios of red or green sensitivity to the blue-violet sensitivity (that is, for those wave lengths absorbed by the silver halide) were "substantially the same" in the two emulsions. The relative spectral sensitivity of an emulsion was, therefore, found to be practically independent of the formation of nuclei. This conclusion is so important that it was tested under a variety of conditions.

It is necessary to present the results in tabular form, because on comparing the undigested and digested portions of an emulsion made with active gelatin there are found differences both in speed and contrast. The same applies to portions of an emulsion made with inert gelatin and digested with and without a nuclear sensitizer.

Table 1 presents the data for an emulsion made with relatively inert gelatin (prepared by digesting an active gelatin with ammonia and then washing very thoroughly). After washing, the emulsion was divided in halves, to one of which was added a trace of sodium thiosulphate. After digestion, each half was divided into four portions; one was left unsensitized and the other three sensitized with three different dyes. It is impossible to express the resulting differences by single numbers. The contrast was changed both by the thiosulphate and the dyes; of the three dyes, pinacyanol produced much the strongest effect. Furthermore, the relation between speed number and time of development was changed by both types of sensitization. However, we may state with confidence that the increase in sensitivity to the longer wave lengths was less than the increase in total sensitivity (to white light). In this emulsion, the sensitivity to wave lengths transmitted by the Minus Blue filter was a

[3] The molecular weights used for this purpose were: Pinacyanol, 412; pinaverdol, 501; pinaflavol, 366; erythrosin, 896. The concentrations are of the order of a few milligrams of dye per liter of emulsion containing 40 to 45 g of silver bromide.

small fraction of the total, so that white light and blue light sensitivities were equivalent within the necessary limits of accuracy.

TABLE 1.—*Effect of nuclear sensitization, by $Na_2S_2O_3$, on sensitization by dyes*

[Neutral type emulsion, 4 per cent AgI (4-103) coated at pH 6.4, [Br-] 9×10-4; half of emulsion sensitized with $Na_2S_2O_3$ sufficient to convert 2 parts AgBr per 100,000 to Ag_2S]

	Dye	Molar ratio dye to AgBr	White light exposure						Minus blue filter exposure						Fog
			Speed			γ			Speed			γ			
			3	6	12	3	6	12	3	6	12	3	6	12	
Control	None	0	35	59	..	0.26	0.33	....	....			....			0.02
	Pinaflavol	6.7×10-4	40	53	65	.34	.35	0.43	4.0	3.5	3.5	0.26	0.57	0.70	.04
	Erythrosin	1.1×10-4	21.5	46	85	.26	.36	.39	3.5	3.1	3.1	.16	.30	.40	.03
	Pinacyanol	5.0×10-4	31.5	30	23	.55	.77	.95	8.0	7.4	6.3	.76	1.14	1.57	.20
Sensitized with $Na_2S_2O_3$	None	0	430	255	230	.51	.91	1.41	....			....			.08
	Pinaflavol	6.7×10-4	315	365	325	.64	.92	1.29	12.3	13.8	13.5	.88	1.07	1.86	.20
	Erythrosin	1.1×10-4	225	235	235	.63	.95	1.40	9.6	9.2	7.0	.57	1.03	1.63	.08
	Pinacyanol	5.9×10-4	120	105	96	.73	1.10	1.66	9.6	9.6	9.1	1.05	1.70	2.42	.41

Similar experiments with other emulsions and two of the dyes are recorded in Table 2. These emulsions had finer grain and lower iodide content, which probably accounts for the much better ratio of red or green sensitivity to blue sensitivity. This ratio was unchanged or somewhat decreased on increasing the total sensitivity by thiocarbamide sensitization.

The emulsions listed in Table 3 were made with active gelatin, and compared with and without digestion after washing. The variation in sensitivity nuclei in this case was obtained by varying the completeness of the reaction with the available sensitizing compounds, instead of by varying the quantity of the latter. The results are very similar to those obtained with the pure sensitizing compounds. The relative sensitivity to longer wave lengths was appreciably decreased by the after-ripening, although the absolute value of sensitivity to red or green increased two or more times.

These last experiments also included the behavior of the dye when present during digestion. Two portions of emulsion were digested under conditions identical except that the dye was added before digestion to one, and after digestion (the usual procedure) to the other. The sensitization by erythrosin was less when it was present during digestion than when it was added afterward (just before coating); pinacyanol was more effective under the former conditions, while pinaflavol was about the same.

TABLE 2.—*Effect of nuclear sensitization, by allyl thiocarbamide, on sensitization by dyes*

[Neutral emulsion (4-48), 1 per cent AgI, coated at pH 7.5, Br=1.3×10⁻⁴; ammonia process emulsion (1-153), 1 per cent AgI, coated at pH 8.3, Br=1.3×10⁻⁴. Half of each emulsion sensitized with allyl thiocarbamide sufficient to convert 8 parts AgBr per million to Ag₂S. "A" red filter used for exposure of pinacyanol sensitized batches; minus blue (yellow) filter for others]

Emulsion	Dye	Allyl thiocarbamide	Blue light exposure ("C" filter)						Yellow or red light exposure						Fog
			Speed			γ			Speed			γ			
			3	6	12	3	6	12	3	6	12	3	6	12	
4-48	Erythrosin......	Without..	14.0	9.5	10.0	0.37	0.61	0.84	7.9	7.9	7.9	0.52	0.74	0.97	0.24
		With......	25	25	23	.54	.76	1.12	11.5	11.0	15.2	.60	.86	1.37	.28
	Pinacyanol......	Without..	11.5	10.0	8.3	.42	.61	.91	15.9	17.4	13.8	.55	.78	1.06	.27
		With......	17	12.5	14	.57	.92	1.18	26.3	26.3	28.1	.63	1.07	1.50	.34
1-153	Erythrosin......	Without ¹..	24.5	24.5	24.5	1.24	1.86	2.21	2.6	2.7	2.8	1.52	2.16	2.36	.23
		With ¹....	166	210	235	1.27	1.75	2.12	20.9	20.9	21.9	1.05	1.80	1.90	.29
	Pinacyanol......	Without..	2.6	2.4	2.6	2.00	2.40	2.75	5.3	5.3	6.4	1.82	2.49	2.73	.28
		With......	9.1	9.0	10.5	1.24	1.86	2.23	17.5	17.5	20.0	1.38	2.12	2.58	.63

¹ White light exposures substituted for blue because of low speeds.

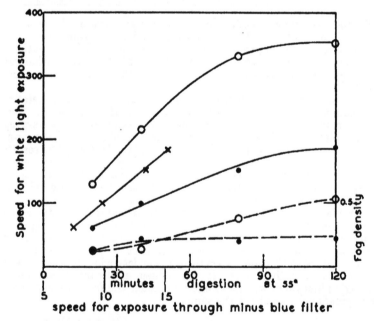

FIGURE 2.—*After-ripening with and without erythrosin in the emulsion*

. With dye.   O Without dye.
——————————— Speed.   ————— Fog.
X Comparison of speed for exposures to white light and through Minus Blue filter at varying stages of digestion.

The data in Table 3 on sensitivities to white light show that the after-ripening was materially retarded by the presence of erythrosin during digestion, while the other two dyes had less effect. Eder (11) observed that some dyes prevented fog when present during ripening.

Lüppo-Cramer (12) has found that in Lippman emulsions erythrosin and other sensitizing dyes can produce a striking retardation of the whole ripening process, and that in normal emulsions erythrosin present during digestion retards after-ripening (13). After-ripening being essentially a surface change in the grains, it is not surprising that it is influenced by the adsorption of a foreign material. Further data were obtained by a slightly different procedure. Emulsions were divided into halves, which were digested at the same temperature and silver ion concentration, one with and one without a sensitizing dye. The results are plotted in Figures 2 and 3 as speeds (at

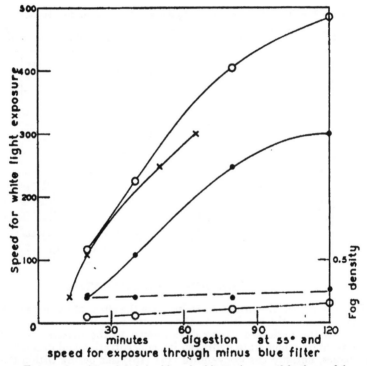

FIGURE 3.—*After-ripening with and without pinacyanol in the emulsion*

. With dye. O Without dye.
——— Speed. — — — — Fog.
X Comparison of speed for exposures to white light and through Minus Blue filter at varying stages of digestion.

$\gamma = 1.5$) for white light exposures against time of digestion. On the same sheet the speeds (at $\gamma = 1.5$) for exposures through the Minus Blue filter are plotted against the white light speeds. The latter curve would be a straight line if the relative spectral sensitivity were unchanged by the after-ripening.[4] The curvature indicates that it is slightly increased. The other curves show that the rate of after-

[4] In both figures, this curve intercepts the horizontal axis at a small positive value of "speed for exposure through Minus Blue filter," as the speed of the unsensitized emulsion did not fall quite to zero on exposure through the Minus Blue filter.

137718—32——6

ripening was divided by about four when erythrosin was present. Pinacyanol approximately halved the rate. Both dyes also retarded the growth of fog. In both figures, the slope of the curve of fog against digestion time is less for the portion digested with dye than for the one digested without it. The effect is less noticeable in Figure 3 because the fog density at any time was increased by the addition of pinacyanol, although the rate of increase with time was less. The data in Table 3 also show that the fog for the portion digested with pinacyanol was less than for the one to which the same amount of dye was added after digestion.

TABLE 3.—*Effect of after-ripening (by digestion) on sensitization by dyes*

[Neutral emulsions 4–50 and 4–51; 2¼ per cent AgI, with subsequent ammonia ripening; coated at pH 8 Br⁻ 1.2×10⁻⁴. Red filter used for exposure of pinacyanol sensitized batches; yellow filter for others]

Dye	Digestion	White light exposure						Exposure through yellow (Minus Blue) or red (A) filter						Fog 12
		Speed			γ			Speed			γ			
		3	6	12	3	6	12	3	6	12	3	6	12	
None	Minimum	50	43	52	0.66	0.94	1.23							0.18
	1.0 hour at 55°	220	302	270	1.01	1.43	2.29							.21
Erythrosin	Minimum	50	40	38	.70	1.02	1.28	3.8	4.0	4.3	0.54	1.17	1.90	.19
	1.0 hour at 55° without dye	265	380	315	.92	1.30	1.83	18	19	22	1.13	1.67	2.29	.30
	1.0 hour at 55° with dye	263	263	240	.78	1.27	1.74	13	13	12	1.20	1.82	2.55	.23
Pinaflavol	Minimum	100	120	115	.67	.85	1.11	6.9	6.6	7.2	1.00	L40	L76	.17
	1.0 hour at 55° without dye	138	220	336	.94	L14	1.71	15.9	15.9	20.0	L10	L52	L97	.79
	1.0 hour at 55° with dye	190	210	288	.86	1.16	1.65	11.5	15.0	16.0	L17	L57	2.25	.58
Pinacyanol	Minimum	50	50	57	.65	.88	1.11	6.6	6.6	6.6	.94	L40	L69	.33
	1.0 hour at 55° without dye	150	178	175	1.07	1.54	2.37	10.5	9.6	10.5	1.41	2.35	2.91	.58
	1.0 hour at 55° with dye	206	209	220	.97	1.47	2.20	19.6	19.0	13.9	1.00	L77	2.71	.40

IV. HYDROGEN ION CONCENTRATION

There is little in the literature to indicate the effect of hydrogen ion concentration of the emulsion on the effectiveness of sensitization by dyes. All the basic sensitizing dyes known to the writers may be decolorized by a moderate hydrogen ion concentration and in some of the classes, notably the isocyanines (14) (15), this occurs within the range which might be encountered in emulsions. The sensitizing properties would obviously be expected to diminish with the color. Approximate tests with the dyes used in these experiments showed that pinaverdol is appreciably decolorized by pH less than 7, and is almost completely decolorized by pH 5. Pinacyanol and pinaflavol did not fade appreciably at pH 5.4; the decolorization was barely perceptible at 4.2, and not complete at 2. Erythrosin has an appreciable shift in hue to the yellow, beginning about pH 5.4; the acid is mostly precipitated at pH 3.

The experimental results are presented in Figures 4 and 5, the logarithm of the speed through the Minus Blue filter (at γ = 1.5) being plotted against pH. For comparison, the logarithm of the speed (to white light) of unsensitized portions of the emulsions sensitized with pinacyanol is also plotted in Figure 4. The emulsions were all digested after washing and before adding the dyes, under the same conditions of temperature, silver ion concentration and pH, and the pH was adjusted just before coating so that it did not affect the after-ripening. With the exception of pinaverdol, the data for each dye (and also for the unsensitized portions) represent two emulsions, one for the range below pH 7 and one for the range above. These fit

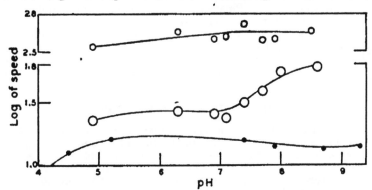

FIGURE 4.—*Variation in sensitivity with hydrogen ion concentration*

o Speed for white light exposure, unsensitized emulsions.
O Speed for exposure through Minus Blue filter, pinacyanol-sensitized emulsions.
. Speed for exposure through Minus Blue filter, pinaverdol-sensitized emulsion.

sufficiently well to draw single curves, except for erythrosin. (Fig. 5.) The discrepancy between the two emulsions in this case is explained by a difference in silver ion concentration, to which erythrosin is especially sensitive. The emulsion used for the acid range had a silver ion concentration at pH 7 of 1.3×10^{-8}, while the concentration for the other was 0.66×10^{-8} at the same pH. The effect of differences in

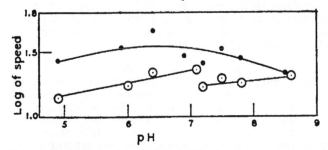

FIGURE 5.—*Variation in sensitivity with hydrogen ion concentration*

All exposures through Minus Blue filter.
. Pinaflavol-sensitized emulsions.
o Erythrosin-sensitized emulsions.

silver ion concentration in other cases was negligible. These emulsions were made with a gelatin giving very little change in sensitivity of the unsensitized emulsions with pH (upper curve in fig. 4), so so that this complication was reduced to a minimum.

The effect of hydrogen ion concentration on sensitization by the dyes was apparently characteristic of the individual dye. At the most, it was less than the effect of silver ion concentration, as may be seen by comparison with the curves of Figures 6, 7, 8, and 9, which are on the same scale. Pinacyanol in the alkaline range (fig. 4) showed the largest effect, with little change between pH 7 and 5. Erythrosin consistently increased in effectiveness with increasing pH; this may possibly be explained by increasing dissociation, as it is the salt of a weak acid. Pinaflavol showed a maximum around the neutral point. (Fig. 5.) Sensitization by pinaverdol (fig. 4), fell off rapidly as the acidity was increased past pH 5, as would be expected from the decolorization of this dye by acid. Sensitivity at pH less than 4.5 became too low to measure accurately. The change from pH 5 to 9 was quite small. No generalization on the basis of these data appears to be justified.

V. SILVER ION CONCENTRATION

The most important of the variables which we have studied is the silver ion concentration of the emulsion. This condition may also be expressed in terms of the excess of soluble bromide or soluble silver salts in the emulsion, as has been the commonest practice in the literature, but our results indicate that the silver ion concentration may be used to simplify the statement and understanding of the condition in a manner quite analogous to the use of hydrogen ion concentration for acidity and alkalinity. As explained in previous communications (4) (5), the emulsion may be considered as saturated with silver bromide, so that the product of silver and bromide ion concentrations is a constant at a given temperature (9×10^{-13} at 30°) and either concentration may be computed from the potential of a silver electrode in the emulsion and used to express its condition. The addition of soluble bromide to an emulsion thus decreases its silver ion concentration; the addition of a soluble silver salt increases the silver ion concentration. In both cases the rate of change is less in the emulsion than it would be in plain aqueous solution. Adsorption of soluble bromide on silver bromide retards the increase in bromide ion concentration on one side of the equivalence point, and formation of unionized silver-gelatin compounds reduces the silver ion concentration on the other side.

As it is essential that the sensitizer should dye the silver halide, sensitization will be dependent on conditions which affect its adsorption to the grain. The sensitizing dyes are known to be ionized, and are adsorbed on the highly polar silver halide lattice, so that it is relatively simple to predict some of the effects of silver ion concentration in the emulsion. In the presence of soluble bromide (bromide ion concentration greatly exceeding silver ion concentration), silver bromide will strongly adsorb bromide ions at the silver ions of the crystal lattice, and will acquire a negative charge. The adsorption of the negatively charged ion of an acid dye, such as erythrosin, will thus be reduced, since it must compete with the bromide ions for the silver ions of the lattice. The presence of an excess of silver will, conversely, increase the adsorption of the color ion of an acid dye on silver bromide, because the silver ions of the lattice will be free of adsorbed bromide and thus more available to the ions of the dye.[5]

---

[5] It should be remembered that in a gelatin emulsion, the adsorption of the gelatin to the silver bromide interferes very materially with the adsorption of either type of dye. This is almost certainly one of the reasons why gelatin emulsions are not sensitized as readily as collodion emulsions.

By the same type of reasoning we may predict that the adsorption of basic dyes should be greater in the presence of an excess of bromide ions, and should be reduced by an excess of silver ions.

The shift in adsorption of dyes with silver ion concentration is used in the titration of silver against halogen with adsorption indicators (16). It is of interest that the best results are obtained with acid dyes (fluorescein and the eosins, including erythrosin) and that the basic dyes which can be used as indicators (17), methyl violet for example, may be used as sensitizers for collodion emulsions. The first observation of the effect was apparently made by von Hübl (18) who discovered a change in color of cyanine on silver bromide with the addition of excess silver or bromide to the solution. Lüppo-Cramer (19) found that, in accordance with the theory, erythrosin was displaced from various insoluble salts by the corresponding anions, while basic dyes were not.

Observations on collodion emulsions (20) have been in accord with the simple theory. Acid sensitizers are adversely affected by the slightest traces of soluble bromide, and are preferably used with the addition of soluble silver salts, while the basic dyes are most effective in the presence of a slight excess of bromide. It should be remembered that collodion emulsions are not sensitized by the basic dyes, such as pinacyanol, until the alcohol has been largely displaced by washing with water, so that we are justified in discussing the ionization on the basis of an aqueous system.

Lüppo-Cramer (21) (22) found that in sensitizing gelatin emulsions by bathing, erythrosin was more affected by soluble bromide in the dye bath than was the basic dye pinachrome. However, the simple theory just given fails to predict that small amounts of soluble bromide will cause any decrease in sensitization by basic dyes, and something further is necessary to account for the fact that this decrease exists and is of such magnitude that most writers have failed to note the difference between the behavior of the acid and basic dyes. In the first place, the adsorption of the basic sensitizing dyes on silver bromide is not as simple as that of erythrosin. Sheppard, Lambert, and Keenan (14) found that at pH 5, pinacyanol was adsorbed much more strongly on silver bromide with excess of silver ions than on the same sample with excess of bromide ions. At pH 7.5 the order was reversed. Adsorption of dichlorofluorescein followed the simple theory. Bokinik (23) also reports stronger adsorption of pinacyanol on silver bromide with excess of silver ions than on an "equivalent" sample; the pH is not given. More data on adsorption of the basic sensitizers is evidently necessary for adequate theoretical treatment. In the second place, there is the possibility of some factor depending on silver ion concentration and affecting acid and basic dyes alike, the effects of changes in adsorption being superposed on this. In a system such as the photographic emulsion where gain or loss of bromine is probably the significant chemical change, the oxidation-reduction potential should depend on the silver (or bromide) ion concentration in a manner exactly analogous to its dependence on hydrogen ion concentration. If we adopt the working hypothesis that sensitization is the result of chemical reduction of silver bromide by the dye, a decrease in sensitization by increasing bromide ion concentration is to be expected. This hypothesis involves the simplest and most definite mechanism of sensitization, and in the absence of proof to the contrary,[6] we believe it is worthy of support (24). It is essentially the converse of the oxidation theory of desensitization. The implied continuous gradation between sensitization and desensitization has been realized by Kögel and Bene (27) and in this laboratory (28) by change in silver ion concentration. Independent evidence of photochemical reaction between silver bromide and sensitizing dyes is given by the accelerated bleaching of the dyes when adsorbed on silver bromide.

As the relative magnitude of the effects of changes in oxidation-reduction potential and in adsorption can not be predicted, the change in sensitization with silver ion concentration, which is the algebraic sum of these effects, may take a variety of forms, but we may at least expect that the basic dyes will be less sensitive to the addition of soluble bromide than the acid dye.

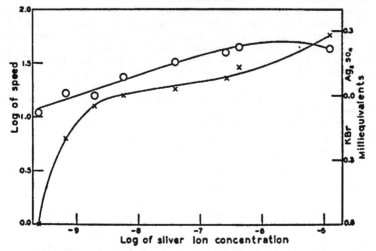

FIGURE 6.—*Variation in sensitivity with silver ion concentration*

O Speed of pinaverdol-sensitized emulsion for exposures through Minus Blue filter.
X Quantities of silver sulphate or potassium bromide added to 175 ml of emulsion, containing 0.040 equivalent of silver bromide, to produce indicated silver ion concentrations.

Our data are presented in Figures 6, 7, 8, and 9. The logarithm of the speed (at $\gamma = 1.0$) for exposures through the Minus Blue filter has been plotted against the logarithm of the silver ion concentration in the emulsions. The logarithmic scale for the speed numbers was adopted as the best means of comparing the data for emulsions which differed considerably in speed. The total (white light) sensitivity of

[6] The results of Leszynski (25) have been widely quoted as proof that sensitization can not depend on photochemical reaction between the dye and the silver halide, since it is practically impossible thus to account for his reported yield of 20 atoms of photo-silver per molecule of erythrosin (without development). The writers are, however, unable to accept his data as adequate to a crucial test. The figure just quoted was obtained by ascribing to the action of the sensitizer all the silver found in an erythrosin-sensitized emulsion after exposure to green light, but Leszynski's own data show that the exposure used to secure the above yield was 10[6] times that necessary to produce a developable density of 1 in the same emulsion without the erythrosin, using exactly the same source of green light (an incandescent light with filter). Granting that green light was many times more effective in producing a developable image in the erythrosin-sensitized portion of the emulsion than in the unsensitized, the absence of control analysis on the unsensitized emulsion after the same exposure renders his conclusion very uncertain, especially as no details are given for his experiments which lead him to believe that the Becquerel effect was not involved. The results of Tollert (26) are open to the same criticism for lack of control experiments.

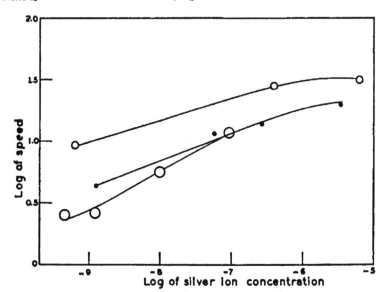

FIGURE 7.—*Variation in sensitivity with silver ion concentration, for erythrosin-sensitized emulsions*

•, o Neutral emulsions.
O   Ammonia-process emulsion.
All speeds for exposures through Minus Blue filter.

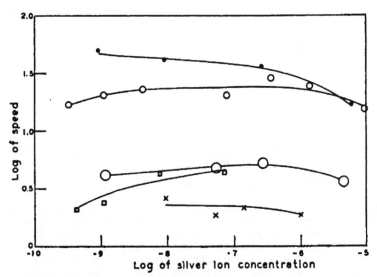

FIGURE 8.—*Variation in sensitivity with silver ion concentration, for pinaflavol sensitized emulsions*

•, o Neutral emulsions.
X, □ Ammonia-process emulsions.
All speeds for exposures through Minus Blue filter.

the emulsions did not vary materially with the silver ion concentration, so that the speed numbers corresponding to the filter exposures indicate the changes in sensitization.

The data were obtained by varying the silver ion concentrations of the emulsions after digestion and addition of the dye and just before coating; the after-ripening of all portions was the same within the limits of error of sensitometry, so that the observed results may be ascribed directly to the effect of environment on the dye-silver halide combination. The silver ion concentration was adjusted by appropriate additions of potassium bromide or silver sulphate. In Figure 6, one curve represents the change in sensitivity of a pinaverdol-sensi-

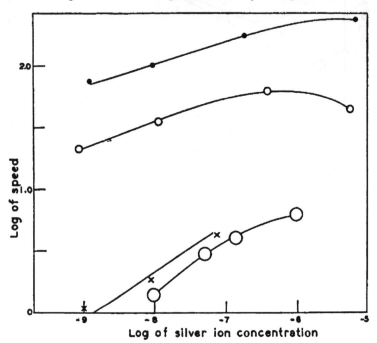

FIGURE 9.—*Variation in sensitivity with silver ion concentration, for pinacyanol-sensitized emulsions*

•, o , Neutral emulsions.
X , O Ammonia-process emulsions.
All speeds for exposures through Minus Blue filter.

tized emulsion with the silver ion concentration; the other shows the amounts of potassium bromide or silver sulphate which were added to the corresponding portions of emulsions to produce these silver ion concentrations. Equivalence of bromide and silver ion concentrations occurs close to −6 on the logarithmic scale, but this corresponds (5) to an excess of silver over bromine in the emulsion. The emulsion after washing (with no addition of bromide or silver) contained only a very slight amount of soluble bromide, but the silver ion concentration was repressed by the gelatin, and, therefore, was much smaller than the bromide ion concentration. The curve obviously corresponds to the change in potential of a silver electrode on passing through the equivalence point of a titration of silver against bromide; in the presence of the gelatin at this pH (about 7), the equivalence point is displaced toward increasing bromide ion concentration and the "break" is not as sharp. The results with other emulsions were very similar.

Two general classes of emulsions were used—the neutral ("C") type, with 4.0 mol per cent AgI, coated at pH 7 ± 0.5, and the ammonia ("A") type, with 1.0 mol per cent AgI, coated at pH 8 to 8.5. No differences which could be ascribed to the emulsion type were detected.

The range of silver ion concentrations which could be used in the experiments was limited by the increasing fog and instability of the batches with increasing excess of silver. The emulsions for this series of experiments were made with partly deactivated gelatin in order to reduce fog. The fog density for 12-minute development in pyro-soda without bromide was normally less than 0.5, and in no case exceeded 0.75.

By centrifuging the experimental emulsions, they could be cleared of silver halide and the concentration of any dye which had not been adsorbed could be observed. The change in adsorption of erythrosin with changing silver ion concentration was readily detected in this way. It could not be measured with any accuracy because the dye changed in hue as well as in concentration, but by comparison with known solutions of the dye made up in a gelatin solution corresponding to the emulsion, it was estimated that in one emulsion, half the dye was adsorbed at a silver ion concentration of 1.2×10^{-9}, while five-sixths was adsorbed at a silver ion concentration of 3×10^{-6}. Pinacyanol and pinaverdol in photographically practicable concentrations were adsorbed too strongly to make any comparison possible, while the color of the gelatin, plus a change in hue of the dye, interfered with measurements of pinaflavol remaining in the emulsion. Quantitative work will obviously have to be done under other conditions, but it is worth recording that the effect could be detected in the emulsion.

The curves for all the dyes show that sensitization decreases with decreasing silver ion concentration in the range 10^{-7} to 10^{-10} which corresponds to concentrations of soluble bromide sufficient to give moderate to good stability to the emulsion. This indicates that the effect common to all dye-emulsion combinations, which we have ascribed to change in oxidation-reduction potential, is the largest factor involved. The differences between the dyes are significant and generally in accordance with the expected change in adsorption. The slopes of the curves for erythrosin-sensitized emulsions (fig. 7) are larger than those of emulsions sensitized with any of the other dyes, since the adsorption of this acid dye increases with increasing silver ion concentration and the sign of the resulting change in sensitization is, therefore, the same as that of the common factor. The curves for sensitization by all three of the basic dyes appear to reach a maximum value, the decrease in sensitization with increasing silver ion concentration being most certainly established for pinaflavol (fig. 8) and least certainly for pinacyanol. (Fig. 9.) The slopes of the curves on the bromide side of the maximum are least for pinaflavol and greatest for pinacyanol, pinaverdol being intermediate. Pinaflavol is by far the most soluble of the three basic dyes, and observations under comparable conditions both in the emulsion and with pure silver bromide indicate that it is less completely adsorbed than the other two. Changes in adsorption caused by change in silver ion concentration should, therefore, have more effect on the sensitization by pinaflavol than on that by the other basic dyes, and it is to be expected that the general trend should be more completely counteracted by the predicted decrease in adsorption with increasing silver ion concentration. The net result is that the change in sensitization by pinaflavol is close to the experimental error over a considerable range of silver ion concentration, although there is an unmistakable decrease at a sufficient excess of silver.

Increase in relative spectral sensitivity by increasing the silver ion concentration in the emulsion is very seriously limited in practice by the instability of the resulting product. If the emulsion must be kept for 6 to 12 months under ordinary conditions without serious deterioration (which might be regarded as the minimum commercial requirement) it is necessary in emulsions of the type used in these experiments to have the silver ion concentration approximately 2×10^{-9} N or less. (This corresponds to 2 to 3 molecules of soluble bromide per 1,000 of silver bromide). Silver ion concentrations up to 10^{-5} N are produced in hypersensitized materials (3), but these correspond to a useful life of a few weeks or even days. Deterioration of the emulsions in which the silver ion concentration was adjusted before coating went on at a rate increasing with this factor; the batches with the highest concentrations indicated behaved like hypersensitized materials.

A number of patents (29) have recently been secured on organic materials which are described as preservatives for emulsions. One of these, nitrobenzimidazol, has also been studied (30) as a substitute for soluble bromide in development. Since its use is patented, any practical investigation by this laboratory is unjustified, but the mechanism of its action is of considerable scientific interest. Nitrobenzimidazol was synthesized in this laboratory and purified by recrystallization from hot water and from alcohol. Its crystals formed the characteristic groups of flesh-colored needles, and melted at 209° C. The silver compound was prepared by precipitation from dilute aqueous solutions of nitrobenzimidazol and silver nitrate. After thorough washing, the silver ion concentration in saturated solutions of the compound was determined by the silver electrode at 30° C. At pH 7.1 (in 0.010 N sodium acetate solution) it was found to be 2.3×10^{-7} N, indicating that the compound is more insoluble than silver bromide. The silver is, however, readily displaced by hydrogen, since at pH 3.5 (0.010 N acetic acid) the silver ion concentration in a saturated solution was 6.4×10^{-5} N. In emulsions at approximate neutrality it produces some decrease in silver ion concentration, but experiments on afterripening with the addition of nitrobenzimidazol indicated that mol for mol, it produces about ten times the effect of soluble bromide in delaying afterripening. Something more than the decrease in silver ion concentration must be involved, and another method of attack would be necessary to decide what this might be.

Experimental emulsions were divided after digestion into portions which were, respectively, coated unsensitized and sensitized with erythrosin, pinaflavol, and pinacyanol, each of these being subdivided nto halves with and without the addition of nitrobenzimidazol. It was found that the preservative selectively depressed the sensitivity conferred by the dyes, the effect being greater for the exposures through the filter than for the white-light exposures of the unsensitized portions. The effect was of the same order of magnitude as would be produced by soluble bromide with the same preservative action, although our data do not exclude the possibility that under other conditions this stabilizer might be superior to soluble bromide in this respect. The effect of the nitrobenzimidazol on the individual dyes was appreciably different from that of soluble bromide. The depression of sensitization increased in the order erythrosin < pinaflavol < pinacyanol. A possible explanation may again be found in terms of adsorption. The insolubility of the silver compound of nitrobenzimidazol indicates that this material should be strongly adsorbed by the silver bromide, most probably by the silver ions. This would bring it into competition with the basic, rather than the acid dyes, so that erythrosin should be the least affected of the three.

VI. Acknowledgment

C. M. Kretchman has rendered valuable assistance in coating and testing the experimental emulsions.

VII. REFERENCES

1. Burka, J. Frank. Inst., vol. 189, p. 25, 1920.
2. Davis and Walters, B. S. Sci. Paper No. 422.
3. Carroll and Hubbard, J. Soc. Motion Picture Eng., vol. 18, p. 600, 1932.
4. Carroll and Hubbard, B. S. Jour. Research, vol. 7 (RP-340), p. 219, 1931.
5. Carroll and Hubbard, B. S. Jour. Research, vol. 7 (RP-376), p. 811, 1931.
6. Bloch and Hamer, Phot. J., vol. 68, p. 21, 1928.
7. Mills and Pope, J. Chem. Soc., vol. 121, p. 946, 1922.
8. Gomberg and Tabern, J. Ind. Eng. Chem., vol. 14, p. 1115, 1922.
9. Eder, Ausführliches Handbuch der Photographie, vol. 3, p. 153, 5th ed., 1903.
10. Sheppard, Colloidal Symposium Monograph 3, p. 1, 1925.
11. Reference 9, p. 659.
12. Lüppo-Cramer, Grundlagen der phot. Negativverfahren, p. 25.
13. Lüppo-Cramer, Phot. Ind., vol. 14, p. 78, 111, 1916.
14. Sheppard, Lambert, and Keenan, J. Phys. Chem., vol. 36, p. 174, 1932.
15. Kolthoff, J. Am. Chem. Soc., vol. 50, p. 1604, 1928.
16. Fajans and Wolff, Z. Anorg. Chem., vol. 137, p. 221, 1924.
17. Hodakow, Z. Physik. Chem., vol. 127, p. 43, 1927.
18. Von Hübl, Jahrbuch für Photographie, p. 189, 1894.
19. Lüppo-Cramer, Jahrbuch für Photographie, p. 54, 1902.
20. Eder, Ausführliches Handbuch der Photographie, vol. II-2, p. 241, 3d ed., 1927.
21. Lüppo-Cramer, Phot. Ind., vol. 15, p. 657, 1917.
22. Lüppo-Cramer, Camera, vol. 2, p. 227.
23. Bokinik, Zeit. f. wiss. Phot., vol. 30, p. 322, 1922.
24. Bancroft, Ackerman, and Gallagher, J. Phys. Chem., vol. 36, p. 154, 1931.
25. Leszynski, Zeit. f. wiss. Phot., vol. 24, p. 261, 1926.
26. Tollert, Zeit. f. Physik. Chem. Abt. A, vol. 140, p. 355, 1929.
27. Kögel and Bene, Phot. Ind., vol. 28, p. 102, 1930.
28. Forthcoming communication on reversal effects with desensitizing dyes.
29. U. S. Patent 1696830, Chem. Abstracts, vol. 23, p. 571, 1929; U. S. Patent 1725934, Chem. Abstracts, vol. 23, p. 4902, 1929; U. S. Patent 1758576, Chem. Abstracts, vol. 24, p. 3186, 1930; U. S. Patent 1756577, Chem. Abstracts, vol. 24, p. 3186, 1930; U. S. Patent 1763989, Chem. Abstracts, vol. 24, p. 3720, 1930; U. S. Patent 1763990, Chem. Abstracts, vol. 24, p. 3720, 1930.
30. Trivelli and Jenson, J. Frank. Inst., vol. 210, p. 287, 1930.

WASHINGTON, June 2, 1932.

RP489

A METHOD FOR THE SEPARATION OF RHODIUM FROM IRIDIUM AND THE GRAVIMETRIC DETERMINATION OF THESE METALS

By Raleigh Gilchrist

ABSTRACT

A method is described for the analytical separation of rhodium from iridium by means of titanous chloride. Two precipitations of the rhodium, as metal, from a solution of its sulphate in sulphuric acid were found to be sufficient to separate it completely from iridium. Following the separation, methods are described for the determination of rhodium by precipitation with hydrogen sulphide, for the elimination of titanium by precipitation with cupferron, and for the determination of iridium by hydrolytic precipitation.

An alternative method is offered in which rhodium and iridium are determined together in one sample by hydrolytic precipitation, and rhodium separately in a second sample.

CONTENTS

I. INTRODUCTION

The separation of rhodium from iridium has long remained one of the most difficult of the analytical problems of the platinum group of metals. Berzelius,[1] in a general scheme for the analysis of crude platinum, used molten potassium bisulphate to separate these two metals from one another. This method is unsatisfactory because of the incomplete removal of rhodium and the partial solution of iridium.

Another method, which is in current use, was employed by Claus,[2] and by Leidié[3] in schemes for the analysis of crude platinum. In these schemes the separation is based upon the precipitation of quadrivalent iridium as ammonium chloroiridate by ammonium chloride. Experience in this laboratory with the purification of the platinum metals has shown that rhodium contaminates iridium, when the latter is precipitated as ammonium chloroiridate, and that it can be eliminated only with great difficulty, if at all, by repeated precipitation of the iridium. Furthermore, since ammonium chloroiridate has a slight solubility, some of the iridium must remain with the rhodium.

---

[1] J. J. Berzelius, Ann. Physik (Pogg.), vol. 13, p. 533, 1828.
[2] C. Claus, Beiträge zur Chemie der Platinmetalle, Dorpat, 1855, p. 55.
[3] E. Leidié, Compt. rend., vol. 131, p. 188, 1900.

In a recent article, Ogburn and Miller [4] proposed a method for the determination of osmium in which strychnine sulphate, $C_{21}H_{22}O_2N_2 - H_2SO_4$, is used as a precipitating reagent. Attempts were made to determine whether this reagent could be used to effect the separation of iridium from rhodium. The reactions of strychnine sulphate with solutions of the quadrivalent platinum metal chlorides were observed. In each case a precipitate was obtained. No precipitate was obtained in chloride solutions of tervalent rhodium, tervalent iridium, or tervalent ruthenium (quadrivalent ruthenium chloride which had been heated with alcohol). In its reaction with the quadrivalent chlorides, the strychnine salt appears to behave like an ammonium salt, but with the production of compounds of much lower solubility.

It was found possible to precipitate iridium completely from solutions of ammonium chloroiridate in diluted hydrochloric acid $(5 + 95)$.[5] The reagent, moreover, caused no precipitation of rhodium from similarly acidified solutions of pure rhodium chloride. However, when strychnine sulphate was added to solutions containing both iridium and rhodium as chlorides, the iridium was found to be contaminated by rhodium to the extent of from 10 to 37 per cent of the rhodium present in the solutions. It was found possible to redissolve the iridium precipitate by treatment with nitric acid or with sodium nitrite, but it was not found possible to treat the solution in such a way as to permit the quantitative reprecipitation of the compound.

Wada and Nakazono,[6] while studying the behavior of titanous sulphate toward various elements, observed that this reagent caused complete precipitation of metallic rhodium from solutions of its chloride, whereas iridium was not precipitated from its chloride solution. Apparently, the effect of the reagent on solutions containing both rhodium and iridium was not studied.

Attempts, in this laboratory,[7] to use the foregoing reactions as a method of separation showed that the precipitated rhodium was contaminated with as much as 10 per cent of the iridium present. In the absence of any known means of redissolving the precipitate and thus preparing it for a second precipitation, the best that could be done was to fuse the ignited metal with pyrosulphate, in order to extract the rhodium. It was found that a small amount of iridium passed into solution and that the residues from the pyrosulphate fusions always retained some rhodium. The method, therefore, could not be made strictly quantitative but, in spite of its shortcomings, it was better than any other known procedure. The objects of the work described in this paper were to develop this method of separation into a strictly quantitative one and to provide suitable methods for the determinations of the two metals after they had been separated.

II. COMPOUNDS USED IN THE EXPERIMENTS

The iridium and the rhodium which were used throughout the investigation were purified by methods previously described [8] and were found to be free from impurities by spectrographic analysis.

[4] S. C. Ogburn, Jr., and L. F. Miller, J. Am. Chem. Soc., vol. 52, p. 42, 1930.
[5] This denotes 5 volumes of hydrochloric acid, specific gravity, 1.18, mixed with 95 volumes of water. This system of designating diluted acids is used throughout this paper. If no dilution is specified, the concentrated reagent is intended. The concentrated sulphuric acid used had a specific gravity of 1.84.
[6] I. Wada and T. Nakazono, Sci. papers Inst. Phys. Chem. Research, Japan, vol. 1, p. 139, 1923.
[7] Unpublished work by W. H. Swanger and by H. A. Buchheit.
[8] E. Wichers, R. Gilchrist, and W. H. Swanger, Trans. Am. Inst. Mining Met. Eng., vol. 76, p. 602, 1928.

In order to avoid errors incident to volumetric measurements, weighed portions of the compounds, ammonium chloroiridate and ammonium chlororhodite, were used. Care was taken to avoid the presence of alkali chloride or of other nonvolatile impurities in the salts, but no attempt was made to prepare compounds of definite composition. The salts were carefully mixed by grinding in an agate mortar and the metallic content determined by direct ignition in hydrogen whenever a series of samples was weighed. It was never calculated from the theoretical composition of the salt.

III. DETERMINATION OF IRIDIUM BY HYDROLYTIC PRECIPITATION

By far the most difficult phase of the problem under consideration was that which dealt with the determination of the iridium after its separation from rhodium by tervalent titanium. Since the solution at this stage contains the titanium which has been added, it is necessary to effect a separation of the iridium from titanium. The most direct way to do this is to precipitate the iridium by hydrogen sulphide.[9] Certain difficulties were encountered in an attempt to use hydrogen sulphide, such as incomplete pr c a n of iridium and partial hydrolysis of titanium compounds, even though considerable quantities of tartaric acid were added to the solutions. The deposition of titanium dioxide was particularly noticeable on the walls of the beaker just above the level of the solutions. The titanium dioxide often formed an adherent mirror on the glassware. It likewise contaminated the sulphide precipitate and could be removed later only by fusion with bisulphate. The recovery of iridium by hydrogen sulphide will undoubtedly have many advantages when the proper conditions of control have been fully established. For the present, however, it seemed expedient to consider some other means by which the iridium could be recovered.

Preliminary experiments showed that the titanium could be entirely eliminated by precipitation with cupferron (ammonium salt of nitrosophenylhydroxylamine, $C_6H_5N \cdot NO \cdot ONH_4$). It was found that the small quantity of iridium, of the order of 1 mg or less, which contaminated the first cupferron precipitate, was completely recovered by reprecipitating the titanium.

Previous work on the determination of ruthenium [10] and of osmium [11] by hydrolytic precipitation suggested that a similar method was feasible for iridium. Sodium bicarbonate was added to boiling aqueous solutions of ammonium chloroiridate until the color became green. Sodium bromate was then added and the solutions were boiled for 25 minutes to coagulate the finely divided precipitate and to insure complete precipitation. The greenish-black precipitates were treated as described in the procedure given later. The acidities of the filtrates from the hydrolytic precipitation were determined approximately by means of indicator solutions. In four of the solutions from which complete recovery of the iridium was obtained, the acidities ranged from pH 4 to pH 4.6. In the fifth filtrate the pH

---

[9] This method of recovery of iridium has certain advantages which make it of importance in the analytical chemistry of the platinum group. While the exact conditions which will allow complete precipitation of iridium have not yet been fully established, further work is being done and will appear later in the description of a general method for the separation of the platinum group.
[10] Gilchrist, B. S. Jour. Research, vol. 3, p. 993, 1929.
[11] Gilchrist, B. S. Jour. Research, vol. 6, p. 421, 1931.

value was somewhere between 1.5 and 4. This filtrate was found to contain a small quantity of iridium.

In order to establish proper conditions for the iridium precipitation, it became necessary to be able to detect small amounts of iridium in the filtrates from the hydrolytic precipitations. A test which has been frequently employed in this laboratory consists in the development of a green or blue color [11] when a concentrated solution of sulphuric acid, containing iridium, is heated with nitric acid. With quantities of iridium less than 0.1 mg, the color developed is green, with larger amounts it is blue. A few precautions, however, must be observed. Chlorides appear to decompose the particular compound responsible for the blue color. The sulphuric acid, therefore, must be heated until heavy vapors are evolved and this treatment repeated, after the addition of a small quantity of nitric acid, until all hydrochloric acid is expelled. Furthermore, the strongly acid solution must not be heated for too long a time after the last addition of nitric acid. If these two precautions are observed, the color will often remain visible for as long as a week. The test is likewise applicable when considerable quantities of alkali sulphates are present. If the composition of the evaporated solution be made to approximate that of sodium bisulphate, with a slight excess of sulphuric acid, the color of the salt on solidification will often remain visible for several weeks, changing slowly from blue to violet with the absorption of water from the air. The limiting quantity of iridium capable of certain detection, under the conditions mentioned above, appeared to be 0.1 mg.

The results of the determination of iridium by hydrolytic precipitation from aqueous solutions of ammonium chloroiridate are given in Table 1. The error in No. 12 is ascribed to the fact that the solution was too acid.

TABLE 1.—*Recovery of iridium from aqueous solutions of ammonium chloroiridate by hydrolytic precipitation*

No	$(NH_4)_2$ IrCl$_6$ taken 42.09 per cent Ir	Iridium present	Iridium recovered	Error	pH of filtrate	H_2SO_4-HNO_3 test of filtrate
	g	*g*	*g*	*g*		
10	0.1550	0.0790	0.0790	0.0000	[1] 4	Very faint. Greenish color.
12	.2210	.0943	.0931	−.0012	1.5–4	Deep blue.
13	.2602	.1111	.1112	+.0001	4.3–4.5	Colorless.
14	.3121	.1332	.1332	.0000	[1] 4	Do.
15	.1869	.0798	.0798	.0000	[1] 4.6	Do.

[1] Approximately.

Since the solution from which the iridium must be recovered after the elimination of titanium will contain a considerable quantity of sulphuric acid, it was necessary to establish the conditions for precipitating iridium from such a solution. Solutions of iridium in 10 ml of sulphuric acid were prepared according to directions given later in the section on recommended procedure. These solutions were diluted with 20 ml of water and 10 ml of hydrochloric acid and boiled for 15 minutes. The solutions were then further diluted, filtered, and

[11] Fresenius, T. W., Qualitative Chemical Analysis (John Wiley & Sons, New York, N. Y.), English translation of 17th German ed. by C. Ainsworth Mitchell, 1921, p. 377.

almost neutralized with a solution of sodium bicarbonate. A considerable quantity of sodium bicarbonate was required (approximately 30 g), but this caused no difficulty in the precipitation. The desired end point (pH 4), in the addition of the sodium bicarbonate solution, was indicated by the appearance of a blue color when a drop of brom phenol blue solution (0.04 per cent) was allowed to run down the stirring rod to which clung a drop of the boiling solution. The presence of bromate made the above manner of testing the only one feasible. The solutions were boiled for 25 minutes and filtered. The precipitates were treated as directed in the section on recommended procedure.

The results of the determination of iridium by hydrolytic precipitation at pH 4 from solutions strongly acid with sulphuric acid and hydrochloric acid are given in Table 2. The incomplete precipitation in Nos. 56 and 57 is ascribed to the fact that the desired end point was not quite reached.

TABLE 2.—*Recovery of iridium from solutions of iridium sulphate in sulphuric acid by hydrolytic precipitation*

No.	$(NH_4)_2IrCl_6$ taken 42.70 per cent Ir	Iridium present	Iridium recovered	Error	H_2SO_4-HNO_3 test of filtrate
	g	*g*	*g*	*g*	
56	0.4139	0.1767	0.1762	−0.0005	Faint bluish color.
57	.3297	.1408	.1403	−.0005	Do.
58	.4029	.1720	.1721	+.0001	No iridium detected.
24	.3510	.1499	.1501	+.0002	Faint greenish color, not more than 0.2 mg Ir.
28	.3410	.1456	.1457	+.0001	Very faint greenish color, not more than 0.1 mg Ir.
32	.4128	.1763	.1764	+.0001	Do.

While it was found that iridium was completely precipitated from solution by hydrolysis at pH 4 in the presence of bromate, it was desirable to learn whether the precipitation was quantitative over a range of pH values and whether the addition of bromate was essential. It was found that precipitation at pH 6 was quantitative and that a shorter time of boiling, 10 minutes, was sufficient, provided bromate was present. Solutions of iridium which had been merely saturated with chlorine failed to yield complete precipitation. Solutions of iridium, which had been evaporated with nitric acid and subsequently with hydrochloric acid, turned green at pH 6 and did not yield complete precipitation without the addition of bromate. In connection with this last-named treatment, it was found that the previous addition of nitric acid did not prevent the precipitation of iridium as it does in the case of ruthenium.[13]

IV. METHOD RECOMMENDED FOR THE SEPARATION OF RHODIUM FROM IRIDIUM AND FOR THE DETERMINATION OF THESE METALS

1. PREPARATION OF THE SOLUTION

Place the solution containing the rhodium and iridium as chlorides in a 500 ml Erlenmeyer flask, closed with a short-stemmed funnel. Add 10 ml of sulphuric acid and 2 to 3 ml of nitric acid and evaporate

[13] See footnote 10, p. 549.

until heavy vapors of sulphuric acid are evolved. Add several portions of nitric acid from time to time and continue to heat over a free flame, keeping the solution in constant motion. Dilute the cooled solution with 20 ml of water and again evaporate until vapors of sulphuric acid appear. This is done to destroy nitroso compounds which may interfere in the precipitation of rhodium by titanous chloride.

2. SEPARATION AND DETERMINATION OF RHODIUM

Transfer the solution to a clean, unetched beaker, dilute it to 200 ml, and heat it to boiling. Add dropwise a solution of titanous chloride (a 20 per cent solution of this reagent may be purchased) until the supernatant liquid appears slightly purple. If the solution is placed over a 100-watt light and stirred, observation of the end point is greatly facilitated. Boil the solution for two minutes and filter it. Wipe the walls of the beaker and also the stirring rod with a piece of ashless filter paper. Wash the filter and precipitated metal thoroughly with cold diluted sulphuric acid (2.5 + 97.5). Place the filter and contents in a 500 ml Erlenmeyer flask, add 10 ml of sulphuric acid, char gently, then heat the solution until vapors of sulphuric acid appear. Destroy the organic matter with nitric acid. Fuming nitric acid is preferable to the ordinary concentrated acid. If solution of the rhodium is not complete, that is, if some black specks remain unattacked, dilute the solution, filter it and return the filter to the flask. Wipe down the walls of the flask with a piece of ashless filter paper. Add 5 ml of fresh sulphuric acid, char the paper and destroy all organic matter with nitric acid. Heat the solution until heavy vapors of sulphuric acid are evolved. This treatment will insure complete solution of any remaining metal and will leave only a slight deposit of colorless silica.

Precipitate the rhodium a second time in the manner described above. Redissolve the rhodium as before, dilute the sulphuric acid solution with 20 ml of water and 10 ml of hydrochloric acid and boil the resulting solution for 15 minutes. This treatment is necessary to convert the rhodium to a form which will allow complete precipitation by hydrogen sulphide. Dilute the solution, now rose colored. Filter it and precipitate the rhodium by hydrogen sulphide from a boiling solution having a volume of from 400 to 500 ml. Filter the solution and wash the sulphide precipitate with cold diluted sulphuric acid (2.5 + 97.5) and finally with water. Place the filter with the sulphide precipitate in a porcelain crucible. Ignite the dried precipitate carefully in air. Ignite the oxidized residue in hydrogen, cool in hydrogen, and weigh as metallic rhodium.

3. RECOVERY AND DETERMINATION OF IRIDIUM

Dilute the combined filtrates from the precipitation of rhodium by titanous chloride to 800 ml. Cool the solution by placing the beaker in crushed ice. Add a chilled, filtered, freshly prepared solution of cupferron (6 per cent) in slight excess. Add some filter paper pulp, filter by suction and wash the titanium precipitate with chilled diluted sulphuric acid (2.5 + 97.5) containing some cupferron. The paper pulp is added to prevent the precipitate from becoming too densely packed. Return the precipitate to the beaker and add nitric acid until the compound is mostly decomposed. Add 20 ml of sulphuric acid and heat until vapors of sulphuric acid appear. Destroy all organic matter with nitric acid. Dilute the resulting solution to 800 ml and repeat the precipitation of the titanium. The first cupferron precipitate will usually be contaminated by about 1 mg or less of iridium. Unite the two filtrates from the cupferron precipitations and evaporate until approximately 10 ml of sulphuric acid remains. Insure the destruction of all organic matter by adding nitric acid. Dilute the sulphuric acid solution with 20 ml of water and 10 ml of hydrochloric acid and boil the resulting solution for 15 minutes. Dilute the solution somewhat, filter it, and precipitate the iridium by hydrolysis. To do this, add a filtered solution of sodium bicarbonate until the acidity of the solution reaches a value of pH 6. Since the color of the solution will, in general, mask the color of the indicator, it is necessary to make the test on a drop of the solution clinging to the stirring rod. If a drop of brom cresol purple indicator solution (0.04 per cent) is colored faintly blue when it comes in contact with the drop of solution on the stirring rod, the solution possesses an acidity which will allow complete precipitation of the iridium. At this point, add 20 ml of a 10 per cent solution of sodium bromate and boil the solution for 20 to 25 minutes. Make sure, however, that sufficient bromate is present to oxidise all of the iridum to the quadrivalent state. Filter the solution and wash the precipitate thoroughly with a hot 1 per cent solution of ammonium chloride. Place the filter and precipitate of hydrated iridium dioxide in a porcelain crucible. Dry somewhat and moisten the contents of the crucible with a saturated solution of ammonium chloride. This is done to prevent deflagration of the hydrated dioxide on subsequent ignition in air. Ignite the filter and precipitate in air and then in hydrogen. Leach the metallic residue with diluted hydrochloric acid $(1+5)$, transfer it to a filter and wash it with hot water. This precaution is taken to insure complete removal of soluble salts. Ignite the metallic residue in air. Finally, ignite the resulting oxidised metal in hydrogen, cool in hydrogen and weigh as metallic iridium.

The results of the determination of rhodium and of iridium, obtained by following the procedure of separation and of recovery which has just been outlined, with the exception that the iridium was precipated at pH 4, are given in Table 3. The result of one experiment, in which rhodium was twice precipitated by titanous chloride in the absence of iridium, is also included.

TABLE 3.—*Results of the analysis of solutions containing both rhodium and iridium*

No.	(NH₄)₃ RhCl₆ taken 30.48 per cent Rh	Rhodium present	Rhodium recovered	Error	(NH₄)₃ IrCl₆ taken 42.70 per cent Ir	Iridium present	Iridium recovered	Error
	g	*g*	*g*	*g*	*g*	*g*	*g*	*g*
25	0.1824	0.0556	0.0556	0.0000				
20–6	.1478	.0451	.0451	.0000	0.1722	0.0735		
60–48	.4444	.1355	.1354	−.0001	.2584	.1103		
66–54	.4732	.1442	.1442	.0000	.3309	.1413	0.1412	−0.0001
67–55	.2991	.1216	.1215	−.0001	.3804	.1624	.1622	−.0002
14–43	.2757	.0840	.0838	−.0002	.3092	.1320	.1320	.0000

NOTE.—A very faint greenish color was produced when the filtrates from the hydrolytic precipitation of iridium in Nos. 66–54 and 67–55 were tested for iridium. No iridium was detected in the filtrate from No. 14–43.

To make certain that the second precipitation by cupferron suffced to separate all of the iridium from the titanium, the second cupferron precipitates were redissolved and the titanium reprecipitated. No trace of iridium was found in the filtrates from the third precipitation of the titanium, nor in the solutions of the third precipitates themselves.

The filtrates from the hydrolytic precipitations of the iridium were likewise tested. A small amount of iridium, not exceeding 0.1 mg, was found in the filtrates from Nos. 66–54 and 67–55.

The chemical examination of the recovered iridium for traces of rhodium is extremely difficult, as no sensitive test for rhodium is known. From the fact that rhodium was found to be completely precipitated by titanous chloride from pure solutions of rhodium and from the fact that the quantities of rhodium recovered from actual separation were identical with the quantities taken, it is reasonable to conclude that no significant amount of rhodium contaminated the recovered iridium.

The chemical examination of the separated rhodium for traces of iridium is simple because a sensitive test for iridium exists. A quantity of iridium, 0.1 to 0.2 mg, is sufficient to give a greenish cast to the yellow solution of rhodium in concentrated sulphuric acid. Since no iridium was detected either in the solutions of the rhodium which had been precipitated twice by titanous chloride or in the filtrates from the hydrogen sulphide precipitation of the rhodium, it was concluded that the recovered rhodium was not contaminated by iridium.

V. DISCUSSION OF THE METHOD

In the early experiments on the separation of rhodium from iridium the first precipitation was made from solutions of rhodium and iridium chlorides in diluted sulphuric acid (5+95) at room temperature. A considerable excess of titanous chloride was added and the solutions were filtered after definite periods of time. The precipitated metal was finely divided and partially passed through the filters. It was very difficult to observe the filtrates for escaped metal because of the intense color of the reagent. The large excess of titanous chloride appeared to cause the deposition of too great a proportion of the iridium. The deposition of iridium was even more pronounced when the solutions were heated in an attempt to coagulate the precipitates. Inasmuch as reprecipitation of the rhodium had to be made from a solution of the metal sulphates in diluted sulphuric acid, because of the method used to redissolve the precipitate, the first precipitation was also made from a similar solution in the later experiments. When both precipitations were made from a boiling solution of the metal sulphates and only a very slight excess of titanous chloride was added, it was found that the second precipitation of the rhodium completely eliminated the last trace of iridium.

Although the other four platinum metals must be absent at this stage of the general analysis, it was, nevertheless, desirable to know how they behaved toward titanous chloride. The reagent was added to solutions containing approximately 50 mg of each metal in 100 ml of diluted hydrochloric acid (5+95) at room temperature. Palladium appeared to be precipitated immediately and platinum almost immediately. The brown color of ruthenium chloride was changed to deep blue, but no precipitation occurred when the solution was allowed to stand for one week. No precipitation occurred with a solution of chloroosmic acid even when it was allowed to stand for three weeks.

When solutions of rhodium, or of rhodium and iridium, are heated to heavy vapors of sulphuric acid, precaution must be taken to prevent the formation of insoluble compounds by local superheating. The formation of such compounds is easily avoided by keeping the hot sulphuric acid solution in constant motion.

The essential feature which makes the scheme proposed by Wada and Nakazono workable is a method by which the precipitated rhodium can be redissolved. A few comments regarding the technic of the operation appear worthy of emphasis. An Erlenmeyer flask should be used in preference to a beaker or to a Kjeldahl flask. With a beaker, there is danger of mechanical loss of material, particularly at the danger of mechanical loss of material, particularly at the stage where the last traces of moisture are being eliminated. The drops of condensed steam on the under side of the cover glass fall into the hot sulphuric acid and' cause violent spattering. With a Kjeldahl flask, it is extremely difficult to wipe down the particles of undissolved metal from the walls. If this latter precaution is not observed, the results for the recoveries of rhodium will invariably be low by amounts up to about 1 mg.

The metallic rhodium, obtained in the second precipitation with titanous chloride, must not be ignited directly because it is always contaminated by silica, introduced through the titanous chloride and possibly from the glassware, and by a small amount of titanium. Consequently, the rhodium is again dissolved and precipitated by hydrogen sulphide. It was not found possible to precipitate the rhodium completely by hydrogen sulphide from solutions of the metal sulphate in sulphuric acid. This same difficulty has often been observed in this laboratory with solutions prepared by dissolving the product of bisulphate fusions of rhodium. A solution of the rhodium can be prepared, however, by adding diluted hydrochloric acid $(1+2)$, equal to three times the volume of the concentrated sulphuric acid solution, and boiling it. The yellow color of the sulphate solution then changes to rose red, which is the same as the characteristic color of rhodium chloride. No difficulty attends the precipitation of rhodium from the solution so treated.

The determination of rhodium and of iridium can be greatly simplified and the titanium precipitations avoided if the solution can be divided conveniently into aliquot parts. Incidental experiments showed that rhodium is completely precipitated as an olive-green hydrated dioxide at pH 6 when bromate is present. Since iridium is precipated quantitatively under the very same conditions, the two can be precipitated together and finally recovered as a metallic mixture. In addition, rhodium must be determined separately on a different portion by precipitation with titanous chloride in the manner described in the recommended procedure.

The method which has been described in this paper is designed for the separation of rhodium from iridium and for the determination of these metals when other elements are absent. It will constitute a part of a general scheme for the analysis of the platinum group and will follow the procedures for the separation of osmium, ruthenium, platinum, and palladium.

WASHINGTON, July 11, 1932.

THE ISOELECTRIC POINT OF SILK

By Milton Harris [1]

ABSTRACT

Suspensions of silk in buffer solutions of different pH were prepared by grinding the dry silk fibroin to a fine powder and shaking it in the buffer solution.

Colloidal solutions of silk were prepared by dissolving the silk fibroin in a 50 per cent lithium bromide solution and dialyzing it to remove the salt. To the solutions were added known buffer mixtures and a small amount of purified quartz powder.

Electrophoretic measurements of the suspensions of silk and of the dissolved silk adsorbed on the quartz particles gave an isoelectric point at pH 2.5.

CONTENTS

I. INTRODUCTION

A colloidal (or other) particle is said to be at its isoelectric point when the particle is electrically neutral with respect to its surrounding medium. This point is usually defined in terms of some property such as the pH of the medium. For an amphoteric substance the acidic and basic ionizations are equal at this point.

The most direct method of determining the isoelectric point is to define the conditions under which the particles do not move in an electric field. Such movement is called electrophoresis.

Electrical mobility measurements have been used in the determination of the isoelectric point of wool.[2] It is the purpose of this paper to present similar measurements of silk.

Satisfactory suspensions of wool could only be prepared mechanically; that is, by grinding the wool to a fine powder, whereas silk may be readily prepared in suspensions and in colloidal solution.[3] Abramson[4] has shown that when quartz particles are suspended in dilute protein solutions, the protein is absorbed on the surface of the quartz particle, and the latter assumes the electrokinetic properties of the protein. This suggested a method for determining the isoelectric point of the dissolved silk.

II. PROCEDURE

The electrophoresis apparatus and methods used in this work were the same as used in the determination of the isoelectric point of wool.[5]

[1] Research associate representing the American Association of Textile Chemists and Colorists.
[2] M. Harris, The Isoelectric Point of Wool, B. S. Jour. Research vol. 8, pp. 779-786, 1932.
[3] M. Harris and T. B. Johnson, Study of Silk Fibroin in the Dispersed State, Ind. & Eng. Chem., vol. 22, p. 965, 1930.
[4] H. A. Abramson, The Adsorption of Serum Proteins by Quartz and Paraffin Oil, J. Gen. Physiol., vol. 13, pp. 169-177, 1929.
[5] See footnote 2.

The silk was purified as follows: A sample of raw China silk was boiled successively in three ½ per cent solutions of soap for 20 minutes each. It was then washed with distilled water, dilute ammonia, dilute hydrochloric acid, and finally repeatedly with distilled water, at 60° C. It was then air dried for 24 hours.

One portion of the purified silk was ground in a Wiley mill. The size of the silk particles varied from 5 μ to 30 μ. A small amount of the ground silk was shaken vigorously with distilled water, and the larger particles allowed to settle. The upper portion was carefully decanted and used for one set of measurements.

A second portion of approximately 5 g of the silk was dissolved in 50 ml of a 50 per cent solution of lithium bromide at about 80° C. The solution was diluted to 200 ml, treated with a few drops of toluene to prevent bacterial action and dialyzed in a cellophane bag against distilled water until it gave no test for LiBr. Approximately 20 ml of the resultant solution of silk was diluted to 200 ml and the purified quartz particles described below were added. The silkcoated quartz thus obtained was used for the second set of measurements.

Quartz particles (size range 5 μ to 20 μ) were purified as described by Abramson.[6] Crude quartz powder was heated with a cleaning solution consisting of concentrated sulphuric acid saturated with sodium dichromate and diluted with water, for about 30 minutes. The particles were allowed to settle and the excess acid carefully decanted. A large amount of distilled water was added and the quartz suspension was filtered. The powder was washed with distilled water and then boiled with an excess of normal hydrochloric acid and allowed to settle. It was again filtered and washed with water for 48 hours.

In making the suspension to be studied, a very small amount of the quartz powder was added to the silk solution, which made it faintly cloudy.

Portions of each of these suspensions were mixed with equal volumes of the following buffer mixtures [7] with double the usual concentrations: pH 1.4 to 2.2, potassium chloride-hydrochloric acid mixture; pH 2.4 to 3.8, acid potassium phthalate-hydrochloric acid mixture; pH 4.0 to 6.0, acid potassium phthalate-sodium hydroxide mixtures.

III. DISCUSSION OF RESULTS

The mobilities of the particles recorded in the following table are the averages calculated from 10 observations.

TABLE 1.—*Electrophoretic mobility of silk (relative velocity)*

pH	Ground silk	Dissolved silk+ quartz	pH	Ground silk	Dissolved silk+ quartz
1.42	+1.20	+1.14	2.86	−0.48	−0.36
1.84	+.73	+.79	3.00	−.57	−.58
2.13	+.48	+.37	3.33	−.88	−.82
2.30	+.24	+.26	3.69	−1.15	−1.27
2.54	−.14	+.11	4.21	−1.37	−1.40

[6] See footnote 4, p. 557.
[7] M. Clark, The Determination of Hydrogen Ions, pp. 106-107, Williams & Wilkins Co.

These data are shown in graphical form in Figure 1. The isoelectric point of silk obtained from the pH-mobility curve is pH 2.5. This value is in good agreement with the values obtained by Hawley and Johnson [8] and by Harris and Johnson [9] but not in agreement with those obtained by Meunier and Rey [10] and Denham and Brash.[11]

Hawley and Johnson employed a U tube electrophoresis cell and obtained values in the range from pH 1.4 to 2.8. Since the diameter of the cell was comparatively large, diffusion and convection currents

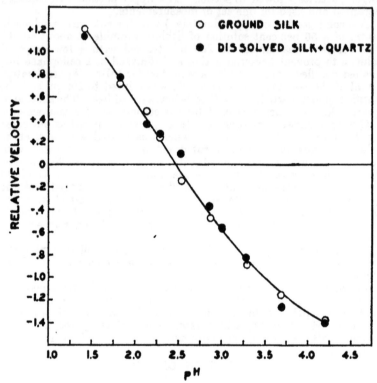

FIGURE 1.—*pH-mobility: curve for silk*

made it difficult to detect small changes in protein concentration in the anode and cathode regions.

Harris and Johnson prepared solutions of silk fibroin by dissolving the degummed silk in concentrated aqueous solutions of lithium bromide and calcium thiocyanate. From solubility and viscosity measurements, values ranging from pH 2.1 to pH 2.4 were obtained.

[8] T. G. Hawle and T. B. Johnson, The Isoelectric Point of Silk Fibroin, Ind. & Eng. Chem., vol. 22, pp. 297–299, 1930.y
[9] M. Harris and T. B. Johnson, Study of the Fibroin from Silk in the Isoelectric Region, Ind. & Eng. Chem., vol. 22, pp. 539–542, 1930.
[10] L. Meunier and G. Rey, Determination du point isoelectrique de la laine et de la soie, Compt. rend, vol. 184, p. 285, 1927.
[11] W. Denham and W. Brash, The Isoelectric Point of Silk Fibroin, J. Textile Inst., vol. 18, p. T520–525, 1927.

Meunier and Rey reported an isoelectric point for silk at pH 4.2. Their results were based on the swelling of the fiber in solutions of varying hydrogen ion concentration. The curve which they obtained is flat in the region from about pH 3 to pH 6 and shows no well defined point of minimum swelling.

Denham and Brash studied the combination of acidic and basic radicals with the silk fibroin when the latter was placed in solutions of acids, bases, and salts. They, too, were unable to obtain a definite isoelectric point and concluded from their experiments that it lies between pH 3.6 and pH 4.0. However, there is ample evidence to show that the combination of proteins with positive and negative ions is not a true measure of the isoelectric point. The isoelectric point is a point of minimum and not zero ionization. Consequently the protein will combine with both positive and negative ions within a certain range on both sides of the isoelectric point.

The low isoelectric point of silk shows it to be one of the most acid proteins known. This is in accordance with some of the properties of silk. Meunier and Rey [12] found silk to contain only 7 parts of amino nitrogen per 10,000. The pronounced affinity of silk for basic dyestuffs is further confirmation of the predominance of its acidic nature.

The shape of the pH-mobility curve is of interest, and shows that the acidic and basic properties are less pronounced than those of wool.[13] Any changes with pH which occur near the isoelectric point are only gradual. This accounts for the large isoelectric range obtained by workers who have used less sensitive methods.

IV. ACKNOWLEDGMENT

This investigation was made possible by a grant to the American Association of Textile Chemists and Colorists by the Textile Foundation. We wish to express our appreciation for the aid given.

Washington, July 8, 1932.

---

[12] See footnote 10, p. 559.
[13] See footnote 2, p. 557.

EFFECTIVE APPLIED VOLTAGE AS AN INDICATOR OF THE RADIATION EMITTED BY AN X-RAY TUBE

By Lauriston S. Taylor, G. Singer, and C. F. Stoneburner

ABSTRACT

A previous study [1] suggested a parallelism between the applied effective voltage and the emission of an X-ray tube. The investigation has been carried further under more carefully controlled conditions. It is found for two mechanical rectifiers and one "constant potential" generator that the X-ray emission per effective (root-mean-square) milliampere of tube current is about the same for all generators operating at the same effective (r. m. s.) voltage (the radiation passing through the same filter in all cases). Thus it becomes possible to express all radiations in terms of that excited by constant potential. This leads to a simplification in the description and reproduction of irradiation conditions in biological work. Although the study has been limited to three generators, it is believed that they are sufficiently typical to warrant generalization.

CONTENTS

I. INTRODUCTION

In a recent study the energy emitted by an X-ray tube, as measured by an ionization chamber, was found to be very different when excited by different types of high voltage generators furnishing the same average tube current and peak voltage (as measured by a sphere gap). But it was further found that for the same average currents and the same effective voltage [2] of the generators, the emission, as measured by the ionization chamber, is approximately the same.

The total energy in an X-ray spectrum devoid of characteristic lines varies as the square of the instantaneous applied voltage, and hence for a known voltage wave form the total energy and its wavelength distribution, for the period of a complete cycle, is calculable. Because of the presence of tungsten line radiation about 70 kv, it is not yet practicable to predict the total X-ray energy by integrating over one cycle of the voltage wave. In the present work the X-ray emission per unit tube current, as it varies with peak voltage, with average voltage, and with effective voltage per unit tube current is given for more carefully controlled and wider ranges of conditions. This is expressed in terms of the X-ray emission per milliampere.

---

[1] L. S. Taylor and K. L. Tucker, B. S. Jour. Research, vol. 9 (R P 475), p. 333, 1932.
[2] Effective voltage is mathematically the same as root-mean-square (r. m. s.) voltage.

II. APPARATUS

Thus far three generators have been used, designated by A, B, and C as in the preceding work.[3] A is a full wave mechanical rectifier having a single high tension transformer, and rectifying over approximately 30 degrees of cycle; B is also a full wave mechanical rectifier but with a divided high tension transformer (two transformers in a single tank), and rectifying over approximately 20 degrees of the cycle; C is a valve-tube-condenser ripple potential[4] generator (so-called "constant potential") having a ripplage of about 2 per cent per milliampere.   The three generators could be interchangeably conconnected to the same X-ray tube.

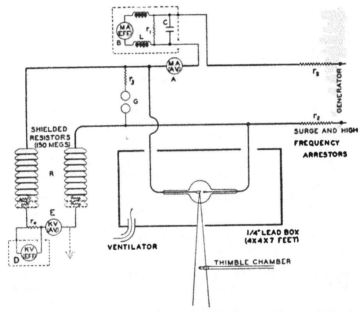

FIGURE 1.—*Apparatus showing voltage and current measuring equipment*

1. CURRENT MEASUREMENTS

The arrangement of the apparatus is indicated in Figure 1.   Average current through the tube is measured in the usual manner by a d. c. milliammeter A; and the effective current by an a. c. milliammeter, B.   For this purpose a thermo milliammeter alone was available. This was protected from high-frequency oscillations by iron core chokes, L, having a d. c. resistance of a fraction of an ohm; a noninductive resistor, r, of about 10,000 ohms, and a 1/20 μf capacitor, C.

[3] See footnote 1, p. 561.

[4] The term "constant potential" has been used loosely to describe the potential supplied by kenotron or other valve tube rectification.   A more accurate designation of a voltage which is not constant in fact but fluctuates slightly about a certain average value is "ripple voltage."   Thus by a "ripple quantity" (potential or current) is meant a simple periodic quantity $y = V_0 + V_1 \sin(\omega t + \alpha_1) + V_2 \sin(2\omega t + \alpha_2) + \cdots$ in which the constant term (V_0) is so large that all values of the quantity are positive (or negative).   The amount of ripple ("ripplage" or "ripplance") in a ripple quantity is the ratio of the difference between the maximum and minimum values of the quantity to the average value.   These definitions are under consideration by the Committee on Electrical Definitions of the American Standards Association under the sponsorship of the A. I. E. E.

Since the resistance of the meter, B, was only about 20 ohms, this
shunt system served as a very effective filter against high-frequency
oscillations in the high-tension system, without affecting the sensitivity.

Excessive high-frequency oscillation when using the mechanical
rectifiers caused difficulty in peak-voltage measurements with the
sphere gap, but this was largely eliminated by placing two 1/2-
megohm inductively wound surge resistors, r_2, immediately next to
the rectifier switches, giving smoother operation of the equipment
and more consistent peak voltage measurements. After raising the
generator voltage sufficiently to compensate for the potential drop
across the resistors, it was found that their presence did not alter the
tube output in any way.

With generator C, the same resistors were used for the purpose of
limiting the current in case of slight tube gassing or when using a
sphere gap for measuring the voltage. It was found that they raised
very materially the voltage limit at which a given tube would operate
satisfactorily.

2. VOLTAGE MEASUREMENTS

Peak voltage of all three generators was measured by the gap
between 12.5 cm spheres, protected by a water resistance r_3, of the
order of 5 megohms to prevent heavy arcing at the gap and consequent tripping of circuit breakers.

Average voltage was measured by means of a d. c. microameter,
E, in a series with a shielded high voltage resistor of 150 mehohms,[5]
the meter being at the center of the resistance so as to be at or near
ground potential.[6]

Effective voltage was measured by means of a 150-volt calibrated
Kelvin multicellular electrostatic voltmeter, D, shunted across a
resistance, r_4, of 75,000 ohms. Since its potential was likely to be
different from ground, it was necessary to shield it by a metal case
having the same potential as one of its terminals.

It should be stated here that the insertion of the high-voltage
resistor, R, across the line has no noticeable influence upon the operation of the high-tension generators because the current it shunts off
is very small. This fact is of course essential, since a wave form
very different from that under normal operating conditions may
otherwise result.

3. RADIATION MEASUREMENTS

The X-ray emission was measured by means of a thimble chamber,
kept in a fixed position with respect to the tube throughout the study.
The beam was in all cases filtered by 0.525 mm of copper, the wall
effect of the chamber being negligible for the transmitted radiation.

No other determination of radiation quality was undertaken, as
the previous study[7] indicated what was to be expected under the
operating conditions employed here, namely, all beams having the
same effective voltage and filtration and giving equal intensities
have approximately the same quality (whether expressed in the half-
value layer of copper, copper absorption coefficient, or effective
wave length).

[5] L. S. Taylor, B. S. Jour. Research, vol. 5 (RP217), p. 609, 1930.
[6] The voltage on the high-tension conductors is seldom divided equally above and below ground potential, although the error in measurement introduced by grounding the resistor at the center is very small.
[7] L. S. Taylor and K. L. Tucker. See footnote 1, p. 561.

III. EXPERIMENTAL RESULTS

In this study, to eliminate the effect of variation in operation, simultaneous values were obtained for (1) average current, (2) effective current, (3) peak voltage, (4) average voltage, (5) effective voltage, and (6) X-ray tube output for a constant filteration of 0.525 mm of copper, the average current (the quantity measured in practice) serving as independent variable.

In Figure 2, groups A and B show the tube emission, expressed in terms of the ionization current per average milliampere (I/ma (average)), as a function of peak voltage for generators A and B, respectively, and for different tube currents, 2, 3, 4, and 5 ma. With generator B it is seen that for a given peak voltage, the value of I/ma (average) varies from 10 to 15 per cent in changing the operating

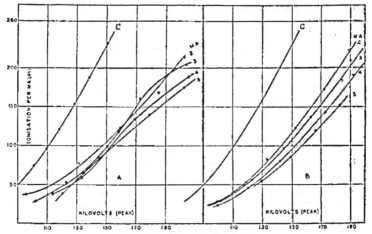

FIGURE 2.—*Relationship between X-ray tube emission per average milliampere, and the applied peak voltage*

A, B, mechanical rectifiers; C, constant potential.

tube current by 1 mA. This is not unexpected since the generator wave form depends to a large extent upon the power drawn from the generator. It is obvious from these curves that it may be very misleading to attempt to deduce the emission of an X-ray tube from current and peak voltages alone. For generator A, I/ma (average) varies erratically with peak voltage under different operating currents.

Figure 2, curve C in both groups is from generator C ("constant potential") for a tube current of 4 ma. Since the peak voltage of this generator does not vary appreciably with tube current (average voltage being kept constant), a single curve embodies the results. It is seen that for the same peak voltage there is a large difference between the X-ray emission of generator C (curve C) on the one hand and that of generators A and B (curves A and B) on the other. To obtain the same X-ray emission from generators A and B as from generator C, the peak voltage of A and B must be very much larger than that of C, depending upon the operating current.

If, for the same radiations plotted in Figure 2, we plot instead the ionization per average milliampere against the average voltage, corresponding curves, *A*, *B*, and *C* in Figure 3 are obtained. It will be noted that I/ma (average) again varies with the tube current. In the case of generator *A* the values of *I*/ma (average) vary from about 2 to 8 per cent in changing the operating current by 1 ma. This percentage change becomes smaller, however, in going to higher average voltages, whereas, when the potential is measured in peak kilovolts, the changes become larger at higher voltages.

In the case of generator *B* the same general results obtain. The variation of *I*/ma (average) with tube current at a given average voltage is, however, much greater at lower potential than for *A* while at higher potentials it is about the same.

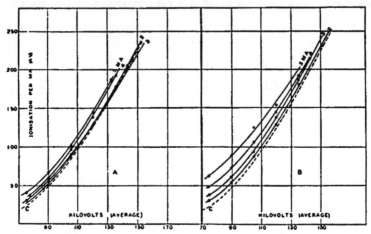

FIGURE 3.—*Relationship between X-ray tube emission per average milliampere, and the applied average voltage*

A, *B*, mechanical rectifiers; *C*,'constant]potential.

It is seen that, while the curves for *A* and *B* differ with tube current, they all lie in the region of curve *C* (dashed) as contrasted to the corresponding curves of Figure 2 for peak voltages. Thus, if the current and voltage for all three generators be measured in terms of average milliamperes and average kilovolts (that is, by using d. c. instruments), the tube emission per milliampere is in all cases of the same order of magnitude.

Using effective milliamperes and effective kilovolts, curves for generators *A*, *B*, and *C* for the X-ray emission per effective milliampere are given in Figure 4.

It should be pointed out that for the comparatively small ripplage present in the voltage of generator *C*, the difference between the average and effective voltage is so small that it may be neglected. Likewise their variation with tube current is negligible; hence, a single curve for *I*/ma (effective) gives a satisfactory basis of comparison.

In the case of generator *A* the change in emission per milliampere, as the operating current is varied, is present and somewhat erratic as in the case of the peak voltage curves. The percentage change, however, is the least of the three cases given. Similarly in the case of generator B the change is present, but again to the least degree.

Of prime importance, however, is the fact that the curves (fig. 4) for both generators A and B lie very closely along the curve for constant potential—much more closely at the higher voltages than those in Figure 3 for average current and voltage measurements. It is concluded that, for equal effective X-ray tube currents and equal effective voltages, the emission per milliampere will be sufficiently near the same with all three generators, for the effective current and voltage to be used as characteristic of the X-ray emission.

Since the emission per milliampere for all strictly constant potential generators is the same, it is logical that it be referred to as a base. By means, then, of the effective current and effective voltage mea-

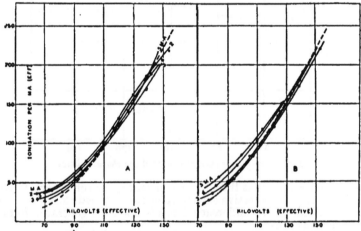

FIGURE 4.—*Relationship between X-ray tube emission per effective milliampere, and the applied effective voltage*

A, B, mechanical rectifiers; broken line curve, constant potential.

surements described, it is possible to relate consistently the X-ray emission excited by the various types of generators.

It is possible to trace the residual variation in X-ray emission per milliampere at lower voltages with the operating current, largely to the current-voltage characteristics of the X-ray tube. This will account also for the divergence from the constant potential base line. Figure 5 shows, for various applied constant potentials, the tube current curves for three values of the filament current, I_f. It is seen that, since the tube current is still increasing about 5 per cent per 30 kilovolts at 120 kilovolts, saturation is not attained. Consequently, with the application of a varying voltage to the tube, the effective current is bound to depend upon the voltage wave form to a varying degree.

This is illustrated in Figure 6 where curves a and b are two hypothetical tube current characteristics, the second of which is ideal in that it reaches saturation at a comparatively low voltage. Curves D, E, and F represent three types of generator voltage wave form

having the same effective voltage, *e*, which, when applied to tubes
having characteristics *a* and *b*, produce the tube current wave forms
a′ and *b′*, respectively, for each associated voltage wave. In each of

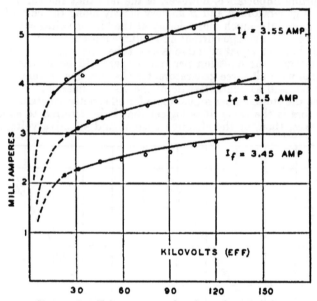

FIGURE 5.—*Tube current—tube voltage characteristics*

the cases it is seen that the effecitve current i_a for tube characteristic
a is lower than the effective current i_b for characteristic *b*. Moreover,
the magnitude of this difference depends upon the generator voltage
wave·form as shown.

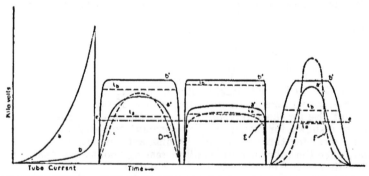

FIGURE 6.—*Effect of tube characteristic and generator wave form on the effective
tube current*

To compare the explanation above with the results found, we note
that the experimentally determined tube characteristics in Figure 5
are similar to *a* in Figure 6; hence, the tube current measured in

effective amperes is less than it would have been had the tube reached saturation at as low voltage as for characteristic *b*.

In Figure 4 the emission per effective milliampere—which is plotted against effective kilovolts—would accordingly have been less with an ideal characteristic such as *b*. In other words, the curves plotted would have fallen more nearly along the constant potential curve *C*. (Fig. 4.)

To show the relationship between peak and effective kilovolts for different tube currents, the curves *A* and *B* in Figure 7 were obtained for the correspondingly designated generators. The straight dashed lines in each set of curves represent the ratio of peak to effective voltage (kv (peak) = 1.41 kv (effective)) for a sine wave. It is seen that there is no simple relationship between the two voltage measurements as in the case of a sine wave.

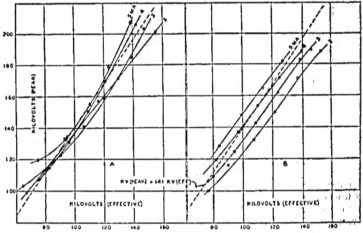

FIGURE 7.—*Relationship between peak and effective kilovolts for two mechanical rectifiers*

Such a set of curves is very useful when interpreting the behavior of a particular generator under different conditions. This was demonstrated in some earlier work [8] when comparing the output of generators *A* and *B*.

The curves also furnish some inference as to the generator wave form, particularly above the point where the tube current characteristic plays an important rôle. Thus, where the points lie below the line kv (peak) = 1.41 kv (effective), they indicate a flat-topped voltage wave form relative to a sine wave; where they lie above that line, a relatively narrow or peaked wave form is indicated. These relations, though only qualitative, are of help in properly adjusting the phase position of the switches on a mechanical rectifier for optimum X-ray tube emission.

---

[8] See Figure 16 with explanation in paper by L. S. Taylor and K. L. Tucker. (See footnote 1, p. 561.)

IV. DISCUSSION

There has been considerable confusion in interpreting the physical magnitudes involved in the therapeutic use of high-voltage X rays. The greatest single factor contributing to this confusion has been the voltage measurement, since it has been sometimes assumed that a given X-ray tube, activated by the same voltage (by sphere gap measurement) and passing equal average currents, should yield the same X-ray emission for all similar generators. We have shown here, however, that the emission may differ by ±20 per cent under supposedly like conditions.

Although a difference existing between mechanical and constant potential rectifiers has been generally recognized, there has been uncertainty as to any exact relationship between them, with the result that few attempts have been made to correlate treatment factors for the two machines. There has been even a belief that the radiation of one could not be clinically reproduced by the other. The experiments described here show that both generators may be made to produce the same emission (as far as may be deduced from absorption measuremen s in copper or paraffin) when the tube potentials are properly adjusted.

The present study has been limited thus far to mechanical and "constant potential" rectifiers. While it is believed by some that mechanical rectifiers will probably be displaced by valve-tube rectifiers, the study is of no less value, since with the great variety of valve-tube circuits in use there is an equal variety of voltage wave forms produced, none of which may be readily correlated as far as they affect the X-ray output of a tube. For example, in addition to "constant potential" there will be half-wave, full-wave, and polyphase rectifiers and voltage doubling circuits, such as the La Tour and Villard systems. In some of these the wave form depends upon load to a greater extent than in others and, as a consequence, measurements of peak voltage alone will lead to similar discordance now encountered in mechanical rectifiers.

This study reveals that, when the generator potential is measured in effective kilovolts, the X-ray emission of a tube is equivalent to that produced by very nearly the same constant potential. It is thus possible to express the emission in terms of that obtainable from a constant potential source.

Expressing tube potential in effective kilovolts simplifies the method of expressing radiation quality. For example, when describing radiation quality in terms of the "half-value layer of copper" it is recognized as necessary to also state the voltage. However, as voltages are now measured, even this is inadequate due to the effect of wave form. To correctly indicate quality by the usual absorption methods, the effective voltage should be stated—not the peak voltage.

The cordial cooperation of the Radiological Research Institute and of the American X-ray equipment manufacturers in this work has been very helpful.

WASHINGTON, June 17, 1932.

COLUMN CURVES AND STRESS-STRAIN DIAGRAMS

By William R. Osgood

ABSTRACT

The only important column formula which rests on a strictly theoretical basis is the Considère-Engesser formula, which includes the Euler formula as a special case. Attempts have been made to underpin by a general theoretical foundation other essentially empirical formulas, notably the Rankine formula; but these attempts have not been successful. It would be very satisfying, however, if these empirical formulas could be shown to be reasonable special forms of the Considère-Engesser formula. Any column formula for centrally loaded columns is a special case of the Considère-Engesser formula provided the compressive stress-strain diagram has a certain definite shape. The present paper examines a few of the commonest types of empirical formulas and determines the shape of the stress-strain diagram in each case which makes them compatible with the Considère-Engesser theory.

The only column formula which rests on a strictly theoretical basis and which at the same time is of practical importance, in the sense that tests confirm the theory, is the Considère-Engesser formula,[1] which includes the Euler formula as a special case. Attempts have been made to underpin by a general theoretical foundation other formulas, essentially empirical, notably the Rankine formula; but no successful attempt has yet been made. It would be very satisfying, however, if these empirical formulas could be shown to be reasonable special forms of the Considère-Engesser formula. Any column formula for centrally loaded columns is a special case of the Considère-Engesser formula, provided the compressive stress-strain diagram of the material has a certain definite shape. It is the purpose of the present paper to examine a few of the commonest empirical formulas and to determine the necessary shape of the stress-strain diagram in order that these formulas may be compatible with the Considère-Engesser theory. A somewhat similar investigation has been carried out by P. M. Frandsen,[2] but he refers the empirical formulas back to the Engesser formula.[3]

---

[1] Developed in 1889 and the years following by A. Considère, Fr. Engesser, and F. Jasinski; re-presented independently by Theo. v. Kármán, Mitteilungen über Forschungsarbeiten, Verein deutscher Ingenieure, Heft 81, Berlin, Julius Springer, 1910, and by R. V. Southwell, Engineering, vol. 94, p. 249, London, Aug. 23, 1912.

[2] Den Teknisk Forenings Tidsskrift, Hæfte 19, p. 139, Copenhagen, Sept. 15, 1920.

[3] Zeitschrift des hannoverischen Architekten- und Ingenieur-Vereins, vol. 35, p. 455, 1889.

The following notation will be used:

P = the load on the column at failure by buckling.

A = the cross-sectional area of the column.

$\sigma = \dfrac{P}{A}$ = the average normal stress on the cross section at failure.

l = the "free length" of the column, the distance between two successive points of inflection of the center line.

i = the least radius of inertia or radius of gyration of the cross section of the column, measured parallel to the plane of bending.

E = the modulus of elasticity of the material of the column.

E' = the "tangent modulus" at the stress σ; that is, E' is the slope of the compressive stress-strain diagram at the stress σ.

$\overline{E} = \dfrac{E'I_1 + EI_2}{I}$, where I_1 is the moment of inertia about the axis of average stress of the part of the cross-sectional area which suffers an increase of stress at the instant of failure of the column, I_2 is the moment of inertia about the axis of average stress of the part of the cross-sectional area which suffers a decrease of stress at the instant of failure of the column, and $I = Ai^2$ is the moment of inertia of the total cross-sectional area of the column about the gravity axis perpendicular to the plane of bending; the position of the axis of average stress is defined by the relation $E'S_1 = ES_2$, where S_1 and S_2 are the statical moments about the axis of average stress, respectively, of the two parts of the cross-sectional area just mentioned in connection with I_1 and I_2.

ϵ = the strain due to the stress σ.

σ_u = the short-column strength of the material, to be taken as the yield point in the case of ductile materials. (See Appendix II.)

$e = \dfrac{E\epsilon}{\sigma_u}$.

$s = \dfrac{\sigma}{\sigma_u}$.

σ_p = the stress above which Euler's formula ceases to apply, strictly the proportional limit of the material, but practically likely to be considerably above the actual proportional limit.

$s_p = \dfrac{\sigma_p}{\sigma_u}$,

$\lambda = \dfrac{1}{\pi}\dfrac{l}{i}\sqrt{\dfrac{\sigma_u}{E}}$.

C_n, C_R, constants.

The assumptions underlying the Considère-Engesser theory are that the material is homogeneous, cross sections remain plane, the stress-strain relation for increasing strain is the same as that given by the stress-strain di gram, the stress-strain relation for decreasing strain is given by a line parallel to the tangent to the stress-strain diagram at the origin, the axis of the column is straight, the cross section of the column is uniform throughout the free length, and the loading is axial.

The Considère-Engesser formula may be written

$$\sigma = \frac{\pi^2 \bar{E}}{\left(\frac{l}{i}\right)^2} \tag{1}$$

In order that any empirical formula expressible as

$$\frac{l}{i} = f(\sigma) \tag{2}$$

give the same relation between σ and $\frac{l}{i}$ as the Considère-Engesser formula, we must have

$$\bar{E} = \frac{\sigma}{\pi^2} \left(\frac{l}{i}\right)^2 \tag{3}$$

where $\frac{l}{i}$ is given by (2). The stress-strain diagram must satisfy this equation.

$\bar{E}$ is a function not only of the slope of the stress-strain diagram at (ϵ, σ) but also of the shape of the cross section, and no general reduction of equation (3) is possible. Fortunately, however, the effect of the shape of the cross section on the value of $\bar{E}$ does not vary greatly for cross sections ordinarily used which have an axis of symmetry perpendicular to the plane of bending.[4] It can be shown that $\bar{E}$ is smaller for the idealized H section (negligible web, flanges thin compared to distance between them) with the plane of bending normal to the flanges than for any other section with the symmetry just mentioned. Consequently, if $\bar{E}$ for the H section is used, equation (1) will give results on the safe side for all symmetrical sections, and equation (3) may be reduced with the assurance that the results obtained represent limiting values for such sections. The value of $\bar{E}$ for the idealized H section is [5]

$$\bar{E} = \frac{2 E E'}{E + E'} \tag{4}$$

which, substituted in equation (3), with $E' = \frac{d\sigma}{d\epsilon}$, gives

$$\sigma = \frac{\pi^2}{\left(\frac{l}{i}\right)^2} \cdot \frac{2}{\frac{d\epsilon}{d\sigma} + \frac{1}{E}} \tag{5}$$

and this is the equation of the stress-strain curve if equation (2) is to represent a theoretically possible column formula.

---

[4] See Appendix I.
[5] Kármán (see footnote 1, p. 571).

Instead of integrating equation (5) for various functions $\frac{l}{i} = f(\sigma)$, it is preferable for comparative purposes to introduce the nondimensional variables [a]

$$e = \frac{E\epsilon}{\sigma_u}, \; s = \frac{\sigma}{\sigma_u}, \; \lambda = \frac{1}{\pi} \frac{l}{i} \sqrt{\frac{\sigma_u}{E}} \tag{6}$$

In terms of these quantities equations (1) and (5) become

$$s = \frac{1}{\lambda^2} \frac{\bar{E}}{E} \tag{7}$$

and

$$s = \frac{1}{\lambda^2} \cdot \frac{2}{\dfrac{de}{ds} + 1} \tag{8}$$

where in the latter equation λ is a given function of s; (7) is the equation of a curve obtained from the column curve by dividing the average stresses by σ_u and the slenderness ratios by $\pi \sqrt{\dfrac{E}{\sigma_u}}$, and (8) is the equation of a curve obtained from the stress-strain curve by dividing the stresses by σ_u and the strains by $\dfrac{\sigma_u}{E}$.

In the Euler range, for which $\bar{E} = E$, equation (7) reduces to

$$s = \frac{1}{\lambda^2} \tag{9}$$

and equation (8) becomes, with the use of equations (6)

$$s = e \tag{10}$$

For values of s above s_p; that is, above the Euler range, for a continuous stress-strain diagram the solution of (8) becomes

$$e = 2 \int_{s_p}^{s} \frac{ds}{s \lambda^2} - s + 2 \, s_p \tag{11}$$

The reduced stress-strain diagram represented by equations (10) and (11) will now be considered for some empirical column formulas.

The parabolic or hyperbolic type of formula.—A common type of formula is one of the form

$$\sigma = \sigma_u \left[1 - C_n \left(\frac{l}{i} \right)^n \right] \tag{12}$$

If

$$C_n = \frac{2}{n \pi^n} \left(\frac{n}{n+2} \right)^{1+\frac{n}{2}} \left(\frac{\sigma_u}{E} \right)^{\frac{n}{2}}$$

the curve represented by (12) is tangent to the Euler curve at $\sigma = \dfrac{n}{n+2} \sigma_u$.

[a] The first of these has been used by K. Hohenemser (Zeitschrift für angewandte Mathematik und Mechanik, vol. 11, p. 15, February, 1931), the second has been used several times before and is obvious, and the third has been used by L. B. Tuckerman, the late S. N. Petrenko, and C. D. Johnson (National Advisory Committee for Aeronautics, Technical Note No. 307, Washington, D. C.).

With this value of C_n, in terms of the variables defined by (6), equation (12) becomes

$$s = 1 - \frac{2}{n}\left(\frac{n}{n+2}\right)^{1+\frac{n}{2}}\lambda^n \qquad (13)$$

which applies for $\frac{n}{n+2} \leq s \leq 1$. Figure 1 shows the curves

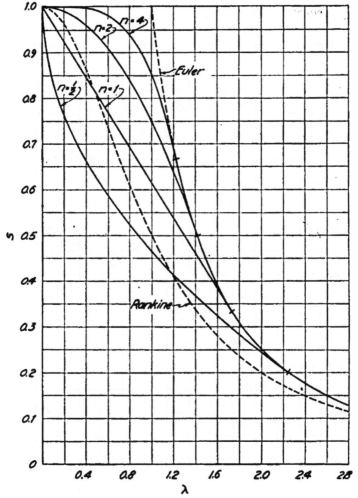

FIGURE 1.—*Reduced column curves of several types*

represented by equation (9) (marked Euler) and equation (13) for different values of n. Equation (13) solved for λ gives

$$\lambda = \left(\frac{n}{2}\right)^{\frac{1}{n}}\left(\frac{n+2}{n}\right)^{\frac{1}{n}+\frac{1}{2}}(1-s)^{\frac{1}{n}} \qquad (14)$$

and substitution of this expression in equation (11) gives finally

$$e = 2\left(\frac{2}{n}\right)^{\frac{2}{n}}\left(\frac{n}{n+2}\right)^{1+\frac{2}{n}}\int_{s_p}^{s}\frac{ds}{s(1-s)^{\frac{2}{n}}}-s+2s_p,\qquad (15)$$

for values of s from $\dfrac{n}{n+2}$ to 1, inclusive. For s less than or equal to $s_p=\dfrac{n}{n+2}$ equation (10) applies.

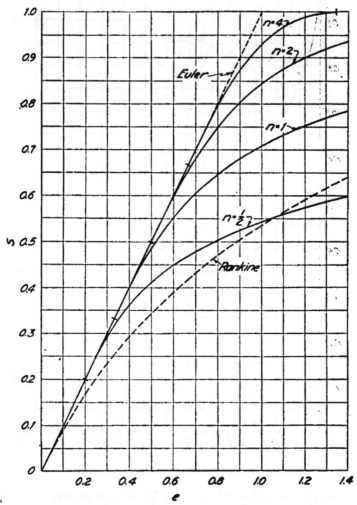

FIGURE 2.—*Reduced compressive stress-strain diagrams corresponding to the column curves of Figure 1*

Figure 2 shows the curves represented by equation (10) (marked Euler and

The case $n=\frac{1}{2}$ represents a column formula which experience has shown to be suitable for brittle materials like cast iron. (See fig. 1.) Equation (15) becomes

$$e=\frac{512}{3125}\int_{\frac{1}{5}}^{s}\frac{ds}{s(1-s)^4}-s+\frac{2}{5} \qquad (16)$$

which reduces to

$$e=0.1638\left[\frac{1}{3(1-s)^3}+\frac{1}{2(1-s)^2}+\frac{1}{1-s}+\log\frac{s}{1-s}\right]-s+0.1877 \qquad (17)$$

applicable for $\frac{1}{5}\leqq s\leqq 1$. (See fig. 2.)

The case $n=1$ represents a column formula suitable for common grade wood. (See fig. 1.) Equation (15) becomes

$$e=\frac{8}{27}\int_{\frac{1}{3}}^{s}\frac{ds}{s(1-s)^2}-s+\frac{2}{3} \qquad (18)$$

which reduces to

$$e=\frac{8}{27}\left(\frac{1}{1-s}+\log\frac{s}{1-s}\right)-s+0.4276 \qquad (19)$$

applicable for $\frac{1}{3}\leqq s\leqq 1$. (See fig. 2.)

The case $n=2$ represents a column formula suitable for ductile materials like mild steel. (See fig. 1.) Equation (15) becomes

$$e=\frac{1}{2}\int_{\frac{1}{2}}^{s}\frac{ds}{s(1-s)}-s+1 \qquad (20)$$

which reduces to

$$e=\frac{1}{2}\log\frac{s}{1-s}-s+1 \qquad (21)$$

applicable for $\frac{1}{2}\leqq s\leqq 1$. (See fig. 2.)

The case $n=4$ represents a column formula suitable for hard materials like hard steel. (See fig. 1.) Equation (15) becomes

$$e=\frac{4\sqrt{3}}{9}\int_{\frac{3}{4}}^{s}\frac{ds}{s(1-s)^{\frac{1}{2}}}-s+\frac{4}{3} \qquad (22)$$

which reduces to

$$e=\frac{4\sqrt{3}}{9}\log\frac{1-\sqrt{1-s}}{1+\sqrt{1-s}}-s+2.347 \qquad (23)$$

applicable for $\frac{2}{3}\leqq s\leqq 1$. (See fig. 2.)

$$1 + C_R\left(\frac{l}{i}\right)^2$$

has been used for all values of $\frac{l}{i}$ with $C_R = \frac{\sigma_u}{\pi^2 E}$, in which case the curve represented by (24) approaches tangency to the Euler curve at infinity. With this value of C_R, in terms of the variables defined by (6), equation (24) becomes

$$s = \frac{1}{1+\lambda^2} \tag{25}$$

which solved for λ gives

$$\lambda = \sqrt{\frac{1-s}{s}} \tag{26}$$

and substitution of this expression in equation (11) gives

$$e = 2\int_0^s \frac{ds}{1-s} - s$$

or

$$e = -2 \log (1-s) - s \tag{27}$$

for values of s from 0 to 1, inclusive.

The curves represented by equations (25) and (27) are shown in Figures 1 and 2, respectively.

Appendix III shows an American Bridge Co. reduced straight-line column curve and the corresponding reduced stress-strain diagram.

All of the curves in Figure 2 except the curve marked $n = 4$ (equation (23)) approach the horizontal line $s = 1$ asymptotically. The curve marked $n = 4$ has a horizontal tangent at $e = 1.347$, $s = 1$. A comparison of the curves of Figure 1 with the corresponding curves of Figure 2 shows the type of compressive stress-strain diagram which is necessary in any particular case in order that a given column curve may represent accurately the strength of columns. The shapes of the reduced stress-strain diagrams are all reasonable except for high values of s and possibly in the case of the Rankine formula. The shape of the diagram for high values of s is relatively unimportant, however, since the corresponding portion of the column curve lies in the region of "short columns," and such columns do not usually fail by buckling. It is doubtful whether any structural material would show a reduced compressive stress-strain diagram like that required for the Rankine formula.

Viewed in another way, any (reduced) compressive stress-strain diagram in Figure 2 will yield the (reduced) column curve corresponding to it in Figure 1. If, therefore, the short-column strength, σ_u, of the material is such that, or can be defined (Appendix III) so that, the (reduced) compressive stress-strain diagram fits a curve of Figure 2, the column strength of the material will be given by the corresponding curve of Figure 1. It will frequently be possible by a suitable choice of σ_u to obtain good agreement between the actual (reduced) compressive stress-strain diagram and one of the curves of Figure 2 up to some limiting value of s, beyond which the agreement will not be good. In that case the corresponding curve of Figure 1 will represent the column strength up to the same value of s. Thus, if representative compressive stress-strain diagrams of a material are available, it may be possible to draw the corresponding column curve without tests. This method is not to be recommended to the exclusion of tests, however, particularly not if the agreement between the actual (reduced) compressive stress-strain diagram and a curve of Figure 2 is not extremely good.

The author wishes to express his indebtedness to Dr. Walter Ramberg, of the engineering mechanics section, Bureau of Standards, for checking the equations.

APPENDIX I

Figure 3 shows the variation of $\dfrac{\bar{E}}{E}$ with $\dfrac{E'}{E}$ for a number of different cross sections, as listed on the figure. The curves for the thin circular-ring section (thickness negligible in comparison with the diameter) and the solid circular section are plotted from formulas by R. V. Southwell.[7]

The formula for the rectangular section is given by T. v. Kármán,[8] and the six points for the I section are taken from a graph by W. Gehler.[9] The two curves for the thin T section (thickness of the web and the flange negligible in comparison with their lengths) have been worked out by the author.

It may be noted that the curve for an unsymmetrical section may lie considerably below the curve for the idealized H section, and the lower limit for such sections is the curve $\dfrac{\bar{E}}{E} = \dfrac{E'}{E}$. (Fig. 3.) Consequently, Frandsen's treatment [10] may be considered as applying unmodified to unsymmetrical sections.

APPENDIX II

In interpreting column tests of ductile materials it is of some importance how σ_u is defined. If σ_u is defined as the stress determined by the intersection of the stress-strain curve with a line through the origin having a slope αE ($0 < \alpha < 1$), the experimental points when plotted as a λ, s-diagram are likely to lie more nearly on a smooth curve than otherwise. Under ideal conditions of test (perfectly straight specimens, perfect centering, etc.) if different materials having affine stress-strain diagrams are tested, and if σ_u is determined as outlined, all points λ, s will lie on one and the same curve. In tests made at the Bureau of Standards the values $\alpha = \dfrac{5}{9}$ and $\alpha = \dfrac{2}{3}$ have

[7] Southwell (see footnote 1, p. 571.)
[8] Kármán (see footnote 1, p. 571.)
[9] Proceedings of the Second International Congress for Applied Mechanics, p. 366, Zurich, Sept. 12–17, 1926.
[10] See foot note 2, p. 571.

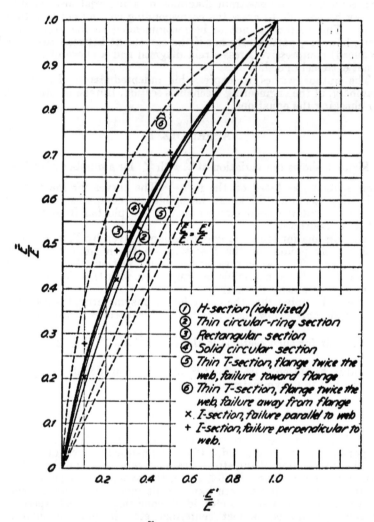

FIGURE 3.—Variation of $\frac{\bar{E}}{E}$ with $\frac{E'}{E}$ for various shapes of cross section

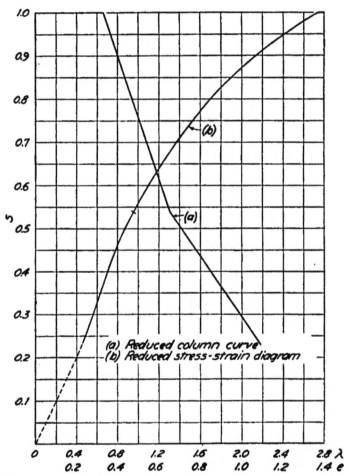

FIGURE 4.—*Reduced American Bridge Co. column curve and corresponding reduced compressive stress-strain diagram*

been found to give satisfactory results for chromium-molybdenum steel and duralumin columns, respectively.

By adjusting the value of α, it may be possible to fit the actual (reduced) compressive stress-strain diagram rather closely to one of the curves of Figure 2 up to a sufficiently high value of s to include all except short columns.

APPENDIX III

It may be of interest to compare a specification column curve with the stress-strain diagram which would make the curve a theoretically correct one. The American Bridge Co. curve given by the following equations has been chosen for this purpose (units are in pounds per square inch):

$$\left.\begin{array}{l} \sigma_s = 13{,}000 \ \ \text{for } 0 \leqq \dfrac{l}{i} \leqq 60, \\[2mm] \sigma_s = 19{,}000 - 100\ \dfrac{l}{i}\ \ \text{for } 60 \leqq \dfrac{l}{i} \leqq 120, \\[2mm] \sigma_s = 13{,}000 - 50\ \dfrac{l}{i}\ \ \text{for } 120 \leqq \dfrac{l}{i} \leqq 200, \end{array}\right\} \qquad (28)$$

where σ_s is the allowable average stress. These equations become, respectively, in terms of λ and s, if we assume $\dfrac{\sigma_u}{E}\left(= \dfrac{\pi^2}{8450} \right) = 0.001168$,

$$\left.\begin{array}{l} s = 1, \\[2mm] s = \dfrac{19}{13} - \dfrac{\sqrt{2}}{2}\ \lambda \ \ \text{for } 1 \geqq s \geqq \dfrac{7}{13} \\[2mm] s = 1 - \dfrac{\sqrt{2}}{4}\ \lambda \ \ \text{for } \dfrac{7}{13} \geqq s \geqq \dfrac{3}{13} \end{array}\right\} \qquad (29)$$

The reduced stress-strain diagram is represented by the following equations:

$$\left.\begin{array}{l} s = 1 \\[2mm] e = \dfrac{169}{361}\left(\dfrac{1}{1 - \dfrac{13}{19}s} + \log \dfrac{s}{\dfrac{19}{13} - s} \right) - s + 0.0913 + C \ \ \text{for } 1 \geqq s \geqq \dfrac{7}{13} \\[2mm] e = \dfrac{1}{4}\left(\dfrac{1}{1 - s} + \log \dfrac{s}{1 - s} \right) - s + C \ \ \text{for } \dfrac{7}{13} \geqq s \geqq \dfrac{3}{13} \end{array}\right\} \qquad (30)$$

where the constant of integration C is arbitrary.

Equations (29) and (30) are represented graphically in Figure 4.

WASHINGTON, July 14, 1932.

O

U. S. DEPARTMENT OF COMMERCE
ROY D. CHAPIN, Secretary
BUREAU OF STANDARDS
LYMAN J. BRIGGS, Acting Director

BUREAU OF STANDARDS
JOURNAL OF RESEARCH

November, 1932

Vol. 9, No. 5

UNITED STATES
GOVERNMENT PRINTING OFFICE
WASHINGTON : 1932

For a partial list of previous research papers appearing in BUREAU OF STANDARDS JOURNAL OF RESEARCH see pages 2 and 3 of the cover

THE VARIATION WITH ANGLE OF EMISSION OF THE RADIATION FROM METALS BOMBARDED BY SLOW ELECTRONS

By C. Boeckner

ABSTRACT

The radiation from Pt and W, with its electric vector parallel to the plane of emission, increases rapidly with increasing emission angle. For the other direction of polarization the radiation varies only slowly with angle. The variation is given by the function $(1-R_\theta)/\cos\theta$ where R_θ is the reflection coefficient of the metal for light incident at angle θ. A discussion of the variation of emission from surface layers with angle is given, based on T. C. Fry's theory of lamellar absorption. The disagreement with the observed variation is in the direction to be accounted for if the radiation is initially emitted anisotropically (in a manner similar to the continuous X rays) and is later modified by refraction effects.

It is shown that the angle variation is somewhat similar to that expected for the inverse vectorial photo-electric effect.

CONTENTS

I. INTRODUCTION

There is considerable evidence showing that metals when bombarded by electrons emit visible and ultra-violet radiation. In 1920 Lillienfeld[1] reported a blue gray luminescence appearing at the focal spot of an X-ray tube anticathode, due apparently to the excitation of the metal atoms by the high speed cathode rays. A similar radiation was reported by Foote, Meggers, and Chenault[2] as emanating from a target bombarded by somewhat slower electrons having energies of the order of 1,000 volts.

Recently F. L. Mohler[3] and the author have described a method for studying the radiation from metals bombarded by very slow electrons having energies of from 2 to 15 volts. The method makes use of the fact that a small metal electrode in a gas discharge draws very intense electron currents when maintained at a potential positive to the surrounding space. It was found that with the large current densities thus obtained (4 or 5 amp. cm⁵), metals emitted radiation

---

[1] Lillienfeld and Rother, Phys. Zeits., vol. 21, p. 49, 1920.
[2] Foote, Meggers, and Chenault, J. Opt. Soc. Am., vol. 9, p. 541, 1924.
[3] F. L. Mohler and C. Boeckner, B. S. Jour. Research, vol. 6 (RP297), p. 673, 1931. F. L. Mohler, B. S. Jour. Research, vol. 8 (RP421), p. 357, 1932.

having a continuous spectrum of rather uniform intensity in the visible and ultra-violet.

Some of the characteristics of this radiation are the approximate equality of the absolute intensities from most metals; the exceptional behavior of silver in ossessing a very intense emission band in the ultra-violet; and the fact that at low electron velocities the spectrum possesses a high frequency limit in the ultra-violet. It may also be mentioned that the radiation is thought to be the analogue of the continuous X-ray spectrum since the absolute intensity is of the order predicted for the "bremsstrahlung" produced by slow electrons.

The present paper deals with the polarization of the radiation and the variation of intensity with the angle of emission. The measurements were undertaken with the thought that the intensity-angle relation might be similar to that of the continuous X rays. It will be shown later that the velocity of the electrons is normal to the metal surface; it might therefore be expected as in the case of the continuous X rays that the radiation would be most intense when observed in a direction perpendicular to the direction of motion of the electrons. By analogy the radiation might also be expected to be polarized with the electric vector parallel to this direction.

II. METHOD

To study the variation of intensity with angle, a cylindrical electrode was observed in a direction perpendicular to its axis. An enlarged image of the cylinder was focused upon the spectograph slit, the axis of the image being in the plane of the slit and perpendicular to it. From a study of the variation of intensity across the image, information could be deduced concerning the variation of intensity with emission angle. It is clear for example that the radiation forming the edges of the image is emitted at a grazing angle while light at the center is emitted normal to the cylinder surface.

It was thought desirable to study the intensity variation for both the significant directions of polarization of the radiation. A quartz double image prism was therefore placed in the optical train so that one of the images had the electric vector of its radiation parallel to the plane of emission (perpendicular to the cylinder axis) and the other perpendicular to the plane. Since the ultra-violet was found to be the most convenient spectral region for the study of the radiation, quartz fluorite achromatic lenses were used to form the cylinder image upon the slit. It was found necessary to place the double image prism on the source side of the lenses to avoid initial depolarization of the radiation by their quartz components.

A discharge in cæsium vapor (several thousandths millimeter pressure) was used as a source of current to the electrodes for numerous experimental reasons. The cylindrical electrode was placed in the positive column of the discharge coaxial with the cylindrical discharge tube. This arrangement was necessary to secure uniform current density over the electrode surface. The measurements were all made with energies of the bombarding electrons of about 7 volts.

It is of interest to mention that the potential drop between the electrode and the discharge occurs in a small sheath several hundredths of a millimeter thick; the electric force is consequently normal to the metal surface. The direction of motion of the electrons is therefore also perpendicular to the metal surface, since their initial energy in he discharge is small compared to the potential drop across the sheath.

The metals studied were latinum, tungsten, and silver. The first two metals could be raised to a bright red heat and outgassed by drawing large discharge currents to them while the discharge tube was connected to the pumps. The surfaces were further cleaned by intense ion bombardment.

It was thought of some significance to compare the variation of intensity with angle for a matte surface and a polished surface. A very good diffusely reflecting surface of platinum was prepared by electroplating a layer of platinum black upon a platinum rod and heating to redness in a flame. An attempt was made to prepare a similar tungsten surface by etching a tungsten rod. In this case, however, the surface was far from being a diffuse reflector and showed marked polarization of light reflected from it.

Densitometer records of the blackening across the cylinder image on the spectograms were made by means of a Kip & Zonen registering microphotometer. The intensities were deduced from these records. A series of differently timed exposures on each spectogram served to give the blackening intensity characteristic of the plate. The wavelength range covered was from 4,200 to about 3,100 A.

The method described has the advantage that the intensities for a range of angles may be obtained from one exposure. It has the disadvantage, however, that the radiation emitted at angles greater than 60° to the surface normal correspond to a region so near the edge of the image that results are falsified by errors in focus. The method was, however, found to be convenient and was thought to be adequate for at least a rough survey of the intensity relations. Due to the long exposure times (six hours with the double image prism in place) a more direct method would have been tedious. One set of measurements, however, was made of the intensity of emission from platinum at 75° by comparing the intensity from a large plane electrode observed at an oblique angle with that obtained when observed in a direction perpendicular to the surface.

III. RESULTS

For both platinum and tungsten it was found that the radiation with the electric vector parallel to the plane of emission increased rapidly with the angle of emission (the angle between the emitted ray and the normal to the surface). For the other direction of polarization the intensity was constant across the image. Figure 1 shows a plot of the intensity variation across the tungsten cylinder image, the tungsten being bombarded by 7-volt electrons. The tungsten in this case had been etched in an attempt to render the surface a diffuse reflector. A polished surface, however, gave the same results showing that the intensity is not sensitive to the degree of polish of the surface. There was no appreciable change in variation with angle between 4,200 and 3,100 A.

By drawing large currents to the cylinder, it was possible to heat it to incandescence, enabling measurements to be made of the variation of the intensity of the thermal radiation with angle. For comparison the results are shown on the lower part of the figure.

Figure 2 shows a plot of the intensity-angle variation for polished platinum (full line). The measurements at 75° were obtained from a large plane electrode which could be observed normally and at an oblique angle. It is seen that in this case the variation with angle changes with wave length. In contrast with tungsten the radiation from the matte platinum surface showed no variation with angle. This is perhaps explained by the fact that the platinum surface is more nearly an ideal diffuse reflector.

Silver showed no variation in intensity across the image of the cylinder. This may, however, be explained by the fact that it was

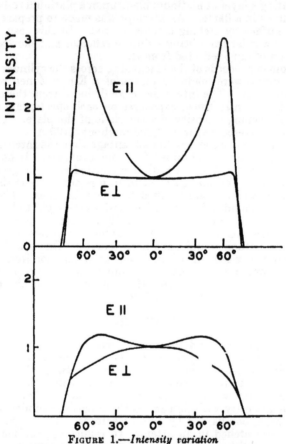

FIGURE 1.—*Intensity variation*

Upper curves, intensity variation across image of tungsten cylinder bombarded by 7-volt electrons. Lower curves, intensity of thermal radiation at 4,200 A

impossible to maintain a polish on the silver electrode in a cæsium discharge, due perhaps to the heating by the large electron currents used.

IV. DISCUSSION

1. RADIATION FROM A THIN LAYER

It is possible to explain the trend of the variation of intensity with angle of emission by assuming that the radiation is emitted by a thin layer under the metal surface (radiating uniformly in all directions) and is weakened by reflection losses on being transmitted through the surface.   The intensity I_θ emitted at angle θ is then given by the relation—

$$I_\theta = \text{constant} \times (1 - R_\theta)/\cos \theta \qquad (1)$$

R_θ is taken to be the reflection coefficient for plane waves incident upon the surface at angle θ.   The factor $1/\cos \theta$ arises from the fact that a given area of the image receives radiation from an area of the surface larger in the ratio $1/\cos \theta$.

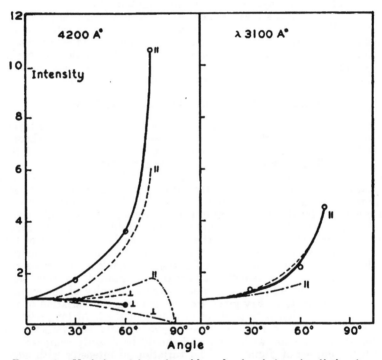

FIGURE 2.—*Variation of intensity with angle of emission of radiation from platinum bombarded by 7-volt electrons, full line*

——— Plot of $(1 - R_\theta)/\cos \theta$
— — — Variation of intensity with angle deduced from the lamellar absorption of platinum.

The dashed lines of Figure 2 are plotted from equation (1).   The reflecting powers are computed from the approximation formula given by Drude[4] using the optical constants of platinum given by the International Critical Tables.

The explanation given above is reasonable only if the concepts of rays and the laws of reflection of plane waves can be applied to waves which have a radius of curvature of a few angstroms at the reflecting surface.   The latter distance is in all probability the depth from which the radiation is emitted.   It is difficult to see, therefore, why the simple considerations used should be correct.   A more reasonable

---

[4] Drude, Ann. d. Phys., vol. 35, p. 523, 1888.

me oα   g   varıa o o   ace emıssıvı y   angıe
will be given below (relation of surface emission to "lamellar absorption").

2. THE INTENSITY VARIATION OF THERMAL RADIATION WITH ANGLE OF EMISSION

The intensity of the thermal radiation emitted by a square centimeter is proportional to $(1-R_\theta) \cos \theta$.[5] The apparent intensity observed at an angle θ is therefore proportional to $(1-R_\theta)$ or

$$I = \text{constant} \ (1-R_\theta) \tag{2}$$

This relation can be derived by assuming an indefinitely thick layer instead of the thin layer used to obtain equation (1).

It is evident that a measurement of the intensity across the image of the cylinder raised to incandescence enables one to estimate the change of reflecting power with angle, R_θ. This was done in order to estimate the reliability of the measurements. The reflection coefficients relative to normal incidence were about 18 per cent lower at 60° than those computed from optical constants of platinum.

3. SURFACE EMISSION AND " LAMELLAR ABSORPTION "

Fry[6] has discussed the absorption of energy by metal surface layers from incident plane light waves using classical electromagnetic theory. He finds the absorption to be somewhat different from that of the metal in bulk and uses the term "lamellar absorption" to describe the phenomenon. The absorption is supposed to be proportional to the square of the electric intensity (radiation energy density) just under the metal surface. Just outside the metal surface the electric intensity is given by the vector sum of electric vectors of the incident and reflected wave. The electric intensity just inside the metal surface can be obtained from the fact that the tangential component of the electric force and the product of the normal component by the amplitude squared of the index of refraction are continuous across the metal surface. The lamellar absorption computed on this basis should by Kirchhoff's law be equal to the emissivity of the surface layers of the metal. Fry gives a curve showing the variation of the lamellar absorption with angle of incidence of the light for platinum at 4,359 A. The curve should also give directly the variation of intensity of emission with angle and should be directly comparable[7] with the curves of Figure 2. The variation of intensity with angle obtained in this way is not much different from that of thermal radiation, and is therefore not adequate to explain the variation of intensity observed in the present work. If the intensity for normal emission is unity, then at 60° one obtains from the lamellar absorption curves 1.4 for parallel light; for "perpendicular" light, 0.50. The

[5] R. H. Fowler, Statistical Mechanics, p. 493.
[6] T. C. Fry, J. Opt. Soc. Am., vol. 22, p. 307, 1932.
[7] Consider a metal surface in equilibrium with radiation at some temperature and let a suitably defined beam of the radiation with unit cross section area and energy flux I be incident upon a metal surface at angle θ. Let aI be the energy absorbed from the beam by a surface layer of definite thickness. Also let eI be the energy emitted by the surface layer in the direction of the incident beam and from an area of the surface equal to $1/\cos \theta$. Equilibrium between the black body radiation and the surface layer demands that e and a be equal. a as a function of angle can be obtained directly from Fry's work. The units of e, radiation from an area proportional to $1/\cos \theta$, are those used in Figure 2. The curves of Fry for the variation of a with angle for platinum at 4,359 A and those of Figure 2 are therefore directly comparable.

observed values at 4,200 A are 3.6 and 0.77. For comparison the variation of emission with angle deduced from the lamellar absorption is also plotted in Figure 2.

4. RELATION TO THE INVERSE PHOTO-ELECTRIC EFFECT

Another explanation of the angle intensity relations is possible if the phenomenon is correlated with the inverse vectorial photoelectric effect. Relations between the angle variations of the direct and inverse effects can be obtained by considering the conditions to be fulfilled in order that a metal surface be in thermal equilibrium with black body radiation and the thermal electrons in the space about the metal.

The following is a brief discussion based on the "principle of detailed balance." Let Idw be the energy of black body radiation of given frequency passing through unit area, s (perpendicular to the beam), and striking the metal surface at angle θ. The directions of the "rays" in the beam are supposed to be in the solid angle dw. Define P_θ such that $IP_\theta dw$ is the energy absorbed from the beam that is used in ejecting photo-electrons from the metal.

Also let J be the thermal electron current [8] striking unit area of the metal surface and define $i_\theta dw$ such that $i_\theta J dw$ is the energy emitted by the metal in the solid angle dw as a result of bombardment by the thermal electrons. Then $Ji_\theta dw/\cos\theta$ is the radiant energy from the metal passing through the unit area s. This latter quantity must by the principle of detailed balance equal the energy absorbed photo-electrically from the beam, or

$$Ji_\theta dw/\cos\theta = P_\theta Idw \tag{3}$$

$i_\theta/\cos\theta$ is the apparent intensity of radiation from the surface observed at angle θ; that is, the quantity measured; if we call it i'_θ

$$i'_\theta = cP_\theta \tag{4}$$

where c is independent of angle.

This latter equation implies that the ordinates of Figures 1 and 2 represent also the variation of photo-electric current with angle, P_θ.

The experimental data for the variation of photo-electric current with angle of incidence of the light is rather meager, at least for metals other than the alkalies.[9] For the alkalies the current with the electric vector parallel to the plane of incidence may be from 2 to 10 times greater at 60° than at normal incidence. For the other direction of polarization the current is usually proportional to the incident energy minus that reflected $(1-R_\theta)$. For metals other than the alkalies the increase with angle for "parallel" radiation is much slower. The variation can be expressed by the relations

$$P_\theta\| = \alpha\|(1-R_\theta\|)$$
$$P_\theta\perp = \alpha\perp(1-R_\theta\perp)$$

If α is made unity at normal incidence then $\alpha\|$ at 60° varies from 1 to 2 or 3; $\alpha\perp$ remains closely unity for all angles. It may be mentioned

---

[8] P_θ should be thought of as referring to the ejection of electrons of some definite velocity and direction and J as the current of electrons having the same velocity and opposite direction.
[9] Photo-electric Phenomena, A. L. Hughes and L. A. Dubridge, McGraw-Hill, 1932.

that $\alpha\| > 1$ is the criterion for the existence of the vectorial photo-electric effect.

If one uses the computed reflecting powers of platinum one obtains from Figure 2 a value for $\alpha\|$ at 60° and 4,200 Å of 2.4, for $\alpha\bot$, 1.3; at 3,100 Å $\alpha\|$ at 60° is 1.6. These values are not inconsistent with the general trend of the photo-electric measurements.

Ives [10] has successfully explained many phases of the vectorial photo-electric effect for the alkalies by assuming that the photo-electric current is proportional to the lamellar absorbing power of the surface layer emitting the photo-electrons. If this is true for platinum then one would predict for platinum at 4,359 Å only a small vectorial effect; that is, $\alpha\| > 1$ and $\alpha\bot$ not much different from unity. The variation of surface emissivity with angle would then be identical with that predicted from the lamellar absorption and would be in conflict with the measurements.

It may be remarked that, if one could use the energy density just above the metal surface instead of that inside as a measure of the absorption or photo-electric current, then one would obtain a very large vectorial effect. If one assumes a gradual transition in energy density between the interior and exterior of the metal, one could explain almost any degree of vectorial effect for the absorption of the first few atom layers. The work of Ives suggests, however, that the discontinuity in energy density is very sharp, occurring within less than an atom diameter.

5. RELATION TO THE VARIATION OF INTENSITY WITH ANGLE OF THE CONTINUOUS X-RAYS

The theory of surface emission given in footnote 7 assumes that the individual atoms radiate uniformly in all directions.[11] The discrepancy between theory and observation may, therefore, be explained by assuming the emitting centers to radiate anisotropically. Such anisotropic radiation is observed for the continuous X rays and is predicted for the radiation from a stream of slow electrons impinging on protons.[12] From analogy to both of these cases the radiation should be most intense when observed in a direction perpendicular to the electron velocity and should be polarized with the electric vector parallel to the direction of motion of the electrons. This is in the right direction to explain the discrepancy since it is found that the "parallel radiation" is much more intense at oblique angles than would be predicted for radiation emitted isotropically. This statement is clear if it is remembered that in the present measurements the electric vector of the "parallel" radiation becomes nearly parallel to the electron velocity at large angles of emission. It is difficult, however, to say quantitatively what should be expected on this view.

In this connection it is of interest to recall the work of Lillienfeld and Rother [13] on the polarization of the visible radiation produced by high speed cathode rays striking a metal target. They observed the polarization for various directions of incidence of the cathode rays and for angles of emission varying from 45° to a very oblique angle.

[10] H. E. Ives and H. B. Briggs. Phys. Rev., vol. 40, p. 802, 1932.
[11] The absorption of a single volume element of the metal is supposed to depend only on the magnitude of the electric vector. This implies that the emission of the element should be independent of direction, its modification by refraction producing the observed angle variation.
[12] A. Sommerfeld, Ann. d. Phys., vol. 11, 257, 1931.
[13] Lillienfeld and Rother, Phys. Zeit., vol. 21, p. 360, 1920.

The radiation was found to be always completely polarized and in the direction to be expected if it were emitted by a classical oscillator vibrating perpendicular to the metal surface. The results can not be explained by a purely optical theory. They suggest, however, that the radiation is produced by the electrons being accelerated through the "work function layer" of the metal. Since such an acceleration is normal to the metal surface, classical electromagnetic theory predicts precisely the type of polarization of the emitted radiation that is observed.

V. SUMMARY

The variation of intensity with angle for both directions of polarization is given approximately by the relation $I(\theta) = (1 - R\theta)/\cos \theta$. The equation is obtained by plausibility arguments and has perhaps not a very sound logical basis.

Angle variation based on sounder reasoning (by use of lamellar absorption theory and Kirchhoff's law) fails to explain the facts. The discrepancy may be due either to the incorrectness of the theory or to the fact that the radiation is initially emitted anisotropically as are the continuous X rays.

WASHINGTON, August 3, 1932.

NITROGEN CONTENT OF SOME STANDARD-SAMPLE STEELS

By J. G. Thompson and E. H. Hamilton

ABSTRACT

The nitrogen contents of four Standard Samples of iron and steel have been determined by both the vacuum fusion and solution-distillation (Allen) methods. These four standard samples, with nitrogen contents ranging from 0.005 to 0.028 per cent, are now available as nitrogen standards for analysts interested in this determination.

Repeated analyses of several standard samples of iron and steel showed that the nitrogen content of these samples did not change over periods as long as 13 years.

CONTENTS

I. INTRODUCTION

Studies of various methods for the determination of nitrogen in ferrous materials have been made at the Bureau of Standards for a number of years. Two methods have been commonly used, the vacuum-fusion method and the solution-distillation method, the latter usually referred to as the Allen method. Both methods have been described in detail in previous publications from the Bureau of Standards [1] [2] and in the recent book by Lundell, Hoffman, and Bright.[3] These and other methods for the determination of nitrogen in steel were summarized in a recent publication [4] by one of the present authors.

II. THE USE OF STANDARD SAMPLES OF IRON AND STEEL AS NITROGEN STANDARDS

During the investigation of methods for the determination of nitrogen, Bureau of Standards standard samples Nos. 8b, 23, and 55 have been in intermittent use as reference standards over periods of from 9 to 13 years. Samples 8b and 23 are Bessemer steels containing 0.1 and 0.8 per cent carbon, respectively; sample 55 is an ingot iron. The uniform and reproducible results recorded in Table 1, show that these three standard samples constitute satisfactory reference standards for the determination of nitrogen. A question might be raised as to the stability or permanence of the nitrides present in irons and steels but it is evident, from the data in Table 1, that the nitrogen content of these samples did not change with time during approximately 10 years.

---

[1] L. Jordan and F. E. Swindells, Gases in Metals. I. The Determination of Combined Nitrogen in Iron and Steel and the Change in Form of Nitrogen by Heat Treatment, B. S. Sci. Paper No. 457, 1922.

[2] H. C. Vacher and L. Jordan, The Determination of Oxygen and Nitrogen in Irons and Steels by the Vacuum Fusion Method, B. S. Jour. Research, vol. 7 (RP345), August, 1931.

[3] G. E. F. Lundell, J. I. Hoffman, and H. A. Bright, Chemical Analysis of Iron and Steel, John Wiley & Sons (Inc.), 1931.

[4] J. G. Thompson, Determination of Oxygen, Nitrogen, and Hydrogen in Steel, Am. Inst. Min. & Met. Eng., Tech. Pub. No. 466, 1932.

III. THE NITROGEN CONTENT OF FOUR CURRENT STANDARD SAMPLES

The available supply of two of these three samples has been exhausted.   In order that nitrogen standards may be readily available to workers interested in this determination, the nitrogen content of some of the current standard samples was determined.   Four standard samples, No. 10d (Bessemer steel, 0.4 per cent C); No. 55 (ingot iron); No. 101 ("18–8" chromium-nickel stainless steel); and No. 106 (nitriding steel) were chosen, representing different types of iron and steel with different nitrogen contents.   Nos. 10d and 55 represent the older, simpler types of ferrous materials with low or medium nitrogen content.   Nos. 101 and 106 represent newer types of steel, with medium or high nitrogen content, in which more importance may be attached to the presence of nitrogen.   Determinations of nitrogen by both the Allen method and the vacuum-fusion method show that sample No. 10d contains 0.008 per cent; No. 55, 0.005 per cent; No. 101, 0.028 per cent; and No. 106, 0.009 per cent, as recorded in Table 2.

TABLE 1.—*Permanence of the nitrogen content of Bureau of Standards standard samples*

No. 8b		No. 23		No. 55	
Year of analysis	Nitrogen	Year of analysis	Nitrogen	Year of analysis	Nitrogen
	Per cent		*Per cent*		*Per cent*
	0. 012		0. 010		0. 006
1919......	. 011		. 010	1923......	. 006
	. 011		. 010		. 004
	. 011		. 010		. 005
	. 012	1922......	. 010	1925......	. 004
1927......	. 011		. 011	1931......	. 005
	. 012		. 011		. 005
	. 011		. 012		¹. 004
1928......	. 011		. 012	1932......	¹. 005
	. 012		. 012		¹. 005
	. 012		¹. 010		
	. 013	1932......	¹. 011		
	. 013		¹. 010		
1932......	. 012				
	¹. 011				
	¹. 010				
	¹. 012				
	¹. 011				

¹ Determinations made by the vacuum-fusion method. All other determinations were made by the Allen method.

Attention is called to the fact that certain precautions must be observed in order to obtain satisfactory results in the analysis of the nitriding steel, No. 106, by the vacuum-fusion method.   The nitrides in this steel are decomposed more slowly in the vacuum-fusion method than are the nitrides of plain carbon steels, and it is necessary to continue the extraction of gases from the sample somewhat longer than usual, in order to insure complete recovery of nitrogen.   If this precaution is observed complete recovery of nitrogen is secured, as is indicated by the agreement between the results from the two methods of analysis recorded in Table 2.

TABLE 2.—Nitrogen content of current standard samples

No. 10d nitrogen	No. 55 nitrogen	No. 101 nitrogen	No. 106 nitrogen	No. 10d nitrogen	No. 55 nitrogen	No. 101 nitrogen	No. 106 nitrogen
Per cent	Per cent	Per cent	Per cent	Per cent	Per cent	Per cent	Per cent
[1] 0.006	[1] 0.004	[1] 0.028	[1] 0.009	0.006	0.005	0.028	[1] 0.009
[1].008	[1].005	[1].028	[1].011	.009	.005	.028	[1].009
[1].009	[1].005	[1].028	[1].010	.008	.005	.028	.010
[1].009	[1].005	[1].028	[1].008	.008		.027	.010
[1].009	.005	[1].028	[1].010			.028	.009
[1].008	.004	[1].026	[1].008				.009
.008	.005	.028	[1].007				.010
.008	.004	.027	[1].009				.010
				Average .006	.005	.028	.009

[1] Determinations made by the vacuum-fusion method. Remaining determinations made by the Allen method.

In view of the numerous requests, which have been received for material to be used as reference standards in the determination of both oxygen and nitrogen, it is unfortunate that these standard samples can not be used as reference standards for oxygen as well as for nitrogen in the vacuum-fusion method. The standard samples are prepared in the form of small, uniform millings, and experience has shown that such samples yield values much too high in the determination of oxygen by the vacuum-fusion method. During the milling process the surface of each chip becomes coated with a thin film of oxide, especially if the chips become heated. In addition to this oxide film, water is usually present as a result of condensation or adsorption during handling and storage. The actual amount of oxygen in the oxide or moisture film on any one chip is very small, but the total from a large number of chips may equal or even exceed the amount of oxygen present in the solid metal of the sample. It is impossible to distinguish between "surface oxygen" and "dissolved oxygen" in the vacuum-fusion method. Consequently, solid specimens with small surface area and correspondingly small contents of "surface oxygen" are preferred for the vacuum-fusion determination of oxygen. The presence of these surface films does not, however, interfere with the determination of nitrogen, and these standard samples are therefore available as nitrogen standards.

WASHINGTON, July 18, 1932.

RP495

ANALYTICAL METHODS
FOR THE DETERMINATION OF LEVULOSE
IN CRUDE PRODUCTS

By R. F. Jackson, J. A. Mathews, and W. D. Chase

ABSTRACT

In sequel to a previous article (RP426) detailed descriptions are presented of analytical methods for the determination of levulose in crude products. The polysaccharides in which levulose is the predominating sugar can be hydrolyzed for analytical purposes in 35 minutes at 80° C. by 0.14 N HCl. After hydrolysis levulose is determined either by the selective copper reduction method or by polarization at two temperatures. The volume of the marc from a 50-g sample of artichokes is found to be 0.8 ml. Rapid methods of titrating reducing sugars in low concentration are suggested in extension of the standard methods of Lane and Eynon. Methods of determining the purity of crude and purified juices are described. Applications of the methods of analysis are made to honey and fruits in most of which levulose is found to be the predominating sugar.

CONTENTS

I. INTRODUCTORY

The wide distribution of levulose in natural and manufactured products makes desirable the formulation of a system of analysis for its quantitative determination. In a previous article [1] some of the fundamental ph sical properties of pure levulose in aqueous solution and improved methods for its selective determination in sugar mixtures were described. In the present article we shall discuss the determination of levulose in crude natural products and particularly in the intermediate products obtained during the process of preparation of levulose from natural sources by the method described by Jackson, Silsbee, and Proffitt. [2]

In natural products levulose occurs in the form of the uncombined sugar or combined in the form of compound sugars or of polysaccharides. When it occurs as the free sugar it is invariably accompanied by dextrose and frequently by sucrose or other sugar. The polysaccharides which contain levulose have properties which make a direct determination of them at present impossible. They are, however, very easily hydrolyzed by dilute acids and may be determined indirectly by analysis of the products of hydrolysis. The reducing sugars formed by hydrolysis consist of levulose mixed with varying proportions of dextrose. The problems involved in analyzing sugar mixtures containing levulose become mainly questions of differentiation between levulose and dextrose since they practically always occur together.

It will be convenient to characterize these mixtures according to the ratio of levulose to total reducing sugar. Various methods of determining this ratio have been described in the previous article. It can be determined rigorously only by direct analysis for reducing sugar and selective analysis for levulose. In the former article a proximate method was described and designated the Mathews formula (RP426, p. 433) in which it was shown that the ratio could be determined conveniently by combining the direct polarization with the titration for total reducing sugar under the assumption that no optically active substances other than dextrose and levulose were present in the mixture. The Mathews ratio will sometimes be found valuable even when it differs from the true ratio, for the difference then becomes a measure of the optically active or reducing materials which are neither dextrose nor levulose.

In previous articles [3] it has been shown that inulin yields upon acid hydrolysis about 5 per cent of a group of nonreducing difructose anhydrides which are of such high stability as to survive the hydrolysis. No analytical method has as yet been devised to determine these substances directly, since it is impossible to hydrolyze them without employing acid of such high concentration as to destroy the reducing sugar in the sample. In the analysis of juices of plants which contain inulin or similar polysaccharides these refractory disaccharides will escape detection under the ordinary procedure. If, however, we assume that they are mainly responsible for the deviations of the Mathews ratio from the true ratio under the assumption that levulose, dextrose, and the disaccharides are the only optically active substances present, we have a means of determining them approximately.

[1] Jackson and Mathews, B. S. Jour. Research, vol. 8 (RP426), p. 403, 1932. Repeated reference will be made to Research Paper No. 426 which for brevity will be designated (RP426).
[2] B. S. Sci. Paper No. 519, vol. 20, p. 604, 1926.
[3] Jackson and Goergen, B. S. Jour. Research, vol. 3 (RP79), p. 27, 1929. Jackson and McDonald, B. S. Jour. Research, vol. 5 (RP251), p. 1151, 1930; vol. 6 (RP299), p. 709, 1931.

II. HYDROLYSIS OF POLYSACCHARIDES IN ARTICHOKE JUICES

The juices contained in the roots of the Jerusalem artichoke, dahlia, chicory, and many other composite flowering plants are rich in polysaccharides in which levulose is the predominating sugar. In order to analyze these juices the conditions of acidity, temperature, and time must be found which will yield complete hydrolysis without destruction of the products of the reaction. The experiments described below were performed with the juices and pulp of the Jerusalem artichoke, but the conclusions are equally applicable to any juice or material which contains similar polysaccharides.

In order to determine the proper conditions for hydrolysis weighed quantities of the expressed juice or comminuted pulp of Jerusalem artichokes were introduced into 200 ml Soxhlet flasks, diluted to about 175 ml, and acidified with 3 or 4 ml of 8 N HCl. The flasks were immersed in a water bath maintained at either 70° or 80° C. and after selected intervals of time were one by one removed and the contents cooled rapidly. To simplify the procedure the solutions were allowed to assume the temperature of the bath without agitation.

TABLE 1.—*Hydrolysis of artichoke pulp with hydrochloric acid*

(3 or 4 ml of 8 N HCl added to 175 ml of solution)

(a) 10-g SAMPLE, 4 ml HCl, 80° C.

Time of hydrolysis	Reducing sugars	Levulose	Ratio: levulose reducing sugar	Mathews ratio: levulose reducing sugar
Minutes	Per cent	Per cent	Per cent	Per cent
10	16.74	....	....	....
20	17.00	....	....	....
30	17.20	12.45	72.4	67.9
40	17.14	12.43	72.5	68.8
60	17.36	12.53	72.2	67.4

(d) 50-g SAMPLE, 4 ml HCl, 70° C.

Time of hydrolysis	Reducing sugars	Levulose	Ratio: levulose reducing sugar	Mathews ratio: levulose reducing sugar
Minutes	Per cent	Per cent	Per cent	Per cent
20	17.25	....	....	....
30	17.28	....	....	....
40	17.26	12.64	73.2	72.1
50	17.23	12.84	74.5	72.5
70	17.18	12.90	75.1	72.6

(b) 10-g SAMPLE, 4 ml HCl, 70° C.

19	16.90	....	....	....
30	17.08	....	....	....
42	17.10	12.45	72.8	70.8
50	17.11	12.23	71.5	70.3
60	17.16	12.54	73.1	69.5

(e) 50-g SAMPLE, 4 ml HCl, 80° C.

7	11.29	....	....	84.0
20	17.10	14.43	84.4	84.1
30	17.04	14.30	83.9	83.7
60	16.96	14.16	83.5	82.4
90	17.19	14.39	83.7	80.7

(c) 50-g SAMPLE, 4 ml HCl, 80° C.

20	17.20	....	....	....
30	17.20	12.80	74.4	71.9
41	17.20	12.72	73.5	71.2
52	17.36	12.71	73.2	70.9
63	17.38	12.69	73.0	69.8

(f) 50-g SAMPLE, 3 ml HCl, 80° C.

10	14.44	....	....	82.9
20	17.30	14.86	85.9	83.7
30	17.31	14.64	84.6	83.9
60	17.41	14.61	83.9	83.9
90	17.42	14.65	84.1	83.1

The samples of pulp were prepared by passing artichoke tubers through a food chopper, the plate of which had perforations 2 mm in diameter. The hydrolysis experiments were conducted with 10-g and with 50-g samples. Ten grams is a sufficient quantity if the sugar analysis is made solely by copper reduction methods, while a 50-g sample is preferable if the procedure includes a polariscopic determination. In the series here described all samples, including the 10-g samples, were polarized, analyzed for total reducing sugar by Lane and Eynon's volumetric method, and selectively analyzed for levulose by Jackson and Mathews's modification of Nyns's method (RP426, p. 425).

The samples after cooling were neutralized and defecated with a saturated solution of normal lead acetate. The addition of alkali was made slowly and with agitation vigorous enough to obviate momentary local alkalinity. The solutions were finally made to volume and filtered. The filtrates were polarized in a 400-mm tube and an aliquot volume was analyzed for total reducing sugar and levulose.

The experimental results are shown in Table 1. Upon examination of the individual series it becomes apparent that regardless of changed conditions of hydrolysis the respective samples arrive at approximately constant compositions after relatively short periods of time. Even when the time is greatly extended there is no evidence of decomposition of the products of hydrolysis. Indeed, in many instances there appears to be a slight increase in total sugar, possibly by gradual hydrolysis of the difructose anhydrides. Table 1(f) shows that 3 ml of HCl is about as effective as the 4 ml used in the other experiments and produces complete hydrolysis in about the same period of time. This can be explained by the fact that the determining factor is the time required to attain the bath temperature and that the hydrolysis itself proceeds very rapidly in either concentration of acid.

The deviations from constancy of the individual determinations are in many cases greater than the experimental error of analysis. These errors arise probably from the difficulty of obtaining complete concentration equilibrium between the solution and the suspended pulp. The divergences diminish, as shown in Table 2, when samples of press juice are hydrolyzed for periods of time between 25 and 60 minutes, although here again there is evidence of gradual hydrolysis of the disaccharides.

The Mathews ratio given in the last column of the tables will be discussed in a later paragraph.

On account of the relative uniformity of all the experimental results the selection of specifications is in a measure arbitrary. Although the hydrolysis can be completed rapidly at 70°, the defecation and filtration of solutions hydrolyzed at 80° were more satisfactory.

Evidently 3 ml of 8 N HCl affords sufficient acidity for the samples under investigation. Conceivably samples of greater dry substance concentration might by buffer action greatly diminish the converting power of the hydrochloric acid. The juices upon which the experiments were made contained by refractometric estimation 17 to 20 per cent dry substance. A 50-g sample therefore contained 8.5 to 10 g of total solids. In order to ascertain how great a variation in dry substance can be tolerated, further experiments were made on a press juice in which the hydrolysis was effected by 3 ml of 8 N HCl for 30 minutes at 80° C., while the weight of sample was varied from 50 to 160 g. In Table 3 are shown the polariscopic readings of the samples diluted after hydrolysis to a uniform concentration. The solutions containing 50 g and 80 g showed identical polarizations, and the one containing 100 g of sample showed but slightly lower polarization. The experiments indicate that the conditions selected for hydrolysis are valid even if the dry substance is increased nearly 100 per cent. Such a range of concentrations is great enough for the analysis of all naturally occurring juices, or more specifically, for samples which contain not in excess of 19 g of dry substance. In case of uncertainty the volume of 8 N HCl can safely be increased to 4 ml.

TABLE 2.—*Hydrolysis of artichoke press juice with hydrochloric acid*

(50 g juice, 4 ml 8 N HCl, 80° C.)

Time of hydrolysis	Reducing sugars	Levulose	Ratio: levulose / reducing sugar	Mathews ratio: levulose / reducing sugar
Minutes	*Per cent*	*Per cent*	*Per cent*	*Per cent*
10	17.73	15.25	86.0	82.4
25	17.78	15.26	85.8	82.6
40	17.82	15.25	85.8	82.7
60	17.83	15.33	86.0	82.7
90	17.91	15.38	85.9	82.7

TABLE 3.—*Buffer action of artichoke juice, 19.8 brix*

(3 ml 8 N HCl, 30 minutes, 80° C.)

Weight of press juice	Polarization at concentration 25 g per 100 ml $\times(-1)$	Relative hydrolysis
g	*°S.*	*Per cent*
50	35.7	100.0
80	35.7	100.0
100	35.6	99.7
125	30.9	86.7
160	18.9	52.9

III. ACCESSORY DATA

1. VOLUME OF MARC AND LEAD PRECIPITATE

When a 50-g sample of pulped artichokes has been digested and defecated in preparation for analysis the residual marc and lead precipitate appear very voluminous. To determine the total volume displaced a sample was hydrolyzed and defecated in a weighed and calibrated 200-ml flask and the total mixture weighed after being made to volume. After thorough mixing, a small "catch" sample, about 25 to 30 g, was weighed, made to 200 ml, and filtered. The remainder of the sample was filtered and a 25.00-ml aliquot sample made to 200 ml. The latter and the filtrate from the weighed out sample were titrated in triplicate for total reducing sugar.

If F is Lane and Eynon's factor, T and T' the titers, respectively, of the aliquot and the "catch" sample, V the calibrated volume of the 200-ml flask, v the volume of the marc and precipitate, W the total weight of the hydrolyzed and defecated sample made to 200 ml, and w the weight of the "catch" sample, then

$$\frac{FV(V-v)}{25T} = \frac{F\left(V-\dfrac{vw}{W}\right)W}{T'w} \tag{1}$$

in which both terms express the total sugar in 50 g of the sample. All quantities are directly observed except v.

Two determinations yielded the values 0.802 and 0.856 ml for the volume of the marc and precipitate, the difference being within the error of analysis. We shall accept the value 0.8 ml for the marc of a 50-g sample. The volume of the marc of a 10-g sample will therefore be about 0.2 ml.

It is of interest to compare this value with the volume of the marc from the hot water digestion of the sugar beet. A 26-g sample of the latter defecated with basic lead acetate yields a volume of 0.6 ml, while a similar sample of artichoke pulp would occupy a volume of 0.43 ml.

2. THE MATHEWS FORMULA

In the previous article (RP426, p. 433) a formula was derived for the calculation of the ratio (R) of levulose to total sugar in pure mixtures from the direct polarization and Lane and Eynon titration. This formula states that

$$R = f\left(\frac{PT}{D}\right) \tag{2}$$

in which P is the direct polarization, T the corrected Lane and Eynon titer, and D the number of volumes to which one volume of the solution polarized is diluted for titration. A graphic solution of the function yielded a single table from which all values of R could be read o interpolated.

The Mathews ratio has been compared with the true ratio in all products which have been analyzed during the present investigation. In some instances because the measurements required for its solution can be made with higher precision, it has proved more serviceable than the ratio determined rigorously. In other instances a considerable discrepancy has appeared.

The discrepancy between the two ratios is significant, for it becomes an indication of the presence of optically active or of reducing substances which are not levulose or dextrose. In special cases the specific rotations of substances accompanying levulose and dextrose in natural products are known and can be employed for quantitative estimation. Thus, if c is the saccharimetric normal weight (derived from the inverse ratio of the specific rotations of sucrose and of the substance in question), V the volume in which g grams of the sample was dissolved for polarization, T the Lane and Eynon titer, D the number of volumes to which one volume of the solution polarized was diluted for titration, and ΔR the difference between the true ratio and the Mathews ratio

$$\frac{VcD\Delta R}{100gT} = \text{per cent optical impurity} \tag{3}$$

In many natural products it is not to be expected that the formula yields an accurate measure of a definite substance. It is rather a measure of all substances which are not dextrose or levulose, arbitrarily grouped together and estimated as a definite substance which is known to be present, but can not be selectively determined. Applications of the Mathews formula will be made in the analyses reported below.

3. TITRATION OF REDUCING SUGAR IN LOW CONCENTRATION

The convenience and precision of Lane and Eynon's volumetric procedure [4] suggest the desirability of extending the range of concentrations of sugar to greater dilutions in order to permit the rapid analysis of sweet waters, exhausted pulp from the diffusion battery, and, in general, any solution of low sugar concentration. Since at the end of Lane and Eynon's titration the concentration of cupric copper is very small, it is evident that the sharpness of the end point is independent of the initial concentration of copper. We can therefore take for titration smaller concentrations of copper than Lane and Eynon specified for their standard methods and still expect the same definiteness of end point, bearing in mind, however, that the slight excess of sugar required to reduce the methylene blue becomes an increasingly large proportion of the total titer as the amount of copper is diminished. It is consequently essential that the volume of indicator be constant for both standardization and analysis.

Lane and Eynon have shown that 5 ml of the alkaline tartrate constituent of Soxhlet's solution is sufficient to keep the cupric copper in solution even for titrations as great as 50 ml. This concentration of tartrate has been retained in the following experiments. The copper solution has been diminished to 2.5 ml and to 1.0 ml, respectively, and water added to make the total volume 10 ml.

The success of the method depends more upon the skill of the analyst than in the case of the standard methods. In order to perceive the end point more precisely we have altered the standard procedure of Lane and Eynon. Ten milliliters of the mixed dilute Soxhlet reagent was transfered to a 150-ml flask. Somewhat less than the total volume of the sugar solution required to reduce the copper was added, and the flask being held with wooden tongs, the reaction mixture was brought to boiling during continuous agitation over a low free flame. Boiling was continued for one-half minute, the indicator was added, and the titration completed in the usual manner. This procedure permits a close perception of the end point.

The titrations against the solutions containing 2.5 ml of copper solution proved to be about as reproducible as those by the standard methods. The factors for varying ratios of levulose to total sugar and varying titers are given in Table 4. The analyst should always standardize his own procedure by titration of a known solution. The deviation of the factor from the tabulated factor found by experiment can safely be considered constant throughout the table.

TABLE 4.—*Lane and Eynon factors for 10-ml Soxhlet solution composed of 2.5 ml copper-sulphate solution, 2.5 ml water, and 5 ml alkaline tartrate*

(Indicator, 1 drop of 1 per cent methylene blue)

R Titer	Dextrose 0	10	20	30	40	Invert sugar 50	60	70	80	90	Levulose 100
ml											
15	26.2	26.3	26.4	26.5	26.6	26.8	26.9	27.0	27.1	27.2	27.3
20	26.4	26.5	26.6	26.7	26.8	26.9	27.0	27.1	27.2	27.3	27.4
25	26.5	26.6	26.7	26.9	27.0	27.1	27.2	27.3	27.4	27.5	27.6
30	26.7	26.8	26.9	27.0	27.1	27.2	27.3	27.4	27.5	27.7	27.8
40	27.0	27.1	27.2	27.3	27.4	27.5	27.7	27.8	27.9	28.0	28.1
50	27.3	27.4	27.5	27.6	27.8	27.9	28.0	28.1	28.2	28.3	28.4

$F = 25.74 + 0.0317\,T + 0.0107\,R$, in which T is titer, and R the ratio of levulose to total sugar expressed in per cent.

[4] J. Soc. Chem. Ind., vol. 42, p. 32, 1923.

The titrations against Soxhlet solutions containing 1 ml of copper-sulphate solution were, as expected, less reproducible than those against the standard Soxhlet solutions. The method can, however, be relied upon probably within 1 or 2 per cent of the quantity measured. Within the errors of titration the factor proved to be sensibly constant for variations of sugar concentration and composition. The mean of 7 titrations (indicator, 1 drop of 0.2 per cent methylene blue) yielded a value of the factor of 11.7.

The employment of solutions containing 2.5 ml of copper sulphate in the Soxhlet solution permits the titration of solutions containing from 55 to 180 mg of hexoses per 100 ml. Soxhlet solutions containing 1.0 ml of copper cover a range of 23 to 78 mg in 100 ml.

4. STEFFEN'S REAGENTS

Many of the intermediate products encountered in the production of levulose by the levulate process require a determination of alkalinity expressed as CaO. For this purpose it is convenient to employ the Steffen reagents which are used in the beet-sugar industry. One milliliter of these solutions which are 1.786 N is equivalent to 0.05 g CaO. Steffen's HNO$_3$ and NaOH are suitable for the determination of total CaO by the usual method. To these are added hydrochloric acid and potassium oxalate of the same normality. For convenience they also are called Steffen's reagents and are useful for the neutralization and removal of calcium in the preparation of samples for sugar analysis. These solutions are all of equivalent concentration and consequently the amount of CaC$_2$O$_4$. H$_2$O which is preci i te is related definitely to the volume of Steffen's reagents added. pRar edch milliliter of Steffen's reagent 0.13 g of calcium oxalate is precipitated, and if the density of the precipitate is considered 2.05 [5] the volume displaced is 0.064 ml. The product of 0.064 and the volume of Steffen's oxalate added is the volume displaced by the precipitate.

Calcium oxalate is occasionally difficult to separate by filtration when precipitated from cold solutions. For removing both calcium and lead a mixture of 77.8 g of K$_2$HPO$_4$ and 82.2 g of K$_2$C$_2$O$_4$. H$_2$O made to 1 liter is particularly satisfactory. This mixture has the same total normality as Steffen's reagents. The mean density of the precipitate is 2.19,[6] and the volume of precipitate from 1 ml of reagent is 0.065 ml.

The volume displaced by the calcium precipitate can be calculated and deducted from the volume of the volumetric flask. It simplifies the subsequent calculation, however, to add by means of a graduated pipette a volume of water above the mark of the flask equal to the volume of the precipitate. This procedure is adopted for the methods suggested below.

IV. ANALYTICAL PROCEDURE

1. ARTICHOKES OR SIMILAR PRODUCTS

The total reducing sugar in these products can be determined by Lane and Eynon titration and the levulose either by the modified Nyns method or by polarization at two temperatures.

---

[5] Calculated from data in Handbook of Chemistry and Physics, 13th ed. Chemical Bulletin Publishing Co., p. 190, 1928.
[6] See footnote 5.

(a) PREPARATION AND HYDROLYSIS OF SAMPLE

Unless the substance is already fluid or finely divided, comminute the material in a food chopper or similar appliance and mix the sample thoroughly. Transfer a weighed sample (the amount depending upon the method of analysis selected) to a 200-ml Soxhlet flask and add water to about 175 ml. Add 3 ml of 8 N HCl, digest in a water bath at 80° C. for 35 minutes. Cool and add a saturated solution of normal lead acetate (about 1 ml for a 7 to 10 g sample and 3 to 6 ml for a 50-g sample). Add 2.9 ml of 8 N NaOH with very vigorous agitation, avoiding any local alkalinity. Make to volume, mix thoroughly, and allow to stand for several minutes. Again mix and filter. Proceed by (b) or (c).

(b) ANALYSIS BY LANE AND EYNON TITRATION AND THE MODIFIED NYNS METHOD

Weigh out an amount of the original sample which contains about 1 to 1.5 g of reducing sugar. This will vary between about 7 and 10 g. The concentration of sugar in the sample can be roughly gaged, if necessary, by refractometric examination of the juice and by assuming a total sugar purity of about 85 per cent. Hydrolyze in a 200-ml flask, defecate, make to volume 0.2 ml above the mark, and filter, as described under (a). Titrate the filtrate directly against 25 ml of mixed Soxhlet reagent according to Lane and Eynon's method.[7] Apply a correction to the titer (T) if titrations of standard solutions fail to show exact correspondence with Lane and Eynon's factors (RP426, p. 439).

Assuming that the ratio of levulose to total sugar is about 75 per cent, calculate from the total sugar titration the volume of the filtrate which contains about 75 mg of levulose. Add this volume to 50 ml of Ost's solution and add sufficient water to make 70 ml. Determine levulose as previously described (RP426, p. 425).

Calculate the ratio (R) of levulose to total sugar by means of Table 25 (RP426, p. 444). Refer to Table 22 (RP426, p. 439) for the proper Lane and Eynon factor. Calculate sugars in the original sample by

$$\frac{F \times 20}{T \times \text{weight sample}} = \text{per cent total sugar}$$

$R \times$ per cent total sugar = per cent levulose.

(c) ANALYSIS BY LANE AND EYNON TITRATION AND TEMPERATURE COEFFICIENT OF POLARIZATION

Weigh out a 50-g sample, transfer to a 200-ml Soxhlet flask, hydrolyze, and defecate as described under (a). Make to volume 0.8 ml above the mark. Remove lead from the filtrate by addition of dry sodium oxalate or a mixture of dry oxalate and phosphate. Polarize at or near 20° C. ($= P_1$) and at about 70° C. in a water-jacketed polariscope tube, noting accurately the respective temperatures of polarization (RP426, p. 418).

Correct the high temperature polarization for the thermal expansion of the solution by

$$P_{obs.}[1 + 0.00043\,(t_2 - t_1)] = P_2$$

and calculate the per cent of levulose by

$$\frac{4(P_2 - P_1)}{0.0344\,(t_2 - t_1)} = \text{per cent levulose}$$

[7] J. Soc. Chem. Ind., vol. 42, p. 32, 1923. J. Assoc. Official Agri. Chem., vol. 9, p. 35, 1926. For correction to titer, see B. S. Jour. Research, vol. 8 (RP426), p. 420, 1932.

valid for a 2-dm polariscope tube. If a 4-dm tube is used divide the result by 2.

Pipette a volume of the filtrate which contains about 1 to 1.5 g of reducing sugar, make to 200 ml and titrate against 25 ml of Soxhlet's solution by Lane and Eynon's method. Calculate total sugar by

$$\frac{F \times 20 \times 200}{T \times v \times 50} = \frac{F \times 80}{T \times v} = \text{per cent total sugar}$$

in which v is the volume diluted to 200 ml for titration. The factor F must be taken from the column in Table 22 (RP426, p. 439) which corresponds to the proper ratio of levulose to total sugar.

(d) PURITY OF DIFFUSION OR PRESS JUICE

The definition of purity of diffusion or press juice is to some extent arbitrary. The dry substance determination is necessarily made before the addition of the hydrolyzing agent, while the sugar analysis is made after such addition. There is therefore a change in the dry substance between the two operations. It would be possible to hydrolyze with some reagent, such as sulphuric acid, and remove it after hydrolysis with barium hydroxide. This procedure would not only be too tedious for routine analysis, but would yield the purity not of the original juice but of the defecated juice, since neutralization of the acid would be accompanied by defecation. It therefore seems advisable for practical purposes to define arbitrarily the purity as the ratio of levulose (or of total sugar) after hydrolysis to the apparent dry substance before hydrolysis. Such purity will of course be "apparent."

Dry substance determination of artichoke juices are most satisfactorily made by refractometric or densimetric estimation. A reference of such observation to Tables 18, 19, or 20 (RP426, pp. 437–438) yields "apparent" dry substance, since the tables are strictly valid only for pure levulose solutions. The dry substance can be determined by desiccation methods, but the latter are difficult because the end point of the drying is never sharply determinable. Moreover, the dry substance is still "apparent" with respect to the sugar analysis, for it consists of polysaccharides which take up water upon hydrolysis.

Having determined the apparent dry substance, weigh out a quantity of juice containing 1 to 1.5 g of total sugar and analyze as described under (b), or 10 to 15 g and proceed as in (c). Divide the percentage of levulose by percentage of dry substance to obtain the apparent purity.

(e) NOTES

The general methods described above can be applied to any plant material which contains polysaccharides similar in nature to those present in the Jerusalem artichoke, dahlia, or chicory. Only when the whole plant material is included in the sample is the allowance made for the volume of the marc.

Juices which have been hydrolyzed and defecated in preparation for the lime precipitation process can be analyzed by the methods described under (b) or (c), omitting the directions for hydrolysis. It is essential that calcium be completely removed by oxalate or oxalate-phosphate mixture, because it interferes seriously with the reducing sugar analysis.

In order to facilitate the introduction of a 50-g sample of pulped plant material into a 200-ml flask it is convenient to make use of a brass funnel, the stem of which extends into and below the neck of the flask, and a plunger consisting of a solid brass cone with a rod attached to the smaller end.

Defecation with normal lead acetate is more complete in neutral or but slightly acid solution. The hydrochloric acid can be neutralized in the manner described if the agitation is sufficiently vigorous to avoid local alkalinity. If preferred an 8 N solution of potassium acetate can be substituted for the NaOH. This solution can be conveniently prepared by neutralizing about 50 g of KOH with about 47 to 50 ml of standardized glacial acetic acid and completing the volume to 100 ml.

An approximate determination of levulose in artichokes can be made by applying the Mathews formula. The filtrate from a 50-g sample is polarized directly and an aliquot volume titrated for reducing sugars. From the observed data total sugar and levulose are calculated in the manner described in RP426, p. 433. The errors in the ratio which result from the application of the Mathews formula to artichoke and dahlia juices are to be ascribed mainly to the presence of the dextrorotatory nonreducing difructose anhydrides. In Tables 1 and 2 are given the true and the Mathews ratio, the deviation of the latter averaging about 3.0 per cent of the quantity measured.

The addition of 3 ml of 8 N HCl is sufficient for hydrolysis if the sample contains not in excess of 19 g of dry substance. If greater weights of dry substance are present add 4 ml.

2. EXHAUSTED PULP AND PULP WATER (RAPID METHOD)

The control of the diffusion process for the extraction of juice from artichokes or similar plants requires a rapid analysis of the pulp and pulp water from the last cell of the battery as soon after its discharge as possible.

Grind the pulp sample through a food chopper and express the juice by means of a tincture press or, preferably, a hydraulic press. Pipette 100 ml into a 200-ml volumetric flask. Add 4 ml of 8 N HCl and immerse the flask in a boiling-water bath for four minutes. Remove from the bath and cool rapidly. Add 1 ml of saturated lead acetate and 4 ml of 8 N NaOH. Make to volume, filter, and titrate by Lane and Eynon's method against 25 or 10 ml of mixed Soxhlet solution according to the concentration of sugar.

$$\frac{F \times 0.2}{T} = \text{grams of reducing sugar in 100 ml}$$

Pipette 175 ml of pulp water into a 200-ml flask and add 4 ml of 8 N HCl. Immerse in boiling water and proceed as described in the previous paragraph.

$$\frac{F \times 0.114}{T} = \text{grams of reducing sugar in 100 ml}$$

If the concentration of sugar is too low for titration by the standard Lane and Eynon methods, titrate against one of the dilute Soxhlet reagents by the method described on page 603.

3. LEVULATE SLUDGE

Because of the instability of the reducing sugars in the alkaline medium in which calcium levulate is precipitated the levulate sludge must be sampled and neutralized without delay.

(a) TOTAL CaO

Weigh out 20 g of the thoroughly mixed sludge and wash into a porcelain casserole. Add 20 ml of Steffen's HNO_3, stir with a glass rod until completely dissolved, and boil for three to five minutes. Cool and titrate to a faint pink with Steffen's NaOH (phenolphthalein)

$$\text{Titer} \div 4 = \text{per cent CaO}$$

(b) TOTAL REDUCING SUGAR

Weigh out 15 g of the sample, wash into a porcelain casserole, and titrate to slight acidity with Steffen's HCl. Add a volume of Steffen's potassium oxalate-phosphate mixture equal to the volume of the Steffen's HCl plus a sufficient quantity to precipitate any neutral calcium which may have been introduced with the defecated juice. Wash into a 200 ml volumetric flask, make to volume, and add by means of a graduated pi e e an excess of water equal to the volume of the precipitate. (See p. 604.) Filter and titrate directly against 25 ml of mixed Soxhlet solution by Lane and Eynon's method for total reducing sugars. Calculate by

$$\frac{4 \times F}{3 \times T} = \text{per cent reducing sugar}$$

in which F is Lane-Eynon's factor and T the corrected titer.

(c) LEVULOSE

In most instances the ratio of levulose to total reducing sugar will have been determined previous to the precipitation process. If so, the known ratio can be applied to the percentage of total reducing sugar to compute the percentage of levulose. The latter can be determined by analysis of the above filtrate by taking a volume containing about 75 mg of levulose and proceeding by the modified Nyns method as described in RP426, page 425. Calculate the ratio, R, as described in RP426, page 444.

$$R \times \text{per cent total sugar} = \text{per cent levulose}$$

4. LEVULATE CAKE

(a) TOTAL CaO

Weigh out 10 g of the sample and wash into a porcelain casserole. Add an excess (usually 25 ml) of Steffen's HNO_3. Stir with a glass rod until dissolved, place over a flame, and allow it to boil from three to five minutes. Add a few drops of phenolphthalein and titrate to a faint pink with Steffen's NaOH.

Titer ÷ 2 = per cent CaO

(b) TOTAL SUGAR AND LEVULOSE

Weigh out 20 g of the sample, wash into a porcelain casserole, and itrate to slight acidity with Steffen's HCl (phenolphthalein). Wash nto a 200 ml volumetric flask. Add a volume of Steffen's oxalate-)hosphate mixture slightly in excess of the determined volume of 5teffen's HNO_3 which was required for the CaO determination. Vlake to volume and add an excess of water equal to the volume of ;he precipitate, that is, 0.065 × milliliters of Steffen's reagents used.

Polarize the filtrate, preferably in a 400 mm tube, at or near 20° C., ·ecording the temperature of observation. The negative polarization)f levulose is enhanced in the presence of KCl, 1 g in 100 ml increasing t in the negative sense by 0.59 per cent. Since there is formed).1332 g of KCl for each milliliter of Steffen's reagent added, diminish the negative polarization by 0.078 per cent of its value for each milliliter of Steffen's reagent used.

Pipette 50 ml of the filtrate into a 200 ml flask, make to volume, and titrate against 25 ml of Soxhlet's solution.

From the corrected polarization and the corrected titer calculate the Mathews ratio in the manner described on p. 440, RP426, remembering that the polarization, P, in the formula refers to the rotation in a 2 dm column. Calculate by

$$\frac{F \times 4}{T} = \text{per cent total sugar,}$$

and

$$R \times \text{per cent total sugar} = \text{per cent levulose}$$

5. WASTE WATER

(a) CaO

Weigh out 100 g of the sample, wash into a porcelain casserole, and titrate with Steffen's HCl (phenolphthalein). Save the solution for (b) and (c).

Titer ÷ 20 = per cent CaO

(b) TOTAL SUGAR

Wash the titrated solution into a 200 ml volumetric flask and add a volume of Steffen's oxalate-phosphate mixture equal to the titer for CaO. If the solution is derived from a juice which has been hydrolyzed with H_2SO_4 and defecated with lime, add an additional 15 ml of the oxalate-phosphate mixture. The calcium must be completely removed. Make to volume and add an excess of water equal to the volume of the precipitate; that is, 0.065 × milliliters of Steffen's reagent. From the filtrate pipette a volume containing about 1 g of total sugar into a 200 ml volumetric flask, make to volume, and titrate against 25 ml of mixed Soxhlet reagent.

(c) LEVULOSE

Assuming a ratio of levulose to total sugar of 35 per cent, compute the volume of the original filtrate which contains about 75 mg of "apparent" levulose, add this volume to 50 ml of Ost's reagent, make to 70 ml, and proceed as directed on page 425, RP426. Calculate the ratio of levulose to total sugar in the usual way by referring to Table 22 (RP426, p. 444) and find the total sugar and levulose by the following formulas:

$$\frac{40 \times F}{T \times v} = \text{per cent total sugar}$$

where F is Lane and Eynon's factor, T is the Lane and Eynon titer, and v is the volume of solution which was diluted to 200 ml for the titration:

$$\text{Per cent total sugar} \times R = \text{per cent levulose}$$

Some uncertainty is introduced by assuming a value for the neutral calcium which is precipitated. The variations from the value given above will, in general, be within the analytical accuracy. If desired, the calcium can be precipitated by dry sodium oxalate and the uncertainty avoided.

6. EVAPORATOR LIQUORS

The levulose solutions which are prepared by carbonation of the levulate cake are of relatively high purity. After filtration from the calcium carbonate they are freed from calcium bicarbonate by evaporation or other means and filtered. At this point an analysis is of significance in determining the completeness of the separation of levulose from its impurities.

It has already been shown that for artichoke juices the Mathews ratio differs but slightly from the true ratio, a fact which indicates that the optically active substances in the juice are mainly levulose and dextrose. It follows, then, that in the purified liquors from which most of the impurities have been removed the Mathews ratio will approach the true ratio very closely. Inasmuch as the analytical operations contributing to the determination of the Mathews ratio are far more precise and reproducible than those depending upon Nyns's method, it is to be recommended that the levulose purity of high purity liquors be determined on the basis of the Mathews ratio solely.

(a) DRY SUBSTANCE

Determine dry substance by measurement of density, referring the observation to Tables 18 or 19 (RP426, p. 437), or by refractometer reading, referring to Table 20 (RP426). If the sirup has too high a concentration for the polarization which is to follow, it must be diluted to below 17.2 per cent for a 20° C. polarization, or below 17.9 per cent for 25° polarization. In this case the dry substance measurement must be repeated for the diluted solution, or a weighed amount of the concentrated solution must be diluted with a weighed amount of water.

(b) DIRECT POLARIZATION AND APPARENT PURITY

Polarize the diluted solution directly, recording the temperature of observation accurately. From this observation and the percentage of dry substance of the solution the "apparent" purity can be calculated by assuming that all optically active material is levulose. Correct the direct polarization to either 20° or 25° C. by assuming

:oefficient of 0.0344° S. for each gram of levulose per degree change)f temperature. Correct for concentration by Table 21 (RP426,). 439). Multiply the normal weight (18.407 g at 20° C. or 19.003 g it 25° C.) by the direct polarization and divide by the number of ;rams of dry substance in 100 ml. (See below.)

(c) TOTAL SUGAR

Compute from the polarization or dry substance analysis the vol-ime· of solution containing about 1 g of total sugar. Transfer this rolume to a 200-ml volumetric flask, make to volume, and titrate igainst 25 ml of mixed Soxhlet reagent.

(d) TRUE PURITY

From the direct polarization and reducing sugar titration compute the product $P \times T/D$ and determine the ratio of levulose to total sugar by reference to Table 24 (RP426, p. 442; see also example on p. 440). The true purity of the sample is

$$\frac{F \times R \times D}{T \times S \times 10} = \text{per cent purity}$$

in which F is Lane and Eynon's factor, R the Mathews ratio, T the corrected titer, D the number of volumes to which the solution polar-ized was diluted for titration, and S the grams of dry substance in 100 ml of the solution polarized. The latter is determined by multi-plying the percentage of dry substance by the density as given in Table 18 (RP426, p. 437). This product is the weight of dry sub-3tance in vacuo and for precise work should be diminished by 0.11 per cent of its value to give its weight in air with brass weights.

V. HONEY AND VARIOUS FRUITS

The methods of selective levulose analysis are applicable to any product containing levulose. In order to illustrate further possible applications, we have analyzed samples of honey and such fruits as were immediately available.

The samples of honey were supplied by the Bee Culture Laboratory, of the United States Department of Agriculture. Dry substance de-terminations were made by C. F. Snyder, of this bureau, by refractom-eter readings referred to Table 20 (RP426, p. 438). The samples were prepared for analysis by weighing out 20 g and washing into 200-ml volumetric flasks. The solutions were clarified satisfactorily by addition of alumina cream. Sucrose was estimated by determin-ing polarizations before and after inversion with invertase. Both the original and inverted solutions were subjected to a complete sugar analysis by the general procedure outlined on pages 604–606. Before calculating the Mathews ratio, the rotation of the sucrose was deducted from the direct polarization.

The analytical data are assembled in Table 5. This group of sam-ples showed a ratio of levulose to total sugar not greatly in excess of 50 per cent; the ratio is therefore not appreciably altered by the inver-sion of the sucrose. The deviation of the Mathews ratio from the true ratio has been solved by formula (3) under the arbitrary assump-tion that the deviation is due to dextrins. If a specific rotation of

+150° C.[3] is assigned to these substances, the constant c in formula (2) acquires a value of 11.5. It is apparent from columns 8 and 13 that these substances occur in all the samples analyzed and that normal nectar honeys contain a mean of about 4.6 per cent. Obviously, if honey were adulterated with any considerable quantity of invert sugar, the deviation of the Mathews ratio from the true ratio will diminish. The suggestion is made that this method of grouping these substances under the designation "apparent dextrins" may be useful in the critical examination of honey. In the honey dew and the melezitose honey the quantity of apparent dextrins becomes very large. The melezitose honey shows no significant change in ratio or in the apparent dextrins before and after inversion, indicating that melezitose is unaffected by invertase. In confirmation of this fact, a sample of pure melezitose showed no change in polarization after treatment with invertase. The total carbohydrates in the melezitose honey fall far short of the total dry substance. This arises from the fact that we ascribed a specific rotation of +150 to the undertermined carbohydrates. If we make the assumption that the total carbohydrate purity is 98.2 per cent (the mean of the remaining honeys) and assign a specific rotation of +150 to the dextrins and +87.8 to melezitose, we can solve simultaneously for the two latter substances. Such a solution yields values of 3.9 per cent dextrins and 20.2 per cent melezitose.

The samples of fruit were obtained by purchase from commercial sources. No attempt was made to secure representative samples, since the purposes of the present article were merely to illustrate the applicability of the methods of analysis. Although some of the fruits were "out of season" at the time of the analysis (March), they were in every case firm and sound.

The analyses refer only to the press juice which was obtained by extraction of the comminuted sample in a small tincture press. Considerable difficulty was encountered in the filtration of the solution after defecation with normal lead acetate, but this was overcome in some cases by sedimentation in a centrifugal machine. The solutions from the deeply colored fruits in some cases remained discolored after defecation, but were completed bleached by one or two drops of a 0.5 N sulphur dioxide solution.

One hundred grams of the sample of press juice was transferred to a 200 ml flask, defecated with normal lead acetate, and filtered or sedimented. Lead was removed by adding a mixture of dry sodium phosphate and sodium oxalate. Sucrose was determined by the invertase method, and total sugars and levulose by the methods previously described. The ratio of levulose to total sugar as determined by the Mathews formula was derived from the titer and from the direct polarization corrected for the rotation of sucrose.[14]

The analytical data are shown in Table 6. It is noteworthy that all of the fruits examined, except the cranberries, showed an excess of levulose, slight in many cases, but rising to great predominance in apples and pears. The deviation of the Mathews ratio from the true ratio as shown in column 8 is small in all cases except the peach and the cranberry in which it assumes large negative values. This indicates the presence of a levorotatory substance.

---

[3] Browne's Handbook of Sugar Analysis, p. 523. John Wiley & Sons, New York, 1912.

TABLE 5.—*Analyses of honeys*

Principal floral source of honey	Original honey						
	Dry sub-stance	Sucrose	Reduc-ing sugars	Levu-lose	Ratio: Levulose / reducing sugar	Mathews ratio: Levulose / reducing sugar	Appar-ent dextrins
	Per cent	Per cent	Per cent	Per cent	Per cent	Per cent	Per cent
Sumac	82.54	3.26	72.13	37.53	51.9	43.2	5.81
White clover	82.42	1.69	76.09	39.26	51.6	46.9	3.32
Buckwheat	81.08	.44	75.53	38.97	51.6	46.3	3.71
Blend No. 1	82.42	1.61	78.69	42.02	53.4	50.3	2.26
Alfalfa	85.68	4.73	76.05	39.63	52.1	44.8	5.15
Sourwood	82.46	1.41	69.17	38.94	56.3	44.8	7.99
Fireweed	84.48	4.54	74.76	40.97	54.8	47.4	5.12
Gallberry	83.24	.83	73.20	41.14	56.2	48.5	5.21
Goldenrod	82.16	.68	77.08	39.93	51.8	47.0	3.43
Blend No. 2	84.84	1.65	77.81	39.92	51.3	45.8	3.97
Honey dew (incense cedar)	86.44	2.90	61.05	28.95	47.4	11.4	20.43
Melezitose honey	83.50	.30	57.61	32.57	56.5	27.0	15.70

Principal floral source of honey	Honey after invertase inversion				
	Reducing sugars	Levulose	Ratio: Levulose / reducing sugar	Mathews ratio: Levulose / reducing sugar	Apparent dextrins
	Per cent	Per cent	Per cent	Per cent	Per cent
Sumac	75.50	38.96	51.6	43.4	5.75
White clover	77.70	40.09	51.6	46.9	3.39
Buckwheat	75.67	39.12	51.7	46.3	3.80
Blend No. 1	79.11	41.37	52.3	50.4	1.40
Alfalfa	80.45	41.11	51.1	45.1	4.49
Sourwood	70.27	39.56	56.3	43.9	8.07
Fireweed	79.01	43.22	54.7	47.5	5.28
Gallberry	74.13	41.74	56.3	48.3	5.49
Goldenrod	77.32	40.52	52.4	47.1	3.80
Blend No. 2	79.23	40.80	51.5	45.8	4.20
Honey dew (incense cedar)	62.78	30.62	48.8	12.5	21.17
Melezitose honey	57.96	33.26	57.4	27.0	16.27

TABLE 6.—*Analyses of fruits*

Fruit	Dry sub-stance by re-fractom-eter	Sucrose	Reduc-ing sugars	Levu-lose	Ratio: Levulose / reducing sugar	Mathews ratio: Levulose / reducing sugar	ΔR	Total sugar purity
	Per cent	Per cent	Per cent	Per cent	Per cent	Per cent	Per cent	Per cent
Peach, "Old Gold"	16.50	8.02	3.17	1.82	57.4	63.8	−6.4	71.5
Apple, "Delicious"	13.14	1.80	9.63	6.58	68.3	69.8	−1.5	87.0
Pear, "D'Anjou"	17.06	1.24	9.99	8.88	88.9	86.7	2.2	85.8
Lemon	8.33	.40	2.03	1.07	52.9	51.8	1.1	29.2
Orange	12.07	4.50	5.07	2.65	52.2	51.7	.5	79.3
Grapefruit	13.84	4.03	6.05	3.17	52.5	51.3	1.2	72.8
Grapes, "Almeria"	18.05	.19	16.28	8.32	51.1	48.6	2.5	91.2
Strawberry	7.63	.82	4.09	2.20	53.8	51.2	2.6	66.4
Cranberry	9.93	.24	5.18	.82	15.8	30.5	−14.7	54.6
Tomato	4.67	0	2.79	1.58	57.7	56.3	1.4	59.7

WASHINGTON, August 13, 1932.

141809—32——3

RP496

ETERMINATION OF ALUMINA AND SILICA IN STEEL BY THE HYDROCHLORIC ACID RESIDUE METHOD

By J. G. Thompson and J. S. Acken

ABSTRACT

The hydrochloric acid residue method, as used at the Bureau of Standards for ıe determination of alumina and silica in steels, is described. Analyses of six .eels indicate that this method yields results for both alumina and silica which ›mpare favorably with those obtained by the bromine and nitric acid residue ıethods. Of the three methods, the hydrochloric acid method is preferred on ıe grounds of speed and simplicity of operation.

Data are presented to show that the presence of aluminum nitride in a steel oes not cause appreciable errors in the determination of alumina by the hydro-ıloric acid residue method.

The recovery of manganese oxide and silica from manganese silicates, by means f the hydrochloric-acid-residue method, is discussed. Data are presented to ιow that satisfactory recovery of silica is obtained only from silicates rich in ilica and that satisfactory recovery of manganese oxide is not obtained by this ıethod.

CONTENTS

I. INTRODUCTION

The use of dilute hydrochloric acid to effect the separation of metallic and nonmetallic constituents of steel was first applied to the determination of alumina in steel by Kichline [1] who reported that not more than 1 or 2 per cent of the alumina (Al_2O_3) present is soluble in dilute hydrochloric acid. He recommended decomposition of the sample in diluted acid (2+3) with subsequent determination of alumina in the residue by the phosphate method.

Oberhoffer and Ammann [2] confirmed Kichline's conclusions regarding the insolubility of alumina in diluted hydrochloric acid and presented comparative analyses made by the bromine method and by the hydrochloric-acid method. The results obtained by the two methods are in satisfactory agreement, although the values obtained by the bromine method usually are slightly higher than those obtained by the hydrochloric-acid method.

---

[1] F. O. Kichline, Note on the Determination of Aluminum Oxide and Total Aluminum in Steel, J. Ind. Eng. Chem., vol. 7, pp. 806–807, 1915.

[2] P. Oberhoffer and E. Ammann, Ein Beitrag zur Bestimmung Oxydischer Einschlüsse in Roheisen und Stahl, Stahl u. Eisen, vol. 47, pp. 1536–1540, 1927.

At the Bureau of Standards it was found that the hydrochloric-acid-residue method could be modified to permit the determination of silica (SiO_2) as well as alumina. The principal modification consists in the use of the 3 per cent sodium carbonate washing solution which Oberhoffer and Ammann employed to separate silica and hydrated silicic acid in the residues obtained in their modified bromine method. The silica in the residue originates in silica and silicate inclusions in the steel. The hydrated silicic acid originates in the decomposition of metallic silicides, and must be separated from the silica before an accurate determination of the latter can be made. The residue which remains after treatment with the acid and alkali solutions may be analyzed by several methods, but it is believed that a complete description of the procedure in use at the Bureau of Standards will be of interest. This procedure is described in the succeeding section.

II. METHODS OF ANALYSIS

1. DETERMINATION OF SILICA AND ALUMINA

Dissolve 50 g of steel chips in 500 ml of diluted hydrochloric acid (1 + 2) on the steam bath. After solution of the iron is complete, filter, with moderate suction, through a tight paper containing a little ashless paper pulp. Wash the residue[3] on the filter several times with cold water or with diluted hydrochloric acid (1 + 20) and then with hot water until the filter paper is free from iron salts. Wash the residue further with 500 ml of hot (80° to 90° C.) 3 per cent sodium carbonate solution, followed by successive washings with hot water, 50 ml of diluted hydrochloric acid (1 + 20) and finally with hot water. The diluted hydrochloric acid removes hydrated iron salts and any sodium silicate which may have formed in dissolving the hydrated silicic acid.

Transfer the paper and residue to a platinum crucible,[4] and heat gently so as to char the paper without allowing it to flame. Finally increase the temperature and ignite, under good oxidizing conditions, but do not let the temperature rise above very dull redness (about 600° C.). Fuse with a small amount of sodium carbonate (2 g will usually suffice). Cool, and dissolve in 50 ml of diluted hydrochloric acid (1 + 10). If a residue remains, filter and wash. Reserve the filtrate designating it (*A*). Ignite paper and residue in a platinum crucible, fuse with 3 g of potassium pyrosulphate, dissolve the cooled melt in a small volume of dilute hydrochloric acid and add to the reserved filtrate (*A*).

Silica.—Evaporate the solution (*A*) to dryness and bake for one hour at 105° C. Cool, drench with 10 ml of hydrochloric acid and digest a few minutes. Add 75 ml of warm water to dissolve salts, filter, and wash several times with cold diluted hydrochloric acid (1 + 10) and then with warm water. Reserve the filtrate, designating it (*B*). Ignite the paper and contents in a platinum crucible, slowly at first, and finally at a temperature of 1,200° C. for about 25 minutes. Cool, weigh, volatilize silica by the usual hydrofluoric-sulphuric acid treatment and reweigh. Any residue remaining in the crucible is dissolved in hydrochloric acid, or fused with bisulphate and added to the reserved filtrate (*B*).

---

[3] The residue may contain carbides of alloying elements, such as chromium and vanadium.
[4] With some steels, such as those which contain appreciable amounts of copper, it is advisable to ignite

Alumina.—Adjust the volume of the filtrate (*B*) to about 100 ml, add 2 ml of strong sulphurous acid,[5] and boil vigorously to expel the excess sulphur dioxide. Add 4 ml of concentrated nitric acid to the hot solution and boil to oxidize ferrous iron. Add sodium hydroxide solution until the acid solution is nearly neutral. Pour the nearly neutral solution slowly and with constant stirring into 80 ml of an 8 per cent solution of sodium hydroxide. Heat to boiling, remove from the source of heat, and allow to stand for at least two hours to permit the precipitate to settle. Filter through a double filter of a strong tight paper (No. 42 Whatman), and wash with a 1 per cent solution of sodium hydroxide.

Acidify the filtrate with hydrochloric acid and concentrate to a volume of about 250 ml. Add 5 ml of concentrated hydrochloric acid and 0.5 g of diammonium phosphate. The addition of a little macerated filter paper will prevent gelatinous aluminum phosphate, formed subsequently, from coagulating in lumps which are difficult to wash free from sodium and ammonium salts. Add two drops of methyl red indicator, make the solution just ammoniacal and restore the pink color with several drops of diluted hydrochloric acid (1+20). Heat the solution to boiling, add 20 ml of a 20 per cent solution of ammonium acetate and boil for five minutes. Let stand for 1 to 2 hours, filter through a tight 9 cm paper (No. 42 Whatman) and wash with hot 5 per cent solution of ammonium nitrate until 10 ml of the washings no longer yield a test for chlorides with acidified silver nitrate. Ignite the residue and paper in platinum under good oxidizing conditions until carbon is gone; cover, and heat at about 1,000° C, to constant weight. The observed weight of aluminum phosphate multiplied by the factor 0.418 is recorded as the weight of Al_2O_3 present in the original residue. Blank determinations are made on all reagents used, and the proper corrections are applied to the results of each analysis.

2. DETERMINATION OF MANGANOUS OXIDE

Any manganese present in the original residue will be found with the iron and chromium precipitated by the sodium hydroxide treatment. If manganese is to be determined, dissolve this precipitate of iron, chromium, and manganese with 50 ml of warm diluted nitric acid (1+3) containing a few milliliters of sulphurous acid. Boil the solution to expel the oxides of nitrogen, allow to cool, and determine manganese by the bismuthate method. If the amount of manganese is very small it is better to use the periodate colorimetric method.[6]

III. EXPERIMENTAL RESULTS

1. BEHAVIOR OF ALUMINUM NITRIDE IN THE DETERMINATION OF ALUMINA

The question arose whether aluminum nitride, if present in the steel, would be found in the acid-insoluble residue and carried through the determination and reported as alumina. To investigate this possi-

---

[5] The sulphurous acid, made by passing sulphur dioxide into distilled water to saturation, is added to reduce any oxidized chromium in order that all of the chromium present may be removed subsequently with the iron. The reduction of chromium by means of hydrogen peroxide was tried, but difficulty was encountered in the subsequent removal of iron from the solution, perhaps due to the formation of small amounts of soluble sodium ferrate.
[6] H. H. Willard and L. H. Greathouse, The Colorimetric Determination of Manganese by Oxidation with periodate. J

bility, a quantity of aluminum nitride, containing 10.1 per cent nitrogen, was obtained through the courtesy of the Fixed Nitrogen Research Laboratory, of the U. S. Department of Agriculture.

Two portions of approximately 50 mg each were treated in separate beakers with 50 ml of diluted hydrochloric acid (1+2). After standing on the steam bath overnight, the solutions were filtered and the residues on the filter papers were washed as in the regular procedure for the separation of acid-insoluble residues. The residues were ignited at a dull red heat to destroy carbonaceous matter, cooled, and transferred together with 50 ml of hydrochloric acid (1+1) to distilling flasks for the determination of nitrogen by the Allen method. Nitrogen is the only constituent of aluminum nitride which was determined, as it was assumed that any attack on the nitrides, by either acid or alkaline solutions, which resulted in solution of the nitrogen, would result in the simultaneous solution of an equivalent amount of aluminum. In other words, if nitrogen dissolves from aluminum nitride, in either acid or alkaline solutions, aluminum likewise will dissolve.

The data obtained from these two samples showed that the insoluble residues contained less than 4 per cent of the nitrogen present in the original samples. This indicates that at least 96 per cent of synthetic aluminum nitride is soluble in the hydrochloric acid (1+2) or in the wash solutions.

In addition to the above, tests were also made on nitrided samples of two commercial steels, one of the aluminum-molybdenum type and the other of the aluminum-molybdenum-chromium type, obtained through the courtesy of Dr. V. O. Homerberg. The composition of these steels prior to nitriding is recorded in Table 1. Specimens for residue analysis consisted of thin nitrided disks, approximately one-sixteenth inch thick and five-eighths inch in diameter, chosen on the assumption that with such thin sections practically all of a specimen would be affected by the nitriding treatment.

Samples of each nitrided steel were subjected to the different operations of a residue analysis, as follows: The samples designated *A* in Table 1 were decomposed in hydrochloric acid (1+2), filtered, and the acid-insoluble residues were washed with the customary solutions. Nitrogen was determined separately in the original filtrates, in the wash waters, and in the unignited residues. The samples designated *B* in Table 1 were decomposed in hydrochloric acid (1+2), filtered without washing, and nitrogen was determined in the filtrates and residues.

The results recorded in Table 1 show that the nitrides present in these two nitrided steels were almost completely decomposed and dissolved by the first solution treatment with hydrochloric acid (1+2) in the usual procedure of residue analysis. The amounts of nitrogen recovered either from the wash waters or from the insoluble residues were of approximately the order of magnitude of the blank correction. However, small but positive traces of nitrogen usually were indicated, as recorded in Table 1. These data show that the amounts of nitrides which survived the solution in acid but dissolved in the wash waters were negligible. The amounts of nitrides retained in the insoluble residues in all cases were an insignificant proportion of the nitrides present in the original sample.

TABLE 1.—*Behavior of nitrides in residue analysis of two nitrided steels*

Sample No.	Weight of sample	Nitrogen recovered		
		Filtrate	Wash waters	Residue
25A	*g* 5.858	*g* 0.1000	*g* 0.00008	*g* 0.0002
257A	4.613	.0720	Nil.	.00014
25B	6.174	.1115	....	.00007
257B	5.247	.0870	....	.0002

COMPOSITION OF STEELS BEFORE NITRIDING

Steel No.	C	Mn	P	S	Si	Al	Mo	Cr
	Per cent	*Per cent*	*Per cent*	*Per cent*	*Per cent*	*Per cent*	*Per cent*	*Per cent*
25	0.19	0.38	0.014	0.012	0.16	1.88	0.83	....
257	.35	.48	.018	.020	.15	.90	.20	0.99

It is concluded, therefore, that the small amounts of aluminum nitride ordinarily present in nonnitrided steels (or in the interior portions of nitrided ones) do not interfere with the determination of alumina by the hydrochloric acid residue procedure. However, in the case of nitrided steels if we assume that all residual nitrogen was present as aluminum nitride and calculate to the equivalent Al_2O_3 we find an appreciable error would occur in the determination of Al_2O_3, particularly if the latter were present in very small amounts.

2. COMPARISON OF RESULTS OBTAINED FROM DIFFERENT RESIDUE METHODS

The data in Table 2 are the results obtained for alumina and silica in comparative analyses of six steels by the nitric acid,[7] the bromine, and the hydrochloric acid residue methods.

The results obtained for alumina are in satisfactory agreement for all three methods. It is believed, however, that the hydrochloric acid residue method is preferable, as it is more rapid and more convenient to operate than either the nitric acid or the bromine methods.

In the determination of silica, the results from the modified hydrochloric acid method and the bromine method are in satisfactory agreement, but some of the results from the nitric acid method are lower than results from either of the other two. The lower results from the nitric acid method probably can be explained on the basis of Oberhoffer and Ammann's statement that the sodium hydroxide wash used in the nitric acid method not only dissolves hydrated silicic acid but also attacks dehydrated silica.

---

[7] J. H. S. Dickenson, A Note on the Distribution of Silicates in Steel Ingots, J. Iron & Steel Inst., vol. 113 (No. 1), pp. 177–196, 1926.

TABLE 2.—*Comparison of results obtained by different residue methods for alumina and silica*

Type of steel	Sample No.	Mold additions for 3 tons of steel	Al_2O_3			SiO_2		
			Hydrochloric acid method	Bromine method [1]	Nitric acid method [1]	Hydrochloric acid method (modified)	Bromine method [1]	Nitric acid method [1]
			Per cent	Per cent	Per cent	Per cent	Per cent	Per cent
Killed steel (0.14 C, 0.44 Mn, 0.15 Si)	113	4 pounds Al	0.026	0.023	0.022	0.001	0.002	<0.001
	119	25 pounds Fe-V	.003	.002	.003	.004	.006	.002
	122	None	.002	.002	.001	.008	.006	.002
Effervescing steel (0.14 C, 0.42 Mn, 0.003 Si)	125	4 pounds Al	.014	.016	.017	<.001	.001	.001
	131	25 pounds Fe-V	.002	.002	<.001	.008	.009	.01
	134	None	.001		<.001	<.001		<.001

[1] Analyses by the bromine method were made by R. J. Krausner, formerly of the Bureau of Standards. Analyses by the nitric-acid method were made by H. A. Bright.

3. RECOVERY OF SILICA AND MANGANOUS OXIDE FROM SYNTHETIC MANGANESE SILICATES

In analyses of acid-insoluble residues by the hydrochloric acid method, it was frequently observed that the sodium carbonate fusions were colored green, indicating the presence of manganese. It is probable that manganous oxide, if present in the steel, would be soluble in hydrochloric acid $(1+2)$, but it is possible that certain manganese silicates, if present, would be insoluble and, therefore, would appear in the residue. In order to investigate this possibility, synthetic manganese silicates containing different proportions of manganous oxide and silica were prepared, and the amounts of manganous oxide and silica recovered by the modified hydrochloric acid residue method were determined.

Manganous oxide was prepared in the laboratory from manganese carbonate by two methods. In one method the carbonate was oxidized to MnO_2 by treatment with nitric acid. The MnO_2 was reduced to MnO by heating in a nickel boat in a stream of hydrogen at 900° C. for about four hours. In the other method manganese carbonate was calcined in air at about 1,200° C. in a graphite crucible in the induction furnace, and was then further heated at 1,500° C. in graphite in a vacuum.

Three manganese silicates, intended to contain 25, 50, and 75 per cent MnO, respectively, were prepared by mixing precipitated manganese carbonate and pulverized quartz in the proper proportions and heating each mixture in a graphite crucible to about 1,500° C. After cooling in the furnace the melt was cleaned on an emery wheel and then crushed and ground. In the course of the investigation it appeared that manganese silicates of additional compositions were needed. Such silicates, containing approximately 30 and 60 per cent MnO, respectively, were supplied by Dr. C. H. Herty, jr., of the Bureau of Mines.

The manganese silicates, manganous oxide, and quartz used in the analyses were ground to pass through a No. 120 sieve. The two silicates received from Doctor Herty passed through a No. 200 sieve. Each sample for residue analysis consisted of about 10 mg of one of the finely ground materials mixed with 50 g of drillings of vacuum-

fused electrolytic iron. Fifty-gram portions of drillings, without
additions, frequently were run as blank determinations.

The results obtained for the recovery of manganous oxide and silica
from these synthetic manganese silicates, by the modified hydrochloric
acid residue method, are presented in Table 3 and are shown graph-

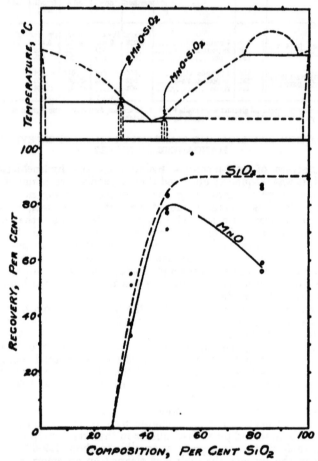

FIGURE 1.—*Recovery of MnO and SiO₂ from synthetic manganese silicates by the hydrochloric acid residue method*

ically in Figure 1. The equilibrium diagram for the system MnO-
SiO₂[8][9] is included in Figure 1 for convenience in interpreting the
results.

⁸ C. Benedicks and H. Löfquist, Nonmetallic Inclusions in Iron and Steel, Chapman & Hall, London,
p. 98, 1930.
⁹ C. H. Herty, jr., and G. R. Fitterer, New Manganese Silicon Alloys for the Deoxidation of Steel, U. S.
Bureau of Mines Report of Investigations, B. I. 3081, 1931.

Table 3.—*Recovery of MnO and SiO$_2$ from synthetic manganese silicates*

Composition of sample		Weight taken		Weight found		Recovery	
MnO	SiO$_2$	MnO	SiO$_2$	MnO	SiO$_2$	MnO	SiO$_2$
Per cent	*Per cent*	*g*	*g*	*g*	*g*	*Per cent*	*Per cent*
0.	100		0.0104		0.0093		89
			.0100		.0090		90
16.4	82.7	0.0016	.0063	0.0009	.0071	56	86
		.0017	.0064	.0010	.0073	59	87
¹ 30.2	¹ 56.3	.0035	.0065	.0027	.0064	77	98
		.0035	.0065	.0027	.0064	77	98
50.7	47.0	.0062	.0058	.0048	.0041	77	71
		.0053	.0049	.0044	.0038	83	78
¹ 58.2	¹ 33.7	.0062	.0036	.0023	.0020	37	55
		.0060	.0035	.0020	.0018	33	51
70.0	26.5	.0070	.0027	.0000	.0000	Nil.	Nil.
		.0073	.0028	.0000	.0000	Nil.	Nil.
100.0	0.	.0109		.0000		Nil.	
		.0101		.0000		Nil.	

¹ 200-mesh samples. Other samples were 120 mesh.

The silicates containing about 50 per cent SiO$_2$, approximately the composition of the compound MnO.SiO$_2$, were the least soluble in dilute hydrochloric acid of all the silicates in this system. As the SiO$_2$ content of the silicates increased above 50 per cent the recovery of SiO$_2$ remained at about 90 per cent, but the recovery of MnO decreased from the maximum. With decreasing SiO$_2$ content, below 50 er cent, the solubility of the silicates increased rapidly, resulting in decreased recovery of both MnO and SiO$_2$ in the insoluble residues. The sample which approximated the compound 2MnO.SiO$_2$ in composition, was completely soluble in dilute hydrochloric acid, as was pure MnO. In view of the limited number of determinations and the difficulties encountered in these analyses, too much emphasis should not be placed on the results of any one determination. However, the general relation between solubility and composition is indicated and it is interesting to note that these curves for the solubility in dilute hydrochloric acid are quite similar to the curves which Herty [10] obtained for the solubility of manganese silicates in dilute nitric acid.

From these data it is concluded that satisfactory recoveries of silica, that is, about 90 per cent, can be obtained by means of the hydrochloric acid residue method, from manganese silicates which contain 50 per cent or more SiO$_2$. Recovery of silica from silicates which contain less than 50 per cent SiO$_2$, is not satisfactory. The data indicate that from 70 to 80 per cent of MnO is recovered from manganese silicates of over 50 per cent SiO$_2$ content. It will be noted that the revovery of MnO in silicates of less than 50 per cent SiO$_2$ is quite low. Obviously, the HCl method is not applicable to the determination of MnO occurring as such (MnO inclusions) and yields only fair recoveries of MnO in manganese silicates of high silica content.

[10] C. H. Herty, Jr., G. R. Fitterer, and J. F. Eckel, The Physical Chemistry of Steel Making: A Study of the Dickenson Method for the Determination of Nonmetallic Inclusions in Steel, U. S. Bureau of Mines, Mining, and Metallurgical Investigations Bull. No. 37, 1928.

IV. SUMMARY

1. The conclusions of previous investigators that the hydrochloric acid residue method yields results for alumina that are comparable with those obtained by the nitric acid and bromine methods, and that the hydrochloric acid method is preferable to the other two on the grounds of speed and simplicity in operation, have been confirmed.

2. The presence of aluminum nitride in a sample ordinarily does not cause appreciable errors in the determination of alumina by the hydrochloric acid residue method. If relatively large amounts of nitrides are present; for example, in a nitrided steel, accompanied by only a small amount of aluminum oxide, a small error due to incomplete decomposition of aluminum nitride might result in a serious error in the determination of a small amount of aluminum oxide.

3. The modified hydrochloric acid residue method yields as satisfactory results for silica as do the nitric acid and bromine methods.

4. If manganese silicates are present in the sample, recoveries of about 90 per cent of the silica may be obtained from silicates which contain more than 50 per cent SiO_2. Recovery of silica from silicates which contain less than 50 per cent SiO_2 is not satisfactory. The hydrochloric acid method recovers about 70 to 80 per cent of MnO from manganese silicates having 50 per cent or more SiO_2 content. The method is not suitable for MnO in silicates of less than 50 per cent SiO_2.

V. ACKNOWLEDGMENT

Acknowledgment is made to H. A. Bright, who is responsible for the section on methods of analysis, and who has contributed helpful advice on other phases of the work reported herein. .

WASHINGTON, August 17, 1932.

RP497

AN ANALYSIS OF LANTHANUM SPECTRA
(LA I, LA II, LA III)

By Henry Norris Russell [1] and William F. Meggers

ABSTRACT

All the available data (wave-length measurements and intensity estimates, temperature classes, Zeeman effects) on lanthanum lines have been correlated and interpreted in an analysis of the successive optical spectra. The total number of lines classified is 540 in the La I spectrum, 728 in the La II spectrum, and 10 in the La III spectrum.

Series-forming terms have been identified in each spectrum and from these the ionization potentials of 5.59 volts for neutral La atoms, 11.38 volts for La+ atoms and 19.1 volts for La++ atoms have been deduced.

Lanthanum is a chemical analogue of scandium and yttrium, but, although the corresponding spectra are strikingly similar, some interesting differences are noted. A doublet-D term (from a d electron) represents the lowest energy (normal state) in the third spectrum of each element and another ^{3}D (from the s^2d configuration) describes the normal state of the neutral atoms in each case. The homologous atoms Sc+, Y+, La+ choose different normal states; (sd) ^{3}D, (s^2) ^{1}S, (d^2) ^{3}F, respectively. In addition, the first two spectra of La exhibit a large number of (odd) middle-set terms ascribed to the binding of an f electron. The La II spectrum is the most completely developed example of a two-electron spectrum which has yet been investigated. All the configuration types, s^2, sp, sd, sf, p^2, pd, pf, d^2, df, f^2, have been identified and almost all of the terms arising from each.

The analyses of all three spectra are supported by measurements of Zeeman effects, which are interpreted with the aid of Landé's theory. The splitting factors (g values) for many levels show marked departure from the theoretical values, but the "g-sum rule" is valid wherever it is tested.

CONTENTS

I. INTRODUCTION

After the analyses of the arc and spark spectra of scandium [2] (Sc, Z=21) and yttrium [3] (Y, Z=39) had been published the authors decided to continue cooperation on the remaining spectra of this type. Lanthanum (La, Z=57) is a chemical analogue of scandium and yttrium; it occupies the same position in the third long period of elements that the former do in the first and second long periods, respectively. The fact that lanthanum occupies a position in the periodic system just preceding the group of 14 elements commonly called "rare earths" makes a complete analysis of its spectra of exceptional interest. Indeed, this analysis shows (vide infra) that the

---

[1] Professor of astronomy, Princeton University.
[2] H. N. Russell and W. F. Meggers, B. S. Sci. Paper No. 558, vol. 22, p. 329, 1927.
[3] W. F. Meggers and H. N. Russell, B. S. Jour. Research, vol. 2 (RP55), p. 733, 1929.

atom building process which accounts for the rare earth elements is actually anticipated in the electron configurations of the lanthanum atom. Furthermore, the exact nature of electron coupling in complex atoms like La, as disclosed by the analysis of spectral structure, is certain to be important in the further development of theory. In the present paper the authors give the results of an extensive analysis of lanthanum spectra in practically the same form as their results for scandium and yttrium; the summary of spectral theory given in the earlier publications applies also to the present case, and the notation is the same except for slight changes which bring it into conformity with the standardized nomenclature [4] and practice.

In 1914, Popow [5] published the first suggestion of regularities in lanthanum spectra. From Zeeman effect observations, he recognized six La lines as comprising a combination of triplet P and triplet D terms. In the same year Paulson [6] published a list of constant differences occurring between wave numbers corresponding to La spark lines, but no attempt was made to interpret the regularities. Even earlier, Rybar [7] had made extensive measurements of the Zeeman effects in La spectra and many complex patterns were published, but there was at that time no satisfactory explanation for most of these. Another decade passed before the theory of complex Zeeman effects began to unfold itself in the work of Landé,[8] thus paving the way for an explanation of the La observations. An attempt was made in 1924 by Goudsmit [9] who identified additional Paulson terms in the La II spectrum with the aid of Rybar's Zeeman effects. From the latter observations the Landé g values and quantum numbers j and l were derived for 20 energy levels accounting for approximately 70 lines.

In the following year the theory of spectral terms as developed by Heisenberg [10] and by Hund [11] gave some important suggestions as to the structures of La spectra and attempts were made by one of us to extend the analysis of the La II spectrum and also to find regularities in the La I spectrum. These efforts succeeded with the availability of new empirical data consisting of a description of La lines with respect to their behavior with temperature in the electric furnace by King and Carter [12] and of unpublished Zeeman effects kindly advanced by Prof. B. E. Moore. The temperature classification gave a reliable separation of La I and La II lines, and the new observations of Zeeman effects, especially in the red portion of the spectrum, gave the first clue to regularities in the La I spectrum. In addition to the 20 levels identified by Goudsmit, 22 more were found for La⁺ atoms and combinations of these 42 levels accounted for about 180 La II lines.[13] For neutral La atoms 48 energy levels were found and their combinations accounted for about 130 lines.[14] In both cases the normal states or lowest energy levels were identified without ambiguity, but in each spectrum many lines remained unclassified, and many theoretical terms were still undiscovered. These preliminary analyses indicated

[4] H. N. Russell, A. G. Shenstone, and L. A. Turner, Phys. Rev., vol. 33, p. 900, 1929.
[5] S. Popow, Ann. d. Physik., vol. 45, p. 147, 1914.
[6] E. Paulson, Ann. d. Physik., vol. 45, p. 1203, 1914.
[7] S. Rybar, Phys. Zeit. vol. 12, p. 889, 1911.
[8] A. Landé, Zeit. f. Physik., vol. 15, p. 189, 1923; vol. 16, p. 391, 1923; vol. 19, p. 112, 1923.
[9] S. Goudsmit, Kon. Akad. Wet. Amsterdam, vol. 33, No. 8, p. 774, 1924.
[10] W. Heisenberg, Zeit. f. Physik., vol. 32, p. 841, 1925.
[11] F. Hund, Zeit. f. Physik., vol. 33, p. 345, 1925. Linienspektren und periodisches system der Elemente, Julius, Springer, Berlin, 1927.
[12] A. S. King and E. Carter, Astrophys. J., vol. 65, p. 86, 1927.
[13] W. F. Meggers, J. Opt. Soc. Am. vol. 14, p. 191, 1927.

that it would be impossible to extend them without still further experimental data,[15] so it was decided to make an entirely new description of La spectra. Such a description was recently completed by one of us,[16] and it serves as a basis for the analyses of La spectra to be detailed in the present paper. In addition to new wave-length determinations for more than 1,500 La lines in the interval 2,100 to 11,000 A, the new data include intensity estimates of arc and spark lines, separating them into three classes, La I, La II, and La III spectra, and improved observations of Zeeman effects for 460 lines ranging from 2,700 to 7,500 A. These data have permitted us to classify almost all of the lines ascribed to lanthanum atoms, to identify a large majority of the spectral terms, and correlate them with electron configurations. In each case it has been possible to recognize series-forming terms, the extrapolation of which lead to calculated ionization potentials of 5.59 volts for neutral La atoms, 11.38 volts for La⁺ atoms, and 19.1 volts for La⁺⁺ atoms. The total numbers of classified lines in the successive spectra are as follows: 540 for La I, 728 for La II, and 10 for La III.

On account of the greater complexity of La spectra, as compared with Sc and Y, and pronounced departures from theoretical interval ratios, line intensities and Zeeman effects, their analysis has been attended by greater difficulties and uncertainties, but patience and perseverance have been rewarded by the final classification of practically all lines without ambiguity. The detailed results will be presented for La III, then for La II, and finally for La I, thus proceeding from the relatively simple (alkali) case of 1-valence electron to the 2-electron spectrum with greatly increased transition possibilities and lastly to the spectrum characteristics of atoms with a full complement of 3-valence electrons.

II. THE SPECTRUM OF DOUBLY IONIZED LANTHANUM
(La III)

Lanthanum belongs to the third long period in which electron orbits of the types $6s$, $6p$, and $5d$ are successively added to the completed xenon shell. Only one valence electron remains in doubly ionized lanthanum and in the normal state this electron is in a $5d$ orbit which produces a ²D term. The next lowest state occurs with the $6s$ orbit, and higher states arise from $6p$, $6d$, $7s$ orbits. Some of these terms were already identified by Gibbs and White [17] and recently Badami [18] classified eight lines of the La III spectrum.

The new description of La spectra yielded 10 lines which are characterized by an enormous intensity difference between arc and spark and are, therefore, ascribed to doubly ionized atoms. Analysis of these data resulted in the identification of spectral terms listed in Table 1; the observed lines and estimated relative intensities appear in Table 2.

The observed and theoretical (Landé) splitting factors (g) are compared in the last column of the term table. Since La spectra possess g values which depart more or less from Landé, the observed Zeeman effects in the table of classified lines are compared with those computed from observed rather than theoretical g's. This procedure shows in

15 W. F. Meggers, J. Wash. Acad. Sci., vol. 17, p. 35, 1927.
16 W. F. Meggers, B. S. Jour. Research, vol. 9 (RP468), p. 239, 1932.
17 R. C. Gibbs and H. E. White, Proc. Nat. Acad. Sci., vol. 12, p. 557, 1926; Phys. Rev., vol. 33, p. 157 1929.
18 J. S. Badami, Proc. Roy. Soc. London, vol. 43, p. 53, 1931.

the first table, the deviations of observed from theoretical g values, and in the second table how closely the observed data are represented by the empirical g values. The observed or computed Zeeman patterns may be compared with the theoretical by reference to "Tables of Theoretical Zeeman Effects" published by Kiess and Meggers.[19] These remarks apply to the La II and La I spectra (vide infra) as well as to La III.

TABLE 1.—*Terms in the La III spectrum*

Electron configuration	Terms	Levels	Separations	g	
				Observed	Landé
5d	5²D1½	0.00	1,603.23		
	5²D1½	1,603.23			
6s	6²S½	13,590.76		2.10	2.000
6p	6²P½	42,014.92	3,095.72	.63	.667
	6²P1½	45,110.64		1.37	1.333
6d	6²D1½	82,378.75	433.76		
	6²D1½	82,812.51			
7s	7²S½	82,345.0			

It will be shown later that the terms arising from the $4f$ electron are higher than those from $6p$ in La I, but lower in La II. This suggested that the La III combinations $5d$ (^{2}D)—$4f$ (^{2}F) might lie in the infra-red. The La spark spectrum in the interval 8,000 to 10,500 A was recently explored with xenocyanine plates, but no La III lines were found although La II lines were recorded all the way to 9,893.8 A.

TABLE 2.—*Classified lines in the La III spectrum*

λ(air) I. A.	Intensity	ν(vac) cm¹	Term combinations	Zeeman effect	
				Observed	Computed
3,517.14	200	28,424.09	6²S½—6²P½	(0.73)1.36	(0.74)1.36
3,171.68	300	31,519.94	6²S½—6²P1½	(0.37)0.99, 1.76	(0.36)1.00, 1.74
2,684.90	50	37,234.29	6²P½—7²S½		
2,682.46	30	37,268.16	6²P½—6²D1½		
2,651.60	300	37,701.87	6²P1½—6²D1½		
2,478.8 ¹	20	40,329.9	6²P1½—7²S½		
2,476.72	100	40,363.79	6²P1½—6²D1½		
2,379.38	200	42,014.92	5²D1½—6²P½		
2,297.75	200	43,507.40	5²D1½—6²P1½		
2,216.08	50	45,110.63	5²D1½—6²P1½		

¹ Near carbon line, 2,478.6 A.

III. THE SPECTRUM OF SINGLY IONIZED LANTHANUM (La II)

The spark spectrum of lanthanum is exceptionally complex; more than 800 lines appearing in spark spectrograms are associated with La⁺ atoms. Superficially, the lanthanum spark spectrum resembles the yttrium spark spectrum; in each case the lines are divided roughly into two classes by comparison of arc and spark spectrograms. One

[19] C. C. Kiess and W. F. Meggers, B. S. Jour. Research, vol. 1 (RP23), p. 64, 1928.

group, including most of the stronger lines, has nearly the same appearance in a 6-ampere, 220-volt arc and in a high voltage condensed discharge, while the second, lying mainly in the ultra-violet, is greatly enhanced upon passing from the arc to the spark and consists largely of hazy and unsymmetrical lines. The former arise from combinations of low energy states with the next higher or middle set, while the latter are practically all identified as combinations of middle terms with a still higher third set.

The lowest energy states which can arise from s and d type electrons are identified with (even) spectral terms as follows:

Electron configuration	Spectral terms		
s^2	^1S		
sd	^1D	^3D	
d^2	$^1S,^1D,^1G$	$^3P\ ^3F$	

All of these have been identified except $(d^2)^1S$. Which particular term will be the lowest energy and represent the normal state of the atom depends on the relative strength of binding of the individual electrons. It is very remarkable that the homologous atoms, Sc^+, Y^+, and La^+, each make a different choice; the normal state of Sc^+ is $(sd)^3D$, of Y^+ $(s^2)^1S$, and of La^+ $(d^2)^3F$.

Substitution of a p electron for an s or a d electron produces the following set of (odd) middle terms:

Electron configuration	Spectral terms	
sp	^1P	^3P
dp	$^1P,^1D,^1F$	$^3P,^3D,^3F$

All of the easily excited lines of Sc^+ and Y^+ are accounted for by the above-mentioned low and middle spectral terms, but in the case of La^+ a large number of otherwise superfluous lines indicate additional middle-set terms to account for which it is necessary to conclude that f-type electrons are present. Thus, the substitution of f- for p-type electrons would yield the following additional (odd) middle-set terms:

Electron configuration	Spectral terms	
sf	^1F	^3F
df	$^1P,^1D,^1F,^1G,^1H$	$^3P,^3D,^3F,^3G,^3H$

All of these terms have been found in the La II spectrum, they increase the number of middle-set levels from 16 to 40 and thus account for the greater complexity of the spectrum.

The terms produced by the $4f$ electrons lie lower than those arising from the $6p$. This is obviously related to the fact that $4f$ electrons are bound into the normal state of the directly following elements, Ce to Lu, while the $6p$ electrons begin to be similarly bound only in Tl. In Ba I the f electron is much more loosely bound than the p. There are numerous high even terms in La II. Those arising from the configurations $5d\ 7s$, $5d\ 6d$, $6s\ 6d$, and $6p^2$ are homologous with similar terms in Sc II and Y II. Two important additional groups evidently arise from $6p\ 4f$ and $4f^2$. The former are the lowest of all the high even terms. Some hazy lines confined to the spark appear to be combinations between these and still higher odd levels which have been denoted by numbers 1° to 8°. There are several configurations (for example, $5d\ 7p$, $5d\ 5f$, $6p\ 6d$, $4f\ 6d$) which may give rise to levels of this sort, and they must be very numerous. They should combine with the ground terms to give lines in the

Schumann region. Observations in this region may detect many more such levels and thus lead to the interpretation of the remaining unclassified lines in the visible and near ultra-violet.

This is the most completely developed example of a 2-electron spectrum which has yet been investigated. All the configuration types, s^2, sp, sd, sf, p^2, pd, pf, d^2, df, f^2, have been identified and almost all of the terms arising from each. The theoretical relations of the terms arising from each configuration have been discussed by Condon and Shortley.[20] The agreement of the observed and computed levels is, in general, good, and their theoretical predictions led to the correct identification of the difficult terms $(df)^1H$ and $(f^2)^1I$.

The terms which have been identified in the La II spectrum are listed in Table 3, in which term symbols, relative values of the levels, level separations, adopted and Landé g values, and combining terms are given in successive columns. The adopted g values are derived from the observed Zeeman effects for La II lines starting with the completely resolved patterns and then applying the formulas of Shenstone and Blair [21] to the unresolved blends. For patterns which though unresolved had perpendicular (n) components distinctly shaded outwards (A^1) or inwards (A^2) the attempt was made to measure the points of maximum intensity corresponding to the strongest components. If x is the observed separation we should then have

$$x = J_1 g_1 - J_2 g_2 \quad (J_1 = J_2 + 1) \tag{1}$$

When no such asymmetry was noticed it was assumed that the settings were on the centroid of the whole pattern and the formula then used was

$$2x = (J_1 + 1)g_1 - J_2 g_2 \tag{2}$$

When $J_1 = J_2$ the equations are

$$2\,x = g_1 + g_2 \quad (n \text{ components}) \tag{3}$$
$$4/3y = (J + 1/2)\,(1 - X)\,(g_1 - g_2) \quad (p \text{ components}) \tag{4}$$

where $X = \dfrac{1}{(2J+1)^2}$ or $\dfrac{1}{4J\,(J+1)}$ according as J is integral or half integral.[22]

The weights assigned to the adopted g's in Table 3 depend on the number and consistency of the derived values. The probable error corresponding to unit weight is ± 0.020.

Most of the g's have nearly the theoretical (Landé) values but marked discrepancies frequently appear which may be attributed to deviations of the actual coupling of the vectors from the ideal SL coupling for narrow multiplets. Some of the largest discordances are clearly due to "g sharing" among neighboring levels with the same J; for example, e^3G_4, e^3F_4; y^3D_1, y^3P_1, z^1P_1. In these cases the intensities of many combinations are also abnormal.

[20] E. U. Condon and C. H. Shortley, Phys. Rev., vol. 37, p. 1025, 1931.
[21] A. G. Shenstone and H. A. Blair, Phil. Mag., vol. 8, p. 765, 1929.
[22] H. N. Russell, Phys. Rev., vol. 36, p. 1590, 1930.

TABLE 3.—*Relative terms in the La II spectrum*

Electron configuration	Term	Level	Level separations	g Adopted weight		Landé	Combinations
5d²	a³F₂	0.00		0.730	7½	0.667	z³F°, z¹F°, z¹G°, y³F°, z³H°, z¹D°, z³G°, z³D°,
	a³F₃	1,016.10	1,016.10	1.092	9	1.083	z³P°, y¹D°, y¹F°, z¹F°, z¹F°, z¹P°, z³P°,
	a³F₄	1,970.70	954.60	1.258	8	1.250	z³F°, z¹F°, z¹P°.
6s5d	a¹D₂	1,394.46		.987	10½	1.000	y¹F°, y¹D°, z¹F°, z¹D°, y³D°, z¹D°, y³P°, y¹P°, z¹P°, z¹F°, z¹P°,
6s5d	a³D₁	1,895.15	696.45	.520	13	.500	z³F°, z¹G°, y³F°, z³H°, z¹D°, z³G°, z³D°, z³P°,
	a³D₂	2,591.60	656.75	1.140	8	1.167	y¹D°, y¹D°, y¹F°, z¹F°, z¹P°, y³P°, y¹P°, y¹P°,
	a³D₃	3,250.35		1.337	11	1.333	z³P°, z¹F°, z¹P°.
	a³P₀	5,249.70		0/0	6	0/0	y³P°, z¹D°, z³D°, z³P°, y¹D°, y¹F°, y³D°, z³F°,
5d²	a³P₁	5,718.12	468.42	1.506	9	1.500	
	a³P₂	6,227.42	509.30	1.488	11½	1.500	
6s²	a¹S₀	7,394.57		0/0	4	0/0	z³D°, y³P°, y¹D°, z¹P°, y¹P°, z¹P°, z¹P°.
5d²	a¹G₄	7,473.32		1.003	2	1.000	z¹G°, y³F°, z³H°, z³G°, z³D°, y³F°, y¹D°, z³F°, z¹H°, z¹F°.
5d²	b¹D₂	10,094.86		1.015	70	1.000	z³D°, z³P°, y¹D°, y³D°, z¹F°, z¹P°, y³P°, y¹P°, z¹P°, y¹P°, z¹P°, z¹P°.
6s4f	z³F₂	14,147.98	227.19	.676	7	.667	a³F, a³D, a³D, e³G, e³F, e¹F, e³D, e³G, f³D,
	z³F₃	14,375.17	323.57	1.087	10	1.083	f¹F, f³G, f³F, g³D, e³H, g³F, h¹D.
	z³F₄	15,608.74		1.258	9	1.250	
6s4f	z¹F₃	15,773.77		1.047	10	1.000	a³F, e¹D, e³G, e³F, e³F, e³D, e³G, e³G, f³D, f³G, f³F, g³D, f³G, e³H, f¹F, h¹D, e³D, f¹D.
5d4f	z¹G₄	16,599.17		.996	4	1.000	a³F, a³D, a³G, e³G, e³F, e³F, e³G, f³F, f³G, f³G, e³H, g³F, g¹G.
5d4f	y³F₂	17,211.93	1,023.63	.757	8	.667	a³F, a³D, a³D, a³P, a³G, e³G, e³F, e³D, e³G, f³P, f³D.
	y³F₃	18,235.56	978.98	1.090	8	1.083	
	y³F₄	19,214.54		1.243	6	1.250	
5d 4f	z³H₄	17,825.62	754.79	.865	2	.800	a³F, a³D, a³G, e³G, e³F, e¹F, e³G, f³F, f³G, e³H, g³F, g¹G, e³I.
	z³H₅	18,580.41	1,169.21	1.042	4	1.033	
	z³H₆	19,749.62		1.17	1	1.167	
5d 4f	z¹D₂	18,895.41		.931	3½	1.000	a³F, a¹D, a³D, a³P, e³G, e³D, e³F, e³F, e³D, e³D, f³D, f³F, f³G, f³F, g³D, e³P, e³S, g³F, h¹D, g³P, h³D, f³P, f³D.
5d 4f	z³G₃	20,402.82	928.78	.775	4	.750	a³F, a³D, a³D, a³G, e³G, e³F, e¹F, e³D, e³D, f³G, f³F, f³G, f³G, e³H, g³F, g³G.
	z³G₄	21,331.60	951.30	1.060	6	1.050	
	z³G₅	22,282.90		1.203	5	1.200	
5d 4f	z³D₁	21,441.73	664.29	.547	8	.500	a³F, a³D, a³P, a³P, a³S, a³G, b¹D, e³G, e³F, e³F, e³D, e³G, e¹D, f³D, f³D, f³G, f³F, g³D, e³P, e³S, e³P, g³D, f³G, e³H, g³P, f³P, f³D.
	z³D₂	22,106.02	431.28	1.186	12	1.167	
	z³D₃	22,537.30		1.317	10	1.333	
5d 4f	z³P₀	22,663.70		0/0	0	0/0	a³F, a³D, a³D, a³P, a³S, b¹D, e³D, e³D, f³D, f³D, f³F, g³D, e³P, e³S, e³P, g³D, g³F, h¹D, g³P, h³D, f³P, f³D.
	z³P₁	22,705.15	21.45	1.458	7	1.500	
	z³P₂	23,246.93	541.78	1.462	7	1.500	
6p 5d	y¹D₂	24,462.66		.897	7½	1.000	a³F, a³D, a³D, a³P, a³P, b¹D, e³G, e³D, f³D, f³D, f³F, f³G, f³F, y¹D, e³P, e³P, h¹D, f³P, f³D.
5d 4f	y¹F₃	24,522.70		1.032	8	1.000	a³F, a¹D, a³D, a³P, a³G, e³G, e³F, e³D, e³G, e³D, f³F, g³D, g¹G, h¹D, g³P, f³P, f³D.
6p 5d	y³D₁	25,973.37	1,414.74	.799	8	.500	a³F, a³D, a³D, a³P, a³S, a³G, b¹D, e³F, e³D, e³D, f³D, f³G, f³P, g³D, e³S, e³S, g³F, e³H, h¹D, g³P, h¹D, f³P, f³D.
	y³D₂	27,388.11	927.14	1.191	10	1.167	
	y³D₃	28,315.25		1.312	7	1.333	
6p 5d	x³F₂	26,414.01	423.65	.836	7½	.667	a³F, a³D, a³D, a³P, a³G, e³G, e³G, e³F, e³F, e¹D, f³D, f³D, f³F, f³G, f³F, g³D, e³P, e³S, g³D, f³G, e³H, h¹D, g³P, f³P, f³D.
	x³F₃	26,837.66	1,727.74	1.081	6	1.083	
	x³F₄	28,565.40		1.246	4	1.250	
5d 4f	z¹P₁	27,423.91		.882	9	1.000	a³F, a³D, a³D, a³P, a³S, b¹D, e³D, f³D, f³D, e³P, e³P, e³S, g³P, g³S, h¹D, f³P, h¹D, f³S.
5d 4f	x¹H₅	28,525.71		1.00	1	1.000	a³G, e³G, e³H, g³G, e³I.
6p 5d	y³P₀	27,545.85		0/0		0/0	
	y³P₁	28,154.55	608.70	1.262	5	1.500	a³F, a³D, a³P, a³P, b¹D, e³D, f³D, f³D, g³D, e³S, e³P, g³S, g³S, h¹D, g³P, h¹D, f³S.
	y³P₂	29,498.05	1,343.50	1.495	5	1.500	
6p 5d	y¹P₁	30,353.33		1.066	6	1.000	e¹D, a³D, a³P, b¹D, e³D, f³D, f³F, e¹P, e³P, g³S, h¹D, f³D, f³S, ¹D?
6s 6p	z³P₀	31,785.82		0/0		0/0	a³F, a³D, a³D, a³P, a³S, b¹D, f³G, g³D, e³P, e³S, e³S, g³P, h¹D, f³P.
	z³P₁	32,160.99	375.17	1.515	6	1.500	
	z³P₂	33,204.41	1,043.42	1.478	4	1.500	
6p 5d	x¹F₃	32,201.06		1.006	7	1.000	a³F, a³D, a³D, a³P, a³G, b¹D, f³D, f³D, f³F, f³G, f³G, e³H, g³F, g¹G, h¹D, h¹D, ¹D?
6p 4f	e³G₃	35,452.66	1,720.13	.873	5	.750	z³F°, z¹F°, z¹G°, y³F°, z³H°, z¹D°, z³G°, z³D°, y¹D°, y¹F°, z¹F°.
	e³G₄	37,172.79	1,845.95	1.146	7	1.050	
	e³G₅	39,018.74		1.208	3	1.200	
6p 4f	e³F₂	35,787.53	1,167.12	.732	5	.667	z³F°, z¹F°, z¹G°, y³F°, z³H°, z¹D°, z³G°, z³D°, y¹D°, z¹F°.
	e³F₃	36,954.65	835.92	1.061	7½	1.083	
	e³F₄	37,790.57		1.141	7	1.250	
6p 4f	e¹F₃	37,209.71		.953	6	1.000	z³F°, z¹F°, z¹G°, z³H°, z¹D°, z³G°, z³D°, y¹F°, z¹F°.
6p 4f	e¹G₄	39,221.65		1.060	6	1.000	z³F°, z¹F°, z¹G°, y³F°, z³H°, z³G°, z³D°, y¹F°, z¹H°.

TABLE 3.—*Relative terms in the La ii spectrum*—Continued

Electron config- uration	Term	Level	Level sep- arations	g Adopted weight	Landé	Combinations
6p 4f	e^3D_1	38,534.11	−312.62	0.495 6	0.500	$z^3F°$, $y^3F°$, $x^3F°$, $z^1D°$, $z^3G°$, $z^3D°$, $z^3P°$, $y^3D°$.
	e^3D_2	38,221.49		1.100 7	1.167	$y^1F°$, $y^3D°$, $x^3F°$, $y^3P°$.
	e^3D_3	39,402.55	1,181.06	1.306 9	1.333	
6p 4f	e^1D_2	40,457.71		1.046 5	1.000	$z^1F°$, $y^1F°$, $z^1D°$, $z^3G°$, $z^3D°$, $z^3P°$, $y^1F°$, $y^3D°$, $z^3F°$, $y^1P°$.
6s 6p	$z^1P_1°$	45,692.17		.97 1	1.000	a^3F, a^1D, a^3D, a^1P, a^1S, b^3D, b^3D.
5d 7s	f^3D_1	49,733.13	151.22	.520 3	.500	$z^3F°$, $y^3D°$, $z^3D°$, $z^3P°$, $y^3D°$, $y^3D°$, $x^3F°$, $z^3P°$,
	f^3D_2	49,884.35	1,344.22	1.141 3½	1.167	$^1yP°$, $z^1F°$.
	f^3D_3	51,228.57		1.331 3	1.333	
5d 7s	f^1D_2	51,523.86		1.02 1½	1.000	$z^1F°$, $y^1F°$, $z^3D°$, $y^1D°$, $x^3F°$, $z^1P°$, $y^1P°$, $y^1P°$, $z^1F°$.
5d 6d	f^3F_2	52,137.67		1.00 1	1.000	$z^3F°$, $z^3G°$, $y^1F°$, $z^1D°$, $z^3P°$, $y^3D°$, $y^1F°$, $x^3F°$, $z^1F°$.
5d 6d	f^3G_3	52,857.88	475.49	.89 1	.750	
	f^3G_4	53,333.37	1,101.28	1.05 1	1.050	$z^3F°$, $z^1F°$, $z^1G°$, $y^3F°$, $z^3G°$, $x^3F°$, $z^1F°$.
	f^3G_5	54,434.65		1.20 1	1.200	
5d 6d	f^3F_2	53,885.24	954.80	.77 2	.667	$z^3F°$, $z^1F°$, $z^3H°$, $z^1D°$, $z^3G°$, $z^3D°$, $y^1D°$, $y^3D°$,
	f^3F_3	54,840.04	481.31	1.14 2	1.083	$z^1F°$.
	f^3F_4	55,321.35		1.16 2	1.250	
5d 6d	g^3D_1	52,169.66	565.15	.67 3	.500	$z^1F°$, $y^3F°$, $z^1D°$, $z^3D°$, $z^3P°$, $y^1D°$, $y^1F°$, $y^3D°$,
	g^3D_2	52,734.81	954.75	1.17 1	1.167	$x^3F°$, $y^1P°$, $z^1F°$.
	g^3D_3	53,689.56		1.21 3	1.333	
5d 6d	e^3S_1	54,365.80		1.42 1	2.000	$z^1D°$, $z^3D°$, $z^3P°$, $y^3D°$, $x^3F°$, $z^1P°$, $y^1P°$, $y^1P°$, $z^1F°$.
5d 6d	e^1P_1	53,302.56				$z^3D°$, $z^3D°$, $z^3P°$, $y^1D°$, $y^3D°$, $z^3F°$, $z^1P°$, $y^1P°$, $z^1F°$.
5d 6d	e^3P_0	54,964.19?	266.14	1.57 1	1.500	$z^1D°$, $z^3P°$, $y^1D°$, $y^3D°$, $z^1P°$, $y^1P°$, $y^1P°$.
	e^3P_1	55,230.33	806.27	1.22 1	1.500	
	e^3P_2	56,036.60		1.03	1.000	
5d 6d	f^1G_4	56,035.70				$z^3F°$, $z^1G°$, $y^3F°$, $z^3H°$, $z^3G°$, $z^1D°$, $y^3D°$, $z^3F°$, $z^1F°$.
5d 6d	$g^1_2)_2$	55,184.05		1.06 1	1.000	$z^3(?°$, $z^3D°$, $z^3P°$, $y^3D°$, $z^3F°$, $y^3P°$.
5d 6d	e^1S_0	54,793.82				$z^1P°$, $y^3P°$.
4f²	e^3H_4	55,107.25	874.84	.94 1	.800	$z^3F°$, $z^3F°$, $z^3G°$, $z^3H°$, $z^3G°$, $z^3D°$, $y^1D°$, $z^3F°$,
	e^3H_5	55,982.09	855.85	1.04 2	1.033	$z^3H°$.
	e^3H_6	56,837.94		1.18 1	1.167	
4f²	g^3F_2	57,399.58	518.92	1.11 1	1.083	z^1F, $z^3F°$, $z^1G°$, $y^3F°$, $z^3H°$, $z^3D°$, $z^3G°$, $z^3D°$,
	g^3F_3	57,918.50	340.91	1.22 1	1.250	$z^3F°$, $z^3F°$.
	g^3F_4	58,259.41				
4f²	g^1G_4	59,527.60		1.07 1	1.000	$z^1G°$, $y^3F°$, $z^3H°$, $z^3G°$, $y^1F°$, $z^3H°$, $z^1F°$.
4f²	h^1D_2	59,900.08		1.00 1	1.000	$z^3F°$, $z^3F°$, $y^3F°$, $z^1D°$, $z^3D°$, $z^3P°$, $y^1D°$, $y^1F°$.
6p²	$f^3P_0,2$	60,094.84		0/0	0/0	$y^1F°$, $z^1D°$, $z^3D°$, $z^3P°$, $y^1D°$, $y^1F°$, $y^3D°$, $z^3P°$,
	f^3P_1	61,128.83	1,033.99	1.47 1½	1.500	$y^1F°$, $z^3F°$.
	f^3P_2	62,506.36	1,377.53	1.45 1½	1.500	
6p²	i^1D_2	62,026.27				$z^1F°$, $y^3F°$, $z^1D°$, $z^3D°$, $z^3P°$, $y^1D°$, $y^1F°$, $z^3F°$, $z^1P°$, $y^3P°$, $y^1P°$, $z^1F°$, $z^1P°$.
4f²	e^1I_6	62,408.40		1.01 1	1.000	$z^3H°$, $z^3H°$.
4f²	g^3P_0	63,463.95	239.23			$z^1D°$, $z^3D°$, $z^3P°$, $y^1F°$, $y^1D°$, $z^3F°$, $z^1P°$, $y^3P°$,
	g^3P_1	63,703.18	575.74			$z^3P°$.
	g^3P_1	64,278.92				
6s 6d	h^3D_1	64,361.28	168.62			$z^1D°$, $z^3P°$, $y^1D°$, $z^1P°$, $y^3P°$, $y^1P°$, $z^3P°$.
	h^3D_2	64,529.90	162.69			
	h^3D_3	64,692.59				
	$^1D_2$?	64,706.76?				$y^1P°$, $z^1P°$.
6p²	f^1S_0	66,591.91				$z^1P°$, $y^3P°$.
4f²	g^1S_0	69,505.06				$z^1P°$, $y^1P°$, $y^1P°$.
6d 6s	i^1D_2	69,233.90				$y^1D°$, $z^1F°$.
	$1°_2$	57,364.12				c^1G, c^3F, c^3D.
	$2°_3$	58,748.90				c^1G, c^3F, c^1F.
	$3°_{1,2}$	59,612.64				c^1G, c^3F, c^3D.
	$4°_3$	60,744.17				c^1G, c^3D.
	$5°_{1,3}$	61,017.66				c^3F.
	$6°_{3,3}$	61,514.46				c^1F, c^1G.
	$7°_{2,3}$	63,598.87				c^1G, c^3G.
	$8°_2$	64,411.17				c^1G, c^3F, c^1D.

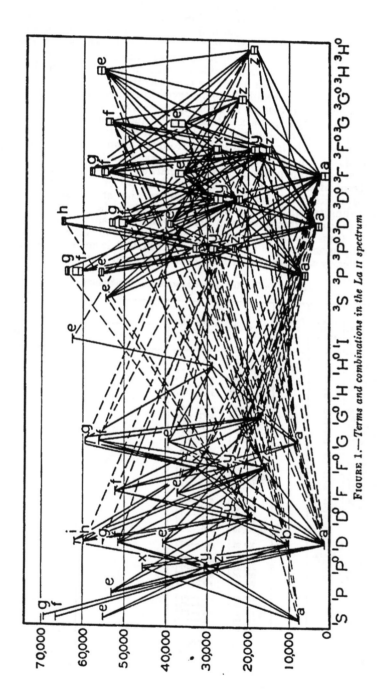

FIGURE 1.—Terms and combinations in the La II spectrum

A diagram of the La II terms and combinations is reproduced in Figure 1. In order to avoid making the figure too confusing some of the combinations have been omitted. The abundance and strength of intersystem connections is very striking in lanthanum spectra. Comparison with the corresponding diagrams for Sc II[23] and Y II[24] show at a glance the remarkable increase in complexity of the La II spectrum due to terms involving the *f*-electron.

Complete data for all of the observed lines characteristic of ionized lanthanum are presented in Table 4, successive columns of which contain wave lengths, spark intensities, furnace classes, vacuum wave numbers, term combinations, and Zeeman effects, both observed and computed. The latter are derived from the adopted *g* values with the formulas given above. When the observed pattern is resolved, the computed components are given separately; when unresolved, the centroid of the blend given by equations (2) or (4), except for the patterns described as A^1 or A^2, where the strongest component according to equation (1) is tabulated. In the latter case the measured position often deviates a little toward the centroid of the group. Apart from this the agreement of the observed and computed values is usually satisfactory. One faint line, 4,193.34 A is entirely discordant, and clearly does not arise from the assigned combination.

TABLE 4.—*The first spark spectrum of lanthanum (La II)*

λ_{air} I. A.	Intensity spark	Temperature class	ν_{vac} cm^{-1}	Term combinations	Zeeman effects	
					Observed	Computed
10, 954. 6	3		9, 126. 09	$a^1G_4-z^1G^o_3$		
10, 186. 5	2		9, 814. 2	$y^1D^o_2-e^1F_3$		
10, 093. 54	1		9, 904. 62	$y^1P^o_1-e^1D_2$		
9, 893. 82	4		10, 104. 55	$y^1P^o_1-e^1D_2$		
9, 672. 94	3		335. 29	$z^1F^o_3-e^1G_4$		
9, 657. 00	20		352. 35	$a^1G_4-z^3H^o_5$		
9, 563. 60	4		453. 45	$z^3F^o_2-e^3G_3$		
9, 346. 69	15		696. 04	$z^1H^o_5-e^1G_4$		
9, 260. 42	3		795. 69	$z^3F^o_3-e^1F_3$		
9, 146. 75	2		929. 86	$z^1F^o_3-e^3G_3$		
9, 127. 5	1		952. 9	$z^3F^o_3-e^3F_4$		
9, 101. 10	2		984. 67	$a^3P_2-y^1F^o_3$		
9, 096. 71	3		10, 989. 97	$y^1D^o_2-e^3G_3$		
9, 016. 80	2		11, 087. 37	$y^3D^o_2-e^1D_2$		
8, 810. 57	3		346. 89	$f^1D_2-z^3D^o_1$		
8, 781. 98	2		383. 83	$z^3F^o_3-e^1D_2$		
8, 650. 82	2		556. 43	$a^3D_2-z^3F^o_2$		
8, 514. 65	3		741. 24	$a^1G_4-y^3F^o_3$		
8, 484. 01	2		11, 783. 65	$a^3D_2-z^1F^o_3$		
8, 323. 35	1		12, 011. 10	$b^1D_2-z^3D^o_1$		
8, 159. 05	10? Spark.		252. 96	$a^1D_1-z^3F^o_2$		
8, 059. 5	3 h		404. 3	$a^3F_4-z^3F^o_3$		
7, 927. 83	2		610. 33	$b^1D_2-z^3P^o_1$		
7, 891. 69	2		666. 08	$a^3P_2-z^1D^o_2$		
79. 93	4		686. 96	$y^1F^o_3-e^1F_3$		
7, 838. 83	1		753. 50	$a^1D_2-z^3F^o_3$		
7, 740. 54	2 h		12, 915. 45	$z^3D^o_1-e^3G_3$		
7, 612. 94	3		13, 131. 92	$a^3F_2-z^3F^o_3$		
7, 489. 15	1		348. 96	$a^3D_2-z^3G_4$		
7, 483. 48	30	IV E	359. 10	$a^3F_2-z^3F^o_3$	(0.00) 1.12	(0.00) 1.09
7, 340. 08	2		620. 08	$z^3F^o_4-e^1D_2$		
7, 297. 99	2		698. 63	$y^1F^o_3-e^1D_2$		
82. 36	150	III E	728. 03	$a^3F_2-z^3F^o_3$	(0.00 w) 1.26	(0.00) 1.26
66. 13			758. 70	$y^1D^o_2-e^1D_2$		
7, 213. 95	2		13, 858. 22	$a^1G_4-z^3G^o_3$		

23 See footnote 2, p. 625.   24 See footnote 3, p. 625.

TABLE 4.—*The first spark spectrum of lanthanum (La II)*—Continued

λ air I. A.	Intensity spark	Temperature class	ν vac cm⁻¹	Term combinations	Zeeman effects Observed	Zeeman effects Computed
7,118.6	2		14,043.8	$z^2F^0_3-z^1D_2$		
16.8	3		047.4	$a^1S_0-z^3D^0_1$		
7,104.7	3 h		071.3	$y^1D^0_2-e^3D_1$		
7,079.74	4		120.92	$z^2G_5-e^3G_5$		
7,066.24	300	III E	147.90	$a^3F_2-z^3F^0_2$	(0.00) 0.70	(0.00) 0.70
6,968.78	25	V E	345.76	$z^3D^0_1-e^3F_2$	(0.00) 0.81	(0.00) 0.82
58.11	100	IV E	367.76	$b^1D_2-y^1D^0_2$	(0.21) 0.96	(0.21) 0.96
54.54	20	IV	375.14	$a^3F_2-z^3F^0_3$		
52.52	10	IV	379.31	$a^1D_2-z^1F^0_3$	(0.00) 1.10	(0.00) 1.11
6,902.08	3	III	484.40	$y^1D^0_2-e^1D_2$		
6,859.03	5	V	575.31	$a^4s_{2}-z^5H^0_3$		
37.91	15	IV ? E	620.32	$a^1D_2-y^1F^0_3$		
34.07	20	IV ? E	628.54	$a^3F_4-z^1G^0_4$	(1.05) ?	(0.87) 1.13
30.83	6	V E	635.48	$z^3D^0_3-e^3G_4$	(0.00) 0.98	(0.00) 0.89
13.66	60	V ? E	672.32	$z^3D^0_3-e^1F_3$	(1.04) 1.29 B	(0.94) 1.14
06.88	30	IV	682.66	$a^3F_2-z^1F^0_3$	(0.00 h) 1.56 h	(0.00) 1.51
6,801.36	5		696.85	$y^1F^0_3-e^1G_4$	(0.00) 1.10	(0.00) 1.10
6,774.29	100	III E	757.65	$a^3F_3-z^1F^0_3$	(0.00 w) 1.10	(0.12) 1.07
50.47	1		809.70	$a^1G_4-z^1G^0_4$		
32.80	40	V E	848.57	$z^3D^0_3-e^3F_3$	(0.00) 0.99	(0.00) 0.94
18.68	60	V E	879.78	$y^1F^0_3-e^3D_2$	(0.69) 1.12 B	(0.70) 1.17
6,714.08	80	V E	889.97	$z^1G^0_4-e^1G_4$	(0.00) 1.38	(0.00) 1.32
6,676.14	3		974.59	$z^1P^0_1-e^3D_2$		
71.41	40	IV ? E	14,985.20	$a^4D_3-y^1F^0_3$	(0.63) 1.17	(0.64) 1.21
42.79	100	V E	15,049.77	$z^3G^0_5-a^1G_4$	(0.33) 0.82	(0.25) 0.82
36.53	5	V	063.96	$a^1G_4-z^3D^0_3$		
6,619.10	2		103.63	$z^3D^0_3-e^1F_3$		
6,570.96	(?)	III	214.28	$a^3P^0_2-z^3D^0_1$		
54.18	(?)	V	253.23	$z^1D^0_2-e^3F_4$		
29.72	4 h	III ?	310.37	$a^1S_0-z^3P^0_1$		
6,526.99	200	III E	316.77	$a^1D_2-y^3F^0_3$	(0.00, 0.22) 0.94 A¹	(0.00, 0.24) 0.52, 0.76, 0.99
6,496.19	250	IV ? E	384.66	$z^3G^0_3-e^3F_2$	(0.00) 0.81	(0.00) 0.82
46.62	200	V E	507.73	$z^3G^0_3-e^3F_4$	(0.00 w) 1.43 A²	(0.00) 1.45 A²
43.05	50 h	V'	516.32	$z^3P^0_1-e^3D_2$	(0.00 w) 0.76 A¹	(0.00) 0.74 A¹
6,415.39	1		583.22	$a^3F_2-z^1G^0_4$		
6,399.04	400	V ? E	621.04	$z^3G^0_3-e^3F_3$	(0.00) 1.04	(0.00) 1.06
90.48	200	III	643.96	$a^1D_2-y^1F^0_3$	(0.00) 1.03	(0.00) 1.04
74.08	30	IV	684.21	$z^3D^0_3-e^1D_2$	(0.00 w) 1.67 A³	(0.00) 1.75 A³
58.12	30	IV	723.58	$a^3P^0_1-z^3D^0_1$	(0.99) 0.49, 1.52	(0.96) 0.54, 1.51
37.85	3 ·	V	773.79	$a^3F_2-z^3F^0_3$		
20.39	200	III	817.44	$a^1D_2-y^3F^0_3$	(0.44) 0.87 B	(0.41) 0.87
15.79	50		828.96	$z^3P^0_1-e^3D_1$	(0.95) 0.49, 1.43	(0.96) 0.50, 1.46
10.91	200	V E	841.20	$z^3G^0_3-e^3G_4$	(0.27) 1.10 B	(0.29) 1.10
07.25	20 h	IV	850.40	$z^1P^0_3-e^3D_1$	(0.00) 0.51	(0.00) 0.50
6,305.46	10	IV	854.90	$a^3F_2-z^3H^0_3$		
6,296.08	300	IV E	878.52	{ $b^1D_2-y^2D^0_1$ / $a^1P_1-z^3D^0_2$ }	} (0.00 w) 1.20 A²	(0.00) 1.23 A²
73.76	100	III ?	935.01	$y^1F^0_3-e^1D_2$	(0.00) 1.04	(0.00) 1.02
62.30	300	III	15,944.17	$a^3D_2-y^3F^0_3$	(0.00 w) 1.10 A¹	(0.00) 0.96 A¹
6,203.51	50 l	V	16,115.46	$z^1D^0_2-e^1D_2$	(0.17) 1.14 h	(0.15) 1.14
6,188.09	100 l	V	155.61	$z^1P^0_1-e^1D_2$	(0.00 w) 1.02 A¹	(0.00) 0.99 A¹
74.15	6	V	192.09	$a^3P_0-z^3D^0_1$	(0.00) 0.63	(0.00) 0.55
72.72	10	V	195.84	$a^3F_2-y^3F^0_3$	(0.00, 0.32, 0.63) 1.69 A²	(0.00, 0.34, 0.67) 1.76 A²
46.53	15	IV	264.85	$a^3F_4-y^3F^0_3$	(0.00 w) 1.69 A³	(0.00) 1.76 A²
1.92	1		303.60	$a^3D_2-z^1D^0_2$		
29.57	50	IV ?	309.85	$a^3P_1-z^3L^0_2$	(0.00 w) 0.99 A¹	(0.00) 0.98 A¹
26.09	50	V ?	319.12	$b^1D_2-z^1F^0_3$	(0.34) 0.92 B	(0.32) 0.93
20.34	1	IV	334.45	$z^1P^0_1-f^1D_2$		
6,100.37	30	V	387.92	$a^3P_2-z^3D^0_3$	(0.00, 0.34) 0.84, 1.16	(0.00, 0.32) 0.86, 1.19, 1.51
6,085.43	10	V	428.15	$z^3D^0_3-e^3D_1$	(0.00, 0.74) 1.20, 1.90	(0.00, 0.69) 0.50, 1.19, 1.88
74.01		III A	459.04	$z^3G^0_3-e^3F_4$	(0.21) 1.12	(0.27) 1.10

TABLE 4.—*The first spark spectrum of lanthanum (La II)*—Continued

λ_air I. A.	Intensity spark	Temperature class	ν_vac cm⁻¹	Term combinations	Zeeman effects Observed	Zeeman effects Computed
67.13	6	V ?	477.71	$a^3P_2 - z^3P_1^o$	(0.00) 1.49	(0.00) 1.50
61.42	2	V ?	483.23			
46.07	2	IV ?	535.10			
0,037.96	2		557.25	$z^1D_2^o - e^1G_3$		
5,991.96	4 h		684.36	$z^1D_2^o - e^1G_4$		
73.52	120 l	V E	735.92	$z^1G_4^o - e^3G_3$	(0.00) 1.20	(0.00) 1.20
71.09	8		742.73	$b^3D_3 - z^3F_3^o$	(0.00) 1.18	(0.00) 1.15
61.43	3		769.86	$z^3G_3^o - e^3G_4$		
57.90	4		770.80	$z^3D_1^o - e^3D_2$		
48.30	20		806.9	$z^3G_3^o - e^1F_3$	(0.52) 0.86 h	(0.46) 0.86
36.22	20	V E	841.08	$a^1D_2 - y^3F_3^o$	(0.00) 1.21	(0.00) 1.19
27.71	30		865.26	$z^3D_1^o - e^3D_2$	(0.00) 1.30	(0.03) 1.31
18.26	4		892.19	$z^1D_2^o - e^3F_3$		
5,901.95	401	IV E	938.87	$z^3G_3^o - e^1G_4$	(0.00 w) 1.60 A²	(0.00) 1.78 A²
5,892.66	4	V	965.58	$a^3P_1 - z^3P_0^o$	(0.00) 1.54	(0.00) 1.51
85.23	1		16,966.99	$a^3P_1 - z^3P_1^o$	(0.00, 0.42) 0.54, 0.96, 1.38	(0.00, 0.41) 0.52, 0.98, 1.34
80.63	50	III E ?	17,000.28	$a^3D_1 - z^1D_2^o$		
74.00	6	IV	019.47	$a^3P_2 - z^3P_2^o$	(0.00) 1.49	(0.05) 1.48
63.70	80	V E	049.36	$a^1G_4 - y^1F_3^o$	(0.00) 0.96	(0.00) 0.96
48.96	20		092.36	$z^3D_1^o - e^1D_2$	(0.00) 0.54	(0.05) 0.52
28.44	2		152.51	$a^3D_1 - z^1G_4^o$		
08.63	8		211.00	$z^1P_1^o - e^1D_2$		
08.31	60	IV ?	211.95	$a^3F_2 - y^3F_3^o$	(0.00) 0.75	(0.05) 0.74
06.56	8		217.14	$y^1F_3^o - e^3G_3$		
5,805.77	120	III E	219.48	$a^3F_3 - y^3F_4^o$	(0.00) 1.09	(0.00) 1.09
5,797.57	150	III E	243.84	$a^3F_4 - y^3F_4^o$	(0.00) 1.24	(0.05) 1.25
81.02	3		293.20	$b^3D_2 - y^3D_2^o$		
79.91	4		296.52	$z^3D_1^o - e^3D_2$	(0.00) 1.38 h	(0.00) 1.43
69.06	60	V E	329.05	$b^3D_2 - z^1P_1^o$	(0.00) 1.09	(0.00) 1.08
49.59	2		387.73	$z^1G_4^o - e^1F_4$		
27.29	20	V E	455.43	$a^3P_0 - z^3P_1^o$	(0.00) 1.45	(0.00) 1.46
12.39	20	III E ?	500.96	$a^1D_2 - z^1D_2^o$	(0.00) 0.98	(0.10) 0.96
5,703.32	20	III	528.80	$a^3P_1 - z^3P_2^o$	(0.00) 1.48	(0.00) 1.44
5,671.54	100	V E	627.02	$z^3H_4^o - e^3G_3$	(0.00) 0.79	(0.00) 0.85
52.3	10 h		687.0	$z^1G_4^o - e^3G_3$		
5,610.53	20		818.69	$z^3G_3^o - e^3D_2$		
5,591.51	1		879.30	$a^3F_3 - z^1D_2^o$		
66.92	40	V E	17,958.28	$y^3F_4^o - e^3G_4$	(0.37) 1.24	(0.32) 1.20
47.56	3 h		18,020.95		(0.00) 1.13	
35.66	80	V E	059.69	$b^3D_2 - y^3P_1^o$	(0.00 w) 0.80 A²	(0.00) 0.77 A²
5,532.17	10	III	071.08	$z^3G_3^o - e^3D_2$	(0.00 w) 0.61 h	(0.00) 0.69
5,493.45	20	V E	198.45	$a^3F_3 - y^3F_2^o$	(0.00 w) 1.57 A²	(0.00) 1.70 A²
84.86	5	V	220.31	$b^1D_2 - y^3D_2^o$		
82.27	40	V E	235.57	$a^3F_4 - y^3F_3^o$	(0.00 w) 1.89 A²	(0.00) 1.81 A²
80.72	25	V E	240.72	$y^3F_3^o - e^3G_3$	(0.00) 1.03 h	(0.00) 0.99
64.37	25	V E	295.30	$a^3P_2 - y^3F_2^o$	(0.00, 0.49, 0.97) 0.00 d, 0.49, 0.97	(0.00, 0.46, 0.91) 0.12, 0.57, 1.03
58.68	50	V E	314.37	$z^1D_2^o - e^1F_3$	(0.00) 1.00	(0.00) 0.96
47.59	10		351.65	$z^1D_2^o - e^1D_2$	(0.22) 1.14	(0.25) 1.12
5,423.82	4		432.08	$a^3F_2 - z^1G_3^o$		
5,381.91	100	V E	575.61	$y^1F_3^o - e^3F_3$	(0.00) 0.74	(0.00) 0.72
81.77	50	V E	576.10	$z^3F_3^o - e^3F_4$		
80.97	100	V E	578.86	$a^1S_0 - y^3D_1^o$	(0.00) 0.77	(0.00) 0.80
77.08	200	V E	592.30	$z^3H_4^o - e^3G_4$	(0.00 w) 0.82 A²	(0.00) 0.83 A²
40.66	100	III E	719.09	$y^3F_3^o - e^1F_3$	(0.00) 1.07	(0.08) 1.06
33.42	2		744.50	$a^1P_1 - y^1D_2^o$		
03.54	100	III E	850.10	$a^3D_2 - z^3D_1^o$	(0.00, 0.61) 0.57, 1.17, 1.78	(0.00, 0.59) 0.55, 1.14, 1.73
02.62	150	V E	853.37	$z^3G_3^o - e^1G_3$	(0.00) 1.16	(0.00) 1.18
5,301.97	200	III E	855.68	$a^3D_3 - z^3D_2^o$	(0.00 w) 1.50 w	(0.00) 1.49
5,290.83	50	III E	895.38	$a^3F_2 - z^3D_1^o$	(0.34) 0.82 B	(0.36) 0.83
79.11	40	V E	18,937.33	$y^3F_3^o - e^3G_4$	(0.00) 1.20	(0.00) 1.23
59.38	50	III E	19,008.37	$a^1D_2 - z^3G_3^o$	(0.00 w) 0.46 A²	(0.00) 0.35 A²
57.28	2		015.96	$z^3D_1^o - e^1D_2$		
26.20	40 l	V E	129.05	$z^3H_4^o - e^3F_3$	(0.00 w) 0.32 w	(0.00) 0.26 A²
22.48	3 h		142.68	$e^3D_2 - 1\frac{1}{2}$	(0.00) 1.15	
21.32	3 h		146.93			

TABLE 4.—*The first spark spectrum of lanthanum (La II)*—Continued

λ_air I. A.	Intensity spark	Temperature class	ν_vac cm⁻¹	Term combinations	Zeeman effects Observed	Zeeman effects Computed
17.83	10 h		159.74			
5,204.14	300	V E	210.14	$z^1H^o_5-e^1F_4$	(0.00) 0.86 A¹?	(0.00) 0.85 A¹
5,191.50	3 h		256.91		(0.00) 1.05	
88.21	500	V E	269.12	$z^1H^o_5-e^1G_4$	(0.00) 1.09	(0.00) 1.09
83.42	400	III E	286.98	$a^1D_2-z^1D^o_3$	(0.00) 1.31	(0.05) 1.33
72.83	25 l	IV ? E	322.67	$z^1F^o_3-f^1D_2$	(0.00) 1.06	(0.00) 0.99
72.80	20 l	V E	326.19	$z^1D^o_2-e^1D_2$	(0.20 1.03	(0.30) 1.02
67.28	10		347.17	$z^2H_4-e^3G_4$		
63.61	40	V ? E	360.92	$e^1F_4-z^3G^o_4$	(0.74) 1.18 B	(0.66) 1.16
62.68	3		364.41	$a^1G_4-z^3F^o_4$		
57.43	150	V E	384.12	$z^1H^o_5-e^1F_3$	(0.00) 0.66	(0.00) 0.73
56.74	40	V E	396.71	$a^1F_3-z^3G^o_4$		
52.31	1		403.38	$b^1D_2-y^1P^o_1$		
22.99	200	III E	514.43	$a^1D_2-z^1D^o_1$	(0.00) 1.16	(0.08) 1.16
14.55	200	III ? E	546.63	$a^1D_1-z^3D^o_1$	(0.00) 0.54	(0.03) 0.53
12.37	2		554.97	$y^1F^o_3-e^1F_4$		
5,107.54	6 h		573.46	$e^1F_4-1^o_3$	(0.00) 1.19	
5,090.56	20 l	V E	638.74	$z^1D^o_2-e^3D_1$	(0.00 w) 1.33 A¹	(0.00) 1.37 A¹
86.71	3 hl		653.61	$z^3F^o_3-f^3G_3$		
80.21	40	V E	678.76	$z^1F^o_3-e^3G_4$	(0.38) 0.98 w	(0.45) 0.96
66.99	20 h		730.10	$e^3G_4-2^o_4$	(0.00) 1.10 h	
63.76	3		742.68	$y^1F^o_3-e^1F_3$		
62.91	20	V E	746.00	$a^3P_2-y^3D^o_1$	(0.00, 0.66) 0.90, 1.56, 2.22	(0.00, 0.69) 0.80, 1.49, 2.18
60.85	3		754.04	$z^1F^o_3-e^3G_3$		
58.56	1 h		762.97			
48.04	30 l	V E	804.16	$y^3F^o_3-3^1G_4$	(0.00) 1.16	(0.00) 1.12
14.45	30 hl		936.82	$z^1F^o_3-f^3F_3$		
5,002.12	40	V E	985.96	$y^3F^o_3-e^1D_2$	(0.00) 1.09	(0.00) 1.08
4,999.46	200	III E	19,996.60	$a^3D_2-z^3P^o_1$	(0.00 w) 1.28 A¹	(0.00) 1.08 A¹
96.82	50	V E	20,007.16	$y^3F^o_3-e^3G_4$		
95.17	· 1		013.77	$z^1F^o_3-e^1F_3$		
91.27	80	IV E	029.41	$a^1S_0-z^1P^o_1$	(0.00) 0.85	(0.00) 0.88
86.82	100	III E	047.28	$a^1D_2-z^3D^o_1$	(0.00, 0.43) 0.57, 1.00, 1.43	(0.00, 0.44) 0.55, 0.99, 1.43
74.20	4 h		008.14	$z^3P^o_2-e^1P_1$		
70.39	100	III E	113.55	$a^1D_2-z^3P^o_1$	(0.00, 0.29) 0.83, 1.12 us	(0.00, 0.32) 0.82, 1.14, 1.46
56.04	2	V	171.79			
52.06	40	V E	188.00	$y^3F^o_3-e^3D_2$	(0.00) 1.17.	(0.00) 1.15
46.47	50	IV? E	210.81	$a^3D_1-z^3D^o_1$	(0.00, 0.67) 1.15, 1.82	(0.00, 0.67) 0.52, 1.19, 1.85
35.61	10	V E	255.28	$a^3P_1-y^3D^o_1$		
34.83	100	V E	258.48	$b^1D_2-y^1P^o_1$	(0.00) 0.97	(0.00) 0.99
21.80	300	III E	312.12	$a^3F_4-z^3G^o_3$	(0.00) 1.11	(0.00) 1.09
20.98	300	III E	315.50	$a^3F_2-z^3G^o_3$	(0.00) 1.00	(0.00) 0.92
11.34	10		355.38	$z^3G^o_3-e^3F_3$		
4,904.43	2 h		384.05	$z^3P^o_3-g^3D_3$		
4,899.92	200	III E	402.82	$a^3F_2-z^3G^o_3$	(0.00) 0.80	(0.00) 0.82
91.43	10		438.23	$z^3H_3-e^3G_3$		
80.20	10 h		485.17	$z^3P^o_2-g^3D_3$		
74.99	1		507.15	$z^1D_3-e^1D_3$		
60.90	80	III E	566.59	$a^3F_4-z^3D^o_3$	(0.00) 1.19	(0.00) 1.17
59.18	5 h		573.87	$z^1G^o_4-e^3G_4$		
50.58	30	V E	610.35	$z^3P^o_1-z^3D_2$; $a^3P_2-z^3F^o_3$		
43.29	5	V	641.37	$z^1G^o_4-e^1F_3$	(0.00) 1.03	(0.00) 1.06
40.02	30	V E	655.32	$a^3D_2-z^3P^o_1$	(0.67) 1.30 ?	(0.58) 1.30
30.51	10	V E	695.98	$a^3P_1-z^3F^o_2$	(0.00) 0.76 R	(0.00, 0.67) 0.17, 0.84 1.51
26.87	20	V E	711.59	$a^1D_2-z^3D^o_2$	(0.33) 1.08 R	(0.36) 1.09
24.05	100	III E	723.70	$a^3P_0-y^3D^o_1$	(0.00) 0.79	(0.00) 0.80
09.00	100	V E	788.55	$a^3D_1-z_2P^o_3$	(0.00) 0.51	(0.00) 0.52
4,804.04	80	V E	810.02	$a_3D_1-z_2P^o_1$	(0.95) 0.49, 1.43	(0.94) 0.52, 1.46
4,796.67	25	V E	841.99	$a^1G_4-y_2G^o_3$		
94.55	3	V E	851.20			

TABLE 4.—*The first spark spectrum of lanthanum (La II)*—Continued

λ_air I. A.	Intensity spark	Temperature class	ν_vac cm⁻¹	Term combinations	Zeeman effects Observed	Zeeman effects Computed
80. 55	2	V	20,912.27			
58. 40	3		21,009.61	$y^3F_2^\circ - c^3D_2$		
48. 73	150	V E	052.39	$a^1G_4 - z^1H_4^\circ$	(0.00) 1.00	(0.00) 0.99
43. 08	250	V E	077.47	$z^3F_3^\circ - c^3G_3$	(0.52 w) 0.95 B	(0.55) 0.98
40. 27	120	III E	089.96	$a^3F_2 - z^3D_2^\circ$	(0.00 w) 0.97 A¹	(0.00) 0.90 A¹
39. 80	15	V E	092.06	$e^1G_4 - z^3F_4$		
30. 73	3 h		132.50	$z^1F_3^\circ - f^3G_3$		
28. 41	100	V E	142.86	$a^1D_2 - z^3D_3^\circ$	(0.00, 0.32) 1.31, 1.61, 1.91	0.00, 0.33, 0.66) —, —, 1.32, 1.65, 1.99
24. 42	40	V E	160.72	$a^3P_2 - y^3D_2^\circ$	(0.65) 0.95 us	(0.30, 0.59) 0.89, 1.19, 1.49, 1.78
22. 14	2 h		170.94	$y^1P_1^\circ - f^1D_2$		
19. 93	150	V E	180.85	$z^1F_3^\circ - c^3F_2$	(0.00) 1.06	(0.00) 1.05
17. 56	50	V E	191.40	$a^3G_3 - c^3F_4$	(0.50) 1.2	(0.48) 1.07
16. 44	80	V E	196.52	$a^3P_2 - z^3P_1^\circ$	(0.00, 0.60) 0.92, 1.52, 2.12	(0.00, 0.61) 0.88, 1.49, 2.09
12. 92	40	V E	212.35	$a^3D_2 - y^3D_3^\circ$	(0.00, 0.47, 0.92) 1.83, 2.26 R	(0.00, 0.44, 0.88) —, —, —, 1.78, 2.22
4,703. 27	150	V E	255.87	$z^3F_3^\circ - c^3F_3$	(0.00 w) 1.70 A¹	(0.00) 1.85 A¹
4,699. 62	50	V E	272.38	$a^3D_2 - y^3F_3^\circ$		
92. 50	200	V E	304.65	$z^3F_2^\circ - c^3G_3$	(0.00 w) 1.21 A¹	(0.00) 1.26 A¹
91. 17	50	V E	310.70	$a^1D_2 - z^3P_1^\circ$	(0.00, 0.46) 0.46 us	(0.00, 0.47) 0.51, 0.99, 1.46
88. 65	40	V E	322.15	$y^3P_2^\circ - c^3D_1$	(0.00) 0.96	(0.00) 0.89
84. 39	2 h		341.54	$c^3D_2 - 4_2^\circ$		
82. 12	5		351.89	$a^3D_1 - z^3P_1^\circ$		
73. 53	1 h		391.14	$c^3D_2 - 31_{,2}$		
71. 82	200	V E	398.96	$z^1F_3^\circ - c^3G_4$	(0.00 w) 1.29 h	(0.00) 1.30
68. 91	250	V E	412.30	$z^3F_3^\circ - c^3F_2$	(0.00, 0.34, 0.68) 0.77, 1.12 1. 47, 1.81 us	(0.00, 0.36, 0.71) —, —, 0.73, 1.09, 1.44, 1.80
63. 76	300	V E	435.94	$z^1F_3^\circ - c^3F_3$	(0.20) 0.99	(0.24) 0.99
62. 51	200	III E	441.69	$a^3F_2 - z^3D_1^\circ$	(0.00 w) 0.89 A¹	(0.00) 0.91 A¹
55. 49	400	V E	474.02	$z^3F_3^\circ - c^3G_4$	(0.43) 1.20 B	(0.37) 1.20.
52. 07	30 hl	I	489.81			
47. 50	100	V E	510.94	$z^3F_3^\circ - c^3F_3$	(0.00, 0.31, 0.61) 1.96 A¹	(0.00, 0.30, 0.61) 2.17 A¹,
45. 28	100	V E	521.22	$a^3F_2 - z^3D_2^\circ$	(0.60) 1.26 B	(0.58) 1.20
41. 40	2 h		539.21	$c^1F_3 - 2_2^\circ$		
36. 42	80	V E	562.35	$z^1D_2^\circ - c^1D_2$	(0.21) 1.00	(0.21) 0.99
34. 95	25 l		569.19	$y^3D_2^\circ - f^1D_2$	(0.00) 1.43	(0.00) 1.46
23. 99	2 h		620.31			
19. 87	300	V E	639.59	$z^3F_3^\circ - c^3F_3$	(0.00) 0.70	(0.10) 0.70
13. 38	200	V E	670.03	$a^3P_1 - y^3D_2^\circ$	(0.00, 0.33) 0.83, 1.16, 1.49	(0.00, 0.32) 0.83, 1.19, 1.51
05. 78	100	V E	705.70	$a^3P_1 - z^1P_1^\circ$	(0.63) 0.89, 1.52	(0.62) 0.88, 1.51
01. 65	3		725.27	$c^3G_3 - 4_3^\circ$		
4,600. 59	5 h		730.27	$y^1P_1^\circ - f^1D_2$		
				$y^1P_1^\circ - f^1D_2$		
4,595. 06	2 h		756.42			
87. 14	2 h		793.99	$c^3F_2 - 2_2^\circ$		
80. 05	150	V E	827.73	$a^3P_1 - y^3P_2^\circ$	(0.00) 1.51	(0.00) 1.51
74. 87	200	III E	852.44	$a^1D_2 - z^3D_2^\circ$	(0.49, 0.98) 0.49, 0.98, 1.46, 1.94	(0.48, 0.95) 0.51, 0.99, 1.46, 1.94
70. 97	10	V E	871.06	$a^3D_2 - y^1D_2^\circ$	(0.43) 1.00 R	(0.44) 1.02
62. 5	5 h		911.7	$c^3G_3 - 1_3^\circ$		
59. 28	100		927.16	$a^3P_2 - y^3P_1^\circ$	(0.00, 0.18) 1.52	(0.00, 0.23) 1.26, 1.49, 1.71
58. 46	200	III E	21,931.10	$a^3D_2 - y^1F_3^\circ$	(0.00 w) 0.86 A¹	(0.00) 0.82 A¹
40. 71	10		22,016.84	$z^1F_3^\circ - c^3F_4$	(0.00) 1.27	(0.00) 1.28
38. 87	8 hl		025.76	$y^3P_2^\circ - f^1D_2$		
30. 54	15	V E	066.26	$b^1D_2 - z^3P_1^\circ$		
26. 12	200	III E	087.81	$a^3P_1 - y^3D_1^\circ$	(0.00 w) 1.01 A¹	(0.00) 0.96 A¹
25. 31	100	V E	091.76	$z^3F_2^\circ - c^3F_2$	(0.46) 1.19 B	(0.39) 1.20
22. 37	400	III E	105.12	$b^1D_2 - z^1F_3^\circ$	(0.00) 1.02	(0.00) 1.00
				$a^3F_2 - z^3D_1^\circ$		
16. 38	5 hl		135.44			
05. 48	10		174.22	$a^3P_2 - z^1P_1^\circ$	(0.00) 0.90	(0.00) 0.88

TABLE 4.—*The first spark spectrum of lanthanum (La II)*—Continued

λ_{air} I. A.	Intensity spark	Temperature class	ν_{vac} cm^{-1}	Term combinations	Zeeman effects	
					Observed	Computed
05.82	3 hl		187.32	$y^3P^\circ_1-f^3D_1$		
4,502.16	10 hl		206.35	$z^3P^\circ_1-e^3S_1$	(0.00) 1.47	(0.00) 1.47
4,498.76	10		222.13	$y^3F^\circ_2-e^1D_2$	(0.00) 1.14	(0.00) 1.13
97.00	2		230.83	$a^3F_3-z^3P^\circ_2$		
84.48	1 h		292.90	$e^1G_4-6^\circ_{4,5}$		
81.21	25 hl		309.16	$z^1P^\circ_1-f^3D_1$	(0.40) 0.50, 0.87	(0.36) 0.52, 0.88
74.03	10		344.96	$y^3D^\circ_2-f^3D_1$	(0.00, 0.66) 0.50, 1.17, 1.83	(0.00, 0.67) 0.52, 1.19, 1.86
59.10	3		419.78	$z^1G^\circ_4-e^3G_4$		
55.79	50	V E	436.43	$a^3P_1-y^3P^\circ_1$	(0.24) 1.41	(0.24) 1.38
43.94	20 hl	I	496.26	$y^3D^\circ_2-f^3D_2$	(0.00) 1.18 h	(0.09) 1.17
35.84	10	IV E	537.34	$a^3F^\circ_3-z^3D^\circ_3$	(0.00, 0.60, 1.19) 1.28, 1.86, 2.44	(0.00, 0.59, 1.17) —, —, 1.32, 1.90, 2.49
32.95	20 ?		552.03	$a^3F_4-y^1F^\circ_3$		
29.90	400	III E	567.56	$a^1D_2-y^1D^\circ_2$	(0.00, 0.40) 0.51, 0.91, 1.31	(0.00, 0.38) 0.52, 0.90, 1.27
27.52	100	V E	579.69	$z^3F^\circ_3-e^3F_3$	(0.00 w) 1.06	(0.07) 1.07
19.16	30	V E	622.41	$z^1G^\circ_4-e^1G_4$	(0.28) 1.04	(0.21) 1.03
17.14	2 h	III	632.75	$z^3P^\circ_1-e^1S_0$		
12.22	2 h		657.98	$e^3F_3-3^\circ_{2,3}$		
11.21	25 hl	V E	663.17	$z^3F^\circ_3-f^3D_3$	(0.00 h) 1.13 h	(0.00) 1.12
4,408.02	2	III A	705.33	$a^3F_3-z^3P^\circ_2$		
4,385.20	40	V E	797.59	$z^3F^\circ_4-e^3G_4$	(0.00) 1.24	(0.00) 1.24
83.44	100	V E	806.75	$z^3F^\circ_3-e^3F_3$	(0.00, 0.40, 0.80) 0.28, 0.68, 1.08, 1.48, 1.88	(0.00, 0.38, 0.77) 0.29, 0.68, 1.06, 1.45, 1.83
78.10	50	IV E	834.56	$z^3F^\circ_3-e^1F_3$	(0.30 w) 0.95	(0.34) 1.02
64.66	100	IV E	904.88	$a^3P_0-y^3P^\circ_1$	(0.00) 1.28	(0.00) 1.26
63.05	50 l	V E	913.33	$y^3D^\circ_3-f^3D_3$	(0.00 h) 1.32 h	(0.03) 1.32
56.18	1		949.46	$y^1P^\circ_1-e^1P_1$		
54.40	200	IV E	22,958.84	$a^1S_0-y^1P^\circ_1$	(0.00) 1.08	(0.00) 1.07
37.78	10 l		23,046.81	$z^1F^\circ_3-f^3D_2$	(0.00) 1.08 h	(0.00) 1.02
34.96	100	V E	061.80	$z^3F^\circ_3-e^1F_3$	(0.00, 0.29, 0.57) 0.69, 0.97, 1.26, 1.54	(0.00, 0.28, 0.55)—0.68, 0.95, 1.23, 1.51
33.76	500	III E	068.19	$a^1D_2-y^1D^\circ_2$	(0.17) 0.94	(0.16) 0.94
22.51	100	III E	128.22	$a^1D_2-y^1F^\circ_3$	(0.00) 1.10	(0.00) 1.08
16.90	30	V E	163.65	$a^3D_2-z^3F^\circ_2$	(0.00, 0.50, 1.00) 0.86, 1.36, 1.86, 2.36	(0.00, 0.50, 1.00)—0.84, 1.34, 1.84, 2.34
04.11	10 hl		227.10	$e^3F_4-5^\circ_{4,5}$	(0.00 h) 1.24 h	
4,300.44	60	IV E	246.92	$a^3F_2-z^3P^\circ_1$	(0.73, 1.48) 0.00, 0.72, 1.45, 2.18	(0.73, 1.46) 0.00, 0.73, 1.46, 2.19
4,296.05	300	IV E	270.67	$a^3P_2-y^3P^\circ_2$	(0.00) 1.48	(0.02) 1.49
86.97	300	V E	319.96	$z^3F^\circ_3-e^1G_4$	(0.00) 1.08	(0.00) 1.09
75.64	100	IV E	381.75	$a^3D_2-y^3D^\circ_2$	(0.00, 0.35) 0.82, 1.17, 1.52	(0.00, 0.34) 0.80, 1.14, 1.48
69.50	300	V E	415.38	$z^3F^\circ_4-e^3F_4$	(0.00) 1.20	(0.00) 1.22
63.59	200	V E	447.83	$z^1F^\circ_3-e^1G_4$	(0.00) 1.12	(0.00) 1.08
59.51	2 h		470.30	$z^3F^\circ_3-f^3D_2$		
56.50	3		486.89			
52.93	4		506.61	$a^3F_3-y^1F^\circ_3$		
49.99	100	V E	522.87	$z^3F^\circ_3-e^1G_4$	(0.74) 0.50, 0.76, 1.02, 1.27, 1.53, 1.76	(—, —, 0.59, 0.79) 0.50, 0.75, 1.00, 1.25, 1.50, 1.75
48.32	2		532.11	$y^1P^\circ_1-f^1F_3$		
41.20	15 hl		571.62	$e^3G_4-4^\circ_4$		
38.38	400	III E	587.30	$a^3D_3-z^3F^\circ_3$	(0.74) 0.56, 0.82, 1.08, 1.34, 1.60, 1.86	(—, 0.51, 0.77) 0.57, 0.83, 1.08, 1.34, 1.59, 1.85
30.95	150	V E	628.72	$z^1F^\circ_3-e^3D_3$	(0.80) 0.65, 0.85, 1.06, 1.26, 1.47, 1.67 ur	(—, 0.52, 0.78) 0.53, 0.79, 1.05, 1.31, 1.56, 1.82
17.56	200	V E	708.74	$z^1F^\circ_3-e^3D_3$	(0.00) 1.22	(0.00) 1.19
10.22	50 hl		745.06		(0.00 h) 1.02 h	
07.61	10 l		759.79	$y^3D^\circ_1-f^3D_1$	(0.27) 0.68 h	(0.28) 0.66
04.03	100	V E	780.02	$a^3P_1-y^3P^\circ_2$	(0.00) 1.48	(0.00) 1.49
4,201.50	6 h		794.34			
4,196.55	250	III E	822.41	$a^3D_2-z^3F^\circ_2$	(0.32, 0.64) 0.51, 0.82, 1.13, 1.44	(0.30, 0.61) 0.53, 0.84, 1.14, 1.44
94.36	30 h		834.85	$z^1F^\circ_3-f^1G_4$	(0.00 h) 1.06 h	(0.00) 1.06
93.34	5		840.64	[$y^3D^\circ_3-f^3D_3$]	(0.25) 0.64	(0.00) 1.47 (?)
92.35	100	V E	846.27	$z^3F^\circ_3-e^3D_2$	(0.00) 1.06	(0.00) 1.07

TABLE 4.—*The first spark spectrum of lanthanum (La II)*—Continued

λ_{air} I. A.	Intensity spark	Temperature class	ν_{vac} cm⁻¹	Term combinations	Zeeman effects	
					Observed	Computed
80.97	12 l		23,911.18	$y^3D_1^o-f^3D_2$	(0.00 h) 1.02	
61.94	8 h		24,020.51			
54.59	2 h		063.00	$e^3F_3-5l_3$		
52.78	100	IV ? E	073.49	$z^3F_2^o-e^3D_2$	(0.43, 0.84) 0.27, 0.67, 1.08, 1.48	(0.42, 0.85) 0.25, 0.65, 1.10, 1.52
51.98	250	III E	078.13	$a^1D_1-y^1D_1^o$	(0.29) 0.50, 0.77	(0.28) 0.52, 0.80
48.2	4 h		100.1	$z^1P_1^o-f^3D_2$		
43.77	15		125.83	$a^3P_2-y^1P_1^o$		
41.73	200	IV E	137.72	$a^3D_2-y^3D_3^o$	(0.00 w) 1.50 w	(0.00) 1.48
37.91	2		160.00	$e^5G_2-3l_2$		
33.33	6 hl		186.77		(0.00) 0.97	
32.50	10 hl		191.63	$y^1P_2^o-g_2D_2$	(0.00 h) 0.83 h	(0.00) 0.92
31.74	5 h		196.08	$e^3D_2-7l_3$	(0.00 h) 1.13 h	
23.23	400	III E	246.01	$a^3D_2-z^3F_3^o$	(0.00) 1.05	(0.00) 1.02
15.35	1 h		292.44	$z^3F_3^o-f^3G_3$		
13.28	40 l		304.66	$e^1F_3-6l_3$	(0.00 h) 1.09 h	
4,101.01	3 h		377.38	$e^1G_4-7l_3$	(0.00) 0.77	(0.00) 0.77
4,099.54	150	V E	386.12	$z_1F_2^o-e^3D_1$		
98.73	5		390.94	$z^3F_3^o-f^3D_3$		
86.72	300	III E	462.62	$a^3F_2-y^1D_2^o$	(0.32) 0.79 w	(0.30) 0.81
77.35	300	III E	518.84	$a^3D_1-z^3F_2^o$	(0.00, 0.32) 0.57, 0.89, 1.20	(0.00, 0.32) 0.52, 0.84, 1.15
76.71	40	IV E	522.68	$a^3F_3-y^1F_3^o$	(0.00, 0.31, 0.62) 0.79, 1.08, 1.38, 1.69	(0.00, 0.30, 0.60) —, 0.73, 1.03, 1.33, 1.64
67.39	100	IV E	578.88	$a^1D_2-y^3D_3^o$	(0.00 w) 1.14 A²	(0.00) 1.17 A²
58.06	5 t		635.26	$a^3P_1-y^1P_1^o$	(0.45) 1.06, 1.47 us	(0.44) 1.07, 1.51
50.06	200	V E	683.92	$z^1F_3^o-e^1D_2$	(0.00) 1.02	(0.00) 1.05
42.91	300	IV E	727.70	$a^1G_4-z^1F_3^o$	(0.00) 0.99	(0.00) 1.00
36.59	15 d	V E	766.41	$a^1S_0-z^1P_1^o$	(0.00) 1.54	(0.00) 1.52
31.68	300	III E	796.57	$a^3D_2-y^3D_3^o$	(0.00) 1.15	(0.09) 1.17
25.87	50	IV E	832.36	$a^3D_2-z^1P_1^o$	(0.00 w) 1.37 A²	(0.00) 1.40 A²
23.58	40	IV E	846.49	$y^3F_3^o-e^1G_4$	(0.00) 1.07	(0.00) 1.02
20.19	2 h		867.44	$y^1P_1^o-e^3S_1$		
4,007.64	7 h		24,945.32		(0.00 w) 1.01 h	
3,995.74	400	III E	25,019.61	$a^1D_2-z^3F_3^o$	(0.30) 0.90 w	(0.27) 0.91
94.50	10		027.37	$z^3F_3^o-e^1D_3$	(0.57 h) 1.25 h	(0.56) 1.20
88.51	500	III E	064.96	$a^3D_2-y^1D_3^o$	(0.00) 1.32	(0.06) 1.32
81.36	10 l		109.97	$z^3F_2^o-f^3D_2$	(0.38) 0.96 h. us	(0.33) 0.93
79.08	8 l		124.35	$z^3F_2^o-g^3D_3$	(0.00) 1.26	(0.00) 1.30
63.04	5 l		226.04		(0.00 w) 0.82 h	
62.03	10 l		232.47		(0.00) 1.10 h	
58.53	2		254.78	$z^3F_3^o-e^3D_3$		
57.25	2		262.95			
56.07	4		270.48	$y^1D_2^o-f^3D_1$	(0.00 h) 1.29 h	(0.00) 1.27
55.21	3 h		275.98		(0.00) 1.15	
53.36	2		287.81		(0.00) 1.06	
51.43	3 h		300.16	$z^3F_3^o-f^1F_3$		
49.10	600	III E	315.09	$a^3D_2-z^3F_3^o$	(0.00) 1.13 w	(0.00) 1.11
44.15	3		346.86	$y^3D_2^o-g^3D_2$	(0.00) 1.18 h	(0.04) 1.18
39.85	20 l		374.52	$y^3D_2^o-g^3D_3$	(0.21) 1.27 h	(0.26) 1.26
36.22	50	IV E	397.92	$a^3F_3-z^3F_3^o$	(0.00, 0.25, 0.48) 0.86, 1.10, 1.34, 1.58,	(0.00, 0.26, 0.51) —, —, 0.84, 1.09, 1.35, 1.60
32.53	10 l		421.75	$y^1D_2^o-f^3D_2$	(0.44) 1.06 B	(0.44) 1.02
36.47	3		435.07		(0.00) 0.92	
29.22	300	III E	443.17	$a^1D_2-z^3F_3^o$	(0.00) 1.20	(0.00) 1.18
25.09	5		469.94	$y^1D_2^o-f^3O_3$		
24.69	3		472.53		(0.00 h) 0.72 h	
21.54	200	III E	492.99	$a^3D_1-y^1D_2^o$	(0.00, 0.66) 0.52, 1.18, 1.84	(0.00, 0.67) 0.52, 1.19, 1.86
16.05	300	III E	528.73	$a^3D_1-z^1P_1^o$	(0.38) 0.51, 0.88	(0.36) 0.52, 0.88
3,910.81	10 l	IV ? E	562.94	$a^3D_2-y^1P_1^o$	(0.00) 1.04	(0.00) 1.06
3,897.43	4		650.70	$a^3D_1-z^1P_1^o$		
92.47	3		683.37	$y^1P_1^o-z^3P_2$	(0.00 h) 1.4 h	(0.00) 1.30
92.05	3		686.15	$y^1P_2^o-g^1D_2$		
86.37	150	III E	723.69	$a^3D_2-y^1D_2^o$	(0.00 w) 1.62 A²	(0.00) 1.66 A²

TABLE 4.—*The first spark spectrum of lanthanum (La II)*—Continued

λ_air I. A.	Intensity spark	Temperature class	ν_{vac} cm^{-1}	Term combinations	Zeeman effects Observed	Zeeman effects Computed
85.09	4		732.16	$y^1P_1^o-e^3P_1$	(0.00) 1.08	
71.64	200	III ? E	821.56	$a^1F_3-z^3F_3^o$		(0.00) 1.09
68.35	3 h		843.52			
64.49	100 l		869.33	$z^3F_4^o-f^3G_5$	(0.00 b) 1.10 h	(0.00) 1.11
63.11	2		878.57	$z^1P_1^o-e^1P_1$		
60.31	2		897.34	$z^3F_3^o-g^3D_2$		
54.91	30		933.62	$z^1P_1^o-z^3P_1^o$	(0.00) 1.50	(0.00) 1.48
49.02	100	III ? E	973.30	$a^3F_2-y^3D_1^o$	(0.00) 0.69	(0.00) 0.70
46.00	20	V ? E	25,993.69	$a^1D_2-y^3D_2^o$	(0.36) 1.11 B	(0.37) 1.09
40.72	60	V ? E	26,029.43	$a^1D_2-z^1P_1^o$	(0.00) 1.04	(0.00) 1.04
36.4	1		058.7	$z^1F_3^o-g^3F_4$		
35.09	50	V E	067.64	$a^1P_1-z^3P_1^o$	(0.00) 1.49.	(0.00) 1.51.
17.24	8 h		189.53	$e^1D_2-8_1^o$	(0.00) 0.82 h.	
16.25	10 h		196.33	$y^3D_1^o-g^3D_1$	(0.00 h) 0.68.	(0.13) 0.73.
14.1	2		211.1	$y^3P_1^o-e^3S_1$		
08.79	15		247.63	$a^1D_2-y^1P_1^o$	(0.00 w) 1.18 h	(0.00) 1.18
07.1	1		259.3	$a^1D_2-y^3P_1^o$		
04.8	2 h		275.2	$z^1F_3^o-f^3F_2$		
3,801.0	1		301.4	$y^3D_3^o-g^3D_3$		
3,798.19	2		320.88	$z^3F_3^o-g^3D_3$		
94.78	400	III E	344.54	$a^3F_4-y^3D_3^o$	(0.00) 1.16	(0.00) 1.18.
90.83	300	III E	371.99	$a^3F_3-y^3D_3^o$	(0.00) 1.00.	(0.00) 0.99.
84.81	15	V E	413.93	$a^3F_2-z^3F_3^o$	(0.17) 0.79.	(0.19) 0.78.
83.06	1		426.15	$e^1G-7_{3,4}^o$		
80.67	50?	V E	442.85	$a^3P_1-z^3P_1^o$	(0.00) 1.54.	(0.01) 1.51.
80.53	50?		443.84	$z^3F_3^o-f^3G_3$	(0.00 h) 0.92 h.	(0.00) 0.94.
73.12	150 l		495.77	$z^3F_4^o-f^3G_4$	(0.00) 1.00 h.	(0.00) 1.00.
68.98	3 h		524.87	$y^3D_3^o-f^3F_2$		
67.05	5 h		538.46	$y^3P_1^o-e^3P_1$	(0.47) 1.24 B.	(0.50) 1.36.
66.58	3 h		541.77	$z^3F_3^o-e^3H_4$		
59.08	300	III E	594.72	$a^3F_4-z^1F_3^o$	(0.00) 1.25.	(0.04) 1.25.
53.04	2 h		637.52	$z^3P_2^o-f^3D_2$		
47.96	5 l		673.63		(0.00 w) 0.90 h.	
44.85	2 h		695.78			
36.41	15 l		756.08	$z^1F_3^o-f^3F_4$	(0.37) 1.20 B.	(0.29) 1.20.
35.85	10	IV E	760.09	$a^1D_2-y^3P_1^o$	(0.00 w) 0.90 d	(0.00) 0.85
35.09	1		765.53	$y^3D_3^o-f^3D_3$		
31.42	8 h		791.86	$y^3D_1^o-e^3H_4$		
28.97	2 h		809.46	$y^3P_1^o-e^1P_2$		
25.06	20	IV E	837.67	$a^3F_2-z^3F_3^o$	(0, 00, 038, 0.75) 0.79, 1.16, 1.53, 1.90	(0, 00, 0.35, 0.70)−, 0.73, 1.08, 1.43, 1.78
20.75	2		868.69	$y^3D_3^o-g^1D_2$		
17.99	2		888.63	$z^3F_3^o-e^1P_1$		
15.53	50	IV E	906.44	$a^1D_2-y^3P_1^o$	(0.40, 0.72) 0.60, 1.13, 1.46, 1.79	(0.36, 0.71) 0.78, 1.14, 1.50, 1.85
14.87	40	V E	911.22	$a^3P_2-z^3P_1^o$	(0.00) 1.50	(0.00) 1.52
13.54	100	IV E	920.85	$a^1D_2-y^3D_3^o$	(0, 00, 0.34, 0.68) 1.00, 1.35, 1.69, 2.03	(0, 00, 0.32, 0.65)−, 0.99, 1.31, 1.64, 1.96
10.61	2		942.11	$z^1P_1^o-e^3S_1$		
05.81	80	V E	26,977.01	$a^3P_2-z^1P_1^o$	(0.00) 1.51	(0.02) 1.48
3,701.81	40 l	V E	27,006.15	$y^3D_3^o-f^3F_4$	(0.00 w) 0.73 A^l	(0.00) 0.70 A^l
3,696.11	2 h		047.80	$z^3F_3^o-f^3F_3$		
95.2	2 h		064.5			
94.27	7 h		061.27	$y^1D_2^o-f^1D_2$	(0.36) 0.88 h	(0.22) 0.96
92.31	2 h		075.64	$y^3P_1^o-e^3P_1$		
78.24	2 h		179.20	$z^3P_1^o-f^3D_2$		
75.22	1		201.54	$e^1F_1-8_3^o$		
70.23	4 h		238.52	$e^3G_4-8_5^o$	(0.30) 1.06	
69.27	3 h		245.65			
65.22	10 l		275.75	$z^1F_3^o-g^3D_2?$	(0.00 d) 1.92 d	(0, 00, 0.37, −)−,−,.1.51, 1.96.
62.06	30	IV E	299.14	$a^3F_2-y^3D_1^o$	(0.66) 1.16 B	(0.57) 1.20
58.40	3		326.60	$z^1F_3^o-g^1G_4$	(0.00) 1.0	(0.00) 1.17
58.04	1		329.29	$y^3D_2^o-e^1P_1$		

TABLE 4.—*The first spark spectrum of lanthanum (La II)*—Continued

λ_air I. A.	Intensity spark	Temperature class	ν_vac cm⁻¹	Term combinations	Zeeman effects Observed	Zeeman effects Computed
56. 62	1 h		369. 84	z¹P₁—e¹S₀		
50. 19	80	V E	388. 06	a³F₂—y³D₂	(0. 48, 0. 91) 0. 29, 0. 73, 1. 18, 1. 62	(0. 46, 0. 92) 0. 27, 0. 73, 1. 18, 1. 65
45. 43	200	IV E	423. 82	a³F₂—z¹P₁	(0. 00) 0. 66	(0. 00) 0. 65
41. 66	80 l		452. 21	y¹D₂—f³F₃	(0. 00) 1. 21	(0. 00) 1. 09
41. 10	2		456. 43	z¹H₁—e⁴H₃	(0. 00) 1. 01	(0. 00) 1. 02
39. 25	3 h		470. 39	z³F₂—f¹G₄		
37. 15	40	V E	486. 25	a³P₁—z³P₂	(0. 00) 1. 51	(0. 00) 1. 46
29. 99	2 hl		540. 46	z¹P₁—e³P₀		
28. 83	60	IV E	549. 27	a³F₂—z¹P₁	(0. 00 w) 1. 68 A²	(0. 00) 1. 71 A²
21. 77	4	V E	602. 97	a³D₁—y³P₂		
20. 16	1		615. 24	y¹F₃—f¹F₃		
18. 50	1		627. 15	z³D₂—f³D₁		
12. 34	60	V E	675. 02	y¹D₂—f³F₂	(0. 00) 1. 11	(0. 00) 1. 10
11. 09	2		684. 60	y³P₂—e³P₁	(0. 00) 1. 57 h	(0. 00) 1. 57
10. 25	30 l		691. 04		(0. 00) 1. 02	(0. 00) 1. 03
09. 22	4		698. 95	z¹F₃—b¹D₂	(0. 00) 1. 00	(0. 00) 1. 01
08. 18	4		708. 93	y¹D₂—g³D₁	(0. 00w) 1. 08 A²	(0. 00) 1. 12 A²
08. 42	4 bl		720. 45	y³D₂—f¹G₄		
3, 601. 07	20 hl	v E	761. 63	a³D₂—y¹P₁	(0. 00) 1. 18	(0. 00) 1. 17
3, 596. 93	3 hl		778. 14	z³D₂—f¹D₂	(0. 00) 1. 15	(0. 08) 1. 16
96. 65	4 hl	v E	795. 75	y³D₂—g¹D₂	(0. 00) 1. 22	(0. 20) 1. 14
93. 29		v E	821. 74			
92. 42	2 h		826. 48			
90. 66	1 h		842. 11	y³D₁—e³P₁		
85. 53	2		861. 95	y³P₁—e³P₂	(0. 00) 1. 25	(0. 00) 1. 20
81. 68	20 hl		911. 92	y³D₁—f³F₂	(0. 00) 0. 74	(0. 00) 0. 76
80. 10	8 h		924. 24	z³P₂—f³P₁	(0. 00) 1. 52	(0. 00) 1. 48
78. 89	5 h		933. 68	z³P₁—f³P₂	(0. 00) 1. 44	(0. 00) 1. 52
76. 56	2 hl		27, 951. 88	z³F₃—e⁴S₁		
70. 10	30 hl		28, 002. 38	z³F₃—f¹F₃	(0. 00) 1. 06	(0. 15) 1. 11
57. 26	8	IV ? E	103. 53	a¹D₂—y³P₂		
50. 82	6	IV ? E	154. 49	a³F₂—y³P₁		
36. 37	3 h		269. 53	z³F₂—e⁴H₄		
33. 67	3 h		291. 13	z³D₂—f³D₁		
30. 67	8	V ? E	315. 17	a³F₂—y¹D₂		
26. 77	2 h		346. 48	z³F₂—g¹D₂	(0. 00) 0. 88	(0. 00) 0. 88
20. 72	10 hl		395. 19	y¹D₂—f³G₂		
14. 87	2 h		442. 45	z³D₂—f³D₂		
12. 93	10	IV E	458. 15	a³D₁—y¹P₁	(0. 53) 0. 83, 1. 10	(0. 55) 0. 82, 1. 07
10. 00	15	IV E	481. 91	a³F₂—y³P₂		
3, 507. 90	4 hl		498. 96	z¹P₁—e³P₂		
3, 493. 97	2 h		612. 58	z³P₂—e³P₂		
84. 39	10 l		691. 25	z⁴O₂—f³D₂	(0. 00) 1. 33	(0. 04) 1. 32
74. 84	8 l		770. 10	z³F₂—g¹D₂		
66. 46	1 h		839. 64	y¹D₂—e³P₁		
62. 32	2 h		874. 13			
60. 31	5 l		890. 90	z⁴P₂—f¹F₃		
53. 17	50	III E	950. 63	a¹D₂—z¹F₃	(0. 93) 0. 37, 0. 68, 1. 06, 1. 31, 1. 63, 1. 94	(—, —, 0. 99) 0. 34, 0. 68, 1. 01, 1. 34, 1. 67, 2. 00
52. 18	40	III E	958. 94	a¹D₂—y¹P₁	(0. 00) 0. 95	(0. 00) 0. 95
51. 12	3 l		28, 967. 83	z³P₁—f³P₁		
32. 81	5		29, 122. 34	z³D₂—f³D₂		
27. 57	8		166. 85	y¹F₃—g³D₂		
23. 9	5		198. 1	z³F₂—f¹G₄		
22. 44	2		210. 57	y¹D₂—g³D₂		
20. 54	5 h		226. 80	y¹D₂—g³D₂		
11. 76	20 hl		302. 01	z³P₂—f³P₂	(0. 00 w) 1. 48	(0. 05) 1. 46
3, 407. 00	8 hl		342. 95	z³P₁—f³P₁	(0. 00) 1. 50	(0. 00) 1. 47
3, 398. 29	2 h		418. 15	z³D₂—f³D₂		
97. 77	40 hl	V E	422. 65	y¹D₂—f³F₂	(0. 24) 0. 85	(0. 23) 0. 83
92. 94	4 h		464. 54	z³P₁—g³D₁		
90. 40	4 h		486. 61 {	z³P₂—g³D₁ z³P₂—g³D₂		
80. 91	300	III E	569. 36	a³D₂—z¹F₃	(0. 00, 0. 37) 0. 75, 1. 11, 1. 47	(0. 00, 0. 38) 0. 76, 1. 14, 1. 52
76. 33	50	III E	609. 49	a³D₂—z¹F₃	(0. 00) 0. 88	(0. 00) 0. 87

TABLE 4.—*The first spark spectrum of lanthanum (La II)*—Continued

λ_air I. A.	Intensity spark	Temperature class	ν_vac cm⁻¹	Term combinations	Zeeman effects	
					Observed	Computed
74.89	3		622.12	z¹F₂—f¹D₂		
51.89	3		825.37			
44.56	200	III E	890.73	a³D₁—z⁴P₁	(0.00) 0.53	(0.00) 0.52
37.49	300	III E	29, 954.05	a³D₂—z⁴P₂	(0.00 w) 1.12 A¹	(0.00) 1.06 A¹
29.07	8		30, 029.81	z⁴P₂—g⁴D₂		
26.21	5		055.63	z⁴P₁—e⁴P₁		
25.23	3		063.59	z⁴D₂—g⁴D₁		
10.62	4		197.16	z⁴D₂—g⁴D₂		
06.98	8	IV E	230.40	a³F₂—z¹F₂	(0.00) 1.77 R	(0.00) 1.03
3, 203.11	150	III E	265.82	a³D₁—z⁴D₁	(1.00) 0.51, I. 51	(1.00) 0.52, 1.52
3, 296.72	5 h		306.09			
97.15	3		320.52	z⁴D₃—f²G₃		
94.44	10		345.46	z⁴P₂—f²P₂		
83.95	8 h		442.39	z⁴P₁—g⁴D₂		
77.83	4		499.23	z⁴P₂—g⁴P₁		
67.31	3		597.43	z⁴P₁—e⁴P₁		
65.67	600	III E	612.79	a³D₂—z⁴P₂	(0.48, 0.80) 0.82, 1.17, 1.51, 1.86	(0.34, 0.88) 0.80. 1.14, 1.48, 1.82
63.96	5		626.64	z⁴D₂—g⁴D₂		
53.41	10 h		728.15	z⁴D₁—g⁴D₁	(0.00) 0.65	(0.12) 0.61
49.35	80	III E	766.54	a¹D₂—z⁴P₂	(0.00, 0.50) 0.46, 0.94, 1.44	(0.00, 0.53) 0.46, 0.99, 1.52
45.13	150	III E	806.55	a¹D₂—z¹F₂	(0.00) 1.04	(0.00) 1.03
26.08	2		30, 988.93	z⁴D₂—f²D₂		
24.71	1		31, 001.62	z¹H₄—g¹G₄		
17.12	8 h		074.76	z⁴P₂—g⁴P₂		
12.56	5		118.86	z⁴P₁—e⁴S₁		
09.13	6		152.12	z⁴D₂—g¹D₂		
06.13	6		161.83			
05.75	4		184.96	a³F₂—z¹F₂		
3, 204.55	3		196.64	z⁴D₂—e¹P₁		
3, 194.70	2		292.83	z⁴D₂—g⁴D₂		
93.02	25	IV E	309.29	a³D₁—z⁴P₂	(0.00, 0.96) 0.61, 1.58, 2.54.	(0.00, 0.96) 0.52, 1.48, 2.44
91.39	10 h		325.26	z⁴P₂—h⁴D₂		
74.86	10 hl		488.15	z⁴P₂—h⁴D₂		
66.26	2		573.90	y⁴D₂—e⁴P₂		
65.19	4		584.57	y⁴D₂—h⁴D₂		
60.56	3		630.84	y⁴P₂—f²P₂		
57.58	2		660.69	z⁴P₁—e⁴S₁		
56.35	2		673.02	y¹P₁—f¹D₂		
45.7	2 h		780.3	z⁴D₂—f²F₂		
42.76	40	IV E	809.96	a¹D₂—z⁴P₂		
32.14	3		917.83	z⁴P₁—g⁴P₁		
30.26	2		937.10	z⁴P₁—g¹D₂		
26.72	4 hl		31, 983.39	z⁴P₁—e⁴P₁		
12.63	8 h		32, 117.88	z⁴P₁—g⁴P₂		
06.46	8	IV E	160.97	a³F₂—z⁴P₁		
3, 104.58	50	IV E	201.16	a³F₂—z¹F₂	(0.00 w) 1.50 A²	(0.00) 1.56 A²
3, 094.76	4 h		303.34			
88.53	4 h		368.49	z⁴P₂—h⁴D₂		
81.42	6 h		443.18	z⁴D₂—f²F₂		
75.51	4 h		505.52	z¹F₂— ¹D₂?		
69.45	3		569.69	z⁴D₂—e⁴H₄		
68.96	4		575.09	z⁴P₂—h⁴D₁		
59.91	8		671.23	z⁴P₁—f²P₀		
54.02	6		734.24	z⁴D₂—f²F₂		
49.39	5		783.94	z⁴D₂—f²F₂		
45.63	1		824.41	z⁴G₂—e⁴H₄		
36.43	2		923.86	z⁴D₂—e⁴S₁		
35.80	1		32, 930.69	z⁴G₂—f²G₄		
28.64	2		33, 008.54	z⁴P₂—f²P₂		
25.86	4 ?		038.65	z⁴G₂—f²F₄		
22.26	5 hl		078.22	z⁴D₂—g⁴D₂		
18.95	6 hl		114.48			

TABLE 4.—*The first spark spectrum of lanthanum (La II)*—Continued

λ_{air} I. A.	Intensity spark	Temperature class	ν_{vac} cm^{-1}	Term combinations	Zeeman effects	
					Observed	Computed
07.32	5		242.54	$z^3D^\circ_2-f^3F_3$		
3,004.68	5 h		271.74			
2,985.76	2		482.57	$z^3G^\circ_3-f^3F_2$		
85.43	5		486.27	$z^3F^\circ_2-h^1D_2$		
84.33	3		498.61	$z^3D^\circ_2-f^1G_4$		
83.44	3		506.61	$z^3G^\circ_3-f^3F_3$		
76.83	3		583.01	$y^3P^\circ_2-f^3P_1$		
71.48	1		643.47	$y^3F^\circ_2-f^3G_3$		
66.55	4		699.38	$z^3G^\circ_5-e^3H_6$		
66.08	2		704.72	$z^1P^\circ_1-f^3P_1$		
62.90	15		740.89	$y^3D^\circ_2-f^3P_1$		
59.85	5		775.65	$z^3G^\circ_4-e^3H_4$		
58.71	1		788.67	$z^3D^\circ_1-e^3P_1$		
51.46	3		871.66	$y^3P^\circ_1-i^1D_2$		
50.50	50	V E	882.68	$z^1H^\circ_5-e^1I_6$	(0.00) 1.04	(0.00) 1.04
48.82	1		901.99	$y^3F^\circ_3-f^1F_3$		
43.56	6 hl		33,962.57	$z^1D^\circ_2-f^3G_3$		
39.64	3 h		34,007.85	$y^1P^\circ_1-h^3D_1$		
29.86	7		121.37	$y^3D^\circ_1-f^3P_0$		
25.15	5 h		176.31	$y^1P^\circ_1-h^3D_2$		
23.90	20		190.92	$y^3D^\circ_3-f^3P_2$	(0.00) 1.23?	(0.00) 1.17
13.60	2		311.78	$y^3F^\circ_3-f^1D_2$		
10.05	1		353.63	$y^1P^\circ_1-{}^1D_2$?		
2,905.53	4 hl		407.07	$z^1D^\circ_2-e^1P_1$		
2,899.80	4 hl		475.06	$y^3F^\circ_2-g^3D_2$		
97.76	5 hl		499.32	$y^3F^\circ_2-g^3D_1$		
93.08	60	V E	555.14	$z^3G^\circ_4-e^3H_5$	(0.00) 1.12	(0.00) 1.12
89.11	1		602.61	$z^1P^\circ_1-i^1D_2$		
85.13	50	V E	650.35	$z^3G^\circ_3-e^3H_4$	(0.00) 1.04	(0.00) 1.00
83.35	1		671.74	$z^3P^\circ_2-g^3F_3$		
80.65	40	V E	704.23	$z^3G^\circ_5-e^3H_4$	(0.00) 1.18	(0.00) 1.19
76.55	1		753.69			
74.28	3		781.14	$z^3G^\circ_3-g^1D_2$		
73.20	2		794.21	$z^1D^\circ_2-g^1D_2$		
67.47	2 h		863.74	$y^3P^\circ_2-h^3D_3$		
62.98	15 hl	V E	918.41	$z^3G^\circ_4-f^3F_4$		
62.37	6		925.85	$y^3F_3-f^3F_3$		
59.76	5		34,957.73	$y^3F^\circ_2-g^3D_1$		
55.90	50 hl	V E	35,004.97	$y^1F^\circ_3-g^1G_4$	(0.00) 1.13	(0.00) 1.13
53.72	4 h		031.71	$y^3P^\circ_2-h^3D_2$		
49.51	2 h		083.47			
48.34	6		097.88	$y^3F^\circ_3-f_3G_4$		
46.67	5		118.47	$y^3D^\circ_2-f^3P_2$		
43.67	4		155.51	$y^3D^\circ_1-f^3P_1$		
40.51	25 hl		194.62	$y^3P^\circ_2-h^3D_3$		
38.45	5 l		220.16	$y^3F^\circ_2-f^3G_3$		
32.53	5		293.77	$z^3D^\circ_2-g^3F_2$		
25.82	1		377.57	$y^1F^\circ_3-h^1D_2$		
25.51	5		381.45	$z^3D^\circ_3-g^3F_3$		
21.03	5		437.64	$y^1D^\circ_2-h^1D_2$		
19.73	2		453.97	$y^3F^\circ_2-g^3D_3$		
18.40	3		470.70	$z^1D^\circ_2-e^3S_1$		
15.36	6		509.00	$z^3F^\circ_2-f^3D_2$		
13.72	5		529.70	$z^3F^\circ_3-f^3D_3$		
13.05	3		538.16	$z^1G^\circ_4-f^1F_3$		
09.35	2		584.96	$z^3F^\circ_2-f^3D_1$		
08.39	150	V E	597.13	$b^1D_2-z^1P^\circ_1$	(0.00) 1.04	(0.00) 1.04
07.20	1		612.22	$z^3F^\circ_2-i^1D_2$		
05.58	5		632.78	$z^3G^\circ_3-f^1G_4$		
2,804.55	4		645.86	$y^3F^\circ_2-f^3G_3$		
2,798.56	40 hl	V E	722.15	$z^3D^\circ_3-g^3F_4$	(0.00) 1.08	(0.00) 1.08
96.40	5		749.75	$z^1F^\circ_3-f^1D_2$		
91.51	25		812.37	$z^3D^\circ_3-g^3F_3$	(0.00) 1.04 R	(0.00) 1.03
81.25	1		944.47	$z^1D^\circ_2-f^3F_3$		
80.23	20		957.66	$z^3D^\circ_1-g^3F_2$		

TABLE 4.—*The first spark spectrum of lanthanum (La II)*—Continued

λ_{air} I. A.	Intensity spark	Temperature class	ν_{vac} cm^{-1}	Term combinations	Zeeman effects Observed	Computed
79.78	10		963.48	$y^3D_?^o-g^3P_2$		
78.76	10		35,976.68	$z^4G_?^o-g^4F_4$		
73.86	1		36,040.23	$z^4P_?^o-g^4P_0$		
67.40	8		124.35	$y^3P_?^o-g^3P_2$		
61.10	5		208.77	$y^3P_?^o-h^3D_1$		
60.51	3		214.51	$y^3D_?^o-h^3D_2$		
59.14	3		232.49	$z^4H_?^o-e^4H_3$		
58.65	3		238.93	$y^1P_?^o-f^1S_0$		
55.57	1		279.43	$z^4P_?^o-g^4P_1$		
52.84	10		315.41	$y^3D_?^o-g^3P_1$		
48.31	8		375.26	$y^3P_?^o-h^3D_2$		
36.90	2		526.90	$z^4H_?^o-e^4H_4$		
36.41	3		533.44	$y^3D_?^o-f^3P_2$		
32.40	10		587.05	$z^3G_?^o-g^3F_4$		
27.5	2		652 8	$z^3P_?^o-h^1D_2$		
26.48	1		666.49	$y^1D_?^o-f^1P_1$		
21.45	2		734.25	$z^1G_?^o-f^1G_4$		
15.43	10 hl		815.68	$y^3P_?^o-h^3D_1$		
12.51	1		855.32	$z^3P_?^o-g^3P_2$		
09.92	3		890.54	$y^3D_?^o-g^3P_3$		
06.49	2		937.29	$z^4P_?^o-h^3D_1$		
2,702.13	8		36,996.88	$z^4G_?^o-g^4F_3$		
2,695.47	35		37,088.29	$z^4H_?^o-e^4H_4$		
94.21	5		105.64	$z^4P_?^o-h^3D_2$		
91.60	1		141.62	$y^3D_?^o-h^3D_2$		
87.75	2		194.81	$z^4P_?^o-h^3D_1$		
81.49	10		281.64	$z^4H_?^o-e^4H_4$		
79.87	4 h		304.17	$y^3D_?^o-h^3D_3$		
75.66	5		362.87	$z^4D_?^o-h^1D_2$		
73.74	3		389.70	$z^4P_?^o-f^3P_0$		
72.90	30		401.45	$z^4H_?^o-e^4H_3$		
72.06	2		413.20			
70.05	2		441.37	$z^3F_?^o-g^3P_3$		
66.54	3		490.65	$y^3D_?^o-g^4P_0$		
66.18	6		495.71	$z^4H_?^o-f^3F_4$		
65.62	4		508.59	$y^1F_?^o-f^1D_2$		
64.75	3		515.83	$z^3G_?^o-g^3F_3$		
62.73	1 h		544.29			
61.66	3		559.38	$z^1F_?^o-f^3G_4$		
61.36	4		563.61	$y^1D_?^o-f^1D_2$		
49.61	1		730.18	$y^3D_?^o-g^3P_1$		
47.36	4		762.25	$z^4F_?^o-f^1F_2$		
44.70	1		800.23	$y^3F_?^o-f^1G_4$		
42.27	1		834.96	$z^1G_?^o-f^1G_3$		
40.15	1		865.37	$z^3F_?^o-g^3P_3$		
39.00	5		881.87	$z^4P_?^o-f^1P_1$		
36.66	1		915.48	$z^1F_?^o-g^1D_2$		
31.94	8		983.47	$y^1P_?^o-f^1P_1$		
31.52	4		3,7089.54	$z^4F_?^o-f^1F_2$		
20.01	7		3,8156.42	$z^4H_?^o-e^4H_3$		
17.29	1		196.07	$z^4G_?^o-g^1G_4$		
16.32	7 hl		210.23	$z^4H_?^o-f^1G_4$		
13.09	4		257.46	$z^4H_?^o-e^4H_3$		
10.34	150		297.76	$e^1S_0-z^1P_?^o$		
04.18	1		388.35	$y^3D_?^o-h^3D_1$		
02.87	1		407.66			
01.79	5		423.60	$z^4P_?^o-f^3P_1$		
00.86	2		437.34	$y^4P_?^o-f^1S_0$		
2,600.33	4		445.18	$z^4P_?^o-f^3P_1$		
2,596.32	3		504.55	$z^1D_?^o-g^3F_2$		
96.08	20		508.11	$z^1G_?^o-e^3H_4$		
92.87	3 h		555.78	$y^3D_?^o-h^3D_2$		
86.35	10		652.97	$z^4D_?^o-f^3P_0$		
82.96	8		703.70	$y^4F_?^o-g^4F_3$		
82.55	6		709.85	$z^3F_?^o-f^3G_3$		

TABLE 4.—*The first spark spectrum of lanthanum (La II)*—Continued

λair I. A.	Intensity spark	Temperature class	ν_vac cm⁻¹	Term combinations	Zeeman effects Observed	Computed
80. 82	8 hl		735. 79	$z^2F_2^2-f^2G_3$		
77. 92	2		779. 36	$z^2P_2^2-f^2D_2$		
73. 47	2 b		846. 42			
66. 09	10 hl		38, 958. 13	$z^2F_2^2-f^2G_4$		
61. 84	20 l		39, 022. 75 {	$z^1D_2^2-g^2F_2$		
				$z^2D_2^2-f^2P_1$		
60. 37	50		045. 15	$y^2F_2^2-g^2F_4$		
58. 99	3		066. 21	$z^1F_2^1-f^2F_3$		
58. 72	2		070. 33			
53. 41	3 h		151. 57	$y^1P_1^2-g^1S_0$		
52. 60	7		164. 00	$y^2F_2^2-g^2F_2$		
52. 36	2		167. 68	$z^1P_1^2-f^1S_0$		
46. 40	20 hl		259. 35	$z^2P_2^2-f^2P_2$		
42. 40	6		321. 11	$z^2P_2^2-f^1D_2$		
41. 60	4		333. 49	$z^2F_2^2-e^2II_4$		
38. 40	2		383. 07	$z^1G_2^2-e^2II_3$		
36. 76	3		406. 53	$z^2F_2^2-e^2II_4$		
34. 98	6		436. 21	$z^1G_2^2-f^2G_4$		
33. 14	15		464. 84	$a^1P_2-z^1P_1^2$		
31. 60	8		488. 84	$z^1D_2^2-f^1D_2$		
30. 26	1		509. 76	$z^2F_2^2-f^2F_2$		
27. 84	3		547. 56	$z^1F_2^2-f^1F_4$		
23. 07	5 hl		622. 34	$z^2F_2^2-f^2F_4$		
19. 22	50		682. 89	$y^1F_2^2-g^1F_2$		
15. 79	4		736. 99	$z^1F_2^2-f^1F_2$		
14. 59	3		755. 95	$y^1F_2^2-g^1F_2$		
2, 501. 18	15 hl		969. 08	$z^2D_2^2-f^2P_2$		
2, 499. 69	1		39, 992. 91			
95. 82	2		40, 054. 91			
94. 90	1		069. 68			
87. 59	40		187. 42	$y^2F_2^2-g^2F_2$		
83. 00	2hl		261. 71	$z^1F_2^2-f^1G_4$		
79. 85	10 l		312. 85	$y^2F_2^2-g^1G_4$		
74. 50	3		400. 00	$z^1D_2^2-f^2P_2$		
72. 44	10		433. 65	$z^1II_2^2-g^1F_4$		
71. 90	20		442. 49	$a^1P_0-z^1P_1^2$		
71. 06	5		456. 23	$z^2P_2^2-g^1P_1$		
70. 55	3		464. 58	$z^2F_2^2-f^2F_3$		
68. 11	1		504. 58			
58. 15	2		668. 69	$z^2F_2^2-f^2D_3$		
55. 88	10		706. 28	$y^1F_2^2-g^1F_5$		
54. 30	1		732. 48	$z^2F_2^2-e^2II_4$		
52. 73	8		758. 55	$z^1P_1^2-g^1P_0$		
51. 59	2		777. 50			
45. 56	10 h		878. 04			
43. 14	2 h		918. 53	$y^2D_2^2-f^2D_2$		
42. 80	3		924. 22			
39. 08	2		965. 63			
38. 42	10		40, 997. 82	$z^1P_1^2-g^1P_1$		
38. 02	20		41, 004. 45	$z^1D_2^2-h^1D_2$		
37. 14	10		019. 26	$z^2P_0^2-g^1P_1$		
36. 42	15		031. 37	$z^2P_2^2-g^1P_2$		
31. 40	6 h		116. 08			
24. 53	2 h		232. 58			
21. 61	5 h		282. 29	$z^1P_1^2-h^1D_2$		
20. 01	5 hl		309. 58			
17. 61	3 h		350. 59	$y^1P_1^2-g^1S_0$		
12. 08	2 h		415. 38	$z^1P_2^2-h^1D_3$		
10. 10	5 hl		479. 43			
07. 79	5 hl		519. 22			
04. 65	6		573. 43	$z^1P_1^2-g^1P_2$		
03. 29	7		596. 96	$z^1D_2^2-g^1P_1$		
2, 401. 45	2 h		628. 66			
2, 399. 64	20 hl		660. 22	$z^1G_4^2-g^1F_4$		
98. 70	3 h		676. 55			
97. 26	7 hl		701. 57	$z^1II_2^2-g^1G_4$		

TABLE 4.—*The first spark spectrum of lanthanum (La II)*—Continued

λ_air I. A.	Intensity spark	Temperature class	ν_vac cm^{-1}	Term combinations	Zeeman effects Observed	Computed
94.98	4		741.27	$z^3D_1-g^3P_2$		
93.27	2 h		771.10			
89.84	3 hl		831.04			
88.96	2		846.45			
86.28	2		893.80			
84.28	3 h		41,928.58			
75.63	2 h		42,061.24	$z^3P_1-g^1S_0$		
70.47	2		172.84	$z^1D_1-g^1P_1$		
69.18	2		195.79			
65.50	3		261.43	$z^3D_1-g^3P_1$		
58.02	3 h		395.48			
56.10	1 h		430.02			
55.81	5 h		435.24			
53.40	2		478.70			
53.03	1		485.48	$z^1F_1-g^3F_4$		
51.93	1		505.24			
48.86	2		560.79	$z^3F_1-g^3F_4$		
41.85	4		688.15	$y^1F_1-A^1D_3$		
25.75	20 hl		42,928.29	$z^1G_1-g^1G_4$		
22.78	3 h		43,038.62			
19.44	20		100.59	$a^3D_7-z^1P_1$		
17.82	20 hl		130.71	$z^1D_1-f^1D_2$		
2,311.45	1		249.56			
2,293.47	2 h		588.59			
92.32	3		610.45	$z^1D_1-f^3P_2$		
80.94	4 h		828.01	$z^1H_1-c^1I_4$		
76.06	1 h		43,921.97			
65.53	3		44,125.90	$z^3F_1-h^1D_2$		
56.77	50		297.36	$a^1D_2-z^1D_1$		
30.74	7		44,811.20	$y^3F_1-f^1D_2$		
07.08	1		45,204.56	$y^3F_1-f^3P_1$		
2,202.72	1		343.38	$z^1D_1-g^3P_2$		
2,195.91	4		524.94	$z^3F_1-A^1D_2$		
90.67	1		633.82	$z^1D_1-A^1D_1$		
87.87	40		45,692.22	$a^3F_2-z^1P_1$		
63.66	20 hl		46,203.12			
61.36	4		252.58	$z^1F_1-f^1D_2$		
42.81	20 hl		46,642.94			

IV. THE SPECTRUM OF NEUTRAL LANTHANUM (La I)

As in Sc and Y, the normal state of the neutral La atom is represented by a doublet-D term associated with the electron configuration ds^2. In each case, the next lowest energy is represented by a quartet-F term arising from d^2s, but in La it is only about one-third as high as in Sc or Y. The remaining metastable low terms also come from d^2s, and are exactly analogous to those in Sc and Y. The ^2G is narrow and inverted in all three spectra, and the predicted ^2S has been found in none. The odd levels are very numerous, and it is difficult to group them into terms. Indeed, parts of the tangle could not have been unraveled without the clues obtained from the new measurements of Zeeman effects.

The triads of quartet terms obtained by the addition of a $6p$ electron to the ^3D, ^3F, and ^3P terms of La II can be securely identified. They all lie lower than the corresponding terms of Y I. Three additional odd quartet terms, $x^4F°$, $y^4G°$, and $w^4D°$, at higher levels can come only from the addition of a $4f$ electron to the same limits. The doublet terms arising from the addition of a $6p$ electron to the various limits have also been certainly or probably identified by the intensities of their combinations and comparison of their levels with Y I. A number of supernumerary doublet terms must also arise from the $4f$ electron, which in this case is less firmly bound than the $6p$.

Many of the numbered odd levels which have not been grouped into terms are probably of similar origin. The high terms, 10° to 15°, probably arise from electrons of higher total quantum number.

The high even terms give much less conspicuous combinations in La I than in Sc I or Y I, and only a few of them have been found—but enough to give the ionization potential. The unassigned even levels have been numbered from 30 onward. If the strong bands of LaO in the yellow and red could be eliminated more such terms could probably be identified, and some fairly strong high temperature arc lines classified.

Table 5 gives the terms of the La I spectrum in the same general form as Table 3 does for La II. The configuration from which each term is believed to arise is given opposite its lowest component. For the odd terms, the probable parent terms in La II are indicated.

TABLE 5.—*Relative terms in the La I spectrum*

Electron configuration	Term	Level	Level separation	*g* Adapted weight	*g* Landé	Combinations
ds^2	$a^2D_{1\frac{1}{2}}$	0.00	1,053.20	0.790 6	0.800	$z^2F°$, $z^2D°$, $z^2D°$, $z^2F°$, $z^2P°$, $y^2P°$, $x^2P°$, $z^2G°$, $y^2D°$, $y^2P°$, $y^2F°$, $z^2F°$, $z^2G°$, $y^2D°$, $3°$, $x^2D°$, $w^2F°$, $v^2F°$, $z^2S°$, $x^2D°$, $x^2F°$, $z^2P°$, $y^2P°$, $y^2G°$, $w^2P°$, $u^2F°$, $t^2F°$, $w^2D°$, $r^2D°$, $w^2D°$, $6°$, $7°$, $8°$, $s^2F°$, $9°$, $r^2F°$, u^2D^d, $s^2P°$, $10°$, $11°$.
	$a^2D_{2\frac{1}{2}}$	1,053.20		1.203 14	1.200	
d^2s	$a^4F_{1\frac{1}{2}}$	2,668.20	341.81	.411 20	.400	$z^4F°$, $z^4D°$, $z^4D°$, $z^2F°$, $z^4P°$, $y^4F°$, $x^2P°$, $z^4G°$, $y^2D°$, $y^4P°$, $y^4F°$, $z^4F°$, $z^4P°$, $y^4G°$, $y^4D°$, $7°$, $2°$, $3°$, $w^4F°$, $x^4D°$, $4°$, $v^4F°$, $z^4S°$, $z^4D°$, $z^4F°$, $z^4P°$, $y^4P°$, $y^4G°$, $y^4G°$, $12°$, $13°$, $14°$, $15°$.
	$a^4F_{2\frac{1}{2}}$	3,010.01	484.57	1.032 11	1.029	
	$a^4F_{3\frac{1}{2}}$	3,494.58	627.03	1.222 10	1.238	
	$a^4F_{4\frac{1}{2}}$	4,121.61		1.327 8	1.333	
d^2s	$a^4P_{\frac{1}{2}}$	7,231.36	259.10	2.706 10	2.667	$z^4P°$, $y^4P°$, $z^4G°$, $y^2D°$, $z^4F°$, $z^4S°$, $z^4D°$, $3°$, $4°$, $w^4F°$, $z^4S°$, $z^4D°$, $z^4F°$, $y^4P°$, $z^4P°$, $w^4D°$, $t^4F°$, $w^4D°$, $r^4D°$, $6°$, $7°$, $8°$, $s^4F°$, $9°$.
	$a^4P_{1\frac{1}{2}}$	7,490.46	189.48	1.69 6	1.733	
	$a^4P_{2\frac{1}{2}}$	7,679.94		1.54 4	1.600	
d^2s	$a^4F_{2\frac{1}{2}}$	7,011.90	1,040.25	.90 3½	.857	$y^4F°$, $z^4G°$, $z^4P°$, $y^4D°$, $y^4F°$, $z^4F°$, $z^4D°$, $y^4D°$, $1°$, $2°$, $w^4F°$, $v^4F°$, $z^4D°$, $z^4F°$, $z^4D°$, $t^4F°$, $y^4G°$, $y^4G°$, $w^4P°$, $u^4F°$, $t^4F°$, $w^4D°$, $r^4D°$, $w^4D°$, $7°$, $s^4F°$, $r^4F°$, $w^4D°$, $10°$, $11°$.
	$a^4F_{3\frac{1}{2}}$	8,052.15		1.13 5	1.143	
d^2s	$b^2D_{1\frac{1}{2}}$	8,446.03	737.74	.89 3	.800	$z^4G°$, $z^4F°$, $y^2D°$, $y^4P°$, $y^4F°$, $z^4S°$, $z^4D°$, $4°$, $s^4F°$, $z^4S°$, $z^4D°$, $z^4F°$, $z^4P°$, $x^4P°$, $w^4P°$, $w^4P°$, $u^4F°$, $5°$, $w^4D°$, $t^4F°$, $v^4D°$, $w^4D°$, $8°$, $s^4F°$, $9°$, $r^4F°$, $w^4D°$.
	$b^2D_{2\frac{1}{2}}$	9,183.77		1.29 3	1.200	
d^2s	$a^2P_{\frac{1}{2}}$	9,044.21	675.23	.67 1	.667	$y^2P°$, $y^4P°$, $y^2F°$, $z^2F°$, $z^2S°$, $z^2D°$, $w^2F°$, $4°$, $x^4F°$, $z^2D°$, $z^4F°$, $z^2P°$, $y^2P°$, $w^2P°$, $5°$, $w^2D°$, $w^2P°$, $v^2D°$, $u^2D°$, $t^2P°$.
	$a^2P_{1\frac{1}{2}}$	9,719.44		1.32 2	1.333	
d^2s	$a^2G_{3\frac{1}{2}}$	9,919.94	−41.02	1.12 1	1.111	$y^2F°$, $z^2F°$, $z^2G°$, $y^2G°$, $w^2F°$, $z^2H°$, $z^2F°$, $y^2G°$, $w^2F°$, $y^2G°$, $w^2D°$, $t^2F°$, $s^2F°$, $9°$, $t^2F°$, $y^2H°$.
	$a^2G_{4\frac{1}{2}}$	9,960.96		.91 1	.880	
$a^2D.p$	$z^4F_{1\frac{1}{2}}$	13,260.36	370.72	1.08	1.029	a^2D, a^4F, 31, 32, 35, e^4F, e^4D.
	$z^4F_{2\frac{1}{2}}$	13,631.08	1,388.47	1.21	1.238	
	$z^4F_{3\frac{1}{2}}$	15,019.55	1,223.70			
	$z^4F_{4\frac{1}{2}}$	16,243.25				
$a^2D.p$	$z^4D_{\frac{1}{2}}$	14,095.70	613.26			a^2D, a^4F, 31, 32, 33, 34, e^4F, e^4D.
	$z^4D_{1\frac{1}{2}}$	14,708.96	794.71			
	$z^4D_{2\frac{1}{2}}$	15,503.67	595.61			
	$z^4D_{3\frac{1}{2}}$	16,099.28				
$a^2D.p$	$z^2D_{1\frac{1}{2}}$	14,804.10	−227.55	1.08	1.200	a^2D, a^4F, 31, e^4F.
	$z^2D_{1\frac{1}{2}}$	15,031.65		.97	.800	
$a^2D.p$	$z^2F_{2\frac{1}{2}}$	15,196.80	1,341.64	.90	.857	a^2D, a^4F, e^4D, e^2F.
	$z^2F_{3\frac{1}{2}}$	16,538.44		1.19	1.143	
$a^2S.p$	$z^2P°_{\frac{1}{2}}$	15,219.90	1,060.30			a^2D, a^4F.
	$z^2P°_{1\frac{1}{2}}$	16,280.20				

TABLE 5.—*Relative terms in the La I spectrum*—Continued

Electron config-uration	Term	Level	Level separa-tion	g Adapted weight	g Landé	Combinations
a³D.p	y³F₃½ y³F₄½	16,856.82 17,910.18	} 1,053.36	0.83 1.08	0.857 1.143	a²D, a⁴F, a²F, a⁴P.
a³D.p	z⁴P°½ z⁴P1½ z⁴P2½	17,567.56 17,797.30 18,157.00	229.74 359.70	1.20	1.600	a²D, a⁴F, a²F, a⁴P, b³D.
a³F.p	z⁴G2½ z⁴G3½ z⁴G4½ z⁴G5½	17,947.16 18,603.95 19,129.34 20,117.40	656.79 525.39 988.06	1.07 1.09 1.15 1.27	.571 .984 1.172 1.273	a²D, a⁴F, a²F, a⁴P, b³D, c⁴F, 35.
a²D.p	y²D1½ y²D2½	18,172.39 19,379.44	} 1,207.05	.83 1.19	.800 1.200	a²D, a⁴F, a²F, a⁴P, b³D, a²P, e⁴F.
a³D.p	y²P½ y²P°1½	20,019.00 20,197.38	} −178.38	1.05 .60	1.333 .667	a²D, a⁴F, b³D, a³P, e⁴F.
a³F.p	y⁴F1½ y⁴F2½ y⁴F3½ y⁴F4½	20,083.02 20,338.30 20,763.31 21,384.06	255.28 425.01 620.75	.73 1.01 1.18 1.28	.400 1.029 1.238 1.333	a²D, a⁴F, a²F, a⁴P, b³D, a³G, c⁴F, e⁴F.
a³F.p	z²F2½ z²F3½	20,972.22 21,662.61	} 690.39	.90 1.13	.857 1.143	a²D, a⁴F, a²F, a⁴P, a³P, a³G, e⁴F.
a³F.p	z²G3½ z²G4½	21,447.92 22,285.85	} 837.93	1.08 1.13	.889 1.111	a²D, a⁴F, a²F, a³G, e²F.
a³F.p	y⁴D°½ y⁴D1½ y⁴D2½ y⁴D3½ 1½ 2½	22,246.64 22,439.37 22,804.26 23,303.31 23,221.16 23,466.85	192.73 364.89 499.05	.03 1.21 1.35 1.22 1.11 1.12	.000 1.200 1.371 1.429 1.111	a²D, a⁴F, a²F. a⁴F, a²F. a⁴F, a²F, a²G.
a⁴P.p	z²S°½ 3½ 4½	23,260.90 23,549.43 24,173.86		1.87 .72	2.000	a⁴P, b³D, a²P. a⁴F, a⁴P. a⁴F, a²P, b³D, a²P.
a²P.p	z⁴D°½ z⁴D1½ z⁴D2½ z⁴D3½	23,528.38 23,704.76 24,046.06 25,083.42	176.38 341.30 1,037.36	.00 1.15 1.22 1.41	.000 1.200 1.371 1.429	a²D, a⁴F, a²F, a⁴P, b³D, a²P.
a³D.f	w²F1½ w²F2½	23,875.00 24,409.70	} 534.70	.94 1.17	.857 1.143	a²D, a⁴F, a²F, a⁴P, a³P, a²G.
	c⁴F1½(?) b³F2½(?)	24,507.89 25,378.46	} 870.57	1.20 1.13	.857 1.143	a²D, a⁴F, a²F, a⁴P, b³D, a²P.
a⁴P.p a⁴F.p	z⁴S1½ z²D1½ z²D2½	24,639.27 24,762.62 25,218.25	} 455.63	1.81 .88 1.25	2.000 .800 1.200	a²D, a⁴F, a⁴P, b³D. a²D, a⁴F, a²F, a⁴P, b³D.
a⁴F.f	z⁴F1½ z⁴F2½ z⁴F3½ z⁴F4½	24,910.39 24,984.33 25,380.33 25,997.27	73.94 396.00 616.94	.73 1.11 1.23 1.36	.400 1.029 1.238 1.333	a²D, a⁴F, a²F, a²G.
a²G.p	z²H1½ z²H2½	25,089.50 25,874.68	} 785.18	.94	.909	a⁴F, a²G.
a³D.p	z²P°½ z²P1½ 31 32½	25,453.92 25,950.39 25,568.49 25,881.53	} 496.47	.92 1.41	.667 1.333	a²D, a⁴F, a⁴P, b³D, a²P. z²D°, z²F°, z²D°. z²D°, z⁴D°, z⁴F°.
a⁴P.p	y⁴P°½ y⁴P1½ y⁴P2½	25,646.90 25,642.02 26,338.90	26.12 695.88	2.29 1.61 1.54	2.667 1.733 1.600	a²D, a⁴F, a²F, a⁴P, b³D, a³P.
a²F.f	y²G2½ y²G3½ y²G4½ y²G5½	27,022.60 27,455.34 28,089.18 28,743.21	432.74 633.84 654.03	.57 .99 1.16 1.29	.571 .984 1.172 1.273	a²D, a⁴F, a²F, a²G.

TABLE 5.—*Relative terms in the La I spectrum*—Continued

Electron configuration	Term	Level	Level separation	g Adapted weight	g Landé	Combinations
a¹G. p	y²G₁½ y²G₁½	27,132.50 27,619.69	487.19	0.95 1.12	0.889 1.111	}a¹F, a²F, a¹G.
b¹D. p b¹D. p	w²P₁½ u²F₁½ u²F₁½	27,225.27 27,393.00 28,039.54	640.54	1.34 .90 1.13	1.333 .857 1.143	a²D, a²F, b¹D, a²P. }a²D, a²F, a¹G.
f	t²F₁½ t²F₁½ 5|½	27,569.38 28,543.10 27,749.05	873.72	.88 1.00	.857 1.143	}a²D, a²F, a¹G. b¹D, a²P.
b¹D. p	w²D₁½ w²D₁½	27,968.53 28,506.39	537.86	.91	.800	}a²D, a²F, a²P, b¹D, a²P.
a²P. f	w⁴D₆½ w⁴D₁½ w⁴D₂½ w⁴D₃½	28,893.47 29,199.53 29,502.17 29,894.91	306.06 302.64 392.74	0.00 1.15 1.20	0.000 1.200 1.429	}a²D, a²F, a²P, b²D.
f	w²D₁½ w²D₁½ 33 6°½ 7|½	28,971.82 29,775.57 29,461.33 29,564.92 29,936.73	803.75	.94 1.13 1.54	.800 1.200	}a²D, a²F, a⁴P, b²D, a²P. z²D°. a²D, a²P, b²D, a²P. a²D, a²F, a²P.
	8|½ 34(?)	30,417.47 29,594.81		1.54		a²D₁ a⁴P₁ b²D. z²D°.
a²F.s	e⁴F₁½ e⁴F₂½ e⁴F₃½ e⁴F₄½	29,874.89 30,354.32 31,059.69 31,923.90	479.43 705.37 864.21			}z²F°, z²D°, y²F°, z²G°.
a¹G.p	s²F₁½ s²F₁½ 9|½	30,783.40 30,064.82 30,896.88	176.42	1.18 1.45	1.143	}a²D, a⁴F, a²F, b²D, a²G. }a²D, a⁴P, b²D, a¹G.
a²F.s	e²F₂½ e²F₃½	31,119.08 32,108.58	989.50			}z²D°, z²F°, y²F°, z²G°.
a¹D.s	e⁴D₃½	31,287.65				z²F°, z²D°, z²F°. .
f	r²F₁½ r²F₁½	31,477.16 32,140.60		.88 1.18	.857 1.143	}a²D, a²F, b²D, a²G.
a²P.p	u²D₁½ u²D₁½	31,751.68 32,492.80	741.12	.82 1.17	.800 1.200	}a²D, a²F, b²D, a²P.
a²P.p	s⁴P°₁½ s²P₁½	32,290.25 33,204.20	913.95	1.26	1.333	}a²D, b²D, a²P.
a¹G.f	y¹I₁₁½ y¹I₁₁½ 35 10° 11° 12°(?) 13° 14° 15°	32,410.76 32,518.12 33,286.50 36,722.38 37,731.90 39,597.58 39,631.27 40,322.45 40,343.40	107.36	.93 1.11	.909 1.091	}a²G. z²F°, z²F°, y²F°, z²G°. a²D, a²F. a²D, a²F. a⁴F. a⁴F. a⁴F. a⁴F.

The spectral terms and combinations for La I are represented diagrammatically in Figure 2, which may be compared with corresponding diagrams for Sc I and Y I spectra in publications already referred to.[25]

Complete details of lines regarded as characteristic of neutral lanthanum atoms are given in Table 6, the arrangement and notation being the same as that in Table 4. Here again the Zeeman patterns computed from the adopted g values are almost always in good agreement with observation.

[25] See footnotes 2 and 3, p. 625.

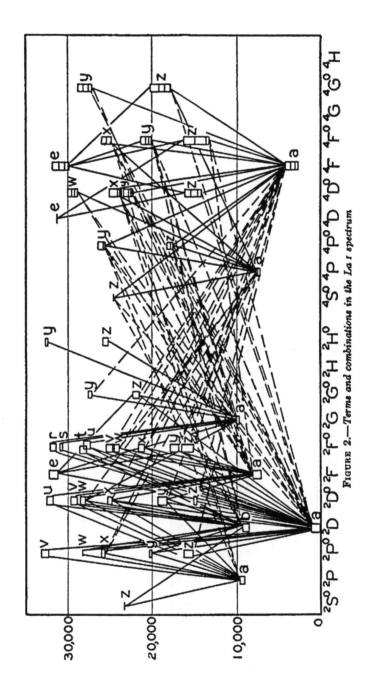

FIGURE 2.—Terms and combinations in the La I spectrum

TABLE 6.—*The arc spectrum of lanthanum (La I)*

λ air I. A.	Intensities, temperature class		ν vac cm⁻¹	Term combination	Zeeman effects	
	B. S.	K. & C.			Observed	Computed
10,952.0	1		9,128.25	$a^2P_{1/2}$—$y^2D_{1 1/2}$		
10,739.66	5		9,308.73			
10,612.86	10		9,420.21	$b^2D_{2 1/2}$—$z^2G_{3 1/2}$		
10,552.41	6		9,473.91			
10,522.09	10		9,501.21	$b^2D_{1 1/2}$—$z^2G_{3 1/2}$		
10,483.0	2		36.6	$y^2F_{3 1/2}$—$e^2F_{1 1/2}$		
61.69	15		56.07			
50.83	20		66.01			
23.4	1		9,591.2	$y^2F_{3 1/2}$—$e^2F_{2 1/2}$		
10,409.55	3		9,602.93			
10,372.4	1		38.4	$z^2F_{1 1/2}$—32		
57.70	20		52.01			
49.08	40		60.05	$a^2P_{1 1/2}$—$y^2D_{1/2}$		
37.20	3		71.15	$z^2G_{3 1/2}$—$e^2F_{3 1/2}$		
32.40	2		75.64	$y^2F_{1 1/2}$—$e^2F_{3 1/2}$		
30.3	1		77.6	$y^2P_{1 1/2}$—$e^2F_{1 1/2}$		
10,318.2	2		9,689.0			
10,294.68	10 d		9,711.09	$b^2D_{1 1/2}$—$z^2P_{3 1/2}$		
85.64	3		21.52			
81.34	10		23.69			
78.52	3		26.37	$b^2D_{1 1/2}$—$y^2D_{1 1/2}$		
74.85	10		29.84			
34.78	2		57.93			
23.76	1		78.46			
19.83	3		82.22	$z^2D_{1/2}$—32		
10,209.85	2		9,791.78	$y^2F_{1 1/2}$—$e^2F_{1 1/2}$		
10,184.60	20		9,816.06			
77.74	6 h		22.67	$z^2G_{3 1/2}$—$e^2F_{1 1/2}$		
54.74	40		44.92	$a^2F_{3 1/2}$—$y^2F_{1 1/2}$		
43.38	2		55.95	$y^2P_{1 1/2}$—$e^2F_{1 1/2}$		
41.20	10		56.06	$a^2F_{3 1/2}$—$y^2F_{3 1/2}$		
30.82	5		68.17			
10,111.9	2 h		9,886.6			
10,083.96	2		9,914.03			
66.77	6		30.95			
58.79	2		34.84			
.54.82	2		42.76			
29.74	2		67.62			
14.45	4		82.84			
10,005.73	50		9,991.54			
9,988.47	10		10,008.80			
81.24	6		016.05	$y^2F_{3 1/2}$—$e^2F_{2 1/2}$		
80.38	10		016.91			
65.70	3		031.67			
32.72	2		064.98	$z^2D_{1/2}$—31		
20.82	150		077.06	$a^2P_{1 1/2}$—$z^2P_{1/2}$		
9,911.06	3		085.95			
9,881.24	100		117.42	$a^2P_{2 1/2}$—$z^2P_{1 1/2}$		
62.60	3		136.54	$a^2F_{1 1/2}$—$z^2F_{1 1/2}$		
52.58	6		146.85	$z^2F_{1 1/2}$—$e^2F_{2 1/2}$		
48.70	4		150.84			
42.0	2		157.8			
33.30	3 h		166.74			
06.2	1 h		195.9	$b^2D_{2 1/2}$—$y^2D_{1 1/2}$		
9,804.20	2		196.92			
9,775.09	8		227.28			
72.24	20		230.26	$a^2P_{2 1/2}$—$y^2F_{1 1/2}$		
66.82	3 h		233.85			
37.09	100		267.20	$a^2P_{2 1/2}$—$z^2G_{3 1/2}$		
13.52	3		292.11			
09.45	10		296.42	$y^2F_{1 1/2}$—$e^2F_{3 1/2}$		
9,706.48	20		299.57	$a^2P_{1 1/2}$—$y^2P_{1 1/2}$		
9,699.64	20		306.84	$a^2P_{1 1/2}$—$z^2P_{1 1/2}$		
96.7	1		310.0			
92.6	2		314.3			

TABLE 6.—*The arc spectrum of lanthanum (La I)*—Continued

λ air I. A.	Intensities, temperature class		ν vac cm⁻¹	Term combination	Zeeman effects	
	B. S.	K. & C.			Observed	Computed
72. 04	8		336. 25	a⁴P₁₄—z⁴P½₄		
46. 47	3		363. 65	a²P₁₁₄—y⁴F½₄		
40. 81	30		369. 73			
23. 72	40		377. 36	a²G₃₁₄—y⁴F½₄		
9, 631. 84	2		379. 38			
9, 570. 36	5		446. 04	z²P½₄—e⁴F₃½₄		
60. 69	10		456. 63	a⁴P₁½₄—z⁴G½₄		
42. 06	80		477. 05	a⁴P₁½₄—z⁴P½₄		
41. 23	20		477. 96	a²P₁½₄—y²P½₄		
9, 528. 0	1 h		492. 5	a⁴P₁½₄—y²D½₄		
9, 485. 15	15		539. 91	y⁴F½₄—e⁴F₀½₄		
84. 2	1		541. 0			
76. 98	3		548. 99	z⁴F½₄—31		
74. 45	5		551. 81	a²F₃½₄—z⁴G½₄		
67. 25	2		559. 84			
61. 82	60		565. 89	a⁴P ½₄—z⁴P½₄		
57. 62	2		570. 59			
41. 7	1		588. 4			
38. 30	100		592. 23	a⁴F₁½₄—z⁴F½₄		
15. 64	3		617. 71			
9, 412. 65	100		621. 08	a⁴F₃½₄—z⁴F½₄		
9, 398. 2	1		637. 4			
90. 56	4		646. 07			
77. 71	3		660. 66	z²G₁½₄—e⁴F₃½₄		
76. 10	3		662. 49			
72. 57	30		666. 51	a⁴P₁½₄—z⁴P½₄		
28. 87	2		716. 47			
24. 5	1		721. 5	y⁴F½₄—e⁴F₃½₄		
9. 321. 9	1 h		734. 5	y⁴F½₄—e²F₃½₄		
9, 293. 3	2 h		757. 5			
87. 5	1 h		764. 2	z²D½₄—31		
54. 70	10		802. 36	a²G₃½₄—y⁴F½₄		
50. 06	20		807. 78			
28. 63	30		835. 22	b²D₃½₄—y²P½₄		
9, 219. 64	10		843. 44	a²G₄½₄—y⁴F½₄		
9, 172. 86	5		898. 71			
72. 39	10		899. 30	b²D₃½₄—y⁴F½₄		
57. 13	10		917. 46			
51. 62	6		924. 03	a⁴P₃½₄—z⁴G½₄		
43. 78	5		933. 40	b²D₁½₄—y²D½₄		
42. 24	6		935. 24	a²D₁½₄—z⁴G½₄		
19. 18	20		962. 89	a⁴F₁½₄—z⁴F½₄		
9, 109. 25	2		10, 974. 84	a²P½₄—y²P½₄		
9, 079. 10	50		11, 011. 29	a²G₄½₄—z²F½₄		
58. 63	2		036. 17	y⁴F½₄—e⁴F₃½₄		
56. 53	5		038. 73	a²P½₄—y⁴F½₄		
46. 97	2		050. 39			
25. 05	4		077. 23	a²F₃½₄—z⁴G½₄		
9, 008. 26	6		097. 88			
8, 977. 39	2		136. 04	z²F½₄—e²F₃½₄		
70. 07	3		145. 13	a²F₃½₄—z⁴P½₄		
65. 41	2		150. 92			
63. 63	10		153. 13	a²P½₄—y²P½₄		
57. 74	50		160. 47	a²F₃½₄—y²D½₄		
48. 89	2		171. 50			
47. 96	1		172. 68			
8, 917. 70	1 h		210. 58			
8, 891. 06	1		244. 17			
84. 24	2		252. 80	a²P₁½₄—z⁴F½₄		
79. 56	3		258. 73			
75. 05	2		264. 45			
71. 00	4		269. 60			
67. 35	3		274. 24			
39. 64	20		309. 57	a⁴F₁½₄—z²D½₄		
26. 86	50		337. 23	a²F₁½₄—y²D½₄		

TABLE 6.—*The arc spectrum of lanthanum (La I)*—Continued

λ air I. A.	Intensities, temperature class		ν vac cm⁻¹	Term combination	Zeeman effects	
	B. S.	K. & C.			Observed	Computed
21.66	1 h		332.62			
8,818.96	20		336.10			
8,797.6	2 h		363.6			
72.02	1		396.76			
67.92	4		402.07			
60.4	1		411.9			
48.42	50		427.50	$a^4F_{1\frac{1}{2}}-z^4D_{\frac{1}{2}}$		
24.12	1		459.33			
20.42	20		464.19	$a^2G_{4\frac{1}{2}}-y^4P_{3\frac{1}{2}}$		
8,703.13	5		486.97	$a^2G_{3\frac{1}{2}}-z^2G_{3\frac{1}{2}}$		
8,674.40	60		525.01	$a^4F_{3\frac{1}{2}}-z^4F_{3\frac{1}{2}}$		
72.10	30		528.07	$a^4G_{4\frac{1}{2}}-z^2G_{3\frac{1}{2}}$		
38.4	10		573.0	$b^3D_{1\frac{1}{2}}-y^4P_{1\frac{1}{2}}$		
24.22	6		592.07	$a^2F_{3\frac{1}{2}}-z^4G_{2\frac{1}{2}}$		
8,621.55	2		595.66			
8,590.97	6		636.94	$b^3D_{1\frac{1}{2}}-y^4F_{1\frac{1}{2}}$		
88.9	1		680.5	$y^2D_{3\frac{1}{2}}-e^4F_{3\frac{1}{2}}$		
45.44	50		698.94	$a^4F_{3\frac{1}{2}}-z^4D_{1\frac{1}{2}}$		
43.46	20		701.65	$a^2G_{3\frac{1}{2}}-z^4F_{3\frac{1}{2}}$		
29.68	3		720.55			
13.55	15		742.76	$a^2G_{4\frac{1}{2}}-x^4F_{3\frac{1}{2}}$		
8,507.37	10		751.29	$b^3D_{1\frac{1}{2}}-y^4P_{0\frac{1}{2}}$		
8,476.48	30		794.12	$a^4F_{1\frac{1}{2}}-z^2D_{1\frac{1}{2}}$		
67.62	15		800.45	$z^4G_{1\frac{1}{2}}-e^4F_{0\frac{1}{2}}$		
8,440.06	3		845.01			
8,379.80	20		930.18	$z^4G_{1\frac{1}{2}}-e^4F_{1\frac{1}{2}}$		
88.50	2 h		960.59			
46.60	100	12, III	977.64	$a^4F_{4\frac{1}{2}}-z^4D_{3\frac{1}{2}}$		
34.44	3	15, III	11,995.11			
24.72	100		12,009.12	$a^4F_{3\frac{1}{2}}-z^4D_{3\frac{1}{2}}$		
24.59	HNR	1, III A	009.30	$a^4F_{3\frac{1}{2}}-z^4F_{1\frac{1}{2}}$		
16.05	10	2, IV A	021.64	$a^4F_{3\frac{1}{2}}-z^4D_{1\frac{1}{2}}$		
8,302.82	4		040.80	$a^4F_{1\frac{1}{2}}-z^4D_{1\frac{1}{2}}$		
8,247.46	60	10, III	121.62	$a^4F_{4\frac{1}{2}}-z^4F_{4\frac{1}{2}}$		
37.90	3		135.68	$a^4F_{1\frac{1}{2}}-z^2D_{2\frac{1}{2}}$		
11.65	2		174.48			
8,203.38	3		186.75	$a^4F_{3\frac{1}{2}}-z^2F_{2\frac{1}{2}}$		
8,086.10	20+M	15, III	363.50	$a^4F_{1\frac{1}{2}}-z^2D_{1\frac{1}{2}}$		
84.53	3		365.92	$a^2G_{4\frac{1}{2}}-z^2G_{3\frac{1}{2}}$		
51.38	10	10, III	416.82	$a^4F_{1\frac{1}{2}}-z^2F_{1\frac{1}{2}}$		
8,001.91	4	4, III A	493.58	$a^4F_{3\frac{1}{2}}-z^4D_{3\frac{1}{2}}$		
7,964.86	5	6, III	551.70	$a^4F_{1\frac{1}{2}}-z^2P_{0\frac{1}{2}}$		
48.30	HNR	10, III	577.85	$a^2D_{2\frac{1}{2}}-z^4F_{1\frac{1}{2}}$		
7,931.18	1		605.00	$a^4F_{1\frac{1}{2}}-z^4D_{1\frac{1}{2}}$		
7,864.96	1		711.10	$a^2F_{3\frac{1}{2}}-y^4F_{1\frac{1}{2}}$		
7,841.76	3		748.74	$a^4F_{1\frac{1}{2}}-z^4F_{1\frac{1}{2}}$		
7,737.74	1 h		12,920.12	$a^2F_{3\frac{1}{2}}-z^2D_{1\frac{1}{2}}$		
7,664.38	4	? III	13,043.78	$a^4F_{1\frac{1}{2}}-z^2F_{1\frac{1}{2}}$		
7,539.24	10	15, II	260.29	$a^2D_{1\frac{1}{2}}-z^4F_{1\frac{1}{2}}$		
33.64	2		270.14	$a^4F_{3\frac{1}{2}}-z^4P_{2\frac{1}{2}}$		
7,501.78	1		326.50	$a^2F_{3\frac{1}{2}}-y^4F_{2\frac{1}{2}}$		
7,498.82	2	5, III A	331.76	$a^2F_{3\frac{1}{2}}-y^4F_{1\frac{1}{2}}$		
7,463.08	5	10, III	395.61	$a^2F_{2\frac{1}{2}}-z^4G_{3\frac{1}{2}}$		
7,382.73	6	5, III A	541.40	$a^2P_{1\frac{1}{2}}-z^2S^o_{\frac{1}{2}}$		
79.71		10, III A	546.94	$a^2G_{4\frac{1}{2}}-y^2G_{3\frac{1}{2}}$		
45.36	25	20, III A	610.30	$z^2F_{1\frac{1}{2}}-z^2F_{3\frac{1}{2}}$	(0.42) 1.18 ?	(0.39) 1.06
44.42	HNR	2, III A	612.03	$a^4F_{1\frac{1}{2}}-z^2P_{1\frac{1}{2}}$		
34.18	50	60, II	631.04	$a^2D_{1\frac{1}{2}}-z^4F_{1\frac{1}{2}}$	(0.00w) 1.30 w	(0.00) 1.30 !
20.90	2		655.77	$a^2D_{1\frac{1}{2}}-z^2D_{1\frac{1}{2}}$		
7,308.46	2 h		679.01	$a^2S_{\frac{1}{2}}-z^4S_{1\frac{1}{2}}$		
7,285.83	1		721.50			
70.30	HNR	5, II A	750.81	$a^2D_{2\frac{1}{2}}-z^2D_{2\frac{1}{2}}$		
70.07	HNR	10, III	751.24	$a^2F_{3\frac{1}{2}}-y^4F_{1\frac{1}{2}}$		
63.66	2		763.34			
62.83	2		764.95			

TABLE 6.—*The arc spectrum of lanthanum (La I)*—Continued

λ air I. A.	Intensities, temperature class		ν vac cm⁻¹	Term combination	Zeeman effects	
	B. S.	K. & C.			Observed	Computed
50.38	1		788.59	$a^4F_{4\frac{1}{2}}-y^2F^\circ_{3\frac{1}{2}}$		
19.92	15	15, II A	846.76	$a^4F_{3\frac{1}{2}}-y^2F^\circ_{3\frac{1}{2}}$		
7,217.16	2		852.06			
7,161.25	40	50, III	960.20	$a^2F_{3\frac{1}{2}}-z^2F^\circ_{3\frac{1}{2}}$	(0.00) 0.91	(0.00) 0.90
58.11	30	50, III A	966.32	$a^2D_{1\frac{1}{2}}-z^4F^\circ_{1\frac{1}{2}}$	(0.00) 1.27	(0.00) 1.22
51.92	HNR	1, III A	978.42	$a^2D_{2\frac{1}{2}}-z^2D^\circ_{1\frac{1}{2}}$		
7,149.76	2	2, III	13,962.63	$a^4P_{2\frac{1}{2}}-z^2F^\circ_{3\frac{1}{2}}$		
7,098.18	2		14,084.24			
92.40	HNR	2, I A	095.72	$a^2D_{1\frac{1}{2}}-z^4D^\circ_{1\frac{1}{2}}$		
76.36	3	3, III	127.67			
68.37	60	100, II	143.64	$a^2D_{2\frac{1}{2}}-z^2F^\circ_{3\frac{1}{2}}$	(0.69)?	(0.66) 1.05
62.4	1 h		155.6	$a^2P_{1\frac{1}{2}}-w^2F^\circ_{2\frac{1}{2}}$		
45.96	200	300, II	188.62	$a^4F_{1\frac{1}{2}}-y^2F^\circ_{2\frac{1}{2}}$	(0.21, 0.65) 1.04, 1.47	(0.21, 0.62) 0.20, 0.63 1.04, 1.46
32.07	25	15, III	216.65	$a^2P_{\frac{1}{2}}-z^2S^\circ_{\frac{1}{2}}$	(0.60) 1.27	(0.60) 1.27
23.67	150	150, III	233.66	$a^2F_{3\frac{1}{2}}-z^2G^\circ_{4\frac{1}{2}}$	(0.00) 1.16	(0.00) 1.16
7,013.15	2		255.00	$z^4D^\circ_{1\frac{1}{2}}-e^4F_{2\frac{1}{2}}$		
6,978.09	2		326.62	$a^2P_{1\frac{1}{2}}-z^4D^\circ_{2\frac{1}{2}}$		
76.87	2	3, III	329.13			
35.03	50	50, III	415.58	$a^4F_{3\frac{1}{2}}-y^2F^\circ_{3\frac{1}{2}}$	(0.48) 1.14	(0.41) 1.15
25.27	80	100, III	435.90	$a^2F_{2\frac{1}{2}}-z^2G^\circ_{3\frac{1}{2}}$	(0.00w) 1.58 A[2]	(0.00) 1.53 A[2]
18.33	8	8, III A	450.38	$a^2D_{2\frac{1}{2}}-z^4D^\circ_{3\frac{1}{2}}$		
17.26	10	10, III A	452.61	$a^4F_{3\frac{1}{2}}-z^4G^\circ_{3\frac{1}{2}}$		
03.08	2	1, III A	482.30	$a^4F_{4\frac{1}{2}}-z^4G^\circ_{4\frac{1}{2}}$		
6,902.08	3	1, III	484.40	$a^2P_{\frac{1}{2}}-z^4D^\circ_{\frac{1}{2}}$		
6,889.63	2 h	tr, V	489.54	$a^2G_{4\frac{1}{2}}-w^2F^\circ_{3\frac{1}{2}}$		
98.41		tr, V	492.10	- - - - - - - -		
23.80	40	15, II A	650.56	$a^2F_{2\frac{1}{2}}-z^2F^\circ_{2\frac{1}{2}}$	(0.00) 1.15	(0.00) 1.12
21.51	5 d	3n, V	655.47			
19.14	1		660.57	$a^2P_{\frac{1}{2}}-z^4D^\circ_{1\frac{1}{2}}$		
6,816.29		1n, V?	666.70			
6,798.73	4	4, II A	708.90	$a^2D_{1\frac{1}{2}}-z^4D^\circ_{1\frac{1}{2}}$		
83.55	1	tr, IV A	737.48			
78.19	1	tr, V	749.14	$z^2F^\circ_{3\frac{1}{2}}-e^4D_{3\frac{1}{2}}$		
76.69	1	1, III A	752.40	$z^4O_{1\frac{1}{2}}-33$		
60.73	HNR	1, III A	787.23	$a^4F_{2\frac{1}{2}}-z^4P^\circ_{1\frac{1}{2}}$		
60.23	1		788.32	$a^2P_{1\frac{1}{2}}-e^2P^\circ_{1\frac{1}{2}}$		
53.05	40	50, I A	804.04	$a^2D_{1\frac{1}{2}}-z^2D^\circ_{2\frac{1}{2}}$	(0.00 w) 1.51 A[2]	(0.00) 1.51 A[2]
48.12	10	10, II A	814.86	$b^2D_{1\frac{1}{2}}-z^2S^\circ_{\frac{1}{2}}$		
41.20	2		830.07			
37.29	1		838.67			
15.96		1, IVA?	885.80	$z^4D^\circ_{1\frac{1}{2}}-34$		
6,709.49	150	200, I	900.16	$a^4F_{2\frac{1}{2}}-y^2F^\circ_{3\frac{1}{2}}$	(0.00) 1.18	(0.00) 1.14
6,699.86	3	4, III A	921.57			
99.26	2	2, III?	922.91			
92.86	20	30, I A	14,937.18	$a^4F_{2\frac{1}{2}}-z^4G^\circ_{3\frac{1}{2}}$	(0.00) 1.06	(0.09) 1.05
64.45	1		15,000.85			
61.40	60	80, I A	007.72	$a^4F_{4\frac{1}{2}}-z^4G^\circ_{4\frac{1}{2}}$	(0.65) 1.23 B?	(0.66) 1.24
58.06	1		015.25	$a^4P_{\frac{1}{2}}-y^4D^\circ_{\frac{1}{2}}$		
50.81	80	100, I A	031.62	$a^2D_{1\frac{1}{2}}-z^2D^\circ_{1\frac{1}{2}}$	(0.00 w) 0.88	(0.25) 0.88
45.15	7	8, IV	044.42	$z^4F^\circ_{1\frac{1}{2}}-e^4D_{1\frac{1}{2}}$		
44.40	30	40, I A	046.12	$a^2D_{2\frac{1}{2}}-z^4D^\circ_{3\frac{1}{2}}$		
31.20	3	3, V?	076.07			
28.4	1	1, IV A	082.4	$b^2D_{1\frac{1}{2}}-z^4D^\circ_{\frac{1}{2}}$		
16.59	60	80, I	109.36	$a^4F_{3\frac{1}{2}}-z^4G^\circ_{4\frac{1}{2}}$	(0.59) 1.12 B	(0.47) 1.14
12.48	3		118.75			
08.26	40	40, II	128.43	$a^2G_{3\frac{1}{2}}-z^2H^\circ_{4\frac{1}{2}}$	(0.00) 1.01	(0.00) 1.00
07.7	1	3, IV	129.7	$a^2P_{\frac{1}{2}}-4^\circ_{\frac{1}{2}}$		
6,600.17	25	50, II A	146.95	$a^4F_{3\frac{1}{2}}-z^4F^\circ_{2\frac{1}{2}}$	(0.35) 1.12	(0.37) 1.12
6,599.48		tr, IV A	148.53			
93.45	40	60, I	162.39	$a^4F_{2\frac{1}{2}}-y^2D^\circ_{1\frac{1}{2}}$	(0.00 w) 1.30 A[2]	(0.00) 1.33 A[2]
90.59	1		168.97	$a^2F_{3\frac{1}{2}}-1_{\frac{1}{2}}$		
82.18	4	6, IV	188.35	$z^4D^\circ_{1\frac{1}{2}}-e^4D_{1\frac{1}{2}}$	(0.00 h) 1.41	
78.51	200	400, I	196.82	$a^2D_{1\frac{1}{2}}-z^2F^\circ_{2\frac{1}{2}}$	(0.00) 1.00	(0.00) 0.98
68.54	1		219.89	$a^2D_{1\frac{1}{2}}-z^2P^\circ_{1\frac{1}{2}}$		
65.45	40	40, I A	227.05	$a^2D_{2\frac{1}{2}}-z^2P^\circ_{1\frac{1}{2}}$		

TABLE 6.—*The arc spectrum of lanthanum (La I)*—Continued

λ air I.A.	Intensities, temperature class		ν vac cm⁻¹	Term combination	Zeeman effects	
	B. S.	K. & C.			Observed	Computed
55. 95	2	3, III A	249. 12	$a^2F_{1\frac12}-y^4D_{1\frac12}$		
55. 11		1, V	251. 07	$b^2D_{1\frac12}-z^2D_{1\frac12}$		
51. 78	1		258. 82			
49. 16	2	2 n, III?	264. 92	$a^2P_{1\frac12}-z^4F_{2\frac12}$		
43. 17	300	500, I	278. 90	$a^4F_{1\frac12}-z^4G_{2\frac12}$	(0. 33, 0. 99) 0. 00, 0. 74, 1. 41, 2. 07	(0. 33, 0. 99) 0. 08, 0. 75, 1. 40, 2. 06
23. 86	2	2, III A?	324. 12	$b^2D_{1\frac12}-z^6F_{1\frac12}$		
20. 70	15	20, IV	331. 55		(0.00 h) 1.10 h	
19. 30	3	4, IV	334. 84	$x^4F_{1\frac12}-e^4F_{1\frac12}$		
17. 35	1		339. 43	$x^4G_{1\frac12}-35$		
6, 506. 25	4	6, IV	365. 59	$x^4D_{1\frac12}-33$		
6, 492. 86	4	5, V	397. 29			
85. 54	20	20, II A	414. 66	$a^2F_{1\frac12}-2i_{1\frac12}$	(0.00) 1.10	(0.00) 1.10
80. 20	1		427. 37	$a^2F_{2\frac12}-y^2D_{1\frac12}$		
68. 44	10	8, II A	455. 42	$b^2D_{1\frac12}-x^4S_{1\frac12}$		
55. 99	250	300, I	488. 22	$a^4D_{1\frac12}-z^2F_{1\frac12}$	(0.00) 1.18	(0.00) 1.15
54. 50	150	200, I	488. 80	$a^4F_{1\frac12}-z^4P_{2\frac12}$	(0.33, 1.11) 0.80, 1.54, 2.34	(0.39, 1.18) 0.02, 0.81, 159, 2.38
50. 34	6	8, II A	498. 78	$a^2P_{1\frac12}-z^2D_{2\frac12}$	(0.00) 1.29	(0.00) 1.20
49. 94	3		499. 74			
48. 25	4	10?, II A	503. 81	$a^2D_{1\frac12}-z^4D_{1\frac12}$		
48. 10	20	50?, II A	504. 17	$a^4F_{1\frac12}-y^2D_{1\frac12}$		
26. 60	2		556. 04	$x^4D_{1\frac12}-e^4F_{1\frac12}$		
20. 90	1		569. 85	$z^4F_{1\frac12}-e^4F_{1\frac12}$		
17. 23	2	3, III? A	578. 85	$b^2D_{2\frac12}-z^2D_{1\frac12}$		
6, 410. 98	200	300, I	593. 94	$z^4F_{1\frac12}-z^4G_{1\frac12}$	(0.00) 1.09	(0.00) 1.10
6, 394. 23	400	600, I	634. 79	$a^4F_{1\frac12}-z^4G_{1\frac12}$	(0.00 W) 1.02 A?	(0.00) 0.90 A?
80. 48	1	1, V	668. 48			
75. 50	2		680. 72	$x^4F_{1\frac12}-e^4F_{1\frac12}$		
75. 11	2	2, V	681. 68			
60. 20	30	30, II	718. 44	$a^4P_{1\frac12}-z^2D_{1\frac12}$	(0.00) 0.89	(0.00) 0.93
56. 38	3	4, III A	727. 88	$b^2D_{1\frac12}-L1_{1\frac12}$		
53. 63	1		734. 69	$a^2P_{1\frac12}-z^2P_{1\frac12}$		
39. 16	2	2, III A	770. 61	$a^4P_{1\frac12}-z^2S_{1\frac12}$		
33. 74	3	2, IV?	784. 10	$z^4D_{2\frac12}-e^4D_{2\frac12}$		
33. 24	2	2, III?	785. 35			
30. 42	3	2, III A	792. 38	$a^2F_{1\frac12}-y^4D_{1\frac12}$		
25. 90	100	150, I	803. 67	$a^4D_{2\frac12}-y^2F_{1\frac12}$	(0. 63, 0. 99) 0. 25, 0. 63, 1. 01, 1. 38, 1. 76	(0. 19, 0. 56, 0. 94) 0. 27, 0. 64, 1. 01, 1. 39, 1. 77
18. 26	5	12, III A	822. 78	$a^2F_{1\frac12}-w^2F_{1\frac12}$		
10. 13	8	12, IV	843. 16			
06. 87	2	2, III	846. 32			
6, 306. 21	2	3, III	847. 98			
6, 293. 57	60	80, II A	884. 85	$a^4F_{1\frac12}-y^2D_{1\frac12}$	(0.00) 1.33	(0.00) 1.26
88. 56	6	7, III A	897. 50	$a^2P_{1\frac12}-y^4P_{1\frac12}$		
87. 73	7	8, III A	899. 60	$b^2D_{2\frac12}-x^4D_{1\frac12}$		
78. 31	1		923. 46	$a^2P_{1\frac12}-y^4P_{1\frac12}$		
66. 00	40	60, III	954. 74	$a^2G_{4\frac12}-z^2H_{1\frac12}$	(0.00) 1.06	(0.00) 1.06
49. 92	300	500, I	15, 995. 79	$a^4F_{4\frac12}-z^4G_{4\frac12}$	(0.00) 1.13	(0.00) 1.14
28. 58	12	15, III A	16, 024. 86	$a^4P_{1\frac12}-z^4D_{1\frac12}$		
36. 76	7	10, III A	029. 54	$a^4P_{1\frac12}-z^2S_{1\frac12}$		
36. 17	8	12, IV	031. 06			
34. 85	10	15, III A	034. 45	$b^2D_{2\frac12}-z^2D_{2\frac12}$	(0.00) 1.26	(0.09) 1.27
33. 51	10	15, III A	037. 90	$a^4P_{1\frac12}-z^4D_{2\frac12}$		
32. 56	1		040. 34	$z^4F_{1\frac12}-e^4F_{2\frac12}$		
25. 33	2 h		058. 97	$a^4P_{1\frac12}-3i_{1\frac12}$		
24. 24	1		061. 78	$b^2D_{1\frac12}-e^6F_{1\frac12}$		
19. 46	2	2, V	074. 13			
18. 19	5	7, III	077. 41	$a^2G_{4\frac12}-x^4F_{1\frac12}$		
14. 35	2		087. 34	$z^4D_{1\frac12}-a^3F_{2\frac12}$		
6, 205. 76	2		107. 02			
6, 173. 74	2	1, IV A	193. 17	$b^2D_{1\frac12}-x^4S_{1\frac12}$		
67. 69	2		209. 05	$a^2F_{1\frac12}-1i_{1\frac12}$		

TABLE 6.—*The arc spectrum of lanthanum (La I)*—Continued

λ_{air} I. A.	Intensities, temperature class		ν_{vac} cm^{-1}	Term combination	Zeeman effects	
	B. S.	K. & C.			Observed	Computed
65.69	30	40, II	214.31	$a^4P_{1½}-z^4D^{\circ}_{1½}$	(0.88) 0.87, 1.43, 1.99	(0.27, 0.81) 0.88, 1.42, 1.96
65.02	3	2, IV	216.07			
54.55	1		243.66	$z^4F^{\circ}_{1½}-e^4F_{1½}$		
46.53	10	5, IV	264.85	$a^2S_{½}-{}^3P_{1½}$		
45.29	3	3, III	268.13	$z^4F^{\circ}_{1½}-e^4D_{1½}$		
42.98	10	10, III	274.25		(0.00) 1.15	
41.71	3		277.61			
36.48	2	2, II A	291.49	$a^2F_{3½}-y^4D^{\circ}_{2½}$		
34.39	20	30, III	297.04	$a^4P_{½}-z^4D^{\circ}_{½}$	(1.34) 1.37	(1.35) 1.35
27.04	8	12, III A	316.59	$b^2D_{1½}-z^2D^{\circ}_{1½}$		
23.75	2	2, IV	325.36			
21.27		1, IV	331.97			
20.34		1, IV	334.45	$z^4F_{1½}-34$		
11.71	20	30, II	357.52	$a^2F_{3½}-w^2F^{\circ}_{3½}$	(0.00) 1.17	(0.12) 1.15
08.47	40	60, II	366.19	$a^4P_{2½}-z^4D^{\circ}_{3½}$	(0.53) 1.34B	(−, 0.48, 0.80) −, 1.06, 1.28, 1.70, −
6,107.26	4	12, II A	369.43	$a^4F_{2½}-y^2D^{\circ}_{1½}$		
6,092.22	2	1, III A	409.85	{ $a^2P_{½}-z^2P^{\circ}_{½}$ / $z^4D^{\circ}_{1½}-e^2F_{2½}$		
88.00	2	1n, IV	421.22			
84.86	5	8, III	429.69	$y^2F^{\circ}_{3½}-35$		
75.24		1, III A	455.71	$a^2F_{3½}-e^2F^{\circ}_{3½}$		
74.01	3	1, III A	459.04	$b^2D_{2½}-y^4P^{\circ}_{2½}$		
72.04	3	4, III A	464.38	$b^2D_{1½}-z^4F^{\circ}_{1½}$		
68.70	20	30, III	473.44	$a^4P_{½}-z^4D^{\circ}_{1½}$	(0.75) 0.40, 1.90	(0.77) 0.38, 1.98
44.8	2	2, III A	538.6	$b^2D_{1½}-z^4F^{\circ}_{2½}$		
41.6	2	2, III	547.3			
38.57	20	25, III A	555.64	$a^4P_{1½}-z^4D^{\circ}_{2½}$		
34.5	2 h	3, V	566.8			
32.38	5		572.63	$a^2P_{½}-y^4P^{\circ}_{½}$		
31.46	5	3n, IV?	575.15			
25.09	2 p?	1, IV?	592.68			
17.16	3	1, III A	614.54	$z^4F^{\circ}_{1½}-e^4F_{1½}$		
6,007.34	50	50, III A	641.70	$a^4F_{4½}-y^4F^{\circ}_{4½}$	(0.00 w) 1.77 A³	(0.00) 1.85 A³
5,992.35	3	2, III A	683.33	$a^4P_{1½}-4^{\circ}_{1½}$		
82.34	10	5, III A	711.25	$a^4F_{1½}-y^2D^{\circ}_{1½}$		
75.75	10	3, III A	729.68	$a^4P_{2½}-w^2F^{\circ}_{2½}$		
70.58	HNR	1, III A	744.16	$a^2D_{2½}-z^4P^{\circ}_{1½}$		
62.30	3	1, III	758.98			
62.59	4	1, III A	766.60	$b^2D_{2½}-z^2P^{\circ}_{1½}$		
60.59	4	3, III A	772.23	$b^2D_{1½}-z^2D^{\circ}_{2½}$		
40.83	3	1, III A	828.01	$a^4P_{2½}-v^2F^{\circ}_{3½}$		
35.29	20	15, II A	843.72	$a^4F_{3½}-y^4F^{\circ}_{3½}$	(0.00 w) 1.72 A³	(0.00) 1.76 A³
30.68	100	}400, I	{ 856.81	$a^2D_{1½}-y^2F^{\circ}_{2½}$		(0.00) 0.86
30.61	200		857.01	$a^2D_{2½}-y^2F^{\circ}_{3½}$	(0.00) 0.87	(0.26) 1.14
28.48	5	4, III A	863.07	$a^2F_{2½}-w^2F^{\circ}_{2½}$		
17.64	20	15, II A	893.96	$a^2D_{2½}-z^4G^{\circ}_{3½}$	(0.27) 1.10 h	(0.29) 1.13
04.28	4	2, III A	932.20	$a^2F_{3½}-z^4F^{\circ}_{3½}$	(0.00) 1.25	(0.00) 1.16
5,900.75	3	1, III A	942.31	$a^4P_{½}-4^{\circ}_{1½}$		
5,894.84	20	8, III A	16,959.30	$a^4P_{2½}-z^4S^{\circ}_{1½}$		
77.96	3	1, III A	17,008.00	$a^2S_{½}-w^2D^{\circ}_{1½}$		
77.62	4	2, III A	008.98	$a^4F_{3½}-y^2P^{\circ}_{1½}$		
74.72	8	8, III A	017.38	$a^4P_{1½}-v^2F^{\circ}_{2½}$		
69.93	2	1, III A	031.27	$a^2F_{3½}-z^4D^{\circ}_{3½}$		
57.44	2		067.58			
55.57	20	15, II A	073.03	$a^4F_{3½}-y^4F^{\circ}_{2½}$	(0.00 w) 1.42 A³	(0.00) 1.41 A³
52.26	6	2, III A	082.69	$a^4P_{1½}-z^4D^{\circ}_{½}$		
48.37	15	8, III A	094.05	$z^4F^{\circ}_{1½}-e^4F_{2½}$	(0.00) 0.90	
45.02	10	6, III A	103.85	$a^2D_{2½}-z^4P^{\circ}_{2½}$	(0.00) 1.18	(0.00) 1.20
39.77	3	2, II A	119.23	$a^2D_{2½}-y^2D^{\circ}_{1½}$		
29.71	20	10, III A	148.77	$a^4P_{1½}-z^4S^{\circ}_{1½}$	(0.00) 1.75	(0.17) 1.75
27.56	8	3, III A	155.10	$b^2D_{1½}-y^4P^{\circ}_{2½}$	(0.59) 1.50 ?	(0.55) 1.42

TABLE 6.—*The arc spectrum of lanthanum (La I)*—Continued

λ_{air} I. A.	Intensities, temperature class		ν_{vac} cm^{-1}	Term combination	Zeeman effects	
	B. S.	K. & C.			Observed	Computed
23.82	15	10, III A	166.11	$a^2F_{3\frac{1}{2}}-z^2D^{\circ}_{2\frac{1}{2}}$	(0.00 w) 0.96 h	(0.00) 0.98
21.98	30	30, III	171.54	$a^2G_{3\frac{1}{2}}-y^2G^{\circ}_{3\frac{1}{2}}$	(0.00) 0.93	(0.12) 0.93
13.44	2	1, III A	196.76	$b^2D_{1\frac{1}{2}}-y^4P^{\circ}_{1\frac{1}{2}}$		
5,802.10	2	1, III A	230.37	$a^4P_{\frac{1}{2}}-z^4F^{\circ}_{1\frac{1}{2}}$		
5,791.32	200	400, I	262.45	$a^4F_{4\frac{1}{2}}-y^4F^{\circ}_{4\frac{1}{2}}$	(0.00 w) 1.31 B	(0.18) 1.30
89.22	150	250, I	268.71	$a^4F_{3\frac{1}{2}}-y^4F^{\circ}_{3\frac{1}{2}}$	(0.15) 1.21	(0.12) 1.20
69.97	25	25, III A	326.32	$a^2F_{3\frac{1}{2}}-v^2F^{\circ}_{3\frac{1}{2}}$		
69.32	80	80, I	328.27	$a^4F_{2\frac{1}{2}}-y^4F^{\circ}_{2\frac{1}{2}}$	(0.00) 1.01	(0.04) 1.02
61.83	50	60, I	350.80	$a^4F_{1\frac{1}{2}}-y^2P^{\circ}_{1\frac{1}{2}}$	(0.94) 0.00, 0.68, 1.34	(0.32, 0.96) 0.09, 0.73, 1.37
44.41	60	80, III	403.41	$a^4P_{2\frac{1}{2}}-z^4D^{\circ}_{2\frac{1}{2}}$	(0.00 w) 1.24	(0.00) 1.25
42.93	4	2, III A	407.90	$a^4P_{\frac{1}{2}}-z^4S^{\circ}_{1\frac{1}{2}}$		
40.65	80	100, I	414.81	$a^4F_{1\frac{1}{2}}-y^4F^{\circ}_{1\frac{1}{2}}$	(0.47) 0.59 B	(0.18, 0.55) 0.33, 0.60, 0.96
34.93	6	5, III A	432.18	$a^2G_{3\frac{1}{2}}-u^2F^{\circ}_{2\frac{1}{2}}$	(0.00) 0.92	(0.00) 0.92
20.01	10	10, III A	477.65	$a^4F_{3\frac{1}{2}}-x^2F^{\circ}_{3\frac{1}{2}}$	(0.16, 0.48, 0.81) 1.93 A[3]	(0.16, 0.48, 0.80) 1.70, 2.03
14.55		1, III A	494.35	$a^2G_{3\frac{1}{2}}-y^4G^{\circ}_{3\frac{1}{2}}$		
14.01	4	5, III A	496.00	$a^2F_{2\frac{1}{2}}-v^2F^{\circ}_{2\frac{1}{2}}$	(0.62) ?	(0.66) 1.05
10.85	2 p?	2, III	505.68	$a^2P_{1\frac{1}{2}}-w^2P^{\circ}_{1\frac{1}{2}}$		
03.32	20	10, III	528.80	$a^4F_{1\frac{1}{2}}-y^2P^{\circ}_{1\frac{1}{2}}$		
02.57		2, III	531.10	$a^4P_{1\frac{1}{2}}-z^2D^{\circ}_{1\frac{1}{2}}$		
5,701.15		1, III	535.47	$a^2G_{4\frac{1}{2}}-y^4G^{\circ}_{3\frac{1}{2}}$		
5,699.32	3	5, III	541.10	$a^4F_{4\frac{1}{2}}-z^2F^{\circ}_{3\frac{1}{2}}$		
96.18	30	40, I	550.77	$a^2D_{2\frac{1}{2}}-z^4G^{\circ}_{2\frac{1}{2}}$	(0.00 w) 0.80 A[1]	(0.00) 0.70 A[1]
61.34	2	1, III A	658.78	$a^2G_{3\frac{1}{2}}-y^2G^{\circ}_{3\frac{1}{2}}$		
57.71	30	50, II	670.10	$a^4F_{1\frac{1}{2}}-y^4F^{\circ}_{2\frac{1}{2}}$	(0.29, 0.88) 0.73, 1.33, 1.94	(0.29, 0.89) 0.12, 0.71, 1.30, 1.90
56.54	2	1, III A	673.76	$a^2P_{1\frac{1}{2}}-w^2F^{\circ}_{2\frac{1}{2}}$		
54.8	20	3, III A	679.2	$a^4P_{\frac{1}{2}}-z^4F^{\circ}_{1\frac{1}{2}}$		
48.24	50	80, III	699.73	$a^2G_{4\frac{1}{2}}-y^2G^{\circ}_{4\frac{1}{2}}$	(0.00) 1.12	(0.00) 1.12
39.31	8	5, III A	727.76	$a^4P_{1\frac{1}{2}}-z^2D^{\circ}_{2\frac{1}{2}}$		
32.02	15	25, II	750.70	$a^2F_{2\frac{1}{2}}-z^2D^{\circ}_{1\frac{1}{2}}$	(0.00) 0.93	(0.00) 0.92
5,631.22	60	100, I	753.22	$a^4F_{2\frac{1}{2}}-y^4F^{\circ}_{3\frac{1}{2}}$	(0.00 W) 1.45 A[2]	(0.00) 1.55 A[2]
5,596.52	3		856.92		(0.00) 0.49	
88.33	100	20, II	889.48	$a^4F_{3\frac{1}{2}}-y^4F^{\circ}_{4\frac{1}{2}}$		
70.37	3	5, II A	947.16	$a^2D_{1\frac{1}{2}}-z^4G^{\circ}_{2\frac{1}{2}}$		
68.45	30	50, II	953.35	$a^4F_{3\frac{1}{2}}-z^2G^{\circ}_{3\frac{1}{2}}$	(0.44) 1.16	(0.41) 1.15
65.70	10	20, II	962.22	$a^4F_{3\frac{1}{2}}-z^2F^{\circ}_{3\frac{1}{2}}$		
65.43	10	20, III	963.09	$a^4P_{2\frac{1}{2}}-y^4P^{\circ}_{1\frac{1}{2}}$	(0.00) 1.52	(0.00) 1.49
62.54	2	2, III	17,972.42	$a^2F_{2\frac{1}{2}}-z^4F^{\circ}_{2\frac{1}{2}}$		
44.90	3	6, III	18,029.60	$a^2P_{1\frac{1}{2}}-Sf_{1\frac{1}{2}}$	(0.28) 1.47	(0.35) 1.44
41.25	15	20, III	041.47	$b^2D_{2\frac{1}{2}}-w^2P^{\circ}_{1\frac{1}{2}}$	(0.00) 1.25	(0.00) 1.25
32.17	4	3, III	071.08	$a^2F_{2\frac{1}{2}}-z^4D^{\circ}_{1\frac{1}{2}}$		
29.86	2	3, III	078.63	$a^2G_{3\frac{1}{2}}-u^2F^{\circ}_{3\frac{1}{2}}$		
26.51		1, V	089.59	$z^2F^{\circ}_{3\frac{1}{2}}-35$		
17.34	20	30, III	119.66	$a^2G_{4\frac{1}{2}}-u^2F^{\circ}_{3\frac{1}{2}}$	(0.00) 1.10	(0.00) 1.10
15.28	5	10, III	126.42	$a^4P_{1\frac{1}{2}}-y^4P^{\circ}_{2\frac{1}{2}}$	(0.25) 1.42	(0.30) 1.39, 2.00
07.33		2, III	152.59	$a^4P_{1\frac{1}{2}}-y^4P^{\circ}_{1\frac{1}{2}}$		
06.00	20	40, II	156.97	$a^2D_{1\frac{1}{2}}-z^4P^{\circ}_{1\frac{1}{2}}$	(0.00 w) 1.54 h	(0.00) 1.50
03.80	40	80, III	164.23	$a^4F_{4\frac{1}{2}}-z^2G^{\circ}_{4\frac{1}{2}}$	(0.70) 1.22 B	(0.69) 1.23
02.66	5	10, III	168.00	$a^4F_{3\frac{1}{2}}-z^2F^{\circ}_{3\frac{1}{2}}$		
02.24	3	4, III	169.38	$a^2G_{4\frac{1}{2}}-y^4G^{\circ}_{4\frac{1}{2}}$		
5,510.34	200	300, I	172.35	$a^2D_{1\frac{1}{2}}-y^2D^{\circ}_{1\frac{1}{2}}$	(0.00) 0.80	(0.06) 0.81
5,498.70	2	2n, III	181.08	$a^4P_{\frac{1}{2}}-w^2P^{\circ}_{1\frac{1}{2}}$		
91.07	5	8, III	206.34	$a^2F_{2\frac{1}{2}}-z^2D^{\circ}_{1\frac{1}{2}}$		
75.17	10	15, III	259.21		(0.00) 1.08	
60.91	2	3, III	286.80	$a^2F_{3\frac{1}{2}}-y^4P^{\circ}_{2\frac{1}{2}}$		
55.14	200	400, I	326.25	$a^2D_{2\frac{1}{2}}-y^2D^{\circ}_{2\frac{1}{2}}$	(0.00) 1.20	(0.02) 1.20
37.55	2	5, III	385.54	$a^4P_{\frac{1}{2}}-y^4P^{\circ}_{\frac{1}{2}}$	(0.00) 2.54	(0.21) 2.50
29.86	6	15, III	411.57	$a^4P_{\frac{1}{2}}-y^4P^{\circ}_{1\frac{1}{2}}$	(0.49) 1.04	(0.55) 1.06, 2.15
22.10		3, III A	437.93	$a^4F_{2\frac{1}{2}}-z^2G^{\circ}_{3\frac{1}{2}}$		
5,415.67	4	8, III	459.82	$a^4P_{1\frac{1}{2}}-z^4P^{\circ}_{1\frac{1}{2}}$	(0.36) 1.56 w	(0.39) 1.56
5,390.63	?		545.56	$a^2G_{3\frac{1}{2}}-w^2D^{\circ}_{2\frac{1}{2}}$		

TABLE 6.—*The arc spectrum of lanthanum (La I)*—Continued

λ air I.A.	Intensities, temperature class		ν vac cm⁻¹	Term combination	Zeeman effects	
	B. S.	K. & C.			Observed	Computed
80.00	3	5, III	582.21	a²G₂½—f²F¹½		
65.87	4	8, III	631.14	a²F₂½—y⁴P¹½		
59.70	2	2, III A	652.59	a⁴F₃½—x²F½		
57.85	25	60, III	659.03	a⁴P½—y⁴P¹½	(0.00) 1.53	(0.00) 1.54
30.64	1		754.27			
28.56	3	3, III	779.22	b²D₁½—x²P¹½		
21.34	1		787.04	a²P₁½—x²D½		
20.14	3	2, III	791.28	a⁴F₃½—z²G¹½		
07.52	3	3, III	835.96			
5,304.01	20	30, III	848.43	a⁴P½—y⁴P½		
5,287.45	1	1, V	907.46			
76.40	5	10, III	947.06	b²D₁½—x²F¹½	(0.00) 0.90	(0.00) 0.90.
71.18	100	150, I	18,965.82	a²D₂½—y²P¹½	(0.00) 1.39 A²	(0.00) 1.44 A²
57.83	4	3, III	19,013.98		(0.00) 1.07	
53.45	100	100, I	029.83	a²D₂½—y⁴F½	(0.23, 0.60) 0.57, 1.03, 1.49, 1.96	(0.21, 0.63) 0.57, 0.99 1.41, 1.83.
40.81	4	2, III	075.72	a²F₃½—y²G½		
39.54	4	4, III	080.35	a²F₃½—y²G¹½		
34.27	150	300, II	099.56	a⁴F₁½—1½	(0.00 W) 2.10 A²	(0.00) 2.10 A².
5,211.85	150	300, II	187.72	a⁴F₁½—y¹D¹½	(0.00 W) 1.73 A²	(0.00) 1.71 A².
5,190.34	8		261.21			
83.91	10	20, II	285.10	a²D₂½—y⁴F₃½		
79.11	2	2, III	302.98	b²D₁½—δ½		
77.30	150	300, II	309.72	a⁴F₁½—y²D½	(0.00) 1.07	(0.00) 1.06.
68.95	2		340.92	a²F₃½—u²F½		
67.79	20	20, III	345.26	a⁴F₁½—2½		
64.03		1, V?	359.34	b²D₁½—f²F½		
61.54	2		368.68			
58.68	40	80, I	379.42	a²D₁½—y⁴D¹½	(0.19, 0.57) 1.05, 1.45, 1.85	(0.20, 0.60) —, 0.98, 1.38, 1.78
52.31	1		403.38	a²F₃½—y⁴G½		
45.42	100	200, II	429.36	a⁴F₃½—y⁴D½	(0.00) 0.89	(0.00) 0.90)
39.16	3		453.03			
35.42	3	2, V	467.20			
34.37	2		471.18			
29.81	3		488.48			
20.87	10	10, III	522.51	b²D₁½—w²D½	(0.00) 0.92	(0.08) 0.90
09.12	3	2, III A	567.40	a²F₂½—y²G¹½		
06.23	100	150, II	578.48	a⁴F₁½—y⁴D₀½	(0.19) 0.62 A²?	(0.19) 0.22, 0.60
5,103.11	3	2, III	590.45			
5,096.59	2	1, IV?	615.51			
86.22	2		655.50	z⁴F½—35		
79.37	4	5, IV	682.01			
78.92	3	2, V?	683.76			
72.10	3	1, III A	710.22	a²D₂½—y⁴F½		
67.90	15	15, III	726.56	a⁴F₃½—1½		
56.46	60	80, II	771.18	a⁴F₁½—y⁴D½	(1.20) 0.66, 0.79, 1.57	(0.39, 1.19) 0.02, 0.81, 1.60
52.10	1		788.25	b²D₂½—z²D½		
50.57	60	80, II	794.24	a⁴F₃½—y⁴D½	(0.80) 1.18 B	(-, -, 0.79) —, 0.88, 1.19, 1.51, —
46.87	30	60, III	808.75	a⁴F₃½—y⁴D½	(0.00 w) 1.24	(0.00) 1.22
37.60	2		845.20	a²F₁½—65½		
33.24	2		862.40			
19.50	10	8, III	916.76			
5,001.78	20	10, III A	19,987.32	a²F₃½—w²F¹½		
4,995.95	1		20,010.64	a²F₁½—y⁴G½		
94.64	2	1, III A	015.90	b²D₃½—w²D¹½		
92.87	15	20, II	018.98	a²D₁½—y²P½		
84.92	3	2, IV?	054.92	a⁴F₁½—3½		
84.63	1		056.09	a²P₁½—c²D½		
83.56	2		060.40	b²D₁½—w²D½		
77.95	8	8, II A	083.01	a²D₁½—y⁴F½		
68.59	4 d	2{III A / V}	120.84	a²F₃½—y³G½		

Table 6.—*The arc spectrum of lanthanum (La I)*—Continued

λ_{air} I. A.	Intensities, temperature class		ν_{vac} cm^{-1}	Term combination	Zeeman effects	
	B. S.	K. & C.			Observed	Computed
64.84	3	4, III A	136.03	$a^4F_{1\frac{1}{2}}-y^4D^{\circ}_{1\frac{1}{2}}$		
57.77	4		164.75			
49.76	50	200, I	197.38	$a^2D_{1\frac{1}{2}}-y^4P^{\circ}_{1\frac{1}{2}}$	(0.00) 0.84	(0.00) 0.84
45.84	3	5, III A	213.39	$a^2F_{2\frac{1}{2}}-w^2P^{\circ}_{1\frac{1}{2}}$		
25.40	3	2, IV	297.27			
16.62	3	1, III? A	333.52			
05.13	4	4, III A	381.15	$a^2F_{3\frac{1}{2}}-u^2F^{\circ}_{3\frac{1}{2}}$		
4,901.87	15	25, I	394.70	$a^2D_{2\frac{1}{2}}-z^2G^{\circ}_{3\frac{1}{2}}$	(0.00) 0.94	(0.00) 0.92
4,894.24	2	1, III	426.49			
87.60	4	5, III	454.24	$a^2F_{2\frac{1}{2}}-w^2D^{\circ}_{2\frac{1}{2}}$		
86.82	3	2, IV	457.51			
81.94	1		477.96	$a^4P_{1\frac{1}{2}}-w^2D^{\circ}_{1\frac{1}{2}}$		
78.86	10	15, III	490.89	$a^2F_{3\frac{1}{2}}-t^2F^{\circ}_{3\frac{1}{2}}$	(0.00) 1.10 R	(0.18) 1.10
70.56	5	5, III	525.81	$b^2D_{1\frac{1}{2}}-v^2D^{\circ}_{1\frac{1}{2}}$		
68.90	3	2, IV?	532.80			
67.37	3		539.26	$a^4F_{3\frac{1}{2}}-3^{\circ}_{3\frac{1}{2}}$		
54.95	8	8, III	591.80	$b^2D_{2\frac{1}{2}}-v^2D^{\circ}_{2\frac{1}{2}}$	(0.00) 1.21 R	(0.35) 1.21
50.81	20	20, I	609.37	$a^2D_{2\frac{1}{2}}-x^2F^{\circ}_{2\frac{1}{2}}$		
39.51	20	25, II	657.49	$a^2F_{2\frac{1}{2}}-t^2F^{\circ}_{2\frac{1}{2}}$	(0.00) 0.89	(0.04) 0.89
17.55	3		751.66			
17.17	10	4, IV?	753.29	$b^2D_{2\frac{1}{2}}-7^{\circ}_{1\frac{1}{2}}$		
4,800.24	9	8, III	826.49	$a^4P_{2\frac{1}{2}}-w^2D^{\circ}_{2\frac{1}{2}}$		
4,799.99	8	8, III	827.57	$a^2G_{3\frac{1}{2}}-s^2F^{\circ}_{3\frac{1}{2}}$		
92.46	1	tr, III A	860.30	$a^4F_{1\frac{1}{2}}-x^4D^{\circ}_{\frac{1}{2}}$		
91.77	1		863.30	$a^4P_{2\frac{1}{2}}-t^2F^{\circ}_{1\frac{1}{2}}$		
91.39	5	5, II	864.96	$a^4F_{2\frac{1}{2}}-w^2F^{\circ}_{2\frac{1}{2}}$		
79.89	4	4, II	915.16	$a^4F_{3\frac{1}{2}}-w^2F^{\circ}_{3\frac{1}{2}}$		
75.14	3	2, V	935.96	$a^2G_{3\frac{1}{2}}-9^{\circ}_{3\frac{1}{2}}$		
70.43	10	15, II	956.63	$a^2F_{3\frac{1}{2}}-w^2D^{\circ}_{2\frac{1}{2}}$	(0.00) 0.93 R	(0.00) 0.89
67.80	1 h		968.19	$a^4F_{4\frac{1}{2}}-z^2H^{\circ}_{4\frac{1}{2}}$		
66.89	60	100, I	20,972.19	$a^2D_{1\frac{1}{2}}-z^2F^{\circ}_{2\frac{1}{2}}$	(0.00) 0.99	(0.00) 0.98
59.71	2	2, IV?	21,003.83	$a^2G_{3\frac{1}{2}}-s^2F^{\circ}_{3\frac{1}{2}}$		
57.14	3	2, IV	015.18			
56.97		1, V	015.93	$a^4P_{1\frac{1}{2}}-w^2D^{\circ}_{2\frac{1}{2}}$		
53.11	2	1, IV	032.99			
52.41	3	3, III	036.09	$a^4F_{3\frac{1}{2}}-x^4D^{\circ}_{2\frac{1}{2}}$		
50.41	10	15, III	044.95	$a^2G_{4\frac{1}{2}}-s^2F^{\circ}_{3\frac{1}{2}}$	(0.00) 1.04 R	(0.00) 1.02
33.82	8	4, V?	118.70	$b^2D_{1\frac{1}{2}}-6^{\circ}_{2\frac{1}{2}}$		
29.09	1	1, V	139.82			
23.72	2	3, II	163.85	$a^4F_{3\frac{1}{2}}-4^{\circ}_{2\frac{1}{2}}$		
14.14	4	5, I	206.86	$a^4F_{1\frac{1}{2}}-w^2F^{\circ}_{2\frac{1}{2}}$		
08.18	8	8, III?	233.71	$b^2D_{2\frac{1}{2}}-8^{\circ}_{1\frac{1}{2}}$	(0.00) 0.97 R	(0.00) 1.15
02.64	8	10, I	258.72	$a^4F_{4\frac{1}{2}}-x^4F^{\circ}_{3\frac{1}{2}}$		
4,700.26	8	8, III? •	289.48			
4,695.30	3		291.95	$a^4P_{2\frac{1}{2}}-v^2D^{\circ}_{1\frac{1}{2}}$		
60.70	8	8, III	450.02	$a^2F_{3\frac{1}{2}}-w^4D^{\circ}_{3\frac{1}{2}}$	(0.00) 0.92 R	(0.00) 0.93
53.90	4		481.36	$a^4P_{1\frac{1}{2}}-v^2D^{\circ}_{1\frac{1}{2}}$		
52.07	15	20, I	489.81	$a^4F_{3\frac{1}{2}}-x^4F^{\circ}_{2\frac{1}{2}}$		
50.32	12	15, I	497.90	$a^4F_{2\frac{1}{2}}-v^4F^{\circ}_{2\frac{1}{2}}$	(0.30) 1.13	(0.37) 1.12
48.64	30	40, I	505.67	$a^4F_{1\frac{1}{2}}-4^{\circ}_{1\frac{1}{2}}$	(0.48)	(0.15, 0.46) 0.26, 0.57, 0.87
46.33	10	12, III	516.36	$a^2G_{3\frac{1}{2}}-r^2F^{\circ}_{2\frac{1}{2}}$	(0.00) 0.89 R	(0.00) 0.95
43.11	5	5, III	531.28	$a^2F_{3\frac{1}{2}}-t^2F^{\circ}_{2\frac{1}{2}}$		
27.35	2		604.61	$b^2D_{2\frac{1}{2}}-s^2F^{\circ}_{2\frac{1}{2}}$		
15.06	8	15, III	662.14	$a^4P_{\frac{1}{2}}-w^4D^{\circ}_{\frac{1}{2}}$	(1.31) 1.37 R	(1.35) 1.35
05.08	6	10, III	709.09	$a^4P_{1\frac{1}{2}}-w^4D^{\circ}_{1\frac{1}{2}}$		
04.24	6	10, III	713.05	$b^2D_{2\frac{1}{2}}-9^{\circ}_{3\frac{1}{2}}$		
4,602.04	10	20, III	723.43	$a^2F_{3\frac{1}{2}}-v^2D^{\circ}_{2\frac{1}{2}}$		
4,596.19	6	10, I	751.08	$a^2D_{2\frac{1}{2}}-y^4D^{\circ}_{2\frac{1}{2}}$		
89.89	5		780.93	$b^2D_{2\frac{1}{2}}-s^2F^{\circ}_{3\frac{1}{2}}$		
81.20	10	12, III	822.25	$a^4P_{2\frac{1}{2}}-w^4D^{\circ}_{2\frac{1}{2}}$		
70.02	60	250, I	875.63	$a^4F_{4\frac{1}{2}}-x^4F^{\circ}_{4\frac{1}{2}}$	(0.00) 1.34	(0.11) 1.34
67.90	50	200, I	885.78	$a^4F_{3\frac{1}{2}}-x^4F^{\circ}_{3\frac{1}{2}}$	(0.00) 1.24	(0.02) 1.23
64.85	6	12, III	900.41	$a^4F_{2\frac{1}{2}}-x^4F^{\circ}_{1\frac{1}{2}}$	(0.00) 1.42 R	(0.15, 0.45) 1.18, 1.48
52.47	8	8, II A	959.96	$a^2F_{2\frac{1}{2}}-v^2D^{\circ}_{1\frac{1}{2}}$	(0.00) 0.87	(0.00) 0.87
50.76	8	10, III A	968.21	$a^4P_{\frac{1}{2}}-w^4D^{\circ}_{1\frac{1}{2}}$	(0.78) 0.38 R	(0.77) 0.38, 1.93

TABLE 6.—*The arc spectrum of lanthanum (La i)*—Continued

λ_air I.A.	Intensities, temperature class		ν_vac cm⁻¹	Term combination	Zeeman effects	
	B. S.	K. & C.			Observed	Computed
50. 16	4	5, III A	971. 11	$a^4F_{1\frac{1}{2}}-r^4S^2_{1\frac{1}{2}}$		
49. 80	40	50, I	21, 974. 30	$a^4F_{1\frac{1}{2}}-z^4F^2_{1\frac{1}{2}}$	(0. 00) 1. 06	(0. 17) 1. 07
41. 78	10	15, III	22, 011. 65	$a^4P_{1\frac{1}{2}}-w^4D^2_{\frac{1}{2}}$		
37. 57	2		032. 07	$a^2P_{1\frac{1}{2}}-u^2D^2_{1\frac{1}{2}}$		
28. 98	3		074. 34	$a^4P_{1\frac{1}{2}}-6^2_{1\frac{1}{2}}$		
07. 4	2		179. 5	$a^2G_{3\frac{1}{2}}-r^2F^2_{3\frac{1}{2}}$		
01. 57	6	10, II A	208. 26	$a^4F_{3\frac{1}{2}}-z^2D^2_{3\frac{1}{2}}$		
4, 500. 21	30	40, II	214. 97	$a^4P_{2\frac{1}{2}}-w^4D^2_{1\frac{1}{2}}$	(0. 00 w) 1. 10 A⁻	(0. 12, 0. 37-) 0. 67,
						0. 92, 1. 17-, -, -
4, 499. 04	10	10, III	220. 75	$a^2G_{4\frac{1}{2}}-r^2F^2_{3\frac{1}{2}}$	(0. 00) 1. 02	(0. 00) 1. 02
94. 71	20	30, I	242. 16	$a^4F_{1\frac{1}{2}}-z^4F^2_{1\frac{1}{2}}$	(0. 51) 0. 27, 0. 58, 0. 91 us	(0. 16, 0. 47) 0. 26, 0. 57, 0. 88
92. 81	5	10, I A	246. 61	$a^2D_{1\frac{1}{2}}-y^2D^2_{1\frac{1}{2}}$	(0. 00) 0. 72 R	(0. 38) 0. 42, 1. 17
93. 11	15	25, I	250. 08	$a^2D_{3\frac{1}{2}}-y^2D^2_{1\frac{1}{2}}$	(0. 00) 1. 16	(0. 04) 1. 21
91. 76	10	15, III	256. 76	$a^4P_{3\frac{1}{2}}-7^2_{1\frac{1}{2}}$	(0. 00) 1. 52	(0. 00) 1. 54
86. 06	10	20, III	285. 04	$a^4P_{1\frac{1}{2}}-w^4D^2_{1\frac{1}{2}}$		
79. 82	6	15, II A	316. 08	$a^4F_{1\frac{1}{2}}-z^4F^2_{1\frac{1}{2}}$		
74. 54	4	5, III A	342. 42	$b^2D_{1\frac{1}{2}}-a^2F^2_{1\frac{1}{2}}$		
68. 97	10	2b, II	370. 26	$a^4F_{1\frac{1}{2}}-z^4F^2_{1\frac{1}{2}}$		
55. 21	3	10, II A	439. 35	$a^2D_{1\frac{1}{2}}-y^2D^2_{1\frac{1}{2}}$		
53. 85		2, IV A	446. 20	$a^4P_{1\frac{1}{2}}-7^2_{1\frac{1}{2}}$		
52. 15	15	30, II	449. 78	$a^2G_{3\frac{1}{2}}-y^2H^2_{3\frac{1}{2}}$	(0. 00) 0. 98	(0. 00) 0. 97
45. 12	2	2, III A	490. 29	$a^2F_{3\frac{1}{2}}-w^4D^2_{1\frac{1}{2}}$		
43. 94	5	10, I	496. 26	$a^2D_{3\frac{1}{2}}-5^2_{1\frac{1}{2}}$		
42. 68	6	12, II	502. 64	$a^4F_{1\frac{1}{2}}-z^4F^2_{1\frac{1}{2}}$		
23. 90	20	30, II	598. 16	$a^2G_{4\frac{1}{2}}-y^2H^2_{4\frac{1}{2}}$	(0. 00) 1. 09	(0. 00) 1. 09
17. 14	2	6n, III	632. 75	$a^4F_{3\frac{1}{2}}-y^4P^2_{1\frac{1}{2}}$		
13. 45		2, III A	651. 67	$a^2D_{3\frac{1}{2}}-z^4D^2_{1\frac{1}{2}}$		
08. 02	2	8, III A	705. 33	$a^4P_{1\frac{1}{2}}-7^2_{1\frac{1}{2}}$		
4, 602. 64	5	15, III	707. 29	$a^2P_{\frac{1}{2}}-w^4D^2_{1\frac{1}{2}}$	(0. 00) 0. 86	(0. 00) 0. 86
4, 597. 04		2, IV A	736. 21	$a^2F_{1\frac{1}{2}}-z^2F^2_{1\frac{1}{2}}$		
96. 79		4, IV A	737. 50	$a^4P_{3\frac{1}{2}}-8^2_{1\frac{1}{2}}$		
96. 31		2, IV	739. 98			
93. 52	2	4, III	754. 42			
89. 87	6	15, III	773. 34	$a^4P_{1\frac{1}{2}}-w^4D^2_{\frac{1}{2}}$	(0. 00) 1. 07	(0. 00) 1. 06
80. 55	4	12, II A	821. 79	$a^2D_{1\frac{1}{2}}-w^4F^2_{3\frac{1}{2}}$		
60. 96	2	2, III A	924. 84	$a^4F_{1\frac{1}{2}}-7^2_{1\frac{1}{2}}$		
60. 49	2	2, III A	926. 78	$a^4P_{1\frac{1}{2}}-8^2_{1\frac{1}{2}}$		
57. 88	2	2, III A	940. 51	$a^4F_{3\frac{1}{2}}-z^2P^2_{1\frac{1}{2}}$		
54. 79	20	25, III	22, 956. 79	$b^2D_{1\frac{1}{2}}-r^2F^2_{3\frac{1}{2}}$		
40. 72	10	15, III	23, 031. 20	$b^2D_{1\frac{1}{2}}-r^2F^2_{3\frac{1}{2}}$	(0. 00) 0. 90	(0. 00) 0. 87
39. 93	5	6, III A	035. 39			
26. 19	2		108. 55	$a^4P_{3\frac{1}{2}}-a^2F^2_{3\frac{1}{2}}$		
11. 73	5	4, III A	186. 05	$a^4P_{\frac{1}{2}}-8^2_{1\frac{1}{2}}$	(0. 59) 0. 96 R	(0. 56) 0. 96, 2. 12
06. 00	6	6, II A	216. 90	$a^4F_{3\frac{1}{2}}-9^2_{1\frac{1}{2}}$	(0. 14) 1. 49	(0. 20) 1. 50
4, 300. 62	HNR	3	245. 94	$a^2P_{\frac{1}{2}}-y^2P^2_{1\frac{1}{2}}$		
4, 291. 00	2		298. 06	$a^4P_{1\frac{1}{2}}-a^2F^2_{1\frac{1}{2}}$		
89. 65	HNR	1	305. 39	$b^2D_{1\frac{1}{2}}-w^2D^2_{1\frac{1}{2}}$		
89. 01	HNR	2	308. 87	$b^2D_{2\frac{1}{2}}-w^2D^2_{1\frac{1}{2}}$		
80. 27	60	100, I	356. 46	$a^2D_{1\frac{1}{2}}-z^4F^2_{1\frac{1}{2}}$	(0. 00) 1. 14	(0. 00) 1. 13
71. 14	4		406. 39	$a^4F_{1\frac{1}{2}}-9^2_{1\frac{1}{2}}$		
67. 74	2		425. 04	$a^2F_{1\frac{1}{2}}-r^2F^2_{1\frac{1}{2}}$		
62. 35	10	15, II A	454. 66	$a^4D_{1\frac{1}{2}}-y^2F^2_{3\frac{1}{2}}$	(0. 00 w) 1. 19	(0. 00) 1. 20
56. 92	6	6, III A	484. 58	$a^2P_{1\frac{1}{2}}-y^2P^2_{1\frac{1}{2}}$	(0. 00) 1. 29 R	(0. 08) 1. 29
38. 59	4	10, III A	586. 13	$a^2D_{3\frac{1}{2}}-z^4S^2_{1\frac{1}{2}}$		
4, 216. 54	4		709. 47	$a^2D_{3\frac{1}{2}}-z^2D^2_{1\frac{1}{2}}$		
4, 192. 72	HNR	2	844. 17	$b^2D_{1\frac{1}{2}}-y^2F^2_{3\frac{1}{2}}$		
87. 31	50	125, I	874. 98	$a^2D_{1\frac{1}{2}}-w^2F^2_{3\frac{1}{2}}$	(0. 00 w) 1. 16 A¹	(0. 00) 1. 16 A¹
77. 48	15	30, I	931. 15	$a^2D_{3\frac{1}{2}}-z^4F^2_{3\frac{1}{2}}$	(0. 00) 1. 50 (?)	(0. 20) 1. 16
72. 32	6	10, III A	9(?). 75	$a^4F_{1\frac{1}{2}}-y^4(i)_{1\frac{1}{2}}$		
71. 13	5	8, III A	23, 967. 58	$a^4F_{1\frac{1}{2}}-y^4G^2_{1\frac{1}{2}}$		
63. 31	5	8, III A	24, 012. 60	$a^4F_{2\frac{1}{2}}-y^4G^2_{3\frac{1}{2}}$		
60. 26	20	30, I	030. 21	$a^2D_{3\frac{1}{2}}-z^4D^2_{1\frac{1}{2}}$	(0. 00 w) 1. 65_w	(0. 00) 1. 67
57. 52	6	10, II A	046. 04	$a^2D_{1\frac{1}{2}}-z^4D^2_{1\frac{1}{2}}$		

TABLE 6.—*The arc spectrum of lanthanum (La I)*—Continued

λ air I. A.	Intensities, temperature class		ν vac cm⁻¹	Term combination	Zeeman effects	
	B. S.	K. & C.			Observed	Computed
50. 24	2		088. 22	$a^4F_{1\frac{1}{2}}$—$r^2F_{1\frac{1}{2}}^2$		
44. 36	4		122. 40	$a^4F_{2\frac{1}{2}}$—$y^2G_{2\frac{1}{2}}^2$		
43. 92	5	12, III A	124. 96	$a^4F_{2\frac{1}{2}}$—$y^2G_{3\frac{1}{2}}^4$		
37. 05	20	40, I	165. 02	$a^2D_{2\frac{1}{2}}$—$z^2D_{2\frac{1}{2}}^2$	(0. 00) 1. 22	(0. 09) 1. 23
17. 67	8	20, III	278. 75		(0. 00) 0. 93	
09. 80	10	20, I A	325. 24	$a^2D_{2\frac{1}{2}}$—$e^2F_{3\frac{1}{2}}^2$	(0. 00) 1. 04	(0. 00) 1. 04
09. 48	6	15, II A	327. 14	$a^2D_{2\frac{1}{2}}$—$x^2F_{1\frac{1}{2}}^2$	(0. 00) 1. 26	(0. 00) 1. 22
4, 104. 87	30	60, I	354. 46	$a^4F_{2\frac{1}{2}}$—$y^2G_{3\frac{1}{2}}^4$	(0. 08, 0. 25) 0. 53, 0. 69, 0. 85 ur	(0. 07, 0. 23) 0. 34, 0. 49, 0. 64, 0. 80
4, 090. 40	2		440. 61	$a^2F_{1\frac{1}{2}}$—$u^2D_{2\frac{1}{2}}^2$		
89. 61	25	50, I	445. 33	$a^4F_{2\frac{1}{2}}$—$y^2G_{2\frac{1}{2}}^2$	(0. 00) 0. 93	(0. 00) 0. 94
79. 17	20	40, I	507. 90	$a^2D_{1\frac{1}{2}}$—$z^2F_{2\frac{1}{2}}^2$		
65. 58	15	30, II	589. 82	$a^2D_{1\frac{1}{2}}$—$y^2P_{1\frac{1}{2}}^2$		
64. 79	25	50, II	594. 60	$a^4F_{1\frac{1}{2}}$—$y^2G_{1\frac{1}{2}}^4$	(0. 00 w) 1. 03 A¹	(0. 00) 0. 94 A¹
60. 33	30	60, II	621. 61	$a^4F_{0\frac{1}{2}}$—$y^4G_{0\frac{1}{2}}^4$	(0. 00 w) 1. 10 A¹	(0. 00) 1. 12 A¹
40. 97	2		739. 57	$a^2F_{0\frac{1}{2}}$—$u^2D_{0\frac{1}{2}}^2$		
37. 21	25	50, I	762. 61	$a^2D_{1\frac{1}{2}}$—$z^2D_{1\frac{1}{2}}^2$	(0. 00) 0. 86	(0. 13) 0. 84
15. 39	25	50, I	897. 17	$a^2D_{2\frac{1}{2}}$—$z^2P_{1\frac{1}{2}}^2$	(0. 10, 0. 31) 0. 94, 1. 10, 1. 30, 1. 48 ur	(0. 16, 0. 30) 0. 94, 1. 10, 1. 30, 1. 51
4, 001. 38	2		24, 984. 34	$a^2D_{1\frac{1}{2}}$—$z^2F_{1\frac{1}{2}}^2$		
3, 953. 67	10	40, II	25, 285. 83	$a^2D_{2\frac{1}{2}}$—$y^2P_{1\frac{1}{2}}^2$		
27. 56	30	80, I	453. 92	$a^2D_{1\frac{1}{2}}$—$z^2P_{1\frac{1}{2}}^2$	(0. 00) 0. 76	(0. 00) 0. 76
3, 902. 57	5	20, II	616. 91	$a^2D_{1\frac{1}{2}}$—$y^2P_{0\frac{1}{2}}^2$		
3, 898. 60	8	40, II	25, 642. 99	$a^2D_{1\frac{1}{2}}$—$y^2P_{1\frac{1}{2}}^2$		
3, 895. 65		87, IV	25, 662. 42			
3, 714. 30	2		26, 915. 34	$a^2D_{2\frac{1}{2}}$—$w^2D_{1\frac{1}{2}}^2$		
3, 704. 54	10	40, II	26, 986. 25	$a^2D_{2\frac{1}{2}}$—$u^2F_{1\frac{1}{2}}^2$	(0. 00) 1. 06	(0. 00) 1. 04
3, 699. 57	4	12, II A	27, 022. 51	$a^2D_{1\frac{1}{2}}$—$y^4G_{1\frac{1}{2}}^2$		
72. 02	8	30, II	225. 24	$a^2D_{1\frac{1}{2}}$—$w^2P_{1\frac{1}{2}}^2$		
49. 55	10	40, II	392. 86	$a^2D_{1\frac{1}{2}}$—$u^2F_{1\frac{1}{2}}^2$	(0. 00) 1. 00	[(0. 00) 0. 96
41. 53	20 ?	100, II	453. 19	$a^2D_{2\frac{1}{2}}$—$w^2D_{2\frac{1}{2}}^2$		
36. 67	6	40, III	489. 88	$a^2D_{1\frac{1}{2}}$—$v^2F_{1\frac{1}{2}}^2$		
3, 613. 08	10	30, II	669. 36	$a^2D_{1\frac{1}{2}}$—$e^2F_{1\frac{1}{2}}^2$	(0. 00) 0. 97	(0. 00) 0. 90
3, 574. 43	20	50, II	27, 968. 53	$a^2D_{1\frac{1}{2}}$—$w^2D_{1\frac{1}{2}}^2$	(0. 00) 0. 83	(0. 17) 0. 85
3, 514. 07	6	20, II A	28, 448. 92	$a^2D_{1\frac{1}{2}}$—$w^2D_{2\frac{1}{2}}^2$		
3, 480. 61	3	8, III A	722. 40	$a^2D_{2\frac{1}{2}}$—$y^2D_{2\frac{1}{2}}^2$		
61. 18	10	25, III A	883. 64	$a^2D_{2\frac{1}{2}}$—$7_{1\frac{1}{2}}$		
50. 65	5	12, III A	28, 971. 78	$a^2D_{1\frac{1}{2}}$—$z^2D_{1\frac{1}{2}}^2$		
3, 404. 53	10	15, III A	29, 364. 24	$a^2D_{2\frac{1}{2}}$—$8_{1\frac{1}{2}}$		
3, 388. 61	6	12, II A	502. 18	$a^2D_{1\frac{1}{2}}$—$w^4D_{1\frac{1}{2}}^4$		
81. 42		15, II A	564. 92	$a^2D_{1\frac{1}{2}}$—$6_{1\frac{1}{2}}$		
68. 36	3		679. 54	$a^2F_{1\frac{1}{2}}$—11°		
64. 88	2		710. 24	$a^4F_{1\frac{1}{2}}$—10°		
62. 04	7	12, III A	735. 33	$a^2D_{1\frac{1}{2}}$—$a^2F_{2\frac{1}{2}}^2$		
57. 50	5	7, III A	775. 54	$a^2D_{1\frac{1}{2}}$—$w^2D_{1\frac{1}{2}}^2$		
49. 82	2	3, III A	843. 80	$a^2D_{1\frac{1}{2}}$—$9_{1\frac{1}{2}}$		
3, 342. 23	10	20, II A	29, 911. 57	$a^2D_{2\frac{1}{2}}$—$a^2F_{2\frac{1}{2}}^2$	(0. 00) 1. 06	(0. 00) 1. 14
3, 256. 60	2		30, 698. 05			
47. 06	5	8, II A	788. 24	$a^2D_{1\frac{1}{2}}$—$a^2F_{1\frac{1}{2}}^2$		
35. 66	3	5, III A	30, 896. 71	$a^2D_{1\frac{1}{2}}$—$9_{1\frac{1}{2}}$		
3, 215. 81	10	15, II A	31, 087. 42	$a^2D_{2\frac{1}{2}}$—$r^2F_{1\frac{1}{2}}^2$	(0.00) 1.13	(0.00) 1.14
3, 179. 78	4	8, III A	439. 65	$a^2D_{1\frac{1}{2}}$—$u^2D_{1\frac{1}{2}}^2$		
75. 99	8	15, II A	477. 17	$a^2D_{1\frac{1}{2}}$—$r^2F_{1\frac{1}{2}}^2$		
48. 51	4	4, III A	31, 751. 89	$a^2D_{1\frac{1}{2}}$—$u^2D_{1\frac{1}{2}}^2$		
3, 109. 42	8	12, II A	32, 151. 04	$a^2D_{2\frac{1}{2}}$—$s^2P_{1\frac{1}{2}}^2$		
3, 096. 02	8	10, II A	32, 290. 19	$a^2D_{1\frac{1}{2}}$—$s^2P_{2\frac{1}{2}}^2$		
10. 78	4	2, IV A	33, 204. 34	$a^2D_{1\frac{1}{2}}$—$s^2P_{1\frac{1}{2}}^2$		
3, 001. 41	2		307. 99			
2, 992. 99	2		33, 401. 69			
09. 65	2		34, 358. 36			
2, 904. 62	1		34, 417. 85			
2, 817. 46	2		35, 482. 54			
2, 794. 03	4		35, 780. 07			
66. 46	4		36, 136. 62	$a^4F_{1\frac{1}{2}}$—13°		
61. 56	7		200. 74	$a^4F_{0\frac{1}{2}}$—14°		

TABLE 6.—*The arc spectrum of lanthanum (La I)*—Continued

λ air I. A.	Intensities, temperature class		ν vac cm⁻¹	Term combination	Zeeman effects	
	B. S.	K. & C.			Observed	Computed
59.54	4		227.24			
56.57	2		266.27			
49.52	2		359.25			
39.25	4		495.56			
37.49	3		519.03			
30.15	2		617.20			
29.85	5		621.22	$a^4F_{1\frac{1}{2}}$—13°		
25.57	15		678.73	$a^2D_{1\frac{1}{2}}$—11°		
22.31	6		722.65	$a^2D_{1\frac{1}{2}}$—10°		
17.33	2		789.90			
15.77	3		811.08			
14.52	8		828.03	$a^4F_{2\frac{1}{2}}$—14°		
10.69	4		880.06			
2,707.07	3		36,929.38	$a^4F_{1\frac{1}{2}}$—12°		
2,684.11	6		37,245.25			
77.77	2		333.43	$a^4F_{2\frac{1}{2}}$—15°		
71.91	2		415.30			
53.48	2		675.16	$a^4F_{1\frac{1}{2}}$—15°		
47.13	2		765.53			

V. IONIZATION POTENTIALS OF LANTHANUM ATOMS

The first two members of series have been identified in the spectra of lanthanum in all three degrees of ionization, and the ionization potentials for three different lanthanum atoms, La, La⁺, La⁺⁺, may thus be deduced spectroscopically from extrapolation of the series to their limits.

To use the Rydberg formula is equivalent to assuming that the denominator n^* changes by exactly 1 from one term to the next in a series. In those spectra for which long series and accurate limits are available, the change Δn^* between the first and second members is usually somewhat greater and it appears to be preferable to assume that the value for lanthanum is comparable with that for similar spectra. For example, in the 1-electron spectra we find the following values of Δn^* for the s electron:

K I	Ca II	Rb I	Sr II	Cs I	Ba II
1.031	1.033	1.040	1.042	1.050	1.052

We may therefore assume $\Delta n^* = 1.050$ for La III. The terms 6^2S and 7^2S then give a limit at 154,630 (above the lowest level $^2D_{1\frac{1}{2}}$) with $n^* = 2.6462, 3.6962$. This corresponds to an ionization potential of 19.07 volts which is the lowest known value for a third ionization. Badami[26] from his analysis of this spectrum derived 20.4 volts for the ionization potential of La⁺⁺.

---

[26] J. S. Badami, Proc. Phys. Soc. London, vol. 43, p. 53, 1931.

For 2-electron spectra we find for the ^{3}S and ^{3}D terms due to "running" s and d electrons:

Term	^{3}S			^{3}D		
Spectrum	Ca I	Sr I	Ba I	Ca I	Sr I	Ba I
Δn^*	1.040	1.042	1.045	1.135	1.181	1.289

The ^{1}S and ^{1}D series in the alkaline earths are too much perturbed [27] to be of use. We therefore assume that in La II $\Delta n^* = 1.05$ for a running s electron; that is, for the terms a^3D, f^3D, and a^1D, f^1D, with limit 5^2D in La III, and 1.15 for a running d electron (a^3D, h^3D with limit 6^2S). The resulting values of n^* (referred to the proper limit for each component) and of the difference between the lowest level in La II (a^3F$_2$) and La III (5^2D$_{1\frac{1}{2}}$) are as follows:

Term	^{3}D$_1$	^{3}D$_2$	^{3}D$_3$	^{1}D$_1$	^{3}D$_1$	^{3}D$_2$	^{3}D$_3$
n^*	{ 2.2232 / 3.2732	2.2333 / 3.2833	2.2206 / 3.2706	2.1826 / 3.2326	2.0384 / 3.1884	2.0453 / 3.1953	2.0518 / 3.2018
Limit	90,704	90,600	90,664	91,924	93,944	93,924	93,916

The mean of these seven values is 92,240, which corresponds to an ionization potential of 11.38 volts for La$^+$.

Finally, for La I, we have the terms a^4F, e^4F and a^2F, e^2F with a running s electron and limit a^3F of La II. With limit a^3D we have only a^4D and e^4D, but these terms, though of different multiplicity, may still be used.[28] The first comes from the configuration $5d\ 6s^2$, the second from $5d\ 6s\ 7s$. Dropping the $5d$ electron we obtain the lowest ^{1}S and ^{3}S terms in Ba I. For these and the corresponding terms in Ca I and Sr I we have the values of n^*.

Spectrum	Ca I	Sr I	Ba I
n^* (^{1}S)	1.442	1.545	1.616
n^* (^{3}S)	2.484	2.548	2.629
Δn^*	1.042	1.003	1.013

We may, therefore, adopt $\Delta n^* = 1.010$ in this case while in the others we use 1.060. We then find

Term	a^2D$_{2\frac{1}{2}}$ / e^2D$_{2\frac{1}{2}}$	a^4F$_{1\frac{1}{2}}$ / d^4F$_{1\frac{1}{2}}$	a^4F$_{2\frac{1}{2}}$ / e^4F$_{2\frac{1}{2}}$	a^4F$_{3\frac{1}{2}}$ / e^4F$_{3\frac{1}{2}}$	a^4F$_{4\frac{1}{2}}$ / e^4F$_{4\frac{1}{2}}$	a^2F$_{2\frac{1}{2}}$ / e^2F$_{2\frac{1}{2}}$	a^2F$_{3\frac{1}{2}}$ / e^2F$_{3\frac{1}{2}}$
n^*	{ 1.5221 / 2.5321	1.6033 / 2.6633	1.6000 / 2.6600	1.5948 / 2.6548	1.5892 / 2.6492	1.6829 / 2.7429	1.6855 / 2.7455
Limit	45,145	45,337	45,856	45,606	45,583	44,721	44,687

The mean is 45,293 for the interval from a^2D$_{1\frac{1}{2}}$ (La I) to a^3F$_2$ (La II). This corresponds to an ionization potential of 5.59 volts.

A change of $+0.01$ in Δn^* alters the computed limits on the average by -152 or -0.019 volts. The average discordance of the seven

---

[27] A. G. Shenstone and H. N. Russell, Phys. Rev., vol. 39, p. 415, 1932.
[28] We owe this suggestion to Prof. A. G. Shenstone.

determinations from the mean is ±334 or ±0.041 volts. We, therefore, adopt 5.59±0.03 volts as the first ionization potential of lanthanum. This is in excellent agreement with the value 5.49 volts
derived by Rolla and Piccardi [29] from observations of the ionization
current and rate of dissociation of La_2O_3 in flames.

For La II a change in Δn^* by ±0.01 affects the ionization potential
by ±0.044 volts, while the average discordance of one determination
is ±0.13 volts.

We may finally adopt for lanthanum atoms:

> First ionization potential   = 5.59±0.03 volts.
> Second ionization potential = 11.38±0.07 volts.
> Third ionization potential = 19.1 ±0.1 volts.

VI. DISCUSSION OF ZEEMAN EFFECTS

The observed Zeeman effects for lanthanum lines in all three
spectra positively identify most of the spectral terms although a
comparison of the observed and theoretical g values discloses many
discrepancies which greatly exceed the experimental errors. Similar
deviations have been observed in other spectra, and have been assigned either to departures from strict SL coupling, or to the sharing
of g values by mutually perturbing terms. If the coupling between
spin moment and orbital moment of each electron is weak individually
as compared with the coupling of the spins and those of the orbital
moments of the electrons mutually we have the so-called SL coupling:

$$\{(s_1\ s_2\text{----})\ (l_1\ l_2\text{----})\} = (SL) = J$$

An atom for which this is true will emit a spectrum in which interval
rules, and Zeeman effects are in accord with the values predicted by
the theory of Landé and others. But if the quantum vectors s and l
are actually compounded to produce a resultant vector J in some other
manner the spectrum will show violations of the interval rules and
deviations from Landé's g values. Such deviations are evident in
the La spectra. For example, the metastable quartet-F term in the
La I spectrum shows only a slight departure from the interval rule:

	Separations	Ratios
Theoretical	$^4F_{1\frac{1}{2}}-^4F_{1\frac{1}{2}}$:$^4F_{1\frac{1}{2}}-^4F_{1\frac{1}{2}}$:$^4F_{3\frac{1}{2}}-^4F_{1\frac{1}{2}}$=	9.00:7.00:5.00
La I	627.0 : 484.6 : 341.8 =	9.00:6.95:4.90

and the observed g values are almost exactly those of Landé

Level	Observed g	Landé g
$^4F_{1\frac{1}{2}}$	0.411	0.400
$^4F_{2\frac{1}{2}}$	1.032	1.029
$^4F_{3\frac{1}{2}}$	1.222	1.238
$^4F_{4\frac{1}{2}}$	1.337	1.333

The low terms in the La II spectrum do greater violence to the interval
rule, and also somewhat larger divergences of observed and theoretical

---

[29] L. Rolla and G. Piccardi, Phil. Mag., vol. 7, p. 286, 1929.

g values occur. Thus, for the normal state represented by a^3F we have the following interval ratios:

	Separations	Ratios
Theoretical........................	$^3F_4-^3F_3:^3F_3-^3F_2$	4.00:3.00
La II........................	954.6 : 1016.1	4.00:4.26

and the following g values:

Level	Observed g	Landé g
$^3F_2$	0.730	0.667
$^3F_3$	1.092	1.083
$^3F_4$	1.258	1.250

Similarly for the term a^3D, the interval ratios are:

	Separations	Ratios
Theoretical....	$^3D_3-^3D_2:^3D_2-^3D_1$	3.00:2.00
La II..........	658.8:696.4	3.00:3.17

and the splitting factors are:

Level	Observed g	Landé
$^3D_1$	0.520	0.500
$^3D_2$	1.140	1.167
$^3D_3$	1.337	1.333

These, and many other departures from theoretical values indicate a considerable deviation from (SL) coupling, probably in the direction of (jj) coupling defined, in the case of two electrons, by

$$\{(s_1l_1)\ (s_2l_2)\} = (j_1j_2) = J$$

in which the interactions between spin and orbital moment of each electron predominate.

No matter what the nature of the coupling may be the so-called g-sum rule of Pauli[30] is expected to be valid. This rule can be stated as follows: If, among all the terms arising through the coupling of one electron, those terms with the same inner quantum number J be grouped, the sum of the g values of the terms in each group must have a value independent of the nature of the coupling, and consequently is, among others, equal to the sum of the Landé g values for a similar group. This rule is tested in Table 7 where the observed and Landé g sums are displayed for all the groups in La II for which the data are complete and for the low terms in La I, although one term has not been found. The sums of observed g values are slightly larger than the theoretical sums, but the differences can perhaps be ascribed to the accumulation of a small systematic error of observation. If the observered g's are diminished by 0.78 per cent, to allow for this, the agreement becomes almost perfect, with an average discordance of only 0.5 per cent.

[30] W. Pauli, Zeits. f. Phys., vol. 16, p. 155, 1923.

TABLE 7.—Zeeman effect: Sums of adopted and Landé g values

La II

Electrons	Terms	6 Adopted	6 Landé	5 Adopted	5 Landé	4 Adopted	4 Landé	3 Adopted	3 Landé	2 Adopted	2 Landé	1 Adopted	1 Landé	Sum Adopted	Sum Landé
ds	a³D							1.337	1.333	1.140	1.167	0.520	0.500	2.997	3.000
s²	a¹D									.967	1.000			.967	3.000
	a³F					1.258	1.250	1.092	1.083	.730	.667			3.080	3.000
	a¹P					1.003	1.000			1.498	1.500	1.506	1.500	2.944	3.000
	a¹G					1.060	1.000							1.003	1.000
	b¹D									1.015	1.000			1.015	1.000
Sum low terms						2.261	2.250	2.429	2.416	5.360	5.334	2.026	2.000	**12.076**	**12.000**
pf	c³G			1.233	1.200	1.146	1.050	.873	.750	.733	.667			3.222	3.000
	c¹F					1.141	1.250	1.061	1.083	1.100	1.107			2.934	3.000
	c³D							1.306	1.333	1.046	1.000	.495	.500	2.901	3.000
	c³G									1.020	1.000			1.060	1.000
	c¹F													.945	1.000
	c¹D													1.045	1.000
ds	f¹D					1.060	1.000	.933	1.000	1.141	1.167	.520	.500	2.942	3.000
	f¹D							1.331	1.333	1.020	1.000			1.020	1.000
Sum all even terms				1.203	1.200	5.608	5.550	7.953	7.916	10.399	10.335	3.041	3.000	**28.204**	**28.000**
sf	z³F	1.17	1.17			1.26	1.25	1.09	1.08	.67	.67			3.02	3.00
	z¹F							1.05	1.00					1.05	1.00
df	z³H			1.08	1.03	.88	.80	.77	.75	.76	.67	.55	.50	3.11	3.00
	z³G			1.20	1.20	1.06	1.05	1.09	1.08	1.19	1.17	1.46	1.50	3.03	3.00
	z³P					1.24	1.25	1.32	1.33	1.46	1.50			3.09	3.00
dp	z³D			1.00	1.00	1.00	1.00	1.03	1.00	.93	1.00	.88	1.00	3.06	3.00
	z³P							1.08	1.06	.84	.67	.80	.50	2.92	3.00
	y³F							1.31	1.33	1.19	1.17	1.26	1.50	1.00	1.00
	y³P							1.06	1.00	1.49	1.50			1.03	1.00
ds	z³D					1.25	1.25			.90	.90	1.07	1.07	.93	1.00
	b³F									1.48	1.50	1.52	1.50	.86	1.00
	b³D											.97	1.00	3.17	3.00
	y³P													2.75	3.00
	z³F													1.08	1.00
	b³D													.90	1.00
	y³P													1.07	1.00
	z³P													3.00	3.00
	z¹P													.97	1.00
Sum all odd terms		1.17	1.17	3.28	3.23	6.67	6.60	9.80	9.66	10.91	10.84	8.51	8.50	**40.84**	**40.00**

TABLE 7.—*Zeeman effect: Sums of adopted and Landé g values*—Continued

La I

Electrons	Terms	4½ Adopted	4½ Landé	3½ Adopted	3½ Landé	2½ Adopted	2½ Landé	1½ Adopted	1½ Landé	½ Adopted	½ Landé	Sum Adopted	Sum Landé
ds^2	a^6D	1.33	1.33	1.22	1.24	1.20	1.20	0.79	0.80			1.99	2.00
d^2s	a^4F					1.03	1.03	.41	.40			3.99	4.00
	a^4P					1.54	1.60	1.69	1.73	2.71	2.67	5.94	6.00
	a^2G	1.12	1.11	.91	.89							2.03	2.00
	a^2F			1.13	1.14	.90	.86					2.18	2.00
	b^2D					1.29	1.20	.89	.80			1.99	2.00
	a^2P							1.32	1.33	.67	.67	'2.00	2.00
	^2S									(2.00)	2.00	(2.00)	2.00
Sum low terms		2.45	2.44	3.26	3.27	5.96	5.89	4.31	4.26	3.38	3.34	20.15	20.00

WASHINGTON, August 6, 1932.

ANALYSIS OF WEIGHTED SILK

By Ralph T. Mease

A generally applicable, rapid, and convenient method for the determination of the amount of pure silk fiber in silk textiles is described. Weighting and finishing materials are removed by repeated extractions with hot water, 2 per cent sodium carbonate solution and a solution containing 2 per cent of hydrochloric and 2 per cent of hydrofluoric acids. Results of analyses of samples of known composition are presented which indicate that the results are correct to within 1 per cent of the weight of the dried finished material. Results obtained by inexperienced analysts working in different laboratories have been in good agreement when samples of the same silk were analyzed. Qualitative methods for the identification of the following weighting materials are given: Aluminum, lead, phosphate, silica, tin, zinc.

CONTENTS

I. INTRODUCTION

Raw silk, as it comes to the manufacturer, contains about 75 per cent fiber (fibroin) and about 25 per cent gum (sericin) exclusive of moisture. At some stage of the processing the gum is removed by "boiling off" in a soap solution. The silk may then be "weighted" by the addition of various metallic salts or other substances, the amount of the weighting being determined by the purposes for which the silk is intended, the selling price, and similar considerations. The amount of weighting added seldom is less than the amount of the gum removed and it may exceed the amount of the silk fibroin. The claim is made that weighted silk has a better "hand" or "feel" than unweighted silk, that it drapes better, and that its lower cost makes possible the use of silk by persons who otherwise could not afford it.

After weighting, the silk may be dyed, or printed. It is finally "finished" by the addition of gums, soaps, waxes, hygroscopic substances, or other "finishing" materials which contribute to the desired properties.

The need for a standard method of analysis for weighted silk arose several years ago when special attention was being given to the problem of overweighting. There was an insistent demand from numerous manufacturers, distributors, and consumers for definite limits on the amount of weighting to be allowed on different types of silk fabrics. This led to the tentative designation of limits by the Silk Association of America in January, 1929.[1] In order to insure satisfactory analyses of weighted silks for conformance with these standards or other standards that might be agreed upon, attention

---

[1] Fifty-seventh Annual Report, Silk Association of America, New York, p. 27, 1929.

was given by a Technical Committee on Weighted Silk [2] to methods of analysis. A proposed standard method was published.[3]

The proposed standard method was the one in common use calling for extraction with hydrofluoric acid, amplified to make possible the successful analysis of silks weighted with materials other than tin-phosphate-silicate. When samples of silk of known weighting content were submitted to cooperating laboratories for analysis by this method, the results were in error in one instance by no less than 32 per cent of the amount actually present and in two other instances by 17 and 20 per cent.[4]

In anticipation of the need for analyses of weighted silks in several projects in the textile section of the Bureau of Standards, and with the expectation that the Federal Trade Commission would require analyses of weighted silks, some attention was given to the development of a generally applicable and satisfactory method. This work resulted in a method which is reasonably rapid, convenient, and reliable. The work is described and the resulting method is given in this paper. Qualitative tests that have been found useful in identifying inorganic weighting materials on silk are also given.

II. EXPERIMENTAL STUDIES

The important determination in the analysis of weighted silk is that of fibroin content. Several methods have been proposed for obtaining this value. In one method [5] the fibroin is evaluated by determining nitrogen, which is a constituent of silk fibroin. This value is in error if dyestuffs containing nitrogen are present in the silk. In another method [6] the fibroin content is indirectly evaluated by making a determination of the ash content of the weighted silk. This is done by igniting a sample in air. The weighting materials are changed during the ignition and the weight of the ash is not a true measure of the weighting materials contained in the material. A third method, which is also indirect, consists in the extraction of inorganic materials with hydrofluoric acid.[7] When lead, zinc, or aluminum salts are present, however, the material is alternately treated with hydrofluoric and hydrochloric acids. Neither of these last two treatments entirely removes the inorganic materials.

It was found in this laboratory that a solution containing 2 per cent of hydrofluoric and 2 per cent of hydrochloric acids was more effective in removing inorganic weighting materials than either acid alone, or the alternate treatments with these two acids. Furthermore, the combination of hydrofluoric and hydrochloric acids was found to be less destructive to silk fibroin than hydrochloric acid alone.

Unweighted silk cloth was purified by boiling it in several changes of neutral soap solution, then extracting with alcohol to remove soap,

---

[2] The Joint Committee of the Textile Industry on Weighted Silk, composed of representatives from national associations representing manufacturers, distributors, and users, established this smaller technical committee to consider technical problems.

[3] Anon, To find silk weighting content, Textile World, vol. 76, p. 1454, Sept. 14, 1929.

[4] Unpublished data of the Technical Committee on Weighted Silk.

[5] Gnehm, R., and Blummer, E., Méthode pour le dosage de la charges des soies noires, Rev. gén. mat. color., vol. 2, pp. 133-134, April, 1898.

[6] Branegan, James A., Practical analysis of methods for determination of artificial weighting of tin weighted silk, Melliand, vol. 1, pp. 735-739, August, 1929. Compares results obtained with several methods of analysis.

[7] Gnehm, R., Eine neue Methode zur quantitativen Bestimmung der Zinn-Phosphat-Silicat-Charge auf Seide, Z. Farben-Textil-Chem., vol. 2, pp. 209-210, June 1, 1903.

and finally with ether. It was then immersed in boiling water and finally dried in air at room temperature. This silk is referred to as "silk fibroin" in the experiments to be described.

The first step in the analysis of weighted silk by extraction with hydrofluoric acid is the removal of soluble finishing materials; that is, materials soluble in water or organic solvents that might be expected to be removed by either wet washing or dry cleaning. The purpose is not only to obtain a value for the amount of such materials, but to expose the insoluble weighting to more ready attack in subsequent treatments. Rinsing of the specimen with alcohol and ether to complete the removal of soluble finishing materials and to expedite the drying was found to be unsatisfactory because silk fibroin may hold these solvents even when dried above their boiling points. This was demonstrated in the following experiments.

Samples of silk fibroin weighing about 2.5 g each were dried to constant weight (± 0.1 per cent) in an air oven at 105° to 110° C. The time required was 1 to 1½ hours. Weighings were made at one-half hour intervals up to five hours. These samples were then soaked in 100 times their weight of distilled water at 65° C. for 20 minutes, rinsed in a fresh portion of distilled water, then in alcohol and finally in ether, a procedure typical of those used for soluble finishing materials. They were then dried to constant weight as before. The dry weight of the samples thus treated was approximately 1.5 per cent more than the weight of the original dry material. The samples were then immersed in boiling water and again dried. After this treatment the samples were found to have returned to their original weight. This procedure was repeated and it was clearly evident that when the silk was rinsed in alcohol and then in ether before drying, the weight of the dry material was considerably higher than when the silk was rinsed finally in boiling water. It was also found by other experiments that the alcohol and ether could be removed by water at temperatures below boiling. On this evidence, the final rinse in alcohol and ether must be followed by immersion of the specimens in water before drying.

After removal of soluble finishing materials, weighting is removed in the hydrofluoric acid method by alternate treatments of the silk with dilute hydrofluoric acid and dilute sodium carbonate solutions. Various concentrations, temperatures, and times of treatment are used in different laboratories. According to the method recommended by the Technical Committee on Weighted Silk, silk weighted with tin phosphate without silicate requires a treatment with dilute hydrochloric acid solution as well as with hydrofluoric acid.

Preliminary experiments indicated that both the concentration and temperature of the reagents are extremely important, too drastic treatment materially affecting the silk fibroin and too mild treatment not removing the weighting completely. Further, a mixture of hydrofluoric and hydrochloric acids [8] was found to be more efficient and less harmful than the separate treatments with hydrochloric and hydrofluoric acids called for in the method of the technical committee.

The treatments which were finally adopted are described in Section III. The effects of these treatments on silk weighted in several differ-

─────────────────────────
[8] Suggested informally several years ago by Dr. Victor Froelicher, chief chemist, Textile Dyeing Co. of America.

ent ways likely to be encountered in practice [9] are shown in Table 1. The amount of silk fibroin found by analysis of silk of known weighting content agreed with the amount present within 1 per cent of the weight of the dry finished silk for tin phosphate, logwood black, and other types of weighting as well as for the usual tin-phosphate-silicate weighted silk. The average loss in weight of eight samples of silk fibroin, when given the treatment for removing weighting, was 0.49 per cent and the maximum loss was less than 1 per cent.

Analyses of over 75 samples of commercial weighted silks have been made according to the method by several analysts, with excellent results. The amount of dry silk fibroin obtained in duplicate determinations agrees well within 1 per cent of the weight of the dried sample. The ash obtained by igniting the sample after treatment for removal of weighting seldom exceeds 0.1 per cent and is usually very much less than this when the silk is weighted with tin-phosphate-silicate.

Results of a few analyses of commercial samples made independently by analysts of two laboratories using different kind of equipment are given in Table 2.[10] Evidently good results may be obtained without the use of special equipment.

TABLE 1.—*Results of analyses of weighted silks*

[Values are in percentages of the weight of the ovendried specimen. Results of duplicate analyses given]

Sample No.	Type of weighting	Extraction with apparatus of Figure 1				Extraction with apparatus of Figure 3		
		Silk content	Soluble finishing materials	Residual ash	Silk fibroin	Soluble finishing materials	Residual ash	Silk fibroin
0	Tin phosphate...............	60.4	4.3 4.5	1.2	61.0 60.8	....	.....	.....
1	Tin phosphate silicate...........	48.3	4.5 3.7	.0	48.6 49.1	3.4 4.2	0.3	50.2 49.2
2	Tin phosphate, lead, silicate...........	50.8	4.5 5.1	.8	50.9 50.8	4.7 4.6	.1	52.8 52.6
3	Tin phosphate logwood black.........	52.1	4.9 5.2	.1	53.0 53.0	2.7 3.4	.1	55.0 54.3
4	Tin phosphate silicate...............	50.1	4.7 4.7	.1	50.5 49.9	....	.....	.....
5	Tin phosphate zinc	60.9	5.2 5.3	.0	60.6 60.8	....	.....	.....
10	Commercial silk, tin phosphate silicate	Unknown.	4.8 4.9	.0	48.4 48.0	4.2 3.9	.0	48.1 48.4
11	.....do.......................	Unknown.	6.2 6.0	.0	50.2 51.0	5.8 5.7	.0	50.9 51.1
12	.....do.......................	Unknown.	5.2 6.6	.0	60.0 59.2	4.6 5.2	.0	59.0 59.0
13	.....do.......................	Unknown.	6.8 6.5	.0	92.5 92.3	3.4 4.9	.0	93.5 93.1
14	.....do.......................	Unknown.	5.5 5.4	.1	49.1 49.4	4.7 5.2	.1	48.3 48.6
15	.....do.......................	Unknown.	5.5 5.2	.0	42.2 42.3	5.3 4.6	.0	41.6 41.2
16	.....do.......................	Unknown.	5.4 5.2	.0	50.5 50.0	4.8 5.9	.1	49.6 49.3
17	.....do.......................	Unknown.	6.8 6.4	.0	53.2 53.2	5.0 5.6	.1	54.9 55.2

[9] The author is indebted to Dr. Victor Froelicher for the samples of known weighting content representative of the different types of weighting.
[10] These results were obtained under the supervision of Prof. Pauline Beery Mack, of the department of home economics, Pennsylvania State College.

TABLE 2.—*Results of analyses by two independent laboratories on commercial silks*

[Values are for silk fibroin expressed as percentages of the weight of the ovendried specimen]

Sample No.	By Bureau of Standards extraction with apparatus of Figure 3	By Pennsylvania State College laboratory inexperienced analyst with no special equipment	Sample No.	By Bureau of Standards extraction with apparatus of Figure 3	By Pennsylvania State College laboratory inexperienced analyst with no special equipment
41	98.8	99.3	51	97.2	98.3
42	91.5	91.8	52	91.8	92.0
43	62.0	61.6	53	65.4	67.2
44	57.7	58.3	54	58.2	58.9
45	49.4	45.7	55	54.8	55.6
46	49.3	48.3	56	49.4	50.8
47	47.5	46.8	57	48.5	50.0
48	43.8	43.3	58	41.4	42.9

III. A METHOD FOR THE ANALYSIS OF WEIGHTED SILK

The specimen for analysis should be representative of the cloth. A strip the full width of the cloth is suggested and should weigh from 3 to 5 g.

The specimen is dried at 105° to 110° C. to constant weight (±0.1 per cent). If the specimen is loosely spread on a watch glass and placed in the usual laboratory electric oven at 105° to 110° C., it should dry within one and one-half hours. It is then placed in a tared weighing bottle, allowed to cool in a desiccator, and weighed. The weight of the dry specimen is called weight A.

Soluble finishing materials are removed by the following treatments. The specimen is immersed for about two minutes each in two 30 ml portions of diethyl ether at room temperature, squeezing by hand after each immersion. Then it is treated similarly with two 30 ml portions of ethyl alcohol at 50° to 60° C. Finally it is immersed in 90 to 100 times its weight of distilled water at 65° to 70° C. for 20 minutes, squeezed by hand, and rinsed by immersion for about one-half minute each in three fresh portions of distilled water at the same temperature, squeezing by hand after each immersion. It is then dried at 105° to 110° C. as before. The weight of the dry specimen at this stage is called weight B.

Weight A−weight B×100/weight A=soluble finishing materials in per cent (based on the weight of the dried specimen).

The more firmly held weighting and finishing materials are now removed by the following series of treatments: (*a*) The specimen is immersed in 90 to 100 times its weight of a solution [11] containing 2 per cent of hydrofluoric acid and 2 per cent of hydrochloric acid maintained at a temperature of 55° C.±1° C. The acid liquid is decanted and the specimen is rinsed with two portions of distilled water at 55° to 60° C. squeezing by hand after each rinse. (*b*) The

[11] This reagent may conveniently be prepared by mixing equal volumes of 4 per cent (±0.1 per cent) hydrofluoric and 4 per cent (±0.1 per cent) hydrochloric acids. High concentrations of mineral acids disintegrate silk. Too low a concentration may not remove inorganic materials sufficiently well from all types of weighted silk fabrics. For this work the concentration of each of the acids was determined by titration with a standard base. The concentrations given mean per cent by weight of actual HF and HCl (not of the ordinary concentrated acids). The reagent is approximately N with respect to HF and N/2 with respect to HCl.

treatment with acid solution (fresh solution) followed by rinsing is repeated. (c) This treatment is again repeated using a 2 per cent solution of sodium carbonate in place of the acid solution. (d and e) The treatment with acid solution is then applied two times more. (f) The specimen is then rinsed several times with distilled water and is dried as previously described. The dry weight is called weight C. Provided all materials other than silk fibroin have been removed

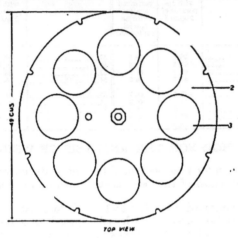

TOP VIEW

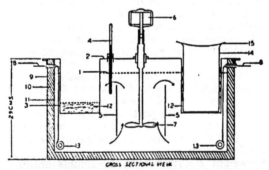

CROSS SECTIONAL VIEW

FIGURE 1.—*Electrically heated oil bath for control of the temperature of the solutions during the extraction process*

1, level of oil in the bath; 2, metal cover; 3, body of water bath container; 4, thermometer; 5, wall of cylinder to guide the circulating oil; 6, electric motor; 7, propeller blade for agitating the oil bath; 8, glass insulated electric leads to heater coil; 9, outer wall of metal container; 10, mineral wool for heat insulation; 11, inner wall of double-jacketed metal container; 12, water bath; 13, nichrome wire heating coil—1 unit of a 3 and 1 of a 4 amp current consumption at 110 v; 14, 800 ml Pyrex beaker, tall form; and 15, watch glass cover.

(see suggestions for ashing the specimen below), then: Weight $C \times 100$/weight A = silk fibroin in per cent (based on the weight of the dried specimen).

Weight A − weight $C \times 100$/weight A = total weighting in per cent (based on the weight of the dried specimen).

The total weighting thus obtained includes all substances other than silk fibroin.

Weighting is adequately removed from some silks when steps (b) and (e) are omitted, but these steps are necessary for the complete removal of weighting from others. It is advisable to determine the ash as a check on the effectiveness of the extraction. Ashing is done preferably by heating the specimen in a well ventilated muffle furnace. If the weight of the ash is more than 0.2 per cent it should be subtracted from the value obtained for weight C in the formulas.

The simple oil bath shown in Figure 1 is a convenience if a number of analyses are to be made. With it the temperature of the solutions can readily be controlled to within 1° C. and eight specimens can be handled simultaneously.

When many samples are to be analyzed time may be saved by using a battery of semiautomatic extraction units. One of these units, constructed of thick walled Pyrex glass, is shown in Figure 2. The specimens to be extracted are placed loosely in the tube, F.

Apparatus of the dimensions indicated will hold five specimens. The liquid is placed in the 1-liter flask, heated to the proper temperature, and the flask is inverted in the position shown at A. The orifice, D, is of such size that 15 to 20 minutes are required for the flask to empty. The extraction is not always quite so thorough with this apparatus as when the specimens are treated individually in beakers in the constant temperature bath.

Concentrated hydrofluoric acid may be measured with a pipette like that shown in Figure 3. Such a pipette may be made of Pyrex glass and coated on the inside by drawing melted ozokerite into the

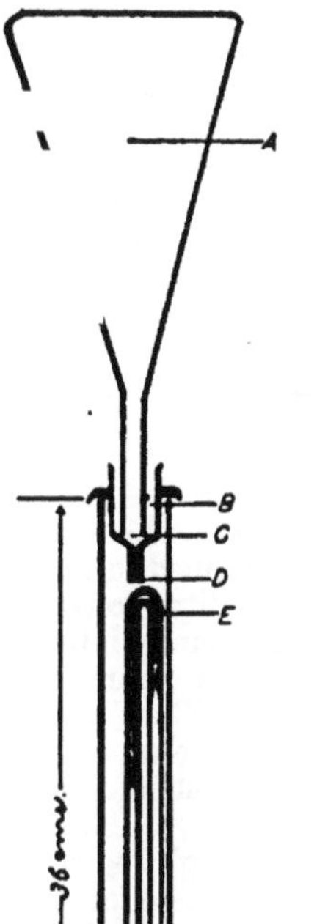

FIGURE 2.—A semiautomatic extraction apparatus for analysis of weighted silk

A, 1,000 ml Erlenmeyer flask; B, trap; C, outlet for liquid and inlet for air to flask A; D, capillary tube controlling flow of liquid; JEG, siphon tube; F, chamber for samples; and H, reservoir for spent liquor.

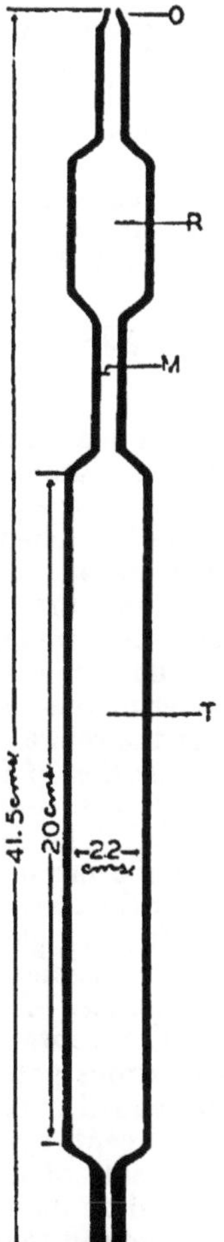

FIGURE 3.—A pipette for transferring and measuring concentrated hydrofluoric acid

T, main reservoir; R, auxiliary reservoir; M, calibration mark; O, suction orifice; and C, capillary discharge tube.

cold pipette and allowing the excess to drain. The acid is drawn into the pipette by means of a rubber tube attached at one end to *O* and at the other to the laboratory vacuum. The acid may be diluted and handled in a Pyrex beaker also coated on the inside with ozokerite. If hydrofluoric acid comes in contact with the skin, it should be removed as soon as possible by washing with a solution of borax to prevent burns. Hydrofluoric acid burns should be treated by keeping the affected parts moist with a saturated solution of borax.

IV. QUALITATIVE TESTS FOR INORGANIC WEIGHTING MATERIALS

The qualitative tests briefly described below provide for the detection of aluminum, lead, phosphate, silicate, tin and zinc when any one or all are present. They are based upon those described in textbooks [12] of chemical analysis and were chosen for reliability, convenience as to length of time required, and simplicity of manipulation. The tests are useful only for the detection of substances commonly used for weighting according to present practices. When small amounts of these substances are to be detected or additional or interfering ones are suspected of being present, the analyst should consult the references given in footnote 12.

Detection of inorganic materials in silk.—Inorganic weighting materials are conveniently detected by burning a small piece of the fabric. The ash will retain the form of the original threads if the material has been weighted with inorganic materials. A sample that is not thus weighted will curl back upon itself as it is burned and leave a soft black globular ash.

Identification of inorganic weighting materials—(a) Preparation of the sample.—A piece about 4 by 10 inches is thoroughly wet with a solution containing 90 g of anhydrous sodium carbonate and 90 g of anhydrous potassium carbonate, dissolved in 400 ml of water. It is suspended over a beaker containing 30 ml of dilute (approximately 10 per cent) hydrochloric acid solution and partially dried by brushing the flame of a hand torch or Bunsen burner across it. It is then ignited by directing the flame on the lower edge. The fused portions are allowed to drop into the acid in the beaker.

(b) Silica.—If silicates are present, a white flocculent precipitate is formed when the acid solution from (a) is boiled. The solution is decanted and filtered and the residue in the beaker digested with about 10 ml of warm concentrated hydrochloric acid for about 15 minutes; this acid is poured through the filter paper and added to the filtrate, which is boiled with 5 ml of concentrated nitric acid, cooled neutralized with sodium hydroxide (25 per cent solution) and enough hydrochloric acid added to produce a clear solution. The solution is used for tests (c) to (g) described below.

(c) Lead.—About 3 ml of the solution is neutralized with sodium hydroxide using phenolphthalein as an indicator. The solution is then made just acid with hydrochloric acid and about 1 ml of water saturated with hydrogen sulphide is added. If lead is present, a gray

[12] Prescott, Albert B., and Johnson, Otis C., Qualitative Chemical Analysis, 7th ed., thoroughly revised. 436 pp., 1916. D. Van Nostrand Co., New York. Scott, Wilfred W., Standard Methods of Chemical Analysis, 4th ed., vol. 1, 745 pp., 1925. D. Van Nostrand Co., New York. Hillebrand, W. F., and Lundell, G. E. F., Applied Inorganic Analysis, 929 pp., 1929. John Wiley & Sons (Inc.), New York

to black coloration is produced. (A yellow precipitate indicates the presence of tin.) If the sample of silk to be tested is light in color, it may be tested directly for lead by moistening it with water and exposing it to a jet of hydrogen sulphide gas. A change to gray or black indicates the presence of lead.

(d) *Aluminum.*—To about 3 ml of the solution, ½ ml of a saturated solution of ammonium chloride, ½ ml of alizarin (prepared by dissolving 0.1 g of the dye, Colour Index No. 1034 in distilled water), and an excess of ammonium hydroxide are added. The mixture is heated to boiling, an excess of glacial acetic acid is added, and the mixture diluted. A red flocculent precipitate denotes the presence of aluminum.

(e) *Tin.*—About 2 ml of the solution is neutralized with 25 per cent sodium hydroxide and about five drops in excess added. The solution is filtered, the filtrate acidified with hydrochloric acid, and a few drops of a 1 per cent water solution of cupferron added. A white precipitate indicates the presence of tin.

(f) *Zinc.*—About 3 ml of the solution is neutralized with 25 per cent sodium hydroxide, enough hydrochloric acid and water are added to produce a clear solution, and about ½ ml of a 2 per cent solution of potassium ferrocyanide is then added. A white precipitate indicates the presence of zinc.

(g) *Phosphate.*—About 2 ml of the solution is neutralized with sodium hydroxide and the resulting solution is acidified with nitric acid. It is heated and then added to an equal volume of hot acid ammonium molybdate. A lemon yellow precipitate is formed on standing if phosphates are present.

WASHINGTON, July 25, 1932.

141809—32——7

RP499

THE HEAT OF FORMATION OF HYDROGEN CHLORIDE AND SOME RELATED THERMODYNAMIC DATA

By Frederick D. Rossini

ABSTRACT

The calorimetric experiments of the present investigation give for the change in heat content associated with the formation of gaseous hydrogen chloride from its elements, at a constant pressure of 1 atmosphere, $\Delta H°_{298.1} = -92.30 \pm 0.05$ international kilojoules mole^{-1}, or $-22,063 \pm 12$ g-cal.$_{15}$ mole^{-1}.

The existing data on the free energy and entropy of formation of gaseous hydrogen chloride are shown to be in accord, within the limits of uncertainty, with the above result. This correlation consists in bringing together (1) data on the emf of certain cells, (2) data on the vapor pressure of hydrogen chloride over its aqueous solution and the activity of aqueous hydrogen chloride, and (3) values of entropy which have been calculated from the data of spectroscopy.

Values of the apparent and partial molal heat capacity, and heat content, for hydrogen chloride, and of the partial molal heat capacity, and heat content, for water, in aqueous solution of hydrogen chloride at 25° C., are given for the entire range of concentration.

The change in heat content associated with the solution of 1 mole of gaseous hydrogen chloride in an infinite amount of water is calculated to be $\Delta H°_{298.1} = -17,880 \pm 40$ g-cal.$_{15}$ mole^{-1}.

The change in heat content for the formation of silver chloride from its elements is given by: $\Delta H°_{298.1} = -30,304 \pm 40$ g-cal.$_{15}$ mole^{-1}. For mercurous chloride the corresponding value is $\Delta H°_{298.1} = -31,580 \pm 45$ g-cal.$_{15}$ mole^{-1}.

The change in heat content for the formation, from its elements, of aqueous hydrogen chloride at infinite dilution is given by: $\Delta H°_{298.1} = -39,943 \pm 40$ g-cal.$_{15}$ mole^{-1}.

The free energy of formation of gaseous hydrogen chloride is calculated to be $\Delta F°_{298.1} = -22,775 \pm 12$ g-cal.$_{15}$ mole^{-1}.

The entropy of aqueous hydrogen chloride, in a hypothetical 1 molal solution, is calculated to be: $S°_{298.1} = 13.40 \pm 0.15$ g-cal.$_{15}$ mole^{-1} °C.$^{-1}$.

CONTENTS

I. INTRODUCTION

It has seemed desirable for a number of years that a new measurement be made of the heat of formation of hydrogen chloride, a fundamentally important chemical substance because it is one of the components of a number of chemical reactions having industrial and scientific importance, and because the value of its heat of formation is used in computing the heats of formation of practically all the metallic chlorides and of numerous other compounds.

The usually accepted value for the heat of formation of gaseous hydrogen chloride is based upon the work of Thomsen [1] in 1873. The pioneer work of Abria,[2] and of Favre and Silbermann,[3] gave values that are now obviously too high. The work of Berthelot [4] is difficult to appraise because of the lack of information concerning his experiments.

In effect, then, there is only one calorimetric investigation of the heat of formation of hydrogen chloride whose detailed data are available for appraisal and recalculation, and that work was performed 60 years ago.

An accurate calorimetric value of the change in heat content, ΔH, for the reaction

$$\tfrac{1}{2}H_2 \text{ (gas)} + \tfrac{1}{2}Cl_2 \text{ (gas)} = HCl \text{ (gas)} \tag{1}$$

is unusually interesting because of the opportunity it affords for correlating the thermodynamic data on this reaction by means of the relation [5]

$$\Delta H° = \Delta F° + T\Delta S° \tag{2}$$

where $\Delta F°$ and $\Delta S°$ are, respectively, the free energy and entropy of formation. The value of $\Delta S°$ is accurately known from the data of spectroscopy; [6] and the value of $\Delta F°$ is obtained by combining data on the emf of cells and the vapor pressure of hydrogen chloride over its aqueous solution.[7] Thus there is the interesting situation where the values of $\Delta H°$, $\Delta F°$, and $\Delta S°$ for the same reaction are obtained from totally independent sources:

    $\Delta H°$, from calorimetric measurements;

    $\Delta S°$, from spectroscopic data; and

    $\Delta F°$, from emf and vapor-pressure data.

1 Thomsen, J., Pogg. Ann., vol. 148, p. 177, 1873.
2 Abria, Compt. rend., vol. 22, p. 372, 1846.
3 Favre, P. A., and Silbermann, J. T., Ann. chim. phys., vol. 34, p. 357, 1852.
4 Berthelot, M., Thermochimie (Gauthier-Villars, Paris), 1897.
5 The thermodynamic symbols used in this paper will follow the nomenclature of Lewis and Randall, Thermodynamics and the Free Energy of Chemical Substances (McGraw-Hill Book Co., New York), 1923.
6 Glauque, W. F., and Overstreet, R., J. Am. Chem. Soc., vol. 54, p. 1741, 1932.
7 Randall, M., and Young, L. E., J. Am. Chem. Soc., vol. 50, p. 989, 1928.

These data should all agree, according to equation (2), within the limits of uncertainty of the weakest link in the chain.

Another interesting correlation can be made on the change in heat content for reaction (1) and the following:

$$\tfrac{1}{2}H_2 \text{ (gas)} + \tfrac{1}{2}Cl_2 \text{ (gas)} = HCl \text{ (aqueous)} \qquad (3)$$
$$HCl \text{ (gas)} = HCl \text{ (aqueous)} \qquad (4)$$

For reaction (3) $\Delta H°$ can be computed from data on the temperature coefficient of the emf of cells, and for reaction (4) it is known from direct calorimetric measurements. The sum of the changes in heat content for reactions (1) and (4) should agree, within the limits of uncertainty, with the value of $\Delta H°$ for reaction (3). For this correlation there are required values of the apparent and partial molal heat content of hydrogen chloride in aqueous solution to infinite dilution, and the heats of formation of silver chloride and mercurous chloride.

With accurate values available for $\Delta H°$ and $\Delta F°$ it becomes possible to compute $\Delta S°$ for reaction (3), and consequently to obtain an accurate value of the entropy of aqueous hydrogen chloride in its standard state of a hypothetical 1 molal solution. This quantity serves as a basis for the calculation of the entropies of aqueous ions according to the methods used by Latimer and his coworkers.[8]

II. UNITS, FACTORS, ATOMIC WEIGHTS, ETC.

The fundamental unit of energy employed in the present calorimetric experiments is the international joule based upon standards of emf and resistance maintained at this bureau.[9]

The conversion factors used throughout this paper are:

1 international joule   = 1.0004 absolute joules
1 g-cal.$_{15}$ = 4.185 absolute joules
R (gas constant)   = 1.9869 g-cal.$_{15}$ °C^{-1}
F (Faraday constant) = 23,067 g-cal.$_{15}$ (international volt-equivalent)$^{-1}$

The atomic weights of hydrogen and chlorine are taken from the 1932 report of the International Committee on Atomic Weights:[10] H, 1.0078; Cl, 35.457.

III. CALORIMETRIC DETERMINATION OF THE HEAT OF FORMATION OF HYDROGEN CHLORIDE

1. METHOD AND APPARATUS

The same calorimetric method and apparatus as were employed in determining the heat of formation of water were used in the present investigation,[11] with the exception that an automatic photo-electric regulator was used to maintain the temperature of the jacket constant

[8] Latimer, W. M., and Ahlberg, J. A., J. Am. Chem. Soc., vol. 54, p. 1900, 1932. References to the earlier publications are given in this paper.
[9] For a report on the international comparison of electrical standards, see Vinal, G. W., B. S. Jour. Research, vol. 8, p. 448, 1932.
[10] J. Am. Chem. Soc., vol. 54, p. 1269, 1932.
[11] Rossini, F. D., B. S. Jour. Research, vol. 6, p. 1, 1931; vol. 7, p. 329, 1931. The flow meters were removed from their previous positions and placed in the waste lines.

to ±0.001° C.[12]   A platinum resistance thermometer, placed under the stirrer and heater in the water of the calorimeter jacket, served as one arm of a Wheatstone bridge whose unbalancing caused a shifting of a beam of light (reflected from a galvanometer mirror) on or off the photo-electric cell, which resulted in actuation of the relay and consequent stopping or starting of the regulatory flow of electrical current through the heater of the calorimeter jacket.

The calorimetric method is substantially the same as that of Thomsen. Chlorine is burned in an atmosphere of hydrogen, and the amount of reaction is determined from the mass of hydrogen chloride formed.

2. CHEMICAL PROCEDURE

(a) CHLORINE

Pure chlorine was obtained by decomposing potassium chloroplatinate which was carefully prepared from pure platinum, potassium chloride, and hydrochloric acid according to the procedure recommended by Noyes and Weber [13] in their work on the atomic weight of chlorine.   The following reactions indicate the nature of the process used in the present investigation:

$$\text{Pt (solid)} + 6 \text{ HCl (aqueous)} = \text{H}_2\text{PtCl}_6 \text{ (aqueous)} + 2 \text{ H}_2 \text{ (gas)} \quad (5)$$

$$\text{H}_2\text{PtCl}_6 \text{ (aqueous)} + \text{KCl (aqueous)} = \text{K}_2\text{PtCl}_6 \text{ (solid)} + 2 \text{ HCl}$$
$$\text{(aqueous)} \quad (6)$$

$$\text{K}_2\text{PtCl}_6 \text{ (solid)} = 2 \text{ KCl (solid)} + \text{Pt (solid)} + 2 \text{ Cl}_2 \text{ (gas)} \quad (7)$$

Each of the above reactions is practically complete and not complicated by any side reaction.   The required reagents, platinum, potassium chloride, and hydrochloric acid, are obtainable in a very pure state.

Pure platinum was obtained from the platinum metals section of this bureau.   It had been prepared according to the methods described by Wichers, Gilchrist, and Swanger.[14]

The potassium chloride was made by recrystallizing twice the best "reagent quality" potassium chloride.   The most probable impurity in this material would be bromine, and tests indicated that this was present to the extent of 1 mole in 50,000 moles of potassium chloride.[15]

Aqueous hydrogen chloride was obtained by treating the best "reagent quality" hydrochloric acid with chlorine (made by the action of hydrogen chloride upon potassium permanganate) and then boiling off the dissolved chlorine.   Tests of this aqueous hydrogen chloride indicated that the bromine present must have been less than 1 mole in 100,000 moles of hydrogen chloride.

The preparation of the chloroplatinic acid was carried out in the electrolytic cell of Weber, according to his directions, except that a current of 6 to 7 amperes instead of 10 was used.[16]

---

[12] This regulator was of the type that has been described quite frequently in the past few years (see for example (a) Southard, J. C., and Andrews, D. H., J. Frank. Inst., vol. 207, p. 323, 1929; (b) Scott, R. B., and Brickwedde, F. G., B. S. Jour. Research, vol. 6, p. 401, 1931; (c) Beattie, J. A., Rev. Sci. Inst., vol. 2, p. 458, 1931).   The author is indebted to E. N. Bunting for assembling the circuit of vacuum tubes, photoelectric cell, and relay.

[13] (a) Noyes, W. A., and Weber, H. C. P., J. Am. Chem. Soc., vol. 30, p. 13, 1908; B. S. Bull., vol. 4, p. 345, 1907-8.   (b) Weber, H. C. P., J. Am. Chem. Soc., vol. 30, p. 29, 1908; B. S. Bull., vol. 4, p. 365, 1907-8.

[14] Wichers, E., Gilchrist, R., and Swanger, W. H., Trans. Am. Inst. Mining Met. Eng., vol. 76, p. 602, 1928.

[15] The tests for bromine were made according to Noyes, A. A., Qualitative Chemical Analysis (Macmillan Co., New York, N. Y.), 1925.

[16] The construction of the electrolytic cell had been improved by Edward Wichers, and the author is greatly indebted to him for his advice with regard to the preparation of the chlorine used in the present investigation.

The potassium chloroplatinate was made according to reaction (6), following the directions of Noyes and Weber except that the salt was washed with water only. The potassium chloroplatinate was then placed in an oven at 110° C. for several days to remove most of the moisture.

For carrying out reaction (7) the apparatus shown in Figure 1 was used. *A* is the furnace with electrical resistance heater; *B* is a silica flask containing the potassium chloroplatinate; *C* is a silica to Pyrex-glass seal; *D* is a trap for catching any potassium or platinum chlorides which might be volatilized; *E* is a condensing tube for collecting the chlorine; *G* is a connection to the vacuum system, which consists of a mercury vapor pump (properly trapped with liquid air) backed by an oil vacuum pump; *K* is a vacuum-tight Pyrex glass-to-copper

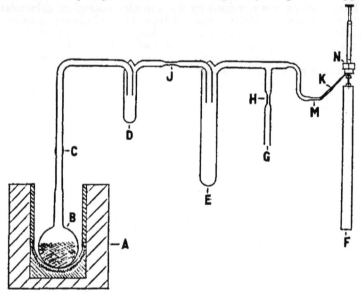

FIGURE 1.—*Apparatus for preparing pure chlorine*
The description is given in the text

seal; *F* is the Monel-metal container, having a Monel-metal diaphragm valve, in which the chlorine was finally stored.

The potassium chloroplatinate was heated to about 300° C. and the system kept evacuated to 0.0001 mm mercury for about three days. This was done to remove the last traces of water and hydrogen chloride from the solid. Then the constriction at *H* was sealed off. The flask *B* was heated to about 750° C. and liquid air placed around the condensing tube *E* where chlorine slowly began to collect as a solid. When chlorine no longer came over, the constriction at *J* was sealed off. The solid chlorine in *E* was liquefied and then distilled into the container *F*, the lower end of which was cooled with liquid air. When all but a small amount of the chlorine was collected in *F*, the valve *N* was closed and the glass system was broken at *M*. The chlorine container *F* was then sealed to the glass line leading to the calorimeter.

A sketch of the chlorine container and valve is shown in Figure 2. A is the cylindrical body of the container. The end plugs B and C are screwed and silver soldered in place. The connecting tube D is screwed and silver soldered to the plug C and tightly connected to the valve F by means of the metal-to-metal connection at E. The valve F, which has a silver diaphragm and in which closures are made with Monel metal to silver seats, is one of the type used in the calorimetric researches on steam by Osborne, Stimson, and Fiock.[17] G is the exit tube which is lead soldered to the body of the valve. The parts A, B, C, D, F, and G are all made of Monel metal. The short copper tube H is lead soldered to the tube G.[18] At J is a vacuum-tight seal of Pyrex glass to copper.

In order to select the material of which the container and valve were to be made, an experiment was made to determine the relative reactivity of gaseous chlorine to various metals. In one end of a Pyrex glass inverted U tube (about 12 by 12 inches) were placed pieces of platinum, silver, Monel metal, steel, brass, and phosphor bronze. To the other end of the U tube was sealed a small quartz tube containing a small amount of potassium chloroplatinate. The entire tube was evacuated to a pressure of about 2 mm of mercury, and the potassium chloroplatinate was then decomposed to give a pressure of 1 to 2 atmospheres of chlorine in the tube (which probably contained also a very small amount of water vapor). After a period of about six months, visual inspection of the tube, which had been at room temperature, showed that the platinum and Monel metal were apparently unaffected by the chlorine, the silver and brass were very slightly coated with chloride, and the phosphor bronze and steel showed considerable evidence of reaction. When at the end of the first series of reaction experiments, which are to be described later, the Monel-metal valve was taken apart, the silver diaphragm and seat, as well as the Monel-metal surface of the interior of the valve, were found to have been unaffected by the chlorine.

G
F
H
E
D
C
A

[17] Osborne, N. S., Stimson, H. F., and Fiock, E. F., B. S. Jour. Research, vol. 5, p. 411, 1930. The success of the present experimental investigation was in large measure due to having a suitable container and valve for handling the chlorine, and the writer is greatly indebted to the heat measurements section for the loan of the Monel-metal diaphragm valve and the Monel-metal tube from which the container was made.

[18] The use of lead solder eliminates the possibility of forming the comparatively volatile stannic chloride.

2.—*Chlorine conith diaphragm valve* ption is given in the text.

(b) HYDROGEN

The hydrogen used in the present investigation was taken from a cylinder of commercial electrolytic hydrogen. Oxygen was removed from this gas by passage through a tube containing asbestos, impregnated with palladium, and quartz, impregnated with platinum, at about 500° C. Before entering the calorimeter, the gas then passed successively through "Ascarite" (a sodium-hydroxide asbestos mixture), "Dehydrite" ($Mg(ClO_4)_2 \cdot 3H_2O$), and phosphorus pentoxide.

Analyses of the hydrogen gas as it would enter the reaction chamber were made by the gas chemistry section. That part of the gas which was noncombustible with oxygen was found to be 0.0 ± 0.1 per cent. A test for the presence of carbon compounds in the hydrogen was made by burning it with a special sample of oxygen[19] and testing the exit gases for carbon dioxide. None was found.

(c) DETERMINATION OF THE AMOUNT OF REACTION

The amount of reaction in each combustion experiment was determined by absorbing the hydrogen chloride in "Ascarite."

With the absorber filled with hydrogen at 1 atmosphere and 25° C., and with brass weights in air, the true mass[20] of hydrogen chloride absorbed is

$$m_{(HCl)} = 0.99991 \Delta m \qquad (8)$$

where Δm is the observed increase in weight of the absorber, corrected for the small amount of unburned chlorine which was also absorbed. The determination of this latter quantity is described on page 686 of this paper.

In its passage from the container to the reaction chamber, the chlorine traversed one 2-way stopcock, which served to permit wasting immediately before the ignition and after the extinction of the flame, and two small ground-glass joints. Phosphoric acid was used to lubricate the stopcock and the two ground joints, the latter being lubricated anew for each experiment. With hydrogen gas flowing through the stopcock and ground joints so lubricated, a carefully weighed absorber connected to the exit end of the reaction chamber showed constancy of weight (± 0.1 mg per experiment, which is 1 part in 50,000).

The two ground joints on the exit side of the reaction chamber were lubricated with the same lubricant as previously used for nonreactive gases. In order to make sure that the lubricant on these two small ground joints would take up no appreciable amount of hydrogen chloride, the lubricant, spread over a large area, was exposed to hydrogen-chloride gas for the time of a calorimetric reaction experiment, and its change in weight determined. It was calculated that under the conditions of a calorimetric reaction experiment, the amount of hydrogen chloride taken up by the lubricant on the two ground joints was about 0.03 mg, or 1 part in 200,000.

Several experiments were made to determine the completeness of the removal, by the "Ascarite," of hydrogen chloride from the gas leaving the reaction vessel. In one experiment a concentrated aqueous solution of silver nitrate was placed beyond the absorber.

[19] See footnote 11, p. 681.
[20] See footnote 11, p. 681.

In another experiment a second absorber, carefully weighed, was placed in series with the first absorber. In both these experiments no hydrogen chloride passed the first absorber with the gas flowing at three or four times the rate actually used in the calorimetric experiments.

3. CALORIMETRIC PROCEDURE

(a) ELECTRICAL ENERGY EXPERIMENTS

These experiments were performed, as nearly as ossible, over the same temperature range and with the same rate ofptemperature rise as the combustion experiments. The description of the procedure for the electrical energy experiments has already been given.[21]

The results of the electrical energy experiments are given in Table 1. The "error," computed [22] as

$$\pm 2\sqrt{\frac{(x-\bar{x})^2}{n(n-1)}}, \text{ is } \pm 0.014 \text{ per cent}$$

TABLE 1.—*Calorimetric results of the electrical energy experiments for hydrogen chloride*

Experiment No.	ΔR	k	u	K	U	Δt corr.	Average temperature	Electrical energy[1]	Mass of calorimeter water[1]	Electrical equivalent of calorimeter system[2]	Deviation from mean
	ohm	*min.⁻¹*	*min.⁻¹*	*ohm*	*ohm*	*°C.*	*°C.*	*Int. joules*	*g*	*Int. joules °C.⁻¹*	*joules °C.⁻¹*
1	0.100338	0.001970	−0.0000084	0.002819	−0.000235	0.97044	30.00	14,988.3	3,620.53	15,359.1	2.7
2	.099976	.001945	−.0000035	.002810	−.00009S	.96557	30.00	14,984.0	3,614.38	15,354.7	−1.7
3	.101019	.001934	.0000024	.002783	.000067	.97455	30.00	15,009.2	3,611.53	15,353.1	−3.3
4	.039543	.001947	.0000015	.002577	.000039	.96225	29.99	14,832.4	3,614.38	15,354.3	−1.9
5	.099308	.001966	−.0000032	.002617	−.000083	.96071	29.99	14,841.5	3,621.98	15,356.8	.6
6	.098320	.001968	.0000003	.002635	.000008	.95975	29.99	14,511.1	3,617.38	15,359.7	3.5
Mean										15,356.3	±2.3

[1] The time of electrical energy input was 1,320.00 seconds in experiments 1, 2, and 3, and 1,200.00 seconds in experiments 4, 5, and 6.
[2] Corrected to 3,600.00 g of water and an average temperature of 30.00° C.

(b) FIRST SERIES OF REACTION EXPERIMENTS

The amount of chlorine unburned during the time of ignition and extinction of the flame, was determined in "correction" experiments which consisted in ignition, burning for one or two seconds, and then extinction of the flame. The following data were required: (1) The increase in mass of the absorber, (2) the total energy imparted to the calorimeter by the reaction and the sparking operation (determined from the corrected temperature rise of the calorimeter and its electrical energy equivalent), and (3) the "spark" energy.

The "spark" energy was determined in two separate experiments to be 10.3 ± 0.1 joules for each sparking time of five seconds. From the energy imparted to the calorimeter by the reaction, which was the total energy less the "spark" energy and the "gas" energy,[23] the amount of hydrogen chloride formed was calculated. The differ-

---

[21] See footnote 11, p. 681.   [22] See footnote 11, p. 681.   [23] See footnote 11, p. 681.

ence between the mass actually taken up by the absorber and the mass of hydrogen chloride formed gave m_{Cl}, the mass of unburned chlorine.

In the first series of reaction experiments the chlorine which was in the burner tube at the time the flow of chlorine was cut off was not burned out, and consequently the value of m_{Cl} for these experiments is relatively large, being about one-half per cent of the total mass absorbed. The results of the "correction" experiments of the first series to determine the value of m_{Cl} are given in Table 2.

TABLE 2.—*Calorimetric results of the correction experiments for hydrogen chloride, first series*

Experiment No.	Number of ignitions and extinctions of flame	Total mass absorbed	Total energy	"Spark" energy	"Gas" energy	Mass of HCl	m_{Cl} per experiment
		g	*int. joules*	*joules*	*joules*	*g*	*g*
a	8	0.3749	473.1	82.4	−2.0	0.1551	0.0275
b	3	.1308	134.3	30.9	.3	.0411	.0299
c	3	.1408	156.5	30.9	.4	.0498	.0303
Mean							.0292

The calorimetric results of the first series of reaction experiments are given in Table 3. All the symbols are defined and explained in an earlier paper.[24] The only change from previous procedure was that the chlorine gas coming from the Monel metal container passed through a coil (about 9 feet long) of glass tubing immersed in a water bath maintained at the temperature of the room.

TABLE 3.—*Calorimetric results of the reaction experiments for hydrogen chloride, first series*

Experiment No.	ΔR	k	u	K	U	$\Delta t_{corr.}$	Average temperature
	ohm	*min*$^{-1}$	*ohm min*$^{-1}$	*ohm*	*ohm*	*° C.*	*° C.*
A	0.05404	0.001963	0.0000038	0.002473	0.000106	0.91157	30.00
B	.093969	.001923	−.0000040	.002461	−.000112	.90954	30.00
C	.095208	.001937	.0000001	.002422	.000003	.92109	30.00
Mean							

Experiment No.	Electrical equivalent of calorimeter system	Total energy	"Gas" energy	"Spark" energy	Mass of HCl formed	$Q_{30°C.}$, 1 atmosphere	Deviation from mean
	Int. joules ° C.$^{-1}$	*Int. joules*	*joules*	*joules*	*mole*	*Int. joules mole*$^{-1}$	*joules mole*$^{-1}$
A	15,418.2	14,054.8	39.2	10.3	0.152679	92,243	−143
B	15,494.4	14,092.8	40.4	10.3	.152770	92,445	59
C	15,432.0	14,214.3	43.8	10.3	.154078	92,471	85
Mean						92,386	±90

[24] See footnote 11, p. 681.

(c) SECOND SERIES OF REACTION EXPERIMENTS

(c) SECOND SERIES OF REACTION EXPERIMENTS

In the second series of reaction experiments the chlorine remaining in the burner tube was all burned out by mixing hydrogen with the chlorine stream before the latter was cut off. This mixing was accomplished by opening the tube connecting the hydrogen and chlorine lines at their entrance into the calorimeter. Because of this improvement in the procedure, the value of m_{Cl} for the second series of reaction experiments was only about one-sixth of that for the first series, or about 0.08 per cent of the total mass absorbed; and a much greater precision was obtained in these experiments.

The results of the "correction" experiments of the second series are given in Table 4.

TABLE 4.—*Calorimetric results of the correction experiments for hydrogen chloride, second series*

Experiment No.	Number of ignitions and extinctions of flame	Total mass absorbed	Total energy	"Spark" energy	"Gas" energy	Mass of hydrogen chloride	m_{Cl} per experiment
		g	*Int. joules*	*joules*	*joules*	*g*	*g*
a	3	0.1652	412.1	30.9	−0.5	0.1508	0.0048
b	3	.1694	423.0	30.9	−0.7	.1552	.0047

The calorimetric results of the reaction experiments of the second series are given in Table 5.

TABLE 5.—*Calorimetric results of the reaction experiments for hydrogen chloride, second series*

Experiment No.	ΔR	k	u	K	U	Δt_{corr}	Average temperature
	Ohm	*Min⁻¹*	*Ohm Min⁻¹*	*Ohm*	*Ohm*	*°C.*	*°C.*
D	0.098361	0.001946	−0.000012	0.002665	−0.000034	0.95034	30.00
E	.098986	.001968	−.0000030	.001771	−.000066	.96575	30.03
F	.089307	.002035	.0000022	.002792	.000062	.85825	30.00
G	.067365	.001927	.0000018	.002753	.000043	.64095	29.85
Mean							

Experiment No.	Electrical equivalent of calorimeter system	Total energy	"Gas" energy	"Spark" energy	Mass of HCl formed	$Q_{30°C}$, 1 atmosphere	Deviation from mean
	Int. joules °C.⁻¹	*Int. joules*	*Joules*	*Joules*	*Mole*	*Int. joules mole⁻¹*	*Joules mole⁻¹*
D	15,415.3	14,649.8	38.5	10.3	0.159049	92,286	−22
E	15,433.8	14,905.2	19.7	10.3	.161479	92,363	+55
F	15,381.0	13,200.7	16.7	10.3	.143083	92,304	−4
G	15,406.4	9,874.8	16.8	10.3	.107079	92,280	−28
Mean						92,308	±27

4. RESULTS OF THE PRESENT INVESTIGATION

The results of the experiments of the present investigation are shown in Figure 3. This plot indicates clearly the increased precision obtained in the second series of reaction experiments and the relation of each result to the average value from the second series.

For the second series of reaction experiments the "error"

$$\pm 2\sqrt{\frac{(x-\bar{x})^2}{n(n-1)}}$$

is computed to be ±0.039 per cent. Combination [25] of this with the "error" of the electrical energy experiments gives a resultant error of 0.041 per cent, or ±38 joules per mole.

The total uncertainty in the average value of the second series is estimated to be not more than ±50 joules per mole.

The present experiments give, then, for the heat evolved in the reaction

$$1/2 \ H_2 \ (gas) + 1/2 \ Cl_2 \ (gas) = HCl \ (gas) \tag{9}$$

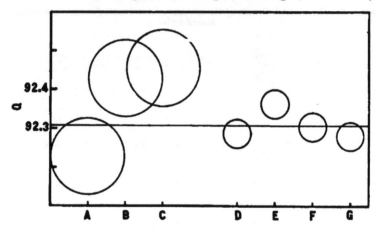

EXPERIMENT NUMBER

FIGURE 3.—*Plot of the experimental results for the heat of formation of gaseous HCl, from the present investigation*

The scale of ordinates gives Q, the heat of formation of gaseous HCl from its elements, at 30° C. and a constant pressure of 1 atmosphere, in international kilojoules mole⁻¹.

On the scale of abscissae are shown the individual values: A, B, C, the first series; and D, E, F, G, the second series. The circles indicate the relative precision for the two series, and the horizontal line, at 92.308, gives the average value from the second series.

at 30° C. and a constant pressure of 1 atmosphere

$$Q_{30°C.} = 92,308 \pm 50 \ \text{international joules mole}^{-1} \tag{10}$$

Since $\Delta C_p = -\dfrac{dQ}{dT} = -1.7$ joules mole⁻¹ °C.⁻¹, the value for 25° C. and 1 atmosphere becomes

$$Q_{25°C.} = 92,300 \pm 50 \ \text{international joules mole}^{-1} \tag{11}$$

Using the factor 1.0004/4.185, this is equivalent to

$$Q_{25°C.} = 22,063 \pm 12 \ \text{g-cal.}_{15} \ \text{mole}^{-1}. \tag{12}$$

5. THE RESULTS OF EARLIER INVESTIGATIONS

As has already been noted, calorimetric determinations of the heat of formation of hydrogen chloride have been made by Abria,[26] Favre and Silbermann,[27] Berthelot,[28] and Thomsen.[29] The results obtained

---

²⁵ See footnote 11, p. 681.   ²⁷ See footnote 3, p. 680.   ²⁹ See footnote 1, p. 680.
²⁶ See footnote 2, p. 680.   ²⁸ See footnote 4, p. 680.

by Abria, and by Favre and Silbermann are, respectively, about 10 and 8 per cent higher than the values obtained by the later investigators, and may be disregarded in view of the more concordant work of later investigators who had the advantage of purer materials and better calorimetric technic. There are no details of Berthelot's experiments, though his value is practically identical with that of Thomsen.

Thomsen published complete details of his four experiments on hydrogen chloride. The writer has recomputed Thomsen's data in

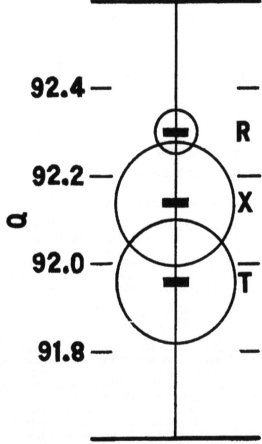

FIGURE 4.—*Plot of the various values for the heat of formation of* HCl

The scale of ordinates gives Q, the heat evolved in the formation of 1 mole of gaseous HCl from its elements, at 25° C. and a constant pressure of 1 atmosphere, in international kilojoules mole⁻¹.

The circles give the estimated uncertainties for values from the following sources: T, calorimetric experiments of Thomsen. X, calculation by way of the free energy and entropy of formation of gaseous HCl. R, calorimetric experiments of the present investigation.

the light of modern atomic weights and units of energy, and obtains as the average value

$$Q_{25°C.} = 91.96 \pm 0.14 \text{ international kilojoules mole}^{-1} \qquad (13)$$

The "error" is estimated from the consistency of Thomsen's data, together with the assumption of a calibration error of 0.10 per cent. There still remains the possibility of there being an unknown systematic error in the above value.

The values from the various investigations are as follows, in international kilojoules per mole at 25° C. and 1 atmosphere:

Abria-- 101
Favre and Silbermann-- 99
Berthelot-- 92
Thomsen--- 91. 96±0. 14
Present investigation-- 92. 30±0. 05

In Figure 4, the value from Thomsen's data is shown in relation to the result obtained in the present investigation. The circles show the estimated uncertainties.

6. CORRELATION OF THE PRESENT DATA ON HEAT CONTENT WITH EXISTING DATA ON THE FREE ENERGY AND ENTROPY OF FORMATION OF HYDROGEN CHLORIDE

For the reaction

$$1/2 \ H_2 \ (gas) + 1/2 \ Cl_2 \ (gas) = HCl \ (gas) \tag{14}$$

the present investigation gives

$$\Delta H_{298.1} = -92,300 \pm 50 \text{ international joules mole}^{-1} \tag{15}$$

or

$$\Delta H_{298.1} = -22,063 \pm 12 \text{ g-cal.}_{15} \text{ mole}^{-1} \tag{16}$$

The reference or standard state [30] used in calculating the values of entropy and free energy of a gas is that of an "ideal" gas where the fugacity is 1 atmosphere. In this state the heat content is the same as that of the real gas at zero pressure. A knowledge of the p-v-T relations suffices to permit computation of ΔH for the above reaction with the gases in their standard states by means of the relation

$$\left(\frac{\partial H}{\partial P}\right)_T = V - T\left(\frac{\partial V}{\partial T}\right)_p \tag{17}$$

But the necessary P-V-T data for chlorine and hydrogen chloride are not available. A calculation employing the Berthelot equation of state, which requires only a knowledge of the critical temperature and pressure for each gas, shows that, although appreciable for chlorine and hydrogen chloride separately, the change in ΔH for reaction (14) from a pressure of 1 atmosphere to a fugacity of 1 atmosphere is only about 0.01 per cent of the value of ΔH. Without additional information this correction can for the present be neglected, as it is less than the uncertainty in the value of ΔH, and one can write

$$\Delta H°_{298.1} = -22,063 \pm 12 \text{ g-cal.}_{15} \text{ mole}^{-1} \tag{18}$$

The values of the entropies of hydrogen, chlorine, and hydrogen chloride (less the entropy of nuclear spin) have been calculated very accurately from the data of spectroscopy by Giauque,[31] and Giauque

---

[30] See footnote 5, p. 580.   [31] Giauque, W. F., J. Am. Chem. Soc., vol. 52, p. 4816, 1930.

and Overstreet.[32]   The values of $S°_{298.1}$, in g-cal.$_{15}$ mole^{-1} °C. $^{-1}$, are: H_2, 31.225; Cl_2, 53.310; HCl, 44.658.   Then for reaction (14)

$$\Delta S°_{298.1} = 2.390 \text{ g-cal.}_{15} \text{ mole}^{-1} \text{ °C}^{-1} \qquad (19)$$

For the present calculations the uncertainty in this value is negligible.
The value of $\Delta F°$ for reaction (14) is best obtained [33] by combining the data for the following reactions:

$$1/2 \ H_2 \text{ (gas)} + AgCl \text{ (solid)} = HCl \text{ (aqueous)} + Ag \text{ (solid)} \qquad (20)$$

$$Ag \text{ (solid)} + 1/2 \ Cl_2 \text{ (gas)} = AgCl \text{ (solid)} \qquad (21)$$

$$HCl \text{ (gas)} = HCl \text{ (aqueous)} \qquad (22)$$

The value for the free energy change in reaction (20) is obtained from the standard emf of the cell in which reaction (20) occurs. The best values are: [34] Randall and Young, 0.2221; Harned and Ehlers, 0.2224; Carmody, 0.2223; Spencer, 0.2222.   These give

$$E°_{298.1} = 0.2222 \text{ international volt} \qquad (23)$$

The uncertainty in this value is about 0.0002 volt. Then, for reaction (20)

$$\Delta F°_{298.1} = -5{,}125 \pm 5 \text{ g-cal.}_{15} \text{ mole}^{-1} \qquad (24)$$

Randall and Young [35] conclude that the best value for the emf of the cell in which reaction (21) occurs is that given by Gerke [36] who finds, with chlorine at a pressure of 1 atmosphere,

$$E°_{298.1} = 1.1362 \text{ international volts} \qquad (25)$$

Correcting the chlorine to a fugacity of 1 atmosphere,[37] one finds

$$E°_{298.1} = 1.1363 \text{ international volts} \qquad (26)$$

The uncertainty in this value can be taken as 0.0004 volt. Therefore, for reaction (21)

$$\Delta F°_{298.1} = -26{,}211 \pm 9 \text{ g-cal.}_{15} \text{ mole}^{-1} \qquad (27)$$

The value of $\Delta F°$ for reaction (22) can be calculated from the ratio of p_2, the vapor pressure of hydrogen chloride over its aqueous solution, to a_2, the activity of hydrogen chloride in that solution:

$$\Delta F° = -RT ln \frac{p_2}{a_2} \qquad (28)$$

Lewis and Randall [38] reviewed the data on the vapor pressure of

[32] Glauque, W. F., and Overstreet, R., J. Am. Chem. Soc., vol. 54, p. 1731, 1932.
[33] See footnote 7, p. 680.
[34] (a) See footnote 7, p. 680; (b) Harned, H. S., and Ehlers, R. W., J. Am. Chem. Soc., vol. 54, p. 1350, 1932; (c) Carmody, W. R., J. Am. Chem. Soc., vol. 54, p. 188, 1932; (d) Spencer, H. M., J. Am. Chem. Soc., vol. 54, p. 3647, 1932.
[35] See footnote 7, p. 680.
[36] Gerke, R. H., J. Am. Chem. Soc., vol. 44, p. 1684, 1922.
[37] See p. 499 of the reference given in footnote 5, p. 680.
[38] See p. 503 of the reference given in footnote 5, p. 680.

hydrogen chloride over its aqueous solution, and the activity of aqueous hydrogen chloride, and concluded that

$$\frac{p_2}{a_2} = 4.41 \times 10^{-7} \tag{29}$$

Later, Randall and Young [20] recomputed this value, having available some few new data on vapor pressure and better values for a_2, and found

$$\frac{p_2}{a_2} = 4.97 \times 10^{-7} \tag{30}$$

This value is 13 per cent higher than the previous one. Assuming the uncertainty in the latest value to be about 5 per cent, then, for reaction (22)

$$\Delta F^{\circ}_{296.1} = 8,598 \pm 30 \text{ g-cal. }_{15} \text{ mole}^{-1} \tag{31}$$

Combination of ΔF° for reactions (20), (21), and (22), gives for reaction (14)

$$\Delta F^{\circ}_{296.1} = -22,738 \pm 33 \text{ g-cal. }_{15} \text{ mole}^{-1} \tag{32}$$

Then from equations (2), (19), and (32) one finds for reaction (14)

$$\Delta H^{\circ}_{296.1} = -22,026 \pm 33 \text{ g-cal. }_{15} \text{ mole}^{-1} \tag{33}$$

or

$$\Delta H^{\circ}_{296.1} = -92.14 \pm 0.14 \text{ international kilojoules mole}^{-1} \tag{34}$$

which agrees, within the limits of uncertainty, with the value obtained from the present calorimetric experiments

$$\Delta H^{\circ}_{296.1} = -92.30 \pm 0.05 \text{ international kilojoules mole}^{-1} \tag{35}$$

The values are shown together in Figure 4.

IV. CALCULATION OF SOME RELATED THERMODYNAMIC DATA

1. THE APPARENT AND PARTIAL MOLAL HEAT CAPACITY OF HYDROGEN CHLORIDE AND THE PARTIAL MOLAL HEAT CAPACITY OF WATER IN AQUEOUS SOLUTION OF HYDROGEN CHLORIDE

In order to calculate the temperature coefficient of reactions involving aqueous hydrogen chloride, it is necessary to know the apparent and partial molal heat capacities. The apparent molal heat capacity, Φ_c, is calculated from the experimental data on the heat capacity of hydrogen chloride solutions by the methods already described.[40]

When Φ_c has been determined as a function of $m^{1/2}$, the square root of the molality,[41] then $\overline{C}_{p_2}$, the partial molal heat capacity of hydrogen chloride, and $\overline{C}_{p_1}$, the partial molal heat capacity of water, are derived from the relations:

[20] See footnote 7, p. 680.
[41] (a) Randall, M., and Rossini, F. D., J. Am. Chem. Soc., vol. 51, p. 323, 1929. (b) Rossini, F. D., B. S. Jour. Research, vol. 4, p. 313, 1930; (c) vol. 7, p. 47, 1931.
[d] *m* is the number of moles of solute per 1,000 g water.

$$\overline{C}_{p_2} = \Phi_c + 1/2m^{1/2}\frac{d\Phi_c}{dm^{1/2}} \tag{36}$$

$$\overline{C}_{p_1} - \overline{C}_{p_1} = -\frac{m}{55.508}\left(1/2m^{1/2}\frac{d\Phi_c}{dm^{1/2}}\right) \tag{37}$$

$$\Phi^\circ_c = \overline{C}_{p_2}{}^\circ \tag{38}$$

For the present compilation of the thermal data on aqueous hydrogen chloride, it is desired to obtain values of Φ_c over as large a range of concentration as possible. In a previous paper,[42] the present author reviewed the existing data from the dilute region of concentration to about 2 molal. To these data can be added the values for the more concentrated solutions and some new data for the dilute region.

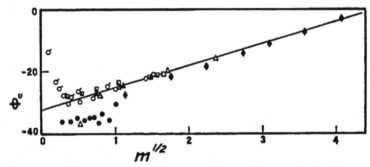

FIGURE 5.—*The apparent molal heat capacity of HCl in aqueous solution of HCl at 25° C.*

The scale of abscissae gives the square root of the molality.
The scale of ordinates gives Φ_c, the apparent molal heat capacity, for HCl in aqueous solution of HCl at 25° C., in g-cal per mole.
The points designate values (corrected to 25° C.) from the data of the following investigators: ○Richards and Rowe; ⊐Richards, Mair, and Hall; △Thomsen; □Thorvaldson, Brown, and Peaker; ◆Wrewsky and Savaritzky; ▲Marignac; ●Randall and Ramage; and, ◁Gucker and Schminke.

In Figure 5 are plotted values of Φ_c for aqueous hydrogen chloride at 25° C. calculated from the data of the following investigators: Thomsen[43] at 18° C.; Marignac[44] at 22° C.; Richards and Rowe[45] at 18° C.; Wrewsky and Kaigorodoff[46] at 20.5° C.; Randall and Ramage[47] at 25° C.; Richards, Mair, and Hall[48] at 18° C.; Thorvaldson, Brown, and Peaker[49] at 18° C.; Gucker and Schminke[50] at 25° C. The last-named investigators reported directly values of Φ_c. Where the data of the above experimenters were obtained at temperatures other than 25° C., they were corrected to 25° C., by means of the temperature coefficient of Φ_c previously given by the present author.[51]

The heavy black line in Figure 5 represents the linear relation between Φ_c and $m^{1/2}$ previously reported by the present author for aqueous hydrogen chloride at 25° C., for the range from infinite dilution to about 2 molal. Inspection of the plot shows that this relation

[42] See footnote 40 (c), p. 693.
[43] Thomsen, J., Ann. Physik, vol. 142, p. 337, 1871.
[44] Marignac, C., Ann. chim. phys., vol. 8, p. 410, 1876.
[45] Richards, T. W., and Rowe, A. S., J. Am. Chem. Soc., vol. 42, p. 162, 1920.
[46] Wrewsky, M., and Kaigorodoff, A., Z. physik. Chem., vol. 112, p. 83, 1924.
[47] Randall, M., and Ramage, L. E., J. Am. Chem. Soc., vol. 49, p. 93, 1927.
[48] Richards, T. W., Mair, B. J., and Hall, L. P., J. Am. Chem. Soc., vol. 51, p. 727, 1929.
[49] Thorvaldson, T., Brown, W. G., and Peaker, C. R., J. Am. Chem. Soc., vol. 52, p. 3927, 1930.
[50] Gucker, F. T., and Schminke, K. H., J. Am. Chem. Soc., vol. 54, p. 1358, 1932.
[51] See footnote 40 (c), p. 693.

represents the data very well over the entire range of concentration, though there is apparently some question as to the extrapolation to infinite dilution.[42]

Then for acqueous hydrogen chloride at 25°C.

$$\Phi_c = -32.5 + 7.2 \ m^{1/2} \text{g-cal. mole}^{-1} \ °\text{C.}^{-1} \tag{39}$$

$$\overline{C}_{p_2} = -32.5 + 10.8 \ m^{1/2} \text{g-cal. mole}^{-1} \ °\text{C.}^{-1} \tag{40}$$

$$\overline{C}_{p_1} - \overline{C}_{p_1}{}^\circ = -0.065 \ m^{3/2} \text{g-cal. mole}^{-1} \ °\text{C.}^{-1} \tag{41}$$

The temperature coefficients of the above quantities can be taken as

$$\frac{d\Phi_c}{dT} = \frac{d\overline{C}_{p_2}}{dT} = \frac{2.0}{7} \ \text{g-cal. mole}^{-1} \ °\text{C.}^{-2} \tag{42}$$

$$\frac{d(\overline{C}_{p_1} - \overline{C}_{p_1}{}^\circ)}{dT} = 0 \tag{43}$$

Equations (42) and (43) are approximations for the temperature range 10° to 35° C.

2. THE APPARENT AND PARTIAL MOLAL HEAT CONTENT OF HYDRO-GEN CHLORIDE AND THE PARTIAL MOLAL HEAT CONTENT OF WATER IN AQUEOUS SOLUTION OF HYDROGEN CHLORIDE AT 25° C.

In order to correlate the existing data on the heat of solution of hydrogen chloride and the heat of formation of aqueous hydrogen chloride with the present data on the heat of formation of gaseous hydrogen chloride, it is necessary to have values of $\Phi_A - \Phi_A{}^\circ$, the relative apparent molal heat content, for aqueous hydrogen chloride. These values can be easily determined from data on heat of dilution by the procedure previously employed by the present author.[53] The method consists simply in plotting the values of the measured heats of dilution as ordinates with the square root of the molality as the scale of abscissas. The intercept at $m^{1/2} = 0$ is made the zero for the scale of ordinates and the curve is then that of $\Phi_A - \Phi_A{}^\circ$ against $m^{1/2}$. Then the relative partial molal heat content of hydrogen chloride is obtained by means of the relation

$$H_2 - H_2{}^\circ = (\Phi_A - \Phi_A{}^\circ) + 1/2 \ m^{1/2} \frac{d(\Phi_A - \Phi_A{}^\circ)}{dm^{1/2}} \tag{44}$$

and the relative partial molal heat content of water is given by

$$\overline{H}_1 - \overline{H}_1{}^\circ = -\frac{m}{55.508} \ 1/2 \ m^{1/2} \frac{d(\Phi_A - \Phi_A{}^\circ)}{dm^{1/2}} \tag{45}$$

The existing data on heats of dilution (with the exception of many data for the dilute region of concentration which appeared in the paper already referred to [54]) are shown in Figure 6, where the ordi-

---

[42] Gucker, F. T., and Schminke, K. H. (see footnote 50, p. 694), found that, for aqueous hydrogen chloride and potassium hydroxide, the values of Φ_c at about 0.04 molal undergo an abrupt change, apparently increasing to very high values at infinite dilution. However, pending the confirmation of this behavior, which Gucker and Schminke found peculiar to hydrogen chloride and potassium hydroxide but not to lithium chloride, one can continue to use the linear extrapolation to infinite dilution.

[53] Rossini, F. D., B. S. Jour. Research, vol. 6, p. 791, 1931.

[54] See footnote 53, p. 695.

.nate scale has already been shifted to make the plot one of $\Phi_A - \Phi_A°$ against $m^{1/2}$. The values plotted in Figure 6 are from the data of Thomsen [55] at 18° C., Berthelot [56] at 18° C., and Wrewsky and Savaritzky [57] at 21.5° C. These data have been corrected to 25° C., using the appropriate values for heat capacity. The curve in the region from $m^{1/2} = 0$ to $m^{1/2} = 1.4$ represents values of $\Phi_A - \Phi_A°$ for 25° C., calculated from those for $\Phi_A - \Phi_A°$ at 18° C. which have already been compiled by the present author. [58]

In the above manner the values given in Tables 6 and 7 have been derived. Table 6 gives $\Phi_A - \Phi_A°$ for aqueous hydrogen chloride

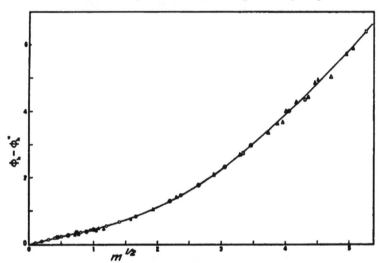

FIGURE 6.—*The relative apparent molal heat content of HCl in aqueous solution of HCl at 25° C.*

The scale of abscissas gives the square root of the molality.
The scale of ordinates gives $\Phi_A - \Phi_A°$, the relative apparent molal heat content, for HCl in aqueous solution of HCl at 25° C., in kg-cal.₁₅ per mole.
The points designate the experimental values (corrected to 25° C.) of the following investigators: ●, Wrewsky and Savaritzky; △, Berthelot; and □, Thomsen.
The points marked ○ are values (corrected to 25° C.) obtained from those for 18° C. previously given by the present author.

at 25° C., with the composition of the solution expressed in terms of the number of moles of water associated with 1 mole of hydrogen chloride. Table 7 gives values for $\Phi_A - \Phi_A°$, $\overline{H}_2 - \overline{H}_2°$, and $\overline{H}_1 - \overline{H}_1°$, at 25° C., for selected values of $m^{1/2}$.

[55] Thomsen, J., Thermochemische Untersuchungen, vol. 2, Barth, Leipzig, 1886.
[56] Berthelot, M., Ann. chim. phys., vol. 4, p. 467, 1875.
[57] Wrewsky, M., and Savaritzky, N., Z. physik. Chem., vol. 112, p. 90, 1924.
[58] See footnote 53, p. 695.

TABLE 6.—*Relative apparent molal heat content of HCl in aqueous solution of HCl, at 25° C.*

[The composition of the solution is given in terms of the number of moles of H_2O associated with 1 mole of HCl]

Solution	$\phi_\text{L}-\phi_\text{L}{}^\circ$	Solution	$\phi_\text{L}-\phi_\text{L}{}^\circ$
	g-cal per mole of HCl		*g-cal per mole of HCl*
$HCl._\infty H_2O$	0	$HCl.18H_2O$	920
$HCl.6400H_2O$	46	$HCl.15H_2O$	1,050
$HCl.3200H_2O$	65	$HCl.12H_2O$	1,250
$HCl.1600H_2O$	90	$HCl.10H_2O$	1,460
$HCl.800H_2O$	128	$HCl.8H_2O$	1,760
$HCl.400H_2O$	181	$HCl.6H_2O$	2,320
$HCl.200H_2O$	249	$HCl.5H_2O$	2,760
$HCl.100H_2O$	343	$HCl.4H_2O$	3,440
$HCl.50H_2O$	483	$HCl.3H_2O$	4,480
$HCl.25H_2O$	730	$HCl.2H_2O$	6,400
$HCl.20H_2O$	850	$HCl.1H_2O$	11,800

TABLE 7.—*Apparent and partial molal heat content of HCl and partial molal heat content of H_2O in aqueous solution of HCl, at 25° C.*

$m^{1/2}$	m moles of HCl per 1,000 g H_2O	$\phi_\text{L}-\phi_\text{L}{}^\circ$ g-cal. per mole of HCl	$\overline{H}_1-\overline{H}_1{}^\circ$ g-cal. per mole of HCl	$\overline{H}_1-\overline{H}_1{}^\circ$ g-cal. per mole of H_2O
0	0	0	0	0
.1	.01	49	74	− .0045
.2	.04	98	146	− .0346
.3	.09	146	219	− .119
.4	.16	193	287	− .271
.5	.25	238	348	− .496
.6	.36	281	412	− .850
.8	.64	367	543	−2.03
1.0	1.00	458	700	−4.35
1.2	1.44	561	884	−8.35
1.4	1.96	675	1,095	−14.8
1.6	2.56	805	1,350	−25.3
1.8	3.24	955	1,650	−40.9
2.0	4.00	1,120	2,000	−63.4
2.5	6.25	1,610	3,040	−161
3.0	9.00	2,260	4,410	−349
3.5	12.25	3,040	5,960	−644
4.0	16.00	3,920	7,580	−1,040
4.5	20.25	4,860	9,170	−1,570
5.0	25.00	5,840	10,840	−2,250
5.5	30.25	6,860	------------	------------

3. THE HEAT OF SOLUTION OF HYDROGEN CHLORIDE IN WATER

The heat of solution of hydrogen chloride in water has been measured by Thomsen,[59] Berthelot and Louguinine,[60] and Wrewsky and Savaritzky.[61] The data of these investigators have been corrected to 25° C. by means of the proper values of heat capacity and are shown in Figure 7. These data are for the reaction

$$HCl \text{ (gas)} + \frac{55.508}{m} H_2O \text{ (liquid)} = HCl.\frac{55.508}{m}H_2O \text{ (solution)} \quad (46)$$

[59] Thomsen, J., Thermochemische Untersuchungen, vol. 2, Barth, Leipzig, 1886.
[60] Berthelot, M., and Louguinine, W., Ann. chim. phys., vol. 6, p. 289, 1875.
[61] See footnote 57, p. 696.

The curve shown in Figure 7 is drawn through the data of Wrewsky and Savaritzky with extrapolation to infinite dilution made with the aid of the values of $\Phi_\Lambda - \Phi_\Lambda^\circ$ given in Table 7.

From these data the value for the heat of solution of hydrogen chloride at infinite dilution in water at 25° C.

$$\text{HCl (gas)} = \text{HCl (aqueous, } m=0) \tag{47}$$

is

$$\Delta H^\circ_{298.1} = -17,880 \pm 40 \text{ g-cal.}_{15} \text{ mole}^{-1} \tag{48}$$

4. THE HEATS OF FORMATION OF SILVER CHLORIDE AND MERCUROUS CHLORIDE

In order to utilize the data on the temperature coefficient of the emf of certain cells for obtaining the heat of formation of aqueous

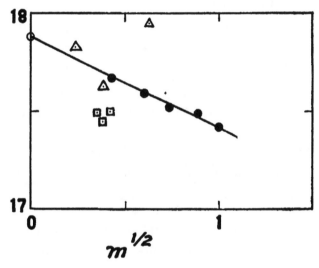

FIGURE 7.—*The heat of solution of HCl $_{(gas)}$ in water at 25° C.*

The scale of abscissae gives the square root of the molality.

The scale of ordinates gives the heat of the reaction $HCl_{(gas)} + \frac{55.508}{m} H_2O_{(liquid)} = HCl \cdot \frac{55.508}{m} H_2O_{(solution)}$ at 25° C., in kg-cal.$_{15}$ per mole of HCl.

The points designate the experimental values (corrected to 25° C.) of the following investigators: ●, Wrewsky and Savaritzky; △, Berthelot and Longuinine; and □, Thomsen.

The point marked ○, at $m_{H}=0$, gives the value for infinite dilution.

hydrogen chloride, it is necessary to have accurate values for the heats of formation of silver chloride and mercurous chloride. Fortunately, data are available for calculating these values with considerable accuracy.

For the reaction

$$\text{Ag (solid)} + 1/2 \text{ Cl}_2 \text{ (gas)} = \text{AgCl (solid)} \tag{49}$$

the free energy change [62] is

$$\Delta F^\circ_{298.1} = -26,211 \pm 9 \text{ g-cal.}_{15} \text{ mole}^{-1} \tag{50}$$

The entropy change for reaction (49) can be calculated from two independent sources: (1) From values of the entropies of silver,

62 See p. 692. The difference in free energy between crystalline and chemically precipitated silver chloride was found by Randall and Young (see footnote 7, p. 680) to be zero within the limits of uncertainty.

chlorine, and silver chloride; and (2) from a knowledge of the temperature coefficient of the emf of the cell in which reaction (49) occurs. New data on the heat capacity of silver and silver chloride down to low temperatures have been obtained recently. From such measurements, Eucken, Clusius, and Woitinek [63] obtained for silver, $S°_{298.1} =$ 10.16; Giauque and Meads [64] reported a preliminary value for silver, $S°_{298.1} = 10.01$; and Eastman and Milner [65] gave for crystalline silver chloride $S°_{298.1} = 22.97 \pm 0.08$. The entropy of silver will be taken as 10.09 ± 0.07. Combining these data with those of Giauque and Overstreet [66] on chlorine one finds for reaction (49)

$$\Delta S°_{298.1} = -13.77 \pm 0.12 \text{ g-cal.}_{15} \text{ mole}^{-1} \tag{51}$$

For the temperature coefficient of the emf of the cell in which reaction (49), with chemically precipitated silver chloride, occurs, Gerke [67] found for 25° C., the value

$$\frac{dE}{dT} = -0.000595 \pm 0.000006 \text{ international volt ° C}^{-1}$$

Correcting the chlorine to unit fugacity, one obtains for reaction (49)

$$\Delta S°_{298.1} = -13.69 \pm 0.14 \text{ g-cal.}_{15} \text{ mole}^{-1} \text{ ° C.}^{-1} \tag{52}$$

The average of (51) and (52) gives for reaction (49)

$$\Delta S°_{298.1} = -13.73 \pm 0.12 \text{ g-cal.}_{15} \text{ mole}^{-1} \text{ ° C.}^{-1} \tag{53}$$

By means of equations (2), (50), and (53) one finds for reaction (49)

$$\Delta H°_{298.1} = -30,304 \pm 40 \text{ g-cal.}_{15} \text{ mole}^{-1} \tag{54}$$

Within the limits of uncertainty this value is apparently that for either crystalline or chemically precipitated silver chloride.

For calculating the heat of formation of mercurous chloride, a slightly different procedure is necessary because an accurate value for the entropy of mercurous chloride is not available.

For the cell in which the reaction was

$$\text{Ag (solid)} + \text{HgCl (solid)} = \text{AgCl (solid)} + \text{Hg (liquid)} \tag{55}$$

Gerke [68] found, at 25° C., with chemically precipitated salts

$$\frac{dE}{dT} = 0.000338 \pm 0.000002 \text{ international volt °C.}^{-1} \tag{56}$$

and

$$E°_{298.1} = 0.0455 \text{ international volt} \tag{57}$$

This latter value is well substantiated [69] by measurements of other investigators, and the uncertainty can be taken as ± 0.0003 volts.

Using the Gibbs-Helmholtz relation

$$\Delta H = NF\left(T\frac{dE}{dT} - E\right) \tag{58}$$

one finds for reaction (55)

$$\Delta H°_{298.1} = 1276 \pm 16 \text{ g-cal.}_{15} \text{ mole}^{-1} \tag{59}$$

[63] Eucken, A., Clusius, K., and Woitinek, H., Z. anorg. Chem., vol. 203, p. 39, 1931.
[64] Giauque, W. F., and Meads, P. F., see footnote 6, p. 680.
[65] Eastman, E. D., and Milner, R. T., personal communication.
[66] See footnote 6, p. 680.
[67] See footnote 36, p. 692.
[68] See footnote 36, p. 692.
[69] See footnote 7, p. 680.

$$Hg \text{ (liquid)} + 1/2 \ Cl_2 \text{ (gas)} = HgCl \text{ (solid)} \tag{60}$$

for which

$$\Delta H^\circ{}_{298.1} = -31,580 \pm 45 \text{ g-cal.}_{15} \text{ mole}^{-1} \tag{61}$$

5. THE HEAT OF FORMATION OF AQUEOUS HYDROGEN CHLORIDE

It is possible to obtain a fairly accurate value for the heat of formation of hydrogen chloride in aqueous solution from data on the temperature coefficient of certain cells and the heats of formation of silver chloride and mercurous chloride.

Lewis and Randall,[70] Ellis,[71] and Harned and Brumbaugh[72] measured at a series of temperatures the emf of cells in which the reaction was

$$1/2 \ H_2 \text{ (gas)} + HgCl \text{ (solid)} = Hg \text{ (liquid)} + HCl \text{ (aqueous, } m) \tag{62}$$

and from these data calculated, by the Gibbs-Helmholtz relation, ΔH for reaction (62). By adding to (62) the reaction for the formation of mercurous chloride (solid) one obtains

$$1/2 \ H_2 \text{ (gas)} + 1/2 \ Cl_2 \text{ (gas)} = HCl \text{ (aqueous, } m) \tag{63}$$

In an analogous manner one can combine the data of Noyes and Ellis,[73] Harned and Brumbaugh,[74] and Butler and Robertson[75] on cells in which the reaction was

$$1/2 \ H_2 \text{ (gas)} + AgCl \text{ (solid)} = Ag \text{ (solid)} + HCl \text{ (aqueous, } m) \tag{64}$$

with the heat of formation of silver chloride to obtain the heat of formation of aqueous hydrogen chloride.

These data give the heat of formation of 1 mole of hydrogen chloride in an infinite amount of aqueous solution of hydrogen chloride of molality m. To the various values of ΔH for reaction (62) have been added the appropriate values of $\overline{H}_2 - \overline{H}_2^\circ$, taken from Table 7, in order to obtain the heat of the reaction

$$1/2 \ H_2 \text{ (gas)} + HgCl \text{ (solid)} = Hg \text{ (liquid)} + HCl \text{ (aqueous, } m = 0) \tag{65}$$

In this manner the following values for reaction (65) were obtained:

Data from—	Number of different molalities	Range of molality	$\Delta H^\circ{}_{298.1}$
Lewis and Randall	1	0.1	−8,420
Ellis	8	4.484–0.0832	−8,370
Harned and Brumbaugh	1	0.1	−8,400

The average of these values is −8,400, exactly the value of the latest determination, and the uncertainty can be taken as ±30. Combination of reactions (60) and (65) gives

$$1/2 \ H_2 \text{ (gas)} + 1/2 \ Cl_2 \text{ (gas)} = HCl \text{ (aqueous, } m = 0) \tag{66}$$

for which

$$\Delta H^\circ{}_{298.1} = -39,980 \pm 55 \text{ g-cal.}_{15} \text{ mole}^{-1} \tag{67}$$

[70] Lewis, G. N., and Randall, M., J. Am. Chem. Soc., vol. 36, p. 1969, 1914.
[71] Ellis, J. H., J. Am. Chem. Soc., vol. 38, p. 737, 1916.
[72] Harned, H. S., and Brumbaugh, N. J., J. Am. Chem. Soc., vol. 44, p. 2729, 1922.
[73] Noyes, A. A., and Ellis, J. H., J. Am. Chem. Soc., vol. 39, p. 2532, 1917.
[74] See footnote 72, p. 700.
[75] Butler, J. A. V., and Robertson, C. M., Proc. Roy. Soc. (London), vol. 125, p. 694, 1929.

In like manner, adding the appropriate values of $H_2 - H_2°$ to convert the data to infinite dilution, one finds for the reaction

$$1/2 \ H_2 \ (gas) + AgCl \ (solid) = Ag \ (solid) + HCl \ (aqueous, \ m = 0) \quad (68)$$

the following values:

Data from—	Number of molalities	Range of molality	$\Delta H°_{298.1}$
Noyes and Ellis...	6	0.1–0.001	−9,552
	1	0.33314	−9,643
Harned and Brumbaugh...	1	0.01	−9,611
	1	0.00	−9,567
Butler and Robertson...	4	0.1–0.01	−9,554
	1	0.3	−9,638

Taking the average of these values, one finds for reaction (68)

$$\Delta H°_{298.1} = -9,594 \pm 50 \ \text{g-cal.}_{15} \ \text{mole}^{-1} \quad (69)$$

Combination of equations (54) and (69) gives for reaction (66)

$$\Delta H°_{298.1} = -39,898 \pm 65 \ \text{g-cal.}_{15} \ \text{mole}^{-1} \quad (70)$$

Taking the average of (67) and (70) one finds for reaction (66)

$$\Delta H°_{298.1} = -39,938 \pm 55 \ \text{g-cal.}_{15} \ \text{mole}^{-1} \quad (71)$$

This value, fortunately, is practically identical with that obtained by combining equations (18) and (48) which give for reaction (66)

$$\Delta H°_{298.1} = -39,943 \pm 40 \ \text{g-cal.}_{15} \ \text{mole}^{-1} \quad (72)$$

which will be taken as the best value for the heat of formation of aqueous hydrogen chloride at infinite dilution.

6. THE FREE ENERGY OF FORMATION OF GASEOUS HYDROGEN CHLORIDE

The free energy of formation of gaseous hydrogen chloride is given by equation (32) as

$$\Delta F°_{298.1} = -22,738 \pm 33 \ \text{g-cal.}_{15} \ \text{mole}^{-1} \quad (32)$$

A value having a smaller uncertainty than the foregoing one can be calculated from equations (2), (18), and (19). This gives

$$\Delta F°_{298.1} = -22,775 \pm 12 \ \text{g-cal.}_{15} \ \text{mole}^{-1} \quad (73)$$

7. THE ENTROPY OF AQUEOUS HYDROGEN CHLORIDE

The entropy of aqueous hydrogen chloride is the basis for the calculation of the entropies of aqueous ions according to the methods of Latimer and coworkers.[76]

For the reaction

$$1/2 H_2 \ (gas) + 1/2 Cl_2 \ (gas) = HCl \ (aqueous) \quad (74)$$

one finds by combining equations (2), (24), (27), and (72)

$$\Delta S°_{298.1} = -28.87 \pm 0.15 \ \text{g-cal.}_{15} \ \text{mole}^{-1} \ °\text{C.}^{-1} \quad (75)$$

---

[76] See footnote 8, p. 681.

Then for aqueous hydrogen chloride at hypothetical 1 molal, where its heat content is the same as that at infinite dilution,

$$S^\circ_{298 \cdot 1} = 13.40 \pm 0.15 \text{ g-cal.}_{15} \text{ mole}^{-1} \text{ }^\circ\text{C.}^{-1} \tag{76}$$

V. ACKNOWLEDGMENT

The author acknowledges the technical advice of E. W. Washburn, under whose direction this work was carried on.

Washington, September 14, 1932.

THERMAL EXPANSION OF LEAD

By Peter Hidnert and W. T. Sweeney

ABSTRACT

Measurements have been made on the linear thermal expansion of three samples of cast lead between room temperature and 300° C. and the results have been correlated with data obtained by other investigators between 1740 and 1931.

A curve has been derived which shows the linear thermal expansion of lead between −253° and +300° C. The summary gives average coefficients of expansion for various temperature ranges between −250° and +300° C.

A comparison of the indirect results by Kopp and Matthiessen with the direct data by other observers, indicates that lead expands the same in all directions.

CONTENTS

I. INTRODUCTION

The thermal expansion of lead has been of considerable interest for nearly 200 years. Its measurement has been the object of more than 25 investigations. A summary of available data obtained by various investigators is given in Table 1.

Between 1740 and 1831, a number of determinations of the coefficient of linear expansion were made for the range from 0° to 100° C. In 1831, Daniell reported the changes in length of a bar of lead heated from 17° to 100° C. and to the point of fusion, respectively. From that time up to 1930, no measurements on the linear thermal expansion of lead above 110° C. have been located in the literature. The present authors appear to have been the first observers after Daniell to report data on the linear thermal expansion of lead above 110° C. Their abstract giving coefficients of linear expansion on heating for various temperature ranges between 20° and 300° C. was published [1] in February, 1930. In October of the same year, Uffelmann [2] published coefficients of linear expansion for various temperatures between 80° and 280° C.

The present investigation was undertaken in order to obtain reliable data on the linear thermal expansion of lead above 100° C. The data obtained between 20° and 300° C. have been correlated with results by other observers.

The authors wish to express appreciation to H. W. Bearce, W. Souder, and H. S. Rawdon for valuable suggestions, and to H. S. Krider for assistance during the preparation of the manuscript.

---

[1] Hidnert and Sweeney. (See Table 1.)
[2] See Table 1.

TABLE 1.—*Summary of expansion data on lead by various observers*

Observers	Date	Reference	Material	Temperature or temperature range	Coefficient of linear expansion per °C	Remarks
				°C.	$\times 10^{-6}$	
Muschenbroeck [1]	1740 (?)	Hist. Acad. Roy. Sci., p. 260, 1715.	Lead	0 to 100	14.2	
Bouguer [1]	1745		do	0 to 100	10.9	
Elliot [1]	Before 1754	Phil. Trans., vol. 48, pt. 2, p. 598, 1754.	do	0 to 100	10.9	
Smeaton	1754		do	0 to 100	28.7	
Herbert [1]	Before 1779		do	0 to 100	26.2	
Berthoud [1]	Before 1803		do	0 to 100	34.5	
Ouyton	1811	Mém. Sci. Math. Phys. Inst. France, 2d pt., p. 89, 1811.	do	0 to 100	27.2	
Lavoisier and Laplace [2]	Before 1831	Phil. Trans., vol. 121 (pt. 1), p. 443, 1831, or Phil. Mag., series 3, vol. 1, pp. 197, 264, 1832.	do	0 to 100	28.5	
Daniell	1831			17 to 100 17 to 327	27.9 29.9	Computed in 1932 from data by Daniell.
Kopp	1852	Ann. Chem. Pharm., vol. 83, p. 1, 1852.	Lead, cast	14 to 99 15 to 42	29.3 27.8	Coefficients of linear expansion were computed by Kopp from coefficients of cubical expansion obtained by density method
Calvert, Johnson, and Lowe	1859	Rep. Brit. Assoc., vol. 23, p. 46, 1859; Proc. Roy. Soc. (London), vol. 10, p. 316, 1859–60 and Chem. News. vol. 3, pp. 315, 357, 371, 1861.	Lead (pure)	0 to 100	30.1	
Matthiessen	1850	Phil. Trans., vol. 156, p. 861, 1866; or Ann. Physik u. Chemie, vol. 130, p. 50, 1867.	Lead, cast in a well-smoked mold and varnished.	14 to 92	¹ 29.0	$L_t=L_o\,(1+10^{-6}\times0.27384+10^{-9}\times0.00742)$, computed by Matthiessen from his data on cubical expansion obtained by weighing 2 samples of lead in water at different temperatures.
Fizeau	1869	Compt. Rend., vol. 68, p. 1125, 1869 or Ann. Physik u. Chemie, vol. 138, p. 26, 1869.	Lead, cast.	40 0 to 100	29.2 29.5 28.5	
Buff	1872	Ann. Physik u. Chemie, vol. 145, p. 636, 1872.	Lead.	0 to 100		Computed in 1932 from coefficient of cubical expansion reported by Buff.
Pfaff	1852	(Erlangen. Berohl. 1872. 8° 1–20), Fortsch. Physik, vol. 29, p. 579, 1873.	Lead wire.	0 to 100	29.1	
Glatzel	1877	Ann. Physik u. Chemie, vol. 160, p. 497, 1877.	Lead rod made from drawn wire.	Room temperature to 100.	29.4	
Rodwell	1877	Proc. Roy. Soc. (London), vol. 25, p. 290, 1875–77.	Lead.	325 (?)	30.2	
Vicentini and Omodei	1888	Atti Accad. Sci. Torino, vol. 23, p. 38, 1887–88.	do	325	29.5	Computed in 1932 from coefficient of cubical expansion reported by Vicentini and Omodei.
Dorsey	1908	Phys. Rev., vol. 27, p. 1, 1908.	Lead, cast.	−160 −120 −80 −40 0 +20 −190 to +20	24.8 25.9 27.2 28.5 28.9 27.1	

Observer	Year	Reference	Material	Temperature (°C)	Coefficient	Remarks
Grüneisen	1910	Ann. Physik, vol. 33, p. 33, 1910	Lead rod	−190 to +17 +17 to 100 18	27.0 29.3 28.8	Determined expansion of lead by comparison with expansion of fused quartz. Lindemann states that he neglected the expansion of the latter for it is extremely small and within the experimental error.
Lindemann	1911	Physik. Z., vol. 12, p. 1197, 1911	Lead, chemically pure	−253 to −192 −190 to −183 −190 to +20	19.8 20.3 26.6	
Raurano and Saarialho	1912	Öfversigt Finska Vetenskaps-Soc. Fördh., vol. 54, 1911-12, Afd. A., N:o 24.	Lead, cast	+15 to 110	29.3	
Friend and Vallance	1924	J. Inst. Metals, vol. 31, p. 75, 1924.	Lead	10 to 100 −253 to 0	29.5 24.7	Computed in 1932 from the data reported by Ebert.
Ebert	1928	Z. Physik, vol. 47, p. 712, 1928.	do	−190 to 0 0 to +100 −253 to −190	26.1 28.9 20.8	
Hidnert and Sweeney	1930	Phys. Rev., vol. 35, p. 296, 1930.	Lead, cast	20 to 60 20 to 100 20 to 200 20 to 300	28.8 29.0 30.0 31.4	Average values obtained on 3 samples from observations on heating.
Uffelmann	1930	Phil. Mag., vol. 10, p. 633, 1930.	Lead, annealed	80 100 120 140 160 170 180 200 220 240 250 260 280	28.9 29.1 29.5 29.9 30.2 30.3 30.5 30.9 31.2 31.6 32.0 32.3 33.0 34.3	
McLennan, Allen, and Wilhelm	1931	Trans. Roy. Soc. Can., III, series 3, vol. 25, p. 1, 1931.	Lead	−269 to −246		Determined relative expansion of lead and fused quartz; the curve of the relative change of length is nearly horizontal in the region of the superconductivity point of lead (−266° C.). The authors state that as there was no discontinuity observed in the thermal expansion curve for lead, it seems fair to conclude that when with the lowering temperature the superconductivity point was passed, no sudden change in the crystal lattice of the metal accompanied the appearance of superconductivity.

1 Cited by Guyton de Morveau (Mém. Sci. Math. Phys. Inst. France, second semestre, p. 1, 1808).
2 Cited by Daniell (Phil. Mag., vol. 10, pp. 191, 268, 350, 1831).
3 Computed in 1932 from Matthiessen's equation.

II. MATERIALS INVESTIGATED

Three samples of cast lead were investigated. The purity and the method of casting of the samples are indicated in Table 2. The length of each sample was about 300 mm and the cross section 10 mm square.

III. APPARATUS

The furnace shown in Figure 1 of Scientific Paper of the Bureau of Standards No. 488, was used for the measurements of the linear thermal expansion of sample 1001 and the white furnace shown at the extreme left of Figure 1 of Scientific Paper of the Bureau of Standards No. 524 was used for samples 1144 and 1215. Figure 4 of the latter paper indicates the method used in mounting samples 1144 and 1215 in the furnace.

Expansion measurements were made by means of micrometer microscopes, which were sighted on fine wires suspended from or in contact with the ends of the specimen. For a detailed description of the apparatus and the methods used the reader should refer to the publications mentioned.

IV. RESULTS

Observations were made on the linear thermal expansion of three cast samples of lead at various temperatures between room temperature and 300° C. and the results obtained are shown in Figure 1. The expansion curves are plotted from different origins to display the individual characteristics of each curve.

The average coefficients of expansion given in Table 2 were derived from the observations on heating and on cooling. This table also gives the differences in length before and after the expansion tests. The plus ($+$) sign indicates an increase in length and the minus ($-$) sign a decrease in length.

TABLE 2.—*Average coefficients of linear expansion of cast lead*

Sample	Lead content [1]	Method of casting	Test No.	Heating or cooling	Average coefficients of expansion per ° C.				Change in length after heating to 300° C. and cooling to room temperature
					20° to 60° C.	20° to 100° C.	20° to 200° C.	20° to 300° C.	
	Per cent				$\times 10^{-4}$	$\times 10^{-4}$	$\times 10^{-4}$	$\times 10^{-4}$	*Per cent*
[3] 1001	......	Preheated steel mold...............	1	Heating...	29.2	29.6	31.2	32.5	+0.02
				Cooling...	......	29.3	30.4	31.6	
[3] 1144	99.9	Sand mold.....................	1	Heating...	28.3	28.6	29.5	31.2	.00
				Cooling...	......	29.5	......	31.2	
[3] 1215	99.8	{Sand mold, from same ingot as sample 1001.	1	Heating...	29.0	28.9	29.7	30.9	—.01
				Cooling...	......	28.8	......	31.1	
			2	Heating...	......	28.8	29.4	31.0	.00
				Cooling...	......	......	28.8	31.0	
		Average......................	......	....	28.8	29.1	30.0	31.3	....

[1] Analysis by H. A. Buchheit, of this bureau.
[2] See sample 1215.
[3] Density for samples 1144 and 1215, 11.310 and 11.329 g/cm³ at 25° C., respectively, determined by Miss E. E. Hill, of this bureau.

The differences obtained in the coefficients of expansion of the three samples of cast lead are probably due to variations in the methods of casting.

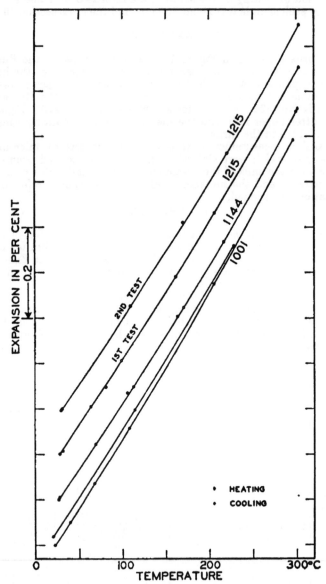

FIGURE 1.—*Linear thermal expansion of three cast samples of lead*

Sample 1001 which was cast in a preheated steel mold, has larger coefficients of expansion than the samples cast in sand molds. From the observations obtained on cooling, it appears that the coefficients of expansion of sample 1001 on the next heating would be less than those obtained on the first heating.

The curve in Figure 2 represents the linear thermal expansion of lead between −253° and +300° C. The portion of the curve between −253° and +20° C. represents the data obtained by previous

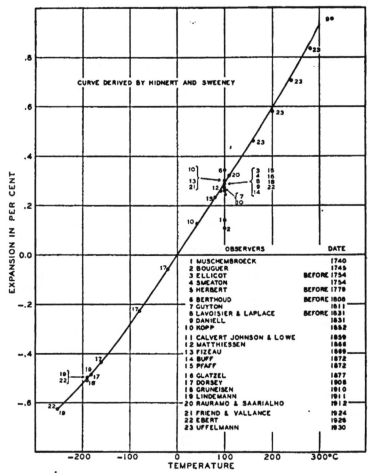

FIGURE 2.—*Comparison of the average expansion curve derived in the present investigation on lead, with data from the observers*

The lower portion of the curve from −253° to +20° C. was derived from data by Lindemann, Ebert, Dorsey, and Grüneisen, and the upper portion of the curve from 20° to 300° C. represents the average of all results obtained in the present investigation on three samples of lead.

investigators, and the portion between 20° and 300° C., the average of all results obtained by the present authors on the three samples of lead. The values obtained by other observers are included in this figure for comparison. Most of these values show good agreement. Bouguer and Muschembroeck obtained the lowest values and Berthoud the highest value at 100° C.

Kopp and Matthiessen's data for linear expansion obtained indirectly from density measurements agree closely with direct measurements made by other observers, and therefore it appears that the expansion of lead is the same in all directions.

The coefficient of linear expansion derived from Vicentini and Omodei's value for the coefficient of cubical expansion of lead near the melting point, appears to be too low.

McLennan, Allen, and Wilhelm published a curve which shows the relative change of length measured against the change of length of fused quartz near absolute zero. Their conclusion relating to the lattice structure of lead, which is based on the assumption that the expansion curve of lead is almost horizontal in the region of the superconductivity point ($-266°$ C.), does not appear to be justified. Since the expansion of fused quartz is not known near absolute zero, it is not possible definitely to determine from their curve the actual expansion of lead in this region. It is possible that when these observers cooled lead and fused quartz in the region of the superconductivity point of lead, the contraction of the lead nearly balanced the expansion of the fused quartz [3] and thus they obtained a nearly horizontal curve for the relative expansion.

V. SUMMARY

Data have been obtained on the linear thermal expansion of three samples of cast lead. Observations were taken at various temperatures between room temperature and 300° C., and the data have been correlated with available data by other investigators.

The average coefficients of linear expansion for various temperature ranges between $-250°$ and $+300°$ C., as derived from the expansion curve in Figure 2 are as follows:

		$\times 10^{-6}$
$-250°$ to $+$ 20° C		25. 1
$-200°$ to 20° C		26. 5
$-100°$ to 20° C		28. 3
$+20°$ to 60° C		28. 8
20° to 100° C		29. 1
20° to 200° C		30. 0
20° to 300° C		31. 3

WASHINGTON, September 12, 1932.

[3] Souder and Hidnert, B. S. Sci. Paper No. 524 (fig. 11).

O

U. S. DEPARTMENT OF COMMERCE
ROY D. CHAPIN, Secretary

BUREAU OF STANDARDS
LYMAN J. BRIGGS, Acting Director

BUREAU OF STANDARDS JOURNAL OF RESEARCH

December, 1932

Vol. 9, No. 6

UNITED STATES
GOVERNMENT PRINTING OFFICE
WASHINGTON : 1932

For sale by the Superintendent of Documents, Washington, D. C. · · Price 25 cents; $2.50 per year on subscription

For a partial list of previous research papers appearing in BUREAU OF STANDARDS JOURNAL OF RESEARCH see pages 2 and 3 of the cover

ISOLATION OF THE THREE XYLENES FROM AN OKLAHOMA PETROLEUM [1]

By Joseph D. White [2] and F. W. Rose, jr. [2]

ABSTRACT

Each of the three xylenes has been isolated in a high state of purity from an Oklahoma petroleum. The relative amounts present are o, 3; m, 3; p, 1, the total xylene content of the crude oil being about 0.3 per cent. The density, refractive index, boiling point, freezing point, and infra-red absorption spectrum have been determined for each xylene.

CONTENTS

I. INTRODUCTION

The presence of xylene in petroleum was first recognized by de La Rue and Müller.[3] Since then many investigators have reported the presence of m-xylene, and two have reported p-xylene in European and Asiatic petroleum,[4] but in no case were the hydrocarbons themselves isolated.

In the first allusion to the occurrence of aromatic hydrocarbons in American petroleum,[5] no reference was made to the presence of xylene. Many years later, Mabery [6] established the presence of m- and p-xylene in the oils of Ohio and Canada and also in those of California [7] by isolating their nitro derivatives from the fraction distilling between 137° and 140° C. Not until recently, however, has o-xylene been detected in any petroleum. Tausz,[8] by oxidizing the xylene fraction of a number of oils (including one from Pennsylvania), succeeded in isolating the three phthalic acids. No one, however, has reported heretofore the separation of each of the xylenes from petroleum.

---

[1] Financial assistance has been received from the research fund of the American Petroleum Institute. This work is part of Project No. 6, The Separation, Identification, and Determination of the Constituents of Petroleum.
[2] Research associate representing the American Petroleum Institute.
[3] Warren de La Rue and Hugo Müller, Proc. Roy. Soc., vol. 8, p. 221, 1856.
[4] Engler-Höfer, Das Erdöl, vol. 1, pp. 361-362, Hirzel, Leipsig, 1913; and the references there cited.
[5] C. Schorlemmer, Chem. News, vol. 7, p. 157, 1863.
[6] C. F. Mabery, Proc. Am. Acad. Arts Sci., vol. 31, p. 35, 1895.
[7] C. F. Mabery and E. J. Hudson, Proc. Am. Acad. Arts Sci., vol. 36, p. 261, 1901.
[8] J. Tausz, Zeit. angew. Chem., vol. 32 (I), p. 361, 1919.

This paper describes the isolation of these three hydrocarbons from an Oklahoma petroleum.[9] Separation of the three isomers from a highly concentrated mixture of the xylenes was accomplished by first extracting with liquid sulphur dioxide the fraction in which the xylenes collected during distillation and then subjecting the extract to systematic crystallization alternated with distillation. *m*-Xylene was also isolated from a eutectic mixture of the *m*- and *p*-isomers by sulphonation and hydrolysis.

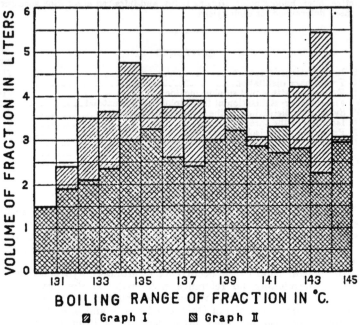

FIGURE 1.—*Distribution of volume of the fractions boiling between 130° and 145° C*

Graph I, before extraction with liquid sulphur dioxide; Graph II, after extraction and subsequent fractional distillation

II. PRELIMINARY DISTILLATION

The petroleum, which had received a preliminary distillation in an oil refinery, was fractionated further with the aid of efficient rectifying columns [10] in the manner described elsewhere.[11] Nearly 45 liters of distillate collected within the temperature range of 127° to 141° C. The distribution by volume of the majority of this fraction with respect to the boiling range is shown by Graph I, in Figure 1. An examination of the 1° cuts of this distillate showed a variation in

[9] For description and properties of the petroleum see E. W. Washburn, J. H. Bruun, and M. M. Hicks, B. S. Jour. Research, vol. 2, p. 469, Table 1, 1929.
[10] (a) E. W. Washburn, J. H. Bruun, and M. M. Hicks, B. S. Jour. Research, vol. 2, p. 470, 1929.
(b) R. T. Leslie and S. T. Schicktanz, B. S. Jour. Research, vol. 6, p. 378, 1931.
(c) J. H. Bruun and S. T. Schicktanz, B. S. Jour. Research, vol. 7, p. 851, 1931.
[11] J. D. White and F. W. Rose, B. S. Jour. Research, vol. 7, p. 907, 1931.

refractive index of $n_D^{20} = 1.420$ to 1.443, with a maximum value for the distillate boiling between 136° and 137° C. The high refractive indices, together with the characteristic odor of the material, strongly suggested the presence of aromatic hydrocarbons. Cooling curves showed that the fraction distilling between 134° and 136° C., which

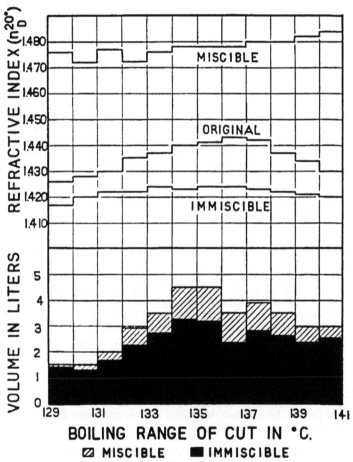

FIGURE 2.—*Change in volume and refractive index of distillation fractions upon extraction with liquid sulphur dioxide*

was the largest, had an initial freezing point of about −80° C. Its behavior on cooling indicated that it might be fractionated further by crystallization. Attempts to fractionate it in this manner resulted in some separation, but not enough to warrant adoption of this method.

III. EXTRACTION WITH LIQUID SULPHUR DIOXIDE

A more promising method for the separation of aromatic hydrocarbons from the distillate was that of extraction with liquid sulphur dioxide.[12] Small samples of a few of the 1° cuts boiling within a range which included the boiling points of ethylbenzene and the xylenes were extracted with liquid sulphur dioxide at −35° to −40° C. The high refractive index of the extract and the low refractive index of the immiscible portion indicated that pronounced separation had taken place. Accordingly, about 40 liters of the distillate boiling between 127° and 141° C. was extracted at an average temperature of −35° C. with the aid of the extraction apparatus designed by Leslie.[13] About 9 liters of extract and 31 liters of immiscible material were obtained. Figure 2 illustrates graphically the results of the extraction. There it may be observed that the major portion of the miscible material was obtained from fractions boiling between 134° and 139° C. It is interesting to note also in Figure 2 that the higher boiling fractions yielded extracts with increasingly higher refractive indices. This may be attributed to a probable greater proportion of *o*-xylene, which has the highest boiling point and the highest refractive index of the three xylenes.

Treating the immiscible portion again with sulphur dioxide resulted in very little additional extraction. Nevertheless it was deemed advisable to redistill this material systematically along with the fractions of the crude oil boiling over the temperature ranges immediately below and above it, and to reextract the resulting distillate. Before redistilling, the sulphur dioxide dissolved in the immiscible fraction was removed, first by sweeping it out with carbon-dioxide gas and then by shaking the gas-swept oil with pulverized soda lime until the odor of sulphur dioxide could no longer be detected.[14] The result of the redistillation is illustrated by Graph II in Figure 1. The refractive indices of the fractions from this distillation were considerably lower than those for the fractions from the initial distillation. In no case was n_D^{20} greater than 1.430. Extraction of the 42-liter portion distilling between 128° and 144° C. yielded only 3.3 liters of extract. As before, the major portion of this material was obtained from the fractions boiling between 134° and 139° C.[15]

The two samples of extract were combined, freed from sulphur-dioxide gas in the manner described above, and were then subjected to fractionation by crystallization.

IV. FRACTIONAL CRYSTALLIZATION OF THE SULPHUR DIOXIDE EXTRACT

A study of the behavior of the extract upon cooling showed that the fractions with the higher refractive indices could be readily

[12] For further study of the merits and technic of extracting petroleum with liquid sulphur dioxide, see
(a) L. Edeleanu, Trans. Am. Inst. Min. Eng., vol. 93, p. 2313, 1914.
(b) S. E. Bowrey, Pet. Rev., vol. 36, p. 351, 1917.
(c) R. L. Brandt, Jour. Ind. Eng. Chem., vol. 22, p. 218, 1918.
(d) R. J. Moore, J. C. Morrell, and G. Egloff, Met. Chem. Eng., vol. 18, p. 396, 1918.
(e) J. Tausz and A. Stüber, Zeit. angew. Chem., vol. 32 (I), p. 175, 1919.
(f) R. T. Leslie, B. S. Jour. Research, vol. 8, p. 591, 1932.
[13] R. T. Leslie, B. S. Jour. Research, vol. 8, p. 591, 1932.
[14] Experience showed that in the presence of even a trace of sulphur dioxide considerable decomposition of the oil took place on heating.
[15] This observation is of interest in view of the fact that *o*-xylene, which was later found to comprise a fourth of the extract, boils at approximately 144° C.

crystallized in a bath cooled with solid carbon dioxide. The fractions with lower refractive indices yielded no crystals at the sublimation temperature of solid carbon dioxide, but did crystallize when cooled further (to about −90° C.) with liquid air. These promising results made it seem advisable to subject the material to further fractionation by equilibrium melting and accordingly this was done by a method described elsewhere.[16] The melted oil was collected in successive cuts which were preserved according to their refractive indices and melting points. For observing the temperature trend during freezing and melting a toluene thermometer, or a single-junction thermocouple of copper constantan, was kept immersed in the oil during operations.

Systematic fractionation in this manner resulted finally in a series of fractions grouped according to Table 1.

TABLE 1.—*Fractions obtained by crystallizing the sulphur-dioxide extract*

Fraction	Volume	n_D^{20}	Crystallizing range
	ml		*°C.*
1	5, 500	1. 495 to 1. 500	−50 to −65
2	1, 425	1. 485 to 1. 495	−65 to −80
3	400	1. 475 to 1. 485	−80 to −90
4	1, 750	1. 470 to 1. 475	−90 to −100

The remainder of the original 12.5-liter portion consisted of mother liquor with $n_D^{20} = 1.470$ and with a crystallizing range from −100° to −125° C. Its fractionation by crystallization could be accomplished only with difficulty and resulted in but little separation. Therefore it was extracted again with liquid sulphur dioxide. The extract had the following properties; $n_D^{20} = 1.483$, B. P. = 136° C., F. P. = −96° C. These properties, as well as its characteristic odor, indicated that the material was impure ethyl benzene. Work is now in progress on its isolation. The immiscible portion has been preserved also. Its physical pro er ies indicate that it is probably a mixture of naphthenes and iso-nonanes. It should be pointed out that fractional crystallization of the original sulphur dioxide extract served to separate the major portion of these impurities from the xylenes, which concentrated in the "crystal fraction."

V. SYSTEMATIC DISTILLATION OF THE CRYSTALLIZATION FRACTIONS

Oxidation of samples of the crystal fractions, with neutral potassium permanganate, yielded the three phthalic acids. This fact, together with the observed refractive index and the crystallization range, and the further observation that the major portion of the material distilled between 138° and 144° C., indicated that it was a mixture of *p*-, *m*-, and *o*-xylene.

With the aid of the rectifying columns used in the earlier distillations, separate portions of the mixture were fractionally distilled under atmospheric pressure and under a pressure of 215 mm Hg. The distribution of the volumes of the fractions with respect to their

[16] R. T. Leslie and S. T. Schicktanz, B. S. Jour. Research, vol. 6, p. 378, 1931. The apparatus used in this work is illustrated on p. 382 of the reference cited.

boiling range indicated that better fractionation resulted under the
reduced pressure. As a consequence the entire mixture of xylenes
was subjected to a systematic distillation at a pressure of 215 mm Hg,
and was finally separated into two fractions. One of these, consisting
of about 2 liters and boiling normally near 144° C., appeared from its
behavior on freezing to be impure o-xylene. The other portion (a 3-
liter fraction) boiling between 138° and 139° C. was found in the same
manner to be chiefly a mixture of p-, and m-xylene.

VI. ISOLATION OF THE THREE XYLENES

The 2-liter fraction of o-xylene was further purified by fractional
crystallization. With the aid of a centrifuge which could be operated
at low temperatures [17] good separation of crystals from mother liquor
was attained. As a result of this fractionation a sample of o-xylene
was isolated, the physical properties of which (see below) indicated it
to be of high purity.

Fractional crystallization of the distillate containing the p- and m-
xylene separated it into a p-xylene fraction and into a mixture of the
two isomers approximating their eutectic in composition. Further
fractionation by crystallization of the material rich in p-xylene
yielded a very pure sample of this hydrocarbon. (See Table 4.)

Systematic crystallization of the impure eutectic mixture resulted
in the separation of a crystal fraction which froze at a constant temperature of −55° C. This freezing point agreed well with the value
recorded by Nakatsuchi [18] for the freezing point of a synthetic eutectic
mixture of p- and m-xylene. The identity of this material was definitely established by comparing its properties with those of a eutectic
mixture of the two isomers, made by mixing them in the proportions
prescribed by Nakatsuchi.

By alternating distillation with crystallization as required, a systematic fractionation of the eutectic mixture, in which cuts were made to
accord with refractive index and freezing point, resulted in the isolation of a small sample of m-xylene. This sample is designated in
Table 3 as "best by physical means." [19]

One other sample of m-xylene was isolated from the eutectic mixture by sulphonating it at 0° C. Twice the equivalent amount of
concentrated sulphuric acid (d=1.84) was slowly added to the oil
over a period of 6 to 12 hours, during which time the mixture was
constantly stirred while surrounded by an ice bath. Reaction was
allowed to continue at this temperature for a total of 24 hours, at the
end of which time the acid layer was separated, mixed with an equal
volume of dilute sulphuric acid (1 : 1 by volume) and steam distilled.
Hydrolysis took place most effectively at 130° to 135° C., although to
some extent below and above this temperature range.[20] Nearly pure

[17] M. M. Hicks-Bruun and J. H. Bruun, B. S. Jour. Research, vol. 8, p. 526, 1932.
[18] A. Nakatsuchi, J. Soc. Chem. Ind., Japan, vol. 29, p. 29, 1926.
[19] The difficulty of isolating m-xylene from its eutectic mixture with p-xylene may be appreciated from the fact that the major portion of the mixture distills at substantially a constant temperature, viz, 138.6° to 138.7° C. The residue remaining in the still at the end of the distillation is, however, somewhat richer in m-xylene. Moreover, the eutectic mixture contains 85 parts by weight of m-xylene and freezes only 7° below the freezing point of that component.
[20] For further study of sulphonation and hydrolysis as a means for separating and analyzing xylene mixtures see: (a) O. Jacobsen, Ber., vol. 10, p. 1009, 1877. (b) J. M. Crafts, Compt. rend., vol. 114, p. 1110, 1892. (c) H. T. Clarke and E. R. Taylor, J. Am. Chem. Soc., vol. 45, p. 830, 1923. (d) T. S. Patterson, A. McMillan, and R. Somerville, J. Chem. Soc., vol. 125, p. 2488, 1924. (e) N. Kishner and G. Vendelshtein, J. Russ. Phys. Chem. Soc., Chem. Part, vol. 57, p. 1, 1926. (f) A. Nakatsuchi, J. Soc. Chem. Ind., Japan, vol. 32, Suppl. Binding, p. 335, 1929.

m-xylene was thus obtained, which upon fractional crystallization yielded a sample of m-xylene of very high purity. This sample is designated in Table 3 and Figure 4 as "best by sulphonation."

VII. PROPERTIES OF THE ISOLATED XYLENES

To establish the identity of the isolated xylenes, the purest sample of each was selected on the basis of its behavior on freezing, and the physical properties of each of these were compared with the physical properties of the corresponding synthetic xylene reported in the literature. In addition, a similar comparison was made in the case of m-xylene and p-xylene, with samples obtained from the Bureau des Étalons Physico-Chimiques in Brussels. These highly pure samples were prepared by J. Timmermans and his coworkers. The results of these comparisons are tabulated in Tables 2, 3, and 4. The density measurements were made by the Division of Weights and Measures of this Bureau. The other properties were determined by the writers. Refractive indices were measured with a calibrated Abbé refractometer (Valentine design) under well-controlled temperature conditions. Readings could be readily made to within two units in the fifth decimal place and the values listed are estimated to be correct to within ±0.00005. Boiling points were determined in a Cottrell boiling-point apparatus. During the process the entire mercury thread of the thermometer was surrounded by the condensing vapors. Freezing points were determined either with a platinum resistance thermometer or a 5-junction thermocouple calibrated against the thermometer. It is believed that the values reported are correct to within ±0.02.

TABLE 2.—*Comparison of the physical constants of o-xylene from petroleum with previously reported constants of synthetic o-xylene*

Sample	$d\,{}^{27}_{4}$	$d\,{}^{20}_{4}$	$n\,{}^{25}_{D}$	Boiling point, 760 mm Hg	Freezing point (in dry air)
				°C.	°C.
From petroleum	[1] 0.87445	[2] 0.88040	1.50301	144.4	[3] −25.30
Synthetic (previously reported)	[4] .874	[4] .880	[4] 1.5033	[5] 144	[6] −25.74

[1] ±0.00001. Determined by the section of capacity and density of this bureau.
[2] Calculated from value determined at 27° C., assuming the same temperature coefficient as for m-xylene.
[3] Determined with platinum resistance thermometer.
[4] K. Von Auwers, Ann., vol. 419, p. 92, 1919. Calculated from values given for 20° C.
[5] Int. Crit. Tables, vol. 1, p. 219 (McGraw-Hill Book Co., 1926).
[6] A. Nakatsuchi, J. Soc. Chem. Ind., Japan, vol. 32, Suppl. Binding, p. 333; 1929.

TABLE 3.—*Comparison of the physical constants of certain samples of m-xylene with previously reported constants of synthetic m-xylene*

Sample	$d\,{}^{27}_{4}$	$d\,{}^{20}_{4}$	$n\,{}^{25}_{D}$	Boiling point 760 mm Hg	Freezing point (in dry air)
From petroleum:				°C.	°C.
Best by sulphonation	[1] 0.85817	[2] 0.86412	1.49467	139.15	[3] −47.89
Best by physical means	[1] .85821	[2] .86416	1.49468	139.15	−49.13
From Bureau des Étalons Physico-Chimiques (Timmermans)	[1] .85821	[2] .86416	1.49467	139.15	[5] −48.05
Synthetic (previously reported)	[4] [5] .85806	[4] [5] .86401	[4] 1.4948	[4] 139.30	[4] [7] −47.55

[1] ±0.00001. Determined by the section of capacity and density of this bureau.
[2] Calculated from values determined at 27° C.
[3] Determined with platinum resistance thermometer.
[4] J. Timmermans, J. chim. phys., vol. 27, p. 402, 1930.
[5] Calculated from value given for 30° C.
[6] Int. Crit. Tables, vol. I, pp. 219, 277 (McGraw-Hill Book Co.), 1926. Calculated from value given for 20° C.
[7] Recent value given in a private communication from Dr. Timmermans.

TABLE 4.—*Comparison of the physical constants of certain samples of p-xylene with previously reported constants of synthetic p-xylene*

Sample	d_4^{27}	d_4^{20}	n_D^{25}	Boiling point 760 mm Hg	Freezing point (in dry air)
				°C.	°C.
From petroleum........................	[1] 0.85498	[2] 0.86107	1.49320	138.4	[3] 13.21
From Bureau des Étalons Physico-Chimiques (Timmermans)........................	[1].85509	[2].86118	1.49320	138.4	[3] 13.21
Synthetic (previously reported).............	[4] [1].85491	[4] [1].86100	[4] [1] 1.49370	[4] 138.40 {	[4] 13.35 [5] 13.19

[1] ±0.00001. Determined by the section of capacity and density of this bureau.
[2] Calculated from values determined at 27° C.
[3] Determined in this bureau by B. J. Mair with a platinum resistance thermometer.
[4] J. Timmermans, J. chim. phys., vol. 23, p. 756, 1926.
[5] Calculated from values given for 15° C.
[6] A. Nakatsuchi, J. Soc. Chem. Ind., Japan, vol. 32, Suppl. Binding, p. 333, 1929.

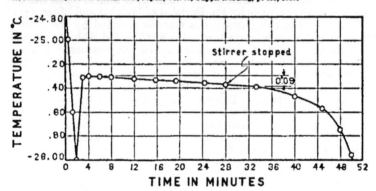

FIGURE 3.—*The time-temperature freezing curve of o-xylene isolated from petroleum*

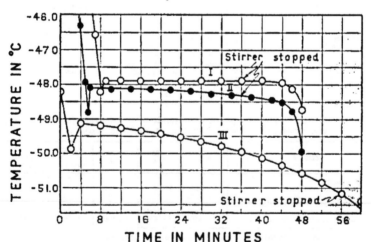

FIGURE 4.—*Freezing curves of m-xylene*

Curve I, from petroleum, best by sulphonation; Curve II, from Bureau des Étalons Physico-Chimiques; and Curve III, from petroleum, best by physical means

As a test of purity of the samples of the isolated hydrocarbons, their behavior during freezing was determined. Figure 3 shows the time-temperature cooling curve of o-xylene isolated from petroleum,

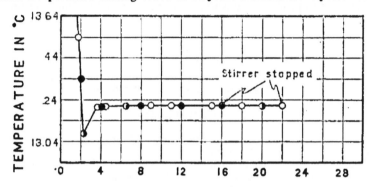

TIME IN MINUTES

O From Petroleum   ● From Bureau d'Étalons

FIGURE 5.—*Freezing curves of p-xylene*

while Figures 4 and 5 show similar curves for the several samples of *m*- and *p*-xylene which include those isolated from e roleum. The very narrow temperature range during the course of pfreezing (shown by the slope of the flattened portion of the curves) indicates that the

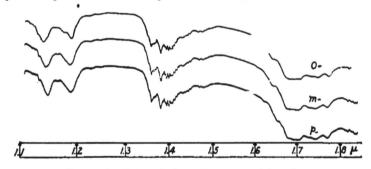

FIGURE 6.—*Infra-red absorption spectra of the xylenes*

Energy transmission curves showing the infra-red absorption spectra of the three isomeric xylenes in 4 mm cell depth from the emission of a tungsten filament lamp. The broad bands from 1.65 μ to 1.76 μ and 1.14 μ to 1.21 μ are the first and second overtones, respectively, of the fundamental hydrocarbon vibration frequency at 3.3 to 3.4 μ. The sharp, peaked absorption bands are due to atmospheric water-vapor in the light-path of the spectrograph. The bands at 1.14 and 1.19 show the clear differentiation in the absorption of the aromatic and aliphatic C ⟷ H vibration; the 1.14 μ band being due to the nuclear C-H absorption, and the 1.19 μ band, the absorption of the side-chain C-H. Slit-width approx. 20 A.

best samples are in a state of high purity. Figure 6 shows the infra-red absorption spectra of the isolated xylenes. The spectra were recorded by U. Liddel, of the Fixed Nitrogen Research Laboratory, Bureau of Chemistry and Soils.

VIII. CONTENT OF THE XYLENES IN THE CRUDE OIL

Analysis of the various xylene fractions by the freezing-point method (thermal analysis) shows that, based upon the total xylene in the crude oil, the three isomers are present in the following proportions by weight:

```
o-xylene_____  3
m-xylene_____  3
p-xylene_____  1
```

Allowing for losses incurred during the various stages of the separation, it is estimated that the total amount of xylenes in the crude oil is 0.3 per cent.

IX. CONCLUSIONS

This investigation confirms the results of previous investigators that the aromatic constituents of petroleum may be readily concentrated by extracting them with liquid sylphur dioxide. This method serves well as an aid in isolating the hydrocarbons themselves from petroleum and is superior to the usual chemical treatment by nitration, which, because of the large number of possible products, would be unsuitable for the determination of each of the xylenes present.

X. ACKNOWLEDGMENTS

The writers desire to acknowledge the technical advice of E. W. Washburn, director of American Petroleum Institute Project No. 6, throughout the course of this investigation.

WASHINGTON, September 17, 1932.

CHARACTERISTIC EQUATIONS OF VACUUM AND GAS-FILLED TUNGSTEN-FILAMENT LAMPS

By L. E. Barbrow and J. Franklin Meyer

ABSTRACT

The manufacture of tungsten-filament incandescent lamps has changed very rapidly during recent years, and methods of lamp photometry have necessarily changed with the changes in the lamps. Photometric measurements have passed very largely from a mean horizontal candlepower basis to a lumen basis. The tables of characteristic relations of vacuum lamps, published in 1914 and based on measurements of horizontal candlepower, are no longer adequate for the newer types and larger sizes of lamps.

Logarithmic equations of the second degree are shown to apply to vacuum lamps and to gas-filled lamps in three steps. The equations describe the characteristics of miniature lamps as well as large lamps. Tables of characteristic relations, based on normal efficiencies of 10.0, 12.5, and 16.0 lumens per watt, computed by means of the characteristic equations, furnish means for ready calculation of light output, current, power input, and operating efficiency over a range of voltages from 55 per cent of normal voltage for vacuum lamps, and 80 per cent of normal voltage for gas-filled lamps, to 120 per cent of normal voltage for all types and sizes up to 150 watts, and to 132 per cent of normal voltage for large gas-filled lamps, sizes 200 watts and up.

CONTENTS

I. INTRODUCTION

In a previous Bureau of Standards publication[1] the derivation of characteristic equations of the form

$$y_n = A_n x^2 + B_n x + C \qquad (1)$$

from data obtained on vacuum tungsten lamps was fully described and applications of the equations to problems of the lamp standardizing and testing laboratory were illustrated. The equations were shown to express the relations of voltage to candlepower, to current, to watts input, and to the "efficiency," expressed in terms of watts per candle, over the range from 3.3 to 0.7 watts per candle. In these equations the variable x is the logarithm of the ratio of any voltage to that voltage at which a lamp operates when it emits light at an assumed

---

[1] Middlekauff and Skogland, Characteristic Equations of Tungsten Filament Lamps, B. S. Sci. Paper No. 238. Bull. B. S., vol. 11, p. 483, 1915.

"normal" efficiency. The variable y_n is the logarithm of the ratios of currents, candlepowers, or watts at these voltages, or the logarithm of "efficiency," expressed in watts per candle, at any voltage. A, B, and C are constants. Tables of the characteristic relations were included in the original paper. These have proved very useful in lamp testing and standardization, but need to be revised and extended in order to be applicable to lamps manufactured now. The necessary revisions and extensions are included in this paper.

II. THE LUMEN BASIS OF INCANDESCENT LAMP MEASUREMENT

1. CHANGES IN LAMP CONSTRUCTION

The earlier paper was published before the integrating sphere photometer had come into general use for measuring the total light output of lamps. Practically all lamps then had "squirrel-cage" filaments in clear bulbs and were rated by the manufacturers at so-called efficiencies expressed as watts per mean horizontal candle. Photometric measurements made on a horizontal bar photometer were therefore all that were required for the rating of lamps.

Within the last few years bulb sizes, shapes, and finishes have been changed, as have also the methods of mounting and supporting the filament. Inside-frosted, diffusing bulbs are used now on all sizes of lamps from 15 to 100 watts, and on many of the larger sizes of lamps. The ring-wound, coiled filament has replaced the "squirrel-cage" type of filament mounting. Besides, practically all lamps of 40 watts and above are now gas-filled instead of vacuum. These changes in the construction of lamps have caused changes in their characteristics which have been taken into account in the equations and tables of this paper.

2. PHOTOMETRY OF VACUUM AND GAS-FILLED LAMPS

It has long been known that horizontal candlepower measurements on gas-filled lamps present great difficulties and are generally unreliable.[2] Besides, for most illuminating engineering purposes a knowledge of the total light output of lamps rather than the candlepower in a particular direction is desired. Horizontal candlepower measurements on all types of lamps therefore have been given up almost entirely in commercial and testing-laboratory practice, and measurements of spherical candlepower or total flux have taken their place. In this paper the watts-per-candle basis of the characteristic equations given in Scientific Paper No. 238 has been replaced by a lumen-per-watt basis.

To make the equations of Scientific Paper No. 238 applicable to vacuum lamps when measured in lumens, conversion of mean horizontal candlepower values to lumen values is necessary. The spherical reduction factor, that is, the ratio of mean spherical candlepower to mean horizontal candlepower, for vacuum tungsten lamps, with "squirrel-cage" filament mountings, such as were used in 1914, is approximately 0.78. The normal basis for photometric measurements of vacuum tungsten lamps was formerly generally accepted as 1.20

---

[2] Middlekauff and Skogland, Photometry of Gas-Filled Lamps, Elec. World, Dec. 26, 1914, p. 1246; and B. S. Sci. Paper No. 264. Bull. B. S., vol. 12, p. 589, 1915-16. C. H. Sharp, Trans. Ill. Eng. Soc., vol. 9, p. 1027, 1914.

watts per mean horizontal candle. This normal value of 1.20 watts per candle becomes 8.17 lumens per watt $\left(\dfrac{0.78 \times 4\pi}{1.20} = 8.17\right)$ when the lamps are measured on a lumen basis. This is a much lower efficiency than that at which most vacuum tungsten lamps are now made, and consequently a value of 10.0 lumens per watt has been arbitrarily chosen, on which basis tabular values for vacuum tungsten lamps are calculated in this paper. The characteristics of clear vacuum tungsten lamps as derived in 1914 have been found to apply also to the inside-frosted vacuum lamps as made to-day, so that the equations and tables in this paper for vacuum lamps are fundamentally the same as those contained in Scientific Paper No. 238, but are modified from the candle to the lumen basis and the normal efficiency is increased from 8.17 to 10.0 lumens per watt. The equations for gas-filled lamps have different constants, but are of the same form as the equations for vacuum lamps.

The data from which were derived the equations for gas-filled lamps of sizes ranging from 200 watts up are the result of a very large number of measurements made over a period of 10 years at the Bureau of Standards and at the laboratories of manufacturers of incandescent lamps. The normal efficiency chosen for these equations is 16.0 lumens per watt. It has been found, however, that these equations do not fit the characteristics of the smaller sizes of gas-filled lamps. Two other sets of equations, one for gas-filled lamps of sizes 60 to 150 watts, and the other for gas-filled lamps of 40 and 50 watts have been derived. Both of these sets of equations are on a normal basis of 12.5 lumens per watt. It has been found, also, that the 200 to 500 watt equations and 60 to 150 watt equations describe the characteristics of 32 and 21 candlepower automobile headlamps, respectively, while the 40 to 50 watt equations are suitable for 3 candlepower, 6 to 8 volt lamps.

III. THE CHARACTERISTIC EQUATIONS OF INCANDESCENT LAMPS

The logarithmic form of characteristic equation as applied to vacuum lamps is completely discussed by Middlekauff and Skogland in their original paper,[3] page 494 and following. The form of equation for vacuum lamps:

$$y_n = A_n x^2 + B_n x + C \qquad (n = 1, 2, 3, 4) \tag{1}$$

is found to apply to gas-filled lamps also. In this equation the independent variable x refers to voltage, and y_n and the corresponding constants A_n, B_n, and C to the efficiency in lumens per watt ($n = 1$), light output ($n = 2$), power ($n = 3$), and current ($n = 4$). There are then four problems to which the general equation is applicable. The variable x is the logarithm of the voltage ratio in each case; that is

$$x = \log \text{(voltage ratio)}$$

and

$y_1 = \log$ (lumens per watt)
$y_2 = \log$ (light output ratio)
$y_3 = \log$ (power ratio)
$y_4 = \log$ (current ratio)

---

[3] See footnote 1.

Constants A_n, B_n, and C have been computed on the selected normal bases of 10, 12.5, 12.5, and 16 lumens per watt, respectively, for vacuum lamps and for the three groups of different sizes of gas-filled lamps. These constants for the vacuum-lamp equations were derived from the constants contained in Bureau of Standards Scientific Paper No. 238, as previously indicated. For the various sizes of gas-filled lamps they were determined by setting up logarithmic curves of the relations, voltage ratio to luminous flux ratio, and voltage ratio to efficiency, using data obtained on several sizes of lamps measured over a wide range of voltages. The coefficients B_1 and B_2 were then determined as the slopes of these curves at the normal efficiency. The coefficients A_1 and A_2 were determined from the rates of change of these slopes. Coefficient C is the logarithm of the normal efficiency. Coefficients A_3, A_4, B_3 and B_4 are not independent of the above coefficients, A_3 and A_4 being equal to $(A_2 - A_1)$, B_3 being equal to $(B_2 - B_1)$, and B_4 being equal to $(B_2 - 1)$.

TABLE 1.—*Constants in the characteristic equations*

General equation:

$$y_n = A_n x^2 + B_n x + C \tag{1}$$

In which x is log (voltage ratio), and y_n is the logarithm of the quantity given in the right-hand column of this table.

CONSTANTS

	Vacuum lamps	Gas-filled lamps				
		Size				
	15–60 w	40–50 w	60–150 w	200–500 w	Used to compute
	Normal lumens per watt					
	10.0	12.5	12.5	16.0	
A_1	−0.918	−1.482	−1.726	−1.690	Lumens per watt.
B_1	1.932	2.162	2.090	1.841	Do.
C	1.00000	1.09691	1.09691	1.20412	Do.
A_2	−.946	−1.425	−1.669	−1.607	Light output (lumens) ratio.
B_2	3.513	3.685	3.613	3.384	Do.
A_3	−.028	.057	.057	.083	Power (watts) ratio.
B_3	1.5805	1.523	1.523	1.543	Do.
A_4	−.028	.057	.057	.083	Current (amperes) ratio.
B_4	.5805	.523	.523	.543	Do.

* $C=0$ for the other relations.

Example.—The lumens-per-watt equation for 500-watt, gas-filled lamps, with normal (100 per cent) value at 16.0 lumens per watt, is

$$y_1 = -1.690x^2 + 1.841x + 1.20412$$

where x is the logarithm of the ratio of any desired voltage to the voltage at normal efficiency (16.0 lumens per watt) and y_1 is the logarithm of the efficiency of the lamps at the desired voltage.

It is of interest to note that the coefficient of x (B_1, B_2, B_3, or B_4) in each case in Table 1 is very closely the exponent of the relationship existing between voltage and any one of the dependent variables listed in the last column of the table. For example, the efficiency of 500-watt lamps varies as the 1.84 power of the voltage; the luminous flux varies as the 3.38 power of the voltage; and the current varies as the 0.54 power of the voltage. These exponents are strictly applicable only for small changes of voltage near normal efficiency, but can be used through a wider range of voltages where only fairly close approximations are required.

There are 16 characteristic equations that can be written by using the constants of Table 1 to cover the four groups of lamp sizes, and the four dependent characteristic variables of each group. It is not convenient, however, to use these equations directly in computations, because one usually does not know the voltage at which a particular lamp is to be operated to give the normal efficiency, and this must be determined before the equations can be used to give the value of any of the dependent variables at any desired voltage. To facilitate computations, Tables 2, 3, 4, and 5, which follow, have been set up by using the 16 characteristic equations, and their use is discussed below.

IV. TABLES OF CHARACTERISTIC RELATIONS OF VACUUM AND GAS-FILLED TUNGSTEN LAMPS

1. CHARACTERISTICS OF VACUUM TUNGSTEN LAMPS

TABLE 2.—*Characteristics of vacuum tungsten lamps, 15 to 60 watts (normal efficiency, 10 lumens per watt)*

$y_1 = -0.918x^2 + 1.932x + 1.000$ (used to obtain column 2)
$y_3 = -0.946x^2 + 3.513x$ (used to obtain column 3)
$y_4 = -0.028x^2 + 0.5805x$ (used to obtain column 4)
$y_5 = -0.028x^2 + 1.5805x$ (used to obtain column 5)

Percentage of normal volts	Lumens per watt	Percentage of normal lumens	Percentage of normal amperes	Percentage of normal watts	Percentage of normal volts	Lumens per watt	Percentage of normal lumens	Percentage of normal amperes	Percentage of normal watts
1	2	3	4	5	1	2	3	4	5
55	2.74	10.6	70.4	38.7	90	8.12	68.7	94.1	84.6
56	2.85	11.4	71.1	39.8	91	8.30	71.5	94.7	86.1
57	2.97	12.2	71.9	41.0	92	8.49	74.4	95.3	87.6
58	3.10	13.1	72.6	42.1	93	8.67	77.3	95.9	89.2
59	3.23	14.0	73.4	43.3	94	8.86	80.3	96.5	90.7
60	3.36	14.9	74.1	44.4	95	9.05	83.4	97.1	92.2
61	3.49	15.9	74.8	45.6	96	9.23	86.6	97.7	93.7
62	3.62	17.0	75.5	46.8	97	9.42	89.8	98.3	95.3
63	3.76	18.1	76.3	48.0	98	9.61	93.1	98.8	96.8
64	3.90	19.2	77.0	49.3	99	9.80	96.5	99.4	98.4
65	4.04	20.4	77.7	50.5	100	10.00	100.0	100.0	100.0
66	4.18	21.6	78.4	51.7	101	10.19	103.5	100.6	101.6
67	4.33	22.9	79.1	53.0	102	10.39	107.2	101.2	103.2
68	4.47	24.3	79.8	54.3	103	10.58	111.0	101.7	104.8
69	4.62	25.7	80.5	55.5	104	10.78	114.8	102.3	106.4
70	4.77	27.1	81.2	56.8	105	10.98	118.6	102.9	108.0
71	4.92	28.6	81.8	58.1	106	11.18	122.5	103.4	109.6
72	5.08	30.2	82.5	59.4	107	11.38	126.5	104.0	111.3
73	5.23	31.8	83.2	60.7	108	11.58	130.6	104.6	112.9
74	5.39	33.5	83.9	62.1	109	11.78	134.8	105.1	114.6
75	5.55	35.2	84.5	63.4	110	11.98	139.2	105.7	116.3
76	5.71	37.0	85.2	64.7	111	12.18	143.6	106.2	117.9
77	5.87	38.8	85.9	66.1	112	12.38	148.1	106.8	119.6
78	6.04	40.7	86.5	67.5	113	12.59	152.7	107.3	121.3
79	6.20	42.7	87.2	68.8	114	12.79	157.4	107.9	123.0
80	6.37	44.7	87.8	70.2	115	13.00	162.1	108.4	124.7
81	6.54	46.8	88.4	71.6	116	13.21	166.9	109.0	126.4
82	6.71	49.0	89.1	73.0	117	13.41	171.8	109.5	128.1
83	6.88	51.2	89.7	74.5	118	13.62	176.8	110.0	129.9
84	7.05	53.5	90.3	75.9	119	13.83	181.9	110.6	131.6
85	7.23	55.9	91.0	77.3	120	14.04	187.2	111.1	133.3
86	7.40	58.3	91.6	78.8					
87	7.58	60.8	92.2	80.2					
88	7.76	63.4	92.8	81.7					
89	7.94	66.0	93.4	83.1					

2. METHOD OF USING THE TABLES OF CHARACTERISTIC RELATIONS

The method of using these tables is as follows:

1. It is assumed that a lamp has been photometered and that at a given voltage V_0, the current A_0, and lumen output L_0 have been directly measured, and from these values, the watts, W_0, and lumens per watt, l_0, have been computed. It is required to find the current, A_1, lumens L_1, watts W_1, and lumens per watt, l_1, at the required voltage V_1.

2. The tables must be entered at the value of lumens per watt, l_0, or if l_0 is not a tabulated value, then at the value of l_0 interpolated into the table.

3. In the line of the table in which l_0 is found or interpolated (column 2 of each table) read or interpolate V_t, A_t, L_t, and W_t, which are the percentages of normal volts, amperes, lumens, and watts, respectively, for the lamp when operated at l_0 lumens per watt. Thus V_0 is V_t per cent of normal volts, A_0 is A_t per cent of normal amperes, L_0 is L_t per cent of normal lumens and W_0 is W_t per cent of normal watts.

4. The voltage V_1 at which it is required to determine the current, watts, lumens, and lumens per watt of the lamp is then $(V_1/V_0) \times V_t$ per cent of normal volts; that is, letting V'_t be the per cent of normal volts of the lamp at V_1 volts

$$V'_t = (V_1/V_0) \times V_t$$

In the line of the table containing V'_t, or in an interpolated line, read or interpolate l'_t, L'_t, A'_t and W'_t, the lumens per watt, per cent of normal lumens, per cent of normal amperes, and per cent of normal watts, respectively, of the lamp when operated at V_1 volts. Then

$$l_1 = l'_t \qquad = \text{lumens per watt at } V_1 \text{ volts}$$

$$L_1 = L_0(L'_t/L_t) = \text{lumens at } V_1 \text{ volts}$$

$$A_1 = A_0(A'_t/A_t) = \text{amperes at } V_1 \text{ volts}$$

$$W_1 = W_0(W'_t/W_t) = \text{watts at } V_1 \text{ volts}$$

also

$$W_1 = V_1 A_1$$

(a) NUMERICAL EXAMPLES ILLUSTRATING THE USE OF TABLES 2 TO 5

Example 1.—Given a vacuum tungsten lamp, which when burned at 115 volts, emits 243 lumens, and takes a current of 0.220 amperes. This lamp will hereafter in this example be referred to as the test lamp.

To find the lumen output, lumens per watt, current and power input of the test lamp at 124 volts.

Solution.—Find first by direct computation the watts and lumens per watt as follows:

Watts$=115\times0.220=25.3$.

Lumens per watt$=243\div25.3=9.61$.

The known values of the lamp are then:

$V_0=115$ volts.

$A_0=0.220$ amperes.

$W_0=25.3$ watts.

$L_0=243$ lumens.

$l_0=9.61$ lumens per watt.

Enter table 2 (column 2) at 9.61 lumens per watt to determine the per cent of normal voltage at this efficiency. From column 1 of the line containing 9.61 lumens per watt we find that the lamp operates at 98 per cent of normal volts.

Since 115 volts is then 98 per cent of normal volts, 124 volts is $(124/115) \times 98 = 105.7$ per cent of normal volts. The line of the table, interpolated (column 1) containing 105.7 per cent of normal volts contains also the lumens per watt and percentages of normal lumens, amperes, and watts of the lamp at 124 volts. By multiplying the known values of lumens, amperes, and watts at 115 volts by the ratios of the percentage values read from lines corresponding to 105.7 per cent and 98 per cent of Table 2, values at 124 volts are obtained,

The tabular statement below contains the complete calculations:

1	Values read directly from table 2, 9.61 lumens per watt corresponding to 98 per cent volts	Values interpolated into table 2, at 105.7 per cent volts	Computation to 124 volts from columns 2 and 3	Computed values at 124 volts
	2	3	4	5
Lumens per watt......................	9.61	11.12	....	11.12 lumens per watt.
	Per cent	*Per cent*		
Lumens.............................	93.1	121.3	$\dfrac{243 \times 121.3}{93.1}$	317 lumens.
Amperes............................	96.8	103.25	$\dfrac{0.220 \times 103.25}{98.8}$	0.230 amperes.
Watts..............................	96.8	109.1	$\dfrac{25.3 \times 109.1}{96.8}$	28.5 watts.

Note.—Column 5 gives the required values.

Example 2.—A 100-watt, gas-filled tungsten-filament lamp is measured at 125 volts and the following values are obtained:

$V_0 = 125$ volts.
$A_0 = 0.912$ amperes.
$W_0 = 114$ watts.
$L_0 = 1,888$ lumens.
$l_0 = 16.56$ lumens per watt.

Find amperes, watts, lumens, and lumens per watt at 100 volts. Table 4, page 728 is applicable in this example. Enter Table 4 to find in column 2, 16.56 lumens per watt. It is not found, but by interpolation between 16.49 and 16.76 we have for 16.56 lumens per watt, the following values:

Per cent normal volts $= 115.26 = V_t$.
Per cent normal lumens $= 164.6 = L_t$.
Per cent normal amperes $= 107.7 = A_t$.
Per cent normal watts $= 124.2 = W_t$.
Lumens per watt $= 16.56 = l_t$.

Then

$$V'_t = V_t/V_0 \times V_t = 100/125 \times 115.26 = 92.2$$

Enter table 4 at 92.2 per cent volts, interpolate and compute by using the ratios from lines 115.26 and 92.2 per cent normal volts in the table. We obtain then:

$$\text{Lumens} = 1,888 \times \frac{74.2}{164.6} = 851.$$

$$\text{Amperes} = 0.912 \times \frac{95.9}{107.7} = 0.812.$$

$$\text{Watts} = 114 \times \frac{88.4}{124.2} = 81.1.$$

Lumens per watt $= 10.49.$

3. CHARACTERISTICS OF GAS-FILLED TUNGSTEN LAMPS

TABLE 3.—*Characteristics of gas-filled lamps, 40 to 50 watts (also 3-candlepower, 6 to 8 volt lamps). Normal efficiency, 12.5 lumens per watt*

$y_1 = 1.482x^2 + 2.162x + 1.09691$ (used to obtain column 2)
$y_1 = 1.425x^2 + 3.685x$ (used to obtain column 3)
$y_1 = 0.057x^2 + 0.523x$ (used to obtain column 4)
$y_1 = 0.057x^2 + 1.523x$ (used to obtain column 5)

Percentage of normal volts	Lumens per watt	Percentage of normal lumens	Percentage of normal amperes	Percentage of normal watts	Percentage of normal volts	Lumens per watt	Percentage of normal lumens	Percentage of normal amperes	Percentage of normal watts
1	2	3	4	5	1	2	3	4	5
80	7.47	42.6	89.1	71.3	100	12.50	100.0	100.0	100.0
81	7.71	44.8	89.7	72.6	101	12.77	103.7	100.5	101.5
82	7.94	47.0	90.2	74.0	102	13.04	107.5	101.0	103.1
83	8.17	49.3	90.8	75.4	103	13.31	111.4	101.6	104.6
84	8.41	51.6	91.4	76.7	104	13.58	115.4	102.1	106.2
85	8.65	54.0	91.9	78.1	105	13.86	119.5	102.6	107.7
86	8.89	56.5	92.5	79.5	106	14.15	123.7	103.1	109.3
87	9.14	59.1	93.0	80.9	107	14.43	128.0	103.6	110.9
88	9.39	61.8	93.6	82.3	108	14.72	132.3	104.1	112.4
89	9.64	64.6	94.1	83.8	109	15.00	136.7	104.6	114.0
90	9.89	67.4	94.7	85.2	110	15.28	141.3	105.1	115.6
91	10.14	70.3	95.2	86.6	111	15.56	145.9	105.6	117.3
92	10.39	73.3	95.8	88.1	112	15.85	150.7	106.1	118.9
93	10.65	76.3	96.3	89.6	113	16.13	155.5	106.6	120.5
94	10.91	79.4	96.8	91.0	114	16.41	160.4	107.1	122.1
95	11.16	82.6	97.4	92.5	115	16.70	165.4	107.6	123.8
96	11.43	86.0	97.9	94.0	116	16.99	170.5	108.1	125.4
97	11.70	89.4	98.4	95.5	117	17.27	175.7	108.6	127.1
98	11.96	92.8	99.0	97.0	118	17.56	180.9	109.1	128.8
99	12.22	96.3	99.5	98.5	119	17.85	186.3	109.6	130.4
					120	18.15	191.8	110.0	132.1

TABLE 4.—*Characteristics of gas-filled lamps, 60 to 150 watts, inclusive (also 21-candlepower 6 to 8 volt lamps). Normal efficiency, 12.5 lumens per watt*

$y_1 = -1.726x^2 + 2.090x + 1.09691$ (used to obtain column 2)
$y_1 = -1.669x^2 + 3.613x$ (used to obtain column 3)
$y_1 = 0.057x^2 + 0.523x$ (used to obtain column 4)
$y_1 = 0.057x^2 + 1.523x$ (used to obtain column 5)

Percentage of normal volts	Lumens per watt	Percentage of normal lumens	Percentage of normal amperes	Percentage of normal watts	Percentage of normal volts	Lumens per watt	Percentage of normal lumens	Percentage of normal amperes	Percentage of normal watts
1	2	3	4	5	1	2	3	4	5
80	7.55	43.1	89.1	71.3	100	12.50	100.0	100.0	100.0
81	7.79	45.2	89.7	72.6	101	12.76	103.6	100.5	101.5
82	8.02	47.4	90.2	74.0	102	13.02	107.4	101.0	103.1
83	8.26	49.7	90.8	75.4	103	13.29	111.2	101.6	104.6
84	8.49	52.1	91.4	76.7	104	13.55	115.1	102.1	106.2
85	8.73	54.5	91.9	78.1	105	13.82	119.1	102.6	107.7
86	8.96	57.0	92.5	79.5	106	14.08	123.1	103.1	109.3
87	9.21	59.6	93.0	80.9	107	14.34	127.3	103.6	110.9
88	9.45	62.3	93.6	82.3	108	14.61	131.5	104.1	112.4
89	9.70	65.0	94.1	83.8	109	14.88	135.8	104.6	114.0
90	9.95	67.8	94.7	85.2	110	15.15	140.2	105.1	115.6
91	10.20	70.7	95.2	86.6	111	15.42	144.7	105.6	117.3
92	10.44	73.6	95.8	88.1	112	15.69	149.2	106.1	118.9
93	10.70	76.6	96.3	89.6	113	15.96	153.8	106.6	120.5
94	10.95	79.7	96.8	91.0	114	16.22	158.6	107.1	122.1
95	11.21	82.9	97.4	92.5	115	16.49	163.4	107.6	123.8
96	11.46	86.2	97.9	94.0	116	16.76	168.2	108.1	125.4
97	11.72	89.5	98.4	95.5	117	17.04	173.2	108.6	127.1
98	11.98	92.9	99.0	97.0	118	17.31	178.3	109.1	128.8
99	12.24	96.4	99.5	98.5	119	17.58	183.4	109.6	130.4
					120	17.84	188.6	110.1	132.1

TABLE 5.—*Characteristics of gas-filled lamps, 200 to 500 watts (also 32-candle-power, 6 to 8 volt, lamps). Normal efficiency, 16.0 lumens per watt*

$$y_1 = -1.690x^2 + 1.841x + 1.20412 \quad \text{(used to obtain column 2)}$$
$$y_3 = -1.607x^2 + 3.384x \quad \text{(used to obtain column 3)}$$
$$y_4 = 0.083x^2 + 0.543x \quad \text{(used to obtain column 4)}$$
$$y_5 = 0.083x^2 + 1.543x \quad \text{(used to obtain column 5)}$$

Percentage of normal volts	Lumens per watt	Percentage of normal lumens	Percentage of normal amperes	Percentage of normal watts	Percentage of normal volts	Lumens per watt	Percentage of normal lumens	Percentage of normal amperes	Percentage of normal watts
1	2	3	4	5	1	2	3	4	5
80	10.23	45.4	88.7	71.0	105	17.49	117.8	102.7	107.8
81	10.51	47.5	89.3	72.4	106	17.78	121.5	103.2	109.4
82	10.79	49.7	89.9	73.7	107	18.07	125.3	103.8	111.0
83	11.07	52.0	90.5	75.1	108	18.36	129.2	104.3	112.6
84	11.35	54.3	91.0	76.5	109	18.66	133.2	104.8	114.2
85	11.64	56.6	91.6	77.9	110	18.95	137.2	105.4	115.9
86	11.92	59.1	92.2	79.3	111	19.23	141.3	105.9	117.5
87	12.21	61.6	92.8	80.7	112	19.52	145.4	106.4	119.2
88	12.50	64.2	93.3	82.1	113	19.81	149.6	106.9	120.8
89	12.78	66.8	93.9	83.6	114	20.10	153.8	107.4	122.5
90	13.07	69.5	94.5	85.0	115	20.4	158	108.0	124.2
91	13.36	72.2	95.0	86.5	116	20.7	162.5	108.5	125.8
92	13.65	75.1	95.6	88.0	117	21.0	167	109.0	127.5
93	13.94	78.0	96.2	89.4	118	21.28	172	109.5	129.2
94	14.24	80.9	96.7	90.9	119	21.59	176.5	110.0	130.9
95	14.53	83.9	97.3	92.4	120	21.84	181	110.5	132.6
96	14.82	87.0	97.8	93.9	121	22.1	186	111.0	134.4
97	15.12	90.2	98.4	95.4	122	22.4	190.5	111.5	136.1
98	15.41	93.4	98.9	96.9	123	22.7	195.5	112.0	137.8
99	15.70	96.6	99.5	98.5	124	23.0	200.5	112.5	139.6
100	16.00	100.0	100.0	100.0	125	23.2	205.5	113.0	141.3
101	16.30	103.4	100.5	101.5	126	23.5	210.5	113.5	143.1
102	16.60	106.9	101.1	103.1	127	23.8	215.5	114.0	144.8
103	16.90	110.5	101.6	104.7	128	24.1	221	114.5	146.6
104	17.20	114.1	102.2	106.2	129	24.4	226	115.0	148.4
					130	24.6	231.5	115.5	150.2
					131	24.9	237	116.0	152.0
					132	25.2	242.5	116.5	153.8

V. VERIFICATION OF THE CHARACTERISTICS

The upper range of efficiencies of 500-watt lamps was investigated several years ago in connection with tests made for the optical division of the Bureau of Standards. Starting with a basic value of 17.6 lumens per watt at 105 volts, the following comparison of observed and computed values is an example of the results obtained.

TABLE 6.—*Efficiency of 500-watt lamps*

Volts	Efficiency in lumens per watt	
	Observed	Computed
90	13.2	13.2
100	16.1	16.1
105	17.6	....
110	19.1	19.1
120	21.9	22.0
125	23.6	23.4
135	26.3	26.2

Values were recorded for 5-volt steps upward to burn out, which occurred within the step from 195 to 200 volts. Departures from computed values were not very extreme even at the higher points; for example, the observed value at 175 volts was 2.5 per cent below the computed value, and at 185 volts the corresponding difference was 2 per cent. A prohibitive amount of work would be required to assign characteristic values to these extreme points, where the observations have to be taken rapidly against great odds of color difference. Consequently, no attempt has been made to include them in the derivation of equations, and the lumens per watt in Table 5 terminate at 25.2.

Extrapolations have been made to points considerably below the limits of the tables for gas-filled lamps in work done on 500-watt lamps. In this work the voltage and light output at color match with and in terms of vacuum tungsten lamps were determined. At this color the 500-watt lamps operated at about 6.7 lumens per watt. Determinations of light output were then made at voltages corresponding to 16.5 lumens per watt in terms of gas-filled standards. Computed from the voltage and light output for color match with vacuum tungsten standards to these upper voltages, the values of light output checked the values obtained by observation to within 0.3 per cent.

The entire range of all the tables has been explored by means of a physical photometer (phototube photometer). A gas-filled cæsium phototube, corrected for color by means of a suitable filter, and attached to a 60-inch integrating sphere, was used. The phototube circuit is similar to that devised by Sharp and Smith.[4] This phototube photometer has been used at the Bureau of Standards for over a year in the rating of lamps for life test. Every month the characteristics of the photometer are checked by reading on it three of each of seven sizes of standard lamps ranging in size from 15 to 100 watts. The color temperature range of the light from these lamps is approximately 2,400° to 2,800° K. Each time there has been a remarkably close check between the readings thus obtained and the values assigned to the lamps by visual photometry using an 88-inch integrating sphere and Lummer-Brodhun photometer head. There have been other checks on the phototube photometer from time to time and these have shown the photometer to be accurate over a large range of color temperature and flux (lumen output).

All regular sizes of lamps from 15 to 500 watts, and also miniature lamps of 3, 21, and 32 candlepower have been photometered at various voltages to cover the range of the tables of this paper. These lamps were selected at random from the lamps received for life test from five different lamp manufacturers. Typical sets of results are given in Table 7.

---

[4] Sharp and Smith, Trans. Ill. Eng. Soc., vol. 23, p. 434, 1928.

TABLE 7.—*Observed and computed values of lumens and amperes on lamps of various sizes*

(a) AVERAGE OF TWO 15-WATT, VACUUM LAMPS (TABLE 2 USED TO OBTAIN COMPUTED VALUES)

Volts	Lumens		Amperes	
	Photo-tube	Com-puted	Potenti-ometer	Com-puted
80	34.7	34.2	0.106	0.107
90	54	54	.114	.114
100	79	80	.122	.122
115	132	....	.132	....
125	176	176	.139	.139
130	202	204	.142	.142

(b) AVERAGE OF TWO 40-WATT GAS-FILLED LAMPS (TABLE 3 USED TO OBTAIN COMPUTED VALUES)

100	252	254	0.332	0.330
110	366	366	.346	.346
120	508	510	.364	.362
125	592	....	.370	....
135	784	784	.384	.385
145	1,005	1,008	.400	.400

(c) AVERAGE OF TWO 3-CANDLE-POWER GAS - FILLED LAMPS (TABLE 3 USED TO OBTAIN COMPUTED VALUES)

6.72₅	34.2	....	0.500	....
7.50	51.2	51.4	.527	0.526
8.00	64.7	64.8	.544	.547
8.5	79.8	80.6	.560	.564

(d) AVERAGE OF TWO 100-WATT, GAS-FILLED LAMPS (TABLE 4 USED TO OBTAIN COMPUTED VALUES)

90	572	570	0.766	0.767
100	852	852	.810	.811
115	1,411	....	.872	....
125	1,888	1,888	.912	.911
130	2,158	2,150	.931	.931

(e) AVERAGE OF TWO 21-CANDLE-POWER GAS - FILLED LAMPS (TABLE 4 USED TO OBTAIN COMPUTED VALUES)

5.00	110	109	2.15	2.15
5.50	156	155	2.26	2.26
6.00	211	211	2.36	2.36
6.35	256	....	2.44	....

(f) AVERAGE OF TWO 200-WATT, GAS-FILLED LAMPS (TABLE 5 USED TO OBTAIN COMPUTED VALUES)

90	1,208	1,216	1.425	1.424
100	1,778	1,786	1.507	1.507
115	2,912	....	1.624	....
125	3,868	3,847	1.699	1.700
129	4,299	4,285	1.728	1.730

(g) AVERAGE OF TWO 32-CANDLE-POWER, GAS-FILLED LAMPS (TABLE 5 USED TO OBTAIN COMPUTED VALUES)

4.00	92	90	2.91	2.89
4.50	142	141	3.08	3.08
5.00	206	206	3.25	3.24
5.50	288	288	3.42	3.42
6.02₅	390	....	3.59	....

VI. CONCLUSION AND ACKNOWLEDGMENTS

The equations and tables embodied in this paper have been used for several years at the Bureau of Standards in connection with the routine testing of incandescent lamps. For the various sizes of lamps, slide rules have been devised and these are used to determine the voltages at which lamps are to be burned in order to operate at any desired efficiency, or the operating efficiency of the lamps when burned at any designated voltage. Although this is the general use of the equations and tables here, they have found other occasional uses, such as for determining the voltages at which to calibrate lamps when a designated lumen output or efficiency of the lamps is required. The results obtained by these computations have been checked many times on a visual photometer by groups of experienced observers.

The authors desire to acknowledge the work of the late J. F. Skogland in connection with this paper. He had much of the data here presented nearly ready for publication when he died. It was felt, however, that the equations should be further checked before publication, and this has been done. We wish to state also that the data from which the 200 to 500 watt lamp equations were derived were obtained by cooperative work between the Nela Park laboratories of the General Electric Co. and the Bureau of Standards.

Washington, August 16, 1932.

RP503

THE CALORIMETRIC DETERMINATION OF THE INTRINSIC ENERGY OF GASES AS A FUNCTION OF THE PRESSURE. DATA ON OXYGEN AND ITS MIXTURES WITH CARBON DIOXIDE TO 40 ATMOSPHERES AT 28° C.

By Frederick D. Rossini and Mikkel Frandsen

ABSTRACT

The intrinsic or internal energy, U, of air, oxygen, and mixtures of oxygen and carbon dioxide (to 37 mole per cent), has been determined calorimetrically as a function of the pressure, to 40 atmospheres, at 28° C.

Within the limits of error of the measurements, $(\partial U/\partial p)_T$ is constant over the given range; and its values for the different gases, at 28° C., 0 to 40 atmospheres, in Joules per atmosphere per mole, are: Air, -6.08; oxygen, -6.51; 0.075 carbon dioxide and 0.925 oxygen, -7.41; 0.175 carbon dioxide and 0.825 oxygen, -8.74; 0.231 carbon dioxide and 0.769 oxygen, -9.58; 0.366 carbon dioxide and 0.634 oxygen, -12.04. These values are estimated to be accurate within $\pm 2\frac{1}{2}$ per cent. The data on the mixtures of oxygen and carbon dioxide can be represented by the equation $(\partial U/\partial p)_{301°K.} = -6.51 - 11.0\ x - 11.0\ x^2$, where x is the mole fraction of carbon dioxide. The value of $(\partial U/\partial p)_T$ for air, oxygen, and the mixtures of oxygen and carbon dioxide at other near temperatures can be computed by means of the temperature coefficient of $(\partial U/\partial p)_T$ given by Washburn, namely, -0.4 per cent per degree.

CONTENTS

I. INTRODUCTION

The theory of determining calorimetrically the intrinsic or internal energy of gases as a function of the pressure has been given in a preceding paper by Washburn.[1] The present paper reports the results of experiments on air, oxygen, and mixtures of oxygen and carbon dioxide (to 36.6 mole per cent), to a pressure of about 40 atmospheres, at 28° C.

II. METHOD

The method employed in these experiments is as follows:

A mass, m, of gas is compressed in a bomb to a pressure of p atmospheres.

---

[1] Washburn, E. W., B. S. Jour. Research, vol. 9 (RP487), p. 521, 1932.

The bomb with its charge of gas is placed in a calorimeter at the temperature, T.

When the entire system has attained the temperature T, the gas is allowed to escape from the bomb to the atmosphere through a coil of tubing immersed in the calorimeter water. At the same time electrical energy is introduced into the calorimeter at such a rate as to maintain the temperature of the calorimeter sensibly constant at T. The amount of the electrical energy will be equal to the work done by the gas as it pushes back the atmosphere plus the change in internal energy which the gas undergoes as it drops from the pressure p to the atmospheric pressure B.

The work done by the gas as it emerges from the bomb is

$$W = n(Bv)_{1\,atm.} - Bv_b \qquad (1)$$

where n is the number of moles of the gas; $(Bv)_{1\,atm.}$ is the product of the pressure and volume for 1 mole of the gas at the temperature T and the pressure B (practically 1 atmosphere): v_b is the internal volume of the bomb at the atmospheric pressure B.

The change in internal energy of the gas is, per mole;

$$\Delta U\Big]_p^B = U_B - U_p \qquad (2)$$

If the amount of electrical energy, Q, that is added maintains the temperature of the calorimeter constant at T, then

$$Q = n\Delta U\Big]_p^B + n(Bv)_{1\,atm.} - Bv_b \qquad (3)$$

and

$$\Delta U\Big]_p^B + (Bv)_{1\,atm.} = \frac{Q + Bv_t}{n} \qquad (4)$$

When $\Delta U\Big]_p^B + (Bv)_{1\,atm.}$ is plotted as ordinate against the pressure p

as abscissa, the curve, at $p = 1$, will pass through the value of $(Bv)_{1\,atm.}$, which is constant and known for the given gas. And the slope of the curve will give the value of $-(\partial U/\partial p)_T$ for the gas.

Because of the fact that the steel of the bomb is not an incompressible substance having a zero temperature coefficient of expansion [2] and because the temperature may not remain absolutely constant, several small corrections to equation (4) need be added.

1. The ideal experiment is one in which the temperature of the calorimeter remains absolutely constant at the initial temperature T throughout the experiment. Actually, the temperature changed by a small amount, $\Delta t_{corr.}$, and the gas left the calorimeter at some average value T_s ($T_s - T$ is about 0.01° C.). The energy required to raise the temperature of the calorimeter from T to $T + \Delta t_{corr.}$ is equal to the product of $\Delta t_{corr.}$ and $C_{p(calor.)}$, the heat capacity of the calorimeter. The energy required to raise the temperature of the gas from T to T_s is equal to $C_{p(gas)}(T_s - T)$. The temperature of the issuing gas at any instant was assumed to be that of the calorimeter.

---

[2] See footnote 1, p. 733.

2. There is a change in the intrinsic energy of the steel bomb itself as the pressure within it changes from p to B atmospheres. This amount of energy is calculable thermodynamically [3] according to the formula

$$\left(\frac{\partial U_b}{\partial p}\right)_T = -T\left(\frac{\partial v}{\partial T}\right)_p - (p-1)\left(\frac{\partial v}{\partial p}\right)_T \qquad (5)$$

For the bomb used in the present experiments

$$v = 1.014 \text{ l} \qquad (6)$$

$$\frac{1}{v}\left(\frac{\partial v}{\partial T}\right)_p = 33 \times 10^{-6} \quad °C.^{-1} \qquad (7)$$

$$\left(\frac{\partial v}{\partial p}\right)_T = 39 \times 10^{-6} \text{ l atm.}^{-1} \qquad (8)$$

Then, at 28° C., the bomb energy, for a drop in pressure from p to 1 atmosphere, is

$$-\Delta U_b\Big]_p^1 = 1.02\ (p-1) + 0.002\ (p-1)^2 \text{ joules} \qquad (9)$$

Equation (9) gives the values shown in Table 1.

At 40 atmospheres, the value of $\Delta U_b\Big]_p^1$ is about 10 per cent of the value of $n\ \Delta U\Big]_p^1$ for air, and about 5 per cent of that for the mixture of 0.634 oxygen and 0.366 carbon dioxide. The uncertainty in the value used for the bomb energy is estimated to be ±10 per cent of itself, or one-half to 1 per cent of the value of $n\Delta U\Big]_p^1$ for the gases in the experiments at 40 atmospheres.

3. The opening of the valve to let the compressed gas escape from the bomb resulted in the transfer of energy to the calorimeter through friction of the stem on the seat of the valve. This amount of energy was found by measurement to be

$$e = 1.5 \pm 0.5 \text{ -joules} \qquad (10)$$

For the present experiments, then

$$\Delta U\Big]_p^B + (Bv)_{1\text{ atm.}} = \frac{Q + Bv_b - C_{p(\text{calor.})}\ (\Delta t_{\text{corr.}}) - C_{p(\text{gas})}\ (T_e - T) + E}{n} \qquad (11)$$

The value of E in equation (11) is given by the sum of equations (9) and (10):

$$E = 1.02\ (p-1) + 0.002\ (p-1)^2 + 1.5 \text{ joules} \qquad (12)$$

The temperature of the experiment was taken as T_e:

---

[1] See footnote 1, p. 733.

TABLE 1

p	$-\Delta U_b \rbrack_p^1$
Atmospheres	*Joules*
10	9.3
20	20.1
30	31.3
40	42.8

III. APPARATUS

The calorimeter, thermometric system, energy measuring devices, etc., were the same as those used in previous investigations [4] with the exception that a more sensitive galvanometer was used with the Mueller bridge and platinum resistance thermometer.

The steel bomb used in the present investigation is shown in place in the calorimeter can in Figure 1. Here A is the calorimeter can; B, the steel bomb; C, the copper coil through which the gas passed before leaving the calorimeter; D, the electrical resistance heater; E, the platinum resistance thermometer; F, the glass tube at the exit end of the copper coil; G, the valve; H and I, wrenches for opening the valve; J, the cover of the calorimeter can; K, the current leads.

The heating coil (D, fig. 1) was made by winding enameled constantan wire (No. 30, B. and S. gage) noninductively over mica on a piece of sheet copper made in the form of an inverted trough with open ends. Current leads of enameled copper (No. 24, B. and S. gage) were soldered to the ends of the constantan wire, and the whole was covered with a thin layer of "Pizein" cement. The heater, which had a resistance of about 22 ohms, was placed in position over the fitting which joined the copper coil to the valve of the bomb. Potential leads of enameled copper (No. 28, B. and S. gage) were soldered to the current leads at a point midway between the calorimeter and the jacket, with which all four leads were in good thermal contact.

The bomb, which was tested to 65 atmospheres, had a wall thickness of 1.65 mm; a mass, with valve, of 745 g; a diameter of 10 cm; a height of 16 cm; an internal volume of 1,014 cm³; and was made of a low carbon (0.12 per cent) steel. The copper coil, whose mass was 1,004 g, was made up of a total length of about 40 feet of ⅛, ³⁄₁₆, and ¼ inch copper tubing.

The assembly of the apparatus used in filling the bomb and for determining the temperature and pressure of the gas in the charged bomb is shown in Figure 2: A is the storage cylinder; B is a heavy steel purifying tube, which contained "ascarite" (sodium hydroxide-asbestos mixture) and "dehydrite" ($Mg(ClO_4)_2 \cdot 3H_2O$) for the experiments on air and oxygen, and "dehydrite" alone for the experiments on the mixtures of oxygen and carbon dioxide; C is a Bourdon gage for reading the pressure; D is the steel bomb; E is a glass jar containing water; and F is the thermometer for reading the temperature. The connecting tubes were made of ⅛-inch copper tubing.

---

[4] Rossini, F. D., B. S. Jour. Research, vol. 6 (RP259), p. 1, 1931.

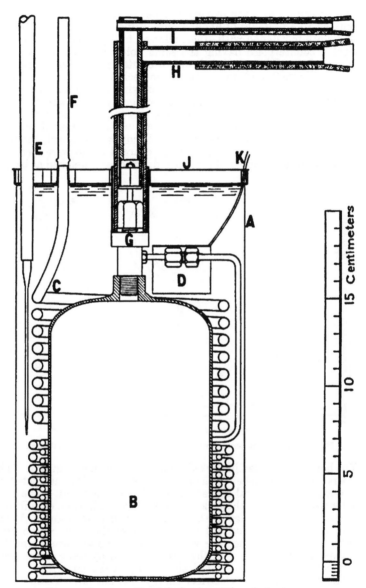

FIGURE 1.—*Assembly of the bomb and the calorimeter can*

The various parts of the assembly, which are drawn to scale, are described in the text

IV. PROCEDURE

The procedure for each experiment was as follows: The bomb was evacuated to a pressure of about 2 mm of mercury, closed, wiped off with a cloth moistened with ether, allowed to dry in air, and weighed against a counterpoise having the same volume as the bomb and a similar surface. Then the bomb was connected to the filling apparatus shown in Figure 2, and immersed to the neck in water maintained at the temperature of the room. The purifying and connecting tubes were flushed out with gas from the storage cylinder, and then the valve of the bomb was opened and gas was permitted to flow slowly through the purifying tube into the bomb until the desired pressure was reached. When temperature equilibrium was attained, the pressure and temperature of the gas in the charged bomb were recorded. The bomb was then closed, disconnected, dried, and weighed as before.

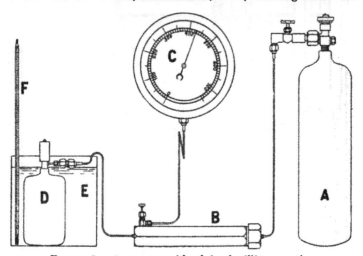

FIGURE 2.—*Apparatus and bomb for the filling operation*

The various parts of the assembly, which are drawn to scale, are described in the text.

The bomb was placed in the calorimeter as shown in Figure 1. The temperature of the jacket of the calorimeter was maintained constant at a temperature several degrees above that of the room. The calorimeter was brought to the jacket temperature, and, when equilibrium was reached, the readings of the "fore" period were begun. At the end of the "fore" period (10 to 20 minutes) the valve was opened slightly and electrical heating was started. The gas was permitted to escape at such a rate that the heat absorbed by the gas was practically equal to the heat evolved electrically in the same time. At a predetermined time the electrical energy was cut off. Temperature equilibrium was established in 5 to 10 minutes and the readings of the "after" period were begun. The lengths of the "fore," "reaction," and "after" periods were usually the same.

After one or two experiments on a given gas it was usually possible to so gage the rate of input of the electrical energy that the net change in the temperature of the calorimeter was not more than 0.01° C.

Readings of the calorimeter temperature, the current, and the voltage were taken every minute and of the jacket temperature every two minutes during the "reaction" perio .

The true mass of the gas in the charged bomb was computed from the apparent increase in weight of the bomb, the residual gas in the bomb before charging (several mg) and the buoyant effect of the air on the brass weights.

The pressure of the gas in the charged bomb at the temperature of the calorimetric experiment was computed by correcting the observed pressure p' at the temperature T' to the calorimetric temperature T_s by means of the perfect gas law, T' and T_s being the same within a few degrees.

V. EXAMINATION OF THE GASES

The air and oxygen and the mixtures of oxygen and carbon dioxide were respectively examined for oxygen and for oxygen and carbon dioxide, and the density of the inert gas in the oxygen was determined by the gas chemistry section of this bureau. The average molecular weights of the oxygen and the mixtures of oxygen and carbon dioxide were computed from the percentages of oxygen, carbon dioxide, and inert gas present, and from their separate molecular weights—the molecular weight of the inert gas being known from its density. These results are given in Table 2.

TABLE 2.—*Composition and molecular weights of the gases*

Gas	O_2	CO_2	N_2+rare gases	Molecular weight
	Mole per cent	*Mole per cent*	*Mole per cent*	*g*
Air	20.99	....	79.01	28.969
"Oxygen"	99.41	....	.59	32.012
"Oxygen"+carbon dioxide, I	91.66	7.51	.82	32.91
"Oxygen"+carbon dioxide, II	81.71	17.46	.84	34.11
"Oxygen"+carbon dioxide, III	76.18	23.08	.78	34.80
"Oxygen"+carbon dioxide, IV	62.80	36.60	.60	36.40

VI. ANALYSIS OF THE MEASURED QUANTITIES

The quantities involved in the calculation of the results of each experiment, exclusive of the corrections discussed in the latter part of Section II are the mass, pressure, and temperature of the gas in the bomb, the temperature change of the calorimeter, the electrical energy input, the atmospheric pressure, and the internal volume of the bomb.

In Table 3 are given, for the experiments on oxygen at 40 atmospheres, the magnitudes of the various quantities, the sensitivity of the measuring devices with respect to these quantities, and the "allowable error" in each that would contribute an absolute error of 1 joule to the result of each experiment.

TABLE 3.—*Analysis of the measured quantities*

[For an experiment on oxygen at 40 atmospheres]

	Magnitude of the quantity	Sensitivity of the measuring device	"Allowable error" in the quantity	Unit
Mass of gas	55	0.003	0.012	Gram.
Pressure of gas	40	.04	.08	Atmosphere.
Temperature of gas	300	.01	.08	°K.
Δt_{corr}	.01	.00002	.0001	°C.
Q	4,700	.2	1	Joule.
B	1	.001	.01	Atmosphere.
v_b	1	.001	.01	Liter.

VII. UNITS AND AUXILIARY DATA

The unit of energy employed in the present work is the international joule.

For computing the value of the product of the pressure and volume for 1 mole of air, oxygen, and carbon dioxide at 28° C. and 1 atmosphere, the following equation was used:

$$(Bv)_{1\ atm.,\ 28\ °C.} = RT_{0\ °C.}(1 + 28\alpha_{28\ °C.})(1 - \lambda_{0\ °C.}) \qquad (13)$$

R was taken as 8.3115 international joules degree^{-1} mole^{-1}, and $T_{0°C.} = 273.10° K$. The values of the constants [5] for air, oxygen, and carbon dioxide, and the computed values of $(Bv)_{1\ atm.,\ 28°C.}$ are given in Table 4. The values of $(Bv)_{1\ atm.,\ 28°C.}$ for the mixtures of oxygen and carbon dioxide were obtained by interpolation.

TABLE 4.—*Values of* $\alpha_{28°C.}$, $\lambda_{0°C.}$, *and* $(Bv)_{1\ atm.,\ 28°C.}$

Gas	$\alpha_{28°C.}$	$\lambda_{0°C.}$	$(Bv)_{1\ atm.\ 28°C.}$
			Int. joules mole^{-1}
Air	0.003672	0.00061	2,501.5
Oxygen	.003675	.00094	2,501.1
Carbon dioxide	.003734	.00706	2,480.6
0.075 carbon dioxide+0.925 oxygen			2,500.2
0.175 carbon dioxide+0.825 oxygen			2,499.1
0.231 carbon dioxide+0.769 oxygen			2,498.4
0.366 carbon dioxide+0.634 oxygen			2,496.9

VIII. EXPERIMENTAL DATA

The experimental data obtained in the present experiments are given in Tables 5 and 6. The column headings give the following information: Column 5, the pressure in the bomb $+(1-B)$, giving the initial pressure in the bomb corresponding to a final pressure of exactly 1 atmosphere; 6, T_s the average temperature at which the gas issued from the calorimeter, which is taken as the temperature of the experiment; 13, the value of E, as given by equation (12); 15, a term which corrects the value of $\Delta U \big]_p^1$ from T_s to 28° C. (computed according to the formula given by Washburn[6]); 17, the value of $(Bv)_{1\ atm.} + \Delta U \big]_p^1$ for 28°C.

---

[5] International Critical Tables, vol. 3, pp. 9, 10, 13. McGraw-Hill Book Co., New York, 1928.
[6] See footnote 1, p. 733. In the last term of that formula $6cT^3$ should read $6c/T^3$.

Table 5.—*Experimental data on air and oxygen*

1	2	3	4	5	6	7	8	9	10	11	12	13	14	15	16	17
Number of experiment	Gas	Mass of gas m	Atmospheric pressure B	Pressure in bomb $+(1-B)$ p	Temperature t_e	Time of electrical energy input Z	Electrical energy input	$\Delta t_{corr.}$	$\Delta t_{corr.} \times C_v'$ cal.	Re	$C_v(e)/(T_e-T)$	E	$n(Be)_{1\,atm} +n\Delta U]_p$ at t_e	Correction of $(Be)_{1\,atm.}$ to 25°C.	Correction of $\Delta U]_p$ to 25°C.	$(Be)_{1\,atm.} +\Delta U]_i$ at 25°C.
		g	$Atm.$	$Atm.$	°C.	$Seconds$	$Int.\,Joules$	°C.	$Joules$	$Joules$	$Joules$	$Joules$	$Int.\,Joules$	$Joules\,mole^{-1}$	$Joules\,mole^{-1}$	$Int.\,Joules\,mole^{-1}$
9		40.254	0.990	41.27	28.51	480.00	4,652.5	0.01150	128.0	101.6	0.2	45.8	4,681.7	-4.2	0.4	2,748.8
10		40.340	.996	41.28	28.11	480.00	4,490.0	-.00234	-26.1	102.2	-.5	45.8	4,664.6	-.9	-.4	2,737.0
11		48.914	.996	40.86	28.48	480.00	4,487.5	-.0003	-7.0	102.2	.8	45.3	4,641.2	-4.0	-.4	2,744.1
12	Air (28.969)	47.043	.997	39.20	27.38	480.00	4,472.9	.01640	184.1	101.3	-.3	43.4	4,433.8	5.1		2,734.0
13		24.430	.987	20.33	26.50	240.00	2,228.0	.01375	155.0	101.8	-.1	21.9	2,200.3	12.5	-1.3	2,628.6
14		37.547	.988	31.25	27.03	300.00	3,344.7	.00218	-24.4	101.4	-.1	34.2	3,504.6	8.0	.6	2,710.4
15		30.308	.996	25.25	27.06	300.00	2,744.7	.00920	103.2	102.4	-.1	37.4	2,771.4	7.6	.5	2,035.2
16		30.393	.995	23.36	27.50	300.00	2,745.9	.00799	89.6	102.2		27.5	2,744.1	4.1	.7	2,665.0
17		36.496	.993	30.30	27.49	300.00	3,284.1	.00391	43.8	101.9	0	33.2	3,375.4	4.3	.3	2,632.3
18		41.972	.992	35.00	28.02	420.00	3,831.3	.00389	43.7	101.8	0	38.5	3,927.9	-24.5	0	2,709.9
45		11.077	.991	10.14	28.97	120.00	1,060.9	.00066	98.4	101.4	0	11.0	1,075.4	-28.7	.6	2,570.8
46		24.107	.988	20.38	31.47	240.00	2,119.4	.00322	35.8	101.4	.1	22.0	2,207.1	-28.7	1.5	2,624.1
53		24.189	.993	20.39	30.49	240.00	2,119.9	.00643	38.3	102.1	.2	22.0	2,204.9	-20.6	1.1	2,630.2
19		26.780	.980	20.30	28.99	240.00	2,185.2	.00900	97.7	100.6	-.2	21.8	2,210.0	-8.2	-.5	2,684.1
20		26.759	.980	20.30	28.53	240.00	2,171.4	.00847	93.0	100.8	-.1	21.9	2,200.6	-4.8	-.2	2,685.6
21	"Oxygen" (32.012)	27.016	.987	20.64	27.61	240.00	2,157.2	.00409	67.3	101.3	-.1	45.3	2,213.2	3.9	-1.6	2,618.6
22		55.133	.991	40.71	27.08	480.00	4,205.0	.02615	-263.1	101.7	-.5	24.1	4,745.7	4.1		2,700.1
23		55.000					4,005.3	.00249	28.0				4,724.1	8.1		2,705.7
24		54.725	.991	40.63	27.51	480.00	4,006.5	.00389	40.1	101.7	.6	45.0	4,712.5	8.0	-.9	2,703.7
25		54.950	.998	40.73	27.53	480.00	4,095.8	.00252	28.2	102.4	.1	45.2	4,724.6	4.1	-.6	2,755.0
26		54.225	.993	40.27	27.53	480.00	4,597.0	.00733	81.9	102.1	.4	44.6	4,661.4	3.9	1.8	2,755.3
50		40.327	.995	30.40	30.48	360.00	3,257.8	.00697	77.6	102.1	-.1	53.2	3,414.8	-30.5		2,692.0
51		20.462	.995	15.55	30.49	180.00	1,591.2	.00302	33.7	102.1	-.1	16.7	1,676.4	-20.5	.9	2,603.1

Table 6.—Experimental data on mixtures of oxygen and carbon dioxide

Number of experiment	Gas	Mass of gas m	Atmospheric pressure B	Pressure in bomb +(1−B) p	Temperature t_i	Time of electrical energy input Z	Electrical energy input	Δt_corr.	Δt_corr. × C_s cal.	B_θ	$\frac{C_s/e}{(T_s-T)}$	E	$s(B_\theta)_i$ atm $+s\Delta U_p$ at t_i	Correction of $(B_\theta)_i$ atm to 28°C	Correction of ΔU_p to 28°C	$(B_\theta)_i$ atm $+\Delta U_i$ at 28°C
1	2	3	4	5	6	7	8	9	10	11	12	13	14	15	16	17
		g	Atm.	Atm.	°C	Seconds	Int. joules	°C	Joules	Joules	Joules	Joules	Int. joules	Joules mol⁻¹	Joules mol⁻¹	Int. joules mol⁻¹
37	0.925 "oxygen" +0.075 CO₂ (32.91).	28.212	0.963	20.51	27.55	240.00	2,372.1	0.02500	208.1	101.9	0.4	22.1	2,227.6	2.7	−0.3	2,601.9
28		27.947	.991	20.37	28.11	240.00	2,184.8	−.00151	−16.9	101.7		21.9	2,335.2	−.9	.1	2,727.3
29		56.525	.991	40.69	28.01	480.00	4,341.8	.00029	−339.0	101.7	.9	44.3	4,837.7	−.1	0	2,706.8
30		56.616	.993	40.84	27.99	480.00	4,312.5	.01271	−141.5	101.9	.7	45.1	4,801.7	−.1	−0	2,791.3
31		49.329	.991	35.64	27.99	480.00	4,012.8	.00252	23.2	101.7	.7	39.1	4,125.8			2,752.7
32		27.838	.989	20.53	28.49	240.00	2,296.9	.01503	167.7	101.5		22.1	2,242.7	−4.0	.3	2,045.7
33		27.890	.987	20.50	28.49	240.00	2,203.1	.00076	76.2	101.3	.1	22.1	2,250.2	−4.1	.3	2,651.4
39	0.825 "oxygen" +0.175 CO₂ (34.11).	58.709	.996	40.56	28.48	480.00	4,717.8	−.00435	−69.0	102.1	.1	45.0	4,914.1	−12.3	2.0	2,840.7
40		60.941	.993	41.55	28.48	480.00	4,706.9	.01764	−196.9	102.1	.6	46.1	5,044.6	−12.8	2.1	2,851.8
41		44.094	.993	30.67	28.48	480.00	3,029.1	.00645	70.8	102.1	.9	33.5	3,388.2	−12.3	1.5	2,704.9
42		30.256	.992	30.53	28.48	240.00	2,197.3	.00097	10.8	101.8	.4	32.1	2,310.8	−16.4	1.4	2,679.2
34	0.769 "oxygen" +0.231 CO₂ (34.50).	30.562	.986	21.00	28.99	240.00	2,299.4	−.00346	38.2	101.2	0	22.6	2,376.0	−8.2	.8	2,606.9
35		61.503	.991	41.15	28.00	480.00	4,698.1	−.00448	−50.2	101.6	.2	44.7	5,096.8	0	0	2,585.8
36		60.746	.990	40.65	28.00	480.00	4,997.5	.00344	27.1	101.6	.4	44.1	5,017.5	0	0	2,574.4
37		45.240	.996	30.64	28.00	480.00	2,946.2	.01652	182.9	101.2	.1	22.6	5,017.6			2,763.4
38		30.617	.996	31.00	28.00	240.00	2,290.6	.00063	40.4	101.3		22.6	2,373.9	−4.1	.4	2,694.5
43	0.634 "oxygen" +0.366 CO₂ (35.40).	65.394	.993	41.27	29.98	480.00	5,279.8	.00417	46.4	101.9	.7	45.8	5,381.8	−16.4	4.1	2,983.3
44		64.768	.997	41.03	30.97	480.00	5,274.0	.00897	99.4	102.3	.2	44.5	5,332.6	−21.3	5.3	2,974.4
45		48.863	.996	31.33	30.98	360.00	3,774.6	.00435	48.2	102.3		34.3	3,862.7	−24.6	4.7	2,857.6
46		31.089	.995	20.21	30.98	240.00	2,470.5	.02171	242.0	102.1	.7	21.8	2,352.3	−24.6	2.9	2,732.5
47		15.415	.992	10.35	30.98	120.00	1,089.2	.00786	88.2	101.8		11.2	1,123.6	−24.6	1.4	2,630.0
53		22.830	1.000	16.10	29.98	180.00	1,773.1	.01871	208.7	102.6	0	16.3	1,688.4	−16.4	1.4	2,077.0
54		15.607	1.000	10.39	29.98	120.00	1,179.7	.01424	159.2	102.6	.1	11.2	1,194.3	−16.4	1.0	2,630.1

IX. RESULTS OF THE PRESENT INVESTIGATION

The values of $\Delta U\big]_p^1 + (Bv)_{1\ atm.}$ that are given in Tables 5 and
6 were plotted against the pressure p; and the experimental values

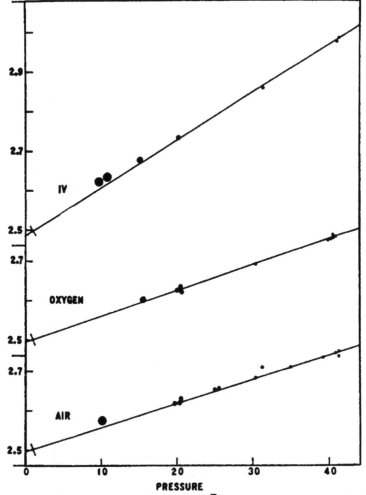

FIGURE 3.—*Plot of the values of* $(Bv)_{1\ atm.} + \Delta U\big]_p^1$ *for air, oxygen, and mixture IV,*
at 28° C.

The scale of ordinates gives the value of $(Bv)_{1\ atm.} + \Delta U\big]_p^1$ for air, oxygen, and mixture IV (0.366 carbon
dioxide and 0.634 oxygen) at 28° C. in international kilojoules mole⁻¹. The scale of abscissas gives the
pressure in atmospheres.

and the resulting curves are shown in Figures 3 and 4. The points
at $p=1$ are the values of $(Bv)_{1\ atm.}$ computed according to equation
(1³) and given in Table 4. In each case the experimental points

are drawn with a radius that is equivalent to an error of 4 joules per experiment, which value, for the experiments on oxygen at 40 atmospheres, is 1 per cent of the value of $n\ \Delta U\big]_p^1$.

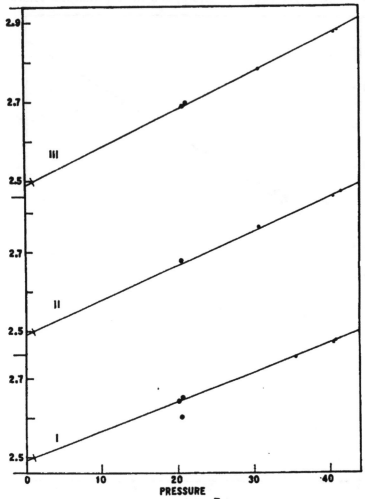

FIGURE 4.—*Plot of the values* $(Bv)_{1\ atm.}+\Delta U\big]_p^1$ *for mixtures I, II, and III of oxygen and carbon.dioxide, at 28° C.*

The scale of ordinates gives the value of $(Bv)_{1\ atm.}+\Delta U\big]_p^1$ for mixture I (0.075 carbon dioxide and 0.925 oxygen), mixture II (0.175 carbon dioxide and 0.825 oxygen), and mixture III (0.231 carbon dioxide and 0.769 oxygen), at 28° C. in international kilojoules mole^{-1}. The scale of abscissas gives the pressure in atmospheres.

Within the limits of error of the measurements, $\Delta U\big]_p^1$ is a linear function of the pressure up to 40 atmospheres, and consequently the value of $(\partial U/\partial p)_T$ is constant. The values of $(\partial U/\partial p)_T$ for the various gases, at 28° C., 0 to 40 atmospheres, in joules per atmosphere per mole, are given in Table 7. These values are estimated to be accurate within $\pm 2\frac{1}{2}$ per cent.

TABLE 7.—*Experimental values of* $(\partial U/\partial p)_T$ *at 28° C., 0 to 40 atmospheres*

Gas	Joules atm^{-1} mole^{-1}
Air	−6.08
Oxygen	−6.51
0.075 carbon dioxide+0.925 oxygen	−7.41
0.175 carbon dioxide+0.825 oxygen	−8.74
0.231 carbon dioxide+0.769 oxygen	−9.56
0.366 carbon dioxide+0.634 oxygen	−12.04

In Figure 5 are plotted the values of $(\partial U/\partial p)_T$, as a function of the mole fraction of carbon dioxide, for oxygen and the mixtures of oxygen and carbon dioxide. The smoothed values of $(\partial U/\partial p)_{301°K.}$ for the mixtures of oxygen and carbon dioxide are given by the equation

$$\left(\frac{\partial U}{\partial p}\right)_{301°K.} = -6.51 - 11.0x - 11.0x^2 \qquad (14)$$

where x is the mole fraction of carbon dioxide.

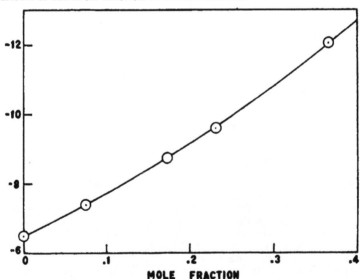

FIGURE 5.—*Plot of the values of* $(\partial U/\partial p)_T$ *against the mole fraction of carbon dioxide, for mixtures of oxygen and carbon dioxide, at 28° C.*

The scale of ordinates gives the value of $(\partial U/\partial p)_T$ at 28° C. in joules atm^{-1} mole^{-1}. The scale of abscissas gives the mole fraction of carbon dioxide. The curve is represented by the equation, $(\partial U/\partial p)_{301°K.} = -6.51 - 11.0x - 11.0x^2$.

The values of $(\partial U/\partial p)_T$ for air, oxygen, and the mixtures of oxygen and carbon dioxide at other near temperatures can be computed by means of the temperature coefficient of $(\partial U/\partial p)_T$ given by Washburn[7], namely, −0.4 per cent per degree.

---

[7] See footnote 6, p. 740.

X. COMPARISON OF THE PRESENT RESULTS WITH VALUES COMPUTED FROM OTHER DATA

Data similar to that reported in the present paper have not previously been obtained. Nevertheless, values of $(\partial U/\partial p)_T$ for some of the gases pertinent to the present work can be computed by means of the proper thermodynamic formulas from data on appropriate other properties.

Given adequate $p-v-T$ data, one can use the formula

$$\left(\frac{\partial U}{\partial p}\right)_T = -T\left(\frac{\partial v}{\partial T}\right)_p - p\left(\frac{\partial v}{\partial p}\right)_T \tag{15}$$

The $p-v-T$ data for nitrogen have been reviewed by Deming and Shupe,[8] and for nitrogen, air, and oxygen by Beattie and Bridgeman.[9]

From these data the values of $(\partial U/\partial p)_{301°K}$ given in Table 8 have been computed for nitrogen, air, and oxygen.

TABLE 8.—*Values of $(\partial U/\partial p)_T$ from various data*

$(\partial U/\partial p)$ 301°K. 0 to 40 atm. (Joules atm.$^{-1}$ mole^{-1})			Reference	Kind of data	Computed by means of formula
Nitrogen	Air	Oxygen			
−5.82 −5.97	------ −5.82	------ −6.65	8------ 9------	} $p-v-T$ measurements------	(15)
------ ------ ------	−5.75 −6.17 −5.92	------ ------ ------	11------ 12------ 13------	} Measurements of Joule-Thomson coefficient.	(16)
------ (−5.97)	−6.03 −6.08	------ −6.51	14------ Present investigation---	Calorimetric measurements of $(\Delta H/\Delta p)_T$. Calorimetric measurements of $(\Delta U/\Delta p)_T$.	(17) $(\partial U/\partial p)_T$

Where measurements of the Joule-Thomson coefficient, $\mu = (\partial T/\partial p)_H$, are available, one can employ the formula

$$\left(\frac{\partial U}{\partial p}\right)_T = -C_p\left(\frac{\partial T}{\partial p}\right)_H - \left(\frac{\partial(pv)}{\partial p}\right)_T \tag{16}[10]$$

The last term in equation (16) is taken from $p-v$ data, and its value, for air, is about one-tenth that of the first term. Measurements of the Joule-Thomson coefficient for air, in the range of pressure and temperature covered in the present investigation, have been made by Roebuck,[11] Noell,[12] and Hausen.[13] The values of $(\partial U/\partial p)_{301°K}$ computed from their data are given in Table 8.

Recently, calorimetric measurements of $(\Delta H/\Delta p)_T$ for air have been reported by Eucken, Clusius, and Berger.[14] The following formula was used to compute from their data the values of $(\partial U/\partial p)_{301°K}$ given in Table 8:

$$\left(\frac{\partial U}{\partial p}\right)_T = \left(\frac{\partial H}{\partial p}\right)_T - \left(\frac{\partial(pv)}{\partial p}\right)_T \tag{17}$$

[8] Deming, W. E., and Shupe, L. E., Phys. Rev., vol. 37, p. 638, 1931.
[9] Beattie, J. A., and Bridgeman, O. C., J. Am. Chem. Soc., vol. 50, p. 3133, 1928.
[10] $H = U + pv$.
[11] Roebuck, J. R., Proc. Am. Acad. Arts Sci., vol. 60, p. 537, 1925.
[12] Noell, F. Mitteilungen Forschungsarbeiten Gebiete Ingenieurwesens. No. 184, 1916. The data are taken from International Critical Tables, vol. 5, p. 144, McGraw-Hill Book Co., New York, 1929.
[13] Hausen, H., Z. tech. Physik. vol. 7, pp. 371, 444, 1926.
[14] Eucken, A., Clusius, K., and Berger, W., Z. tech. Physik. vol. 13, p. 267, 1932.

In the bottom row of Table 8 are given the values of $(\partial U/\partial p)_{301^\circ \mathrm{K}}$. for air and oxygen as determined from the experiments of the present investigation. The value for nitrogen given in the bottom row was obtained by extrapolation from the values for oxygen and air, assuming the additivity of $(\partial U/p\partial)_T$ for the two components of air.

Inspection of Table 8 shows that, given an estimated error of $\pm 2\frac{1}{2}$ per cent in each of the various values, all are in substantial agreement with the values obtained in the present investigation.

XI. ACKNOWLEDGMENT

The authors acknowledge the technical advice of E. W. Washburn, under whose direction this investigation was carried out.

WASHINGTON, October 1, 1932.

RP504

SHEAR TESTS OF REINFORCED BRICK MASONRY BEAMS

By D. E. Parsons, A. H. Stang, and J. W. McBurney

ABSTRACT

Eighteen beams of reinforced brick masonry were tested to determine the resistance of such beams to failure by diagonal tension. The beams were 14 feet long and about 1 foot square in cross section. Beams of three different types of construction were made, an equal number with each of two kinds of brick. A 1:3 Portland cement mortar, with addition of lime equal to 15 per cent of the volume of the cement, was used in all beams. Each beam contained six ½-inch square steel bars as tensile reinforcement. Tensile and shear tests of the bond between mortar and brick and pull-out tests of steel bars embedded in brick masonry were made to supplement the data from the beam tests.

Positions of the neutral axes in the beams varied with kind of brick, arrangement of bricks in the beams, and loads. The ratio of depth to neutral axis to depth of the tensile reinforcement increased with an increase in the number and total thickness of mortar joints in the masonry. The position of the neutral axis corresponded to that calculated by means of the design formulas applying to beams of reinforced concrete, with an assumed modulus of elasticity of the masonry equal to 50 to 70 per cent of that of the masonry piers.

The failures of all beams were accompanied by cracks near the ends of the beams, between a support, and the nearer load. The cracks were evidence of failures by diagonal tension.

Maximum shearing stresses for the different types of beams ranged from 65 to 159 lbs./in.^{2} Resistance to diagonal tension increased with an increase in the proportion of bricks laid with staggered joints. Shearing strengths of the beams were in the same order as shearing and tensile bond strengths of small masonry specimens.

With the relatively absorbent bricks used, tensile and shearing strengths of the masonry were much greater when bricks were wetted before laying than when laid dry.

Bond strengths as determined by pull-out tests of ½-inch square deformed bars embedded about 8 inches in brick masonry ranged from 870 to 950 lbs./in.^{2} Differences in the kinds of brick and curing conditions did not cause significant changes in bond strength.

CONTENTS

I. INTRODUCTION

Brick masonry containing steel reinforcement in the form of bars bands or straps, wires or mesh apparently is one of the oldest forms of reinforced masonry construction, and there already exists a fund of information relating to its structural value.[1] Resistance of brickwork to compressive stresses has been extensively investigated, and the information obtained is available in ublishe reports[2] of tests on piers and walls. Data on resistance ofpbrickwork beams to shearing stresses are not as complete as desired and the chief purpose of tests described herein was to obtain information on resistance of reinforced brickwork beams to diagonal tension failures. The compressive strength of masonry having a like arrangement of the bricks, with reference to the direction of stress to that in the beams was also determined, compressive tests being made on piers resembling short lengths of the beams. Tests of the adhesion of mortar to the bricks and to steel were made to supplement the data from the beam tests.

The tests were made by the Bureau of Standards with the cooperation of the Common Brick Manufacturers Association. The association paid for the materials and labor for the construction of all specimens. The beams were built under supervision of and were tested by members of the bureau staff.

The authors are indebted to Robert Hamilton, Judson Vogdes, and Hugo Filippi, representing the association, for their assistance in planning the investigation; to L. R. Sweetman, A. U. Theuer, D. A. Parsons, E. E. W. Bowen, and W. W. Harrison for assisting in making the tests and to S. E. Wade for making the drawings and some of the computations.

---

[1] Edw. E. Krauss and Judson Vogdes, Reinforced Brick Masonry: History, Summary of Tests. Structures Erected and Bibliography to Date, Report No. 5 of Committee on Reinforced Brick Masonry, National Brick Manufacturers Research Foundation, February, 1932. Reinforced Brickwork: A New Construction Material, Engineering News-Record, vol. 109, p. 71, July 21, 1932.

[2] The Building Code Committee of the Department of Commerce has prepared a mimeographed circular giving test data obtained prior to 1926. B. S. RP106, "Compressive Strength of Clay Brick Walls" describes tests of 297 masonry specimens, completed in 1928.

II. DESCRIPTION OF SPECIMENS AND THE TESTING METHODS

1. BRICKS

Bricks from two localities, Chicago and Philadelphia, were used. Both kinds of bricks were made from surface clays and were formed by the stiff-mud and end-cut process. The Chicago bricks were irregular in shape, contained lime nodules, and were considerably laminated. The Philadelphia bricks were somewhat more regular in shape, quite free from lime nodules, but also laminated. In so far as applicable, methods of specification C67–31 of the American Society for Testing Materials [3] were followed in making the tests.

2. MORTAR

The mortar was proportioned by weight to give a mixture approximately equivalent by volume to 1 part of Portland cement to 3 parts of loose, damp sand, with an addition of hydrated lime equal to 15 per cent of the volume of cement. The weights used were 94 pounds of cement and 6 pounds of hydrated lime to 220 pounds of dry sand, water being added in the amounts desired by the masons.

On each day during construction of the beams a sample of mortar was taken from a mason's board and tested for consistency on the 10-inch flow table,[4] and six 2 by 4 inch cylinders were cast from the same mortar. After remaining in the mold for one day three of each set of cylinders were immersed in water at 70° F. and the other three were stored with the beams. All cylinders were tested for compressive strength at the age of 28 days.

3. REINFORCEMENT

The steel reinforcement consisted of deformed $\frac{1}{2}$-inch square bars. Specimens cut from five bars were subjected to tensile tests following the methods (in so far as applicable) of specification A15–14 of the American Society for Testing Materials.[5]

4. BEAMS

(a) TYPES

Eighteen beams were tested, each being 14 feet long and roughly 1 foot square in cross section. Nine beams were built with common bricks from Chicago and the other nine with common bricks from Philadelphia. The specimens were of three different types of construction, there being three beams of each type with each kind of brick. As illustrated in Figure 1, the bricks in beams of type A were laid in common American bond as for a wall $12\frac{1}{2}$ inches in thickness; in those of type B the bonding was similar to that in a lintel beam with exposed soldier courses. Beams of type C were built on end in which position they resembled portions of a $12\frac{1}{2}$-inch wall with all bricks laid as stretchers and having vertical reinforcement near one face as if designed to resist lateral pressures.

Each beam contained six $\frac{1}{2}$-inch square deformed bars as tensile reinforcement. The ends of all bars were bent into hooks in order

---

[3] 1931 Supplement to Book of A. S. T. M. Standards, p. 5.
[4] The flow tests were made by the method of Federal specification SS-C-181 for cement; masonry (Jan. 6, 1931).
[5] Book of A. S. T. M. Standards, Pt. I, p. 132, 1930.

to minimize the probability of beam failures resulting from slipping of the bars. The piers contained no reinforcement, but each one was similar otherwise to a short length (about 34 inches) of a beam.

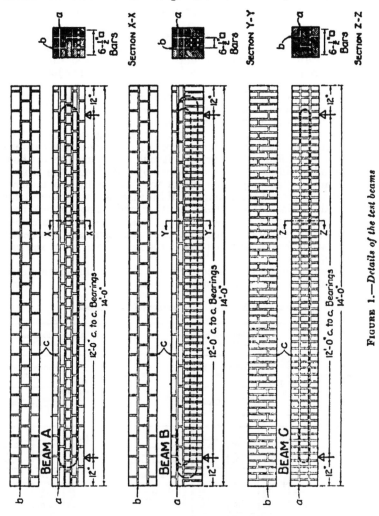

FIGURE 1.—*Details of the test beams*

(b) IDENTIFICATION SYMBOLS

The following symbols, in the order given, are used to identify the beams:

Bricks	C = Chicago.
	P = Philadelphia.
Type of beam	A as shown in Figure 1.
	B as shown in Figure 1.
	C as shown in Figure 1.

Numbers 1, 2, and 3 indicate individual beams.

(c) CONSTRUCTION

The beams were built in the laboratories of the Bureau of Standards by a mason contractor. Bids were obtained from three contractors and the work of constructing the specimens was awarded to the lowest bidder.

The bricks for the beams were dumped from the delivery trucks on an outdoor concrete pavement. On the day before using the bricks they were sprinkled in the pile until the water flowed from every portion. They were again sprinkled in the same manner just before laying.

Beams of types A and B were built on horizontal wooden forms and those of type C were built on end on the laboratory floor. Views of one beam of each type are shown in Figure 2.

The beams were built by two masons working together. The masons were instructed to produce masonry having the spaces between the bricks well filled with mortar, without specifying the method. The method of filling the vertical joints by "slushing" the mortar rather than "shoving" the bricks was chosen by the masons. During the building of the beams one of the masons was cautioned not to furrow the horizontal beds while the other made smooth spread beds without special effort. Although these masons had no previous experience with reinforced brickwork, the two masons and one helper built the last 12 beams and 12 piers at the rate of one beam and one pier per 3¼ hours.

The average thicknesses of mortar joints as determined from measurements of the beams are given in Table 1.

TABLE 1.—*Thickness of the mortar joints in beams*

Beam No.	Average thickness of mortar joints [1]			Beam No.	Average thickness of mortar joints [1]		
	Hori- zontal *a*	Vertical in top course			Hori- zontal *a*	Vertical in top course	
		Longi- tudinal *b*	Cross *c*			Longi- tudinal *b*	Cross *c*
	Inches	*Inches*	*Inch*		*Inches*	*Inches*	[*Inch*
CA-1	0.66	0.90	0.52	PA-1	0.79	0.85	0.54
CA-2	.66	.90	.69	PA-2	.78	.80	.54
CA-3	.66	.94	.60	PA-3	.69	.70	.54
Average	.66	.91	.60	Average	.72	.78	.54
CB-1	.68	.66	.47	PB-1	.85	.87	.50
CB-2	.70	.68	.53	PB-2	.80	.74	.53
CB-3	.65	.87	.63	PB-3	.80	1.03	.58
Average	.68	.74	.54	Average	.82	.88	.54
CC-1	.99	1.25	.52	PC-1	1.12	1.40	.52
CC-2	1.01	1.25	.44	PC-2	1.02	1.33	.48
CC-3	1.01	1.15	.44	PC-3	1.01	1.12	.48
Average	1.00	1.22	.47	Average	1.06	1.29	.49

[1] With the beam in the same position as during testing. The location of joints *a*, *b*, and *c* are shown in Figure 1.

(d) AGING

The beams of types A and B were lifted from the forms when one week old and were stored, until tested, in stacks which allowed free circulation of air around each specimen. Those of type C were placed in a horizontal position in the same stacks. All beams were tested at ages ranging from 27 to 29 days.

(e) METHOD OF TEST

The beams were tested in a vertical screw-testing machine having a capacity of 600,000 pounds. They were supported over a span of 12 feet and loaded along two lines, each 3 feet from mid span. The type of supports and method of transmitting the loads to the beam are illustrated in Figure 3.

Measurements were made of deflections at mid span of the beams and of deformations in the masonry and steel along longitudinal gage lines near mid span. Deflections were measured by means of micrometer dials which indicated the vertical displacement of the upper surface of the beam at mid span relative to a metal frame supported at each end on steel spheres directly over the beam supports. (Shown in fig. 3.) Compressive longitudinal deformations in the upper surfaces of the beams were measured on 20-inch gage lengths by means of a hand strain gage. Tensile longitudinal deformations in the steel over 8-inch gage lengths were measured in one of the bars near each lateral face of the beams by means of gages clamped to the under sides of the bars. Small strap-iron bars, set in vertical joints, were used to support micrometer dial gages and spacing bars for making measurements over longer gage lengths. With this apparatus the deformations in gage lengths of about 40 inches were measured at each lateral face about 1 inch below the upper surface and about 1 inch above the lower surface of a beam.

In making a test the load was increased by increments until deformations indicated stresses in the steel of about 15,000 lbs./in.^{2}. The total load on the beam was then reduced to 1,000 pounds, and finally the load was increased by increments until failure occurred. Usually the devices for measuring strains were removed before the beams failed.

5. PIERS

Eighteen piers, corresponding to the 18 beams, were also built. The piers contained no reinforcement, but each was similar otherwise to a short length (about 34 inches) of a beam and was built in a manner similar to that of the corresponding beam, identified by the same symbol and aged under the same conditions. A view of an unfinished pier is shown in Figure 2.

The piers were tested in compression under central loading in a 10,000,000-pound-capacity testing machine.[6] The lower bearing surface of the pier was bedded in plaster of Paris (calcined gypsum) on the lower platen of the machine. After tilting this platen until the upper surface of the pier was parallel with the upper platen, the upper surface was capped with plaster of Paris.

Vertical compressometers having gage lengths usually between 20 and 30 inches were attached to two opposite sides of the pier. Read-

⁶ Described in Bureau of Standards Research Paper No. 108, p. 526.

FIGURE 2.—*Test beams under construction*
Note the type C beam on end at the left.

B. S. Journal of Research, RP504

FIGURE 3.—*Beam PC-3 in testing machine*

FIGURE 7.—*Beam PA-3 after diagonal tension failure*

ings were taken at equal increments of load usually obtaining about eight readings as the stress increased from 0 to 1,000 lbs./in.², the load being held constant while each set of readings was made. The compressometers were then removed and the machine run at constant speed until failure occurred.

6. BRICK-MORTAR TENSILE SPECIMENS

As the tensile strength of masonry in beams without web reinforcement is the chief source of resistance to diagonal tension, the factors which affect tensile strengths of brick and mortar specimens probably are the same as those governing web resistance of beams. Failures of specimens of brick and mortar when subjected to tensile stresses might conceivably occur by failures of the bricks, of the mortar, of the adhesion between mortar and brick, or of any combination of these. The purpose of tests described was to compare tensile strengths of specimens made with the same types of bricks and the same mortar as were used in the beams. Variations in moisture content of the bricks when laid and in curing conditions were included in the study in order to obtain an estimate of their influence on the strengths of the masonry.

Tests of strength of bond in tension of brick-mortar joints were made. The specimens consisted of two bricks laid flatwise, separated by mortar, the cross section of the joint being 30 square inches for standard size bricks. Twenty test specimens were made for each of the following conditions: Two makes of bricks (Chicago and Philadelphia), dry bricks and dry cure, dry bricks and damp cure, wet bricks and dry cure, and wet bricks and damp cure.

Bricks were considered as dry after one week's storage in a steam heated room. Some bricks were used after 48 hours in a drying oven at 220° F., followed by 24 hours' cooling in the laboratory.

For wetting bricks, the procedure was to totally immerse previously dried bricks for one hour, stand on end in air for one-half hour, and then make the test specimens within the next half hour.

All construction was done in a room kept at a temperature of 70° ±1° F. and a relative humidity of 40 to 60 per cent. The "dry cure" specimens were left in this room for 48 hours after construction and then removed to a laboratory at "room temperature." The "damp cure" specimens were removed at the end of the half hour construction period to the damp storage room of the concrete laboratory, which is kept at a temperature of 70° ±1° F. and a relative humidity of over 90 per cent.

The mortar used was a cement-lime-sand mixture of the same proportions as that used in constructing the beams. An attempt was made to have the flow immediately after mixing between 110 and 120 per cent. The mortar was proportioned and mixed dry. Time was counted from the moment of adding the water to the mortar mix. A flow test was made before starting construction of the test specimens. Mortar was thrown on the flat of the bottom brick. The top brick was quickly put in place with a shoving motion, using considerable pressure. Excess mortar was cut off with a trowel. Since comparison of the two makes of bricks was considered the main purpose of the investigation, the individual specimens were constructed alternately of Chicago and Philadelphia bricks.

The action of the dry bricks of both makes was to suck water out of the mortar, hence the utmost possible speed was used in getting the bricks in place. At best, a number of joints in the dry brick specimens were imperfectly filled.

Care was taken to avoid jarring specimens after construction. In spite of care used in handling, a considerable number of the dry brick specimens separated before testing.

It was the intention to make all tests at 28 days, hence the specimens were removed from damp storage at the end of 24 days, exposed in the laboratory for two days and then capped. Circumstances required that some of the tests be delayed, but the damp cure was restricted in all cases to 24 days.

The method of determining the tensile strength was essentially that used for testing whole bricks in tension described in another paper.[7] Palmer and Hall[8] further describe the apparatus and method.

7. BRICK-MORTAR SHEARING SPECIMENS

The shearing specimens were equal in number and made with the same mortar mixture and with the same procedure as the brick mortar tensile specimens. The specimens for shear were made by laying three bricks flat, the top and bottom bricks having their ends in line while the center brick was displaced lengthwise from one-half to three-quarters of an inch.

The projecting ends of the two outer bricks of each specimen were capped with plaster of Paris, the surfaces of both caps being in one plane. The projecting end of the center brick was also capped, its surface being as far as possible parallel to the surfaces of the two other caps.

The specimens after capping were loaded in compression, the load being applied to the center brick and the specimen resting on the ends of the two outside brick. This is the method of Douty and Gibson.[9]

8. PULL-OUT TESTS

Specimens for pull-out tests consisted of a ½-inch square deformed steel bar imbedded lengthwise in a mortar joint of a small brick pier. The same variables in brick and curing were used as for the brick-mortar tensile and shearing specimens, but the brick were wetted before laying. A deformed steel bar (one-half inch square) was held vertically by clamps, the lower end resting on oiled paper. Around this bar was built a small brick pier approximately 8 by 8 inches in cross section and three bricks high. The middle course of bricks were laid as headers with respect to the top and bottom courses. Care was taken to secure imbedding of the bar in mortar and to avoid contact of brick with the bar. All joints were filled with mortar. No dry bricks were used in this series, but both kinds of cure were employed. The mortar was the same as that used for the tensile and shearing specimens.

Before testing, the specimens were capped with plaster of Paris on top of the bricks, care being taken to have the capped surface smooth and normal to the projecting steel bar.

---

[7] J. W. McBurney, Strength of Brick in Tension, J. Am. Ceramic Soc., vol. 11 (2), pp. 114–117, 1928.
[8] L. A. Palmer and J. V. Hall, Durability and Strength of Bond Between Mortar and Brick, B. S. Jour. Research vol. 6 (3), pp. 473–492, 1931.
[9] R. F. Douty and H. C. Gibson, Influence of the Absorptive Capacity of Brick Upon the Adhesion of Mortar, Proc. Am. Soc. Testing Materials, vol. 8, pp. 513–530, 1908.

During a test the specimen was inverted, resting on a steel plate supported by the top head of the testing machine. The bar projected downward through a ⅜-inch hole in the steel plate and was gripped by wedges in the movable head of the testing machine.

III. RESULTS OF THE AUXILIARY TESTS

1. BRICKS

Results of tests of the bricks are given in Table 2, in which each average value is the mean from 25 tests. As a measure of the dispersion of the results of single tests about their corresponding averages, there are given in Table 2 values for the standard deviations, which were calculated by means of the following formula:

$$\sigma = \sqrt{\frac{\Sigma v^2}{n-2}}$$

where

σ = standard deviation.
n = number of individual values.
Σv^2 = sum of the squares of the deviations of the single values from their mean.

The values of standard deviations given in other tables were calculated by means of the same formula.

2. MORTAR IN THE BEAMS

Results of tests of the specimens representing the mortar in the beams are given in Table 3.

TABLE 2.—*Properties of the bricks*

Each value was derived from results of tests of 25 bricks

Property	Kind of bricks			
	Chicago		Philadelphia	
	Mean value	Standard deviation	Mean value	Standard deviation
Length_____inches__	8.00	0.08	8.15	0.14
Width_____do____	3.64	.06	3.72	.06
Thickness_____do____	2.23	.03	2.29	.02
Compressive strength flatwise_____lbs./in.²__	3,910	860	4,510	840
Compressive strength edgewise_____do____	4,290	950	5,240	950
Compressive strength endwise_____do____	7,030	1,330	4,200	1,730
Modulus of rupture flatwise_____do____	1,530	570	650	340
Absorption by 5 hours immersion____per cent__	8.8	2.7	11.1	1.6
Absorption by 48 hours immersion_____do____	10.8	2.8	12.7	1.9
Absorption by 5 hours boiling_____do____	14.7	2.6	16.1	1.8

TABLE 3.—*Properties of mortar for the beams*

Each value was derived from results of 21 tests

Property	Mean value	Standard deviation
Flow_____per cent__	108	8.2
Compressive strengths of 2 by 4 inch cylinders, age 28 days:		
Dry storage_____lbs./in.²__	2,340	440
Damp storage_____do____	3,740	530

3. REINFORCEMENT

Tensile properties of the ½-inch square deformed steel bars are given in Table 4.

TABLE 4.—*Properties of reinforcement for the beams*

Each value was derived from the results of five tests

Property	Mean value	Standard deviation
Cross sectional area..inches ²..	0. 240	0. 002
Proportional limit..lbs./in.²..	45, 200	6, 700
Yield point...do....	51, 100	1, 600
Tensile strength..do....	81, 600	1, 400
Modulus of elasticity...do....	29, 400, 000	1, 100, 000
Elongation in 8 inches...per cent..	22. 4	2. 0

4. BRICK-MORTAR TENSILE SPECIMENS

Results of tests of the brick-mortar tensile specimens are given in Table 6.

TABLE 5.—*Properties of mortar for the tensile and shear test specimens*

Each value was derived from results of from 40 to 51 tests. Strength tests at ages ranging from 28 to 60 days

Property	Mean value	Standard deviation
Initial flow ....per cent..	116	9.8
Flow one-half hour after mixing...................................do....	92	14. 1
Tensile strengths of briquettes with:		
Dry storage..lbs./in.²..	290	25
Damp storage..do....	470	34
Compressive strengths of 2 by 4 inch cylinders with:		
Dry storage..lbs./in.²..	3, 410	400
Damp storage..do....	4, 010	250

TABLE 6.—*Results of tensile tests of bond between mortar and brick*

Age at test 28 to 60 days

Brick		Storage	Number of specimens	Average strength	Standard deviation
Kind	Condition when laid				
				Lbs./in.³	Lbs./in.³
Chicago................	{Dry........	{Dry................	16	38	17
		{Damp.............	16	38	13
	{Wet.......	{Dry................	20	55	17
		{Damp.............	20	61	16
Philadelphia...........	{Dry........	{Dry................	16	18	10
		{Damp.............	17	27	12
	{Wet.......	{Dry................	20	53	18
		{Damp.............	20	44	13

Mortar used in brick-mortar tensile specimens and in brick-mortar shearing specimens had the same proportions of dry constituents as the mortar in the beams and piers. In the latter, water was added as desired by the masons; while in the brick-mortar specimens, made under more careful laboratory conditions, the amount of water to give a flow of approximately 110 per cent was used. For this reason, properties of the mortar, given in Table 5, for brick-mortar specimens differ somewhat from values of Table 3 for the mortar of the beams.

Where test specimens represented wetted bricks, the characteristic failure was not at the junction of brick and mortar but was a failure in the brick. Chicago bricks left a "skin" adhering to the mortar, and Philadelphia bricks frequently pulled off or sheared off their flats to a depth of one-eighth of an inch. In other words, when bricks had been wetted, bond between brick and mortar exceeded the strength of the brick. On the other hand, a few of the "dry brick" specimens showed separation in the mortar, mortar adhering to both bricks. The difference between number of tests indicated and the 20 test specimens originally constructed represents failures of bond occurring by handling the specimens. The number of tests is too few and the variation too great to permit much weight to be given either to averages or distributions but several conclusions are evident: First, wetting Chicago and Philadelphia bricks much increased strength of bond; second, the Chicago brick tended to give stronger bonds than the Philadelphia brick.

5. BRICK-MORTAR SHEARING SPECIMENS

Results of tests of brick-mortar shearing specimens are given in Table 7. In general, results of these tests are similar to the results of tests of the brick-mortar tensile specimens.

TABLE 7.—*Results of shear tests of bond between mortar and brick*

Age at test, 28 to 60 days

Brick		Storage	Number of specimens	Average strength	Standard deviation
Kind	Condition when laid				
				Lbs./in.²	Lbs./in.²
Chicago	Dry	Dry	16	100	50
		Damp	12	115	47
	Wet	Dry	20	275	59
		Damp	19	245	101
Philadelphia	Dry	Dry	10	91	35
		Damp	11	120	38
	Wet	Dry	19	231	86
		Damp	20	173	70

6. PULL-OUT TESTS

Results of the pull-out tests are given in Table 8. As shown by the data of this table there was not a significant difference in bond strength of specimens made with Chicago or Philadelphia bricks. Furthermore, curing conditions did not affect results significantly. Failure was by splitting of the brickwork usually into halves but in some cases into quarters. Specimens giving highest individual loads had strengths which exceeded the proportional limit of the steel.

TABLE 8.—*Results of pull-out tests*

Specimens were deformed steel bars, one-half inch square, embedded about 8 inches in 8 by 8 inch square brick masonry piers

Brick		Storage	Number of specimens	Age at test	Average bond strength	Standard deviation
Kind	Condition when laid					
				Days	*Lbs./in.²*	*Lbs./in.²*
Chicago	Wet	Dry	5	32-48	880	140
Philadelphia	do	do	4	32-48	950	140
Chicago	do	Damp	5	41-57	870	190
Philadelphia	do	do	5	41-57	920	150

The maximum bond stresses were slightly greater than those reported by Abrams [10] for deformed bars in 1:2:4 concrete two years old. It has been shown that for smooth bars embedded in concrete the bond strength is more dependent on length of embedment than on other size and shape factors.[11] The lengths of embedment in these two series were about the same (8 inches). Hence, if the effects of size and shape of specimens are approximately the same with deformed bars as with smooth bars, it may be concluded that bond strengths in the brick masonry specimens were about the same as in 1:2:4 concrete specimens tested by Abrams.

7. PIERS

Results of tests of the piers are given in Table 9. There was not a marked departure from a linear relation between loads on the piers and their compressive deformations until 25 per cent of the maximum load was reached, after which there was a tendency for the deformations to increase more rapidly than the loads increased.

Table 9 gives values of the secant modulus of elasticity at a stress of 250 lbs./in.², obtained by dividing this stress by the corresponding compressive strain. Values of the secant modulus of elasticity for any stresses less than one-fourth of the maximum did not differ significantly from those given.

TABLE 9.—*Results of pier tests*

Pier	Secant modulus of elasticity to 250 lbs./in.²	Compressive strength	Pier	Secant modulus of elasticity to 250 lbs./in.²	Compressive strength
	Lbs./in.²	*Lbs./in.²*		*Lbs./in.²*	*Lbs./in.²*
CA-1	2,690,000	1,735	PA-1	1,050,000	1,438
CA-2	2,400,000	1,955	PA-2	1,800,000	1,577
CA-3	2,340,000	1,879	PA-3	1,770,000	2,039
Average	2,480,000	1,856	Average	1,540,000	1,686
CB-1	2,120,000	1,956	PB-1	980,000	1,210
CB-2	2,140,000	2,458	PB-2	1,020,000	1,421
CB-3	2,380,000	1,661	PB-3	1,220,000	1,535
Average	2,210,000	2,025	Average	1,070,000	1,389
CC-1	1,570,000	1,235	PC-1	740,000	1,072
CC-2	960,000	810	PC-2	1,070,000	1,423
CC-3	1,020,000	1,019	PC-3	800,000	1,112
Average	1,180,000	1,021	Average	870,000	1,202

[10] Duff A. Abrams, Tests of Bond Between Concrete and Steel, Bul. No. 71, University of Illinois.
[11] W. H. Glanville, Studies in Reinforced Concrete. I. Bond Resistance, Building Research Tech. Paper No. 10, Dept. of Sci. and Ind. Res., London, 1930.

Both the moduli of elasticity and compressive strengths of the piers were greater for piers of types A and B, having some bricks on end, than for those of type C having all bricks flatwise. As indicated by the data of Table 9 on compressive strength of the piers and that of Table 2 on physical properties of the bricks, there were not large differences in strength properties of either bricks or piers.

IV. RESULTS OF BEAM TESTS

1. DEFORMATIONS IN THE BEAMS

Load deformation curves for the brickwork and the steel and load-deflection curves for the beams are shown in Figures 4 and 5. Values shown for the strains in the brickwork and steel are averages calculated from data obtained with the 40-inch gage length extensometer attached to the sides of beams and the readings of the strain gages. The curves are similar in shape to those representing like data from tests of concrete beams containing heavy reinforcement. Disregarding the portion due to the partial release of the load, the curves tend to consist of a short straight portion, corresponding to the first stage of loading, while the brickwork was still effective in resisting tensile stresses; then a curved portion as the failure of the brickwork in tension took place and then a nearly straight portion. Usually this was followed by another curved portion similar to that of a stress strain diagram representing the later stages in a compressive test of brick masonry.

Some bricks under the reinforcement bars at mid span were removed, prior to testing the beams, in order to facilitate attachment of strain gages on the bars. Due partly to their removal and partly to lack of sensitivity of the long and the short gage tensometers, reliable values for extensibility of the brickwork were not obtained.

When loads on the beams were reduced to 1,000 pounds after the first loading, strain and deflection indicators did not return to the positions taken under the first application of a load of 1,000 pounds. The differences, which are measures of sets produced by the loadings, were equal usually to from one-seventh to one-fifth of the deformations under the greatest loading which preceded.

2. POSITIONS OF THE NEUTRAL AXIS

Average positions of the neutral axes for each type of beam at several loads as determined from measured deformations are indicated in Figure 6, k being the ratio of depth of neutral axis to depth of centroid of the steel. Each plotted point represents a value obtained by averaging the data for three like beams. On the same diagram are shown horizontal lines indicating positions of the neutral axes corresponding to several values of n, calculated by means of the following well known formula:

$$k = \sqrt{2pn + (pn)^2} - pn$$

where k = ratio of depth to neutral axis to depth of the centroid of the tensile reinforcement.

    p = ratio of area of tensile reinforcement to area of masonry above the centroid of the tensile reinforcement.

    $n = \dfrac{E_s}{E_m}$ = ratio of the modulus of elasticity of steel to that of the masonry.

As loads were increased there was a tendency for the neutral axis
to rise in the beams of types A and B, indicating a gradual lessening of
tensile resistance of the brickwork. The tendency of the neutral

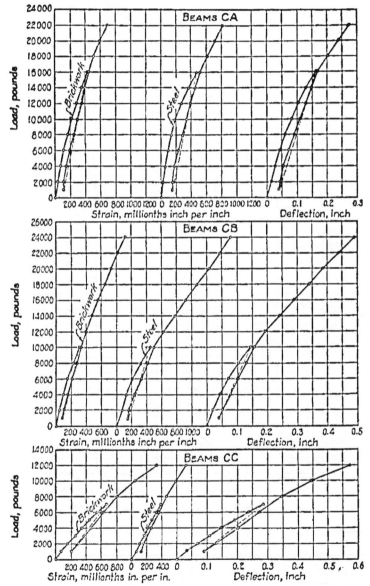

Figure 4.—*Load-deformation curves for beams with Chicago brick*

axis to rise continued until the compressive stresses in the brickwork
were so great that the rate of change of strain with compressive stress
in the brickwork had become great enough to either partially or com-

pletely overcome this tendency.   Changes in the depths to the neutral
axis with increased loads on these beams resembled those in reinforced concrete beams of similar shape and reinforcement, except that
the reduction of tensile resistance of the brickwork in beams of types
A and B seemed to take place more gradually than is common in
beams of concrete.   As the modulus of elasticity of the masonry
E_m varies with stress and as the effects of tensile resistance of the

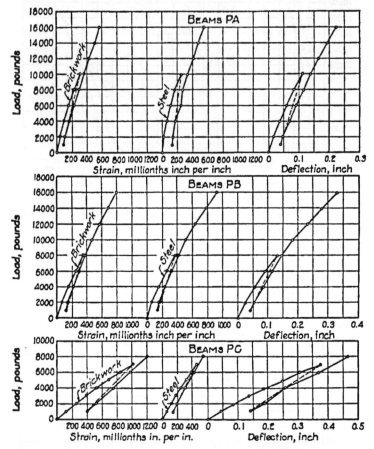

FIGURE 5.—*Load-deformation curves for beams with Philadelphia brick*

masonry are neglected in the derivation of the design formulas, it is
desirable to determine what values of n correspond in the formula
$k = \sqrt{2pn + (pn)^2} - pn$ to the observed values of k.   For this purpose
it is best to use data obtained while conditions in the beams most
nearly correspond to those assumed in the derivation of the design
formulas.   The minimum values of the ratio k usually were obtained
when loads on the beams were sufficiently large to have caused tensile
cracks in the masonry below the neutral axis and yet not large enough
to have produced a rapid plastic yielding of the masonry in com-

pression.   The minimum values of k observed for two or more loads are given in Table 10, together with values of n calculated by means of the foregoing formula.

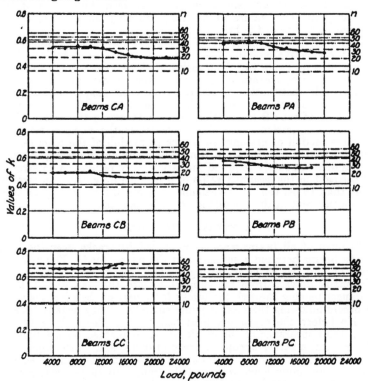

Figure 6.—*Position of the neutral axes*

The circles indicate values of k determined from the measured deformations. Calculated values of k corresponding to the values of n, shown on the right-hand scale of each graph, are indicated by the horizonal dash-dot lines

TABLE 10.—*Values of n calculated from dimensions of beams and positions of neutral axes during tests.   Values of k were determined from the long gage deformations*

Beam	p	k	n	Beam	p	k	n
	Per cent				*Per cent*		
CA-1	1.03	0.43	16	PA-1	0.97	0.54	33
CA-2	1.01	.44	17	PA-2	.97	.48	23
CA-3	1.00	.45	18	PA-3	1.02	.53	29
CB-1	1.20	.44	15	PB-1	1.07	.48	21
CB-2	1.17	.45	16	PB-2	1.10	.57	34
CB-3	1.16	.45	16	PB-3	1.05	.56	34
CC-1	1.31	.66	50	PC-1	1.24	.70	66
CC-2	1.31	.62	38	PC-2	1.25	.69	62
CC-3	1.30	.68	56	PC-3	1.27	.69	60

These calculated values of n were greater for beams of Philadelphia bricks than with those of Chicago bricks.   They were greatest with beams of type C and least with those of types A and B, the values corresponding in order of magnitude to the average number and total thickness of joints per unit length of beam in the upper two courses of bricks. However, no general relations between dimensions of the mortar joints and values of n exist, probably on account of the laminar structure of the bricks.

Secant moduli of elasticity of the piers were in all cases larger than the effective moduli of the masonry of beams as indicated by values of n in Table 10. The ratios of these values range from 0.5 to 0.9. A close correspondence would not be expected because of difference in distribution of stresses over their unsymmetrical cross sections. The entire section of piers A and C resembled more nearly than those of type B the portion of the masonry in the beams which were in compression. For these two types the average ratios of effective moduli of masonry in beams to moduli of the piers ranged from 0.5 to 0.7. Fortunately, from the standpoint of design, accurate values of n are rarely required for satisfactory results, as an error of 50 per cent in the assumed value of n would rarely cause an error of more than 15 per cent in the calculated stress in the masonry or more than 4 per cent in the calculated stress in the steel.

3. TYPES OF FAILURES

Failures of all beams were accompanied by cracks near the ends of the beams between a support and the nearer load. Figure 7 is a view of two beams after testing. These cracks tended to extend diagonally upward from the support toward the load line, in some instances passing through the bricks for a part of their lengths while in others following the mortar joints entirely. They usually became visible before the maximum load had been applied. In a typical case, the appearance of the crack was accompanied by a falling off of load and a rather abrupt increase in deflection. After this, with the machine running at constant speed, load increased more slowly, but the maximum loads were usually from 5 to 10 per cent greater than when the crack was first observed.

All beams except CA–2 and CC–1 failed by diagonal tension. The tensile reinforcement in beam CA–2 began to yield before the maximum load had been applied. Using the ordinary formulas for working stresses in concrete beams and considering the position of the neutral axis as observed, the calculated maximum stresses under maximum load in this beam were 54,000 lbs./in.² in the steel and 2,400 lbs./in.² in the masonry. Strain-gage readings indicated yielding of masonry in the upper surface of beam CC–1 and spalling was observed prior to the maximum load. Calculated stresses under the maximum load were 34,000 lbs./in.² in the steel and 1,340 lbs./in.² in the masonry. The beams were so conservatively designed against failures by slipping of the bars that the bond stresses developed in the beam tests have no significance.

4. RESISTANCE TO DIAGONAL TENSION

(a) EFFECT OF ARRANGEMENT AND BONDING OF BRICKS IN BEAMS

Resistance of the beams to failures by diagonal tension was affected markedly by the bonding of the bricks as is shown by the data of Table 11. With the Chicago bricks the ratios of the average maximum shearing stress to that for type C beams were, respectively, 1.69, 1.38, and 1.00 for beams of types A, B, and C; corresponding ratios for beams of Philadelphia bricks were 1.49, 1.42, and 1.00.

TABLE 11.—*Results of beam tests*

Beam No.	Width b	Depth d	j	Maximum load W	Maximum shearing stress $v = \dfrac{Vf}{bjd}$
	Inches	*Inches*		*Pounds*	*Lbs./in.²*
CA-1	12.73	11.00	0.85	36,700	154
CA-2	12.73	11.25	.85	41,750	171
CA-3	12.80	11.30	.85	37,600	153
Average	12.75	11.18	.85	38,680	159
CB-1	12.23	9.84	.85	28,500	140
CB-2	12.57	9.84	.85	26,000	123
CB-3	12.67	9.84	.85	27,600	130
Average	12.49	9.84	.85	27,370	131
CC-1	12.90	8.50	.78	18,000	105
CC-2	12.90	8.50	.78	14,900	87
CC-3	12.80	8.65	.78	15,900	91
Average	12.87	8.55	.78	16,270	94
PA-1	12.87	11.95	.83	23,450	92
PA-2	12.77	11.65	.83	22,100	89
PA-3	12.57	11.20	.83	26,000	111
Average	12.74	11.60	.83	23,850	97
PB-1	12.90	10.40	.83	20,000	90
PB-2	12.63	10.40	.83	17,400	80
PB-3	13.23	10.40	.83	24,000	105
Average	12.92	10.40	.83	20,470	92
PC-1	13.27	8.75	.77	8,650	48
PC-2	13.20	8.75	.77	14,400	81
PC-3	13.00	8.75	.77	11,700	67
Average	13.16	8.75	.77	11,580	65

NOTE.—$j = 1 - \frac{k}{3}$, where k is the average of experimentally determined values of ratio of depth of neutral axis to effective depth. The values of k used were taken from Figure 6.

Maximum shearing stresses were in the same order as the proportion of bricks laid with staggered vertical joints. As shown in Figure 1, joints which were vertical in the type C beams during a test were not staggered; they extended over the full width and from the lower to upper surfaces of the beams. In type B beams all vertical joints were staggered, but those between the soldier bricks of the outer wythes had an unbroken vertical length about equal to the height of three courses of bricks laid flatwise. The vertical joints in the A beams were broken at each course. The proportion of the staggered joints in the different beams my be expressed by approximate numerical values as:

15 out of 15 or 100 per cent in the A beams.
9 out of 15 or 60 per cent in the B beams.
0 out of 9 or 0 per cent in the C beams.

(b) EFFECT OF STRENGTH AND ABSORPTION OF THE BRICKS

Resistance of beams to shearing stresses (Table 11) were in reverse order to compressive strengths flatwise and edgewise of the bricks (Table 2), the strengths of Chicago bricks in these tests being less than that of Philadelphia bricks. Chicago bricks were stronger endwise and when laid as in the top course of beams A and B were subjected to compressive stresses in the direction of greatest strength. However, strengths of the C beams were not in the same order as compressive strengths of bricks in the direction of the axis of the beams. It appears, therefore, that there was no consistent relation between compressive strength of bricks and shearing resistance of masonry.

Shearing resistance of the beams was in the same order as moduli of rupture of the bricks. The moduli of the bricks were determined, however, only in the direction of their lengths. Bricks from both sources were laminated and their moduli in other directions may not have been in the same order. Therefore, the data are not conclusive as to the effect of the moduli of rupture of the bricks on shearing strength of the masonry.

Absorption properties of the two kinds of bricks did not differ greatly, and it does not seem likely that the small differences in absorptions had an important effect on the relative strengths of beams made with them.

(c) EFFECT OF BOND STRENGTH BETWEEN MORTAR AND BRICKS

As noted in the description of tensile and shearing tests of brick-mortar specimens, some specimens were made with wet and the rest with dry bricks. Some were aged in damp storage and others in dry storage. Bricks for the beams were wetted before use. Although the beams were aged in the laboratory where some drying would take pl ce, the rate of drying was probably much less than for the smaller bond test specimens stored in air. Hence, it would be expected that the moisture content during fabrication and storage of the beams was not the same as for any of the bond-test specimens but was intermediate between the dry and wet storage bond-test specimens made with wet bricks. Moreover, as conditions during the first few days of storage have a greater effect upon the properties of Portland cement mortars than those during a similar equal period, it seems likely that the properties of the mortars in the beams were more nearly like those of the damp-cured bond specimens.

Table 12 gives ratios of the average maximum shearing stress of beams of Chicago bricks to those of Philadelphia bricks for each type of beam. These were calculated from the data of Table 11. Similar ratios calculated from data of Tables 6 and 7 for bond tests are given also. The close agreement between ratios for the beams and for the damp-cured bond-test specimens indicates, as would be expected,[12] that tensile and shearing strengths of the masonry were closely related to the resistance to diagonal tension of the masonry beams.

---

[12] Because the appearance of the cracks indicated failures due to diagonal tension.

TABLE 12.—*Effect of tensile and shearing strengths of masonry on shearing resistance of beams*

[Ratio of strength with Chicago bricks to strength with Philadelphia bricks]

Maximum shearing stress beams:

A_____ 1. 64
B_____ 1. 43
C_____ 1. 45
Bond strength in tension, wet bricks, damp storage_____ 1. 39
Bond strength in shear, wet bricks, damp storage_____ 1. 41
Bond strength in tension, wet bricks, air storage_____ 1. 04
Bond strength in shear, wet bricks, air storage_____ 1. 19

V. CONCLUSIONS

1. Maximum shearing stresses for the different types of beams, which are taken as measures of their resistances to diagonal tension, ranged from 65 to 159 lbs./in.^{2}.

2. Resistance to diagonal tension increased with an increase in the proportion of the bricks laid with staggered joints. Maximum shearing stresses in the beams having all bricks laid flatwise with staggered joints were about 60 per cent greater than in beams with continuous (not staggered) joints.

3. Maximum shearing stresses in beams with one kind of brick were from 40 to 60 per cent greater than in beams made with the other kind. These stresses were in the same order as the tensile and shearing strengths of small masonry specimens made with the same kinds of brick and mortar. The shearing strength of the beams appeared to be independent of the compressive strengths of the bricks. The only qualities of the bricks which appeared to have a major influence were those affecting tensile and shearing strength of the masonry.

4. With the rapidly absorbing bricks used in these tests tensile and shearing strengths of the masonry were much greater when the bricks were wetted before laying than when laid dry.· Curing conditions had a relatively small effect; specimens made with dry brick were made stronger by damp curing, while those made with wet bricks were stronger when air cured.

5. The neutral axes in the beams rose during the early stages of the loading, but during the later stages usually became lower. The positions as indicated by minimum observed values of k corresponded to values of $n = \frac{E_s}{E_m}$ ranging from 15 to 66. The effective n varied with the kind of bricks in the beam and increased with an increase in the average number and total thickness of mortar joints per unit of length of beam. The effective E_m in the beams ranged from 0.5 to 0.7 of the secant modulus of elasticity of the piers, when the bonding of the bricks in the piers was similar to that in the upper half of the beams.

6. The bond strength at an age of from one to two months as determined by pull-out tests of ¾-inch-square deformed bars embedded 8 inches in brickwork ranged from 870 to 950 lbs./in.^{2}. Differences in the kinds of bricks and of curing conditions did not cause significant changes in these bond strengths.

WASHINGTON, September 17, 1932.

RP505

THE MEASUREMENT OF LOW-VOLTAGE X-RAY INTENSITIES [1]

By Lauriston S. Taylor and C. F. Stoneburner

ABSTRACT

A special type of chamber has been designed to measure the ionization produced by 3 to 12 kv X rays in a known volume of unrestricted air, thus making it possible to express the intensity of the X-ray beam in roentgens per minute. By the use of a very small guarded field ionization chamber (5 by 5 cm), the air-absorption correction is reduced several fold, and hence errors in its determination are not so serious in the final result sought. It is shown that air-absorption corrections must be determined separately for each beam of radiation used, and the chamber is so designed that the necessary corrections may be determined without recourse to other special apparatus. Examples are given showing in detail the methods of making measurements with the chamber described. Compared at higher voltages (40 to 90 kv) with a large primary standard ionization chamber, the small type here described shows a divergence of about ±0.5 per cent which is believed to be the over-all error present in its use. In order to obtain this close agreement it is necessary to take special precautions in the construction and measurement of the limiting diaphragm and the collector electrode.

CONTENTS

I. INTRODUCTION

With the increasing clinical use of soft X rays, excited by potentials of 3 to 12 kv (peak), arises the necessity for their more accurate measurement such as obtains with the higher voltage radiations. This is particularly important in view of the fact that such radiation may furnish an erythema dose in one or two minutes. In fact errors in measurement assume even greater importance than for high-voltage X rays.

Although several investigators have reported on the measurement of soft X rays (so-called grenz rays), they have given but few essential details of the methods they employed.[2][3][4] This obscurity led us to construct an open-air ionization chamber capable of measuring soft

[1] Presented at annual meeting of Radiological Society of North America, St. Louis, December, 1931.
[2] H. Kustner, Strahlentherapie, vol. 27, p. 124, 1928.
[3] B. Rajewsky and G. Gabriel, Strahlentherapie, vol. 30, p. 20, 1928.
[4] Otto Glasser, Chap. III of Grenz Ray Therapy, by Gustav Bucky, Macmillan, 1929. See also Radiology.

X rays, and to compare it with the standard used for measuring high-voltage X rays.[5] A chamber has, therefore, been designed which measures reliably X rays excited by voltages ranging from 3 to 80 kv (peak), covering not only the "grenz-ray therapy" (3 to 12 kv) but also the lower "diagnostic X-ray range" (60 to 80 kv).

Open-air chambers lacking a guarded field are not readily applicable to the measurement of soft X rays because the large air absorption correction involved can not be determined with desired accuracy. The principle of the guarded-field air-ionization chamber [6] renders it, however, particularly adaptable by permitting the air path between diaphragm and collector plates to be reduced to a minimum. The points essential to its successful application to low-voltage X rays are worked out in the present investigation.

II. DESCRIPTION OF CHAMBER

Described briefly, the guarded-field ionization chamber furnishes adequate field uniformity over the width of the collector electrode of a parallel plate ionization chamber by means of properly charged guard wires stretched across the open faces of chamber, the potentials of the wires being divided in equal successive steps between zero (at collector) and the plate potential. It has been shown that such a system permits the use of grounded metal parts (case, diaphragm, etc.) much closer to the electrodes than would otherwise, be possible. In the present case the shielding and the limiting diaphragm may be brought to within 5 cm of the center of the collector pla e, and air-absorption correction need be made only for this small distance.

A cross section of this special ionization chamber for soft X rays is shown in Figure 1. The electrode system consists of the high-potential plate, A, the two grounded guard plates, B, and the collector electrode, C, all mounted as a unit, similar to the larger type of guarded field chamber. The 4 rectangular guard wire loops, a, b, c, d, made of 10-mil aluminum, are maintained at equal potential steps, respectively, between earth and the potential of A by means of 5 half-megohm resistors, r, r, mounted on a hard rubber supporting frame. The entire electrode system can be slid the length of the box, D, on two guide rods, and its position read on the scale, S, which also serves as a "handle" with which to move the system from the outside.

The box, D, is of ¼-inch brass, which provides sufficient radiation protection for lower-voltage X rays. For higher-voltage X rays it is further shielded with about 15 mm of lead. The rod, H, is connected to a source of sufficiently high potential to produce saturation current in the chamber and makes contact with the plate, A, through a sliding spring. Similarly the collector electrode, C, is connected by a sliding contact to E, which leads to the electrometer system. The rod, E, is shielded over its entire length, as shown by the section X, for the purpose of preventing any ions not formed within the electrode system from reaching the electrometer system. The adequacy of this shielding was tested experimentally.

The diaphragm, n, has an 8-mm orifice with the limiting edges cylindrical and about 1 mm thick. The chamber as a whole is

⁵ L. S. Taylor and G. Singer, B. S. Jour. Research, vol. 5 (RP211), p. 507, 1930.

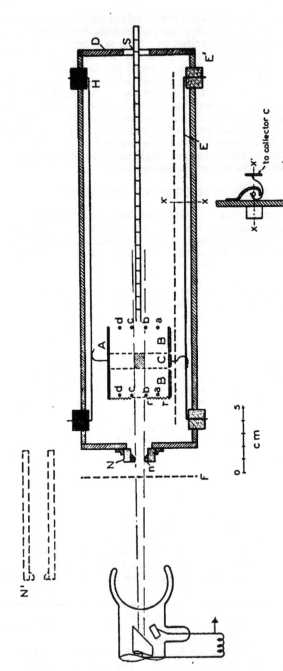

FIGURE 1.—Cross section of special guarded-field ionization chamber for measuring low-voltage X rays or grenz rays

mounted on a base that slides in a direction parallel to the X-ray beam axis. A series of aluminum filters, *F*, about 3 cm in diameter, are mounted on a celluloid disk, the axis of which is independently movable parallel to the axis.

III. CALIBRATION MEASUREMENTS

1. COLLECTOR ELECTRODE

Since the chamber is so small, it is imperative that its essential dimensions be known with high accuracy. The collector electrode being only about 1 cm wide, along the length of the beam, an error of $\frac{1}{10}$ mm in determining its effective width (true width plus one-half the width of both air gaps between collector and guard plate) would introduce an error of 1 per cent in the effective air volume. The necessity for determining this with accuracy is obvious. Accordingly, through the help of the gage section of this bureau, the following typical measurements were obtained:

mm

Width of collector electrode_____ 9.492±0.005
Separation between guard plates_____ 9.904± .005
Effective collector width_____ 9.698± .005

Furthermore, to avoid field distortion at the edges, it is essential that the faces of the collector and guard plate lie exactly in the same plane. Field distortion at these edges presents one of the most serious obstacles to accurate current measurements in case, during the compensation of current, it is difficult to hold the collector at zero potential.

2. CHAMBER-LIMITING DIAPHRAGM

This diaphragm consists of a coin-gold sleeve, *n*, held in a Pb-Bi alloy ring, *N*. Its orifice was made and measured in two ways—the first by lapping the inner surface and then measuring by the plug-gage method; the second by forcing a standard steel ball through an orifice which had been previously turned to a diameter of one or two thousandths of an inch less than the ball diameter. The latter method burnished the surface and at the same time determined its size. The measured diameters of two such diaphragms were:

mm at 68° F.

Diaphragm 1 (plug-gage method)_____ 7.8608±0.0001
Diaphragm 2 (steel-ball method)_____ 7.9375± .0001

It is seen that the accuracy in determining the diaphragm area by either method is about 1 part in 9,000.

3. SATURATION VOLTAGE

Voltage to saturate the ionization chamber was supplied from a 1,300-volt d. c. generator. Curve *A* in Figure 2 shows that saturation was reached at about 300 volts; hence, for normal operation, 500 volts was applied.

4. EFFECT OF SCATTERED RADIATION

In chambers for higher-voltage X rays the limiting diaphragm must be placed at least 10 to 15 cm from the collector to avoid the influence of X rays and *β* rays scattered from the diaphragm edges. Since the radiation from low-voltage X-ray tubes]must, in general, be measured so close to their source that a comparatively divergent beam enters the chamber, the customary type of chamber diaphragm is not satisfactory.

The diaphragm placed as shown in N by Figure 1 is satisfactory, provided the scattering from its walls is negligible. This was tested by measuring the ionization with the diaphragm in place, then again after shifting it to the end of a 10 cm extension tube, N', the chamber being shifted so as to maintain the distance from N to N' to the tube target the same. Correction was made as shown later for absorption in the 10 cm additional air path. Measurements at 40 kv (eff) gave the following results:

Intensity, diaphragm at normal position_____ 0. 964±0. 0033
Intensity, diaphragm shifted to 10 cm_____ . 970± . 0027

Difference_____ . 006

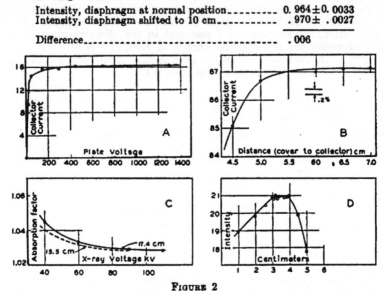

FIGURE 2

a, Saturation curve for special ionization chamber; b, variation of measured ionization with distance between diaphragm and ionization chamber; c, air absorption factor for calibration of special ionization chamber; d, X-ray intensity along cross section of beam used in calibrating ionization chamber.

The observed difference being of the same magnitude as the error of observation but in the direction opposite to that to be expected,[6] it is assumed that the two arrangements give identical results for the comparatively narrow beams used. In other words, the scattering effect from the walls of the diaphragm orifice is negligible.

Use of the divergent beam exposes the guard wires to the direct radiation, and this may have a scattering effect. To test for scattering from this source, two strands of 20-mil aluminum wire were suspended, insulated, in the beam between the 10-mil guard wires b and c, hence did not influence the field. The following measurements were obtained at 40 kv (eff), with no filtration:

Intensity, wires in beam_____ 0. 295±0. 0016
Intensity, wires out of beam_____ . 294± . 0008

Difference_____ . 001

[6] This is probably due to error in the air-absorption correction.

The difference being of the magnitude of the experimental error, no appreciable effect arises from permitting the beam to strike the guard wires.

5. FIELD DISTORTION

To determine how close the electrode system may be brought to the front end of the box without introducing appreciable field distortion, X-ray intensities were measured at each of several successive positions of the electrodes with respect to the box front. Curve *B*, Figure 2, shows the observed increase in ionization as the distance from collector electrode to the diaphragm is increased. Beyond 5 ⸱ cm, further change is only about 0.2 to 0.3 per cent, a magnitude smaller than the experimental error in these particular measurements.

IV. COMPARISON WITH STANDARD CHAMBER

The reliability of the special ionization chamber was finally proved by comparing it with the primary standard on the higher voltage radiation. Since it was not feasible to do this directly, the special chamber was compared with a guarded field secondary chamber and this, in turn, with the guarded field primary standard. For the same beam of 70 kv unfiltered radiation, the following measurements show an agreement between the primary and secondary chambers closer than the experimental error:

Primary standard ionization current_____ 0. 1315±0. 00026
Secondary standard ionization current_____ . 1316± . 00030
                                                 ――――――
                                                 . 0001

Comparison of the special chamber with the secondary standard was made with unfiltered radiation excited by 40 to 90 kv (peak). To eliminate as many variables as possible, the same diaphragm was used interchangeably on both chambers. Likewise, the collector electrode of each chamber was connected to the same electrometer. Saturation voltages were supplied separately. Both chambers were placed in a fixed position and the tube shifted on a track, so that the same beam could enter either chamber as desired. To have this beam moderately well defined, a diaphragm 16 mm in diameter was placed about 2 cm from the wall of the 200 kv thin-walled Coolidge tube used.[7]

Ionization currents were measured by means of a null electrostatic compensator, such as was used in previous work,[8] wherein the ionization charge was compensated over a known interval of time. However, instead of exposing the chamber for a given time by means of an electrically operated shutter in the X-ray beam, the electrometer was connected to the chamber for a definite period by means of an electrically operated switch [9] located at *E′* (fig. 1). The chamber and electrometer were connected by a shielded rubber cable.

The effective length along the beam of the collector of the secondary standard was 4.975 cm, so that, for the same diaphragms, the effective air volumes ionized in the special and secondary chambers were in the ratio 0.9698 to 4.975. All readings with the secondary chamber were

[7] L. S. Taylor, B. S. Jour. Research, vol. 2 (RP56), p. 771, 1929.
[8] L. S. Taylor, B. S. Jour. Research, vol. 6 (RP306), p. 807, 1931.
[9] L. S. Taylor, B. S. Jour. Research, vol. 8 (RP397), p. 9, 1932.

therefore divided by 5.129 to reduce to the same effective air volume as the special chamber.

The distance between the limiting diaphragm and center of the collector electrode on the secondary chamber was 22.5 cm, as compared with 7 cm for the corresponding distance on the special chamber. Absorption in the air-path difference of 15.5 cm will, for the same beam, lower the measured current of the secondary chamber relative to that of the special chamber by a corresponding amount. Since this absorption depends upon such factors as voltage wave form, tube characteristic, and tube wall thickness, it must be determined for each set of conditions employed.

At voltages ranging from 40 to 100 kv, where the X rays pass initially through moderately thick tube walls, the radiation is sufficiently hardened so that its quality undergoes no appreciable further change in passing through 50 cm of air. In such case the air absorption is approximately proportional to the length of the air path.[10] But, as shown later, this is not true for the 4 to 12 kv radiation for which the special chamber is designed.

Air-absorption measurements to be applied in the present comparison were made with the large open air standard ionization chamber,[11] using the limiting diaphragm in a fixed position relative to the target so as to define a narrow beam passing between the chamber plates. X-ray intensities, I_1 and I_2, were then measured with the plates at the distances 110.6 and 128.0 cm, respectively, from the target. The ratio $\dfrac{I_1}{I_2}$, giving the air absorption correction factor for an air path of 17.4 cm, is plotted against applied voltage in curve C of Figure 2. The absorption being practically linear with distance, it is simple to correct it to fit any path length as, for example, 15.5 cm, required in the present set-up and shown by the broken curve.

To insure, in the comparison, proper alignment of the special chamber, measurements were made of the intensity of the X-ray beam over its horizontal cross section in the plane of the diaphragm, N. As shown by curve D in Figure 2, the area of effective uniformity, about 1 cm wide, indicates the range of proper alignment.[12]

Typical results of the comparison of the special with the secondary chambers, with unfiltered radiation from a thin wall, deep therapy Coolidge tube, are given in Table 1.

TABLE 1.—*Comparisons of special ionization chamber with secondary chamber*

kv	I_s	I_s	Absorption factor $(I_s/I_s)_{15.5}$	$\dfrac{I_s}{I_s}\left(\dfrac{I_1}{I_2}\right)_{15.5}$
40	4.63	4.46	1.040	0.998
50	2.30	2.205	1.033	1.008
70	2.88	2.79	1.028	1.003
90	6.08	6.08	1.027	.968

Column 2 gives the intensities I_s in Röntgens per minute as measured with the special chamber; column 3, the intensities, I_s, at the

[10] L. S. Taylor and G. Singer, B. S. Jour. Research, vol. 6 (RP271), p. 219, 1931.
[11] L. S. Taylor, B. S. Jour. Research, vol. 2 (RP56), p. 771, 1929.
[12] L. S. Taylor, B. S. Jour. Research. vol. 3 (RP119), p. 807, 1929.

middle of the collector, as determined with the secondary standard but uncorrected for air absorption in the extra 15.5 cm of path; column 4, the corresponding air-absorption factor as obtained from the broken curve of Figure 3 (C), and by which I_x is multiplied in order to obtain the intensity 15.5 cm nearer the diaphragm, at which point the special chamber is used; and the last column gives the ratio of the intensities furnished by the two chambers.

At 90 kv (peak) the small chamber indicates definitely too little ionization. This is attributed to the close spacing (5 cm) of the collector, C, and high-potential plate, A, which does not permit full utilization of the range of the photoelectrons emitted along the path of the X-ray beam. This sets a definite X-ray voltage limit above which the small chamber should not be used. The safe working range for the present chamber includes, however, the lower part of the diagnostic X-ray region and may be used where, due to "wall effect," measurements with most thimble ionization chambers become inaccurate.

V. MEASUREMENTS OF "GRENZ RAYS"

Having found agreement between the indications of the special and the standard chambers at voltages ranging from 40 to 80 kv (peak), the use of the special chamber may be extended down to any X-ray voltage for which adequate air absorption correction can be made.

For measuring so-called "grenz rays," 4 to 12 kv (peak), the necessary corrections will be indicated and the necessity for accurately measuring the air absorption for each set of experimental conditions under which the radiation is used will be brought out.

Very soft X rays (4 to 12 kv) are absorbed by air to such a degree that the radiation quality changes rapidly even in a distance of only 5 cm. The beam intensity should, therefore, be measured at the exact position, with respect to both tube and filter, where the radiation is to be applied, because it is unsafe to apply the inverse square law for computing intensitites at points other than where measured. Furthermore, it is essential to determine the air absorption between diaphragm and collector, C, for an air column having exactly the same position with respect to the target as the column between diaphragm and collector when the chamber is in its working position, if the point of application is taken at the position of the limiting diaphragm.

The requirements are illustrated in Figure 3, which gives a plot of the intensity of an X-ray beam as a function of the distance, e, from the target, T. At the lower part are indicated several possible positions of an ionization chamber with respect to the tube. Suppose, for example, the intensity, I_b, at the position, b, is to be measured. The chamber diaphragm would be placed at position, $n_1 = b$, and the middle of the collector electrode at e_1, a distance, d, from n_1, where d is the minimum working distance between collector and diaphragm. Owing to air absorption over the path, $b - c$, the intensity measured at e_1 must be corrected to give the value as if at the position, n_1. This correction can not be obtained by simply keeping n_1 fixed, and moving e_1 away an equal distance, d, to the position e_1' because, due to the filtering action of the air, the radiation quality at g is different from that at c.

To obtain the air absorption factor to be applied in the chamber
working position, it is necessary to move the chamber, including
diaphragm, toward the tube a distance, d, to the position, n_2, and then
measure the intensities, I_b and I_c, obtained, respectively, at the two

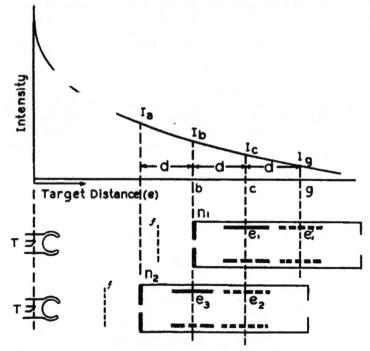

FIGURE 3.—*Diagram of working positions of special ionization chamber in
determining air absorption factors*

positions of the collector plate, $e_3 = n_1$, and $e_2 = e_1$. The ratio $\dfrac{I_b}{I_c}$ gives
the air absorption factor to be applied to the readings made with the
chamber at the position n_1. Corrections made in this manner are

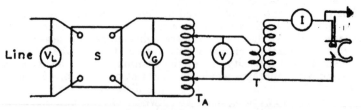

FIGURE 4.—*Power circuit for grenz-ray generator*

limited to distances, d, or greater, from the X-ray tube. The special
chamber described, owing to its small size, is particularly well adapted
to the problem; an unguarded field type can not be made sufficiently
small without introducing too great field distortion.

The observed change in the air absorption factor, as the distance from the tube is changed, is given in Table 2.

TABLE 2.—*Air absorption factors 8 kv (peak) (Slack window, Grenz ray tube)* [1]

Diaphragm position	Absorption path distance from target	Absorption factor
	cm	
1 (a-b)	12-18	1.635
2 (b-c)	18-24	1.593
3 (c-d)	24-30	1.570

[1] C. M. Black, J. Opt. Soc. Am., vol. 18, p. 123, 1929.

The letters in parentheses refer to the positions in Figure 4. The correction factor changes between positions 1 and 2 by 4.2 per cent; between 2 and 3, due to hardening of the radiation, by only 1.7 per cent. This proves that air-absorption corrections made in one position can not be applied to another position without introducing a corresponding error.

It is also evident from these measurements that, as the distance, d (fig. 3), of the middle of the collector electrode increases, the absorption correction correspondingly increases, and may become so large that its accuracy of determination is insufficient. It is obvious that d should be kept as small as possible. With the chamber used here, d may be as small as 5 cm for radiation up to 80 kv (peak); while, for the same range, a simple parallel plate or cylindrical ionization chamber would have a minimum distance $d = 13$ cm and, therefore, at least double the absorption.

Measurements were made using a commercial grenz ray generator and a tube with a thin indrawn-bubble type of window [13] of such size and position that the ionization chamber diaphragm could be brought to about 5 cm of the target. The filament side of the tube was grounded (fig. 4) and the voltage regulated by means of a slide wire auto transformer T_A in the primary circuit of the high-voltage transformer. Peak voltage was determined from the manufacturer's calibration in terms of the voltage, V, on the primary of the transformer. (Since accurate values of the tube voltage were of no great significance in this particular work, the voltage calibration was not checked.)

Air absorption, expressed in terms of an absorption factor, I_b/I_c (see fig. 3) multiplied by the ionization reading, I_c, gives the intensity in Röntgens per minute at the point, b, the position of the limiting diaphragm. The absorption factor, I_b/I_c, for $d = 6$ cm is plotted in Figure 5 as a function of the thickness of aluminum filter used in the beam. It is seen that for lightly filtered radiations air absorption correction may be, with 6 kv radiation, as large as 40 per cent. The need for accuracy of this correction is obvious.

Figure 6 gives a series of curves showing the intensity, I_b, in Röntgens per minute at the position $b = 12$ cm as a function of the aluminum filtration. The lower curve of each pair gives the measured

---

[13] See footnote to Table 2.

intensity while the upper curve gives the corresponding calculated
value at the diaphragm—obtained by multiplying by the appropriate
absorption factor from Figure 6. It is interesting to note in this
connection that a lateral shift, corresponding to 0.0055 mm thickness
of aluminum, of the first curve of any pair brings the two curves
together, within experimental error. This means that the absorption
of 6 cm of air is equivalent to 0.00055 cm of aluminum for the range

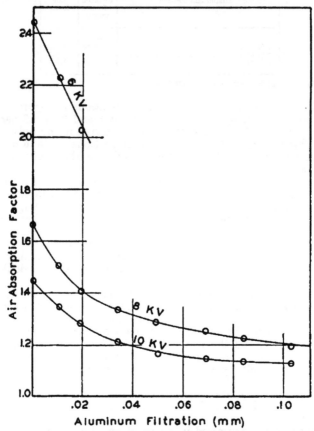

FIGURE 5.—*Air absorption factor for different voltage X rays necessary in the use of special ionization chamber*

of voltages and qualities covered in Figures 5 and 6. The aluminum
equivalence of air is from this 9.2×10^{-5} mm Al per mm of air; which
agrees favorably with the value 9×10^{-5} reported by Siegbahm,[14]
but is appreciably higher than Glasser's value of 7.9×10^{-5} mm.

The logarithm of the intensity plotted against thickness of filter
shows a distinct curvature up to 0.1 mm/Al filtration and indicates
that there is no "homogeneity filter"[15] beyond which the quality
does not change. Typical examples of intensity and absorption

[14] M. Siegbahm, The Spectroscopy of X Rays (Oxford), 1925.
[15] H. Behnken, Zeit. fur. Tech. Phys., vol. 2, p. 153, 1921.

curves for grenz rays have been given by Jacobson [16] and others. Such curves have value only for a particular tube and should be · determined separately for each type of generator, as brought out by using the system indicated in Figure 4. Here tube emission was measured for a constant voltage, V, applied to the transformer, T, and constant current, I, through the tube, the input voltage V_G, to the control autotransformer being stabilized in one case and unstabilized in the other. [17] The results in Table 3 show a 50 per cent decrease in emission with the particular type of stabilization used. Since the effect of wave form on emission varies between tubes of different

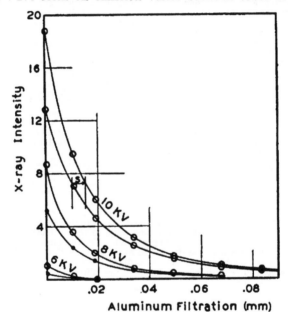

FIGURE 6.—*X-ray intensities measured with the special ionization chamber before and after application of the air absorption correction*

types, there is little to be gained by further investigation of the variation; it simply emphasizes the necessity for individual measurements with each type of generator used.

TABLE 3.—*Grenz-ray emissions for different voltage wave forms*

Line voltage V_L	Stabilizer	Input voltage V_G	Emission
109	Out	109	16.07
104	......do	104	16.24
109	In	106.5	8.83
104	......do	106.5	8.52

WASHINGTON, August 1, 1932.

---

[16] L. Jacobson, Am. J. Roent., vol. 22, p. 547, 1929.
[17] All stabilizers of the general type here used produce a distorted wave form depending on the load power factor. This does not affect the usefulness of the instrument for its intended purpose.

RP506

A MULTIRANGE POTENTIOMETER AND ITS APPLICATION TO THE MEASUREMENT OF SMALL TEMPERATURE DIFFERENCES

By H. B. Brooks and A. W. Spinks

ABSTRACT

In determining, as a criterion of purity, the boiling range or the freezing range of a liquid which is being purified by fractionation, it is convenient to be able to determine accurately the small difference between the boiling points or the freezing points of the two end fractions. The occasional measurement of relatively large temperature differences between two fractions also requires that the total range of measurement be large in comparison with the smallest measurable temperature difference.

Thermocouples are very suitable for measurements of temperature difference. To increase the electromotive force per degree it was decided in a particular case to use a group of 10 copper-constantan thermocouples in series. Even with these 10 couples, the electromotive force corresponding to a difference of 0.001° C. would be only 0.4 μv, and its measurement to the nearest 0.0001° would require the measurement of the electromotive force to 0.04 μv. Measurements accurate to such a small quantity require not only that the potentiometer used shall be extraordinarily free from parasitic electromotive force, but also that provision should be made for readily detecting and eliminating any such intruding electromotive force in the galvanometer circuit.

The relatively little-used "second method" of Poggendorff, as first realized in practical form by Lindeck and Rothe, appeared to offer the most desirable basis for the design of a multirange potentiometer suitable for the above purposes. The potentiometer which will be described has 6 ranges, and when used with the group of 10 thermocouples provides 6 ranges of temperature difference; namely, from 0° to 0.1°, 0.2°, 0.5°, 1°, 2°, and 5°, respectively, corresponding in each case to a deflection of the pointer of 100 divisions, readable by estimation to 0.1 division. The potentiometer is readily adaptable to read directly in microvolts or in temperature difference, as may be preferred. For the particular application for which it was designed, it was desired to avoid entirely the use of the curves or tables which are necessary when the readings are in microvolts, and it was therefore arranged to read directly in temperature difference.

CONTENTS

I. INTRODUCTION

In the work of isolating particular hydrocarbons from petroleum by fractionation it became necessary to measure the difference between the boiling points or the freezing points of two nearly identical liquids with a precision which could be very moderate when the boiling points differed by several degrees, but which had to be much greater as the two boiling points approached equality. For example, if the difference in boiling points was only 0.001° C., it was desired to determine it within, say, 0.0001° C. If the boiling points differed by 0.1° C., a measurement of the difference to about 0.001° C. would suffice, and so on for still larger differences. The requirements were such that some form of electric thermometer was evidently needed, and the thermocouple was chosen as being well suited to the purpose. To obtain an adequate value of thermal emf with the very small temperature differences to be measured a group of 10 copper-constantan couples in series was used. It was decided to cover a range of temperature differences up to 5° C., and to be able to detect a change of 0.0001° C. when measuring temperature differences of 0.1° C. or lower. This required that the parasitic emf in the potentiometer to be used should be well below 0.04 μv, even under relatively severe thermal conditions. It was felt that no potentiometer was commercially available in which the extreme precautions necessary to secure this result had been taken; furthermore, it was desired that the potentiometer should be capable of ready adaptation to make it read directly in temperature difference rather than in a unit of electromotive force. For these reasons the design of a special potentiometer was undertaken.

II. PRINCIPLE OF OPERATION

In announcing his development of the "compensation" process which forms the basis of the art of potentiometry, Poggendorff[1] described what he called the "first method" and the "second method." The first method was taken up with great interest by other workers, and has resulted in a long line of potentiometers, embodying many ingenious ideas to extend the range and increase the precision of measurements. The underlying idea of the first method is the balancing of the unknown emf by an equal potential difference which is adjusted by changing the value of resistance between two tap-off points on a circuit in which a current is maintained constant at a preassigned standard value.

During all of this active development of the first method, Poggendorff's second method was almost forgotten. It was used in 1896 by Holman,[2] who merely assembled stock pieces of apparatus for the purpose. The first recorded development of Poggendorff's second method into a definite instrument for specific applications appears to have been made by Lindeck and Rothe[3] at the Reichsanstalt in 1899. The instrument was developed with the cooperation of Siemens and Halske,[4] who placed it on the market.

[1] J. C. Poggendorff, Ann. der Physik und Chemie, vol. 54, p. 161, 1841.
[2] Holman, Lawrence, and Barr, Proc. Am. Acad. Arts and Sci. vol. 31, pp. 218–233, 1895–96.
[3] Lindeck and Rothe, Zeitschrift für Instrumentenkunde, vol. 20, pp. 293–299, 1900; foreshadowed in the annual report of the Reichsanstalt, ibid., vol. 19, p. 249, 1899. Hoffmann and Rothe, ibid., vol. 25, pp. 271–248, 1905, described a recording potentiometer developed from the Lindeck-Rothe potentiometer, using the deflection-potentiometer principle.
[4] Keinath, Elektrische Temperaturmessgeräte, 1923 edition, pp. 30–31.

In Poggendorff's second method the adjustment of the controllable
potential difference was effected by changing the value of the current
through a circuit while the resistance between the two tap-off points
remained constant. This method requires the use of an instrument
which will measure a direct current of any value between zero and a
given maximum value; that is, ordinarily, an ammeter. Herein lies
the chief limitation of the method, for an ordinary ammeter can be
read to a precision of only about 0.001 of the full-scale value, and
the total error resulting from inaccuracy of marking of the scale,
imperfect elasticity of the springs, etc., in a good instrument may be
assumed to be several times this value. At the start, therefore,
one must reckon with this limitation, and apply the second method
only where its limited absolute accuracy is sufficient. When this is the
case the second method offers the following noteworthy advantages:
(a) With a given ammeter, any number of ranges in any desired
relation to each other may be readily obtained by providing a corresponding number of tap-off points; (b) sliding contacts being absent
from the measurement circuit, parasitic emf can be excluded to
almost any desired degree; and (c) since one has liberty of choice of both
the I and the R which by their product give the desired potential
difference the "internal resistance" of the potentiometer may in
principle be made as low or as high as desired by choosing an ammeter
of corresponding range.

Two possible misconceptions regarding potentiometers of the
Lindeck-Rothe type should be avoided. The first is that they are
"deflection potentiometers." This is not correct because in the
deflection potentiometer, properly so called, an unbalanced part of the
unknown emf occasions the observed deflection of the indicating
instrument, which is directly in the measurement circuit. In the
Lindeck-Rothe potentiometer the deflection is all occasioned by a
current from an auxiliary source, and all of the unknown emf is balanced
by the IR drop in the potentiometer, hence no current flows through
the source of the unknown emf. The other possible erroneous idea
is that the apparatus is not a potentiometer, because the entire result
of the measurement depends on a deflection, and no standard cell is
used. A potentiometer may be broadly defined as an instrument
for measuring an unknown potential difference by balancing it, in
whole or in part, against a known potential difference produced
by the passage of known currents through a network of known
electrical constants. Where the usual types of potentiometer
produce this adjustable potential difference by means of an adjustable
value of resistance and a current of fixed value which is frequently
checked by reference to a standard cell, the Lindeck-Rothe potentiometer utilizes a fixed resistance and an adjustable current which is
checked less exactly and much less frequently, namely, when the
accuracy of the milliammeter is checked by reference to a resistance
standard and a standard cell.

III. BASIS FOR DESIGN

In practice the auxiliary current for a Lindeck-Rothe potentiometer
will be supplied by some form of battery, and should therefore be
kept down to moderate values. While storage cells are much used
with potentiometers, they have some objectionable features which

make the use of dry cells attractive. This suggests the use of currents not exceeding a few hundredths of an ampere, a suggestion which is fortified by the fact that the accuracy of direct-current milliammeters in which the entire current to be measured flows through the moving coil is almost independent of changes of instrument temperature.[5]

In the present case the current to give full-scale deflection of the milliammeter could be conveniently given the value 10 milliamperes. To produce a potential difference of 2,000 μv,[6] this current must flow through a resistance of 0.2 ohm, a value which is negligible in comparison with the resistance of the group of thermocouples, namely, about 50 ohms. Considerations of speed of working required that the galvanometer to be used should be critically (or slightly under) damped with this latter resistance across its terminals.[7]

The galvanometer selected has a nominal coil resistance of 12 ohms, is critically damped with an external resistance of 40 ohms, has a sensitivity of 5 mm (at 1 m) per microvolt with this external resistance and a complete period of five seconds. It is constructed in such a way as to be relatively free from parasitic emf, but because of the very stringent requirements of the problem, the potentiometer was provided with means for readily detecting any parasitic emf in the galvanometer and for neutralizing its effect. Although the parasitic emf compensator can also be used to neutralize the effect of parasitic emf within the potentiometer, it was considered much better to simplify the manipulation by keeping such internal parasitic emf below the limit of detection, even at the expense of extraordinary precautions in design, materials, and construction.

IV. CONSTRUCTION

1. PLAN OF CIRCUITS

Figure 1 illustrates the principle of operation of the potentiometer and Figure 2 shows the plan of circuits used. In Figure 2, a resistor R_{13} of manganin strip [8] is provided with nine taps, of which the two outside ones are used as potential leads. A current from the dry cell B_2, controlled by the coarse rheostat R_5 and the fine rheostat R_4, flows through the slide rheostat R_{11}, the shunted milliammeter MA, the current-limiting resistor R_6, the reversing switch S_1, then through the manganin resistor R_{13}, entering at the common point C and leaving at one of the six taps connected to the studs of the range-changing switch S_2. The values of the 4-terminal resistance of R_{13} for the six positions of this switch are (from left to right) 0.004, 0.008, 0.02, 0.04, 0.08, and 0.2 ohm.

---

[5] This independence of temperature changes depends on the facts (1) that no change in distribution of the measured current between the moving coil and a shunt circuit can occur as a result of temperature change; (2) that the small temperature coefficient of magnetic flux density in the air gap (about 0.01 to 0.03 per cent per degree C.) tends to offset a temperature coefficient of rigidity of the springs, which is about 0.04 per cent per degree C.

Although the requirements of the present case made it necessary to use a shunt around the moving coil, it was possible to give the shunt a value of temperature coefficient which reduced the temperature coefficient of the instrument (as a milliammeter) to zero.

[6] These nominal values of current and potential difference would correspond to the use of the actual group of thermocouples in baths at a mean temperature of about 22° C. As explained further on, the milliammeter was provided with a continuously adjustable Ayrton-Mather shunt which increased the total current for full-scale deflection to values between the limits 10.57 and 14.76 milliamperes, corresponding to the rate of change of thermal emf with temperature at mean bath temperatures of 50° and 300° C., respectively.

[7] This criterion is the one to be used in selecting a moving-coil galvanometer for a specific application, rather than the old rule, still quoted occasionally, that the galvanometer resistance and the resistance of the circuit to which it is connected should be as nearly equal as practicable.

[8] This strip was three-sixteenths inch wide, 25 mils thick, and was about 6 feet long.

The potential difference between the ends of the resistor R_{12} is opposed, through a galvanometer, to the emf of the thermocouples, the terminals of which are brought to the binding posts A and B. In the lead from the left-hand end of the resistor R_{12} to the binding post A are included three keys, two having the protective resistors R_1 and R_2. The lead from the right-hand end of the manganin resistor to one of the galvanometer binding posts includes the copper rheostat

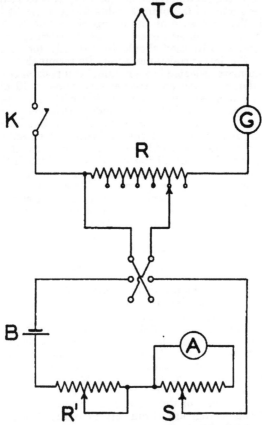

FIGURE 1.—*Diagram illustrating the principle of operation of the multirange potentiometer*

The emf developed in the thermocouple *TC* is balanced by the drop of potential produced in the resistor *R* by the current from a battery *B*. This current is adjusted by the rheostat *R'* and measured by the shunted milliammeter *A*.

R_3 forming part of the parasitic emf compensator. A very small current from a flashlight cell B_1 may be made to flow in either direction through any part of R_3 from zero up to one-half of R_3. Depression of the key marked "Shunt off" opens a shunt circuit across the galvanometer, namely, a copper resistor, R_{12} with taps at 30, 40, 60, and 80 ohms. As much of this coil is used as will make the motion of the galvanometer coil critically damped when the keys R_1, R_2, 0 are open.

The function of the reversing switch S_1 is to change the sign of the potential difference between the ends of the resistor R_{13} to take care of the fact that either of the two baths of liquid may on occasion be the hotter. The fixed resistance R_7 and the slide rheostat R_8 constitute the continuously adjustable Ayrton-Mather shunt to the milliammeter.

2. MAIN RESISTOR

The junctions of the copper potential leads with the ends of the manganin strip constitute a possible source of disturbing thermal emf.[9] It was necessary to take such precautions in the design and construction of this resistor as would insure that the two end junctions of manganin with copper would not differ in temperature by more than a few thousandths of a degree. This severe requirement was met by the combination of a number of expedients. As shown in Figure 3, the manganin strip was reflexed in circular fashion around two circles

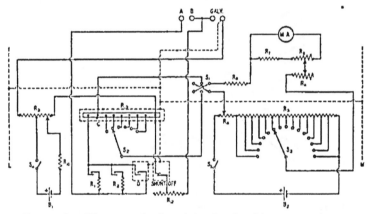

FIGURE 2.—*Diagrammatic plan of circuits of multirange potentiometer*

of bakelite pegs set in a bakelite plate; several thicknesses of $\frac{1}{16}$-inch felt separated the strip from the plate to increase the thermal insulation of the strip. The ends of the strip were brought in radially to the center of the circle, electrically insulated from each other by mica, but in good thermal contact. A layer of thin felt was placed over the strip, and the nine copper tap wires were wound several times around the outer circle of bakelite pegs,[10] then covered with two additional layers of felt and a second bakelite plate which was then secured by screws through the corners to the first plate. The space between the two bakelite plates was then filled in by winding a narrow strip of felt over the outer circle of bakelite pegs.

Before the main resistor was mounted in the structure just described, a similar resistor of constantan, made and mounted in this manner, was used as a check. Tests, such as placing a can of water

---

[9] For a copper-manganin junction at ordinary room temperature, the thermal emf changes at the rate of about 1 to 2 μv per °C.

[10] The object of this procedure was to increase the length of path along which heat from other parts of the potentiometer must travel in order to reach the main resistor, and at the same time to permit heat exchange between these wires. The flow of heat uniformly to (or from) all parts of the main resistor would have no bad effect because it would not cause a difference of temperature between the end junctions.

at 40° C. on the upper bakelite plate, produced a thermal emf of the
order of a microvolt. This showed that only a few hundredths of a
microvolt would be developed with a manganin resistor similarly
treated. To be on the safe side, however, the complete structure,
with the manganin resistor in it, was mounted below the bakelite
top of the potentiometer, midway between the bakelite and the bottom
of the box, on four bakelite posts symmetrically placed, and was
covered with a box of sheet aluminum 0.03 inch thick. In Figure 4
the bottom of this box has been removed to show the bakelite structure containing the main resistor.

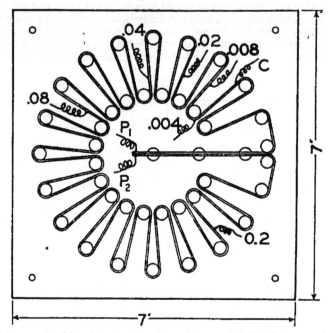

FIGURE 3.—*Plan view of main resistor with upper bakelite plate removed*

3. MILLIAMMETER

This instrument is a fan-shaped d. c. milliammeter of a type primarily intended for switchboard use, modified by the use of a knife-
edge pointer, a parallax mirror, and a fine-line 100-division scale of
the quality used in standard portable instruments. The resistance
of the instrument was about 5 ohms and the current for full-scale
deflection (before applying the shunt) was 10 milliamperes.

4. BATH-TEMPERATURE SCALE

In the ordinary procedure for measuring temperatures or temperature differences by means of thermocouples and a potentiometer, the
latter is usually graduated to read in terms of a unit of emf (millivolt
or microvolt) and its readings are subsequently translated into tem-

peratures (or temperature differences) with the help of curves or tables for the particular couples used. In the present case, it was desired to be able to adapt the potentiometer to read directly in temperature difference for a given group of couples. To determine the electrical conditions which must be satisfied to obtain this result, one must start with the temperature-emf relation of the couples used. A sample couple made of the same wires as the 10 couples to be used was certified as having the following relation:

$$E = 38.062T + 0.04457T^2 - 0.0000288T^3$$

where E is the emf in microvolts and T is the temperature of one of the junctions in degrees C., the other junction being at 0° C. It was assumed that a similar equation, with numerical coefficients ten times greater, applied with sufficient accuracy to the group of couples to be used. If the curve of the above equation be plotted between $T = 0°$ and $T = 300°$, its slope, dE/dT, will be seen to increase with T throughout this range. In order that the potentiometer shall indicate small temperature differences directly for mean bath temperatures in the region 50° to 300° C., some means of adjustment must obviously be incorporated which, when set to a given mean bath temperature, will make the current through the manganin resistor for any given reading of the indicator proportional to dE/dT. Differentiating the expression for E gives the equation

$$\frac{dE}{dT} = 38.062 + 0.08914T - 0.0000864T^2$$

Using this formula, values of dE/dT were computed for $T = 50°$, 60°, ... 300°. They ranged from 42.303 microvolts per degree at 50° to 57.028 microvolts at 300°.

Taking the lowest range, 0 to 0.1°, for example, for which the 4-terminal resistance of the corresponding section of the main resistor R_{13} is 0.004 ohm; if the full-scale current of the unshunted milliammeter is passed through this section the potential difference at the potential terminals of R_{13} will be $0.004 \times 0.01 = 0.00004$ volt $= 40$ μv. This is equal to the thermal emf of the particular group of thermocouples when the temperature difference between the two sets of junctions is 0.1° C. and their mean temperature is 22.2° C., and consequently for these conditions the full-scale deflection of the milliammeter corresponds to 0.1° C. temperature difference. As the mean bath temperature increases from 22.2° to 300° C. the value of dE/dT increases continuously from 40 μv per degree to 57.028 μv per degree. Therefore, the current which must be passed through any section of the main resistor R_{13} to give a difference of potential equal to the thermal emf of the couples, for a given temperature difference, must be greater than 10 milliamperes by a factor which is equal to dE/dT for the given temperature divided by that for 22.2° C., namely 40 μv per degree. Thus for a mean bath temperature of 50° a shunt must be connected across the milliammeter to increase the full-scale current of the shunted instrument by the factor 42.303/40; that is, to 10.576 milliamperes; for a mean bath temperature of 300° the current for full-scale deflection must be 10 milliamperes multiplied by the factor 57.028/40; that is, 14.257 milliamperes; and similarly for intermediate values of mean bath temperature. This object was

B. S. Journal of Research, RP506

FIGURE 4.—*Structure of multirange potentiometer as seen from below the bakelite top*

Part of the aluminum thermal shield has been removed from the main resistor (center) and from the key marked 0 (near lower right-hand corner of main resistor).

FIGURE 5.—*Plan view of multirange potentiometer*

B. S. Journal of Research, RP506

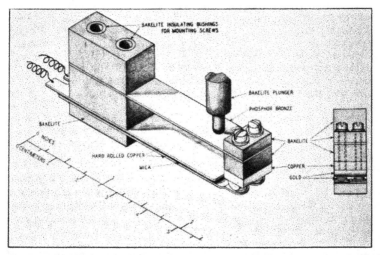

FIGURE 6.—*Galvanometer key embodying refinements to minimize thermal emf*

readily attained by a continuously adjustable shunt of the Ayrton-Mather type [11] connected to the terminals of the milliammeter. This shunt is shown diagrammatically in Figure 2 as consisting of a fixed resistor R_7 in series with the winding of a circular slide rheostat R_8, the two being connected to the terminals of the milliammeter. The current enters by the slider of R_8 and divides between the two parallel paths. As the slider is moved toward the left, the total current for full-scale deflection of the milliammeter increases. A pointer attached to the slider may therefore indicate on a fixed scale either the corresponding value of dE/dT in microvolts per degree or the mean bath temperature. The former would be preferable in general because it would facilitate the use of the potentiometer with any thermocouple for which the value of dE/dT under the conditions of use fell within the range of the scale. In the present instrument, however, it was desired to avoid entirely the use of tables and curves, and the scale was accordingly marked in terms of mean bath temperature for the particular set of copper-constantan thermocouples which had already been made. Attention is called to the fact that the scale divisions were engraved on a disk which is an integral part of the knob of the circular slide rheostat which performs the function of R_8 in Figure 2, and only a fiducial mark was engraved in the bakelite top. (See figure 5.) If for any reason the original set of couples is to be replaced by another, a new scale may be engraved on the disk, which may be readily removed for the purpose. If desired, the scale of bath temperature could be replaced by another scale reading values of dE/dT in microvolts per degree.

If the circular slide rheostat R_8 were of the ordinary type in which the rate of change of resistance with respect to rotation is approximately constant, the bath-temperature scale would be open at one end and crowded at the other because of the quadratic form of the relation between dE/dT and T. It was possible to obtain a nearly linear scale by using a circular slide rheostat in which the resistance wire is wound on a tapered strip of sheet insulating material, giving a quadratic relation between the resistance and the angle of rotation. The close approach to a linear bath-temperature scale which was obtained may be seen from Figure 5, in which the dial at the upper left-hand corner of the potentiometer is the one to be set to the mean bath temperature. The scale was laid out by adjusting the current through the shunted milliammeter successively to the computed values of current for full-scale deflection for bath temperatures of 50°, 60°, ... 300°, and turning the knob of the bath-temperature rheostat until the pointer of the milliammeter came to the full-scale mark. For each such adjustment a mark was drawn, opposite the fiducial mark, on the scale disk which moves with this knob. The 10° lines were then engraved, with 5° lines spaced midway between them. The results of a test to check the correctness of the resulting scale are given in a later section of this paper.

The current-regulating rheostats R_4 and R_5 operate properly only if the resistance of the circuit beyond them (that is, connected between the slider of R_4 and that of R_5) is substantially constant. The effect of moving the slider of R_8 to the left is to increase the resultant resistance of the milliammeter and its shunt, and the tapered slide

[11] Ayrton and Mather, J. Inst. Elecl. Engrs. (London), vol. 23, p. 314, 1894.

rheostat R_{11}, connected as shown, compensates for this change and maintains a substantially constant resistance in the circuit beyond the sliders of R_4 and R_5.

5. GALVANOMETER KEYS

In the effort to reduce parasitic emf in the "measurement circuit" to a value below the limit of detection, particular attention was given to some of the keys which had to be used. These keys must be operated frequently; they must contain contact points of metals or alloys nobler than copper because the latter is subject to corrosion; heat from the observer's hand will flow to the contact points through the key spindle; and heat is developed at the contact points by their mutual friction. In many ways the problem of avoiding parasitic emf in the keys is much more difficult than in the main resistor, which has no moving parts and can be thermally shielded to any desired degree.

Referring to Figure 2, it will be seen that there are four keys in the galvanometer circuit. Of these, the ones marked R_1 and R_2 may be of ordinary construction because they are used only in obtaining an approximate balance between the emf to be measured and the adjustable potential difference developed within the potentiometer. The "working microvolt sensitivity"[12] with either R_1 or R_2 in circuit is too low to permit the effect of parasitic emf to be detected. The keys marked "0" and "Shunt off," on the contrary, are in circuit with the galvanometer when measurements are being made; the total resistance in the galvanometer circuit is then low and the working sensitivity correspondingly high; consequently any appreciable parasitic emf in these keys will cause an error in the measurement. The conditions in these two keys are somewhat different. The key marked "0," normally open, is closed just before a reading is taken; that marked "Shunt off" is closed until just after the key marked "0" is closed. Any parasitic emf in the "Shunt-off" key would cause an error in the zero reading of the galvanometer while any such emf in the "0" key would cause an error in the deflection. It was obviously necessary to make these keys so nearly "thermofree"[13] that no perceptible error could be introduced by them. Their design, materials, and construction are based upon the results of a series of experiments which will not be detailed here. If a galvanometer key be constructed of random materials, with consideration given only to its mechanical functioning, it is liable to be the source of objectionable thermal emf under many conditions encountered in practice. The production of a good thermofree key[14] requires that attention be given to three important matters, each of which acts in its own way to minimize the thermal emf. These three matters are:

---

[12] This term, ordinarily to be abbreviated to "working sensitivity," has been proposed by one of the authors (B. S. Jour. Research, vol. 4, p. 299, 1930) to denote a constant of importance to the galvanometer user, defined as follows: "The working sensitivity applies only to the user's particular problem at the moment, and may be defined as the response of the galvanometer for unit electromotive force in a circuit which includes the galvanometer and a particular external circuit."

[13] This term is offered as an arbitrary short equivalent of complete but unwieldy expressions, such as "free from thermoelectric forces," "thermal emf free," "thermoelectrically neutral," etc. While it was suggested by the well-known German adjective "thermokraftfrei," it was felt unnecessary to include the English equivalent of the syllable "kraft."

[14] Some makers of potentiometers have been using keys intended primarily for telephone purposes. Some telephone keys contain pairs of springs of very dissimilar alloys (evidently phosphor bronze and nickel silver) and in consequence should not be used in the construction of potentiometers. There should be no difficulty in having telephone keys assembled for the purpose with none but bronze springs.

1. The choice of materials for the key springs and the contact points which are very close to copper in their thermoelectric properties.

2. The design of the key to be "thermoelectrically astatic" with respect to heat flow; that is, so that all junctions of dissimilar metals occur in adjacent pairs so arranged that a flow of heat in any direction will set up nearly equal thermal emf's of opposite sign in the circuit containing the key.

3. The inclosure of the key within a shield which tends to maintain uniformity of temperature of all parts of the key in spite of heat radiation and conduction to or from the shield.

Such a shield may be a jacket of heat-insulating material, or a metal shield of high thermal conductivity, or a multi-layer thermal shield embodying both these features repeated as often as necessary. It is clearly useless to incur expense in reducing the parasitic emf much below a value which will produce a barely perceptible deflection (for example, 0.1 mm) of the galvanometer to be used.

The plunger for operating the key provides a path along which heat may travel into the key structure. The plunger should therefore be preferably of material of low thermal conductivity and be rounded or pointed where it bears on the part of the key to be depressed; furthermore, special care should be taken to make this part of the key "astatic." If the use of metal for the plunger is considered essential, it may be noted that either manganin or constantan is preferable to brass because of their very much lower heat conductivity.

Keys meeting the exacting needs of the present case have been made by caring for the three principal matters as follows: (*a*) The use of hard-rolled copper for the key springs and United States coin gold [15] for the contact points, (*b*) using an astatic arrangement which will be described below, and (*c*) inclosing the key in a box of sheet aluminum 0.03 inch thick.

These two keys, one with part of the aluminum shield removed to show the interior, are shown in Figure 4 near the lower edge' of the bakelite top, to the right of the center. To the right of them are the two ordinary keys, R_1 and R_2.

Figure 6 shows the construction of the key marked 0. Two springs of hard-rolled copper, separated by a sheet of mica, are clamped at one end between bakelite blocks and carry gold contact disks at their free ends. Above these contacts is a transverse half-cylindrical gold contact piece soldered to the lower surface of a block of copper which is secured to a flat phosphor-bronze spring by blocks of bakelite used for thermal rather than electrical insulation. The plunger through which the bronze spring is depressed is of bakelite to retard the flow of heat from the observer's hand. However, even if an appreciable amount of heat did flow through the plunger, the construction of the key is thermally astatic to such a high degree that the development of an appreciable thermal emf would be almost impossible. Heat entering the bronze spring by way of the plunger will flow in part to the left, where a thick block of bakelite impedes its flow to the upper copper strap. Although the temperature of this strap will be slightly raised above that of the lower strap, the adjacent junctions to connecting wires are of hard-rolled copper to soft copper,

---

[15] This alloy is composed of gold 90 parts, copper 10 parts, and has a thermal emf against copper, at room temperature, of 1.2 μv per ° C.

for which combination the thermal emf is very small, namely, about 0.25 μv per °C.

Heat flowing to the right through the bronze spring is retarded by the block of bakelite between the spring and the massive copper block. Because of the symmetrical construction there will be no tendency for unequal heating of the copper block; and even if heat did enter the upper surface of the copper block in a nonuniform manner, the high thermal conductivity of the copper and the small rate of heat transfer would maintain a very high degree of temperature uniformity in the copper and therefore in the transverse gold contact piece soldered to it. Heat will flow from this contact piece to the gold contact disks in the free ends of the hard-rolled copper springs. However, this part of the key is astatic to heat flow in the vertical direction because a current in the circuit which is closed by the key flows up to the transverse gold contact piece on one side and down on the other. Any thermal emf in one of the copper-gold junctions will thus be balanced by an equal and opposing emf in the other junction.

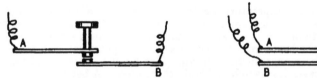

FIGURE 7.—*Keys illustrating poor design (left) and mediocre design (right) as regards the liability to thermal emf*

In contrast to this key, Figure 7 shows diagrammatically two constructions which are not thermally astatic. The key at the left has the junctions A and B as far apart as possible, and is consequently not astatic as to lateral heat flow. The key at the right is free from this objection, but neither key is astatic as to vertical heat flow.

6. PARASITIC EMF COMPENSATOR

The parasitic emf compensator, shown diagrammatically as R_3 in Figure 2, is shown in Figure 4 just above the two ordinary keys. It is merely a circular slide rheostat as used in radio apparatus but with the resistance-alloy winding replaced by one of copper wire. The material [16] of which the sliding contact lever is made is of no consequence thermoelectrically.

The manner of using the parasitic emf compensator may be understood by reference to Figure 2, which shows that when none of the four keys is depressed the tapped copper coil R_{12} is across the terminals of the galvanometer. Assuming that the slider of R_{12} is at the central point of its winding (or that the switch in series with the cell B_1 is open), any parasitic emf in the galvanometer will maintain a deflection. When the "Shunt off" key is depressed, the galvanometer coil will come to its true zero position, revealing by its motion the fact that an unwanted emf exists in the galvanometer circuit. With the switch S_4 closed, the slider of R_3 may be moved until a position is

---

[16] An interesting experience illustrates the necessity for great caution in the construction of highly thermo-free apparatus. The complete potentiometer, exposed to heat radiation from a slide rheostat 10 inches away, dissipating 300 watts, showed a parasitic emf of 0.2 μv. Careful search revealed that a strip of metal, forming a lug to which one end of the copper winding of the compensator had been soldered, was of bronze instead of copper as had been assumed. After this strip had been replaced with one of copper, the parasitic emf with the radiation test above described dropped to about one-tenth of its previous value.

found for which the galvanometer coil does not move when the "Shunt off" key is opened or closed. In this condition the parasitic emf in the galvanometer is just annulled by an equal and opposite potential difference in the copper winding of the compensator.

The resistor R_{10} in series with the cell B_1 has a resistance of 50,000 ohms. It is of the "wire-wound" type developed in recent years for radio purposes, and may be easily removed from holding clips if a change in the value of this resistance is desired. The greater the parasitic emf found in the galvanometer by actual use, the lower this resistance must be, but it is desirable not to make it much lower than necessary. The current taken from the flashlight cell is so small as to have no appreciable effect on the life of the cell, and the only reason for providing the switch S_4 for opening the circuit of this cell is to provide a ready means of annulling the small compensating emf set up by the parasitic emf compensator, without altering the position of the slider.

7. MISCELLANEOUS DETAILS

When the temperature difference between the baths A and B changes sign the direction of the thermal emf to be measured reverses. One way to meet this situation would be to provide a reversing switch between the leads from the couples and the potential taps from the ends of the manganin resistor R_{13}. (See fig. 2.) This would require that the reversing switch be highly thermofree. A better method was used, which is shown in Figure 2, namely, the reversing switch S_1 is placed in the wires which carry the adjusted current from R_4 and R_6 to the main resistor. Parasitic emf in the switch can do no harm whatever with this arrangement. Above the bakelite top the two positions of this switch are marked "A higher" and "B higher," respectively, A and B being the designations of the two baths of liquids.

The dry cells B_1 and B_2 are contained in metal receptacles within the potentiometer. There is a door in the back of the potentiometer box which does not appear in Figure 5 and which gives access to the cells and to four copper binding posts. These posts are supported on a small bakelite panel attached to the bakelite top, as shown near the top of Figure 4. As a precaution against possible differences of temperature among the four wires which come from the couples and the galvanometer to these binding posts, the two insulated wires from the couples, twisted together, enter the rear wall of the box and pass through a copper tube which extends to a point near two of the binding posts; and the pair of wires from the galvanometer pass through a similar tube. Any difference of temperature between the two wires of either pair will be equalized by their thermal contact with each other and with the tube.

Near the middle of the small bakelite panel are mounted two pairs of jacks for disconnecting the dry-cell leads when the bakelite top is to be removed from the box. Below the binding-post panel (fig. 4) are the two single-pole switches S_4 and S_5 of Figure 2, for opening the circuits of the two dry cells.

When the potentiometer was under test for freedom from thermal emf during the winter some difficulty was experienced because of electrostatic effects. For example, the galvanometer was found to deflect when the observer rubbed his shoes on the dry floor while his hands were on the galvanometer keys. These difficulties were overcome by shielding as indicated in Figure 2, in which the dotted lines L and M denote two brass angle plates for supporting some of the resistors at the sides of the potentiometer (see fig. 4) and the dotted outlines around R_{12} and the keys marked 0 and "Shunt off" denote the aluminum thermal shields. These five metal structures were connected together and to the right-hand galvanometer binding post as shown by the dotted lines. It was also necessary to shield the leads from the potentiometer to the galvanometer by inclosing them in a metal tube which was electrically connected to the same binding post.

V. TESTS

1. PARASITIC EMF IN COMPONENT PARTS

In testing experimental keys and other parts to determine their thermoelectric behavior under thermal exposure, a sensitive reflecting galvanometer was used in series with a parasitic emf compensator [17] and a device (a small Lindeck-Rothe potentiometer) by means of which a measured emf from 0 to 10 μv could be injected into the galvanometer circuit. To prepare for the measurement of parasitic emf in a key, for example, the galvanometer circuit as above described was first closed by directly joining two copper wires. No current flowed in the compensator circuit and no emf was injected into the galvanometer circuit by the Lindeck-Rothe potentiometer. The resulting deflection, if any, was caused by parasitic emf in the galvanometer, and was then reduced to zero by using the compensator. Then a small emf, say 1 μv, was injected, and the resulting deflection noted as the basis for reducing to microvolts the subsequent observations with the key in ciruit and the injected emf removed.

While ordinary keys, not designed to be specially thermofree, showed appreciable values of parasitic emf, the combination of good materials and thermally astatic design reduced the thermal emf to values so small as to be almost inappreciable. Two expedients assisted in the measurements of emf of only a few hundredths of a microvolt. The first was to replace a part made of good material (for example, the hard-rolled copper spring of a key, or the manganin main resistor) by a similar part made of a very unsuitable material. Constantan was used for the latter. Since its thermal emf per degree with respect to copper is about 20 times the thermal emf per degree for the combination manganin-copper, the values of parasitic emf observed with constantan, divided by 20, gave the values of thermal emf which would have been obtained for the same structure made of good materials; that is, with copper for the key springs and manganin for the main resistor.

The other expedient, used in final tests of parts (and later of the complete potentiometer) which had been made of very suitable materials, was to increase very greatly the sensitivity of reading the galvanometer deflection. A special optical arrangement was devised, usable only for very small (but nevertheless adequate) angular deflections of the galvanometer coil, which gave a sensitivity equal to that which would have been obtained if the galvanometer had been

[17] This compensator was similar to the one in the potentiometer. It was a small, self-contained device, complete with dry cell, and has been very convenient for purposes other than the present development.

used in the ordinary way, but with the scale 13 m from the galvanometer mirror.

2. MAIN RESISTOR

The 4-terminal resistances of the six sections of the main resistor were determined after final adjustment and found to have the desired values within 0.06 per cent. The manganin strip composing this resistor had been formed on an iron plate with two circles of iron pegs, which was a replica of the bakelite housing in which the strip was finally mounted. The manganin was annealed while on the iron plate at about 550° C. to relieve internal stresses. It is believed that this treatment insures the permanency of the resistance values to a degree much higher than the needs of the case require.

3. BATH-TEMPERATURE SCALE

The second of the two factors which jointly determine the accuracy of measurements with the potentiometer is the accuracy with which the current for full-scale deflection of the milliammeter approaches the ideal values, calculated from the emf-temperature relation of the couples, as the bath-temperature dial is turned to various readings over its range of 50° to 300° C. This matter was checked by measuring the full-scale current, using a resistance standard and Wolff potentiometer, at each of the 10° divisions over the entire range. The relative differences between these measured currents and the currents calculated from the emf-temperature relation for the thermocouples averaged 3 parts, maximum 8 parts, in 10,000. The errors in temperature-difference measurements from this cause will therefore bepbelow the usual limit of reading of the scale.

4. OVER-ALL ACCURACY

The final test of the potentiometer was a direct check of the accuracy of the potential difference which it develops at the terminals marked A and B (fig. 2) for full-scale deflection of the milliammeter and various settings of the bath-temperature scale. This potential difference was measured with a Wolff potentiometer, with the bath-temperature dial set in succession at each 10° point over its whole range. The check was made with the range-changing switch S_2 set at the extreme position on the right, corresponding to a difference of 5° C. between the temperatures of the two baths for full-scale deflection of the milliammeter. The average relative difference between observed and computed values of potential difference in microvolts was 0.1 per cent, the differences ranging from 0 to 0.2 per cent at various points on the bath-temperature scale.

5. PARASITIC EMF IN COMPLETE POTENTIOMETER

In addition to the heat-radiation tests of separate portions of the potentiometer, main resistor, keys, etc., as constructed, to determine how much parasitic emf would be set up in them, a similar test was made on the complete potentiometer. The source of heat was a small tubular slide rheostat dissipating 300 watts. It was placed on the table 10 inches from the front of the potentiometer; that is, in about the usual position of the observer when measurements are in progress. With the heat source in this position, the heat passing through the wooden box would be communicated to the keys and the main resistor. (See fig. 4.) That these parts were properly designed and well protected by their aluminum shields was shown by the fact that even after prolonged exposure the thermal emf in the potentiometer was so minute that it would have produced an error corresponding to about 0.1 division on the indicator scale with the lowest range (0 to 0.1° C.); that is, corresponding to an error of 0.0001° C. in the result. Although, it would doubtless be possible to reduce parasitic emf in potentiometers to a lower amount, by more refined methods of shielding the "thermoelectrically vulnerable" parts, the refinement would not usually be worth the extra cost because even high-grade moving-coil galvanometers would not be capable of showing the difference in performance.

VI. PROCEDURE IN USE

The complete procedure to be followed in measuring the temperature difference of the two baths is as follows:

1. With both battery switches open, depression of the "Shunt off" key will show whether any appreciable deflection of the galvanometer has been existing because of parasitic emf. If the spot of light moves when this key is depressed, the left-hand battery switch is to be closed and the knob of the parasitic emf compensator is to be turned by trial to a position such that the spot of light does not move when the "Shunt off" key is depressed.

2. The bath-temperature dial is to be set to the mean temperature of the two baths, as determined by independent means. This temperature does not need to be known very accurately because an error of 5° to 10° C. in T will produce an error of only 1 per cent in the value of dE/dT.

3. The right-hand battery switch is to be closed, the range-selecting switch set to a value of full-scale temperature difference which includes the estimated or expected value, and the reversing switch set to "A higher" or "B higher" as may be thought necessary; then by means of the coarse rheostat and the fine rheostat at the right of the "indicator" (milliammeter) the pointer is brought to a reading corresponding to the estimated temperature difference. Key R_1 is now closed, the direction of the galvanometer deflection noted, and the reading of the indicator changed by using the coarse rheostat until the direction of the galvanometer deflection becomes reversed. With the aid of the fine rheostat the deflection is to be reduced to zero; the key R_2 is then closed, increasing the working sensitivity; the resulting deflection is again reduced to zero; and similarly with both the keys 0 and "Shunt off" depressed. The difference between the temperatures of the baths is then indicated by the pointer, the reading in divisions being interpreted in degrees C. by noting the setting of the range-selecting switch. If the actual temperature difference is greater than the anticipated value, it may be impossible to reduce the galvanometer deflection to zero until the range-selecting switch has been advanced to a higher setting. If the temperature of bath B is actually higher than that of bath A, and the knob has been set to "A higher," the fact will be shown by the impossibility of reducing the galvanometer deflection to zero, this deflection increasing con-

tinually as the reading of the indicator increases and as the range-selecting switch is advanced to higher settings.

The check for parasitic emf in the galvanometer, made as the first step in the procedure, should be repeated occasionally, the intervals depending on the degree of freedom of the galvanometer from such emf, its environment as regards temperature uniformity, air currents, etc., and the magnitude of the temperature difference under measurement. The smaller this difference the more carefully must parasitic emf be detected and compensated.

By setting the bath-temperature scale at 158°, at which temperature dE/dT is 50 μv per degree, the potentiometer may be made to indicate the potential difference between the terminals A and B (fig. 2) in microvolts, the full-scale deflection of 100 divisions then corresponding to 50, 100, 250, 500, 1,000, and 2,500 μv when the range switch is set at 0.1°, 0.2°, 0.5°, 1°, 2°, and 5°, respectively. This procedure adapts the potentiometer to the measurement of very small values of emf for any purpose where the limitations on the relative accuracy of Poggendorff's second method do not interfere. A good illustration is the determination of the emf of a standard cell in terms of the emf of another cell taken as a reference cell, the two emfs being opposed and their difference being measured by the potentiometer. Even if the two cells differed in emf by as much as 1,000 μv and the relative error of measuring this quantity were 0.5 per cent, the value of the unknown cell in terms of the reference cell would be obtained with an error of only 5 parts in 1,000,000. Reading directly in microvolts, the potentiometer may be used with thermocouples of characteristics different from those for which it was designed, and might also be used, with the couples for which it was designed, for temperatures lower than 50° or higher than 300°. However, since this procedure would involve inconvenient calculations or reference to curves to obtain values of temperature difference, a more convenient method would be as follows:

For any given mean bath temperature, set the bath-temperature dial to a corresponding value, taken from a curve or table, such that all observed values of temperature difference need only multiplication by a convenient factor to give actual temperature difference. For example, in the present case, the range +50° C. down to −40° C. can be covered by setting the bath-temperature dial at 208° for actual bath temperature of 50°; at 57° for actual bath temperature of −40°, and multiplying observed temperature differences by 1.25. The use of the factor 1.5 makes it possible to cover the range −40° to −100° C., and the factor 2 continues this down to −160°. These factors, increasing as the bath temperature decreases, represent the fact that dE/dT decreases with the temperature.

It would be possible to design the rheostat R_8 (see fig. 2) to provide a scale of dE/dT (or of mean bath temperature) covering a very much wider range of bath temperature, but this might make the scale more crowded than is desirable. It is probably preferable, as a rule, to provide an open scale covering the temperature range for which the greater part of the work is to be done, and to use expedients, such as those of the preceding paragraph, for the very occasional work outside this range.

Although particular attention has been given in this paper to the specific application which occasioned the design and construction of this potentiometer, it is obvious that it, or others of similar design, may readily be adapted to many other purposes involving the rapid measurement of small electromotive forces, particularly in the lower ranges where it is necessary to avoid errors arising from parasitic electromotive forces in the measurement circuit.

WASHINGTON, August 16, 1932.

RP507

THE DENSITY OF SOME SODA-LIME-SILICA GLASSES AS A FUNCTION OF THE COMPOSITION

By F. W. Glaze, J. C. Young, and A. N. Finn

ABSTRACT

Thirty-seven soda-silica and 22 soda-lime-silica glasses were made in platinum crucibles and the density and chemical composition of each glass were determined. From the data obtained equations were derived and a diagram was prepared showing the relations between density and composition of these glasses. Within the range of compositions considered, the diagram makes it possible to predict with considerable accuracy (1) the density of any glass from the composition and (2) the compositions of the various glasses having equal densities. Some evidence is presented indicating that the density of the soda-silica glasses is a simple function of certain soda-silica compounds which may be present in the glass.

CONTENTS

I. INTRODUCTION

Studies of the relations between chemical composition and certain physical properties of glasses have received considerable attention in the past and a report on The Index of Refraction of Some Soda-Lime-Silica Glasses as a Function of the Composition, by C. A. Faick and A. N. Finn, was recently published.[1] In that report the purpose of this general study was given, the method of making the glasses was detailed, essential parts of the methods of chemical analysis were outlined, and the condition of annealing was described.

Since that report was published some additional glasses were made, and the present paper gives the results obtained with respect to density.

II. THE MEASUREMENT OF DENSITY

The samples prepared for the work on index of refraction were also used for density determinations and weighed approximately 20 g each. Since no samples were used that contained gaseous inclusions of sufficient magnitude to affect results, no corrections for "seeds" were necessary.

Density was determined by weighing the samples in air and then weighing them suspended in kerosene by means of a platinum wire basket which was connected with the balance beam by means of a single platinum wire. The suspending wire was covered with unbur-

---

[1] B. S. Jour. Research, vol. 6 (RP320), p. 993, June, 1931; also J. Am. Cer. Soc., vol. 14 (7), p. 518, 1931.

nished electroplated gold. The density of the kerosene was determined at first by means of a 25 ml pycnometer; later it was determined by means of a plummet whose volume was 35.0300 ml at 20° C. and whose weight (corrected for buoyancy of air) was 43.7584 g.

The density of the kerosene (average of 24 determinations) was 0.80870; the maximum change in density of the kerosene, based on observations extending over a period of 14 months, was 0.00028. Measured densities of kerosene were computed to the density at 20° C., using the coefficient 0.0007 per degree, which is sufficiently accurate if the temperature of observation does not differ from 20° C. by more than 5°. When the densities of a large number of glasses were determined during the same day, the density of the kerosene was determined before and after the other determinations; the maximum observed change in any one day, probably resulting largely from unavoidable temperature variations, was 0.00014; the average change was 0.00005.

The average variation in density observations, as determined from the results on six stable glasses, was 0.0003, but since some of the other glasses were decidedly hygroscopic, the reported values for the latter may be in error by as much as 0.001.

All weights were corrected for the buoyancy of air and results were computed to the density of water at 4° C. The data obtained are given in Table 1, as are also values computed by equation (3), differences between computed and observed values and temperatures at which the various glasses were annealed.

TABLE 1.—*Composition, observed and computed densities, and annealing temperatures used for 59 soda-lime-silica glasses*

Glass No.[1]	Composition			Density			Annealing temperature used[a]
	SiO_2	Na_2O	CaO	Observed	Computed	Difference $\times 10^4$	
	Per cent	Per cent	Per cent				°C.
1	50.22	49.78		2.5650	2.5686	−10	450
2	50.51	49.49		2.5640	2.5654	−14	450
3	51.61	48.39		2.5607	2.5602	+5	440
4	52.60	47.40		2.5537	2.5555	−18	460
5	53.62	46.38		2.5515	2.5508	+7	450
6 (4)	54.14	45.86		2.5475	2.5484	−9	420
7	56.56	43.44		2.5383	2.5372	+11	450
8 (5)	57.45	42.55		2.5318	2.5330	−12	430
9	58.98	41.02		2.5266	2.5261	+5	460
10	59.69	40.31		2.5240	2.5226	+14	450
11	59.71	40.29		2.5228	2.5224	+4	470
12 (6)	59.97	40.03		2.5208	2.5210	−2	430
13	62.77	37.23		2.5071	2.5062	+9	470
14 (7)	62.89	37.11		2.5044	2.5058	−14	440
15 (8)	63.06	36.94		2.5038	2.5047	−9	440
16 (9)	65.30	34.70		2.4890	2.4930	−40	450
17 (10)	65.32	34.68		2.4924	2.4928	−4	480
18	66.52	33.48		2.4865	2.4862	+3	480
19 (11)	67.14	32.86		2.4807	2.4819	−12	500
20 (12)	69.65	30.35		2.4644	2.4650	−6	515
21 (13)	70.21	29.79		2.4612	2.4612	0	523
22 (14)	70.44	29.56		2.4603	2.4597	+6	525
23 (15)	72.15	27.85		2.4488	2.4483	+5	525
24	72.33	27.67		2.4479	2.4472	+7	500
25	74.16	25.84		2.4343	2.4351	−8	505

[1] Figures in parentheses are the numbers of the same glasses in the refractivity paper.

TABLE 1.—*Composition, observed and computed densities, and annealing temperatures used for 59 soda-lime-silica glasses*—Continued

Glass No.[1]	Composition			Density			Annealing temperature used
	SiO$_2$	Na$_2$O	CaO	Observed	Computed	Difference ×10⁴	
	Per cent	*Per cent*	*Per cent*				°C.
26	74.69	25.31	....	2.4305	2.4323	−18	505
27 (16)	75.29	24.71	....	2.4260	2.4264	−4	560
28	76.60	23.40	....	2.4140	2.4133	+7	510
29	76.65	23.35	....	2.4133	2.4128	+5	510
30	76.70	23.30	....	2.4126	2.4124	+2	550
31 (17)	77.85	22.15	....	2.4007	2.4011	−4	575
32 (18)	78.61	21.39	....	2.3938	2.3936	+2	575
33 (19)	79.73	20.27	....	2.3813	2.3829	−16	575
34	82.76	17.24	....	2.3545	2.3541	+4	550
35 [2] (20)	82.86	17.14	....	2.3536	2.3531	+5	600
36	85.15	14.85	....	2.3307	2.3319	−12	540
37 (21)	86.41	13.59	....	2.3204	2.3203	+1	600
38 [2]	100	....	....	2.2033	2.2026	+7	....
39	50.18	37.80	12.02	2.6413	2.6400	+13	460
40	50.30	43.04	6.66	2.6076	2.6082	−6	450
41 (23)	54.37	32.85	12.78	2.6236	2.6232	+4	530
42 (24)	56.20	34.00	9.80	2.5976	2.5978	−2	480
43 (25)	56.76	34.48	5.76	2.5719	2.5718	+1	470
44 (26)	58.41	38.54	3.05	2.5474	2.5476	−2	480
45 (27)	60.32	24.50	15.18	2.6074	2.6068	+6	530
46 (28)	63.34	24.39	12.27	2.5757	2.5755	+2	525
47 (29)	64.14	21.22	14.64	2.5851	2.5845	+6	565
48 (30)	64.70	26.84	8.46	2.5460	2.5460	0	525
49 (31)	65.71	28.79	5.50	2.5229	2.5234	−5	520
50 (32)	66.47	21.74	11.79	2.5564	2.5555	+9	565
51 (33)	67.30	29.43	3.27	2.4998	2.5009	−11	520
52 (34)	67.98	22.50	9.52	2.5331	2.5333	−2	565
53 (35)	70.50	23.00	6.50	2.4980	2.4980	0	530
54 (36)	72.08	14.21	13.71	2.5276	2.5269	+7	580
55 (37)	72.61	24.24	3.15	2.4641	2.4640	+1	525
56 (38)	74.09	15.23	10.68	2.4935	2.4932	+3	545
57 [2] (39)	74.69	12.28	13.03	2.4961	2.4964	−3	590
58 (40)	75.48	15.26	9.26	2.4734	2.4724	+10	590
59 (41)	78.77	16.33	4.90	2.4190	2.4184	+6	545
60 (42)	80.59	16.17	3.24	2.3886	2.3921	−35	565

[1] New analysis made since report on refractivity.
[2] This sample was not made at the Bureau of Standards, nor was it analyzed.

III. DATA OBTAINED AND RESULTS

If the densities of the soda-silica glasses are plotted against the silica or soda content of the glasses, an approximately smooth curve will be obtained for those glasses containing less than 80 per cent silica; the best equation derived to represent that portion of the curve was

$$D = 2.4756 + 0.00301B - \frac{2.988}{B-1} \qquad (1)$$

in which D is the density and B is the percentage of soda. This equation is not very satisfactory because of its limited range and also because it could not easily be adapted to glasses containing lime.

Since, in the paper on index of refraction, it was pointed out that more satisfactory results could be obtained by drawing straight lines between certain points on a composition-refractivity diagram for the soda-silica glasses, the same procedure was applied to the density data. In this case the need for straight lines rather than a continuous curve again became evident, but the changes in slope of the straight lines drawn through the plotted data were not decidedly pronounced.

When, however, specific volumes (reciprocal of density) were plotted against silica, a very satisfactory series of straight lines could be drawn through the plotted data. (Fig. 1.) The intersections of these lines occurred at approximately 59.4, 66.3, and 74.9 per cent silica. These values correspond closely to three simple molecular ratios of soda to silica, namely, 4:6, 3:6, and 2:6. Of these the only

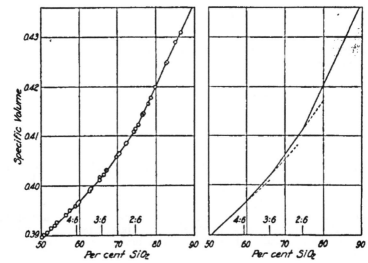

FIGURE 1.—*Diagram showing relation between composition and specific volume of some soda-silica glasses; three molecular ratios of soda to silica are also indicated*

All reported observations are plotted in the graph at the left and are omitted in the graph at the right; the lines in both these graphs are identical and the dotted extensions are added, in the right-hand graph, to emphasize the change in slope of the various sections

one obtained in a crystalline state is the 3:6 compound[2] (ordinarily written 1:2).

Equations of the form $\frac{1}{D} = a + bA$, in which A is the percentage of

silica, can be used to express the four straight lines in Figure 1. Since such equations involve only one constituent of the binary glasses they could not be readily applied to more complex glasses; consequently the initial equations were converted, through the relation $A + B = 100$, to the form

$$\frac{1}{D} = \alpha A + \beta B \tag{2}$$

in which α and β are empirical constants having different values for the different ranges of composition, as shown in Table 2. A simul-

[2] Morey, G. W., and Bowen, N. L., The Binary System Sodium Metasilicate-silica, J. Phys. Chem., vol. 28, No. 11, pp. 1167-1179, 1924.

taneous solution of each adjacent pair of equations gives the indicated silica limits.

In general the data on the soda-lime glasses were treated as if the combined amount of soda and lime were all soda, and a value for $\frac{1}{D}$ for each glass was computed from equation (2), using appropriate constants. The differences between these computed values and the observed specific volumes were plotted as functions of the percentage of lime, C, and four equations of the form

$$f(C) = kC + \gamma'C^2$$

were obtained, k and γ' being empirical constants. Each of these equations, when added to its corresponding soda-silica equation, gives the completed form

$$\frac{1}{D} = \alpha A + \beta(B + C) + kC + \gamma C^2$$

or

$$\frac{1}{D} = \alpha A + \beta B + \gamma C + \gamma'C^2 \tag{3}$$

in which A, B, and C represent the percentages of silica, soda and lime, respectively, and α, β, γ, and γ' are empirical constants having the values indicated in Table 2.

It is evident that this procedure assumes initially that the lines of demarcation (aa', bb', and cc', in fig 2) will lie along constant silica lines, but a simultaneous solution of each adjacent pair of equations (3) gives the following values of the silica limits for the glasses containing lime:

$$A_a = 59.4 - 0.23C + 0.032C^2 \text{ (indicated by } aa' \text{ in fig. 2)}$$

$$A_b = 66.3 + .15C - .004C^2 \text{ (indicated by } bb' \text{ in fig. 2)}$$

$$A_c = 74.9 \qquad - .017C^2 \text{ (indicated by } cc' \text{ in fig. 2)}$$

in which A_a, A_b, and A_c represent, respectively, the percentages of silica for the lines aa', bb', and cc'.

In arriving at the best values for γ and γ', therefore, it was necessary with some glasses near the limiting silica values to depart from the general procedure outlined by using an adjacent set of values of α and β in anticipation of the final positions of the lines of demarcation. The proper selection of the values of α and β could only be determined by trial. These curved lines are preferred to the straight lines indicated in the refractivity paper because density changes with composition are approximately seven times greater than corresponding refractivity changes, and, hence, give a better criterion of the location of these lines. Equation (3) and the constants in Table 2 were used in computing data to draw the lines of equal density shown in Figure 2 in which all the experimental glasses are indicated. As additional data are obtained, however, it may be necessary to make other changes in the location of these lines of demarcation.

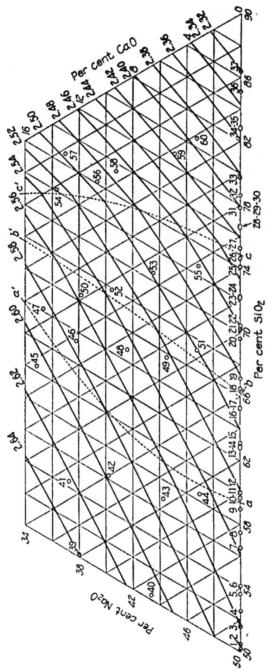

FIGURE 2.—*Diagram showing the relation between composition and density of some soda-lime-silica glasses*

TABLE 2.—*Silica limits and values of corresponding constants for equations (2) and (3)*

Silica limits, per cent silica	α	β	γ	γ'
50 (?) to A_a	0.0042520	0.0035370	0.002545	0.0000080
A_a to A_b	.0043028	.0034628	.002500	.0000040
A_b to A_c	.0043922	.0032872	.002285	.0000050
A_c to 100.0	.0045400	.0028460	.001844	.0000148

IV. CONCLUSION

Considering the data on the soda-silica glasses only (and disregarding the trend of specific volume changes in the soda-lime glasses), it is difficult to avoid the conclusion that the increment in specific volume of any of these glasses between the end members of the group (indicated in Table 2) in which that particular glass lies, is directly proportional to the increment in silica. In other words, the data suggest that the soda-silica glasses, in the range studied, should be regarded either as simple compounds or as simple mixtures of two adjacent compounds.

WASHINGTON, August 3, 1932.

145879—32——7

COILED FILAMENT RESISTANCE THERMOMETERS

By C. H. Meyers

ABSTRACT

Platinum resistance thermometers with strain free windings which are smaller than those previously used are described, and the method of construction is given. By double helical winding the coil of a 25 ohm thermometer is reduced to about 5 mm diameter and 2 cm length, and the outside diameter of the thermometer is about the same as that of the ordinary mercury-in-glass thermometer. These thermometers are hermetically sealed, and thermometric lag is reduced by filling them with helium.

CONTENTS

I. INTRODUCTION

In a previous paper, T. S. Sligh [1] has discussed the properties of platinum resistance thermometers in gener l, explained the method used for calculating temperatures, and described the construction of resistance thermometersprdinarily used at this bureau.

Strain-free thermometers with smaller dimensions have sometimes been found desirable for use in special apparatus, and to fill this need several such thermometers have been constructed. When these thermometers are mounted in a glass tube the over-all diameter is about 7 mm, or about the size of an ordinary mercury-in-glass thermometer. Since the length of the coil is only about 2 cm this design of resistance thermometer is suitable for any location where a mercury thermometer may be used.

A detailed description of this special design of thermometer coil and its construction will be given as well as a description of modifications in the design of leads and thermometer head.

II. THERMOMETER COIL

The design of the thermometer coil is illustrated in Figure 1, which is a photograph of a model built for illustrative purposes. In Figure 1, (A) shows this model mounted on the quartered mandrel upon which it was wound, and (B) shows the coil removed from the mandrel. The size of the actual thermometer coil is illustrated by the scale drawn in (A).

In the construction of the thermometer coil, the platinum winding was mounted upon a mica cross of the type mentioned in Sligh's

---

[1] B. S. Sci. Papers, vol. 17 (8407), p. 49, 1922.

paper. The various steps in the construction of this cross and the coil are given in Figure 2. The dimensions given are those actually used and are given in millimeters unless otherwise stated in the figure.

Mica was cut into strips by clamping the part to be used under a bar of the proper width and trimming away the portions extending on each side by a repeated back and forth motion of a sharp-edged steel tool. As mentioned in the paper already referred to, a thin safety-razor blade forms a convenient source of material for such a

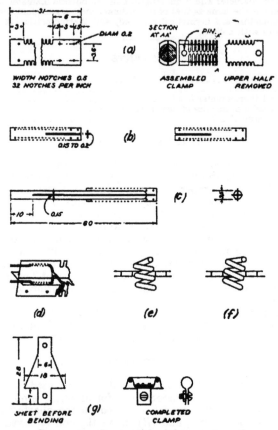

FIGURE 2.—*Construction of thermometer coil*

tool. The mica strip may be split to the desired thickness (about 0.1 mm), or teeth may cut in the strip before splitting. For cutting the teeth the mica was placed in a toothed clamp, such as shown to the right in Figure 2 (a), the halves of which were kept in alignment by pins. This clamp was gripped between vise jaws. Teeth were cut in the mica with a special saw consisting of 0.5 mm steel wire mounted upon a light jeweler's hack-saw frame. The saw cut equally well when the wire was charged with carborundum by rolling between two plates of glass or when the wire was merely roughened by rub-

B. S. Journal of Research. RP508

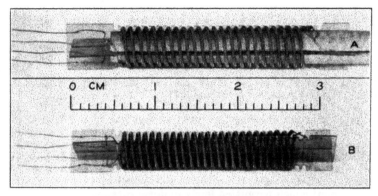

FIGURE 1.—*Large model of thermometer coil*
A, Mounted upon mandrel; B, removed from mandrel.

bing transversely with a carborundum stone. This saw produced a smooth-edged notch with parallel sides and a rounded bottom. Both steel piano wire and nickel wire have been used successfully in cutting notches from 0.1 to 0.5 mm wide. Copper wire was not found as satisfactory on account of excessive stretching of the wire. After the mica had been notched it was removed from the clamp, and the six holes shown in Figure 2 (b) were drilled. Piano wire was ground to form a flat drill for this purpose. The slots shown were cut and the slotted pieces of mica were pushed together to form a cross in which the teeth were arranged to accommodate a bifilar winding of uniform pitch. This cross was inserted in the quartered steel mandrel illustrated in Figure 2 (c).

Platinum wire nominally 0.1 mm in diameter was used in these thermometers, and the length of wire required was 1.8 to 2.4 m, depending upon the actual diameter of the lot of wire used. The requisite amount of wire was cut into two pieces, each of which was wound into a helix with an outer diameter of about 0.45 mm. For this purpose a steel wire (0.23 mm diameter) was used as a mandrel, one end being mounted in a lathe and the other end attached to a ball bearing, an arrangement which permitted fairly uniform rotation of the taut steel wire. When hard-drawn platinum wire was available the following procedure was successful: One end of the wire was attached to the mandrel and the other end to a thread which led over a pulley to a weight of about 30 g. The pulley was placed in such position relative to the lathe that the consecutive turns of the helix were wound in contact; it was far enough from the lathe (about 3 m) so that the angle of winding remained substantially constant. When the winding was completed both ends of the helix were released and one end was kept from rotating while the lathe was rotated in such direction as would tend to unwind the helix until the other end began to slip on the mandrel. The mandrel was then removed from the lathe, its end freed from any roughness which might scratch the platinum wire and the helix pushed off. The helix was then stretched until its length under no tension was approximately doubled, the result being an almost uniform spacing of the wire.

The method just described was not successful for annealed platinum wire. When the use of such wire was necessary, it was either wound along with a spacing wire so that the turns were in contact and the spacing wire later unwound, or the platinum wire was fed from the moving tool carriage of a thread-cutting lathe. For the latter method a lathe with two heads geared together is desirable since the irregularities in the motion of the ball bearing cause irregular spacing of the wire. After winding, the mandrel was dissolved in hot hydrochloric acid.

Short leads of platinum were welded to form a T on one end of each of the two helices. A small oxy-gas flame was used in the present work, although the arc method described by Sligh might have been used with due precaution to arrange the work so that the electric current would not anneal the helix. It was found that the helices stretched less during winding and that the coil suffered less from accidental distortions if the wire was not annealed until the winding upon the mica cross was completed. Both pairs of leads were then threaded into the mica cross as shown for one pair in Figure 2 (d).

The choice of the direction of winding upon the cross was determined by the direction of winding of the helices which had been wound with a left-hand pitch. Figure 2 (e) and (f) show a magnified view of the edge of the mica with such a helix crossing it. The direction of winding upon the cross corresponding to that at (e); that is, a winding with a right-hand pitch was chosen as the one in which two consecutive turns of the helix would be less likely to straddle the mica. The helices were wound upon the cross under the tension caused by a weight of 2 or 3 g attached to each of their free ends. The clamp of thin sheet copper designed as shown in Figure 2 (g) for both right and left hand windings, was slipped over the coil to hold the winding in place while the ends of the helices were joined to the platinum tie wire visible in Figure 1, and tied to the mica. After removing the clamp and the quartered mandrel from the coil, the latter was annealed in an electric furnace for two or three hours at 600° to 650° C.

During this annealing the resistance of the 25 ohm thermometers decreased about 0.5 ohm, an amount which had to be allowed for in choosing the initial length of wire.

After the preliminary annealing the resistance of the coil was adjusted by cutting out the proper amount of wire and rewelding, the mandrel and clamp being replaced during the operation. The coil was then again annealed at the same temperature for about five hours.

III. LEADS AND THERMOMETER HEAD

As in the thermometers described by Sligh, mica disks served the double purpose of separating the internal leads and breaking up convection currents in the stem of the thermometer. The mica disks were cut with a punch and die of the type ordinarily used for blanking and perforating sheet metal. An attempt to use Sligh's method of keeping the mica disks normal to the axis of the thermometer by staggering the holes through which the leads were threaded was unsuccessful for the small size of washers used.

Better results were obtained by tapering the wire so that disks with holes which fitted the wire snugly could be used. Although tapering the full length of the leads gave the best appearing result, it was found that with care the mica disks could be slid along the uniform wires without serious enlargement of the holes, and that part or all of the tapered portion could be discarded. Gold wire 0.2 mm diameter was tapered in hot acqua regia by moving the wire up and down thus varying the time of immersion along the length of the wire. The holes in the disks used with these leads were made with drills formed from 0.16-mm steel wire. A pile of the disks was placed in the jig illustrated in Figure 3 (a) and four equally spaced holes were drilled. The insertion of a piece of the wire used for drills in one hole before removal of the disks from the jig made it possible to thread the whole pile of disks onto the ends of the leads as a unit. After the requisite number of disks had been threaded on, the leads were put under tension and each disk slid singly into the desired position, the spacing used being about 5 to 10 mm. A short copper wire was silver soldered or welded to one end of each of the gold leads for convenience in later sealing the thermometer head with soft solder, and one of the short platinum leads on the thermometer coils was welded to the other end.

Since some of these thermometers were to be used in an inverted position and would probably come in contact with oil, it was desirable to avoid the use of cement or wax for sealing. The method of bringing the internal leads out through the Pyrex tube is shown in Figure 3 (b). Four platinum disks 0.02 or 0.03 mm thick were beveled at the edge

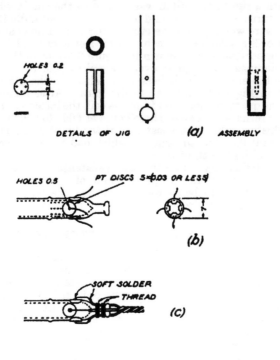

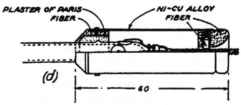

FIGURE 3.—*Construction of leads and head*

by running through rolls and curled to fit the glass tube. These were temporarily held in place by a clamp of nichrome wire and fused to the glass as described by Housekeeper.[2] It was found that the formation of bubbles under the platinum could be minimized by keeping the work far out in the flame. The addition of some oxygen to the air supplied to the torch was also found advantageous for Pyrex. Holes

² J. Am. Inst. Elect. Eng., vol. 42, p. 954, 1923.

through the platinum and glass were made at the centers of the disks with a hot tungsten wire.   Benton [3] used a hot platinum wire for perforating glass, but tungsten has the advantage of being stiffer and forming an oxide which does not readily stick to the glass.

The short copper wires at the end of the gold leads were threaded through the holes, which were then closed with soft solder.   Caution must be used in soldering; a hot iron must not be applied to the cold glass lest the platinum be loosened.   The external leads were then attached as in Figure 3 (c).   The assembled head is shown in Figure 3 (d).

The thermometers were filled through the bulb end with dry helium at various pressures less than 1 atmosphere and sealed.

IV. CHARACTERISTIC PROPERTIES OF THE THER-MOMETERS

As is to be expected, the values for the resistance in ice, steam, and sulphur have the same proportions as for previous thermometers in which platinum of the same purity was used.

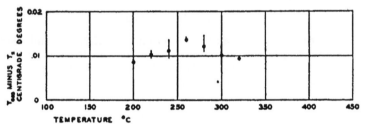

FIGURE 4.—*Comparison of coiled-filament thermometer with ordinary platinum resistance thermometer*

To determine whether the readings of a coiled-filament ther-mometer deviated appreciably from those of thermometers previously used, such a thermometer (No. 2) was compared in the range 200° to 320° C. with C_{23}, a thermometer of the customary design as de-scribed by Sligh.   During this comparison the thermometers were immersed in a stirred oil bath of uniform temperature.

In Figure 4 the difference in centigrade degrees obtained by sub-tracting the readings of the coiled-filament thermometer from those of C_{23} are plotted as ordinate with the temperature as abscissa.   The differences are no greater than might be expected in this region between two thermometers of the customary design.   Each circle represents the mean of the observations at a given temperature while the vertical line through the circle represents the spread of the observations.   A curve of differences should by definition pass through zero at 100° and at 444° C., but experimental errors in calibration, especially at the latter point, probably introduce considerable differences on the scale chosen in Figure 4, hence no attempt has been made to draw such a curve.   It may be concluded that the coiled-filament winding intro-duces no appreciable peculiarity in the readings of the thermometer.

Since winding the wire in a helix reduces the effective area for heat

[3] J. Ind. Eng. Chem., vol. 11, p. 623, 1919

transfer between the thermometer and its surroundings, it is to be expected that the thermometric lag will be thereby increased, as well as the temperature difference between the thermometer coil and its surroundings necessary to dissipate the heat generated by the measuring electric current.

Measurements of the apparent resistance in an ice bath for several values of measuring current show that this temperature difference is proportional to the square of the current, and for 1.5 ma through a 25 ohm coil 5 mm in diameter and 2 cm long is about 0.004°C. when air filled and 0.001° C. when helium filled. In the paper by Sligh already mentioned the temperature difference for the larger type of thermometer filled with air is given as 0.001° C. for a current of 1.5 ma.

With the use of currents up to 5 ma the ratio between this temperature difference and the power input in the thermometer may be determined with sufficient accuracy so that variations with time in this ratio may be used as a measure of air leakage into the thermometer tube. One thermometer of the type described filled with helium at about 40 mm absolute pressure has shown no appreciable leakage over a period of two years.

V. ACKNOWLEDGMENTS

The author wishes to thank H. F. Stimson, F. R. Caldwell, and others for valuable suggestions and assistance in the construction of these thermometers, and E. F. Fiock for the data given in Figure 4.

WASHINGTON, September 1, 1932.

RP509

DEFLECTION OF COSMIC RAYS BY A MAGNETIC FIELD

By L. F. Curtiss

ABSTRACT

A continuation of earlier experiments on this subject (Phys. Rev., vol. 34, p. 1931, 1929; vol. 35, p, 1433, 1930), gives decisive evidence that a considerable fraction of an unfiltered cosmic-ray beam can be deviated by a magnetic field of 7,000 gauss. Using four Geiger-Müller tube counters, coincidences were observed with and without a magnetic field. The arrangement was such that electrons in the beam, possessing energies corresponding to 10^9 volts or less, could be prevented from producing coincidences. Observations were made for a total of 615 hours without the field and 508 hours with the field, in alternating periods of about 48 hours. The average number of coincidences without the magnetic field was found to be 1.31 ± 0.09 per hour and with the magnetic field 0.88 ± 0.14 per hour. This indicates that, for approximately 70 per cent of the cosmic-ray particles, the deviation is less than that for 10^9 volt electrons. This harder portion of the beam may contain, in addition to electrons of higher energy than 10^9 electron volts, protons of 5×10^9 electron volts and neutrons. These data indicate that the results obtained by Rossi (Nature, vol. 128, p. 300; 1931) and Motth-Smith (Phys. Rev., vol. 37, p. 1001; 1931), showing no effect of the magnetic field, were affected by scattering or absorption in the iron core in which the magnetic field was produced.

CONTENTS

I. INTRODUCTION

In a letter[1] to the Physical Review for November, 1929, a report was made of experiments with Geiger-Müller tube counters which seemed to show that cosmic rays consisted, in part at least, of electrically charged particles which could be deflected by a magnetic field. These observations were made with a pair of tube counters, the coincidences between which were observed by making simultaneous records on a paper tape with dotting pens actuated by the counters through amplifiers. Although qualitatively there was indication of a decrease on applying a magnetic field between the counters, the method was too crude to serve as a basis for a definite decision of the point in question.

The experiments were, therefore, repeated, using the vacuum tube circuit developed by Bothe[2] for resolving coincidences electrically and following the suggestion by Tuve[3] of using more than two counters so that a more sharply defined beam of cosmic rays may be studied.

[1] Phys. Rev., vol. 34, p. 1391, 1929.
[2] W. Bothe, ZS. f. Phys., vol. 59, p. 1, 1929.
[3] M. Tuve, Phys. Rev., vol. 35, p. 651, 1930.

These experiments were made with three counters, the results of which were reported to the Physical Society[4] in April, 1930. These data also pointed to a magnetic deviation of a part of the cosmic-ray beam. A decrease of the order of 25 per cent in coincidence was observed on applying a magnetic field of 7,000 gauss to the path of the rays between two of the counters. These observations were made with considerable difficulty. The counters were necessarily of a small size, since they could be no longer than the air gap of the electromagnet (5 cm). This reduced the sensitivity of the arrangement, requiring, therefore, long periods of observation. Much trouble was also encountered with the Bothe screen grid vacuum tube circuit as modified for use with three counters. An even more serious source of trouble and possible error was discovered when it was noticed that the counters themselves are very considerably affected by changes of temperature. Sealed counters, described elsewhere,[5] were used for these experiments. The rate of counting at a given voltage falls off rapidly as the temperature is increased. Some counters have been found to cease counting entirely when their temperature is raised 10° or 15°. This effect has not been investigated, but it is presumably caused by the increase of vapor pressure of nonpermanent gases and more volatile substances present in the partially exhausted counters. It could easily falsify observations made with the magnetic field, since the temperature of the windings of the electromagnet rises considerably during the prolonged periods of observations, causing the temperature of some of the counters to rise also.

Another disturbing condition in the former experiments was the presence in the building of several grams of radium which, although stored at some distance in a safe with lead walls 6 inches thick, nevertheless increased the rate of counting. This penetrating radiation could not be screened off readily and was responsible for a large number of accidental coincidences, these being proportional to the rate of counting of each counter.

Interest in the question of the magnetic deviation of cosmic rays was increased by the reports made by Rossi[6] and by Mott-Smith[7] that no deflection was observed when cosmic ray beams pass through magnetized iron. It is of importance to make certain of the earlier results. Therefore the investigation has been continued under improved conditions designed to eliminate as far as possible difficulties and sources of error encountered in the earlier experiments. This paper gives an account of the results which have been obtained.

II. EXPERIMENTAL ARRANGEMENT

The most annoying difficulties encountered in the previous experiments arose from the presence of a relatively large number of accidental coincidences and from changes in counting rate due to temperature changes. Particular attention was given to the elimination of these disturbances.

Considerable reduction in the number of accidental coincidences was effected by the removal of the apparatus from the building in which

[4] Phys. Rev., vol. 35, p. 1433, 1930.
[5] B. S. Jour. Research, vol. 4 (RP 165), p. 593, 1930.
[6] B. Rossi, Nature, vol. 128, p. 300, 1931.
[7] L. M. Mott-Smith, Phys. Rev., vol. 37, p. 1001, 1931

radium is stored. Since the accidentals are proportional to the product of the rate of the individual counters, the importance of this change, which reduces simultaneously the rate of all counters, is obvious. It seemed desirable, however, to reduce these accidentals even further, in view of the long periods covered by the observations. At most, only a few genuine coincidences per hour could be expected so that it was desirable to reduce as far as possible the probability of any accidental coincidences.

The most effective method of reducing accidentals is to increase the number of counters. Accordingly a fourth counter, added to the previous arrangement, had the desired effect of reducing the accidental coincidences practically to zero. The advanatge of additional counters can best be shown by computing the accidentals for the case of two and four counters under otherwise similar conditions. For two counters the expected number, E, of accidentals is given by

$$E = 2 \, N_1 \, N_2 \, \tau$$

where N_1 is the average rate of counting of the first counter, N_2 the rate for the second, and τ is the "resolving power" of the apparatus used to detect coincidences. Putting $N_1 = N_2 = 30$ per minute and $\tau = 10^{-3}$ sec., E becomes 1.7 per hour. For the case of four counters

$$E = 2^3 \, N_1 \, N_2 \, N_3 \, N_4 \, \tau^3$$

Putting $N_1 = N_2 = N_3 = N_4 = 30$ per minute and $\tau = 10^{-3}$ sec., E drops in this case to 1 coincidence in about 70 years. Consequently it is safe to assume in general that a circuit using four counters only responds to the passage of an ionizing particle through all four counters. A possible exception will be mentioned in the discussion of the results.

The vacuum tube circuit devised by Rossi [8] was adopted, since it is simpler and easier to operate when extended to four counters than any other arrangement which has been tried. The details of the circuit are shown in Figure 1. The output was connected to a telegraph relay which actuated electrically a dial counter. Switches were arranged to shunt out any three counters by applying a negative voltage to the grids of the three vacuum tubes connected with these counters. In this way the rate of counting of the fourth alone could be observed directly, making possible a frequent check of the rate and sensitivity of each counter.

To eliminate errors arising from changes in temperature each counter was inclosed in a thermally insulated box containing a heating coil and thermostat capable of maintaining the temperature constant to within a few tenths of a degree. The electromagnet and counters were also inclosed so that their temperature could be kept uniform by rapid circulation of the air by means of an electric fan. A small radium preparation inclosed in lead was used to test the sensitivity to penetrating γ-rays.

The geometrical arrangement of the counters with respect to the poles of the electromagnet is shown in Figure 2, which is drawn to scale. The air gap between the poles, 5 cm long, determined the useful length of the counters. Observations were made with a magnetic

[8] B. Rossi, Nature, vol. 125, p. 136, 1930.

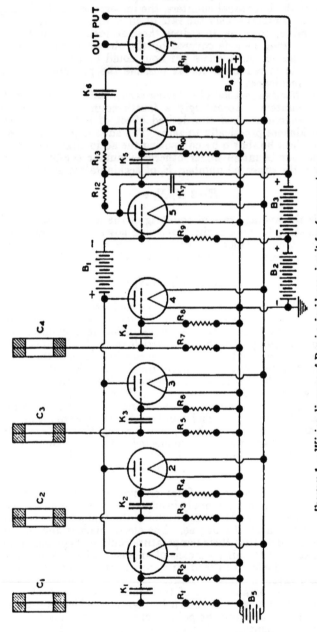

FIGURE 1.—*Wiring diagram of Rossi coincidence circuit for four counters*

C_1, C_2, C_3, C_4—Geiger-Müller tube counters. K_1, K_2, K_3, K_4—0.00006 μf condensers. K_5, K_6—0.1 μf condenser. K_7—0.5 μf condenser. R_1, R_2, R_3, R_7—10^6 ohms.
R_8, R_4, R_5, R_6—5×10^5 ohms. R_9, R_{10}, R_{11}—10^5 ohms. R_{12}, R_{13}—0.25×10^6 ohms. B_1—16 volts. B_2—6 volts. B_3—90 volts.

field in the air gap of 7,000 gauss, determined from a calibration curve. As indicated by the dotted lines across the end view of the pole face, a beam of 10^9 volt electrons would, with this arrangement, be deviated just sufficiently to miss the lower counter.

The experiments were carried out under a thin wooden roof with no other filtering, so that the softer components of the cosmic radiation are included in the observations. Since the walls of the counters were of steel, approximately 1 mm thick, and a particle to produce one coincidence would have to traverse 7 mm of steel, beta-radiation of radioactive origin could produce very few systematic coincidences.

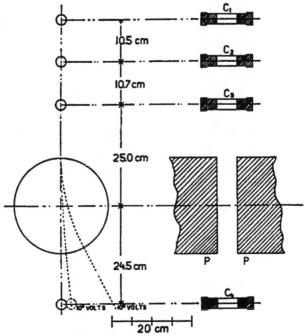

FIGURE 2.—*Plan of arrangement of counters with respect to poles of electromagnet*

C_1, C_2, C_3, C_4=Geiger-Müller tube counters. PP=poles of electromagnet.

Actual observations were preceded by a careful study of the operating characteristics of each counter, making it possible to select for each an operating temperature such that all four could be run at the same voltage. A small 110-volt alternating-current generator driven by a synchronous motor, having its output stepped up by a transformer to about 1,100 volts, supplied this voltage. All parts of this circuit were inclosed and kept at a constant temperature.

III. OBSERVATIONS

The experiments were undertaken primarily to determine whether or not cosmic-ray particles could be deflected by a magnetic field. Accordingly, the procedure adopted in taking observations was to record the number of coincidences for several hours in the absence of the magnetic field and then to take a similar record with the field applied. This method does not permit a study of the direction of deflections to determine the sign of the charge carried by the particles. The small size of the counters made long perio s of observation necessary so that it was decided to settle first, if possible, the question of the existence of a deflection.

A summary of observations is given in the following table. Although in some cases uninterrupted series of observations of a week or more were possible, interruptions frequently occurred. A failure of any part of the complicated arrangements was cause for the rejection of the record taken since the last time it was known to be functioning properly. The chief difficulty arose from temperature variations of the room which was in an unheated building. These variations were sometimes too sudden or too great for the temperature controls to cope with.

As the table shows, observations were made for a total of 636.5 hours with no magnetic field and for a total of 508 hours with a magnetic field of 7,000 gauss. The average number of coincidences without the field was 1.31 ± 0.09 per hour and with the field 0.88 ± 0.14 per hour. Although, as to be expected under the circumstances, there was a considerable fluctuation in the number of coincidences per hour between individual periods of observation, there can be very little doubt that there were fewer coincidences when the field was applied. The amount of this decrease, as shown by the above data, was approximately 30 per cent.

TABLE 1.—*Record of observations*

Dates	Duration (hours)	Number of coincidences—		Coincidences per hour—	
		Without field	With field	Without field	With field
June 23-25	48.75	107		2.19	
June 25-29	100.00		141		1.41
July 5-6	20.00	14		.70	
July 7-8	23.75		7		.30
July 15-16	24.00	45		1.87	
July 16-18	45.25		37		.82
July 18-21	71.25	71		1.00	
July 21-23	45.50		104		2.29
July 23-25	50.25	79		1.57	
July 26-28	41.75		24		.56
July 28-Aug. 1	95.25	117		1.23	
Aug. 1-4	70.50		62		.88
Aug. 5-8	72.00	105		1.46	
Aug. 9-11	41.50		32		.77
Aug. 12-13	27.50	36		1.31	
Aug. 13-17	92.50		48		.52
Aug. 17-20	72.75	64		.88	
Aug. 20-22	47.25		19		.40
Aug. 25-26	23.75	15		.63	
Aug. 26-27	25.25	42		1.66	
Sept. 2-6	95.75	113		1.18	
Average number of coincidences per hour				1.31±0.09	0.88±0.14

IV. DISCUSSION

Although the figures given represent only a preliminary approach to the study of the magnetic deviation of cosmic-ray particles, certain conclusions concerning the energy and character of the cosmic-ray particles can be drawn from them. Referring to Figure 2, it is seen that a beam of electrified particles consisting of electrons of energy greater than 10^9 electron volts or protons of energy greater than 5×10^8 electron volts will be detected in the undeflected beam when the magnetic field is applied. This represents 70 per cent of the cosmic rays according to the figures given above. If these undeflected particles have no charge, these experiments reveal nothing concerning their energy. The results could then be explained quite simply by assuming that the primary cosmic-ray beam consists of neutrons. One of the characteristics of the neutron, known at present, is its ability to transmit high velocities to protons and, to a lesser extent, to electrons. Thus, a beam of neutrons penetrating our atmosphere should be accompanied by protons and electrons with high velocities. From the results obtained by Dee [9] and by Feather,[10] one would expect the protons to predominate. Thus, if it could be shown that, of the 30 per cent of the beam that shows a deflection, the protons greatly exceed the electrons in number, it could be taken as an indication of the existence of neutrons in the cosmic-ray beam. A more direct attack would be to determine the loss of particles for various strengths of magnetic fields up to the values greatly in excess of that used in the present experiments. It seems reasonable to suppose that under such conditions a point should be found where an increase of strength of the magnetic field produced no further loss of particles. The deviability to be expected for a neutron moving in a magnetic field depends upon the structure assumed for the neutron. As pointed out by Huff,[11] the model proposed by Pauli may be deflected in a nonhomogeneous magnetic field. However, this deflection is of observable magnitude only when the neutron is traveling at low speeds when it would probably not ionize at all. Consequently, a beam of neutrons, having the high velocities necessary to account for cosmic-ray phenomena, could not be distinguished by a magnetic experiment from a beam of photons.

In making an attempt to interpret the apparent existence of an undeflected beam of cosmic rays amounting to approximately 70 per cent of the total radiation, the possibility of psuedo coincidences must be considered. Such coincidences are caused by the simultaneous ejection of several secondary particles by a single cosmic ray in such a way that separate secondary particles traverse different counters at the same time. Such an effect has been demonstrated experimentally by Rossi [12] and by Johnson and Street.[13] If an appreciable number of the coincidences observed in the present experiments were produced in this way, it would be difficult to arrive at any satisfactory interpretation of the results obtained. It is easy to see in a general way that this could not be true, since the effect is only readily detectable when masses of absorbing material of high atomic number (that is,

[9] P. I. Dee, Proc. Roy. Soc., vol. 136, p. 727, 1932.
[10] N. Feather, Proc. Roy. Soc., vol. 136, p. 709, 1932.
[11] Huff, L. D., Phys. Rev., vol. 38, p. 2292, 1932.
[12] Rossi, B., Phys. ZS, vol. 33, p. 304, 1932.
[13] Johnson, T. H., and Street, J. C., Phys. Rev., vol. 40, p. 638, 1932.

lead) are placed slightly above and in the immediate neighborhood of the counters. The secondary particles which cause the pseudo coincidences are projected mainly in the direction of the primary radiation which has its maximum of intensity in the vertical direction. In the present experiments the only large mass of heavy material, the iron of the electromagnet, was below all except one of the counters. There was no absorbing material of high atomic number above the counters. Nevertheless it is important to make certain of this by an experimental determination of the probable number of coincidences which might be caused by this effect. Observations extending over several weeks were made in which one of the counters was placed entirely out of line from the others but near to its original position. The average number of coincidences under these conditions was found to be 0.15 per hour. Although representing about 10 per cent of the total number of coincidences observed with no magnetic field, this figure is but slightly greater than the experimental error. Thus it seems that the effect under discussion presumably did not contribute appreciably to the coincidences observed when all four counters were in line.

The fact that a large proportion of the cosmic rays appear to be less deviable than 10^9-volt electrons seems to offer, when considered in connection with the suggestion made by Johnson,[14] and explanation for the failure of Rossi,[15] and of Mott-Smith [16] to obtain evidence of a deflection when allowing the cosmic rays to pass through magnetized iron. The results given above show that only the softer 30 per cent of the cosmic rays could be expected to show an effect. This is the portion of the beam most easily absorbed or scattered. The scattering of particles in this portion of the beam may be sufficient, when passing through several centimeters of iron, to prevent any of these particles from producing a systematic coincidence.

It is of interest to compare the results obtained here with some of the more recent investigations which offer a basis of comparsion. The remarkable photographs obtained by Anderson [17] of cosmic ray particles in a Wilson chamber, in the presence of a magnetic field, exhibit particles showing a wide range of magnetic deviations. From the analysis of these photographs which he gives it is seen that of a total of 64 observed particles, 58 have energies over 10×10^6 electron volts and could be expected to produce coincidences through 7 mm of steel. Only 13 of these 58 particles have energies greater than 400×10^6 electron volts and might, therefore, be considered to form a part of the undeflected beam in the present experiments. These particles are of the energy range where accurate measurements of the Wilson photographs are difficult and, therefore, there may be considerable error in this estimate. On this basis only about 25 per cent of the cosmic-ray beam should have been undeflected in the present experiments as opposed to about 70 per cent observed. Even allowing for the low precision of the observations of coincidences, this discrepancy appears to be genuine, and points to the existence of high-energy cosmic-ray particles of an ionizing power sufficient to actuate a tube counter but difficult to distinguish in a Wilson photograph. This assumption is also supported to some extent by the experiments of

[14] Johnson, T. H., Phys. Rev., vol. 40, p. 468, 1932.
[15] See footnote 6, p. 816.
[16] See footnote 7, p. 816.
[17] Anderson, C. D., Phys. Rev., vol. 41, p. 405, 1932.

Johnson, Fleisher, and Street,[18] in which Wilson photographs were taken only when coincidences occurred between the two tube counters. Although, in the absence of exact information regarding geometrical arrangements and number of possible accidental coincidences, no definite conclusion can be drawn, it is significant that a considerable number of photographs showed no tracks.

V. ACKNOWLEDGMENTS

The writer wishes to thank B. W. Brown and L. L. Stockman for help in constructing the apparatus and recording the observations. He is also indebted to Dr. A. V. Astin for helpful discussions of various phases of the experiment.

WASHINGTON, October 12, 1932.

---

[18] Johnson, T. H., Fleisher, W., and Street, J. C., Phys. Rev. vol. 40, p. 1048, 1932.

THE SYSTEM: CaO-B₂O₃

By Elmer T. Carlson

ABSTRACT

The phase equilibrium diagram for the system $CaO-B_2O_3$ has been partially worked out by means of heating curves. Completion of the diagram was impossible because the melting point of B_2O_3 could not be determined, while that of CaO was beyond the range of the apparatus. The following compounds were identified and their optical properties determined with the petrographic microscope: $CaO \cdot 2B_2O_3$, melting at $986° \pm 5°$ C.; $CaO \cdot B_2O_3$, melting at $1,154° \pm 5°$ C.; $2CaO \cdot B_2O_3$, melting at $1,298° \pm 5°$ C.; and $3CaO \cdot B_2O_3$, melting at $1,479° \pm 5°$ C. Mixtures containing less than 23 per cent CaO were found to separate, on fusion, into two liquid layers, one of which contained 23 per cent CaO, while the other was nearly pure B_2O_3. This immiscibility region extended above 1,500° C.

CONTENTS

I. INTRODUCTION

The study of this system was undertaken as a preliminary to a study of part of the ternary system, $CaO-B_2O_3-SiO_2$, the purpose of the latter being to determine the effect of the addition of small amounts of B_2O_3 to the calcium silicates found in Portland cement. It is hoped, however, that the results of the investigation of the binary system, $CaO-B_2O_3$, may also prove of interest in connection with the production of glazes and other ceramic materials.

Very little work has been done on the anhydrous calcium borates in recent years, and much of the earlier data is conflicting or inconclusive. The first comprehensive study of the system $CaO-B_2O_3$ was made by W. Guertler,[1] who constructed the liquidus diagram through the use of cooling curves. He found maxima at 1,030°,[2] 1,095°, and 1,225°, corresponding to the compounds $CaO \cdot 2B_2O_3$, $CaO \cdot B_2O_3$, and $2CaO \cdot B_2O_3$, respectively. Due to the limitations of his apparatus, he was unable to go much beyond the last of these, although he argued from analogy with the system $BaO-B_2O_3$ that the compound $3CaO \cdot B_2O_3$ should exist. Due to the excessive super-

---

[1] Guertler, On the Melting Points of Mixtures of the Alkaline Earths with Boric Anhydride, Z. anorg. Chem., vol. 40, pp. 337-354, 1904.
[2] All temperatures given in °C. unless otherwise stated.

825

cooling which occurs in this system, the accuracy of his results is somewhat doubtful.

The four borates mentioned have since been prepared by R. Griveau,[3] who also studied their heats of formation. He gives the values 1,025°, 1,100°, and 1,215° as the approximate melting points of the compounds $CaO \cdot 2 B_2O_3$, $CaO \cdot B_2O_3$, and $2CaO \cdot B_2O_3$, respectively. Roberts[4] reported a melting point of 1,304° for $2CaO \cdot B_2O_3$. Optical properties for the metaborate, $CaO \cdot B_2O_3$, are given by A. N. Winchell.[5]

Burgess and Holt[6] reported that CaO is insoluble in B_2O_3 in small quantities, but if present in larger amounts the mixture fuses to a clear glass. Guertler[7] found that a fused mixture of B_2O_3 and a small amount of CaO separated into two liquid layers which were immiscible even at 1,400°.

The melting point of pure CaO was placed at 2,572° by Kanolt,[8] and at 2,849° K. (2,576° C.) by Schumacher.[9] The melting point of boric oxide is unknown, as this compound has never been obtained in crystalline form.

Because of the absence of optical data for most of the calcium borates, as well as the conflicting nature of the melting-point data, it seemed desirable to make a study of the system CaO-B_2O_3, tracing the liquidus as far as possible and determining the optical properties of the various phases appearing in the system.

The general plan followed was to prepare mixtures of CaO and B_2O_3 in varying proportions and to determine their melting points by means of heating curves. The phases existing in equilibrium with the melt at the liquidus were identified by petrographic examination of samples quenched from various temperatures near the liquidus.

II. EXPERIMENTAL WORK

1. PREPARATION OF MIXTURES

Calcium carbonate ($CaCO_3$) and boric acid (H_3BO_3) were used in the preparation of the mixtures. The percentage composition of the calcium carbonate was as follows: Ignition loss, 43.65; SiO_2, 0.07; R_2O_3, 0.04; MgO, 0.04; CaO (by difference), 56.20. The boric acid was analyzed only for impurities (analysis by chemistry division of this bureau), which were reported as follows: Fe, < 0.001 per cent; SO_3, < 0.01 per cent; Cl, < 0.001 per cent. This material was recrystallized in order to render it finer, as it was difficult to pulverize the flakes by grinding.

For each mixture the calculated amount of calcium carbonate was first calcined and then thoroughly mixed with the required quantity of boric acid and heated over a burner (Meker type). The materials combined readily to form a mixture sufficiently homogeneous for the

[3] Griveau, Heats of Formation of the Anhydrous Calcium Borates, Compt. rend., vol. 166, pp. 993–995, 1918.
[4] Roberts, Some New Standard Melting Points at High Temperatures, Phys. Rev., vol. 23, pp. 386–395, 1924.
[5] A. N. Winchell, The Optic and Microscopic Characters of Artificial Minerals. Univ. of Wisconsin Studies in Science, 1924.
[6] Burgess and Holt, The Behavior of Metallic Oxides Toward Fused Boric Anhydride, Proc. Chem. Soc., vol. 19, pp. 221–222, 1903.
[7] Guertler, On the Limits of Miscibility of Boric Anhydride and Borates in the Fused State, Z. anorg. Chem., vol. 40, pp. 225–53, 1904.
[8] Kanolt, Melting Points of Some Refractory Oxides, B. S. Bull., vol. 10, pp. 295–313, 1914.
[9] Schumacher, Melting Points of Barium, Strontium, and Calcium Oxides, J. Am. Chem. Soc., vol. 48, pp. 396–405, 1926.

purpose for which it was intended. The product varied in appearance from a glass to a soft clinker, depending on the compostion. In some instances a test for free lime was made by the ammonium acetate titration method,[10] but no free lime was found except in one mixture containing CaO in excess of $3CaO \cdot B_2O_3$.

2. ANALYSIS OF MIXTURES

Due to the loss of boric acid by volatilization during the preparation of the mixtures, it was necessary to analyze the product in each case.

CaO was determined by titration with half-normal hydrocbloric acid, using methyl red as indicator. The acid also served as a solvent for the sample, as the calcium borates are only slightly soluble in water. The B_2O_3 was then determined on the same sample by adding mannite and titrating with half-normal sodium hydroxide to the end point indicated by phenolphthalein.

In general, it was found that mixtures containing an excess of either uncombined lime or boric oxide were hygroscopic, while the others showed low ignition losses. The moisture present was of no significance, since it was expelled when the samples were heated in the furnace. Consequently the percentages of CaO and B_2O_3 were in all cases adjusted so as to total 100 per cent on a nonvolatile basis.

Although free boric oxide is somewhat volatile at high temperatures, it was found by analysis that no appreciable loss of B_2O_3 occurred when the mixtures, prepared as described above, were subsequently used in obtaining heating curves.

3. EXPERIMENTAL PROCEDURE

The melting points of the various mixtures were determined by means of heating curves. For this purpose a platinum-wound, vertical-tube furnace, equipped with automatic temperature control,[11] was used. Temperatures were measured by means of platinum:platinum-rhodium thermocouples.

In obtaining a heating curve the sample was placed in one compartment of a small platinum thimble which was divided vertically by a platinum septum welded in place. The other half of the thimble was filled with artificial corundum as a reference medium. The two hot junctions of a differential thermocouple were imbedded in the two materials and the thimble suspended in the furnace. The temperature was raised at a rate of 5° per minute, and readings of the actual and differential temperatures were taken every minute, or, when necessary, every half minute.

Due to pronounced supercooling, no satisfactory cooling curves could be obtained.

In order to determine the solid phase present at the liquidus, a small sample of each mixture was wrapped in platinum foil, suspended in the furnace, and held at a constant temperature near the melting point for 15 minutes to allow it to come to equilibrium. It was then quenched by dropping it into water or mercury, and the phases present identified with the aid of the microscope. This process was then repeated at a temperature 2° or 3° higher or lower until a point was found, above which the primary solid phase did not appear.

[10] Lerch and Bogue, Revised Procedure for the Determination of Uncombined Lime in Portland Cement, Ind. Eng. Chem., Anal. ed., vol. 2, pp. 296-298, 1930.
[11] Adams, J. Opt. Soc. Am., vol. 9, p. 509, 1924.

Unfortunately, mixtures containing more than about 45 per cent CaO were found to crystallize during quenching, thus limiting the use of this method.

4. INTERPRETATION OF HEATING CURVES

A few words regarding the interpretation of the heating curves may be helpful.

The curves fall into two general classes. In the first class are those which resemble the typical heating curve for a pure compound or a

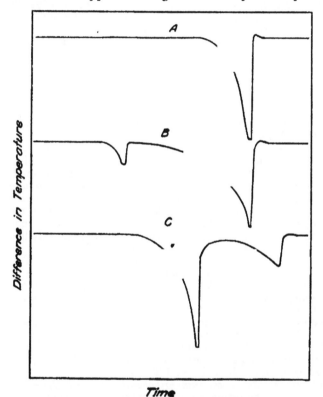

FIGURE 1.—*Types of differential heating curves*

eutectic mixture (curve *A*, fig. 1). Due to the temperature gradient in the sample, the outer edges fuse first, gradually reducing the amount of heat conducted to the thermocouple, which causes the curve of the differential temperature to dip downward. As the melting proceeds toward the center, the curve drops more and more sharply, reaching a minimum at the instant at which fusion is complete. Recovery is then very rapid, resulting in an almost vertical rise in the curve. The point at which the curve reaches the minimum was taken as the melting point of the sample. Heating curves for a pure reference material, potassium sulphate (m. p. = 1,069° [12]), confirmed this interpretation.

---

[12] See footnote 4, p. 826.

Mixtures lying between a eutectic and a compound give curves of the second type, represented by B (fig. 1). The first break represents the melting point of the eutectic; that is, the last portion of the mixture to solidify on cooling. Continued heating causes a gradual fusion (or solution) of the solid phase until it has completely disappeared. This is illustrated by the gradual drop in the curve, culminating as before in a sudden turn upward.

Curve C is of the same type as curve B, but in this case the composition is closer to that of the eutectic, causing the first break to be very pronounced, whereas the upper break, representing the true melting point, is relatively faint. For several of the mixtures whose compositions were close to those of the various eutectics the upper break was obscured entirely, making an absolute determination of the composition of the eutectic by this method impossible.

III. RESULTS

1. BINARY COMPOUNDS

Four binary compounds were found in the system: $CaO \cdot 2B_2O_3$, $CaO \cdot B_2O_3$, $2CaO \cdot B_2O_3$, and $3CaO \cdot B_2O_3$.

Calcium diborate, or calcium tetraborate, $CaO \cdot 2B_2O_3$.—This compound occurred as the primary phase in all the mixtures containing up to about 28 per cent CaO. As ordinarily prepared it remained in the form of a glass; but by heating for a few minutes at a temperature slightly below the melting point, this could be devitrified readily. It appears as irregular grains, highly birefringent, uniaxial, negative; $\omega = 1.638$, $\epsilon = 1.568$. Whether or not it melts congruently could not be determined with certainty, as its composition apparently lies at the break in the liquidus. The melting point is $986° \pm 5°$.

Monocalcium borate or calcium metaborate, $CaO \cdot B_2O_3$.—This compound occurs as the primary phase in mixtures containing approximately 29 to 51 per cent CaO. It appears as long, flat plates, highly birefringent, biaxial, negative, with a large optic axial angle; $\alpha = 1.550$, $\beta = 1.660$, $\gamma = 1.680$. It melts congruently at $1,154° \pm 5°$.

Dicalcium borate or calcium pyroborate, $2CaO \cdot B_2O_3$.—This compound occurs as the primary phase in mixtures containing approximately 51 to 64 per cent CaO. It appears as irregular grains, highly birefringent, biaxial, negative, with small optic axial angle; $\alpha = 1.585$, $\beta = 1.662$, $\gamma = 1.667$. It melts at $1,298° \pm 5°$.

Tricalcium borate or calcium orthoborate, $3CaO \cdot B_2O_3$.—This compound occurs as the primary phase in mixtures containing from 64 to at least 71 per cent CaO, and probably somewhat higher. It appears as irregular grains, highly birefringent, uniaxial, negative; $\omega = 1.728$, $\epsilon = 1.630$. It melts at $1,479° \pm 5°$.

2. THE PHASE EQUILIBRIUM DIAGRAM

The phase equilibrium diagram is shown in Figure 2.

Mixtures containing less than 24 per cent CaO were found to separate into two immiscible liquids (which for convenience may be designated as A and B) on fusion. The heavier of the two liquids (A) on cooling formed a clear glass with a refractive index of about 1.56. It was nearly insoluble in water but soluble in hydrochloric acid. The other liquid (B) formed a cloudy glass, which was seen under the microscope to be full of minute globules, presumably of the other glass. These appeared to separate during cooling, since in quenched samples both glasses were clear. The index of the lighter glass was about 1.45. It was soluble in water, thus affording a simple method of separating it from the other glass for analysis. The water-soluble glass was found to be practically pure B_2O_3, while the other contained about 23 per cent CaO.

An attempt was made to determine the temperature above which the components would be completely miscible, but little evidence of a change in composition was found up to 1500°. For this experiment a small amount of one of the mixtures was placed in a platinum thimble and held at a definite temperature long enough to permit

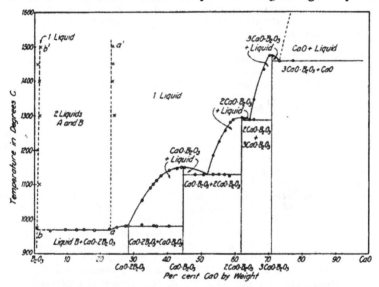

FIGURE 2.—*Phase equilibrium diagram for the system* $CaO–B_2O_3$

the liquids to separate into two layers. The charge was then cooled quickly, the two glasses separated by solution of the upper one in water, and the ratio of CaO to B_2O_3, determined in each. The results are given in Table 1.

TABLE 1.—*Composition of the immiscible liquids in the system* $CaO–B_2O_3$, *at various temperatures*

Temperature in °C.	Composition of glass A		Composition of glass B		Temperature in °C.	Composition of glass A		Composition of glass B	
	CaO	B_2O_3	CaO	B_2O_3		CaO	B_2O_3	CaO	B_2O_3
	Per cent	*Per cent*	*Per cent*	*Per cent*		*Per cent*	*Per cent*	*Per cent*	*Per cent*
900–950	23.0	77.0	0.6	99.4	1,200			.9	99.1
	23.1	76.9	.6	99.4	1,300	24.2	75.8	1.5	98.5
	24.9	75.1	.2	99.8	1,400	23.5	76.5	1.4	98.6
975			.7	99.3	1,450	23.9	76.1	.9	99.1
1,100			1.5	98.5	1,500	23.7	76.3	1.7	98.3

It will be seen that the data are not entirely consistent, but this is not surprising considering the slowness of attainment of equilibrium, the possibility of p r icles of one glass remaining dispersed throughout the other, and the rather inexact method of separation. In considering the values for the composition of glass *A* it should be borne in mind that this glass represents the lower CaO limit of stable liquid mixtures in this system. Hence it seems reasonable to take the minimum value obtained, 23.0 per cent CaO and 77.0 per cent B_2O_3, as the most reliable figure for the composition of this glass.

The data seem to indicate a slight increase in the CaO content of glass *B* with increasing temperature, but this is not definitely proved. The low CaO values are more reliable because of the probability that glass *B* was contaminated with glass *A*. This supposition is supported by the observation that glass *B* was always cloudy, due to the suspension in it of minute droplets of the other phase.

It is probable that the compositions of these two liquids approach each other, and that there is some temperature above which only one liquid phase can exist. Data obtained at 1,600° showed a marked tendency in this direction, but they can not be considered reliable, as the problem is complicated by the high volatility of B_2O_3 at this temperature, as well as by the failure of the two glasses to separate completely.

The region of liquid immiscibility up to 1,500° is represented approximately by the area between the dotted lines *a a'* and *b b'* in Figure 2. Any mixture in this area must exist as two liquid phases. On cooling, it will solidify to a mixture of two glasses; but if held for a few minutes at a temperature slightly below 970°, the compound $CaO \cdot 2B_2O_3$ starts to crystallize out from liquid *A*. As this will tend to increase the proportion of B_2O_3 remaining in *A*, it is evident that a small amount of liquid *B* must also separate out, and this will continue until the mixture consists entirely of the compound and liquid *B*, the latter solidifying to a glass on cooling. A heating curve of the partially crystallized mixture will show a break at about 971° corresponding to the melting of $CaO \cdot 2B_2O_3$ in the presence of B_2O_3 glass. The melting oin for any mixture in this region is represented by the horizontablinetin Figure 2, extending from 0.2 to 23 per cent CaO.

From *a* the liquidus rises gradually to 986°, the melting point of the diborate, $CaO \cdot 2B_2O_3$ (composition, CaO, 28.71 per cent; B_2O_3, 71.29 per cent). The curve near this point is nearly horizontal, and it could not be determined with certainty whether the compound melts congruently or not. A mixture consisting of 28.6 per cent CaO and 71.4 per cent B_2O_3 showed a trace of monocalcium borate as the primary phase. This would indicate that calcium diborate is unstable at the melting point. However, the composition of the mixture in question was so close to that of $CaO \cdot 2B_2O_3$ that if allowance is made for a slight error in chemical analysis the above conclusion is unwarranted.

From the melting point of $CaO \cdot 2B_2O_3$, the liquidus rises sharply, reaches a maximum of 1,154° at the composition $CaO \cdot B_2O_3$ (CaO, 44.61 per cent; B_2O_3, 55.39 per cent), and descends to 1,132°, the melting point of a eutectic containing approximately 51.5 per cent CaO and 48.5 per cent B_2O_3. As previously mentioned, the true melting points near the eutectic could not be determined from the heating curves. The two points to the left of the eutectic in the diagram, therefore, represent the melting point of the eutectic itself, rather than those of the two mixtures.

The curve then rises to another maximum of 1,298°, corresponding to the melting point of $2CaO \cdot B_2O_3$ (CaO, 61.70 per cent; B_2O_3, 38.30 per cent), and drops to 1,291°, the melting point of a second eutectic containing approximately 64 per cent CaO. A maximum of 1,479° corresponds to the melting point of $3CaO \cdot B_2O_3$ (CaO, 70.73 per cent; B_2O_3, 29.27 per cent), which is followed by a drop in the curve to another eutectic melting at 1,460° and consisting of about 73 per cent CaO and 27 per cent B_2O_3. Heating curves on two mixtures still higher in lime indicated the melting of this eutectic, but showed no further breaks up to 1,600°, the maximum temperature attained. Petrographic examination of these samples after fusion showed that they consisted of two phases, $3CaO \cdot B_2O_3$ and CaO. Hence it is assumed that no compounds are formed containing more CaO than the orthoborate, $3CaO \cdot B_2O_3$, and that the liquidus rises continuously to the melting point of CaO at 2,572°.

IV. SUMMARY

The phase equilibrium diagram for the system $CaO \cdot B_2O_3$ has been partially worked out. A region of liquid immiscibility was found, extending from 0.2 to 23 per cent CaO, and from 971° to above 1,500° C. The optic l properties of the four compounds in this system were determined. The melting points of these compounds are given in Table 2, together with the melting points determined by previous investigators.

TABLE 2.—*Melting points of the calcium borates*

Compound	Melting point, determined by—			
	Guertler (1904)	Griveau (1918)	Roberts (1924)	Carlson
	°C.	°C.	°C.	°C.
$CaO \cdot 2B_2O_3$	1,030	ca. 1,025		986±5
$CaO \cdot 2B_2O_3$	1,095	ca. 1,100		1,154±5
$2CaO \cdot B_2O_3$	1,225	ca. 1,215	1,304±5	1,298±5
$3CaO \cdot B_2O_3$				1,479±5

WASHINGTON, October 10, 1932.

A METAL-CONNECTED GLASS ELECTRODE

By M. R. Thompson

ABSTRACT

Glass electrodes have come into extensive use for the measurement of hydrogen-ion activity in special cases in which other kinds of electrodes are unsuitable. Previous methods of construction have produced glass electrodes which were fragile, had an uncertain length of life, and required connecting in circuit with two standard electrodes.

In the work reported in this paper a glass electrode was developed which has a comparatively thick wall and a direct metallic connection to the glass. This makes a very durable apparatus and eliminates one standard electrode from the circuit. It is shown that the accuracy of such a metal-connected glass electrode is comparable with that of other electrodes used in pH measurement.

CONTENTS

I. INTRODUCTION

The glass "electrode," or, more strictly speaking, the glass half-cell, has come into extensive use recently to measure hydrogen-ion concentration in certain kinds of solutions for which other types of electrodes are unsuitable. Its advantages include freedom from disturbance by oxidation-reduction potentials and obviation of added reagents, such as gaseous hydrogen or solid quinhydrone. Its disadvantages are the inherent high resistance (at present greater than 1 megohm), requiring the use of delicate or special apparatus for detecting potential changes; and the fragility, which makes the useful life short or uncertain. The purpose of this investigation was to remove or reduce these disadvantages.

II. HISTORICAL

The discovery of a reaction between glass and hydrogen ions goes back at least to Cremer (1).[1] Haber and Klemensiewicz (5), who acknowledged Cremer's work, appear to have been the first to actually apply the principle to laboratory measurements. Their systematic and extensive treatise is so well known that they are usually credited with the development of the glass electrode in something like its present form. As often happens, over a decade elapsed before further notice appeared, and about 20 years passed before general interest in the subject was aroused.

Numerous theories have been proposed to account for this peculiar action of glass. These theories may be roughly classified into three groups, based on phase boundary, ion adsorption (or exchange), and liquid junction (or diffusion) potentials. All of these seem to imply that the glass contains or attracts water and hydrogen ions (protons), and some of them imply that only hydrogen ions can penetrate into or through the glass. The situation is complicated by the fact that glass is not unique in this respect. Thus, similar effects have been detected at interfaces of water with quartz (41), paraffin, zeolite minerals (19, p. 252), and benzene (5). Borelius (8, p. 447) summarize concisely three possible conditions for electrode action. Much information on the theoretical aspects will be found in recent papers by MacInnes and Belcher (34), and by Dole (35) (44).

The composition of the glass used is important and has been studied by Hughes (23), MacInnes and Dole (27) (31), Elder (29), and others. Most investigators agree that a soft glass containing about 22 per cent of Na_2O, 6 per cent of CaO, and 72 per cent of SiO_2[2] gives the best results. Means to effect further improvement in the electrode action of glass are not now apparent, but it is conceivable that the abnormalities shown in very acid and alkaline solutions might be decreased by the addition of other substances to the glass.

Small bulbs with thin walls have been used by most investigators, including Haber and Klemensiewicz (5), Hughes (9) (23), Elder and Wright (25), Voegtlin, De Eds and Kahler (32), Robertson (36), and others. Kerridge (20) modified this construction by blowing a local reentrant thin area (0.025 to 0.030 mm thick) in the bulb wall to serve as the active surface. Horovitz (15, p. 345) sealed a relatively thick (0.05 to 0.1 mm) plate of glass on the end of a glass tube. MacInnes and Dole (27) used an electrode of similar shape, but with an extremely thin (0.001 mm) active surface, about 4 mm in diameter, obtained from the wall of a glass bubble. This might be called a "membrane" electrode. The forms described may be conveniently classified into the two general types of bulb and membrane, although there appears to be no essential difference in their action. Figure 1 shows several types of electrodes and methods of connection.

The characteristic arrangement for these active glass surfaces has been with a solution on each side of the bulb or membrane wall. One solution is of known or fixed hydrogen ion activity, while the other is of unknown or variable activity. This in effect makes a hydrogen

---

[1] Figures in parenthesis here and throughout the text refer to the numbers used in the bibliography at the end of this paper.

[2] Commercial glass of this composition is obtainable, for instance, No. 015 glass of the Corning Glass Co., Corning, N. Y. Tubing with an outside diameter of 7 mm, and 1 mm wall is convenient for blowing membrane electrodes.

electrode on each side of the glass. A standard electrode of some sort is immersed in each solution to complete the circuit. These standard electrodes need not be identical but, if so, their potentials are opposed and cancel each other in the measurement. It is important to note that in this arrangement only electrolytic contact is made with the glass.

Maximum sensitivity is associated with minimum resistance, and consequently the tendency has been to make the glass as thin as possible. For small electrodes, the practical lower limit of resistance thus reached lies between 2 and 50 megohms (27) (34) (36), as determined by direct-current measurement (27) (34). The best electrodes thus become rather difficult to construct, excessively fragile, and of uncertain life even with careful handling.

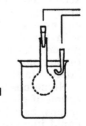

Where extreme compactness is not required, more durable, although not necessarily as sensitive, electrodes of the same resistance could obviously be prepared by using a thicker wall with a larger area. The bulb type is better suited to this modification than the membrane type, as it is difficult to properly seal a large area of membrane. Hazel and Sorum (39), however, have recently increased the membrane diameter to about 10 mm. On the other hand, the results obtained by Robertson (36) indicate that it is possible to form a comparatively small bulb having a resistance as low as that of a membrane.

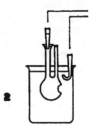

The situation described limits the widespread use of glass electrodes, especially in industrial work, where rugged and permanent apparatus is desirable.

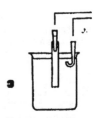

III. THE METAL-CONNECTED GLASS ELECTRODE

1. PRINCIPLE

In connection with the mechanical advantages of thick-walled electrodes, it appeared possible to greatly simplify the mounting and eliminate the extra (inside) standard electrode by substituting a direct metallic connection to the glass wall. With glass of sufficient thickness to withstand ordinary handling, for instance, 0.1 to 0.5 mm, such a metallic connection could be easily applied and would furnish additional protection to the glass. This was tried (March 25, 1930) by wrapping tinfoil around a soft-glass tube, and the results were promising enough to warrant further trial and development, as reported in this paper.

FIGURE 1.—*Forms of the ordinary glass electrode*

(The active surface portion is shown in dotted line. Each glass electrode is connected to two calomel electrodes, of which only the tips are shown.) 1, plain bulb; 2, reentrant bulb; 3, membrane

No publication has been noted which describes the use of a metal to glass connection for the systematic accurate measurement of hydrogen-ion concentration. Previous investigators have doubtless made incidental trials of a simple metallic connection, such as mercury within a glass bulb, analogous to the mercury cups so often used in making electrical contact with sealed-in wires on glass apparatus. B. von Lengyel (41, pp. 426 and 431) tried a mercury to.quartz connection, but found it to be unsatisfactory. He also used a Wood's metal to quartz connection for checking an electrode chain. Haber and Klemensiewicz (5, p. 424) used a connection, not with glass, but with a layer of benzene, by means of an immersed disk of amalgamated brass. Many experiments have also been published describing electrolysis through glass, or determinations of the conductivity of glass, in which electrodes of metal, amalgam, or powdered graphite were applied.

The exact function of the metal coating on a "metal-connected" glass electrode has not yet been determined. If, as is very generally believed, conduction through glass is entirely electrolytic; that is, due to the movement of ions, there may be set up a metal-metal oxide or metal-metal silicate electrode with a comparatively steady potential. If the passage of current through the glass results from electronic—that is, solid conduction—the metal serves, of course, merely as a convenient terminal. Besides being a conductor (or nonconductor according to the point of view) glass is also a dielectric; that is, it transfers charges by induction. The apparatus used for detecting a change of potential in the circuit is capable of measuring static charges. This point will be mentioned subsequently in connection with the details of construction. It should be emphasized, however, that further work will be required to furnish a satisfactory explanation of the complicated phenomena involved.

2. CONSTRUCTION

There are obviously many ways of arranging a metal-glass electrode. Thus, wires, rods, sheets, and tubes of suitable metals might be coated with a film of the proper glass; or, conversely, tubes, beakers, flasks, and other vessels of the proper glass might be coated with a film of an appropriate metal. The latter might be applied by pressure, spraying, casting, sputtering, or chemical precipitation from a solution or a gas, and subsequently thickened and protected by electroplating. Various possible arrangements are shown in Figure 2.

The metal-connected glass electrodes first employed in this investigation were constructed from various stocks of laboratory tubing, using tinfoil on those containing the solution to be tested, and mercury or coiled wire inside those dipping into the solution. Subsequent work showed that glass having the composition ordinarily used for membranes also gave the best results for metal-connected electrodes, and that silvering followed by copper plating gave a more durable and satisfactory coating than tinfoil. Electrodes of the dipping type were found to be less accurate than those of the containing type. Further investigation is needed, but this difference may be the result of greater dehydration of the outer surface during the working of the hot glass.

For the final exploratory work, three containing and three dipping type electrodes were prepared from Corning No. 015 glass tubing by blowing it into the shape of test tubes with walls about 0.5 mm thick. These tubes had an outside diameter of 1.2 cm and a length of 12 cm, and contained (or displaced) about 10 ml of solution. Silvering was applied on either the inside or outside surface by the Rochelle salts process (37), and copper [3] was then deposited to a thickness of about 0.008 mm. The glass tubes were coated with paraffin for about 1 cm on the outside at the open end to diminish surface leakage. A copper

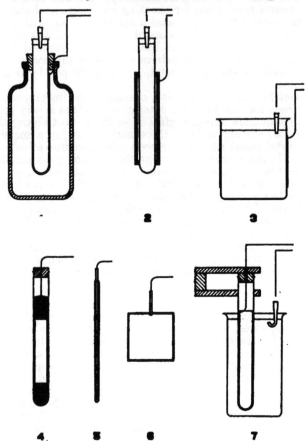

FIGURE 2.—*Arrangements of the metal-connected glass electrode*

(Only the tips of calomel electrodes are shown.) 1, Containing type; tube form, metal coated; 2, containing type; tube form, with metal portion insulated from the glass by means of mica wedges at top and bottom, and an air gap. For special study; 3, containing type; beaker form, metal coated; 4, dipping type; tube form, mercury filled; 5, dipping type; metal rod or wire form, coated with glass; 6, dipping type; metal sheet form, coated with glass; 7, dipping type; tube form, metal coated and supported by bakelite clamp.

wire was wrapped or pressed against the metal coating and the completed electrode was mounted in a suitable support. This consisted

---

[3] Plating on the outside was done with a solution containing 100 g/l of $CuSO_4.5H_2O$. The coating was applied by "flashing" at a current density of about 5.5 amp/dm² for a few minutes, after which plating was continued at 0.6 amp/dm² for one-half hour. Plating on the inside was found difficult on account of the small bore and consequent poor circulation. Satisfactory results were finally obtained by using a more concentrated solution containing 250 g/l of $CuSO_4.5H_2O$, and an accurately centered copper wire as anode. Because of the necessarily excessive anode current density, the 110-volt circuit with a rheostat in series was required to overcome anode polarization. A flash of copper was applied at about 5 amp/dm² for one minute, after which plating was continued for one and one-half hours at 0.36 amp/dm². All plating was done at room temperature. The copper coating was lacquered, except at the pressure contact with the connecting wire.

of a rubber-stoppered glass bottle for a containing electrode and a bakelite cl m on a stand for a dipping electrode. (See fig. 2, Nos. 1 and 7.) a p

Rough measurements of the resistance of these electrodes, in contact with 0.1 *N* HCl into which a copper wire was dipped, were made using a high-resistance bridge and direct current at 1,000 volts.[4] The capacity [5] at 1,000 cycles was also determined, by means of a capacity bridge. The data on resistance and capacity are given in Table 1. Because of their greater accuracy, the results presented subsequently in this paper were obtained by means of electrodes of the containing type.

TABLE 1.—*Resistance and capacity of metal-connected glass electrodes*

Electrode No.	Type	Resistance		Capacity	
		Initial	After 8 months	Initial	After 8 months
		megohm	*megohm*	*μμf*	*μμf*
1 [1]	Containing	43	53	590	625
2	do	37	39	660	695
3	do	48	55	535	585
4	Dipping	57	58	460	505
5	do	53	48	530	525
6	do	53	60	510	485

[1] The resistance of this electrode when previously coated with tinfoil was 70 megohms.

In actual use, the electrode was rinsed once with the solution to be measured and was then filled, and the potential reading obtained at once. Then the electrode was rinsed with distilled water and kept filled with distilled water until the next measurement. When working with solutions difficult to remove by rinsing, it was necessary to clean the electrodes with chromic acid solution, or to immerse them in water for a long time. Recovery from ordinary disturbances was usually complete within one hour and almost always on standing over night.

The special electrode shown in Figure 2, No. 2, was constructed with the metal side insulated from the glass, mainly by an air gap. The resistance was thus raised to 5,000 megohms, with a corresponding reduction in capacity to 45 *μμf*. Readings on buffer solutions were unsteady, but an accuracy of about ±0.5 pH was obtained even under these very unpromising conditions. This experiment indicated that electrode action does not require actual contact between

[4] This scheme was used merely for convenience and is not suggested for general use. The application of a high voltage tends to cause the passage of appreciable current, with the possibility of polarization and disturbance to readings. These electrodes were affected temporarily, as shown at *A* in Figure 7; but after standing in contact with water for a few weeks, they returned to the original condition.

It has been pointed out by MacInnes and Belcher (34) that resistance measurements on glass electrodes made with direct current are conventional and that alternating current gives much lower and more accurate results. They concluded, however, that: "It does not appear desirable to change the current practice of reporting the apparent direct-current resistances of electrodes in megohms."

In measurements using direct current, MacInnes and Dole (27), (31) applied much lower voltages, either 3.7 or 1.2 volts; MacInnes and Belcher (34) applied less than 1 volt. It may be pointed out that the surface ratio of a metal-connected electrode to a membrane electrode, each of the usual size, is about 3,300 to 12 mm². The application of 1,000 volts on the former produces the same current density as does 3.6 volts on the latter, if both have the same total resistance. As a matter of fact, the direct-current resistances obtained for the metal-connected glass electrodes are of the expected order of magnitude, when compared with the direct-current resistances of membrane electrodes as measured by the author, using the method of MacInnes and Dole (27) (31).

[5] Residual noise in the telephones indicated a wide phase angle. This is characteristic of low dielectric resistance, or of a "leaky" condenser.

metal and glass. Completion of the circuit may have been effected through surface leakage, or possibly by induction. In this connection, Borelius (6) (7) (8) has published many measurements including electrolytic potentials at glass or paraffin surfaces, where an air gap was used in the circuit.

IV. EXPERIMENTAL WORK AND CONCLUSIONS

1. APPARATUS AND GENERAL METHODS

Potential measurements were made with a "student type" potentiometer calibrated to ±0.5 mv. A Compton quadrant electrometer[*] was used as the indicating instrument in the glass electrode measurements, by the null-point method. The saturated potassium chloride-calomel reference electrode was employed. For standardization of buffer solutions and for comparative readings, other electrodes were used, including the hydrogen (with platinum black), the quinhydrone (with gold), and the glass membrane (with Corning No. 015 glass).

The glass-membrane electrodes were mounted or connected, not through an inside silver-silver chloride electrode as used by MacInnes and Dole (27) but through a saturated calomel electrode according to the system of Kerridge (20) and others. The electrode was filled with any suitable buffer solution, into which dipped the plugged tip of the calomel electrode. The plug consisted of agar jelly saturated with potassium chloride, and this device was used for its mechanical convenience. Such devices are, of course, possible sources of error by the introduction of an additional potential into the circuit. The error in this case appeared to be negligible within the limits of accuracy sought. An alternative procedure would involve the use of an extremely small bent tip, or else a ground-glass seal (20). The second or reference calomel electrode had a bent tip, according to the usual scheme. (See fig. 1, No. 3.) Silver-silver chloride electrodes are more compact, but somewhat less convenient to prepare and maintain than calomel electrodes.

A saturated calomel electrode, having a tip plugged with agar jelly saturated with potassium chloride, was also used for connection with the containing type of metal-connected glass electrodes. (See fig. 2, No. 1.)

Potential readings were made at room temperature, usually about 25° C. As the saturated calomel electrode was used for reference and most of the solutions were buffered, the accuracy of the measurements was about ±0.1 pH (±6 mv). This was sufficient for the exploratory purposes of this investigation. The readings, including those with the metal-connected glass electrode, were made, however, to the nearest 0.02 pH (±1 mv) and occasionally to 0.01 pH (±0.5 mv). The observations indicated that with closer control of temperature and a decreased wall thickness the accuracy with this electrode could probably be made equal to the above precision.

Switches in the circuit of the electrometer and glass electrode were made of paraffin blocks with mercury-filled copper cups as terminals.

[*] Where portability is required, the Lindemann electrometer may be substituted. Numerous circuits have been published in which the delicate and relatively expensive electrometer was replaced by vacuum tubes. A recent paper by Compton and Haring (47) described a compensated thermionic electrometer which is stated to have possibilities for use with thick glass electrodes. Robertson (36) reported the use of a galvanometer of high sensitivity with electrodes of very low resistance.

Fused sulphur would probably have been better than paraffin, but it is less easy to handle. The electrometer leads were insulated with mica and were run in grounded brass pipes, although bakelite insulation was used in a few places for convenience. The electrometer switches were placed in a grounded copper gauze cabinet, one compartment of which also served to hold and partially shield the glass-calomel

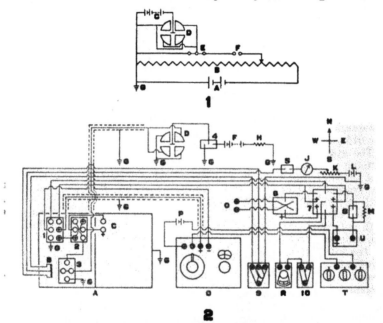

FIGURE 3.—*Wiring diagram of apparatus*

1. Simplified diagram for the potentiometer using an electrometer by the null-point method. *A*, Battery for potentiometer; *B*, potentiometer; *C*, battery for electrometer needle; *D*, electrometer; *E*, single pole double throw switch; *F*, binding posts for connecting glass electrode-standard electrode cell; *G*, ground. 2. Complete diagram (student type potentiometer). *A*, shielded cabinet holding special paraffin base switches and the glass electrode-standard electrode cell; *B*, electromagnets (1 ohm each) for tripping electrometer switch; *C*, terminals for connecting glass electrode-standard electrode cell; *D*, electrometer; *F*, battery for electrometer needle (it makes no difference which pole is connected to the needle); *G*, ground; *H*, protective resistance (0.1 megohm) for *F*: *J*, ammeter (1 ampere scale) for reading the electromagnet current; *K*, variable resistance (3 ohms) for controlling the electromagnet current; *L*, battery (3 volts) for supplying the electromagnet current; *M*, protective resistance (20,000 ohms) for standard cell; *O*, terminals for connecting hydrogen or quinhydrone electrodes; *P*, battery for potentiometer; *Q*, potentiometer (student type); *R*, galvanometer (desk type); *T*, variable resistance (1,000 ohms) for adjusting potentiometer; *U*, standard cell; *1*, paraffin base main D. P. D. T. switch; *2*, paraffin base reversing D. P. D. T. switch; *3*, paraffin base electrometer S. P. D. T. switch; *4*, porcelain base electrometer needle S. P. D. T. switch; *5*, porcelain base S. P. S. T. switch for *B* circuit; *6*, porcelain base D. P. D. T. reversing switch; *7*, porcelain base T. P. D. T. switch; *8*, porcelain base S. P. S. T. shorting switch; *9*, hard rubber base three point key; *10*, hard rubber base two point key. Switching directions, adjusting potentiometer, (*1*) north and (*7*) east. Set potentiometer by means of *U*, *R*, and (*10*). Using hydrogen or quinhydrone electrode, (*1*) north and (*7*) west. Hydrogen or quinhydrone electrode connected to *O* at north reads on potentiometer, using *R* and (*10*). (*6*) west for hydrogen electrode. (*6*) east for quinhydrone electrode and pH below 7.67. (*6*) west for quinhydrone electrode and pH above 7.67 (using saturated calomel electrode). Using glass electrode, (*1*) south, glass electrode reads on potentiometer, using the electrometer, close (*5*) to operate (*3*), use (*9*) for remote control of (*3*), set (*2*) as may be necessary.

cell. The main switch connecting the electrometer was tripped by an electromagnet operated by remote control from a key placed outside the cabinet. This device proved very convenient and eliminated any disturbance due to body capacity. The other paraffin switches needed only occasional shifting, which could be done manually without difficulty.

Electrometer disturbances could usually be traced to either too much or too little mercury, or to dirty mercury, in the switch cups. A quick, positive make or break is necessary and contact electromotive forces must be prevented. In view of occasional difficulties, however, the elimination of mercury is being considered, by means of a switch with solid contacts of the brush or spring type. It was preferable not to operate the electrometer at maximum sensitivity, on account of its instability. Best results were obtained with the particular instrument used by keeping the needle voltage as low as possible, about 70 volts usually being satisfactory. A scale deflection of 3 mm/mv at 1 m scale distance was then obtained by lift and tilt adjustment of the quadrants. This sensitivity was adequate for the present work, although a much higher sensitivity is possible and has been used at times by others (34).

It is occasionally stated that an electrometer can not be used in very humid weather. In these experiments, suitable measurements could be obtained up to 80 per cent humidity of the room atmosphere. This favorable result may have been due to special care in insulating and shielding the circuit. It was sometimes found advantageous to ground the glass electrode just before connecting it to the electrometer for a reading. This removed stray charges from the system.

The potentiometer circuit was so wired that either hydrogen, quinhydrone, or glass electrodes could be connected in the appropriate position and all necessary measurements could be obtained by means of permanently connected switches and keys. The wiring diagram is shown in Figure 3. When using a more accurate potentiometer, such as the "type K" which has more of the accessory apparatus located internally, certain changes would be required in the wiring (34, p. 3316). The equations applying to the respective electrodes when used at 25° C. in combination with a saturated calomel reference electrode are given in Table 2, in both an extended and a somewhat simplified form. The extended form shows the factors which control the first term on the right-hand side of the simplified equation used in the actual measurements. E is the measured cell potential, expressed in volts, from which the desired pH value is computed.

TABLE 2.—*Electrode equations at 25° C.*

Electrode	Extended equation	Simplified equation
Hydrogen (24, p. 672)................................	$pH = \dfrac{-E-0.2458}{0.05912}$	$pH = -4.16 - \dfrac{E}{0.05912}$
Quinhydrone (24, p. 672)........................	$pH = \dfrac{0.6992-E-0.2458}{0.5912}$	$pH = 7.57 - \dfrac{E}{0.05912}$
Glass (a), mounted with silver-silver chloride electrode and 0.1 N HCl (28).	$pH = \dfrac{0.3524-E-0.2458}{0.05912}$	$pH = 1.80 - \dfrac{E}{0.05912}$
Glass (b), mounted with saturated calomel electrode and 0.1 N HCl (24, p. 672).	$pH = \dfrac{-E+0.0636}{0.05912}$	$pH = 1.08 - \dfrac{E}{0.05912}$
Glass (c), mounted with saturated calomel electrode and 0.05 M acid potassium phthalate (20).	$pH = \dfrac{-E+0.2349}{0.05912}$	$pH = 3.97 - \dfrac{E}{0.05912}$
Glass (d), mounted with saturated calomel electrode and any buffer solution of pH = K.	$pH = \dfrac{-E-E_K}{0.05912}$	$pH = K - \dfrac{E}{0.05912}$
Glass (e), metal-connected; X (determined by calibration) is the pH of the solution for which $E = 0$.		$pH = X - \dfrac{E}{0.05912}$

In the equations of Table 2, liquid junction potentials and uncertainties in pH values have been disregarded, but in most cases these are too small to need consideration for the present purposes. Any appreciable potential due to the glass itself, variously known as the self, strain, or asymmetry potential, must be determined, however, and a corresponding correction applied to the over-all potential. The asymmetry potential is readily measured with the ordinary glass electrode by having the same kind of solution and auxiliary electrode on each side. (With the metal-connected glass electrode it is not possible to determine the asymmetry potential, which is included in the calibration.) With good electrodes, this potential is seldom more than a few millivolts at the start and tends to decrease to zero in a short time. It is less confusing to omit such a term from the above equations, with the understanding that it must be subtracted (algebraically) from E when necessary, as will be illustrated below.

The equations are given in algebraic form, and potential values must be inserted with proper signs. In general, the sign of any electrode is defined as that of the charge on the outside metal terminal. As the saturated calomel electrode is used here as the standard half cell for reference, the sign of the measured cell potential (E) is that of the other electrode with respect to the calomel, or the same as that on the outside terminal of the other electrode. A little consideration of the circuit conditions also shows that the sign of a glass bulb or membrane electrode is that of the inside connecting electrode (whether silver-silver chloride, calomel, etc.) in the solution the pH of which is kept constant. This is brought out in Figure 4, in which the relative number of + or − signs indicates conventionally the tendency for current to flow to the external circuit in the ordinary sense. With the metal-connected glass electrode, the sign is that of the terminal of the metal coating, wherever this is located; that is, whether or not the electrode contains, or is immersed in, the solution to be tested.

The following calculations are typical for a glass bulb or membrane electrode filled with 0.05 M acid potassium phthalate and mounted between two saturated calomel electrodes.

With 0.05 M acid potassium phthalate also placed on the outside—

$E_1 = -0.002$ volt = asymmetry potential (measured).
(Inner calomel negative to outer calomel. In other instances the sign might be opposite.)
With solution of unknown pH placed on the outside:
$E_2 = -0.126$ volt (measured).
(Inner calomel negative to outer calomel.)
$E = E_2 - E_1 = -0.126 - (-0.002) = -0.124$ volt.

$$pH = 3.97 - \frac{-0.124}{0.0591} = 3.97 + 2.10 = 6.07$$

As most publications on this subject omit any explanation of the signs, it is believed that the above brief treatment may be helpful. The conventional straight line graphs for the various equations listed in Table 2 are shown for illustrative purposes in Figure 5. Such a diagram is useful for bringing out the relations involved and for quickly plotting the position of an ordinary glass electrode when the conditions are changed. The following relations should be noted:

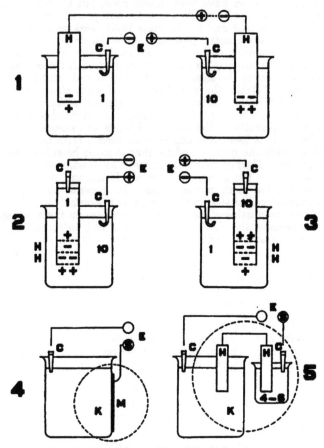

FIGURE 4.—*Sign of electrodes*

(Only the tips of calomel electrodes are shown). *1*, example of two hydrogen electrodes connected with two calomel electrodes. *H*=hydrogen electrode. *C*=calomel electrode. pH of left solution=1. pH of right solution=10. *E*=measured potential. The hydrogen electrode in the solution of lower pH is + to the other hydrogen electrode. The calomel electrode in the solution of higher pH is + to the other calomel electrode. Taking one calomel electrode as the standard of reference, in a solution of fixed pH, the sign of *E* is that of this electrode; *2*, example of a membrane glass electrode filled with a solution of relatively low and constant pH. The membrane is imagined to be divided into two hydrogen electrodes connected in series as in (*1*). *H*, *C*, and *E* as before. pH of solution within membrane electrode=1. pH of solution in beaker=10. The calomel electrode within the glass electrode is the standard of reference and is negative, which is taken as the sign of *E*. This results from the lower hydrogen electrode being negative to the upper hydrogen electrode, just as in (*1*) the right hydrogen electrode is negative to the left hydrogen electrode. The relation is similar if a silver-silver chloride electrode is substituted for the inside calomel electrode; *3*, this is like (*2*), with all conditions reversed. As before, the sign of *E* is that on the inside calomel electrode, or + in this case; *4*, example of a containing type metal-connected glass electrode. pH of the solution to be measured=*K*. The sign of *E* is that on the terminal *S*, which is connected to the metal, as is explained in (*5*); *5*, the area within the dotted circle represents in more detail the supposed conditions within the dotted circle of (*4*). The glass-metal section can be substituted by two hydrogen electrodes in series, as in (*1*). One hydrogen electrode is immersed in the solution of pH=*K* and the other hydrogen electrode is immersed in a solution of constant pH, in these experiments somewhere between 4 and 6. The circuit is completed by another calomel electrode in the latter solution. This electrode becomes the standard of reference, as in (*1*). The sign of *E* is that on terminal *S* as in (*4*)

(a) All lines are parallel and have the same (negative) slope of 0.05912 volt per unit pH at 25° C., corresponding to the denominator of the last right-hand term of the equations.

(b) Each line intersects the zero axis of abscissae (that is, each electrode reads zero to a saturated calomel reference electrode) at a pH value equal to the first right-hand term of its equation, or the "pH constant" for the electrode.

(c) When the electrode mounting or connection is symmetrical like that of Kerridge, the pH constant is equal to the pH of the inside

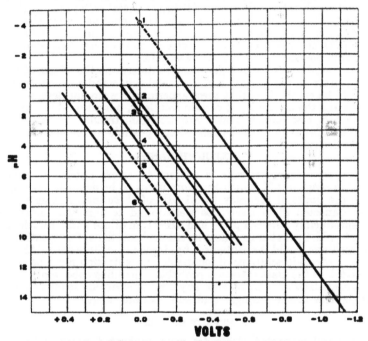

FIGURE 5.—*Standard graphs of electrode equations*

pH and potential in volts v. saturated calomel electrode at 25° C. (See Table 2.) *1*, Hydrogen electrode; *2*, glass electrode filled with 0.1 N HCl and connected with an inside saturated calomel electrode; *3*, glass electrode filled with 0.1 N HCl and connected with an inside silver-silver chloride electrode; *4*, glass electrode filled with 0.05 M acid potassium phthalate and connected with an inside saturated calomel electrode; *5*, metal-connected glass electrode. This line is shown dotted to indicate that the position is only approximate and must be determined by calibration; *6*, quinhydrone electrode

solution. Hence, a line can be immediately drawn in for a given value of pH for the inside buffer solution.

(d) The sign of the electrode reverses when the pH of the solution being measured passes the 'electrode constant, being positive for lower pH values and vice versa.

(e) An electrode (glass or other) will be positive to any other electrode placed in the same solution, if the first electrode line lies to the left, and vice versa.

(f) The approximate useful pH range for each electrode is indicated by the positions of the ends of its line.

The various kinds of standard and glass electrodes used in this investigation were calibrated or checked by means of a series of buffer solutions that were either 0.05 or 0.1 M and prepared according to the directions of Clark (24). These solutions included phthalate, phosphate, borate, and carbonate and were standardized by either the hydrogen or quinhydrone electrode.

The suitability of an electrode for measuring hydrogen ion concentration depends upon the extent to which the accuracy and reproducibility are affected by various factors. Among these are: (a) Useful pH range; (b) age; that is, any change in the calibration with time; (c) alkali salts (salt error); (d) heavy metal salts; (e) proteins; and (f) oxidants and reductants. Certain "poisons" may also be found in some of these categories. Thus, copper sulphate (a heavy metal salt) poisons a hydrogen electrode and sodium bisulphite (a reductant) poisons a quinhydrone electrode.

A brief study was accordingly made to determine the effects of the above factors on the operation of the metal-connected glass electrode. It should be emphasized that these experiments were not designed to redetermine the constants of certain solutions, nor to establish exact limits for the electrode, but merely to determine trends. In this way, the general behavior could be defined and possible means of improvement indicated.

2. USEFUL pH RANGE

Every secondary electrode is subject to errors outside of some favorable range of pH, which therefore needs to be defined. Briefly, within the limits of accuracy (± 0.1 pH) of these measurements the metal-connected glass electrodes were found to function linearly[7] and reproducibly between pH-1 and pH-11, in dilute buffer solutions. Down to pH-0; that is in concentrated acid solutions, fairly satisfactory readings could be obtained by frequent calibration. Above pH-11, hydroxide solutions of increasing concentration must be used and attack of the glass surface is probable. Readings above pH-11.5 were unsatisfactory. In general, any kind of electrode tends to show discrepancies in the strongly acid and alkaline ranges.

3. EFFECT OF AGE

The metal-connected electrodes were calibrated occasionally by means of buffer solutions, using the technique previously described. Typical calibrations are shown for illustrative purposes in Figure 6[7]. The approximate variation in the constant thus determined for the same electrodes over a period of about eight months[8] is shown in Figure 7.

The change in the constant is evidently due to a change in the initial potential of (or in) the glass. This is usually accompanied by a slight increase in resistance (Table 1), but without any apparent decrease in accuracy. The change can not, therefore, be definitely ascribed either to increased surface leakage or to deterioration of the glass. Except in rare cases, the normal drift or rate of change was probably less than 0.01 pH per day. In two cases the constant decreased, and at an increasing rate (fig. 7, Nos. 1 and 3), while in the third case (fig. 7, No. 2) it is remarkable that very little change occurred, although this electrode was most frequently used in the experiments.

The electrodes were actually used to a much greater extent than is indicated in Figure 7, but were not always calibrated for other than a required short range of pH, which did not necessarily include the pH constant. Figure 7 merely records observed limits and was not used for purposes of measurement. Like other types of electrodes,

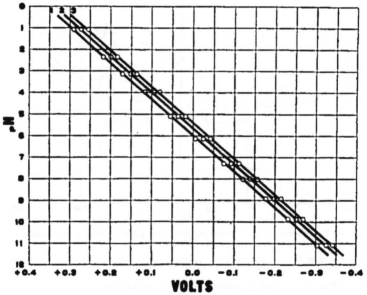

FIGURE 6.—*Actual calibrations of three m etal-connected glass electrodes of the containing type*

pH and potential in volts versus saturated calomel electrode at 25° C.

the metal-connected glass electrode required frequent checking, depending on the accuracy desired and the kind of solution under study. It was found advantageous to check an electrode in use several times daily.

4. EFFECT OF ALKALI SALTS (SALT ERROR)

At a given pH, the total concentration and kind of neutral salt present affect the potential of an electrode, causing what is known as the salt error. For the hydrogen electrode this error is zero by definition and for the quinhydrone electrode it is relatively small. To determine the approximate salt error for the metal-connected glass electrode, buffer solutions were prepared containing a large amount of potassium chloride or sodium chloride. The apparent change in pH was then determined.

The results obtained are shown in Tables 3 and 4.

TABLE 3.—*Effect of addition of normal potassium chloride to buffer solutions*

Buffer solution		Change in pH after addition			Salt error	
Composition	pH without addition	Hydrogen electrode	Quinhy- drone electrode	Metal- connected glass electrode	Quinhy- drone electrode	Metal- connected glass electrode
0.1 *N* HCl	1.10	−0.20	−0.18	−0.24	+0.02	−0.04
0.05 *M* KH₂PO₄+K₂HPO₄	7.00	−.41	−.37	−.44	+.04	−.03
0.1 *N* K₂CO₃	11.20	−.28	....	−.32	....	−.04

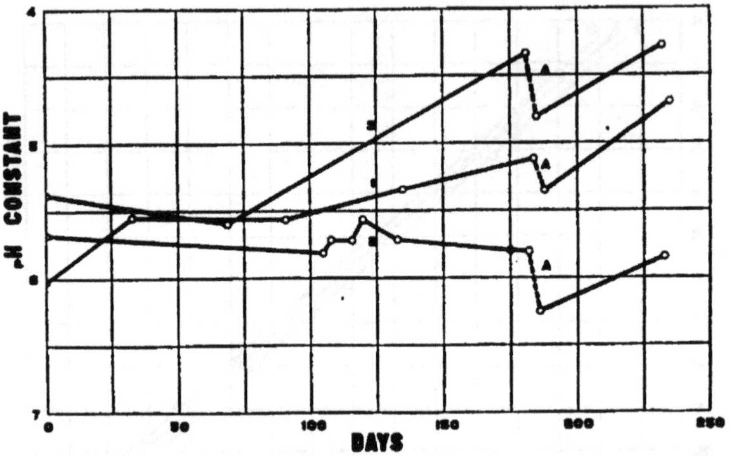

FIGURE 7.—*Variation of pH constant with time for three metal-connected glass electrodes of the containing type*

pH and potential in volts versus saturated calomel electrode at 25° C. *A*, Effect of applying 1,000 volts d. c. for a measurement of resistance

TABLE 4.—*Effect of addition of sodium chloride to buffer solutions*

Buffer solution composition.	NaCl added *N*	Change in pH after addition			Salt error	
		Hydrogen electrode	Quinhy- drone electrode	Metal- connected glass electrode	Quinhy- drone electrode	Metal- connected glass electrode
0.1 *N* HCl	1	−0.26	−0.28	−0.24	−0.02	+0.02
	2	−.50	−.43	−.41	+.07	+.09
	4	−.90	−.79	−.66	+.11	+.24
0.05 *M* KH₂PO₄+K₂HPO₄	1	−.71	−.62	−.70	+.09	+.01
0.1 *N* K₂CO₃	1	−.46	....	−.76	....	−.30
	2	−.66	....	−1.03	....	−.34
	4	−.81	....	−1.39	....	−.58

The addition of salts caused a decrease in pH (an increase in hydrogen ion activity) in every case. As mentioned above, the change for the hydrogen electrode is attributed to an actual change in the solution and the differential change between the hydrogen and any other electrode then becomes the salt error for that electrode.

Table 3 shows that the salt error caused by potassium chloride up to 1 *N* is only a few hundredths pH throughout the useful range of the metal-connected glass electrode, or about the same as for the quin-hydrone electrode. The salt error caused by sodium chloride (Table 4) is also low in the acid and neutral ranges, but is quite appreciable in the alkaline range.

While these data are not complete, it seems safe to conclude that for best results the total salt concentration should be kept below 1 N if possible, and that potassium salts are preferable in alkaline solutions. When high salt concentrations are unavoidable it may be possible to secure more accurate measurements by calibrating the electrode in concentrated buffer solutions.

Similar results have been reported by Hughes (23), Kahler and De Eds (38) and others. Various investigators have noted that divalent ions, such as calcium, have a smaller effect than sodium or potassium ions.

5. EFFECT OF HEAVY METAL SALTS

The presence of a heavy metal salt often makes electrometric pH determinations difficult. Specific applications of an inappropriate nature are, of course, readily excluded. These include cases where the reagent used precipitates metal from solution, as gaseous hydrogen does with copper and quinhydrone with silver. In addition, difficulties may arise from the presence of impurities, such as colloids (2), from oxidation-reduction potentials including those possibly caused by reduction of metal ions to subvalent states (3) (4) (26), side reactions (18), and adsorption effects. It must also be remembered that these solutions are unbuffered unless a solid phase of hydroxide is present.

The measurements on heavy metal salt solutions were made to compare results by the different electrodes and not to redetermine hydrolysis points. Only the nickel ammonium sulphate and the copper sulphate were recrystallized salts and none of the solutions was treated with the corresponding metal hydroxide. The salts used were, however, all of high quality and there is reason to believe that the values obtained are typical, except possibly for manganese.

Colorimetric measurements of pH (uncorrected) could be obtained conveniently with these solutions and were therefore made by the Gillespie drop ratio method (24).

The results obtained with a few typical heavy metal solutions are given in Table 5.

TABLE 5.—*Effect of heavy metal salts on electrodes*

pH of solutions

Solution 0.05 M	Hydrogen electrode	Quinhy-drone electrode	Mem-brane glass electrode	Metal-connected glass electrode	Colorimetric (uncorrected) [1]
ZnSO₄	5.2–5.6	5.1–5.8	6.15	5.80	6.55 B. c. p. / 6.55 B. t. b.
MnCl₂	5.5–5.6	4.8–5.2	6.05	5.95	6.80 B. c. p. / 7.00 B. t. b.
NiSO [2]		5.20	5.40	5.45	5.40	5.70 M. r. / 5.80 B. c. p.
CuSO₄			4.20	4.40	4.30	4.75 B. p. b. / 4.50 M. r.
HgCl₂				3.60	3.60	3.80 B. p. b.
AgNO₃ [3]					5.07	6.55 B. c. p. / 6.80 B. t. b.

[1] B. p. b., brom phenol blue; M. r., methyl red; B. c. p., brom cresol purple; B. t. b., brom thymol blue.
[2] As NiSO₄. (NH₄)₂SO₄·6H₂O.
[3] 0.1 M.

The two types of the glass electrode showed good agreement in all the solutions except that of zinc sulphate. In this case the metal-connected electrode gave a value that was nearer than that of the membrane electrode to the hydrogen and quinhydrone values, although these were unsteady. Britton's (14, p. 2120) result for a more dilute (0.025 M) solution was 5.2 by the hydrogen electrode. Kolthoff and Kameda (42) found for 0.1 M solution a value of 5.67 by means of a specially prepared hydrogen electrode (with a bright platinum deposit) and colorimetrically by methyl red a value of 5.66, practically the same. Britton and Dodd (45) have recently found about 6.3 (interpolated from their fig. 1, p. 1943) for a 0.01 M solution, by the glass electrode. Britton and R°binson (46) have discussed the status of this subject. Some previous investigators (2) (21) (22) (26) have also reported a drift of the hydrogen electrode in zinc solutions.

The two types of the glass electrode agreed closely with the quinhydrone electrode in the nickel ammonium sulphate solution. Such solutions have, of course, a somewhat lower pH than do those of nickel sulphate. Better agreement between the quinhydrone and hydrogen electrodes is usually obtained in nickel solutions, particularly at higher concentrations or when buffered (30).

In the copper sulphate solution the glass electrodes also showed good agreement with the quinhydrone. Values for 0.05 M CuSO$_4$ at 18° C have been reported by O'Sullivan (17) using the quinhydrone and by Hughes (23) using the glass electrode, as 4.14 and 4.24, respectively, which are in good agreement with the present results. Britton's (14, p. 2148) value for 0.02 M copper sulphate was 5.4 by the oxygen electrode.

In the manganese solution readings were somewhat unsteady with all electrodes used. Those made with the glass electrode were considerably higher than those with the hydrogen or quinhydrone. The latter both showed drifts and did not agree closely. Rideal (18) has pointed out that manganese salts tend to catalyze the auto-oxidation of quinhydrone, so that the result by the quinhydrone electrode is doubtful. The accepted value for manganese solutions in the literature is 8.5 to 8.8 (11) (14, p. 2110) by the hydrogen electrode, indicating that these solutions at equilibrium conditions are slightly alkaline, a conclusion that seems anomalous in view of the weakly basic character of divalent manganese. Possibly an oxidation-reduction potential interferes.

The value obtained for the mercuric chloride solution is close to that of Britton and Dodd (45), who found about 3.7 (interpolated from their fig. 1, p. 1943) for a 0.01 M solution, by the glass electrode. This replaces an earlier and much higher value of 7.3, by the oxygen electrode, reported by Britton (14, p. 2148).

The value for the silver nitrate solution agrees with that of about 5.0 (interpolated from their fig. 1, p. 1943) found for a more dilute (0.01 M) solution by Britton and Dodd (45), by the glass electrode. Britton (14, p. 2148) previousl found 5.7 for 0.02 M silver nitrate, by the oxygen electrode. He then adopted "pH 9 (?)" tentatively, however, basing this figure on calculations of the solubility product of silver hydroxide. Evidently, this value was too high and the experimental one was nearer correct. Further data and discussion have been given by Britton and Robinson (46). Horovitz (10, p. 389) thought that he detected a silver electrode function of glass in silver solutions, but this was not confirmed by Hughes (23). Any direct effect of the silver (or copper) coating of the metal-connected electrode is unlikely, as the glass used contained no heavy metals and was comparatively thick.

In general, the results by the colorimetric method (uncorrected) show somewhat higher salt errors for the indicators than would be expected for such dilute solutions. Combination between metal ion and indicator may have occurred in some cases.

6. EFFECT OF PROTEIN

It is well known that proteins are likely to disturb electrode readings. Gelatin was selected for trial, as it is a substance of special interest in electrochemical work, and a few experiments were made on solutions containing various amounts at low, intermediate and high pH values. The data obtained are presented in Table 6.

TABLE 6.—*Effect of gelatin*

Buffer solution	Gelatin g/l	pH Hydrogen electrode	pH Quinhydrone electrode	pH Metal-connected glass electrode
0.1 N HCl	0		1.18	1.19
	1		1.19	1.17
	5		1.19	1.17
	10		1.17	1.14
0.05 M KH$_2$PO$_4$+K$_2$HPO$_4$	0		6.96	7.01
	1		6.99	7.01
	5		6.96	7.02
	10		6.96	6.96
0.1 N K$_2$CO$_3$	0	11.00		11.03
	1	10.94		10.90
	5	10.77		10.88
	10	10.59		10.73

Large amounts of gelatin did not appreciably affect the readings of the metal-connected glass electrode at low and intermediate pH values, as compared with the quinhydrone electrode. It is well known that the quinhydrone electrode is itself affected by some proteins. The metal-connected electrode agreed closely with the hydrogen electrode at the high pH when only small amounts of gelatin were present. The pH of the carbonate solution actually decreased when gelatin was dissolved in it. This was accompanied by a slight precipitation.

7. EFFECT OF OXIDANTS AND REDUCTANTS

Strongly oxidizing or reducing compounds cause a potential separate from that of hydrogen ions and usually great enough to prevent the use of the hydrogen or quinhydrone electrode. The latter can be used occasionally, for instance, in dilute nitric acid. Sometimes a metal-metal oxide electrode can be substituted, as that of mercury, manganese, or antimony, which may be convenient even if empirical. Indicators are usually unsatisfactory.

Previous types of the glass electrode have been shown by Hughes (9) MacInnes and Dole (27) and others to be quite unaffected by oxidation-reduction potentials. To determine if this relation holds also for the metal-connected electrode, comparative measurements with this and the membrane glass electrode were made on a few typical solutions. The results are presented in Table 7.

TABLE 7.—*Effect of oxidants and reductants*

Solution	pH		Solution	pH	
	Membrane electrode	Metal-connected electrode		Membrane electrode	Metal-connected electrode
0.1 M HNO$_3$ [1]	1.10	1.20	0.1 M K$_2$Cr$_2$O$_7$	3.75	3.85
1 M CrO$_3$	.45	.65	0.2 M K$_2$CrO$_4$	9.25	9.30
0.45 M K$_2$Cr$_2$O$_7$ [2]	3.60	3.80	0.05 M As$_2$O$_3$	5.60	5.55
1 M K$_2$CrO$_4$	9.45	9.55	0.25 M Na$_2$S$_2$O$_3$	4.90	4.80
0.2 M CrO$_3$	.90	1.05	0.1 M NaCN [3]		11.0

[1] 1.14 by quinhydrone electrode.
[2] Saturated solution.
[3] 11.5 (maximum) by hydrogen electrode.

The two types of glass electrodes gave about the same results. The lower values by the membrane electrode, particularly in the more concentrated oxidizing solutions, are probably more accurate, as this type seems less subject to disturbances, possibly of the nature of adsorption effects.

The results for the chromic acid and chromate solutions agree roughly with the end points of the titration curves published by Hughes (9) and by Britton (12) (13) (43, p. 73, p. 162). Hughes used a glass electrode of the bulb type and 0.1 M chromic acid, while Britton used both oxygen and hydrogen electrodes in very dilute (about 0.025 M) chromic acid.

It is of interest to note the great difference in strength between dichromic and chromic acid, on account of which chromate solutions are distinctly alkaline because of hydrolysis. (The acidity of the dichromate solutions may be attributed to secondary dissociation of the dichromate ion.)

Readings in the bisulphite solution were often unsteady, probably because of absorption of oxygen from the air and a resulting change in acidity. The hydrogen (43, p. 73), quinhydrone and oxygen (43, p. 73) electrodes all fail in bisulphite or sulphite solutions.

The value for sodium cyanide is consistent with the hydrolysis measurements of Harman and Worley (16) and with a measurement by Britton and Robinson (40) on dilute potassium cyanide solution (0.04 M), with the antimony electrode. Britton and Dodd (45) later found the glass electrode to be more accurate than the antimony electrode, in cyanide solutions. The hydrolysis of alkali cyanide is evidently about the same as that of alkali carbonate.

8. APPLICATIONS

Applications of the metal-connected glass electrode in research have been made by the author in connection with pH measurements on chromic acid and chromium chromate solutions, cyanide silver-plating solutions (preferably diluted), and ammoniacal silver bromide-gelatin photographic emulsions. In general, consistent and useful results were obtained. These will be published in separate papers by other investigators.

V. SUMMARY

1. Gl ss electrodes with a direct metallic connection may be constructed having comparatively good accuracy and sensitivity.

2. This shows that an equilibrium between hydrogen ions on opposite sides of a glass wall is not essential to the functioning of glass as an electrode, and suggests that deep penetration of hydrogen ions into the glass may not occur.

3. The metal-connected electrodes may be constructed with walls thick enough to make breakage unlikely with ordinary care in handling.

4. Their useful life is indefinite and measured by months or years.

5. Only small errors are caused by alkali salts in moderate concentration, by heavy metal salts in low concentration and by gelatin.

6. Oxidation-reduction potentials have little, if any, effect with this or other types of glass electrodes.

VI. ACKNOWLEDGMENTS

The writer acquired his first experience with the technique of glass electrodes during March and April, 1929, as a guest at the Rockefeller Institute for Medical Research in New York City. Grateful acknowledgment is herewith made to Dr. D. A. MacInnes of that institution and to Dr. M. Dole, formerly there but now at Northwestern University, Evanston, Ill.

Acknowledgment is also made to Dr. W. Blum, who directed this research and whose advice has been most helpful.

VII. BIBLIOGRAPHY

1. Cremer, M., Zeits. Biologie, vol. 47, p. 562. 1906.
2. Denham, H. G., J. Chem. Soc., vol. 93, p. 41. 1908.
3. Denham, H. G. and Allmand, A. J., J. Chem. Soc., vol. 93, p. 424. 1908.
4. Denham, H. G., J. Chem. Soc., vol. 93, p. 833. 1908.
5. Haber, F. and Klemensiewicz, Z., Zeits. Phys. Chemie, vol. 67, p. 385. 1909.
6. Borelius, G., Ann. Physik (4), vol. 42, p. 1129. 1913.
7. Borelius, G., Ann. Physik. (4), vol. 45, p. 929. 1914.
8. Borelius, G., Ann. Physik (4), vol. 50, p. 447. 1916.
9. Hughes, W. S., J. Am. Chem. Soc., vol. 44, p. 2860. 1922.
10. Horovitz, K., Zeits Physik., vol. 15, p. 369. 1923.
11. Allmand, A. J. and Campbell, A. N., Trans. Faraday Soc., vol. 20, p. 379. 1924.
12. Britton, H. T. S., J. Chem. Soc., vol. 125, p. 1572. 1924.
13. Britton, H. T. S., J. Chem. Soc., vol. 127, p. 1896. 1925.
14. Britton, H. T. S., J. Chem. Soc., vol. 127, pp. 2110, 2120, 2142, and 2148. 1925.
15. Horovitz, K., Sitz. Akad. Wiss. Wien, Abt. IIa, vol. 134, p. 335. 1925.
16. Harman, R. W., and Worley, F. P., Trans. Faraday Soc., vol. 20, p. 502. 1925.
17. O'Sullivan, J. B., Trans. Faraday Soc., vol. 21, p. 319. 1925.
18. Rideal, E. K., Trans. Faraday Soc., vol. 21, p. 325. 1925.
19. Rideal, E. K., an introduction to Surface Chemistry, Cambridge University Press. London, 1926.
20. Kerridge, P. M. T., J. Sci. Instruments, vol. 3, p. 404, 1926.

21. Frölich, Per K., Trans. Am. Electrochem. Soc., vol. 49, p. 395. 1926.
22. Thompson, M. R., Trans. Am. Electrochem. Soc., vol. 50, p. 193. 1926.
23. Hughes, W. S., J. Chem. Soc., p. 491. 1928.
24. Clark, W. M., The Determination of Hydrogen Ions, 3d ed. 1928. Williams & W kins Co., Baltimore, Md.
25. Elder, L. W., jr., and Wright, W. H., Proc. Nat. Acad. Sci., vol. 14, p. 936. 1928.
26. Denham, H. G., and Marris, N. A., Trans. Faraday Soc., vol. 24, p. 510. 1928.
27. MacInnes, D. A., and Dole, M., Ind. & Eng. Chem., Anal. ed., vol. 1, p 57. 1929.
28. MacInnes, D. A., and Dole, M., J. Gen. Physiol., vol. 12, p. 805. 1929.
29. Elder, L. W., jr., J. Am. Chem. Soc., vol. 51, p. 3266. 1929.
30. Blum, W., and Bekkedahl, N., Trans. Am. Electrochem. Soc., vol. 56, p. 291. 1929.
31. MacInnes, D. A., and Dole, M., J. Am. Chem. Soc., vol. 52, p. 29. 1930.
32. Voegtlin, C., De Eds, F., and Kahler, H., Reprint No. 1413, Public Health Reports, vol. 45, p. 2223. 1930.
33. Elder, L. W., jr., Trans. Am. Electrochem. Soc., vol. 57, p. 383. 1930.
34. MacInnes, D. A., and Belcher, D., J. Am. Chem. Soc., vol. 53, p. 3315. 1931.
35. Dole, M., J. Am. Chem. Soc., vol. 53, p. 4260. 1931.
36. Robertson, G. R., Ind. & Eng. Chem., Anal. ed., vol. 3, p. 5. 1931.
37. The Making of Mirrors by the Deposition of Metal on Glass, circular of the Bureau of Standards, No. 389. 1931.
38. Kahler, H., and De Eds, F., J. Am. Chem. Soc., vol. 53, p. 2998. 1931.
39. Hazel, F., and Sorum, C. H., J. Am. Chem. Soc., vol. 53, p. 49. 1931.
40. Britton, H. T. S., and Robinson, R. A., J. Chem. Soc., 458. 1931.
41. Lengyel, B. v., Zeits. Physik. Chem., Abt. A, vol. 153, p. 425. 1931.
42. Kolthoff, I. M., and Kameda, T., J. Am. Chem. Soc., vol. 53, p. 832. 1931.
43. Britton, H. T. S., Hydrogen Ions, 2d ed., D. Van Nostrand Co., New York City. 1932.
44. Dole, M., J. Am. Chem. Soc., vol. 54, p. 3095. 1932.
45. Britton, H. T. S., and Dodd, E. N., J. Chem. Soc., p. 1940. 1932.
46. Britton, H. T. S., and Robinson, R. A., Trans. Faraday Soc., vol. 28, p. 531. 1932.
47. Compton, K. G., and Haring, H. E., Trans. Electrochem. Soc., vol. 62. 1932 (preprint).

WASHINGTON October 11, 1932.

145879—32——10

Page

Page

858 Bureau of Standards Journal of Research

Page

SUPERINTENDENT OF DOCUMENTS,
Government Printing Office,
Washington, D. C.

(Cut here)

DEAR SIR: Send me the Bureau of Standards publications marked X below. I inclose remittance ¹ to cover the cost.

Mark X	Title of publication	Price		Amount inclosed
		United States, Canada, Cuba, Newfoundland, Republic of Panama	Foreign countries	
	Commercial Standards Monthly (one year)	$1.00	$1.60	
	Bureau of Standards Journal of Research ² (one year)	2.50	3.25	
	Blue buckram bound volumes 1 to 8 of above journal are now available..each volume	³2.75	³3.50	
	Technical News Bulletin (one year)	.50	.70	
	Standards Yearbook, 1932	1.00	1.20	
	Total remittance			

Send to_____

Number and Street_____

City and State (or County)_____

¹ Remittances should be either in the form of a post-office money order, coupons (issued for the specific purpose of purchasing Government publications), express money order, New York draft, or cash, at the sender's risk.

² This issue completes volume 9. Volume 10 will commence with the January (1933) issue. Your subscription should be submitted at once to insure starting with No. 1 of volume 10.

³ These prices are for volumes 1 to 7. Commencing with volume 8 the price will be $2.50 and 3.25 (foreign).

Lightning Source UK Ltd.
Milton Keynes UK
UKHW020634011218
333024UK00012B/1970/P